CAMBRIDGE STUDIES IN ADVANCED MATHEMATICS 151

Foundations of Ergodic Theory

Rich with examples and applications, this textbook provides a coherent and self-contained introduction to ergodic theory suitable for a variety of one- or two-semester courses. The authors' clear and fluent exposition helps the reader to grasp quickly the most important ideas of the theory, and their use of concrete examples illustrates these ideas and puts the results into perspective.

The book requires few prerequisites, with background material supplied in the appendix. The first four chapters cover elementary material suitable for undergraduate students – invariance, recurrence and ergodicity – as well as some of the main examples. The authors then gradually build up to more sophisticated topics, including correlations, equivalent systems, entropy, the variational principle, and thermodynamic formalism. The 400 exercises increase in difficulty through the text and test the reader's understanding of the whole theory. Hints and solutions are provided at the end of the book.

Marcelo Viana is Professor of Mathematics at Instituto Nacional de Matemática Pura e Aplicada (IMPA), Rio de Janeiro and a leading research expert in ergodic theory and dynamical systems. He has served in several academic organizations, such as the International Mathematical Union (Vice-president 2011–2014), the Brazilian Mathematical Society (President 2013–2015), the Latin American Mathematical Union (Scientific Coordinator, 2001–2008) and the newly founded Mathematical Council of the Americas. He is also a member of the academies of science of Brazil, Portugal, Chile and the Developing World and he has received several academic distinctions, including the Grand Croix of Scientific Merit, granted by the President of Brazil, in 2000, and the Ramanujan Prize of ICTP and IMU, in 2005. He was an invited speaker at the international congress of mathematicians ICM 1994, in Zurich, a plenary speaker at the International Congress of Mathematical Physics ICMP 1994, and a Plenary Speaker at the ICM 1998, in Berlin. To date, he has supervised 32 doctoral theses. Currently, he leads the organization of the ICM 2018 in Rio de Janeiro and is also involved in initiatives to improve mathematical education in his country.

Krerley Oliveira is Associate Professor at the Universidade Federal de Alagoas (UFAL), Brazil, where he founded the graduate program in mathematics and the State of Alagoas Math Olympiad program, where he promotes young talents in mathematics. He was a medalist at the Brazilian Mathematical Olympiad (1996) and twice at the Iberoamerican Mathematical Olympiad for university students (1999 and 2000). He was elected an Affiliate Member of the Brazilian Academy of Sciences (2007–2012) and his research is focused on dynamical systems and ergodic theory.

CAMBRIDGE STUDIES IN ADVANCED MATHEMATICS

All the titles listed below can be obtained from good booksellers or from Cambridge University Press.
For a complete series listing visit www.cambridge.org/mathematics.

Already published

112 D. W. Stroock *Partial differential equations for probabilists*
113 A. Kirillov, Jr *An introduction to Lie groups and Lie algebras*
114 F. Gesztesy *et al. Soliton equations and their algebro-geometric solutions, II*
115 E. de Faria & W. de Melo *Mathematical tools for one-dimensional dynamics*
116 D. Applebaum *Lévy processes and stochastic calculus (2nd edition)*
117 T. Szamuely *Galois groups and fundamental groups*
118 G. W. Anderson, A. Guionnet & O. Zeitouni *An introduction to random matrices*
119 C. Perez-Garcia & W. H. Schikh of *Locally convex spaces over non-Archimedean valued fields*
120 P. K. Friz & N. B. Victoir *Multidimensional stochastic processes as rough paths*
121 T. Ceccherini-Silberstein, F. Scarabotti & F. Tolli *Representation theory of the symmetric groups*
122 S. Kalikow & R. McCutcheon *An outline of ergodic theory*
123 G. F. Lawler & V. Limic *Random walk: A modern introduction*
124 K. Lux & H. Pahlings *Representations of groups*
125 K. S. Kedlaya *p-adic differential equations*
126 R. Beals & R. Wong *Special functions*
127 E. de Faria & W. de Melo *Mathematical aspects of quantum field theory*
128 A. Terras *Zeta functions of graphs*
129 D. Goldfeld & J. Hundley *Automorphic representations and L-functions for the general lineargroup, I*
130 D. Goldfeld & J. Hundley *Automorphic representations and L-functions for the general lineargroup, II*
131 D. A. Craven *The theory of fusion systems*
132 J. Väänänen *Models and games*
133 G. Malle & D. Testerman *Linear algebraic groups and finite groups of Lie type*
134 P. Li *Geometric analysis*
135 F. Maggi *Sets of finite perimeter and geometric variational problems*
136 M. Brodmann & R. Y. Sharp *Local cohomology (2nd Edition)*
137 C. Muscalu & W. Schlag *Classical and multilinear harmonic analysis, I*
138 C. Muscalu & W. Schlag *Classical and multilinear harmonic analysis, II*
139 B. Helffer *Spectral theory and its applications*
140 R. Pemantle & M. C. Wilson *Analytic combinatorics in several variables*
141 B. Branner & N. Fagella *Quasiconformalsurgery in holomorphic dynamics*
142 R. M. Dudley *Uniform central limit theorems (2nd Edition)*
143 T. Leinster *Basic category theory*
144 I. Arzhantsev, U. Derenthal, J. Hausen & A. Laface *Cox rings*
145 M. Viana *Lectures on Lyapunov exponents*
146 J.-H.Evertse & K. Győry *Unit equations in Diophantine number theory*
147 A. Prasad *Representation theory*
148 S. R. Garcia, J. Mashreghi & W. T. Ross *Introduction to model spaces and their operators*
149 C. Godsil & K. Meagher *Erdős–Ko–Rado theorems: Algebraic approaches*
150 P. Mattila *Fourier analysis and Hausdorff dimension*
151 M. Viana & K. Oliveira *Foundations of ergodic theory*

Foundations of Ergodic Theory

MARCELO VIANA
IMPA, Rio de Janeiro

KRERLEY OLIVEIRA
Universidade Federal de Alagoas, Brazil

CAMBRIDGE
UNIVERSITY PRESS

University Printing House, Cambridge CB2 8BS, United Kingdom

One Liberty Plaza, 20th Floor, New York, NY 10006, USA

477 Williamstown Road, Port Melbourne, VIC 3207, Australia

4843/24, 2nd Floor, Ansari Road, Daryaganj, Delhi - 110002, India

79 Anson Road, #06-04/06, Singapore 079906

Cambridge University Press is part of the University of Cambridge.

It furthers the University's mission by disseminating knowledge in the pursuit of education, learning and research at the highest international levels of excellence.

www.cambridge.org
Information on this title: www.cambridge.org/9781107126961

First published 2016

A catalogue record for this publication is available from the British Library

Library of Congress Cataloging in Publication data
Viana, Marcelo.
Foundations of ergodic theory / Marcelo Viana, IMPA, Rio de Janeiro, Krerley Oliveira, Universidade Federal de Alagoas, Brazil.
pages cm. – (Cambridge studies in advanced mathematics ; 151)
Includes bibliographical references and index.
ISBN 978-1-107-12696-1 (hbk)
1. Ergodic theory–Textbooks. 2. Topology–Textbooks. I. Oliveira, Krerley.
II. Title.
QA611.5.V53 2016
515′.48–dc23 2015022583

ISBN 978-1-107-12696-1 Hardback

Contents

Preface

In short terms, ergodic theory is the mathematical discipline that deals with dynamical systems endowed with invariant measures. Let us begin by explaining what we mean by this and why these mathematical objects are so worth studying. Next, we highlight some of the major achievements in this field, whose roots go back to the physics of the late 19th century. Near the end of the preface, we outline the content of this book, its structure and its prerequisites.

What is a dynamical system?

There are several definitions of what a dynamical system is some more general than others. We restrict ourselves to two main models.

The first one, to which we refer most of the time, is a transformation $f : M \to M$ in some space M. Heuristically, we think of M as the space of all possible states of a given system. Then f is the evolution law, associating with each state $x \in M$ the one state $f(x) \in M$ the system will be in a unit of time later. Thus, time is a discrete parameter in this model.

We also consider models of dynamical systems with continuous time, namely flows. Recall that a *flow* in a space M is a family $f^t : M \to M$, $t \in \mathbb{R}$ of transformations satisfying

$$f^0 = \text{identity} \quad \text{and} \quad f^t \circ f^s = f^{t+s} \text{ for all } t,s \in \mathbb{R}. \tag{0.0.1}$$

Flows appear, most notably, in connection with differential equations: take f^t to be the transformation associating with each $x \in M$ the value at time t of the solution of the equation that passes through x at time zero.

We always assume that the dynamical system is measurable, that is, that the space M carries a σ-algebra of *measurable subsets* that is preserved by the dynamics, in the sense that the pre-image of any measurable subset is still a measurable subset. Often, we take M to be a topological space, or even a metric space, endowed with the Borel σ-algebra, that is, the smallest σ-algebra that contains all open sets. Even more, in many of the situations we consider in this book, M is a smooth manifold and the dynamical system is taken to be differentiable.

What is an invariant measure?

A *measure* in M is a non-negative function μ defined on the σ-algebra of M, such that $\mu(\emptyset) = 0$ and

$$\mu\left(\cup_n A_n\right) = \sum_n \mu(A_n)$$

for any countable family $\{A_n\}$ of pairwise disjoint measurable subsets. We call μ a probability measure if $\mu(M) = 1$. In most cases, we deal with finite measures, that is, such that $\mu(M) < \infty$. Then we can easily turn μ into a probability ν: just define

$$\nu(E) = \frac{\mu(E)}{\mu(M)} \quad \text{for every measurable set } E \subset M.$$

In general, we say that a measure μ is invariant under a transformation f if

$$\mu(E) = \mu(f^{-1}(E)) \quad \text{for every measurable set } E \subset M. \tag{0.0.2}$$

Heuristically, this may be read as follows: the probability that a point is in any given measurable set is the same as the probability that its image is in that set. For flows, we replace (0.0.2) by

$$\mu(E) = \mu(f^{-t}(E)) \quad \text{for every measurable set } E \subset M \text{ and } t \in \mathbb{R}. \tag{0.0.3}$$

Notice that (0.0.2)–(0.0.3) do make sense since, by assumption, the pre-image of a measurable set is also a measurable set.

Why study invariant measures?

As in any other branch of mathematics, an important part of the motivation is intrinsic and aesthetical: as we will see, these mathematical structures have deep and surprising properties, which are expressed through beautiful theorems. Equally fascinating, ideas and results from ergodic theory can be applied in many other areas of mathematics, including some that do not seem to have anything to do with probabilistic concepts, such as combinatorics and number theory.

Another key motivation is that many problems in the experimental sciences, including many complicated natural phenomena, can be modelled by dynamical systems that leave some interesting measure invariant. Historically, the most important example came from physics: Hamiltonian systems, which describe the evolution of conservative systems in Newtonian mechanics, are described by certain flows that preserve a natural measure, the so-called Liouville measure. Actually, we will see that very general dynamical systems do possess invariant measures.

Yet another fundamental reason to be interested in invariant measures is that their study may yield important information on the dynamical system's behavior that would be difficult to obtain otherwise. Poincaré's recurrence theorem, one of the first results we analyze in this book, is a great illustration of this: it asserts that, relative to any finite invariant measure, almost every orbit returns arbitrarily close to its initial state.

Brief historic survey

The word *ergodic* is a concatenation of two Greek words, $\epsilon\rho\gamma\,o\nu$ (*ergon*) = work and $o\delta o\sigma$ (*odos*) = way, and was introduced in the 19th century by the Austrian physicist L. Boltzmann. The systems that interested Boltzmann, J. C. Maxwell and J. C. Gibbs, the founders of the kinetic theory of gases, can be described by a Hamiltonian flow, associated with a differential equation of the form

$$\left(\frac{dq_1}{dt},\ldots,\frac{dq_n}{dt},\frac{dp_1}{dt},\ldots,\frac{dp_n}{dt}\right)=\left(\frac{\partial H}{\partial p_1},\ldots,\frac{\partial H}{\partial p_n},-\frac{\partial H}{\partial q_1},\ldots,-\frac{\partial H}{\partial q_n}\right).$$

Boltzmann believed that typical orbits of such a flow fill in the whole energy surface $H^{-1}(c)$ that contains them. Starting from this *ergodic hypothesis*, he deduced that the (time) averages of observable quantities along typical orbits coincide with the (space) averages of such quantities on the energy surface, which was crucial for his formulation of the kinetic theory of gases.

In fact, the way it was formulated originally by Boltzmann, this hypothesis is clearly false. So, the denomination *ergodic hypothesis* was gradually displaced to what would have been a consequence, namely, the claim that time averages and space averages coincide. Systems for which this is true were called *ergodic*. And it is fair to say that a great part of the progress experienced by ergodic theory in the 20th century was motivated by the quest to understand whether most Hamiltonian systems, especially those that appear in connection with the kinetic theory of gases, are ergodic or not.

The foundations were set in the 1930's, when J. von Neumann and G. D. Birkhoff proved that time averages are indeed well defined for almost every orbit. However, in the mid 1950's, the great Russian mathematician A. N. Kolmogorov observed that many Hamiltonian systems are actually *not* ergodic. This spectacular discovery was much expanded by V. Arnold and J. Moser, in what came to be called KAM (Kolmogorov–Arnold–Moser) theory.

On the other hand, still in the 1930's, E. Hopf had given the first important examples of Hamiltonian systems that *are* ergodic, namely, the geodesic flows on surfaces with negative curvature. His result was generalized to geodesic flows on manifolds of any dimension by D. Anosov, in the 1960's. In fact, Anosov proved ergodicity for a much more general class of systems, both with discrete time and in continuous time, which are now called Anosov systems.

An even broader class, called uniformly hyperbolic systems, was introduced by S. Smale and became a major focus for the theory of dynamical systems through the last half a century or so. In the 1970's, Ya. Sinai developed the theory of Gibbs measures for Anosov systems, conservative or dissipative, which D. Ruelle and R. Bowen rapidly extended to uniformly hyperbolic systems. This certainly ranks among the greatest achievements of smooth ergodic theory.

Two other major contributions must also be mentioned in this brief survey. One is the introduction of the notion of *entropy*, by Kolmogorov and Sinai, near the end of the 1950's. Another is the proof that the entropy is a complete invariant for Bernoulli shifts (two Bernoulli shifts are equivalent if and only if they have the same entropy), by D. Ornstein, some ten years later.

By then, the theory of non-uniformly hyperbolic systems was being initiated by V. I. Oseledets, Ya. Pesin and others. But that would take us beyond the scope of the present book.

How this book came to be

This book grew from lecture notes we wrote for the participants of mini-courses we taught at the Department of Mathematics of the Universidade Federal de Pernambuco (Recife, Brazil), in January 2003, and at the meeting *Novos Talentos em Matemática* held by Fundação Calouste Gulbenkian (Lisbon, Portugal), in September 2004.

In both cases, most of the audience consisted of young undergraduates with little previous contact with measure theory, let alone ergodic theory. Thus, it was necessary to provide very friendly material that allowed such students to follow the main ideas to be presented. Still at that stage, our text was used by other colleagues, such as Vanderlei Horita (São José do Rio Preto, Brazil), for teaching mini-courses to audiences with a similar profile.

As the text evolved, we have tried to preserve this elementary character of the early chapters, especially Chapters 1 and 2, so that they can used independently of the rest of the book, with as few prerequisites as possible.

Starting from the mini-course we gave at the 2005 *Colóquio Brasileiro de Matemática* (IMPA, Rio de Janeiro), this project acquired a broader purpose. Gradually, we evolved towards trying to present in a consistent textbook format the material that, in our view, constitutes the core of ergodic theory. Inspired by our own research experience in this area, we endeavored to assemble in a unified presentation the ideas and facts upon which is built the remarkable development this field experienced over the last decades.

A main concern was to try and keep the text as self-contained as possible. Ergodic theory is based on several other mathematical disciplines, especially measure theory, topology and analysis. In the appendix, we have collected the main material from those disciplines that is used throughout the text. As a rule, proofs are omitted, since they can easily be found in many of the excellent references we provide. However, we do assume that the reader is familiar with the main tools of linear algebra, such as the canonical Jordan form.

Structure of the book

The main part of this book consists of 12 chapters, divided into sections and subsections, and one appendix, also divided into section and subsections. A list of exercises is given at the end of every section, appendix included. Statements (theorems, propositions, lemmas, corollaries, etc.), exercises and formulas are numbered by section and chapter: for instance, (2.3.7) is the seventh formula in the third section of the second chapter and Exercise A.5.1 is the first exercise in the fifth section of the appendix. Hints for selected exercises are given in special chapter after the appendix. At the end, we provide a list of references and an index.

Chapters 1 through 12 organized as follows:

- Chapters 1 through 4 constitute a kind of introductory cycle, in which we present the basic notions and facts in ergodic theory—invariance, recurrence and ergodicity—as well as some main examples. Chapter 3 introduces the fundamental results (ergodic theorems) upon which the whole theory is built.
- Chapter 4, where we introduce the key notion of ergodicity, is a turning point in our text. The next two chapters (Chapters 5 and 6) develop a couple of important related topics: decomposition of invariant measures into ergodic measures and systems admitting a unique, necessarily ergodic, invariant measure.
- Chapters 7 through 9 deal with very diverse subjects—loss of memory, the isomorphism problem and entropy—but they also form a coherent structure, built around the idea of considering increasingly chaotic systems: mixing, Lebesgue spectrum, Kolmogorov and Bernoulli systems.
- Chapter 9 is another turning point. As we introduce the fundamental concept of entropy, we take our time to present it to the reader from several different viewpoints. This is naturally articulated with the content of Chapter 10, where we develop the topological version of entropy, including an important generalization called pressure.
- In the two final chapters, 11 and 12, we focus on a specific class of dynamical systems, called expanding transformations, that allows us to exhibit a concrete (and spectacular!) application of many of the general ideas presented the text. This includes Ruelle's theorem and its applications, which we view as a natural climax of the book.

Appendices A.1 through A.2 cover several basic topics of measure and integration. Appendix A.3 deals with the special case of Borel measures in metric spaces. In Appendix A.4 we recall some basic facts from the theory of manifolds and smooth maps. Similarly, Appendices A.5 and A.6 cover some useful basic material about Banach spaces and Hilbert spaces. Finally, Appendix A.7 is devoted to the spectral theorem.

Examples and applications have a key part in any mathematical discipline and, perhaps, even more so in ergodic theory. For this reason, we devote special attention to presenting concrete situations that illustrate and put in perspective the general results. Such examples and constructions are introduced gradually, whenever the context seems better suited to highlight their relevance. They often return later in the text, to illustrate new fundamental concepts as we introduce them.

The exercises at the end of each section have a threefold purpose. There are routine exercises meant to help the reader become acquainted with the concepts and the results presented in the text. Also, we leave as exercises certain arguments and proofs that are not used in the sequel or belong to more elementary related areas, such as topology or measure theory. Finally, more sophisticated exercises test the reader's global understanding of the theory. For the reader's convenience, hints for selected exercises are given in a special chapter following the appendix.

How to use this book?

These comments are meant, primarily, for the reader who plans to use this book to teach a course. Appendices A.1 through A.7 provide quick references to background material. In principle, they are not meant to be presented in class.

The content of Chapters 1 through 12 is suitable for a one-year course, or a sequence of two one-semester courses. In either case, the reader should be able to cover most of the material, possibly reserving some topics for seminars given by the students. The following sections are especially suited for that:

Section 1.5, Section 2.5, Section 3.4, Section 4.4, Section 6.4, Section 7.3, Section 7.4, Section 8.3, Section 8.4, Section 8.5, Section 9.5, Section 9.7, Section 10.4, Section 10.5, Section 11.1, Section 11.3, Section 12.3 and Section 12.4.

In this format, Ruelle's theorem (Theorem 12.1) and its applications are a natural closure for the course.

In case only one semester is available, some selection of topics will be necessary. The authors' suggestion is to try and cover the following program:

Chapter 1: Sections 1.1, 1.2 and 1.3.
Chapter 2: Sections 2.1 and 2.2.
Chapter 3: Sections 3.1, 3.2 and 3.3.
Chapter 4: Sections 4.1, 4.2 and 4.3.
Chapter 5: Section 5.1 (mention Rokhlin's theorem).

Chapter 6: Sections 6.1, 6.2 and 6.3.
Chapter 7: Sections 7.1 and 7.2.
Chapter 8: Section 8.1 and 8.2 (mention Ornstein's theorem).
Chapter 9: Sections 9.1, 9.2, 9.3 and 9.4.
Chapter 10: Sections 10.1 and 10.2.
Chapter 11: Section 11.1.

In this format, the course could close either with the proof of the variational principle for the entropy (Theorem 10.1) or with the construction of absolutely continuous invariant measures for expanding maps on manifolds (Theorem 11.1.2).

We have designed the text in such a way as to make it feasible for the lecturer to focus on presenting the central ideas, leaving it to the student to study in detail many of the proofs and complementary results. Indeed, we devoted considerable effort to making the explanations as friendly as possible, detailing the arguments and including plenty of cross-references to previous related results as well to the definitions of the relevant notions.

In addition to the regular appearance of examples, we have often chosen to approach the same notion more than once, from different points of view, if that seemed useful for its in-depth understanding. The special chapter containing the hints for selected exercises is also part of that effort to encourage and facilitate the autonomous use of this book by the student.

Acknowledgments

The writing of this book extended for over a decade. During this period we benefitted from constructive criticism from several colleagues and students.

Many colleagues used different preliminary versions of the book to teach courses and shared their experiences with us. Besides Vanderlei Horita (São José do Rio Preto, Brazil), Nivaldo Muniz (São Luis, Brazil) and Meysam Nassiri (Teheran, Iran), we would like to thank Vítor Araújo (Salvador, Brazil) for an extended list of suggestions that influenced significantly the way the text evolved from then on. François Ledrappier (Paris, France) helped us with questions about substitution systems.

We also had the chance to test the material in a number of regular graduate courses at IMPA (Instituto de Matemática Pura e Aplicada) and at UFAL (Universidade Federal de Alagoas). Feedback from graduate students Adriana Sánchez, Aline Gomes Cerqueira, El Hadji Yaya Tall, Ermerson Araujo, Ignacio Atal, Rafael Lucena, Raphaël Cyna and Xiao-Chuan Liu allowed us to correct many of the weaknesses in earlier versions.

The first draft of Appendices A.1–A.2 was written by João Gouveia, Vítor Saraiva and Ricardo Andrade, who acted as assistants for the course in *Novos Talentos em Matemática 2004* mentioned previously. IMPA students Edileno de Almeida Santos, Felippe Soares Guimarães, Fernando Nera Lenarduzzi,

Ítalo Dowell Lira Melo, Marco Vinicius Bahi Aymone and Renan Henrique Finder wrote many of the hints for the exercises in Chapters 1 through 8 and the appendix.

The original Portuguese version of this book, *Fundamentos da Teoria Ergódica* [VO14], was published in 2014 by SBM (Sociedade Brasileira de Matemática). Feedback from colleagues who used that back to teach graduate courses in different places helped eliminate some of the remaining shortcomings. The extended list of remarks by Bernardo Lima (Belo Horizonte, Brazil) and his student Leonardo Guerini was particularly useful in this regard.

Several other changes and corrections were made in the course of the translation to English. We are grateful to the many colleagues and students who agreed to revise different parts of the translated text, especially Cristina Lizana, Elaís C. Malheiro, Fernando Nera Lenarduzzi, Jiagang Yang, Karina Marin, Lucas Backes, Maria João Resende, Mauricio Poletti, Paulo Varandas, Ricardo Turolla, Sina Türelli, Vanessa Ramos, Vítor Araújo and Xiao-Chuan Liu.

Marcelo Viana[1] and Krerley Oliveira[2]
Rio de Janeiro and Maceió, March 31, 2015

[1] IMPA, Estrada D. Castorina 110, 22460-320 Rio de Janeiro, Brasil. viana@impa.br.

[2] Departamento de Matemática, Universidade Federal de Alagoas, Campus A. C. Simões s/n, 57072-090 Maceió, Brasil. krerley@mat.ufal.br.

1
Recurrence

Ergodic theory studies the behavior of dynamical systems with respect to measures that remain invariant under time evolution. Indeed, it aims to describe those properties that are valid for the trajectories of almost all initial states of the system, that is, all but a subset that has zero weight for the invariant measure. Our first task, in Section 1.1, will be to explain what we mean by 'dynamical system' and 'invariant measure'.

The roots of the theory date back to the first half of the 19th century. By 1838, the French mathematician Joseph Liouville observed that every energy-preserving system in classical (Newtonian) mechanics admits a natural invariant volume measure in the space of configurations. Just a bit later, in 1845, the great German mathematician Carl Friedrich Gauss pointed out that the transformation

$$(0,1] \to \mathbb{R}, \quad x \mapsto \text{fractional part of } \frac{1}{x},$$

which has an important role in number theory, admits an invariant measure equivalent to the Lebesgue measure (in the sense that the two have the same zero measure sets). These are two of the examples of applications of ergodic theory that we discuss in Section 1.3. Many others are introduced throughout this book.

The first important result was found by the great French mathematician Henri Poincaré by the end of the 19th century. Poincaré was particularly interested in the motion of celestial bodies, such as planets and comets, which is described by certain differential equations originating from Newton's law of universal gravitation. Starting from Liouville's observation, Poincaré realized that for almost every initial state of the system, that is, almost every value of the initial position and velocity, the solution of the differential equation comes back arbitrarily close to that initial state, unless it goes to infinity. Even more, this *recurrence* property is not specific to (celestial) mechanics: it is shared by any dynamical system that admits a finite invariant measure. That is the theme of Section 1.2.

The same theme reappears in Section 1.5, in a more elaborate context: there, we deal with any finite number of dynamical systems commuting with each other, and we seek *simultaneous* returns of the orbits of all those systems to the neighborhood of the initial state. This kind of result has important applications in combinatorics and number theory, as we will see.

The recurrence phenomenon is also behind the constructions that we present in Section 1.4. The basic idea is to fix some positive measure subset of the domain and to consider the first return to that subset. This first-return transformation is often easier to analyze, and it may be used to shed much light on the behavior of the original transformation.

1.1 Invariant measures

Let $(M, \mathcal{B}, \mu)$ be a measure space and $f : M \to M$ be a measurable transformation. We say that the measure μ is *invariant* under f if

$$\mu(E) = \mu(f^{-1}(E)) \quad \text{for every measurable set } E \subset M. \tag{1.1.1}$$

We also say that μ is *f-invariant*, or that f *preserves* μ, to mean just the same. Notice that the definition (1.1.1) makes sense, since the pre-image of a measurable set under a measurable transformation is still a measurable set. Heuristically, the definition means that the probability that a point picked "at random" is in a given subset is equal to the probability that its image is in that subset.

It is possible, and convenient, to extend this definition to other types of dynamical systems, beyond transformations. We are especially interested in *flows*, that is, families of transformations $f^t : M \to M$, with $t \in \mathbb{R}$, satisfying the following conditions:

$$f^0 = \mathrm{id} \quad \text{and} \quad f^{s+t} = f^s \circ f^t \text{ for every } s, t \in \mathbb{R}. \tag{1.1.2}$$

In particular, each transformation f^t is invertible and the inverse is f^{-t}. Flows arise naturally in connection with differential equations of the form

$$\frac{d\gamma}{dt}(t) = X(\gamma(t))$$

in the following way: under suitable conditions on the vector field X, for each point x in the domain M there exists exactly one solution $t \mapsto \gamma_x(t)$ of the differential equation with $\gamma_x(0) = x$; then $f^t(x) = \gamma_x(t)$ defines a flow in M.

We say that a measure μ is *invariant* under a flow $(f^t)_t$ if it is invariant under each one of the transformations f^t, that is, if

$$\mu(E) = \mu(f^{-t}(E)) \quad \text{for every measurable set } E \subset M \text{ and } t \in \mathbb{R}. \tag{1.1.3}$$

Proposition 1.1.1. *Let $f : M \to M$ be a measurable transformation and μ be a measure on M. Then f preserves μ if and only if*

$$\int \phi \, d\mu = \int \phi \circ f \, d\mu \tag{1.1.4}$$

for every μ-integrable function $\phi : M \to \mathbb{R}$.

Proof. Suppose that the measure μ is invariant under f. We are going to show that the relation (1.1.4) is valid for increasingly broader classes of functions. Let $\mathcal{X}_B$ denote the characteristic function of any measurable subset B. Then

$$\mu(B) = \int \mathcal{X}_B \, d\mu \quad \text{and} \quad \mu(f^{-1}(B)) = \int \mathcal{X}_{f^{-1}(B)} \, d\mu = \int (\mathcal{X}_B \circ f) \, d\mu.$$

Thus, the hypothesis $\mu(B) = \mu(f^{-1}(B))$ means that (1.1.4) is valid for characteristic functions. Then, by linearity of the integral, (1.1.4) is valid for all simple functions. Next, given any integrable $\phi : M \to \mathbb{R}$, consider a sequence $(s_n)_n$ of simple functions, converging to ϕ and such that $|s_n| \le |\phi|$ for every n. That such a sequence exists is guaranteed by Proposition A.1.33. Then, using the dominated convergence theorem (Theorem A.2.11) twice:

$$\int \phi \, d\mu = \lim_n \int s_n \, d\mu = \lim_n \int (s_n \circ f) \, d\mu = \int (\phi \circ f) \, d\mu.$$

This shows that (1.1.4) holds for every integrable function if μ is invariant. The converse is also contained in the arguments we just presented.

1.1.1 Exercises

1.1.1. Let $f : M \to M$ be a measurable transformation. Show that a Dirac measure δ_p is invariant under f if and only if p is a fixed point of f. More generally, a probability measure $\delta_{p,k} = k^{-1}\big(\delta_p + \delta_{f(p)} + \cdots + \delta_{f^{k-1}(p)}\big)$ is invariant under f if and only if $f^k(p) = p$.

1.1.2. Prove the following version of Proposition 1.1.1. Let M be a metric space, $f : M \to M$ be a measurable transformation and μ be a measure on M. Show that f preserves μ if and only if

$$\int \phi \, d\mu = \int \phi \circ f \, d\mu$$

for every bounded continuous function $\phi : M \to \mathbb{R}$.

1.1.3. Prove that if $f : M \to M$ preserves a measure μ then, given any $k \ge 2$, the iterate f^k also preserves μ. Is the converse true?

1.1.4. Suppose that $f : M \to M$ preserves a probability measure μ. Let $B \subset M$ be a measurable set satisfying any one of the following conditions:
(a) $\mu(B \setminus f^{-1}(B)) = 0$;
(b) $\mu(f^{-1}(B) \setminus B) = 0$;
(c) $\mu(B \Delta f^{-1}(B)) = 0$;
(d) $f(B) \subset B$.
Show that there exists $C \subset M$ such that $f^{-1}(C) = C$ and $\mu(B \Delta C) = 0$.

1.1.5. Let $f : U \to U$ be a C^1 diffeomorphism on an open set $U \subset \mathbb{R}^d$. Show that the Lebesgue measure m is invariant under f if and only if $|\det Df| \equiv 1$.

1.2 Poincaré recurrence theorem

We are going to study two versions of Poincaré's theorem. The first one (Section 1.2.1) is formulated in the context of (finite) measure spaces. The theorem of Kač, that we state and prove in Section 1.2.2, provides a quantitative complement to that statement. The second version of the recurrence theorem (Section 1.2.3) assumes that the ambient is a topological space with certain additional properties. We will also prove a third version of the recurrence theorem, due to Birkhoff, whose statement is purely topological.

1.2.1 Measurable version

Our first result asserts that, given any *finite* invariant measure, almost every point in any positive measure set E returns to E an infinite number of times:

Theorem 1.2.1 (Poincaré recurrence). *Let $f : M \to M$ be a measurable transformation and μ be a finite measure invariant under f. Let $E \subset M$ be any measurable set with $\mu(E) > 0$. Then, for μ-almost every point $x \in E$ there exist infinitely many values of n for which $f^n(x)$ is also in E.*

Proof. Denote by E_0 the set of points $x \in E$ that never return to E. As a first step, let us prove that E_0 has zero measure. To this end, let us observe that the pre-images $f^{-n}(E_0)$ are pairwise disjoint. Indeed, suppose there exist $m > n \geq 1$ such that $f^{-m}(E_0)$ intersects $f^{-n}(E_0)$. Let x be a point in the intersection and $y = f^n(x)$. Then $y \in E_0$ and $f^{m-n}(y) = f^m(x) \in E_0$. Since $E_0 \subset E$, this means that y returns to E at least once, which contradicts the definition of E_0. This contradiction proves that the pre-images are pairwise disjoint, as claimed.

Since μ is invariant, we also have that $\mu(f^{-n}(E_0)) = \mu(E_0)$ for all $n \geq 1$. It follows that

$$\mu\left(\bigcup_{n=1}^{\infty} f^{-n}(E_0)\right) = \sum_{n=1}^{\infty} \mu(f^{-n}(E_0)) = \sum_{n=1}^{\infty} \mu(E_0).$$

The expression on the left-hand side is finite, since the measure μ is assumed to be finite. On the right-hand side we have a sum of infinitely many terms that are all equal. The only way such a sum can be finite is if the terms vanish. So, $\mu(E_0) = 0$ as claimed.

Now let us denote by F the set of points $x \in E$ that return to E a finite number of times. It is clear from the definition that every point $x \in F$ has some iterate

$f^k(x)$ in E_0. In other words,

$$F \subset \bigcup_{k=0}^{\infty} f^{-k}(E_0).$$

Since $\mu(E_0) = 0$ and μ is invariant, it follows that

$$\mu(F) \le \mu\Big(\bigcup_{k=0}^{\infty} f^{-k}(E_0)\Big) \le \sum_{k=0}^{\infty} \mu(f^{-k}(E_0)) = \sum_{k=0}^{\infty} \mu(E_0) = 0.$$

Thus, $\mu(F) = 0$ as we wanted to prove.

Theorem 1.2.1 implies an analogous result for continuous time systems: *if μ is a finite invariant measure of a flow $(f^t)_t$ then for every measurable set $E \subset M$ with positive measure and for μ-almost every $x \in E$ there exist times $t_j \to +\infty$ such that $f^{t_j}(x) \in E$.* Indeed, if μ is invariant under the flow then, in particular, it is invariant under the so-called *time-1 map* f^1. So, the statement we just made follows immediately from Theorem 1.2.1 applied to f^1 (the times t_j one finds in this way are integers). Similar observations apply to the other versions of the recurrence theorem that we present in the sequel.

On the other hand, the theorem in the next section is specific to discrete time systems.

1.2.2 Kač theorem

Let $f : M \to M$ be a measurable transformation and μ be a finite measure invariant under f. Let $E \subset M$ be any measurable set with $\mu(E) > 0$. Consider the *first-return time* function $\rho_E : E \to \mathbb{N} \cup \{\infty\}$, defined by

$$\rho_E(x) = \min\{n \ge 1 : f^n(x) \in E\} \tag{1.2.1}$$

if the set on the right-hand side is non-empty and $\rho_E(x) = \infty$ if, on the contrary, x has no iterate in E. According to Theorem 1.2.1, the second alternative occurs only on a set with zero measure.

The next result shows that this function is integrable and even provides the value of the integral. For the statement we need the following notation:

$$E_0 = \{x \in E : f^n(x) \notin E \text{ for every } n \ge 1\} \quad \text{and}$$
$$E_0^* = \{x \in M : f^n(x) \notin E \text{ for every } n \ge 0\}.$$

In other words, E_0 is the set of points in E that never *return* to E and E_0^* is the set of points in M that never *enter* E. We have seen in Theorem 1.2.1 that $\mu(E_0) = 0$.

Theorem 1.2.2 (Kač). *Let $f : M \to M$ be a measurable transformation, μ be a finite invariant measure and $E \subset M$ be a positive measure set. Then the function*

ρ_E is integrable and

$$\int_E \rho_E \, d\mu = \mu(M) - \mu(E_0^*).$$

Proof. For each $n \ge 1$, define

$$E_n = \{x \in E : f(x) \notin E, \dots, f^{n-1}(x) \notin E, \text{ but } f^n(x) \in E\} \quad \text{and}$$
$$E_n^* = \{x \in M : x \notin E, f(x) \notin E, \dots, f^{n-1}(x) \notin E, \text{ but } f^n(x) \in E\}.$$

That is, E_n is the set of points of E that return to E for the first time exactly at time n,

$$E_n = \{x \in E : \rho_E(x) = n\},$$

and E_n^* is the set points that are not in E and enter E for the first time exactly at time n. It is clear that these sets are measurable and, hence, ρ_E is a measurable function. Moreover, the sets E_n, E_n^*, $n \ge 0$ constitute a *partition* of the ambient space: they are pairwise disjoint and their union is the whole of M. So,

$$\mu(M) = \sum_{n=0}^{\infty} \big(\mu(E_n) + \mu(E_n^*)\big) = \mu(E_0^*) + \sum_{n=1}^{\infty} \big(\mu(E_n) + \mu(E_n^*)\big). \tag{1.2.2}$$

Now observe that

$$f^{-1}(E_n^*) = E_{n+1}^* \cup E_{n+1} \quad \text{for every } n. \tag{1.2.3}$$

Indeed, $f(y) \in E_n^*$ means that the first iterate of $f(y)$ that belongs to E is $f^n(f(y)) = f^{n+1}(y)$ and that occurs if and only if $y \in E_{n+1}^*$ or else $y \in E_{n+1}$. This proves the equality (1.2.3). So, given that μ is invariant,

$$\mu(E_n^*) = \mu(f^{-1}(E_n^*)) = \mu(E_{n+1}^*) + \mu(E_{n+1}) \quad \text{for every } n.$$

Applying this relation successively, we find that

$$\mu(E_n^*) = \mu(E_m^*) + \sum_{i=n+1}^{m} \mu(E_i) \quad \text{for every } m > n. \tag{1.2.4}$$

The relation (1.2.2) implies that $\mu(E_m^*) \to 0$ when $m \to \infty$. So, taking the limit as $m \to \infty$ in the equality (1.2.4), we find that

$$\mu(E_n^*) = \sum_{i=n+1}^{\infty} \mu(E_i). \tag{1.2.5}$$

To complete the proof, replace (1.2.5) in the equality (1.2.2). In this way we find that

$$\mu(M) - \mu(E_0^*) = \sum_{n=1}^{\infty} \Big(\sum_{i=n}^{\infty} \mu(E_i)\Big) = \sum_{n=1}^{\infty} n\mu(E_n) = \int_E \rho_E \, d\mu,$$

as we wanted to prove.

In some cases, for example when the system (f,μ) is *ergodic* (this property will be defined and studied later, starting from Chapter 4), the set E_0^* has zero measure. Then the conclusion of the Kač theorem means that

$$\frac{1}{\mu(E)}\int_E \rho_E\, d\mu = \frac{\mu(M)}{\mu(E)} \tag{1.2.6}$$

for every measurable set E with positive measure. The left-hand side of this expression is the *mean return time* to E. So, (1.2.6) asserts that *the mean return time is inversely proportional to the measure of E.*

Remark 1.2.3. By definition, $E_n^* = f^{-n}(E) \setminus \bigcup_{k=0}^{n-1} f^{-k}(E)$. So, the fact that the sum (1.2.2) is finite implies that the measure of E_n^* converges to zero when $n \to \infty$. This fact will be useful later.

1.2.3 Topological version

Now let us suppose that M is a topological space, endowed with its Borel σ-algebra $\mathcal{B}$. A point $x \in M$ is *recurrent* for a transformation $f : M \to M$ if there exists a sequence $n_j \to \infty$ of natural numbers such that $f^{n_j}(x) \to x$. Analogously, we say that $x \in M$ is recurrent for a flow $(f^t)_t$ if there exists a sequence $t_j \to +\infty$ of real numbers such that $f^{t_j}(x) \to x$ when $j \to \infty$.

In the next theorem we assume that the topological space M admits a countable basis of open sets, that is, there exists a countable family $\{U_k : k \in \mathbb{N}\}$ of open sets such that every open subset of M may be written as a union of elements U_k of this family. This condition holds in most interesting examples.

Theorem 1.2.4 (Poincaré recurrence)**.** *Suppose that M admits a countable basis of open sets. Let $f : M \to M$ be a measurable transformation and μ be a finite measure on M invariant under f. Then, μ-almost every $x \in M$ is recurrent for f.*

Proof. For each k, denote by $\tilde{U}_k$ the set of points $x \in U_k$ that never return to U_k. According to Theorem 1.2.1, every $\tilde{U}_k$ has zero measure. Consequently, the countable union

$$\tilde{U} = \bigcup_{k\in\mathbb{N}} \tilde{U}_k$$

also has zero measure. Hence, to prove the theorem it suffices to check that every point x that is not in $\tilde{U}$ is recurrent. That is easy, as we are going to see. Consider $x \in M \setminus \tilde{U}$ and let U be any neighborhood of x. By definition, there exists some element U_k of the basis of open sets such that $x \in U_k$ and $U_k \subset U$. Since x is not in $\tilde{U}$, we also have that $x \notin \tilde{U}_k$. In other words, there exists $n \geq 1$ such that $f^n(x)$ is in U_k. In particular, $f^n(x)$ is also in U. Since the neighborhood U is arbitrary, this proves that x is a recurrent point.

Let us point out that the conclusions of Theorems 1.2.1 and 1.2.4 are false, in general, if the measure μ is not finite:

Example 1.2.5. Let $f : \mathbb{R} \to \mathbb{R}$ be the translation by 1, that is, the transformation defined by $f(x) = x + 1$ for every $x \in \mathbb{R}$. It is easy to check that f preserves the Lebesgue measure on $\mathbb{R}$ (which is infinite). On the other hand, no point $x \in \mathbb{R}$ is recurrent for f. According to the recurrence theorem, this last observation implies that f can not admit any finite invariant measure.

However, it is possible to extend these statements for certain cases of infinite measures: see Exercise 1.2.2.

To conclude, we present a purely topological version of Theorem 1.2.4, called the Birkhoff recurrence theorem, that makes no reference at all to invariant measures:

Theorem 1.2.6 (Birkhoff recurrence). *If $f : M \to M$ is a continuous transformation on a compact metric space M then there exists some point $x \in X$ that is recurrent for f.*

Proof. Consider the family $\mathcal{I}$ of all non-empty closed sets $X \subset M$ that are invariant under f, in the sense that $f(X) \subset X$. This family is non-empty, since $M \in \mathcal{I}$. We claim that an element $X \in \mathcal{I}$ is minimal for the inclusion relation if and only if the orbit of every $x \in X$ is dense in X. Indeed, it is clear that if X is a closed invariant subset then X contains the closure of the orbit of each one of its elements. Hence, in order to be minimal, X must coincide with every one of these closures. Conversely, for the same reason, if X coincides with the orbit closure of each one of its points then it has no proper subset that is closed and invariant. That is, X is minimal. This proves our claim. In particular, every point x in a minimal set is recurrent. Therefore, to prove the theorem it suffices to prove that there exists some minimal set.

We claim that every totally ordered set $\{X_\alpha\} \subset \mathcal{I}$ admits a lower bound. Indeed, consider $X = \bigcap_\alpha X_\alpha$. Observe that X is non-empty, since the X_α are compact and they form a totally ordered family. It is clear that X is closed and invariant under f and it is equally clear that X is a lower bound for the set $\{X_\alpha\}$. That proves our claim. Now it follows from Zorn's lemma that $\mathcal{I}$ does contain minimal elements.

Theorem 1.2.6 can also be deduced from Theorem 1.2.4 together with the fact, which we will prove later (in Chapter 2), that every continuous transformation on a compact metric space admits some invariant probability measure.

1.2.4 Exercises

1.2.1. Show that the following statement is equivalent to Theorem 1.2.1, meaning that each one of them can be obtained from the other. Let $f : M \to M$ be a measurable transformation and μ be a finite invariant measure. Let $E \subset M$ be any measurable set with $\mu(E) > 0$. Then there exists $N \geq 1$ and a positive measure set $D \subset E$ such that $f^N(x) \in E$ for every $x \in D$.

1.2.2. Let $f : M \to M$ be an invertible transformation and suppose that μ is an invariant measure, not necessarily finite. Let $B \subset M$ be a set with finite measure. Prove that, given any measurable set $E \subset M$ with positive measure, μ-almost every point $x \in E$ either returns to E an infinite number of times or has only a finite number of iterates in B.

1.2.3. Let $f : M \to M$ be an invertible transformation and suppose that μ is a *σ-finite* invariant measure: there exists an increasing sequence of measurable subsets M_k with $\mu(M_k) < \infty$ for every k and $\bigcup_k M_k = M$. We say that a point x *goes to infinity* if, for every k, there exists only a finite number of iterates of x that are in M_k. Show that, given any $E \subset M$ with positive measure, μ-almost every point $x \in E$ returns to E an infinite number of times or else goes to infinity.

1.2.4. Let $f : M \to M$ be a measurable transformation, not necessarily invertible, μ be an invariant probability measure and $D \subset M$ be a set with positive measure. Prove that almost every point of D spends a *positive fraction of time* in D:

$$\limsup_n \frac{1}{n}\#\{0 \leq j \leq n-1 : f^j(x) \in D\} > 0$$

for μ-almost every $x \in D$. [Note: One may replace lim sup by lim inf in the statement, but the proof of that fact will have to wait until Chapter 3.]

1.2.5. Let $f : M \to M$ be a measurable transformation preserving a finite measure μ. Given any measurable set $A \subset M$ with $\mu(A) > 0$, let $n_1 < n_2 < \cdots$ be the sequence of values of n such that $\mu(f^{-n}(A) \cap A) > 0$. The goal of this exercise is to prove that $V_A = \{n_1, n_2, \ldots\}$ is a *syndetic*, that is, that there exists $C > 0$ such that $n_{i+1} - n_i \leq C$ for every i.

(a) Show that for any increasing sequence $k_1 < k_2 < \cdots$ there exist $j > i \geq 1$ such that $\mu(A \cap f^{-(k_j - k_i)}(A)) > 0$.

(b) Given any infinite sequence $\ell = (l_j)_j$ of natural numbers, denote by $S(\ell)$ the set of all finite sums of consecutive elements of ℓ. Show that V_A intersects $S(\ell)$ for every ℓ.

(c) Deduce that the set V_A is syndetic.

[Note: Exercise 3.1.2 provides a different proof of this fact.]

1.2.6. Show that if $f : [0,1] \to [0,1]$ is a measurable transformation preserving the Lebesgue measure m then m-almost every point $x \in [0,1]$ satisfies

$$\liminf_n n|f^n(x) - x| \leq 1.$$

[Note: Boshernitzan [Bos93] proved a much more general result, namely that $\liminf_n n^{1/d} d(f^n(x), x) < \infty$ for μ-almost every point and every probability measure μ invariant under $f : M \to M$, assuming M is a separable metric whose d-dimensional Hausdorff measure is σ-finite.]

1.2.7. Define $f : [0,1] \to [0,1]$ by $f(x) = (x + \omega) - [x + \omega]$, where ω represents the *golden ratio* $(1 + \sqrt{5})/2$. Given $x \in [0,1]$, check that $n|f^n(x) - x| = n^2|\omega - q_n|$ for every n, where $(q_n)_n \to \omega$ is the sequence of rational numbers given by $q_n = [x + n\omega]/n$. Using that the roots of the polynomial $R(z) = z^2 - z - 1$ are precisely ω and $\omega - \sqrt{5}$, prove that $\liminf_n n^2|\omega - q_n| \geq 1/\sqrt{5}$. [Note: This shows that the constant 1 in Exercise 1.2.6 cannot be replaced by any constant smaller than

$1/\sqrt{5}$. It is not known whether 1 is the smallest constant such that the statement holds for *every* transformation on the interval.]

1.3 Examples

Next, we describe some simple examples of invariant measures for transformations and flows that help us interpret the significance of the Poincaré recurrence theorems and also lead to some interesting conclusions.

1.3.1 Decimal expansion

Our first example is the transformation defined on the interval $[0,1]$ in the following way:

$$f : [0,1] \to [0,1], \quad f(x) = 10x - [10x].$$

Here and in what follows, we use $[y]$ as the *integer part* of a real number y, that is, the largest integer smaller than or equal y. So, f is the map sending each $x \in [0,1]$ to the *fractional part* of $10x$. Figure 1.1 represents the graph of f.

We claim that the Lebesgue measure μ on the interval is invariant under the transformation f, that is, it satisfies

$$\mu(E) = \mu(f^{-1}(E)) \quad \text{for every measurable set } E \subset M. \tag{1.3.1}$$

This can be checked as follows. Let us begin by supposing that E is an interval. Then, as illustrated in Figure 1.1, its pre-image $f^{-1}(E)$ consists of ten intervals, each of which is ten times shorter than E. Hence, the Lebesgue measure of $f^{-1}(E)$ is equal to the Lebesgue measure of E. This proves that (1.3.1) does hold in the case of intervals. As a consequence, it also holds when E is a finite

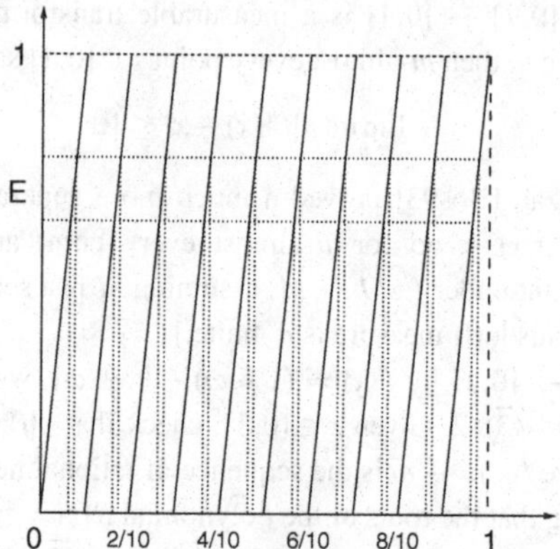

Figure 1.1. Fractional part of $10x$

union of intervals. Now, the family of all finite unions of intervals is an algebra that generates the Borel σ-algebra of $[0,1]$. Hence, to conclude the proof it is enough to use the following general fact:

Lemma 1.3.1. *Let $f : M \to M$ be a measurable transformation and μ be a finite measure on M. Suppose that there exists some algebra $\mathcal{A}$ of measurable subsets of M such that $\mathcal{A}$ generates the σ-algebra $\mathcal{B}$ of M and $\mu(E) = \mu(f^{-1}(E))$ for every $E \in \mathcal{A}$. Then the latter remains true for every set $E \in \mathcal{B}$, that is, the measure μ is invariant under f.*

Proof. We start by proving that $\mathcal{C} = \{E \in \mathcal{B} : \mu(E) = \mu(f^{-1}(E))\}$ is a monotone class. Let $E_1 \subset E_2 \subset \cdots$ be any increasing sequence of elements of $\mathcal{C}$ and let $E = \bigcup_{i=1}^{\infty} E_i$. By Theorem A.1.14 (see Exercise A.1.9),

$$\mu(E) = \lim_i \mu(E_i) \quad \text{and} \quad \mu(f^{-1}(E)) = \lim_i \mu(f^{-1}(E_i)).$$

So, using the fact that $E_i \in \mathcal{C}$,

$$\mu(E) = \lim_i \mu(E_i) = \lim_i \mu(f^{-1}(E_i)) = \mu(f^{-1}(E)).$$

Hence, $E \in \mathcal{C}$. In precisely the same way, one gets that the intersection of any decreasing sequence of elements of $\mathcal{C}$ is in $\mathcal{C}$. This proves that $\mathcal{C}$ is indeed a monotone class.

Now it is easy to deduce the conclusion of the lemma. Indeed, since $\mathcal{C}$ is assumed to contain $\mathcal{A}$, we may use the monotone class theorem (Theorem A.1.18), to conclude that $\mathcal{C}$ contains the σ-algebra $\mathcal{B}$ generated by $\mathcal{A}$. That is precisely what we wanted to prove.

Now we explain how one may use the fact that the Lebesgue measure is invariant under f, together with the Poincaré recurrence theorem, to reach some interesting conclusions. The transformation f is directly related to the usual decimal expansion of a real number: if x is given by

$$x = 0.a_0 a_1 a_2 a_3 \cdots$$

with $a_i \in \{0,1,2,3,4,5,6,7,8,9\}$ and $a_i \neq 9$ for infinitely many values of i, then its image under f is given by

$$f(x) = 0.a_1 a_2 a_3 \cdots .$$

Thus, more generally, the n-th iterate of f can be expressed in the following way, for every $n \geq 1$:

$$f^n(x) = 0.a_n a_{n+1} a_{n+2} \cdots \tag{1.3.2}$$

Let E be the subset of points $x \in [0,1]$ whose decimal expansion starts with the digit 7, that is, such that $a_0 = 7$. According to Theorem 1.2.1, almost every element in E has infinitely many iterates that are also in E. By the expression (1.3.2), this means that there are infinitely many values of n such that $a_n = 7$.

So, we have shown that *almost every number x whose decimal expansion starts with 7 has infinitely many digits equal to* 7.

Of course, instead of 7 we may consider any other digit. Even more, there is a similar result (see Exercise 1.3.2) when, instead of a single digit, one considers a block of $k \geq 1$ consecutive digits. Later on, in Chapter 3, we will prove a much stronger fact: for almost every number $x \in [0,1]$, every digit occurs with frequency $1/10$ (more generally, every block of $k \geq 1$ digits occurs with frequency $1/10^k$) in the decimal expansion of x.

1.3.2 Gauss map

The system we present in this section is related to another important algorithm in number theory, the *continued fraction expansion*, which plays a central role in the problem of finding the best rational approximation to any real number. Let us start with a brief presentation of this algorithm.

Given any number $x_0 \in (0,1)$, let

$$a_1 = \left[\frac{1}{x_0}\right] \quad \text{and} \quad x_1 = \frac{1}{x_0} - a_1.$$

Note that a_1 is a natural number, $x_1 \in [0,1)$ and

$$x_0 = \frac{1}{a_1 + x_1}.$$

Supposing that x_1 is different from zero, we may repeat this procedure, defining

$$a_2 = \left[\frac{1}{x_1}\right] \quad \text{and} \quad x_2 = \frac{1}{x_1} - a_2.$$

Then

$$x_1 = \frac{1}{a_1 + x_2} \quad \text{and so} \quad x_0 = \cfrac{1}{a_1 + \cfrac{1}{a_2 + x_2}}.$$

Now we may proceed by induction: for each $n \geq 1$ such that $x_{n-1} \in (0,1)$, define

$$a_n = \left[\frac{1}{x_{n-1}}\right] \quad \text{and} \quad x_n = \frac{1}{x_{n-1}} - a_n = G(x_{n-1}),$$

and observe that

$$x_0 = \cfrac{1}{a_1 + \cfrac{1}{a_2 + \cfrac{1}{\cdots + \cfrac{1}{a_n + x_n}}}}. \tag{1.3.3}$$

It can be shown that the sequence

$$z_n = \cfrac{1}{a_1 + \cfrac{1}{a_2 + \cfrac{1}{\cdots + \cfrac{1}{a_n}}}} \tag{1.3.4}$$

converges to x_0 when $n \to \infty$. This is usually expressed through the expression

$$x_0 = \cfrac{1}{a_1 + \cfrac{1}{a_2 + \cfrac{1}{\cdots + \cfrac{1}{a_n + \cfrac{1}{\cdots}}}}}, \tag{1.3.5}$$

which is called *continued fraction expansion* of x_0.

Note that the sequence $(z_n)_n$ defined by the relation (1.3.4) consists of rational numbers. Indeed, one can show that these are the *best rational approximations* of the number x_0, in the sense that each z_n is closer to x_0 than any other rational number whose denominator is smaller than or equal to the denominator of z_n (written in irreducible form). Observe also that to obtain (1.3.5) we had to assume that $x_n \in (0,1)$ for every $n \in \mathbb{N}$. If in the course of the process one encounters some $x_n = 0$, then the algorithm halts and we consider (1.3.3) to be the continued fraction expansion of x_0. It is clear that this can happen only if x_0 itself is a rational number.

This continued fraction algorithm is intimately related to a certain dynamical system on the interval $[0,1]$ that we describe in the following. The *Gauss map* $G : [0,1] \to [0,1]$ is defined by

$$G(x) = \frac{1}{x} - \left[\frac{1}{x}\right] = \text{fractional part of } 1/x,$$

if $x \in (0,1]$ and $G(0) = 0$. The graph of G can be easily sketched (see Figure 1.2), starting from the following observation: for every x in each interval $I_k = (1/(k+1), 1/k]$, the integer part of $1/x$ is equal to k and so $G(x) = 1/x - k$.

The continued fraction expansion of any number $x_0 \in (0,1)$ can be obtained from the Gauss map in the following way: for each $n \geq 1$, the natural number a_n is determined by

$$G^{n-1}(x_0) \in I_{a_n},$$

and the real number x_n is simply the n-th iterate $G^n(x_0)$ of the point x_0. This process halts whenever we encounter some $x_n = 0$; as we explained previously, this can only happen if x_0 is a rational number (see Exercise 1.3.4). In particular, there exists a full Lebesgue measure subset of $(0,1)$ such that all the iterates of G are defined for all the points in that subset.

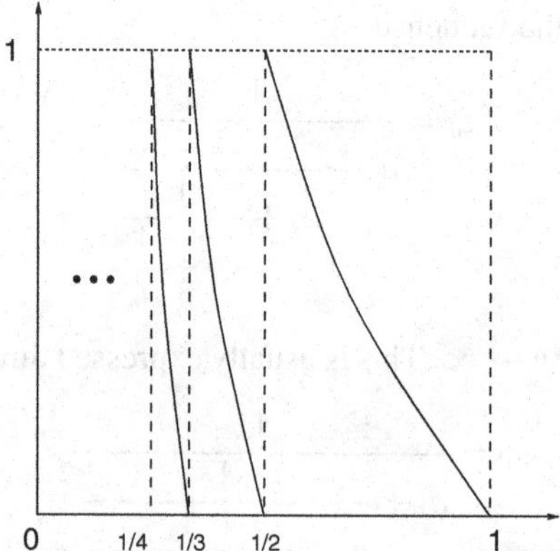

Figure 1.2. Gauss map

A remarkable fact that makes this transformation interesting from the point of view of ergodic theory is that G admits an invariant probability measure that, in addition, is equivalent to the Lebesgue measure on the interval. Indeed, consider the measure defined by

$$\mu(E) = \int_E \frac{c}{1+x}\, dx \quad \text{for every measurable set } E \subset [0,1], \tag{1.3.6}$$

where c is a positive constant. The integral is well defined, since the function in the integral is continuous on the interval $[0,1]$. Moreover, this function takes values inside the interval $[c/2, c]$ and that implies

$$\frac{c}{2}\, m(E) \le \mu(E) \le c\, m(E) \quad \text{for every measurable set } E \subset [0,1]. \tag{1.3.7}$$

In particular, μ is indeed equivalent to the Lebesgue measure m.

Proposition 1.3.2. *The measure μ is invariant under G. Moreover, if we choose $c = 1/\log 2$ then μ is a probability measure.*

Proof. We are going to use the following lemma:

Lemma 1.3.3. *Let $f : [0,1] \to [0,1]$ be a transformation such that there exist pairwise disjoint open intervals $I_1, I_2, \dots$ satisfying*

1. *the union $\bigcup_k I_k$ has full Lebesgue measure in $[0,1]$ and*
2. *the restriction $f_k = f \mid I_k$ to each I_k is a diffeomorphism onto $(0,1)$.*

Let $\rho : [0,1] \to [0,\infty)$ be an integrable function (relative to the Lebesgue measure) satisfying

$$\rho(y) = \sum_{x \in f^{-1}(y)} \frac{\rho(x)}{|f'(x)|} \tag{1.3.8}$$

for almost every $y \in [0,1]$. Then the measure $\mu = \rho dx$ is invariant under f.

Proof. Let $\phi = \chi_E$ be the characteristic function of an arbitrary measurable set $E \subset [0,1]$. Changing variables in the integral,

$$\int_{I_k} \phi(f(x))\rho(x)\,dx = \int_0^1 \phi(y)\rho(f_k^{-1}(y))|(f_k^{-1})'(y)|\,dy.$$

Note that $(f_k^{-1})'(y) = 1/f'(f_k^{-1}(y))$. So, the previous relation implies that

$$\begin{aligned}\int_0^1 \phi(f(x))\rho(x)\,dx &= \sum_{k=1}^{\infty} \int_{I_k} \phi(f(x))\rho(x)\,dx \\ &= \sum_{k=1}^{\infty} \int_0^1 \phi(y) \frac{\rho(f_k^{-1}(y))}{|f'(f_k^{-1}(y))|}\,dy.\end{aligned} \tag{1.3.9}$$

Using the monotone convergence theorem (Theorem A.2.9) and the hypothesis (1.3.8), we see that the last expression in (1.3.9) is equal to

$$\int_0^1 \phi(y) \sum_{k=1}^{\infty} \frac{\rho(f_k^{-1}(y))}{|f'(f_k^{-1}(y))|}\,dy = \int_0^1 \phi(y)\rho(y)\,dy.$$

In this way we find that $\int_0^1 \phi(f(x))\rho(x)\,dx = \int_0^1 \phi(y)\rho(y)\,dy$. Since $\mu = \rho dx$ and $\phi = \mathcal{X}_E$, this means that $\mu(f^{-1}(E)) = \mu(E)$ for every measurable set $E \subset [0,1]$. In other words, μ is invariant under f.

To conclude the proof of Proposition 1.3.2 we must show that the condition (1.3.8) holds for $\rho(x) = c/(1+x)$ and $f = G$. Let G_k denote the restriction of G to the interval $I_k = (1/(k+1), 1/k)$, for $k \geq 1$. Note that $G_k^{-1}(y) = 1/(y+k)$ for every k. Note also that $G'(x) = (1/x)' = -1/x^2$ for every $x \neq 0$. Therefore,

$$\sum_{k=1}^{\infty} \frac{\rho(G_k^{-1}(y))}{|G'(G_k^{-1}(y))|} = \sum_{k=1}^{\infty} \frac{c(y+k)}{y+k+1}\left(\frac{1}{y+k}\right)^2 = \sum_{k=1}^{\infty} \frac{c}{(y+k)(y+k+1)}. \tag{1.3.10}$$

Observing that

$$\frac{1}{(y+k)(y+k+1)} = \frac{1}{y+k} - \frac{1}{y+k+1},$$

we see that the last sum in (1.3.10) has a telescopic structure: except for the first one, all the terms occur twice, with opposite signs, and so they cancel out. This means that the sum is equal to the first term:

$$\sum_{k=1}^{\infty} \frac{c}{(y+k)(y+k+1)} = \frac{c}{y+1} = \rho(y).$$

This proves that the equality (1.3.8) is indeed satisfied and, hence, we may use Lemma 1.3.1 to conclude that μ is invariant under f.

Finally, observing that $c\log(1+x)$ is a primitive of the function $\rho(x)$, we find that

$$\mu([0,1]) = \int_0^1 \frac{c}{1+x}\,dx = c\log 2.$$

So, picking $c = 1/\log 2$ ensures that μ is a probability measure.

This proposition allows us to use ideas from ergodic theory, applied to the Gauss map, to obtain interesting conclusions in number theory. For example (see Exercise 1.3.3), the natural number 7 occurs infinitely many times in the continued fraction expansion of almost every number $x_0 \in (1/8, 1/7)$, that is, one has $a_n = 7$ for infinitely many values of $n \in \mathbb{N}$. Later on, in Chapter 3, we will prove a much more precise statement, that contains the following conclusion: for almost every $x_0 \in (0,1)$ the number 7 occurs with frequency

$$\frac{1}{\log 2}\log\frac{64}{63}$$

in the continued fraction expansion of x_0. Try to guess right away where this number comes from!

1.3.3 Circle rotations

Let us consider on the real line $\mathbb{R}$ the equivalence relation $\sim$ that identifies any numbers whose difference is an integer number:

$$x \sim y \quad \Leftrightarrow \quad x - y \in \mathbb{Z}.$$

We represent by $[x] \in \mathbb{R}/\mathbb{Z}$ the equivalence class of each $x \in \mathbb{R}$ and denote by $\mathbb{R}/\mathbb{Z}$ the space of all equivalence classes. This space is called the *circle* and is also denoted by S^1. The reason for this terminology is that $\mathbb{R}/\mathbb{Z}$ can be identified in a natural way with the unit circle $\{z \in \mathbb{C} : |z| = 1\}$ on the complex plane, by means of the map

$$\phi : \mathbb{R}/\mathbb{Z} \to \{z \in \mathbb{C} : |z| = 1\}, \quad [x] \mapsto e^{2\pi x i}. \tag{1.3.11}$$

Note that ϕ is well defined: since the function $x \mapsto e^{2\pi x i}$ is periodic of period 1, the expression $e^{2\pi x i}$ does not depend on the choice of a representative x for the class $[x]$. Moreover, ϕ is a bijection.

The circle $\mathbb{R}/\mathbb{Z}$ inherits from the real line $\mathbb{R}$ the structure of an abelian group, given by the operation

$$[x] + [y] = [x + y].$$

Observe that this is well defined: the equivalence class on the right-hand side does not depend on the choice of representatives x and y for the classes on the left-hand side. Given $\theta \in \mathbb{R}$, we call *rotation* of angle θ the transformation

$$R_\theta : \mathbb{R}/\mathbb{Z} \to \mathbb{R}/\mathbb{Z}, \quad [x] \mapsto [x + \theta] = [x] + [\theta].$$

Note that R_θ corresponds, via the identification (1.3.11), to the transformation $z \mapsto e^{2\pi\theta i}z$ on $\{z \in \mathbb{C} : |z| = 1\}$. The latter is just the restriction to the unit circle of the rotation of angle $2\pi\theta$ around the origin in the complex plane. It is clear from the definition that R_0 is the identity map and $R_\theta \circ R_\tau = R_{\theta+\tau}$ for every θ and τ. In particular, every R_θ is invertible and the inverse is $R_{-\theta}$.

We can also endow S^1 with a natural structure of a probability space, as follows. Let $\pi : \mathbb{R} \to S^1$ be the canonical projection, that assigns to each $x \in \mathbb{R}$ its equivalence class $[x]$. We say that a set $E \subset S^1$ is measurable if $\pi^{-1}(E)$ is a measurable subset of the real line. Next, let m be the Lebesgue measure on the real line. We define the *Lebesgue measure* μ on the circle to be given by

$$\mu(E) = m\big(\pi^{-1}(E) \cap [k, k+1)\big) \quad \text{for every } k \in \mathbb{Z}.$$

Note that the left-hand side of this equality does not depend on k, since, by definition, $\pi^{-1}(E) \cap [k, k+1) = \big(\pi^{-1}(E) \cap [0, 1)\big) + k$ and the measure m is invariant under translations.

It is clear from the definition that μ is a probability. Moreover, μ is invariant under every rotation R_θ (according to Exercise 1.3.8, it is the only probability measure with this property). This can be shown as follows. By definition, $\pi^{-1}(R_\theta^{-1}(E)) = \pi^{-1}(E) - \theta$ for every measurable set $E \subset S^1$. Let k be the integer part of θ. Since m is invariant under all the translations,

$$\begin{aligned} m\big((\pi^{-1}(E) - \theta) \cap [0, 1)\big) &= m\big(\pi^{-1}(E) \cap [\theta, \theta+1)\big) \\ &= m\big(\pi^{-1}(E) \cap [\theta, k+1)\big) + m\big(\pi^{-1}(E) \cap [k+1, \theta+1)\big). \end{aligned}$$

Note that $\pi^{-1}(E) \cap [k+1, \theta+1) = \big(\pi^{-1}(E) \cap [k, \theta)\big) + 1$. So, the expression on the right-hand side of the previous equality may be written as

$$m\big(\pi^{-1}(E) \cap [\theta, k+1)\big) + m\big(\pi^{-1}(E) \cap [k, \theta)\big) = m\big(\pi^{-1}(E) \cap [k, k+1)\big).$$

Combining these two relations we find that

$$\mu\big(R_\theta^{-1}(E)\big) = m\big(\pi^{-1}(R_\theta^{-1}(E) \cap [0, 1))\big) = m\big(\pi^{-1}(E) \cap [k, k+1)\big) = \mu(E)$$

for every measurable set $E \subset S^1$.

The rotations $R_\theta : S^1 \to S^1$ exhibit two very different types of dynamical behavior, depending on the value of θ. If θ is rational, say $\theta = p/q$ with $p \in \mathbb{Z}$ and $q \in \mathbb{N}$, then

$$R_\theta^q([x]) = [x + q\theta] = [x] \quad \text{for every } [x].$$

Consequently, in this case every point $x \in S^1$ is periodic with period q. In the opposite case we have:

Proposition 1.3.4. *If θ is irrational then $\mathcal{O}([x]) = \{R_\theta^n([x]) : n \in \mathbb{N}\}$ is a dense subset of the circle $\mathbb{R}/\mathbb{Z}$ for every $[x]$.*

Proof. We claim that the set $\mathcal{D} = \{m + n\theta : m \in \mathbb{Z}, n \in \mathbb{N}\}$ is dense in $\mathbb{R}$. Indeed, consider any number $r \in \mathbb{R}$. Given any $\varepsilon > 0$, we may choose $p \in \mathbb{Z}$ and $q \in \mathbb{N}$

such that $|q\theta - p| < \varepsilon$. Note that the number $a = q\theta - p$ is necessarily different from zero, since θ is irrational. Let us suppose that a is positive (the case when a is negative is analogous). Subdividing the real line into intervals of length a, we see that there exists an integer l such that $0 \le r - la < a$. This implies that

$$|r - (lq\theta - lp)| = |r - la| < a < \varepsilon.$$

As $m = lq$ and $n = -lq$ are integers and ε is arbitrary, this proves that r is in the closure of the set $\mathcal{D}$, for every $r \in \mathbb{R}$.

Now, given $y \in \mathbb{R}$ and $\varepsilon > 0$, we may take $r = y - x$ and, using the previous paragraph, we may find $m, n \in \mathbb{Z}$ such that $|m + n\theta - (y - x)| < \varepsilon$. This is equivalent to saying that the distance from $[y]$ to the iterate $R_\theta^n([x])$ is less than ε. Since x, y and ε are arbitrary, this shows that every orbit $\mathcal{O}([x])$ is dense in S^1.

In particular, it follows that *every* point on the circle is recurrent for R_θ (this is also true when θ is rational). The previous proposition also leads to some interesting conclusions in the study of the invariant measures of R_θ. Among other things, we will learn later (in Chapter 6) that if θ is irrational then the Lebesgue measure is the unique probability measure that is preserved by R_θ. Related to this, we will see that the orbits of R_θ are uniformly distributed subsets of S^1.

1.3.4 Rotations on tori

The notions we just presented can be generalized to arbitrary dimension, as we are going to explain. For each $d \ge 1$, consider the equivalence relation on $\mathbb{R}^d$ that identifies any two vectors whose difference is an integer vector:

$$(x_1, \dots, x_d) \sim (y_1, \dots, y_d) \quad \Leftrightarrow \quad (x_1 - y_1, \dots, x_d - y_d) \in \mathbb{Z}^d.$$

We denote by $[x]$ or $[(x_1, \dots, x_d)]$ the equivalence class of any $x = (x_1, \dots, x_d)$. Then we call the *d-dimensional torus*, or simply the *d-torus*, the space

$$\mathbb{T}^d = \mathbb{R}^d / \mathbb{Z}^d = (\mathbb{R}/\mathbb{Z})^d$$

formed by those equivalence classes. Let m be the Lebesgue measure on $\mathbb{R}^d$. The operation

$$[(x_1, \dots, x_d)] + [(y_1, \dots, y_d)] = [(x_1 + y_1, \dots, x_d + y_d)]$$

is well defined and turns $\mathbb{T}^d$ into an abelian group. Given $\theta = (\theta_1, \dots, \theta_d) \in \mathbb{R}^d$, we call

$$R_\theta : \mathbb{T}^d \to \mathbb{T}^d, \quad R_\theta([x]) = [x] + [\theta]$$

the *rotation* by θ (sometimes, R_θ is also called the *translation* by θ). The map

$$\phi : [0,1]^d \to \mathbb{T}^d, \quad (x_1, \dots, x_d) \mapsto [(x_1, \dots, x_d)]$$

is surjective and allows us to define a Lebesgue probability measure μ on the d-torus, through the following formula:

$$\mu(B) = m\big(\phi^{-1}(B)\big) \quad \text{for every } B \subset \mathbb{T}^d \text{ such that } \phi^{-1}(B) \text{ is measurable.}$$

This measure μ is invariant under R_θ for every θ.

We say that a vector $\theta = (\theta_1, \dots, \theta_d) \in \mathbb{R}^d$ is *rationally independent* if, for any integer numbers $n_0, n_1, \dots, n_d$, we have that

$$n_0 + n_1\theta_1 + \cdots + n_d\theta_d = 0 \quad \Rightarrow \quad n_0 = n_1 = \cdots = n_d = 0.$$

Otherwise, we say that θ is rationally dependent. One can show that θ is rationally independent if and only if the rotation R_θ is *minimal*, meaning that the orbit $\mathcal{O}([x]) = \{R_\theta^n([x]) : n \in \mathbb{N}\}$ of every $[x] \in \mathbb{T}^d$ is a dense subset of $\mathbb{T}^d$. In this regard, see Exercises 1.3.9–1.3.10 and also Corollary 4.2.3.

1.3.5 Conservative maps

Let M be an open subset of the Euclidian space $\mathbb{R}^d$ and $f : M \to M$ be a C^1 diffeomorphism. This means that f is a bijection, both f and its inverse f^{-1} are differentiable and the two derivatives are continuous. Denote by vol the restriction to M of the Lebesgue measure (volume measure) on $\mathbb{R}^d$. The formula of change of variables asserts that, for any measurable set $B \subset M$,

$$\operatorname{vol}(f(B)) = \int_B |\det Df|\, dx. \tag{1.3.12}$$

One can easily deduce the following consequence:

Lemma 1.3.5. *A C^1 diffeomorphism $f : M \to M$ preserves the volume measure* vol *if and only if the absolute value* $|\det Df|$ *of its Jacobian is equal to* 1 *at every point.*

Proof. Suppose that the absolute value $|\det Df|$ of its Jacobian is equal to 1 at every point. Let E be any measurable set E and $B = f^{-1}(E)$. The formula (1.3.12) yields

$$\operatorname{vol}(E) = \int_B 1\, dx = \operatorname{vol}(B) = \operatorname{vol}(f^{-1}(E)).$$

This means that f preserves the measure vol and so we proved the "if" part of the statement.

To prove the "only if," suppose that $|\det Df(x)| > 1$ for some point $x \in M$. Then, since the Jacobian is continuous, there exists a neighborhood U of x and some number $\sigma > 1$ such that

$$|\det Df(y)| \geq \sigma \quad \text{for all } y \in U.$$

Then, applying (1.3.12) to $B = U$, we get that

$$\operatorname{vol}(f(U)) \geq \int_U \sigma\, dx \geq \sigma \operatorname{vol}(U).$$

Denote $E = f(U)$. Since $\mathrm{vol}(U) > 0$, the previous inequality implies that $\mathrm{vol}(E) > \mathrm{vol}(f^{-1}(E))$. Hence, f does not leave vol invariant. In precisely the same way, one shows that if $|\det Df(x)| < 1$ for some point $x \in M$ then f does not leave the measure vol invariant.

1.3.6 Conservative flows

Now we discuss the invariance of the volume measure in the setting of flows $f^t : M \to M$, $t \in \mathbb{R}$. As before, take M to be an open subset of the Euclidean space $\mathbb{R}^d$. Let us suppose that the flow is C^1, in the sense that the map $(t,x) \mapsto f^t(x)$ is differentiable and all the derivatives are continuous. Then, in particular, every flow transformation $f^t : M \to M$ is a C^1 diffeomorphism: the inverse is f^{-t}. Since f^0 is the identity map and the Jacobian varies continuously, we have that $\det Df^t(x) > 0$ at every point.

Applying Lemma 1.3.5 in this context, we find that the flow preserves the volume measure if and only if

$$\det Df^t(x) = 1 \quad \text{for every } x \in U \text{ and every } t \in \mathbb{R}. \tag{1.3.13}$$

However, this is not very useful in practice because most of the time we do not have an explicit expression for f^t and, hence, it is not clear how to check the condition (1.3.13). Fortunately, there is a reasonably explicit expression for the Jacobian of the flow that can be used in some interesting situations. Let us explain this.

Let us suppose that the flow $f^t : M \to M$ corresponds to the trajectories of a C^1 vector field $F : M \to \mathbb{R}^d$. In other words, each $t \mapsto f^t(x)$ is the solution of the differential equation

$$\frac{dy}{dt} = F(y) \tag{1.3.14}$$

that has x as the initial condition (when dealing with differential equations we always assume that their solutions are defined for all times).

The *Liouville formula* relates the Jacobian of f^t to the *divergence* $\operatorname{div} F$ of the vector field:

$$\det Df^t(x) = \exp\left(\int_0^t \operatorname{div} F(f^s(x))\, ds\right) \quad \text{for every } x \text{ and every } t.$$

Recall that the divergence of a vector field F is the trace of its Jacobian matrix, that is

$$\operatorname{div} F = \frac{\partial F_1}{\partial x_1} + \cdots + \frac{\partial F_d}{\partial x_d}. \tag{1.3.15}$$

Combining the Liouville formula with (1.3.13), we obtain:

Lemma 1.3.6 (Liouville). *The flow $(f^t)_t$ associated with a C^1 vector field F preserves the volume measure if and only if the divergence of F is identically zero.*

We can extend this discussion to the case when M is any Riemannian manifold of dimension $d \geq 2$. The reader who is unfamiliar with this notion may wish to check Appendix A.4.5 before proceeding.

For simplicity, let us suppose that the manifold is orientable. Then the volume measure on M is given by a differentiable d-form ω, called the *volume form* (this remains true in the non-orientable case, except that the form ω is defined up to sign only). What this means is that the volume of any measurable set B contained in the domain of local coordinates $(x_1, \dots, x_d)$ is given by

$$\operatorname{vol}(B) = \int_B \rho(x_1, \dots, x_d)\, dx_1 \cdots dx_d,$$

where $\omega = \rho dx_1 \cdots dx_d$ is the expression of the volume form in those local coordinates. Let F be a C^1 vector field on M. Writing

$$F(x_1, \dots, x_d) = (F_1(x_1, \dots, x_d), \dots, F_d(x_1, \dots, x_d)),$$

we may express the divergence as

$$\operatorname{div} F = \frac{\partial(\rho F)}{\partial x_1} + \cdots + \frac{\partial(\rho F)}{\partial x_d}$$

(it can be shown that the right-hand side does not depend on the choice of the local coordinates). A proof of the following generalization of Lemma 1.3.6 can be found in Sternberg [Ste58]:

Theorem 1.3.7 (Liouville). *The flow $(f^t)_t$ associated with a C^1 vector field F on a Riemannian manifold preserves the volume measure on the manifold if and only if* $\operatorname{div} F = 0$ *at every point.*

Then, it follows from the recurrence theorem for flows that, assuming that the manifold has finite volume (for example, if M is compact) and $\operatorname{div} F = 0$, then almost every point is recurrent for the flow of F.

1.3.7 Exercises

1.3.1. Use Lemma 1.3.3 to give another proof of the fact that the decimal expansion transformation $f(x) = 10x - [10x]$ preserves the Lebesgue measure on the interval.

1.3.2. Prove that, for any number $x \in [0,1]$ whose decimal expansion contains the block 617 (for instance, $x = 0.3375\mathbf{617}264\cdots$), that block occurs infinitely many times in the decimal expansion of x. Even more, the block 617 occurs infinitely many times in the decimal expansion of almost every $x \in [0,1]$.

1.3.3. Prove that the number 617 appears infinitely many times in the continued fraction expression of almost every number $x_0 \in (1/618, 1/617)$, that is, one has $a_n = 617$ for infinitely many values of $n \in \mathbb{N}$.

1.3.4. Let G be the Gauss map. Show that a number $x \in (0,1)$ is rational if and only if there exists $n \geq 1$ such that $G^n(x) = 0$.

1.3.5. Consider the sequence $1, 2, 4, 8, \ldots, a_n = 2^n, \ldots$ of all the powers of 2. Prove that, given any digit $i \in \{1, \ldots, 9\}$, there exist infinitely many values of n for which a_n starts with that digit.

1.3.6. Prove the following extension of Lemma 1.3.3. Let $f : M \to M$ be a C^1 local diffeomorphism on a compact Riemannian manifold M. Let vol be the volume measure on M and $\rho : M \to [0, \infty)$ be a continuous function. Then f preserves the measure $\mu = \rho$ vol if and only if

$$\sum_{x \in f^{-1}(y)} \frac{\rho(x)}{|\det Df(x)|} = \rho(y) \quad \text{for every } y \in M.$$

When f is invertible this means that f preserves the measure μ if and only if $\rho(x) = \rho(f(x)) |\det Df(x)|$ for every $x \in M$.

1.3.7. Check that if A is a $d \times d$ matrix with integer coefficients and determinant different from zero then the transformation $f_A : \mathbb{T}^d \to \mathbb{T}^d$ defined on the torus by $f_A([x]) = [A(x)]$ preserves the Lebesgue measure on $\mathbb{T}^d$.

1.3.8. Show that the Lebesgue measure on S^1 is the only probability measure invariant under all the rotations of S^1, even if we restrict to *rational* rotations. [Note: We will see in Chapter 6 that, for any *irrational* θ, the Lebesgue measure is the unique probability measure invariant under R_θ.]

1.3.9. Suppose that $\theta = (\theta_1, \ldots, \theta_d)$ is rationally dependent. Show that there exists a continuous non-constant function $\varphi : \mathbb{T}^d \to \mathbb{C}$ such that $\varphi \circ R_\theta = \varphi$. Conclude that there exist non-empty open subsets U and V of $\mathbb{T}^d$ that are disjoint and invariant under R_θ, in the sense that $R_\theta(U) = U$ and $R_\theta(V) = V$. Deduce that no orbit $\mathcal{O}([x])$ of the rotation R_θ is dense in $\mathbb{T}^d$.

1.3.10. Suppose that $\theta = (\theta_1, \ldots, \theta_d)$ is rationally independent. Prove that if V is a non-empty open subset of $\mathbb{T}^d$ invariant under R_θ, then V is dense in $\mathbb{T}^d$. Conclude that $\bigcup_{n \in \mathbb{Z}} R_\theta^n(U)$ is dense in the torus, for every non-empty open subset U. Deduce that there exists $[x]$ whose orbit $\mathcal{O}([x])$ under the rotation R_θ is dense in $\mathbb{T}^d$. Conclude that $\mathcal{O}([y])$ is dense in $\mathbb{T}^d$ for every $[y]$.

1.3.11. Let U be an open subset of $\mathbb{R}^{2d}$ and $H : U \to \mathbb{R}$ be a C^2 function. Denote by $(p_1, \ldots, p_d, q_1, \ldots, q_d)$ the coordinate variables in $\mathbb{R}^{2d}$. The *Hamiltonian vector field* associated with H is defined by

$$F(p_1, \ldots, p_d, q_1, \ldots, q_d) = \left(\frac{\partial H}{\partial q_1}, \ldots, \frac{\partial H}{\partial q_d}, -\frac{\partial H}{\partial p_1}, \ldots, -\frac{\partial H}{\partial p_d} \right).$$

Check that the flow defined by F preserves the volume measure.

1.3.12. Let $f : U \to U$ be a C^1 diffeomorphism preserving the volume measure on an open subset U of $\mathbb{R}^d$. Let $H : U \to \mathbb{R}$ be a first integral of f, that is, a C^1 function such that $H \circ f = H$. Let c be a regular value of H and ds be the volume measure defined on the hypersurface $H_c = H^{-1}(c)$ by the restriction of the Riemannian metric of $\mathbb{R}^d$. Prove that the restriction of f to the hypersurface H_c preserves the measure $ds / \| \operatorname{grad} H \|$.

1.4 Induction

In this section we describe a general method, based on the Poincaré recurrence theorem, to construct from a given system (f,μ) other systems, that we refer to as *systems induced* by (f,μ). The reason this is interesting is the following. On the one hand, it is often the case that an induced system is easier to analyze, because it has better global properties than the original one. On the other hand, interesting conclusions about the original system can often be obtained from analyzing the induced one. Examples will appear in a while.

1.4.1 First-return map

Let $f : M \to M$ be a measurable transformation and μ be an invariant probability measure. Let $E \subset M$ be a measurable set with $\mu(E) > 0$ and $\rho(x) = \rho_E(x)$ be the first-return time of x to E, as given by (1.2.1). The *first-return map* to the domain E is the map g given by

$$g(x) = f^{\rho(x)}(x)$$

whenever $\rho(x)$ is finite. The Poincaré recurrence theorem ensures that this is the case for μ-almost every $x \in E$ and so g is defined on a full measure subset of E. We also denote by μ_E the restriction of μ to the measurable subsets E.

Proposition 1.4.1. *The measure μ_E is invariant under the map $g : E \to E$.*

Proof. For every $k \geq 1$, denote by E_k the subset of points $x \in E$ such that $\rho(x) = k$. By definition, $g(x) = f^k(x)$ for every $x \in E_k$. Let B be any measurable subset of E. Then

$$\mu(g^{-1}(B)) = \sum_{k=1}^{\infty} \mu\left(f^{-k}(B) \cap E_k\right). \tag{1.4.1}$$

On the other hand, since μ is f-invariant,

$$\mu(B) = \mu(f^{-1}(B)) = \mu(f^{-1}(B) \cap E_1) + \mu(f^{-1}(B) \setminus E). \tag{1.4.2}$$

Analogously,

$$\begin{aligned} \mu(f^{-1}(B) \setminus E) &= \mu(f^{-2}(B) \setminus f^{-1}(E)) \\ &= \mu(f^{-2}(B) \cap E_2) + \mu(f^{-2}(B) \setminus (E \cup f^{-1}(E))). \end{aligned}$$

Replacing this expression in (1.4.2), we find that

$$\mu(B) = \sum_{k=1}^{2} \mu(f^{-k}(B) \cap E_k) + \mu\left(f^{-2}(B) \setminus \bigcup_{k=0}^{1} f^{-k}(E)\right).$$

Repeating this argument successively, we obtain

$$\mu(B) = \sum_{k=1}^{n} \mu(f^{-k}(B) \cap E_k) + \mu\left(f^{-n}(B) \setminus \bigcup_{k=0}^{n-1} f^{-k}(E)\right). \tag{1.4.3}$$

Now let us go to the limit when $n \to \infty$. It is clear that the last term is bounded above by $\mu\big(f^{-n}(E) \setminus \bigcup_{k=0}^{n-1} f^{-k}(E)\big)$. So, using Remark 1.2.3, that term converges to zero when $n \to \infty$. In this way we conclude that

$$\mu(B) = \sum_{k=1}^{\infty} \mu\big(f^{-k}(B) \cap E_k\big).$$

Together with (1.4.1), this shows that $\mu(g^{-1}(B)) = \mu(B)$ for every measurable subset B of E. That is to say, the measure μ_E is invariant under g.

Example 1.4.2. Consider the transformation $f : [0,\infty) \to [0,\infty)$ defined by

$$f(0) = 0 \quad \text{and} \quad f(x) = 1/x \text{ if } x \in (0,1) \quad \text{and} \quad f(x) = x-1 \text{ if } x \geq 1.$$

Let $E = [0,1]$. The time ρ of first return to E is given by

$$\rho(0) = 1 \quad \text{and} \quad \rho(x) = k+1 \text{ if } x \in (1/(k+1), 1/k] \text{ with } k \geq 1.$$

So, the first-return map to E is given by

$$g(0) = 0 \quad \text{and} \quad g(x) = 1/x - k \text{ if } x \in (1/(k+1), 1/k] \text{ with } k \geq 1.$$

In other words, g is the Gauss map. We saw in Section 1.3.2 that the Gauss map admits an invariant probability measure equivalent to the Lebesgue measure on $[0,1)$. From this, one can draw some interesting conclusions about the original map f. For instance, using the ideas in the next section one finds that f admits an (infinite) invariant measure equivalent to the Lebesgue measure on $[0,\infty)$.

1.4.2 Induced transformations

In an opposite direction, given any measure ν invariant under $g : E \to E$, we may construct a certain related measure ν_ρ that is invariant under $f : M \to M$. For this, g does not even have to be a first-return map: the construction that we present below is valid for any map *induced* from f, that is, any map of the form

$$g : E \to E, \quad g(x) = f^{\rho(x)}(x), \tag{1.4.4}$$

where $\rho : E \to \mathbb{N}$ is a measurable function (it suffices that ρ is defined on some full measure subset of E). As before, we denote by E_k the subset of points $x \in E$ such that $\rho(x) = k$. Then we define

$$\nu_\rho(B) = \sum_{n=0}^{\infty} \sum_{k>n} \nu\big(f^{-n}(B) \cap E_k\big), \tag{1.4.5}$$

for every measurable set $B \subset M$.

Proposition 1.4.3. *The measure ν_ρ defined in* (1.4.5) *is invariant under f and satisfies $\nu_\rho(M) = \int_E \rho \, d\nu$. In particular, ν_ρ is finite if and only if the function ρ is integrable with respect to ν.*

Proof. First, let us prove that ν_ρ is invariant. By the definition (1.4.5),

$$\nu_\rho\big(f^{-1}(B)\big) = \sum_{n=0}^{\infty}\sum_{k>n} \nu\big(f^{-(n+1)}(B)\cap E_k\big) = \sum_{n=1}^{\infty}\sum_{k\geq n} \nu\big(f^{-n}(B)\cap E_k\big).$$

We may rewrite this expression as follows:

$$\nu_\rho\big(f^{-1}(B)\big) = \sum_{n=1}^{\infty}\sum_{k>n} \nu\big(f^{-n}(B)\cap E_k\big) + \sum_{k=1}^{\infty} \nu\big(f^{-k}(B)\cap E_k\big). \tag{1.4.6}$$

Concerning the last term, observe that

$$\sum_{k=1}^{\infty} \nu\big(f^{-k}(B)\cap E_k\big) = \nu\big(g^{-1}(B)\big) = \nu\big(B\big) = \sum_{k=1}^{\infty} \nu\big(B\cap E_k\big),$$

since ν is invariant under g. Replacing this in (1.4.6), we see that

$$\nu_\rho\big(f^{-1}(B)\big) = \sum_{n=1}^{\infty}\sum_{k>n} \nu\big(f^{-n}(B)\cap E_k\big) + \sum_{k=1}^{\infty} \nu\big(B\cap E_k\big) = \nu_\rho\big(B\big)$$

for every measurable set $B \subset E$. The second claim is a direct consequence of the definitions:

$$\nu_\rho(M) = \sum_{n=0}^{\infty}\sum_{k>n} \nu\big(f^{-n}(M)\cap E_k\big) = \sum_{n=0}^{\infty}\sum_{k>n} \nu(E_k) = \sum_{k=1}^{\infty} k\nu(E_k) = \int_E \rho\, d\nu.$$

This completes the proof.

It is interesting to analyze how this construction relates to the one in the previous section when g is a first-return map of f and the measure ν is the restriction $\mu \mid E$ of some invariant measure μ of f:

Corollary 1.4.4. *If g is the first-return map of f to a measurable subset E and $\nu = \mu \mid E$, then*

1. *$\nu_\rho(B) = \nu(B) = \mu(B)$ for every measurable set $B \subset E$.*
2. *$\nu_\rho(B) \leq \mu(B)$ for every measurable set $B \subset M$.*

Proof. By definition, $f^{-n}(E) \cap E_k = \emptyset$ for every $0 < n < k$. This implies that, given any measurable set $B \subset E$, all the terms with $n > 0$ in the definition (1.4.5) are zero. Hence, $\nu_\rho(B) = \sum_{k>0} \nu(B\cap E_k) = \nu(B)$ as claimed in the first part of the statement.

Consider any measurable set $B \subset M$. Then,

$$\begin{aligned} \mu\big(B\big) &= \mu\big(B\cap E\big) + \mu\big(B\cap E^c\big) = \nu\big(B\cap E\big) + \mu\big(B\cap E^c\big) \\ &= \sum_{k=1}^{\infty} \nu\big(B\cap E_k\big) + \mu\big(B\cap E^c\big). \end{aligned} \tag{1.4.7}$$

Since μ is invariant, $\mu(B \cap E^c) = \mu(f^{-1}(B) \cap f^{-1}(E^c))$. Then, as in the previous equality,

$$\begin{aligned}\mu(B \cap E^c) &= \mu(f^{-1}(B) \cap E \cap f^{-1}(E^c)) + \mu(f^{-1}(B) \cap E^c \cap f^{-1}(E^c)) \\ &= \sum_{k=2}^{\infty} \nu(f^{-1}(B) \cap E_k) + \mu(f^{-1}(B) \cap E^c \cap f^{-1}(E^c)).\end{aligned}$$

Replacing this in (1.4.7), we find that

$$\mu(B) = \sum_{n=0}^{1} \sum_{k>n} \nu(f^{-n}(B) \cap E_k) + \mu\left(f^{-1}(B) \cap \bigcap_{n=0}^{1} f^{-n}(E^c)\right).$$

Repeating this argument successively, we get that

$$\begin{aligned}\mu(B) &= \sum_{n=0}^{N} \sum_{k>n} \nu(f^{-n}(B) \cap E_k) + \mu\left(f^{-N}(B) \cap \bigcap_{k=0}^{N} f^{-n}(E^c)\right) \\ &\geq \sum_{n=0}^{N} \sum_{k>n} \nu(f^{-n}(B) \cap E_k) \quad \text{for every } N \geq 1.\end{aligned}$$

Taking the limit as $N \to \infty$, we conclude that $\mu(B) \geq \nu_\rho(B)$.

We also have from the Kač theorem (Theorem 1.2.2) that

$$\nu_\rho(M) = \int_E \rho \, d\nu = \int_E \rho \, d\mu = \mu(M) - \mu(E_0^*).$$

So, it follows from Corollary 1.4.4 that $\nu_\rho = \mu$ if and only if $\mu(E_0^*) = 0$.

Example 1.4.5 (Manneville–Pomeau)**.** Given $d > 0$, let a be the only number in $(0,1)$ such that $a(1+a^d) = 1$. Then define $f\colon [0,1] \to [0,1]$ as follows:

$$f(x) = x(1+x^d) \quad \text{if} \quad x \in [0,a] \quad \text{and} \quad f(x) = \frac{x-a}{1-a} \quad \text{if} \quad x \in (a,1].$$

The graph of f is depicted on the left-hand side of Figure 1.3. Observe that $|f'(x)| \geq 1$ at every point, and the inequality is strict at every $x > 0$. Let $(a_n)_n$ be the sequence on the interval $[0,a]$ defined by $a_1 = a$ and $f(a_{n+1}) = a_n$ for $n \geq 1$. We also write $a_0 = 1$. Some properties of this sequence are studied in Exercise 1.4.2.

Now consider the map $g(x) = f^{\rho(x)}(x)$, where

$$\rho : [0,1] \to \mathbb{N}, \quad \rho(x) = 1 + \min\{n \geq 0 : f^n(x) \in (a,1]\}.$$

In other words, $\rho(x) = k$ and so $g(x) = f^k(x)$ for every $x \in (a_k, a_{k-1}]$. The graph of g is represented on the right-hand side of Figure 1.3. Note that the restriction to each interval $(a_k, a_{k-1}]$ is a bijection onto $(0,1]$. A key point is that the induced map g is *expanding*:

$$|g'(x)| \geq \frac{1}{1-a} > 1 \quad \text{for every } x \in [0,1].$$

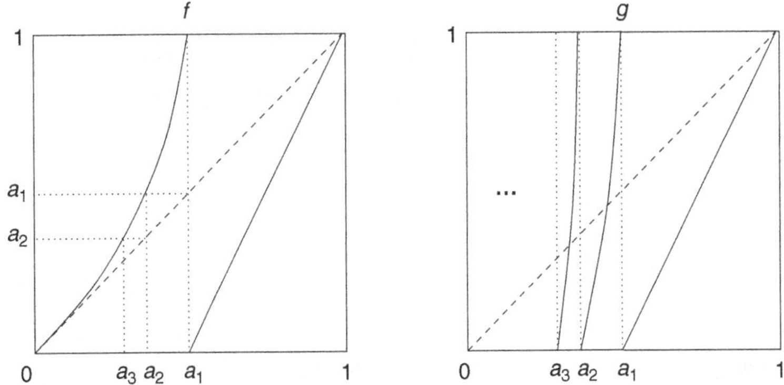

Figure 1.3. Construction of an induced transformation

Using the ideas that will be developed in Chapter 11, one can show that g admits a unique invariant probability measure ν equivalent to the Lebesgue measure on $(0,1]$. In fact, the density (Radon–Nikodym derivative) of ν with respect to the Lebesgue measure is bounded from zero and infinity. Then, the f-invariant measure ν_ρ in (1.4.5) is equivalent to Lebesgue measure. It follows (see Exercise 1.4.2) that this measure is finite if and only if $d \in (0,1)$.

1.4.3 Kakutani–Rokhlin towers

It is possible, and useful, to generalize the previous constructions even further, by omitting the initial transformation $f : M \to M$ altogether. More precisely, given a transformation $g : E \to E$, a measure ν on E invariant under g and a measurable function $\rho : E \to \mathbb{N}$, we are going to *construct* a transformation $f : M \to M$ and a measure ν_ρ invariant under f such that E can be identified with a subset of M, g is the first-return map of f to E, with first-return time given by ρ, and the restriction of ν_ρ to E coincides with ν.

This transformation f is called the *Kakutani–Rokhlin tower* of g with time ρ. The measure ν_ρ is finite if and only if ρ is integrable with respect to ν. They are constructed as follows. Begin by defining

$$M = \{(x,n) : x \in E \text{ and } 0 \le n < \rho(x)\} \\ = \bigcup_{k=1}^{\infty} \bigcup_{n=0}^{k-1} E_k \times \{n\}.$$

In other words, M consists of k copies of each set $E_k = \{x \in E : \rho(x) = k\}$, "piled up" on top of each other. We call each $\bigcup_{k>n} E_k \times \{n\}$ the *n-th floor* of M. See Figure 1.4.

Next, define $f : M \to M$ as follows:

$$f(x,n) = \begin{cases} (x,n+1) & \text{if } n < \rho(x) - 1 \\ (g(x),0) & \text{if } n = \rho(x) - 1 \end{cases}.$$

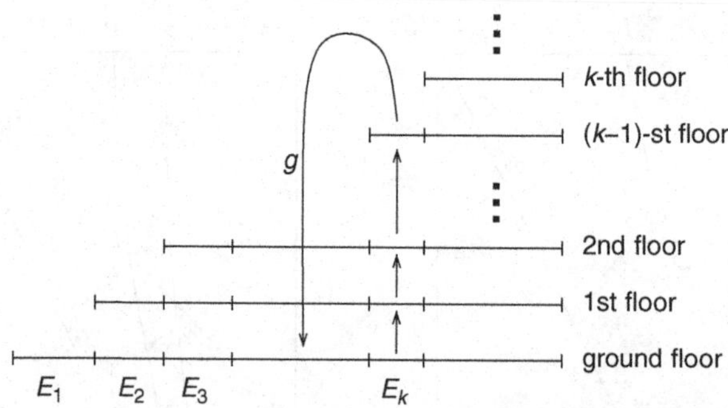

Figure 1.4. Kakutani–Rokhlin tower of g with time ρ

In other words, each point (x,n) is "lifted" one floor at a time, until reaching the floor $\rho(x)-1$; at that stage, the point "falls" directly to $(g(x),0)$ on the ground (zero-th) floor. The ground floor $E\times\{0\}$ is naturally identified with the set E. Besides, the first-return map to $E\times\{0\}$ corresponds precisely to $g:E\to E$.

Finally, the measure ν_ρ is defined by

$$\nu_\rho \mid (E_k\times\{n\}) = \nu \mid E_k$$

for every $0\le n<k$. It is clear that the restriction of ν_ρ to the ground floor coincides with ν. Moreover, ν_ρ is invariant under f and

$$\nu_\rho(M)=\sum_{k=1}^{\infty} k\nu(E_k)=\int_E \rho\, d\nu.$$

This completes the construction of the Kakutani–Rokhlin tower.

1.4.4 Exercises

1.4.1. Let $f:S^1\to S^1$ be the transformation $f(x)=2x \mod \mathbb{Z}$. Show that the function $\tau(x)=\min\{k\ge 0: f^k(x)\in(1/2,1)\}$ is integrable with respect to the Lebesgue measure. State and prove a corresponding result for any C^1 transformation $g: S^1\to S^1$ that is close to f, in the sense that $\sup_x\{\|g(x)-f(x)\|,\|g'(x)-f'(x)\|\}$ is sufficiently small.

1.4.2. Consider the measure ν_ρ and the sequence $(a_n)_n$ defined in Example 1.4.5. Check that ν_ρ is always σ-finite. Show that $(a_n)_n$ is decreasing and converges to zero. Moreover, there exist $c_1,c_2,c_3,c_4>0$ such that

$$c_1\le a_j j^{1/d}\le c_2 \quad\text{and}\quad c_3\le (a_j-a_{j+1})j^{1+1/d}\le c_4 \quad \text{for every } j. \tag{1.4.8}$$

Deduce that the g-invariant measure ν_ρ is finite if and only if $d\in(0,1)$.

1.4.3. Let $\sigma:\Sigma\to\Sigma$ be the map defined on the space $\Sigma=\{1,\dots,d\}^{\mathbb{Z}}$ by $\sigma((x_n)_n)=(x_{n+1})_n$. Describe the first-return map g to the subset $\{(x_n)_n\in\Sigma: x_0=1\}$.

1.4.4. [Kakutani–Rokhlin lemma] Let $f:M\to M$ be an invertible transformation and μ be an invariant probability measure without atoms and such that $\mu(\bigcup_{n\in\mathbb{N}} f^n(E))=1$ for every $E\subset M$ with $\mu(E)>0$. Show that for every

$n \geq 1$ and $\varepsilon > 0$ there exists a measurable set $B \subset M$ such that the iterates $B, f(B), \ldots, f^{n-1}(B)$ are pairwise disjoint and the complement of their union has measure less than ε. In particular, this holds for every invertible system that is *aperiodic*, that is, whose periodic points form a zero measure set.

1.4.5. Let $f : M \to M$ be a transformation and $(H_j)_{j\geq 1}$ be a collection of subsets of M such that if $x \in H_n$ then $f^j(x) \in H_{n-j}$ for every $0 \leq j < n$. Let H be the set of points that belong to H_j for infinitely many values of j, that is, $H = \bigcap_{k=1}^{\infty} \bigcup_{j=k}^{\infty} H_j$. For $y \in H$, define $\tau(y) = \min\{j \geq 1 : y \in H_j\}$ and $T(y) = f^{\tau(y)}(y)$. Observe that T maps H inside H. Moreover, show that

$$\limsup_n \frac{1}{n} \#\{1 \leq j \leq n : x \in H_j\} \geq \theta > 0 \quad \Rightarrow \quad \liminf_k \frac{1}{k} \sum_{i=0}^{k-1} \tau(T^i(x)) \leq \frac{1}{\theta}.$$

1.4.6. Let $f : M \to M$ be a transformation preserving a measure μ. Let $(H_j)_{j\geq 1}$ and $\tau : M \to \mathbb{N}$ be as in Exercise 1.4.5. Consider the sequence of functions $(\tau_n)_n$ defined by $\tau_1(x) = \tau(x)$ and $\tau_n(x) = \tau(f^{\tau_{n-1}(x)}(x)) + \tau_{n-1}(x)$ for $n > 1$. Suppose that

$$\limsup_n \frac{1}{n} \#\{1 \leq j \leq n : x \in H_j\} \geq \theta > 0 \quad \text{for } \mu\text{-almost every } x \in M.$$

Show that $\tau_{n+1}(x)/\tau_n(x) \to 1$ for μ-almost every $x \in M$. [Note: Sequences with this property are called *non-lacunary*.]

1.5 Multiple recurrence theorems

Now we consider finite families of *commuting maps* $f_i : M \to M$, $i = 1, \ldots, q$, that is, such that

$$f_i \circ f_j = f_j \circ f_i \quad \text{for every } i, j \in \{1, \ldots, q\}.$$

Our goal is to explain that the results in Section 1.2 extend to this setting: we find points that are *simultaneously recurrent* for these transformations.

The first result in this direction generalizes the Birkhoff recurrence theorem (Theorem 1.2.6):

Theorem 1.5.1 (Birkhoff multiple recurrence). *Let M be a compact metric space and $f_1, \ldots, f_q : M \to M$ be continuous commuting maps. Then there exists $a \in M$ and a sequence $(n_k)_k \to \infty$ such that*

$$\lim_k f_i^{n_k}(a) = a \quad \textit{for every } i = 1, \ldots, q. \tag{1.5.1}$$

The key point here is that the sequence $(n_k)_k$ does not depend on i: we say that the point a is *simultaneously recurrent* for all the maps f_i, $i = 1, \ldots, q$. A proof of Theorem 1.5.1 is given in Section 1.5.1. Next, we discuss the following generalization of the Poincaré recurrence theorem (Theorem 1.2.1):

Theorem 1.5.2 (Poincaré multiple recurrence). *Let $(M, \mathcal{B}, \mu)$ be a probability space and $f_i : M \to M$, $i = 1, \ldots, q$ be measurable commuting maps that* *preserve the measure μ. Then, given any set $E \subset M$ with positive measure, there exists $n \geq 1$ such that*

$$\mu\big(E \cap f_1^{-n}(E) \cap \cdots \cap f_q^{-n}(E)\big) > 0.$$

In other words, for a positive measure subset of points $x \in E$, their orbits under all the maps f_i, $i = 1,\ldots,q$ return to E *simultaneously* at time n (we say that n is a *simultaneous return* of x to E): once more, the crucial point with the statement is that n does not depend on i.

The proof of Theorem 1.5.2 will not be presented here; we refer the interested reader to the book of Furstenberg [Fur77]. We are just going to mention some direct consequences and, in Chapter 2, we will use this theorem to prove the Szemerédi theorem on the existence of arithmetic progressions inside "dense" subsets of integer numbers.

To begin with, observe that the set of simultaneous returns is always infinite. Indeed, let n be as in the statement of Theorem 1.5.2. Applying the theorem to the set $F = E \cap f_1^{-n}(E) \cap \cdots \cap f_q^{-n}(E)$, we find $m \geq 1$ such that

$$\mu\big(E \cap f_1^{-(m+n)}(E) \cap \cdots \cap f_q^{-(m+n)}(E)\big) \geq \mu\big(F \cap f_1^{-m}(F) \cap \cdots \cap f_q^{-m}(F)\big) > 0.$$

Thus, $m+n$ is also a simultaneous return to E, for all the points in some subset of E with positive measure.

It follows that, for any set $E \subset M$ with $\mu(E) > 0$ and for μ-almost every point $x \in E$, there exist infinitely many *simultaneous returns* of x to E. Indeed, suppose there is a positive measure set $F \subset E$ such that every point of F has a finite number of simultaneous returns to E. On the one hand, up to replacing F by a suitable subset, we may suppose that the simultaneous returns to E of all the points of F are bounded by some $k \geq 1$. On the other hand, using the previous paragraph, there exists $n > k$ such that $G = F \cap f_1^{-n}(F) \cap \cdots \cap f_q^{-n}(F)$ has positive measure. Now, it is clear from the definition that n is a simultaneous return to E of every $x \in G$. This contradicts the choice of F, thus proving our claim.

Another direct corollary is the Birkhoff multiple recurrence theorem (Theorem 1.5.1). Indeed, if $f_i : M \to M$, $i = 1,\ldots,q$ are continuous commuting transformations on a compact metric space then there exists some probability measure μ that is invariant under all these transformations (this fact will be checked in the next chapter, see Exercise 2.2.2). From this point on, we may argue exactly as in the proof of Theorem 1.2.4. More precisely, consider any countable basis $\{U_k\}$ for the topology of M. According to the previous paragraph, for every k there exists a set $\tilde{U}_k \subset U_k$ with zero measure such that every point in $U_k \setminus \tilde{U}_k$ has infinitely many simultaneous returns to U_k. Then $\tilde{U} = \bigcup_k \tilde{U}_k$ has measure zero and every point in its complement is simultaneously recurrent, in the sense of Theorem 1.5.1.

1.5.1 Birkhoff multiple recurrence theorem

In this section we prove Theorem 1.5.1 in the case when the transformations $f_1,\dots,f_q$ are homeomorphisms of M, which suffices for all our purposes in the present chapter. The general case may be deduced easily (see Exercise 2.4.7) using the concept of natural extension, which we will present in the next chapter.

The theorem may be reformulated in the following useful way. Consider the transformation $F : M^q \to M^q$ defined on the product space $M^q = M \times \cdots \times M$ by $F(x_1,\dots,x_q) = (f_1(x_1),\dots,f_q(x_q))$. Denote by Δ_q the *diagonal* of M^q, that is, the subset of points of the form $\tilde{x} = (x,\dots,x)$. Theorem 1.5.1 claims, precisely, that there exist $\tilde{a} \in \Delta_q$ and $(n_k)_k \to \infty$ such that

$$\lim_k F^{n_k}(\tilde{a}) = \tilde{a}. \tag{1.5.2}$$

The proof of Theorem 1.5.1 is by induction on the number q of transformations. The case $q = 1$ is contained in Theorem 1.2.6. Consider any $q \geq 2$ and suppose that the statement is true for every family of $q-1$ commuting homeomorphisms. We are going to prove that it is true for the family $f_1,\dots,f_q$.

Let $\mathcal{G}$ be the (abelian) group generated by the homeomorphisms $f_1,\dots,f_q$. We say that a set $X \subset M$ is *$\mathcal{G}$-invariant* if $g(X) \subset X$ for every $g \in \mathcal{G}$. Observing that the inverse g^{-1} is also in $\mathcal{G}$, we see that this implies $g(X) = X$ for every $g \in \mathcal{G}$. Just as we did in Theorem 1.2.6, we may use Zorn's lemma to conclude that there exists some minimal, non-empty, closed, $\mathcal{G}$-invariant set $X \subset M$ (this is Exercise 1.5.2). The statement of the theorem is not affected if we replace M by X. Thus, it is no restriction to assume that the ambient space M is minimal. This assumption is used as follows:

Lemma 1.5.3. *If M is minimal then for every non-empty open set $U \subset M$ there exists a finite subset $\mathcal{H} \subset \mathcal{G}$ such that*

$$\bigcup_{h\in\mathcal{H}} h^{-1}(U) = M.$$

Proof. For any $x \in M$, the closure of the orbit $\mathcal{G}(x) = \{g(x) : g \in \mathcal{G}\}$ is a non-empty, closed, $\mathcal{G}$-invariant subset of M. So, the hypothesis that M is minimal implies that every orbit $\mathcal{G}(x)$ is dense in M. In particular, there is $g \in \mathcal{G}$ such that $g(x) \in U$. This proves that $\{g^{-1}(U) : g \in \mathcal{G}\}$ is an open cover of M. By compactness, it follows that there exists a finite subcover, as claimed.

Consider the product M^q endowed with the distance function

$$d\big((x_1,\dots,x_q),(y_1,\dots,y_q)\big) = \max\{d(x_i,y_i) : 1 \leq i \leq q\}.$$

Note that the map $M \to \Delta_q$, $x \mapsto \tilde{x} = (x,\dots,x)$ is a homeomorphism, and even an isometry for this choice of a distance. Every open set $U \subset M$ corresponds to an open set $\widetilde{U} \subset \Delta_q$ through this homeomorphism. Given any $g \in \mathcal{G}$, we denote by $\tilde{g} : M^q \to M^q$ the homeomorphism defined by $\tilde{g}(x_1,\dots,x_q) = (g(x_1),\dots,g(x_q))$. The fact that the group $\mathcal{G}$ is abelian implies that $\tilde{g}$ commutes with F; note also that every $\tilde{g}$ preserves the diagonal Δ_q. Then the conclusion of Lemma 1.5.3 may be rewritten in the following form:

$$\bigcup_{h\in\mathcal{H}} \tilde{h}^{-1}(\widetilde{U}) = \Delta_q. \tag{1.5.3}$$

Lemma 1.5.4. *Given $\varepsilon > 0$ there exist $\tilde{x} \in \Delta_q$, $\tilde{y} \in \Delta_q$ and $n \geq 1$ such that $d(F^n(\tilde{x}),\tilde{y}) < \varepsilon$.*

Proof. Define $g_i = f_i \circ f_q^{-1}$ for each $i = 1,\dots,q-1$. Since the maps f_i commute with each other, so do the maps g_i. Then, by induction, there exist $y \in M$ and $(n_k)_k \to \infty$ such that

$$\lim_k g_i^{n_k}(y) = y \quad \text{for every } i = 1,\dots,q-1.$$

Denote $x_k = f_q^{-n_k}(y)$ and consider $\tilde{x}_k = (x_k,\dots,x_k) \in \Delta_q$. Then,

$$\begin{aligned} F^{n_k}(\tilde{x}_k) &= (f_1^{n_k} f_q^{-n_k}(y),\dots,f_{q-1}^{n_k} f_q^{-n_k}(y), f_q^{n_k} f_q^{-n_k}(y)) \\ &= (g_1^{n_k}(y),\dots,g_{q-1}^{n_k}(y),y) \end{aligned}$$

converges to $(y,\dots,y,y)$ when $k \to \infty$. This proves the lemma, with $\tilde{x} = \tilde{x}_k$, $\tilde{y} = (y,\dots,y,y)$ and $n = n_k$ for every k sufficiently large.

The next step is to show that the point $\tilde{y}$ in Lemma 1.5.4 is arbitrary:

Lemma 1.5.5. *Given $\varepsilon > 0$ and $\tilde{z} \in \Delta_q$ there exist $\tilde{w} \in \Delta_q$ and $m \geq 1$ such that $d(F^m(\tilde{w}),\tilde{z}) < \varepsilon$.*

Proof. Given $\varepsilon > 0$ and $\tilde{z} \in \Delta_q$, consider $\widetilde{U}$ = open ball of center $\tilde{z}$ and radius $\varepsilon/2$. By Lemma 1.5.3 and the observation (1.5.3), we may find a finite set $\mathcal{H} \subset \mathcal{G}$ such that the sets $\tilde{h}^{-1}(\widetilde{U})$, $h \in \mathcal{H}$ cover Δ_q. Since the elements of $\mathcal{G}$ are (uniformly) continuous functions, there exists $\delta > 0$ such that

$$d(\tilde{x}_1,\tilde{x}_2) < \delta \quad \Rightarrow \quad d(\tilde{h}(\tilde{x}_1),\tilde{h}(\tilde{x}_2)) < \varepsilon/2 \quad \text{for every } h \in \mathcal{H}.$$

By Lemma 1.5.4 there exist $\tilde{x},\tilde{y} \in \Delta_q$ and $n \geq 1$ such that $d(F^n(\tilde{x}),\tilde{y}) < \delta$. Fix $h \in \mathcal{H}$ such that $\tilde{y} \in \tilde{h}^{-1}(\widetilde{U})$. Then,

$$d\big(\tilde{h}(F^n(\tilde{x})),\tilde{z}\big) \leq d\big(\tilde{h}(F^n(\tilde{x})),\tilde{h}(\tilde{y})\big) + d\big(\tilde{h}(\tilde{y}),\tilde{z}\big) < \varepsilon/2 + \varepsilon/2.$$

Take $\tilde{w} = \tilde{h}(\tilde{x})$. Since $\tilde{h}$ commutes with F^n, the previous inequality implies that $d(F^n(\tilde{w}),\tilde{z}) < \varepsilon$, as we wanted to prove.

Next, we prove that one may take $\tilde{x} = \tilde{y}$ in Lemma 1.5.4:

Lemma 1.5.6 (Bowen). *Given $\varepsilon > 0$ there exist $\tilde{v} \in \Delta_q$ and $k \geq 1$ with $d(F^k(\tilde{v}),\tilde{v}) < \varepsilon$.*

Proof. Given $\varepsilon > 0$ and $\tilde{z}_0 \in \Delta_q$, consider the sequences ε_j, m_j and $\tilde{z}_j$, $j \geq 1$ defined by recurrence as follows. Initially, take $\varepsilon_1 = \varepsilon/2$.

- By Lemma 1.5.5 there are $\tilde{z}_1 \in \Delta_q$ and $m_1 \geq 1$ with $d(F^{m_1}(\tilde{z}_1), \tilde{z}_0) < \varepsilon_1$.
- By the continuity of F^{m_1}, there exists $\varepsilon_2 < \varepsilon_1$ such that $d(\tilde{z}, \tilde{z}_1) < \varepsilon_2$ implies $d(F^{m_1}(\tilde{z}), \tilde{z}_0) < \varepsilon_1$.

Next, given any $j \geq 2$:

- By Lemma 1.5.5 there are $\tilde{z}_j \in \Delta_q$ and $m_j \geq 1$ with $d(F^{m_j}(\tilde{z}_j), \tilde{z}_{j-1}) < \varepsilon_j$.
- By the continuity of F^{m_j}, there exists $\varepsilon_{j+1} < \varepsilon_j$ such that $d(\tilde{z}, \tilde{z}_j) < \varepsilon_{j+1}$ implies $d(F^{m_j}(\tilde{z}), \tilde{z}_{j-1}) < \varepsilon_j$.

In particular, for any $i < j$,

$$d(F^{m_{i+1}+\cdots+m_j}(\tilde{z}_j), \tilde{z}_i) < \varepsilon_{i+1} \leq \frac{\varepsilon}{2}.$$

Since Δ_q is compact, we can find i,j with $i < j$ such that $d(\tilde{z}_i, \tilde{z}_j) < \varepsilon/2$. Take $k = m_{i+1} + \cdots + m_j$. Then,

$$d(F^k(\tilde{z}_j), \tilde{z}_j) \leq d(F^k(\tilde{z}_j), \tilde{z}_i) + d(\tilde{z}_i, \tilde{z}_j) < \varepsilon.$$

This completes the proof of the lemma.

Now we are ready to conclude the proof of Theorem 1.5.1. For that, let us consider the function

$$\phi : \Delta_q \to [0, \infty), \quad \phi(\tilde{x}) = \inf\{d(F^n(\tilde{x}), \tilde{x}) : n \geq 1\}.$$

Observe that ϕ is upper semi-continuous: given any $\varepsilon > 0$, every point $\tilde{x}$ admits some neighborhood V such that $\phi(\tilde{y}) < \phi(\tilde{x}) + \varepsilon$ for every $y \in V$. This is an immediate consequence of the fact that ϕ is given by the infimum of a family of continuous functions. Then (Exercise 1.5.4), ϕ admits some continuity point $\tilde{a}$. We are going to show that this point satisfies the conclusion of Theorem 1.5.1.

Let us begin by observing that $\phi(\tilde{a}) = 0$. Indeed, suppose that $\phi(\tilde{a})$ is positive. Then, by continuity, there exist $\beta > 0$ and a neighborhood V of $\tilde{a}$ such that $\phi(\tilde{y}) \geq \beta > 0$ for every $\tilde{y} \in V$. Then,

$$d(F^n(\tilde{y}), \tilde{y}) \geq \beta \quad \text{for every } y \in V \text{ and } n \geq 1. \tag{1.5.4}$$

On the other hand, according to (1.5.3), for every $\tilde{x} \in \Delta_q$ there exists $h \in \mathcal{H}$ such that $\tilde{h}(\tilde{x}) \in V$. Since the transformations h are uniformly continuous, we may fix $\alpha > 0$ such that

$$d(\tilde{z}, \tilde{w}) < \alpha \quad \Rightarrow \quad d\big(\tilde{h}(\tilde{z}), \tilde{h}(\tilde{w})\big) < \beta \quad \text{for every } h \in \mathcal{H}. \tag{1.5.5}$$

By Lemma 1.5.6, there exists $n \geq 1$ such that $d(\tilde{x}, F^n(\tilde{x})) < \alpha$. Then, using (1.5.5) and recalling that F commutes with every $\tilde{h}$,

$$d\big(\tilde{h}(\tilde{x}), F^n(\tilde{h}(\tilde{x}))\big) < \beta.$$

This contradicts (1.5.4). This contradiction proves that $\phi(\tilde{a}) = 0$, as claimed.

In other words, there exists $(n_k)_k \to \infty$ such that $d(F^{n_k}(\tilde{a}), \tilde{a}) \to 0$ when $k \to \infty$. This means that (1.5.2) is satisfied and, hence, the proof of Theorem 1.5.1 is complete.

1.5.2 Exercises

1.5.1. Show, by means of examples, that the conclusion of Theorem 1.5.1 is generally false if the transformations f_i do not commute with each other.

1.5.2. Let $\mathcal{G}$ be the abelian group generated by commuting homeomorphisms $f_1, \ldots, f_q : M \to M$ on a compact metric space. Prove that there exists some minimal element $X \subset M$ for the inclusion relation in the family of non-empty, closed, $\mathcal{G}$-invariant subsets of M.

1.5.3. Show that if $\varphi : M \to \mathbb{R}$ is an upper semi-continuous function on a compact metric space then φ attains its maximum, that is, there exists $p \in M$ such that $\varphi(p) \geq \varphi(x)$ for every $x \in M$.

1.5.4. Show that if $\varphi : M \to \mathbb{R}$ is an (upper or lower) semi-continuous function on a compact metric space then the set of continuity points of φ contains a countable intersection of open and dense subsets of M. In particular, the set of continuity points is dense in M.

1.5.5. Let $f : M \to M$ be a measurable transformation preserving a finite measure μ. Given $k \geq 1$ and a positive measure set $A \subset M$, show that for almost every $x \in A$ there exists $n \geq 1$ such that $f^{jn}(x) \in A$ for every $1 \leq j \leq k$.

1.5.6. Let $f_1, \ldots, f_q : M \to M$ be commuting homeomorphisms on a compact metric space. A point $x \in M$ is called *non-wandering* if for every neighborhood U of x there exist $n_1, \ldots, n_q \geq 1$ such that $f_1^{n_1} \cdots f_q^{n_q}(U)$ intersects U. The *non-wandering set* is the set $\Omega(f_1, \ldots, f_q)$ of all non-wandering points. Prove that $\Omega(f_1, \ldots, f_q)$ is non-empty and compact.

2

Existence of invariant measures

In this chapter we prove the following result, which guarantees the existence of invariant measures for a broad class of transformations:

Theorem 2.1 (Existence of invariant measures). *Let $f : M \to M$ be a continuous transformation on a compact metric space. Then there exists some probability measure on M invariant under f.*

The main point in the proof is to introduce a certain topology in the set $\mathcal{M}_1(M)$ of probability measures on M, that we call weak* topology. The idea is that two measures are close, with respect to this topology, if the integrals they assign to (many) bounded continuous functions are close. The precise definition and some of the properties of the weak* topology are presented in Section 2.1. The crucial property, that makes this topology so useful for proving the existence theorem, is that it turns $\mathcal{M}_1(M)$ into a compact space (Theorem 2.1.5).

The proof of Theorem 2.1 is given in Section 2.2. We will also see, through examples, that the hypotheses of continuity and compactness cannot be omitted.

In Section 2.3 we insert the construction of the weak* topology into a broader framework from functional analysis and we also take the opportunity to introduce the notion of the Koopman operator of a transformation, which will be very useful in the sequel. In particular, as we are going to see, it allows us to give an alternative proof of Theorem 2.1, based on tools from functional analysis.

In Section 2.4 we describe certain explicit constructions of invariant measures for two important classes of systems: skew-products and natural extensions (or inverse limits) of non-invertible transformations.

Finally, in Section 2.5 we discuss some important applications of the idea of multiple recurrence (Section 1.5) in the context of combinatorial arithmetics. Theorem 2.1.5 has an important role in the arguments, which is the reason why this discussion was postponed to the present chapter.

2.1 Weak* topology

In this section M will always be a metric space. Our goal is to define the so-called weak* topology in the set $\mathcal{M}_1(M)$ of Borel probability measures on M and to discuss its main properties.

Let $d(\cdot,\cdot)$ be the distance function on M and $B(x,\delta)$ denote the ball of center $x \in M$ and radius $\delta > 0$. Given $B \subset M$, we define

$$d(x,B) = \inf\{d(x,y) : y \in B\}$$

and we call the *δ-neighborhood* of B the set B^δ of points $x \in M$ with $d(x,B) < \delta$.

2.1.1 Definition and properties of the weak* topology

Given a measure $\mu \in \mathcal{M}_1(M)$, a finite set $\Phi = \{\phi_1,\dots,\phi_N\}$ of bounded continuous functions $\phi_i : M \to \mathbb{R}$ and a number $\varepsilon > 0$, we define

$$V(\mu,\Phi,\varepsilon) = \{\nu \in \mathcal{M}_1(M) : \left|\int \phi_i d\nu - \int \phi_i d\mu\right| < \varepsilon \text{ for every } i\}. \quad (2.1.1)$$

Note that the intersection of any two such sets contains some set of this form. Thus, the family $\{V(\mu,\Phi,\varepsilon) : \Phi,\varepsilon\}$ may be taken as a basis of neighborhoods of each $\mu \in \mathcal{M}_1(M)$.

The *weak* topology* is the topology defined by these bases of neighborhoods. In other words, the open sets in the weak* topology are the sets $\mathcal{A} \subset \mathcal{M}_1(M)$ such that for every $\mu \in \mathcal{A}$ there exists some $V(\mu,\Phi,\varepsilon)$ contained in $\mathcal{A}$. Observe that the definition depends only on the topology of M, not on its distance. Furthermore, this topology is Hausdorff: Proposition A.3.3 implies that if μ and ν are distinct probabilities then there exist $\varepsilon > 0$ and some bounded continuous function $\phi : M \to \mathbb{R}$ such that $V(\mu,\{\phi\},\varepsilon) \cap V(\nu,\{\phi\},\varepsilon) = \emptyset$.

Lemma 2.1.1. *A sequence $(\mu_n)_{n\in\mathbb{N}}$ converges to a measure $\mu \in \mathcal{M}_1(M)$ in the weak* topology if and only if*

$$\int \phi\, d\mu_n \to \int \phi\, d\mu \quad \textit{for every bounded continuous function } \phi : M \to \mathbb{R}.$$

Proof. To prove the "only if" claim, consider any set $\Phi = \{\phi\}$ consisting of a single bounded continuous function ϕ. Since $(\mu_n)_n \to \mu$, for any $\varepsilon > 0$ there exists $\bar{n} \geq 1$ such that $\mu_n \in V(\mu,\Phi,\varepsilon)$ for every $n \geq \bar{n}$. This means, precisely, that

$$\left|\int \phi\, d\mu_n - \int \phi\, d\mu\right| < \varepsilon \quad \text{for every } n \geq \bar{n}.$$

In other words, the sequence $(\int \phi\, d\mu_n)_n$ converges to $\int \phi\, d\mu$.

The converse asserts that if $(\int \phi\, d\mu_n)_n$ converges to $\int \phi\, d\mu$ for every bounded continuous function ϕ then, given any $\Phi = \{\phi_1,\dots,\phi_N\}$ and $\varepsilon > 0$, there exists $\bar{n} \geq 1$ such that $\mu_n \in V(\mu,\Phi,\varepsilon)$ for $n \geq \bar{n}$. To check that this is so, let $\Phi = \{\phi_1, \dots, \phi_N\}$. The hypothesis ensures that for every i there exists $\bar{n}_i$ such that

$$\left| \int \phi_i \, d\mu_n - \int \phi_i \, d\mu \right| < \varepsilon \quad \text{for every } n \ge \bar{n}_i .$$

Taking $\bar{n} = \max\{\bar{n}_1, \dots, \bar{n}_N\}$ we get that $\mu_n \in V(\mu, \Phi, \varepsilon)$ for every $n \ge \bar{n}$.

2.1.2 Portmanteau theorem

Now let us discuss other useful ways of defining the weak* topology. Indeed, the relations (2.1.2), (2.1.3), (2.1.4) and (2.1.5) below introduce other natural choices for neighborhoods of a probability measure $\mu \in \mathcal{M}_1$. In Theorem 2.1.2 we prove that all these choices give rise to the same topology in $\mathcal{M}_1(M)$, which coincides with the weak* topology.

A direct variation of the definition of weak* topology is obtained by taking as the basis of neighborhoods the family of sets

$$V(\mu, \Psi, \varepsilon) = \{\eta \in \mathcal{M}_1(M) : \left| \int \psi_i \, d\eta - \int \psi_i \, d\mu \right| < \varepsilon \text{ for every } i\}, \quad (2.1.2)$$

where $\varepsilon > 0$ and $\Psi = \{\psi_1, \dots, \psi_N\}$ is a family of Lipschitz functions. The next definition is formulated in terms of closed subsets. Given any finite family $\mathcal{F} = \{F_1, \dots, F_N\}$ of closed subsets of M and given any $\varepsilon > 0$, consider

$$V_f(\mu, \mathcal{F}, \varepsilon) = \{\nu \in \mathcal{M}_1 : \nu(F_i) < \mu(F_i) + \varepsilon \text{ for every } i\}. \quad (2.1.3)$$

The next construction is analogous, just with open subsets instead of closed subsets. Given any finite family $\mathcal{A} = \{A_1, \dots, A_N\}$ of open subsets of M and given any $\varepsilon > 0$, consider

$$V_a(\mu, \mathcal{A}, \varepsilon) = \{\nu \in \mathcal{M}_1 : \nu(A_i) > \mu(A_i) - \varepsilon \text{ for every } i\}. \quad (2.1.4)$$

We call a *continuity set* of a measure μ any Borel subset B of M whose boundary ∂B has zero measure for μ. Given any finite family $\mathcal{B} = \{B_1, \dots, B_N\}$ of continuity sets of μ and given any $\varepsilon > 0$, consider

$$V_c(\mu, \mathcal{B}, \varepsilon) = \{\nu \in \mathcal{M}_1 : |\mu(B_i) - \nu(B_i)| < \varepsilon \text{ for every } i\}. \quad (2.1.5)$$

Given any two topologies $\mathcal{T}_1$ and $\mathcal{T}_2$ in the same set, we say that $\mathcal{T}_1$ is *weaker* than $\mathcal{T}_2$ (or, equivalently, that $\mathcal{T}_2$ is *stronger* than $\mathcal{T}_1$) if every subset that is open for $\mathcal{T}_1$ is also open for $\mathcal{T}_2$. We say that the two topologies are equivalent if they have exactly the same open sets.

Theorem 2.1.2. *The topologies defined by the bases of neighborhoods* (2.1.1), (2.1.2), (2.1.3), (2.1.4) *and* (2.1.5) *are all equivalent.*

Proof. Since every Lipschitz function is continuous, it is clear that the topology (2.1.2) is weaker than the topology (2.1.1).

To show that the topology (2.1.3) is weaker than the topology (2.1.2), consider any finite family $\mathcal{F} = \{F_1, \dots, F_N\}$ of closed subsets of M. According to Lemma A.3.4, for each $\delta > 0$ and each i there exists a Lipschitz function $\psi_i : M \to [0,1]$ such that $\mathcal{X}_{F_i} \le \psi_i \le \mathcal{X}_{F_i^\delta}$. Observe that $\bigcap_\delta F_i^\delta = F_i$, because F_i is closed, and so $\mu(F_i^\delta) \to \mu(F_i)$ when $\delta \to 0$. Fix $\delta > 0$ small enough so that $\mu(F_i^\delta) - \mu(F_i) < \varepsilon/2$ for every i. Let Ψ be the set of functions $\psi_1, \dots, \psi_N$ obtained in this way. Observe that

$$\left| \int \psi_i \, d\nu - \int \psi_i \, d\mu \right| < \varepsilon/2 \Rightarrow \nu(F_i) - \mu(F_i^\delta) < \varepsilon/2 \Rightarrow \nu(F_i) \le \mu(F_i) + \varepsilon$$

for every i. In other words, $V(\mu, \Psi, \varepsilon/2)$ is contained in $V_f(\mu, \mathcal{F}, \varepsilon)$.

It is easy to see that the topologies (2.1.3) and (2.1.4) are equivalent. Indeed, let $\mathcal{F} = \{F_1, \dots, F_n\}$ be any finite family of closed subsets and let $\mathcal{A} = \{A_1, \dots, A_N\}$, where each A_i is the complement of F_i. Clearly,

$$\begin{aligned} V_f(\mu, \mathcal{F}, \varepsilon) &= \{\nu \in \mathcal{M}_1 : \nu(F_i) < \mu(F_i) + \varepsilon \text{ for every } i\} \\ &= \{\nu \in \mathcal{M}_1 : \nu(A_i) > \mu(A_i) - \varepsilon \text{ for every } i\} = V_a(\mu, \mathcal{A}, \varepsilon). \end{aligned}$$

Next, let us show that the topology (2.1.5) is weaker than these equivalent topologies (2.1.3) and (2.1.4). Given any finite family $\mathcal{B} = \{B_1, \dots, B_N\}$ of continuity sets of μ, let F_i be the closure and A_i be the interior of each B_i. Denote $\mathcal{F} = \{F_1, \dots, F_N\}$ and $\mathcal{A} = \{A_1, \dots, A_N\}$. Since $\mu(F_i) = \mu(B_i) = \mu(A_i)$,

$$\nu(F_i) < \mu(F_i) + \varepsilon \Rightarrow \nu(B_i) < \mu(B_i) + \varepsilon$$

$$\nu(A_i) > \mu(A_i) - \varepsilon \Rightarrow \nu(B_i) > \mu(B_i) - \varepsilon$$

for every i. This means that $V_f(\mu, \mathcal{F}, \varepsilon) \cap V_a(\mu, \mathcal{A}, \varepsilon)$ is contained in $V_c(\mu, \mathcal{B}, \varepsilon)$.

Finally, let us prove that the topology (2.1.1) is weaker than the topology (2.1.5). Let $\Phi = \{\phi_1, \dots, \phi_N\}$ be a finite family of bounded continuous functions. Fix an integer number ℓ such that $\sup |\phi_i(x)| < \ell$ for every i. For each i, the pre-images $\phi_i^{-1}(s)$, $s \in [-\ell, \ell]$ are pairwise disjoint. Hence, $\mu(\phi_i^{-1}(s)) = 0$ except for a countable set of values of s. In particular, we may choose $k \in \mathbb{N}$ and points $-\ell = t_0 < t_1 < \cdots < t_{k-1} < t_k = \ell$ such that $t_j - t_{j-1} < \varepsilon/2$ and $\mu(\{\phi_i^{-1}(t_j)\}) = 0$ for every j. Then, each

$$B_{i,j} = \phi_i^{-1}((t_{j-1}, t_j])$$

is a continuity set of μ. Moreover,

$$\sum_{j=1}^{k} t_j \, \mu(B_{i,j}) \ge \int \phi_i \, d\mu \ge \sum_{j=1}^{k} t_{j-1} \, \mu(B_{i,j}) > \sum_{j=1}^{k} t_j \, \mu(B_{i,j}) - \varepsilon/2,$$

and we also have similar inequalities for the integrals relative to ν. It follows that

$$\left| \int \phi_i \, d\mu - \int \phi_i \, d\nu \right| \le \sum_{j=1}^{k} \ell \, |\mu(B_{i,j}) - \nu(B_{i,j})| + \varepsilon/2 \tag{2.1.6}$$

for every i. Denote $\mathcal{B} = \{B_{i,j} : i = 1, \dots, N \text{ and } j = 1, \dots, k\}$. Then the relation (2.1.6) implies that $V_c(\mu, \mathcal{B}, \varepsilon/(2k\ell))$ is contained in $V(\mu, \Phi, \varepsilon)$.

2.1.3 The weak* topology is metrizable

Now assume that the metric space M is separable. We will see in Exercise 2.1.3 that the weak* topology on $\mathcal{M}_1(M)$ is separable. Here we show that it is also metrizable: we exhibit a distance function on $\mathcal{M}_1(M)$ that induces the weak* topology.

Given $\mu, \nu \in \mathcal{M}_1(M)$, define $D(\mu, \nu)$ to be the infimum of all numbers $\delta > 0$ such that

$$\mu(B) < \nu(B^\delta) + \delta \quad \text{and} \quad \nu(B) < \mu(B^\delta) + \delta \tag{2.1.7}$$

for every Borel set $B \subset M$.

Lemma 2.1.3. *The function D is a distance on $\mathcal{M}_1(M)$.*

Proof. Let us start by showing that $D(\mu, \nu) = 0$ implies $\mu = \nu$. Indeed, the hypothesis implies

$$\mu(B) \le \nu(\bar{B}) \quad \text{and} \quad \nu(B) \le \mu(\bar{B})$$

for every Borel set $B \subset M$, where $\bar{B}$ denotes the closure of B. When B is closed, these inequalities mean that $\mu(B) = \nu(B)$. As we have seen previously, any two measures that coincide on the closed subsets are necessarily the same.

We leave it to the reader to check all the other conditions in the definition of a distance (Exercise 2.1.5).

This distance D is called the *Levy–Prohorov metric* on $\mathcal{M}_1(M)$. In what follows we denote by $B_D(\mu, r)$ the ball of radius $r > 0$ around any $\mu \in \mathcal{M}_1(M)$.

Proposition 2.1.4. *If M is a separable metric space then the topology induced by the Levy–Prohorov distance D coincides with the weak* topology on $\mathcal{M}_1(M)$.*

Proof. Let $\varepsilon > 0$ and $\mathcal{F} = \{F_1, \dots, F_N\}$ be a finite family of closed subsets of M. Fix $\delta \in (0, \varepsilon/2)$ such that $\mu(F_i^\delta) < \mu(F_i) + \varepsilon/2$ for every i. If $\nu \in B_D(\mu, \delta)$ then

$$\nu(F_i) < \mu(F_i^\delta) + \delta < \mu(F_i) + \varepsilon \quad \text{for every } i,$$

which means that $\nu \in V_f(\mu, \mathcal{F}, \varepsilon)$. This shows that the topology induced by the distance D is stronger than the topology (2.1.3) which, as we have seen, is equivalent to the weak* topology.

We are left to prove that if M is separable then the weak* topology is stronger than the topology induced by D. For that, let $\{p_1, p_2, \dots\}$ be any countable dense subset of M. Given $\varepsilon > 0$, let us fix $\delta \in (0, \varepsilon/3)$. For each j, the spheres $\partial B(p_j, r) = \{x : d(x, p_j) = r\}$, $r > 0$ are pairwise disjoint. So, we may find $r > 0$ arbitrarily small such that $\mu(\partial B(p_j, r)) = 0$ for every j. Fix any such r, with $r \in (0, \delta/3)$. The family $\{B(p_j, r) : j = 1, 2, \dots\}$ is a countable cover of M by continuity sets of μ. Fix $k \ge 1$ such that the set $U = \bigcup_{j=1}^k B(p_j, r)$ satisfies

$$\mu(U) > 1 - \delta. \tag{2.1.8}$$

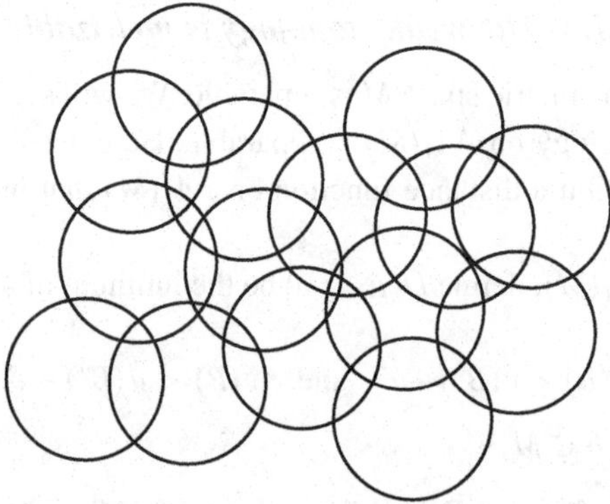

Figure 2.1. Partition defined by a finite cover

Next, let us consider the (finite) partition $\mathcal{P}$ of U defined by the family of balls

$$\{B(p_j,r) : j = 1,\dots,k\}.$$

That is, the elements of $\mathcal{P}$ are the maximal sets $P \subset U$ such that, for each j, either P is contained in $B(p_j,r)$ or P is disjoint from $B(p_j,r)$. See Figure 2.1. Now let $\mathcal{E}$ be the family of all finite unions of elements of $\mathcal{P}$. Note that the boundary of every element of $\mathcal{E}$ has measure zero, since it is contained in the union of the boundaries of the balls $B(p_j,r)$, $1 \le j \le k$. That is, every element of $\mathcal{E}$ is a continuity set of μ.

If $\nu \in V_c(\mu,\mathcal{E},\delta)$ then

$$|\mu(E) - \nu(E)| < \delta \quad \text{for every } E \in \mathcal{E}. \tag{2.1.9}$$

In particular, (2.1.8) together with (2.1.9) imply that

$$\nu(U) > 1 - 2\delta. \tag{2.1.10}$$

Now, given any Borel subset B, denote by E_B the union of all the elements of $\mathcal{P}$ that intersect B. Then $E_B \in \mathcal{E}$ and so the relation (2.1.9) yields

$$|\mu(E_B) - \nu(E_B)| < \delta.$$

Observe that B is contained in $E_B \bigcup U^c$. Moreover, $E_B \subset B^\delta$ because every element of $\mathcal{P}$ has diameter less than $2r < \delta$. These facts, together with (2.1.8) and (2.1.10), imply that

$$\mu(B) \le \mu(E_B) + \delta < \nu(E_B) + 2\delta \le \nu(B^\delta) + 2\delta$$

$$\nu(B) \le \nu(E_B) + 2\delta < \mu(E_B) + 3\delta \le \mu(B^\delta) + 3\delta.$$

Since $3\delta < \varepsilon$, these relations imply that $\nu \in B_D(\mu,\varepsilon)$.

One can show that if M is a complete separable metric space then the Levy–Prohorov metric on $\mathcal{M}_1(M)$ is complete (and separable, by Exercise 2.1.3). See, for example, Theorem 6.8 in Billingsley [Bil68].

2.1.4 The weak* topology is compact

In this section we take the metric space M to be compact. We are going to prove the following fundamental result:

Theorem 2.1.5. *The space $\mathcal{M}_1(M)$ is compact for the weak* topology.*

Since we already know that $\mathcal{M}_1(M)$ is metrizable, it suffices to prove:

Proposition 2.1.6. *Every sequence $(\mu_k)_{k\in\mathbb{N}}$ in $\mathcal{M}_1(M)$ has some subsequence that converges in the weak* topology.*

Proof. Let $\{\phi_n : n \in \mathbb{N}\}$ be a countable dense subset of the unit ball of $C^0(M)$ (recall Theorem A.3.13). For each $n \in \mathbb{N}$, the sequence of real numbers $\int \phi_n \, d\mu_k$, $k \in \mathbb{N}$ is bounded by 1. Hence, for each $n \in \mathbb{N}$ there exists a sequence $(k_j^n)_{j\in\mathbb{N}}$ such that

$$\int \phi_n \, d\mu_{k_j^n} \text{ converges to some number } \Phi_n \in \mathbb{R} \text{ when } j \to \infty.$$

Moreover, each sequence $(k_j^{n+1})_{j\in\mathbb{N}}$ may be chosen to be a subsequence of the previous $(k_j^n)_{j\in\mathbb{N}}$. Define $\ell_j = k_j^j$ for each $j \in \mathbb{N}$. By construction, $(\ell_j)_{j\in\mathbb{N}}$ is a subsequence of every $(k_j^n)_{j\in\mathbb{N}}$, up to finitely many terms. Hence,

$$\left(\int \phi_n \, d\mu_{\ell_j}\right)_j \to \Phi_n \quad \text{for every } n \in \mathbb{N}.$$

One can easily deduce that

$$\Phi(\varphi) = \lim_j \int \varphi \, d\mu_{\ell_j} \tag{2.1.11}$$

exists, for every function $\varphi \in C^0(M)$. Indeed, suppose first that φ is in the unit ball of $C^0(M)$. Given any $\varepsilon > 0$, we may find $n \in \mathbb{N}$ such that $\|\varphi - \phi_n\| \le \varepsilon$. Then,

$$\left|\int \varphi \, d\mu_{\ell_j} - \int \phi_n \, d\mu_{\ell_j}\right| \le \varepsilon$$

for every j. Since $\int \phi_n \, d\mu_{\ell_j}$ converges (to Φ_n), it follows that

$$\limsup_j \int \varphi \, d\mu_{\ell_j} - \liminf_j \int \varphi \, d\mu_{\ell_j} \le 2\varepsilon.$$

Since ε is arbitrary, we find that $\lim_j \int \varphi \, d\mu_{\ell_j}$ exists. This proves (2.1.11) when the function is in the unit ball. The general case reduces immediately to this one, just replacing φ with $\varphi/\|\varphi\|$. In this way, we have completed the proof of (2.1.11).

Finally, it is clear that the operator $\Phi : C^0(M) \to \mathbb{R}$ defined by (2.1.11) is linear and positive: $\Phi(\varphi) \ge \min \varphi \ge 0$ whenever $\varphi \ge 0$ at all points. Moreover, $\Phi(1) = 1$. Thus, by Theorem A.3.11, there exists some Borel probability measure μ on M such that $\Phi(\varphi) = \int \varphi \, d\mu$ for every continuous function φ. Now, the equality in (2.1.11) may be rewritten as

$$\int \varphi \, d\mu = \lim_j \int \varphi \, d\mu_{\ell_j} \quad \text{for every } \varphi \in C^0(M).$$

According to Lemma 2.1.1, this means that the subsequence $(\mu_{\ell_j})_{j\in\mathbb{N}}$ converges to μ in the weak* topology.

As we observed previously, Theorem 2.1.5 is an immediate consequence of the proposition we have just proved.

2.1.5 Theorem of Prohorov

The theorem that we are going to state in this section provides a very general criterion for a family of probability measures to be compact. Indeed, the class of metric spaces M to which it applies includes virtually all interesting examples.

Definition 2.1.7. A set $\mathcal{M}$ of Borel measures in a topological space is *tight* if for every $\varepsilon > 0$ there exists a compact set $K \subset M$ such that $\mu(K^c) < \varepsilon$ for every measure $\mu \in \mathcal{M}$.

Note that when $\mathcal{M}$ consists of a single measure this definition corresponds exactly to Definition A.3.6. Clearly, tightness is a hereditary property: if a set is tight then all its subsets are also tight. Note also that if M is a compact metric space then the space $\mathcal{M}_1(M)$ of all probability measures is a tight set. So, the next result is an extension of Theorem 2.1.5:

Theorem 2.1.8 (Prohorov). *Let M be a complete separable metric space. A set $\mathcal{K} \subset \mathcal{M}_1(M)$ is tight if and only if every sequence in $\mathcal{K}$ admits some subsequence that is convergent in the weak* topology of $\mathcal{M}_1(M)$.*

Proof. We only prove the necessary condition, which is the most useful part of the statement. Then, in Exercise 2.1.8, we invite the reader to prove the converse.

Suppose that $\mathcal{K}$ is tight. Consider an increasing sequence $(K_l)_l$ of compact subsets of M such that $\eta(K_l^c) \le 1/l$ for every l and every $\eta \in \mathcal{K}$. Fix any sequence $(\mu_n)_n$ in $\mathcal{K}$. To begin with, we claim that for every l there exists a subsequence $(n_j)_j$ and there exists a measure ν_l on M such that $\nu_l(K_l^c) = 0$ and $(\mu_{n_j} \mid K_l)_j$ converges to ν_l, in the sense that

$$\int_{K_l} \psi \, d\mu_{n_j} \to \int_{K_l} \psi \, d\nu_l \quad \text{for every continuous function } \psi : K_l \to \mathbb{R}. \tag{2.1.12}$$

Indeed, that is a simple consequence of Theorem 2.1.5: up to restricting to a subsequence, we may suppose that the limit $b_l = \lim_n \mu_n(K_l)$ exists (note that

$1 \geq b_l \geq 1 - 1/l$); it follows from the theorem that the sequence of normalized restrictions

$$\big((\mu_n \mid K_l)/\mu_n(K_l)\big)_n$$

admits a subsequence converging to some probability measure $\eta_l \in \mathcal{M}_1(K_l)$; to conclude the proof of the claim it suffices to take η_l to be a probability measure on M with $\eta_l(K_l^c) = 0$ and to choose $\nu_l = b_l \eta_l$.

Next, using a diagonal argument analogous to the one in Proposition 2.1.6, we may choose a subsequence $(n_j)_j$ in such a way that (2.1.12) holds, simultaneously, for every $l \geq 1$. Observe that the sequence $(\nu_l)_l$ is non-decreasing: given $k > l$ and any continuous function $\phi : M \to [0,1]$,

$$\int \phi \, d\nu_l = \lim_j \int_{K_l} \phi \, d\mu_{n_j} \leq \lim_j \int_{K_k} \phi \, d\mu_{n_j} = \int \phi \, d\nu_k.$$

Analogously, for any $k > l$ and any continuous function $\phi : M \to [0,1]$,

$$\int \phi \, d\nu_k - \int \phi \, d\nu_l = \lim_j \int_{K_k \setminus K_l} \phi \, d\mu_{n_j} \leq \limsup_j \mu_{n_j}(K_l^c) \leq 1/l.$$

Using Exercise A.3.5, we may translate this in terms of measures of sets (rather than integrals of functions): for every $k > l$ and every Borel set $E \subset M$,

$$\nu_l(E) \leq \nu_k(E) \leq \nu_l(E) + 1/l. \tag{2.1.13}$$

Define $\nu(E) = \lim_l \nu_l(E)$ for each Borel set E. We claim that ν is a probability measure on M. It is immediate from the definition that $\nu(\emptyset) = 0$ and that ν is additive. Furthermore, $\nu(M) = \lim_l \nu(K_l) = \lim_l b_l = 1$. To show that ν is countably additive (σ-additive), we use the criterion of continuity at the empty set (Theorem A.1.14). Consider any decreasing sequence $(B_n)_n$ of Borel subsets of M with $\bigcap_n B_n = \emptyset$. Given $\varepsilon > 0$, choose l such that $1/l < \varepsilon$. Since ν_l is countably additive, Theorem A.1.14 shows that $\nu_l(B_n) < \varepsilon$ for every n sufficiently large. Hence, $\nu(B_n) \leq \nu_l(B_n) + 1/l < 2\varepsilon$ for every n sufficiently large. This proves that $(\nu(B_n))_n$ converges to zero and, by Theorem A.1.14, it follows that ν is indeed countably additive.

The definition of ν implies (see Exercise 2.1.1 or Exercise 2.1.4) that $(\nu_l)_l$ converges to ν in the weak* topology. So, given $\varepsilon > 0$ and any bounded continuous function $\varphi : M \to \mathbb{R}$, we have that $|\int \varphi \, d\nu_l - \int \varphi \, d\nu| < \varepsilon$ for every l sufficiently large. Fix l such that, in addition, $\sup |\varphi|/l < \varepsilon$. Then,

$$\left| \int \varphi \, d\mu_{n_j} - \int \varphi \, d\nu_l \right| \leq \left| \int_{K_l^c} \varphi \, d\mu_{n_j} \right| + \left| \int_{K_l} \varphi \, d\mu_{n_j} - \int_{K_l} \varphi \, d\nu_l \right| \leq 2\varepsilon$$

for every j sufficiently large. This shows that $|\int \varphi \, d\mu_{n_j} - \int \varphi \, d\nu| < 3\varepsilon$ whenever j is large enough. Thus, $(\mu_{n_j})_j$ converges to ν in the weak* topology.

2.1.6 Exercises

2.1.1. Let M be a metric space and $(\mu_n)_n$ be a sequence in $\mathcal{M}_1(M)$. Show that the following conditions are all equivalent:
1. $(\mu_n)_n$ converges to a probability measure μ in the weak* topology.
2. $\limsup_n \mu_n(F) \leq \mu(F)$ for every closed set $F \subset M$.
3. $\liminf_n \mu_n(A) \geq \mu(A)$ for every open set $A \subset M$.
4. $\lim_n \mu_n(B) = \mu(B)$ for every continuity set B of μ.
5. $\lim_n \int \psi \, d\mu_n = \int \psi \, d\mu$ for every Lipschitz function $\psi : M \to \mathbb{R}$.

2.1.2. Fix any dense subset $\mathcal{F}$ of the unit ball of $C^0(M)$. Show that a sequence $(\mu_n)_{n\in\mathbb{N}}$ of probability measures on M converges to some $\mu \in \mathcal{M}_1(M)$ in the weak* topology if and only if

$$\int \phi \, d\mu_n \text{ converges to } \int \phi \, d\mu, \quad \text{for every } \phi \in \mathcal{F}.$$

2.1.3. Show that the subset formed by the measures with finite support is dense in $\mathcal{M}_1(M)$, relative to the weak* topology. Assuming that the metric space M is separable, conclude that $\mathcal{M}_1(M)$ is also separable.

2.1.4. The *uniform topology* in $\mathcal{M}_1(M)$ is defined by the basis of neighborhoods

$$V_u(\mu,\varepsilon) = \{\nu \in \mathcal{M}_1(M) : |\mu(B) - \nu(B)| < \varepsilon \text{ for every } B \in \mathcal{B}\},$$

and the *pointwise topology* is defined by the basis of neighborhoods

$$V_p(\mu,\mathcal{B},\varepsilon) = \{\nu \in \mathcal{M}_1(M) : |\mu(B_i) - \nu(B_i)| < \varepsilon \text{ for every } i\},$$

where $\varepsilon > 0$, $n \geq 1$ and $\mathcal{B} = \{B_1,\ldots,B_N\}$ is a finite family of measurable sets. Check that the uniform topology is stronger than the pointwise topology and the latter is stronger than the weak* topology. Show, by means of examples, that these relations may be strict.

2.1.5. Complete the proof of Lemma 2.1.3.

2.1.6. Let V_k, $k = 1,2,\ldots$ be random variables with real values, that is, measurable functions $V_k : (X,\mathcal{B},\mu) \to \mathbb{R}$ defined in some probability space $(X,\mathcal{B},\mu)$. The *distribution function* of V_k is the monotone function $F_k : \mathbb{R} \to [0,1]$ defined by $F_k(a) = \mu(\{x \in X : V_k(x) \leq a\})$. We say that $(V_k)_k$ *converges in distribution* to some random variable V if $\lim_k F_k(a) = F(a)$ for every continuity point a of the distribution function F of the random variable V. What does this have to do with the weak* topology?

2.1.7. Let $(\mu_n)_{n\in\mathbb{N}}$ be a sequence of probability measures converging to some μ in the weak* topology. Let B be a continuity set of μ with $\mu(B) > 0$. Prove that the normalized restrictions $(\mu_n \mid B)/\mu_n(B)$ converge to the normalized restriction $(\mu \mid B)/\mu(B)$ when $n \to \infty$. What can be said if we replace continuity sets by closed sets or by open sets?

2.1.8. (Converse to the theorem of Prohorov) Prove that if $\mathcal{K} \subset \mathcal{M}_1(M)$ is such that every sequence in $\mathcal{K}$ admits some convergent subsequence then $\mathcal{K}$ is tight.

2.2 Proof of the existence theorem

Given any $f: M \to M$ and any measure η on M, we denote by $f_*\eta$ and call the *iterate (or image) of η under f* the measure defined by

$$f_*\eta(B) = \eta(f^{-1}(B))$$

for each measurable set $B \subset M$. Note that the measure η is invariant under f if and only if $f_*\eta = \eta$.

Lemma 2.2.1. *Let η be a measure and ϕ be a bounded measurable function. Then*

$$\int \phi \, df_*\eta = \int \phi \circ f \, d\eta. \tag{2.2.1}$$

Proof. If ϕ is the characteristic function of a measurable set B then the relation (2.2.1) means that $f_*\eta(B) = \eta(f^{-1}(B))$, which holds by definition. By linearity of the integral, it follows that (2.2.1) holds whenever ϕ is a simple function. Finally, since every bounded measurable function can be approximated by simple functions (see Proposition A.1.33), it follows that the claim in the lemma is true in general.

Proposition 2.2.2. *If $f: M \to M$ is continuous then $f_*: \mathcal{M}_1(M) \to \mathcal{M}_1(M)$ is continuous relative to the weak* topology.*

Proof. Let $\varepsilon > 0$ and $\Phi = \{\phi_1, \dots, \phi_n\}$ be any family of bounded continuous functions. Since f is continuous, the family $\Psi = \{\phi_1 \circ f, \dots, \phi_n \circ f\}$ also consists of bounded continuous functions. By the previous lemma,

$$\left| \int \phi_i \, d(f_*\mu) - \int \phi_i \, d(f_*\nu) \right| = \left| \int (\phi_i \circ f) \, d\mu - \int (\phi_i \circ f) \, d\nu \right|$$

and so the left-hand side is smaller than ε if the right-hand side is smaller than ε. That means that

$$f_*\big(V(\mu, \Psi, \varepsilon)\big) \subset V(f_*\mu, \Phi, \varepsilon)) \quad \text{for every } \mu, \Phi \text{ and } \varepsilon,$$

and this last fact shows that f_* is continuous.

At this point, Theorem 2.1 could be deduced from the classical Schauder–Tychonoff fixed point theorem for continuous operators in topological vector spaces. A *topological vector space* is a vector space V endowed with a topology relative to which both operations of V (addition and multiplication by a scalar) are continuous. A set $K \subset V$ is said to be *convex* if $(1-t)x + ty \in K$ for every $x, y \in K$ and every $t \in [0,1]$.

Theorem 2.2.3 (Schauder–Tychonoff). *Let $F: V \to V$ be a continuous transformation on a topological vector space V. Suppose that there exists a compact convex set $K \subset V$ such that $F(K) \subset K$. Then $F(v) = v$ for some $v \in K$.*

Theorem 2.1 corresponds to the special case when $V = \mathcal{M}(M)$ is the space of complex measures, $K = \mathcal{M}_1(M)$ is the space of probability measures on M and $F = f_*$ is the action of f on $\mathcal{M}(M)$.

However, the situation of Theorem 2.1 is a lot simpler than the general case of the Schauder–Tychonoff theorem because the operator f_* besides being continuous is also *linear*. This allows for a direct and elementary proof of Theorem 2.1 that also provides some additional information about the invariant measure.

To that end, let ν be any probability measure on M: for example, ν could be the Dirac mass at any point. Form the sequence of probability measures

$$\mu_n = \frac{1}{n}\sum_{j=0}^{n-1} f_*^j \nu, \tag{2.2.2}$$

where $f_*^j\nu$ is the image of ν under the iterate f^j. By Theorem 2.1.5, this sequence has some accumulation point, that is, there exists some subsequence $(n_k)_{k\in\mathbb{N}}$ and some probability measure $\mu \in \mathcal{M}_1(M)$ such that

$$\frac{1}{n_k}\sum_{j=0}^{n_k-1} f_*^j \nu \to \mu \tag{2.2.3}$$

in the weak* topology. Now we only need to prove:

Lemma 2.2.4. *Every accumulation point of a sequence $(\mu_n)_{n\in\mathbb{N}}$ of the form (2.2.2) is a probability measure invariant under f.*

Proof. The relation (2.2.3) asserts that, given any family $\Phi = \{\phi_1,\dots,\phi_N\}$ of bounded continuous functions and given any $\varepsilon > 0$, we have

$$\left|\frac{1}{n_k}\sum_{j=0}^{n_k-1}\int (\phi_i \circ f^j)\, d\nu - \int \phi_i\, d\mu\right| < \varepsilon/2 \tag{2.2.4}$$

for every i and every k sufficiently large. By Proposition 2.2.2,

$$f_*\mu = f_*\left(\lim_k \frac{1}{n_k}\sum_{j=0}^{n_k-1} f_*^j\nu\right) = \lim_k \frac{1}{n_k}\sum_{j=1}^{n_k} f_*^j\nu. \tag{2.2.5}$$

Now observe that

$$\begin{aligned}\left|\frac{1}{n_k}\sum_{j=0}^{n_k-1}\int (\phi_i\circ f^j)\,d\nu - \frac{1}{n_k}\sum_{j=1}^{n_k}\int (\phi_i\circ f^j)\,d\nu\right| \\ = \frac{1}{n_k}\left|\int \phi_i\,d\nu - \int(\phi_i\circ f^{n_k})\,d\nu\right| \le \frac{2}{n_k}\sup|\phi_i|,\end{aligned}$$

and the latter expression is smaller than $\varepsilon/2$ for every i and every k sufficiently large. Combining this fact with (2.2.4), we conclude that

$$\left|\frac{1}{n_k}\sum_{j=1}^{n_k}\int(\phi_i \circ f^j)\,d\nu - \int \phi_i\,d\mu\right| < \varepsilon \tag{2.2.6}$$

for every i and every k sufficiently large. This means that

$$\frac{1}{n_k}\sum_{j=1}^{n_k} f_*^j \nu \to \mu$$

when $k \to \infty$. However, (2.2.5) means that this sequence converges to $f_*\mu$. By uniqueness of the limit, it follows that $f_*\mu = \mu$.

Now the proof of Theorem 2.1 is complete. The examples that follow show that neither of the two hypotheses in the theorem, continuity and compactness, may be omitted.

Example 2.2.5. Consider $f : (0,1] \to (0,1]$ given by $f(x) = x/2$. Suppose that f admits some invariant probability measure: we are going to show that this is actually not true. By the recurrence theorem (Theorem 1.2.4), relative to that probability measure almost every point of $(0,1]$ is recurrent. However, it is clear that there are no recurrent points: the orbit of every $x \in (0,1]$ converges to zero and, in particular, does not accumulate on the initial point x. Hence, f is an example of a continuous transformation (on a non-compact space) that does not have any invariant probability measure.

Example 2.2.6. Modifying a little the previous construction, we see that the same phenomenon may occur in compact spaces, if the transformation is not continuous. Consider $f : [0,1] \to [0,1]$ given by $f(x) = x/2$ if $x \neq 0$ and $f(0) = 1$. For the same reason as before, no point $x \in (0,1]$ is recurrent. So, if there exists some invariant probability measure μ then it must give full weight to the sole recurrent point $x = 0$. In other words, μ must be the Dirac mass supported at zero, that is, the measure δ_0 defined by

$$\delta_0(E) = 1 \text{ if } 0 \in E \quad \text{and} \quad \delta_0(E) = 0 \text{ if } 0 \notin E.$$

However, the measure δ_0 *is not* invariant under f: for example, the measurable set $E = \{0\}$ has measure 1 and yet its pre-image $f^{-1}(E)$ is the empty set, which has measure zero. Thus, this transformation f has no invariant probability measures.

Our third example is of a different nature. We include it to stress the limitations of Theorem 2.1 (which are inherent to its great generality): the measures whose existence is ensured by the theorem may be completely trivial; for example, in the situation that we are going to describe "almost every point" just means the point $x = 0$. For this reason, an important objective in ergodic theory is to construct more sophisticated invariant measures, with additional interesting properties such as, for instance, being equivalent to the Lebesgue measure.

Example 2.2.7. Consider $f : [0,1] \to [0,1]$ given by $f(x) = x/2$. This is a continuous transformation on a compact space. So, by Theorem 2.1, f admits some invariant probability measure. Using the same arguments as in the previous example, we find that there exists a unique invariant probability measure, namely, the Dirac mass δ_0 at the origin. Note that in this case the measure δ_0 is indeed invariant.

As an immediate application of Theorem 2.1, we have the following alternative proof of the Birkhoff recurrence theorem (Theorem 1.2.6). Suppose that $f : M \to M$ is a continuous transformation on a compact metric space. By Theorem 2.1, there exists some f-invariant probability measure μ. Every compact metric space admits a countable basis of open sets. So, we may apply Theorem 1.2.4 to conclude that μ-almost every point is recurrent. In particular, the set of recurrent points is non-empty, as stated by Theorem 1.2.6.

2.2.1 Exercises

2.2.1. Prove the following generalization of Lemma 2.2.4. Let $f : M \to M$ be a continuous transformation on a compact metric space, ν be a probability measure on M and $(I_n)_n$ be a sequence of intervals of natural numbers such that $\#I_n$ converges to infinity when n goes to infinity. Then every accumulation point of the sequence

$$\mu_n = \frac{1}{\#I_n} \sum_{j \in I_n} f_*^j \nu$$

is an f-invariant probability measure.

2.2.2. Let $f_1, \dots, f_q : M \to M$ be any finite family of *commuting* continuous transformations on a compact metric space. Prove that there exists some probability measure μ that is invariant under f_i for every $i \in \{1, \dots, q\}$. In fact, the conclusion remains true for any countable family $\{f_j : j \in \mathbb{N}\}$ of commuting continuous transformations on a compact metric space.

2.2.3. Let $f : [0,1] \to [0,1]$ be the decimal expansion transformation. Show that for every $k \geq 1$ there exists some invariant probability measure whose support is formed by exactly k points (in particular, f admits infinitely many invariant probability measures). Determine whether there are invariant probability measures μ such that

(a) the support of μ is infinite countable;
(b) the support of μ is non-countable but has empty interior;
(c) the support of μ has non-empty interior but μ is singular with respect to the Lebesgue measure m.

2.2.4. Prove the theorem of existence of invariant measures for continuous flows: every continuous flow $(f^t)_{t\in\mathbb{R}}$ on a compact metric space admits some invariant probability measure.

2.2.5. Show that the transformation $f : [-1,1] \to [-1,1]$, $f(x) = 1 - 2x^2$ has some invariant probability measure equivalent to the Lebesgue on the interval.

2.2.6. Let $f : M \to M$ be an invertible measurable transformation and m be a probability measure on M such that $m(A) = 0$ if and only if $m(f(A)) = 0$. We say that the pair (f,m) is *totally dissipative* if there exists a measurable set $W \subset M$ whose iterates $f^j(W), j \in \mathbb{Z}$ are pairwise disjoint and such that their union has full measure. Prove that if (f,m) is totally dissipative then f admits some σ-finite invariant measure equivalent to Lebesgue measure m. This measure is necessarily infinite.

2.2.7. Let $f : M \to M$ be an invertible measurable transformation and m be a probability measure on M such that $m(A) = 0$ if and only if $m(f(A)) = 0$. We say that the pair (f,m) is *conservative* if there is no measurable set $W \subset M$ with positive measure whose iterates $f^j(W), j \in \mathbb{Z}$ are pairwise disjoint. Show that if (f,m) is conservative then, for every measurable set $X \subset M$, m-almost every point of X returns to X infinitely times.

2.2.8. Suppose that (f,m) is conservative. Show that f admits a σ-finite invariant measure μ equivalent to m if and only if there exist sets $X_1 \subset \cdots \subset X_n \subset \cdots$ with $M = \bigcup_n X_n$ and $m(X_n) < \infty$ for every n, such that the first-return map f_n to each X_n admits a finite invariant measure μ_n absolutely continuous with respect to the restriction of m to X_n.

2.2.9. Find conservative pairs (f,m) such that f has no finite invariant measures equivalent to m. [Observation: Ornstein [Orn60] gave examples such that f does not even have σ-finite invariant measures equivalent to m.]

2.3 Comments in functional analysis

The definition of weak* topology in the space of probability measures is a special case of a construction from functional analysis that is worthwhile recalling here. It leads us to introducing a certain linear isometry U_f in the space $L^1(\mu)$, called the *Koopman operator* of the system (f,μ). These operators have an important role in ergodic theory because they allow for powerful tools from Analysis to be used in the study of invariant measures. To illustrate this fact, we present an alternative proof of Theorem 2.1 based on the spectral properties of the Koopman operator.

2.3.1 Duality and weak topologies

Let E be a Banach space, that is, a vector space endowed with a complete norm. The *dual* of E is the space E^* of all continuous linear functionals defined on E. This is also a Banach space, with the norm

$$\|g\| = \sup\left\{\frac{|g(v)|}{\|v\|} : v \in E \setminus \{0\}\right\}. \tag{2.3.1}$$

The *weak topology* in the space E is the topology defined by the following basis of neighborhoods:

$$V(v,\{g_1,\dots,g_N\},\varepsilon) = \{w \in E : |g_i(v) - g_i(w)| < \varepsilon \text{ for every } i\}, \qquad (2.3.2)$$

where $g_1,\dots,g_N \in E^*$. In terms of sequences, it satisfies

$$(v_n)_n \to v \quad \Rightarrow \quad (g(v_n))_n \to g(v) \quad \text{for every } g \in E^*.$$

The *weak* topology* in the dual space E^* is the topology defined by the following basis of neighborhoods:

$$V^*(g,\{v_1,\dots,v_N\},\varepsilon) = \{h \in E^* : |g(v_i) - h(v_i)| < \varepsilon \text{ for every } i\}, \qquad (2.3.3)$$

where $v_1,\dots,v_N \in E$. It satisfies

$$(g_n)_n \to g \quad \Rightarrow \quad (g_n(v))_n \to g(v) \quad \text{for every } v \in E.$$

The weak* topology has the following remarkable property:

Theorem 2.3.1 (Banach–Alaoglu). *The closed unit ball of E^* is compact for the weak* topology.*

The construction carried out in the previous sections corresponds to the situation where E is the space $C^0(M)$ of (complex) continuous functions and E^* is the space $\mathcal{M}(M)$ of complex measures on a compact metric space M: according to the theorem of Riesz–Markov (Theorem A.3.12), $\mathcal{M}(M)$ corresponds to the dual of $C^0(M)$ when we identify each measure $\mu \in \mathcal{M}(M)$ with the linear functional $I_\mu(\phi) = \int \phi\, d\mu$. Note that the definition of the norm (2.3.1) implies that

$$\|\mu\| = \sup\left\{\frac{|\int \phi\, d\mu|}{\sup|\phi|} : \phi \in C^0(M) \setminus \{0\}\right\}.$$

In particular, the set $\mathcal{M}_1(M)$ of probability measures is contained in the unit ball of $\mathcal{M}(M)$. Since this set is closed for the weak* topology, we conclude that Theorem 2.1.5 is also a direct consequence of the theorem of Banach–Alaoglu.

Now consider any continuous transformation $f : M \to M$ and the corresponding action $f_* : \mathcal{M}(M) \to \mathcal{M}(M)$, $\mu \mapsto f_*\mu$ in the space of complex measures. Then f_* is a linear operator on $\mathcal{M}(M)$ and it is continuous with respect to the weak* topology. There exists another continuous linear operator naturally associated with f, namely $U_f : C^0(M) \to C^0(M)$, $\phi \mapsto \phi \circ f$. Observe that these two operators are dual, in the following sense (remember Lemma 2.2.1):

$$\int U_f(\phi)\, d\mu = \int (\phi \circ f)\, d\mu = \int \phi\, d(f_*\mu). \qquad (2.3.4)$$

These observations motivate the important notion that we are going to introduce in the next section.

2.3.2 Koopman operator

Let $(M,\mathcal{B})$ be a measurable space, $f: M \to M$ be a measurable transformation and μ be an f-invariant measure. The *Koopman operator* of (f,μ) is the linear operator

$$U_f : L^1(\mu) \to L^1(\mu), \quad U_f(\phi) = \phi \circ f.$$

Note that U_f is well defined and is an *isometry*, that is, it preserves the norm in the Banach space $L^1(\mu)$: since μ is invariant under f,

$$\|U_f(\phi)\|_1 = \int |U_f(\phi)|\,d\mu = \int |\phi| \circ f\,d\mu = \int |\phi|\,d\mu = \|\phi\|_1. \tag{2.3.5}$$

Moreover, U_f is a *positive* linear operator: $U_f(\phi) \geq 0$ at μ-almost every point whenever $\phi \geq 0$ at μ-almost every point. For future reference, we summarize these facts in the following proposition:

Proposition 2.3.2. *The Koopman operator $U_f : L^1(M) \to L^1(M)$ of any system (f,μ) is a positive linear isometry.*

The property (2.3.5) implies that the operator U_f is injective. In general, U_f is not surjective (see Exercise 2.3.5). It is clear that if f is invertible then U_f is an isomorphism: the inverse is just the Koopman operator $U_{f^{-1}}$ of the inverse of f.

We may also consider versions of the Koopman operator defined on the spaces $L^p(\mu)$,

$$U_f : L^p(\mu) \to L^p(\mu), \quad U_f(\phi) = \phi \circ f$$

for each $p \in [1,\infty]$. Proposition 2.3.2 remains valid in all these cases: all these operators are positive linear isometries.

When M is a compact metric space and f is continuous, it is particularly interesting to investigate the action of U_f restricted to the space $C^0(M)$ of continuous functions:

$$U_f : C^0(M) \to C^0(M).$$

It is clear that this operator is continuous relative to the norm of uniform convergence. As we have seen previously, the dual space of $C^0(M)$ is naturally identified with the space $\mathcal{M}(M)$ of complex measures on M. Moreover, the relation (2.3.4) shows that, under that identification, the dual operator

$$U_f^* : C^0(M)^* \to C^0(M)^*$$

corresponds precisely to the action $f_* : \mathcal{M}(M) \to \mathcal{M}(M)$ of the transformation f on $\mathcal{M}(M)$. This fact allows us to give an alternative proof of Theorem 2.1, based on certain facts from spectral theory.

For that, we need to recall some notions from the theory of positive linear operators. The reader can find a lot more details in Deimling [Dei85], including the proofs of the results quoted in the following.

Let E be a Banach space. A closed convex subset C is called a *cone* of E if it satisfies:

$$\lambda C \subset C \text{ for every } \lambda \geq 0 \quad \text{and} \quad C \cap (-C) = \{0\}. \tag{2.3.6}$$

We call the cone C *normal* if

$$\inf\{\|x+y\| : x, y \in C \text{ such that } \|x\| = \|y\| = 1\} > 0.$$

Let us fix a cone C of E. Given any continuous linear operator $T: E \to E$, we say that T is *positive over* C if the image $T(C) \subset C$. Given a continuous linear functional $\phi: E \to \mathbb{R}$, we say that ϕ is *positive over* C if $\phi(v) \geq 0$ for every $v \in C$. By definition, the *dual cone* C^* is the cone of E^* formed by all the linear functionals positive over C.

Example 2.3.3. The cone $C^0_+(M) = \{\varphi \in C^0(M) : \varphi \geq 0\}$ is a normal cone of $C^0(M)$ (Exercise 2.3.3). By the Riesz–Markov theorem (Theorem A.3.11), the dual cone is naturally identified with the space of finite positive measures on M.

Denote by $r(T)$ the *spectral radius* of the continuous linear operator T:

$$r(T) = \lim_n \sqrt[n]{\|T^n\|}.$$

Then $r(T) = r(T^*)$, where $T^* : E^* \to E^*$ represents the linear operator dual to T. The next result is a consequence of the theorem of Banach–Mazur; see Proposition 7.2 in Deimling [Dei85]:

Theorem 2.3.4. *Let C be a normal cone of a Banach space E and $T: E \to E$ be a linear operator positive over C. Then $r(T^*)$ is an eigenvalue of the dual operator $T^* : E^* \to E^*$ and it admits some eigenvector $v^* \in C^*$.*

As an application, let us give an alternative proof of the existence of invariant measures for continuous transformations on compact spaces. Consider the cone $C = C^0_+(M)$ of $E = C^0(M)$. As we observed before, the dual cone C^* is the space of finite positive measures on M. It is clear from the definition that the operator $T = U_f$ is positive over C. Also, its spectral radius is equal to 1, since $\sup|T(\varphi)| \leq \sup|\varphi|$ for every $\varphi \in C^0(M)$ and $T(1) = 1$. So, by Theorem 2.3.4, there exists some finite positive measure μ on M that is an eigenvalue of the dual operator $T^* = f_*$ associated with the eigenvalue 1. In other words, the measure μ is invariant. Multiplying by a suitable constant, we may suppose that μ is a probability measure.

2.3.3 Exercises

2.3.1. Let ℓ^1 be the space of summable sequences of complex numbers, endowed with the norm $\|(a_n)_n\|_1 = \sum_{n=0}^{\infty} |a_n|$. Let ℓ^∞ be the space of bounded sequences and c_0 be the space of sequences converging to zero, both endowed with the norm $\|(a_n)_n\|_\infty = \sup_{n\geq 0} |a_n|$.

(a) Check that ℓ^∞, ℓ^1 and c_0 are Banach spaces.

(b) Show that the map $(a_n)_n \mapsto [(b_n)_n \mapsto \sum_n a_n b_n]$ defines norm-preserving isomorphisms from ℓ^∞ to the dual space $(\ell^1)^*$ and from ℓ^1 to the dual space $(c_0)^*$.

2.3.2. Show that a sequence $(x^k)_k$ in ℓ^1 (write $x^k = (x_n^k)_n$ for each k) converges in the topology defined by the norm if and only if it converges in the weak topology, that is, if and only if $(\sum_n a_n x_n^k)_k$ converges for every $(a_n)_n \in \ell^\infty$. [Observation: This does not imply that the two topologies are the same. Why not?] Show that this is no longer true if we replace the weak topology by the weak* topology.

2.3.3. Check that $C_+^0(M)$ is a normal cone.

2.3.4. Let $R_\theta : S^1 \to S^1$ be an irrational rotation and m be the Lebesgue measure on the circle. Calculate the eigenvalues and the eigenvectors of the Koopman operator $U_\theta : L^2(m) \to L^2(m)$. Show that the spectrum of U_θ coincides with the unit circle $\{z \in \mathbb{C} : |z| = 1\}$.

2.3.5. Show, through examples, that the Koopman operator U_f need not be surjective.

2.3.6. Let $U : H \to H$ be an isometry of a Hilbert space H. By Exercise A.6.8, the image of U is a closed subspace of H. Deduce that there exist closed subspaces V and W such that $U(V) = V$, the iterates of W are pairwise orthogonal and orthogonal to V, and

$$H = V \oplus \bigoplus_{n=0}^{\infty} U^n(W).$$

Furthermore, U is an isomorphism if and only if $W = \{0\}$.

2.3.7. Let $\phi : E \to \mathbb{R}$ be a continuous convex functional on a separable Banach space E. Assume that ϕ is differentiable in all directions at some point $u \in E$. Prove that there exists at most one bounded linear functional $T : E \to \mathbb{R}$ *tangent* to ϕ at u, that is, such that $T(v) \le \phi(u+v) - \phi(u)$ for every $v \in E$. If ϕ is differentiable at u then the derivative $D\phi(u)$ is a linear functional tangent to ϕ at u. [Observation: The smoothness theorem of Mazur (Theorem 1.20 in Phelps [Phe93]) states that the set of points where ϕ is differentiable and, consequently, there exists a unique linear functional tangent to ϕ is a residual subset of E.]

2.4 Skew-products and natural extensions

In this section we describe two general constructions that are quite useful in ergodic theory. The first one is a basic model for the situation where two dynamical systems are coupled in the following way: the first system is autonomous but the second one is not, because its evolution depends on the evolution of the former. The second construction associates an invertible system with any given system, in such a way that their invariant measures are in one-to-one correspondence. This permits reduction to the invertible case for many statements about general, not necessarily invertible systems.

2.4.1 Measures on skew-products

Let $(X, \mathcal{A})$ and $(Y, \mathcal{B})$ be measurable spaces. We call a *skew-product* any measurable transformation $F : X \times Y \to X \times Y$ of the form $F(x,y) =$

$(f(x), g(x,y))$. Represent by $\pi : X \times Y \to X$ the canonical projection to the first coordinate. By definition,

$$\pi \circ F = f \circ \pi. \tag{2.4.1}$$

Let m be a probability measure on $X \times Y$ invariant under F and let $\mu = \pi_* m$ be its projection to X. Using that m is invariant under F, we get that

$$f_* \mu = f_* \pi_* m = \pi_* F_* m = \pi_* m = \mu,$$

that is, μ is invariant under f. The next proposition provides a partial converse to this conclusion: under suitable hypotheses, every f-invariant measure is the projection of some F-invariant measure.

Proposition 2.4.1. *Let X be a complete separable metric space, Y be a compact metric space and F be continuous. Then, for every probability measure μ on X invariant under f there exists some probability measure m on $X \times Y$ invariant under F and such that $\pi_* m = \mu$.*

Proof. Given any f-invariant probability measure invariant μ on X, let $\mathcal{K} \subset \mathcal{M}_1(X \times Y)$ be the set of measures η on $X \times Y$ such that $\pi_* \eta = \mu$. Consider any $\eta \in \mathcal{K}$. Then, $\pi_* F_* \eta = f_* \pi_* \eta = f_* \mu = \mu$. This proves that $\mathcal{K}$ is invariant under F_*. Next, note that the projection $\pi : X \times Y \to X$ is continuous and, thus, the operator π_* is continuous relative to the weak* topology. So, $\mathcal{K}$ is closed in $\mathcal{M}_1(X \times Y)$. By Proposition A.3.7, given any $\varepsilon > 0$ there exists a compact set $K \subset X$ such that $\mu(K^c) < \varepsilon$. Then $K \times Y$ is compact and $\eta\big((K \times Y)^c\big) = \mu(K^c) < \varepsilon$ for every $\eta \in \mathcal{K}$. This proves that the set $\mathcal{K}$ is tight. Consider any $\eta \in \mathcal{K}$. By the theorem of Prohorov (Theorem 2.1.8), the sequence

$$\frac{1}{n} \sum_{j=0}^{n-1} F_*^j \eta$$

has some accumulation point $m \in \mathcal{K}$. Arguing as in the proof of Lemma 2.2.4, we conclude that m is invariant under F.

2.4.2 Natural extensions

We are going to see that, given any surjective transformation $f : M \to M$, one can always find an extension $\hat{f} : \hat{M} \to \hat{M}$ that is invertible. By *extension* we mean that there exists a surjective map $\pi : \hat{M} \to M$ such that $\pi \circ \hat{f} = f \circ \pi$. This fact is very useful, for it makes it possible to reduce to the invertible case the proofs of many statements about general systems. We comment on the surjective hypothesis in Example 2.4.2: we will see that this hypothesis can be omitted in many interesting cases.

To begin with, take $\hat{M}$ to be the set of all *pre-orbits* of f, that is, all sequences $(x_n)_{n \le 0}$ indexed by the non-positive integers and satisfying $f(x_n) = x_{n+1}$ for every $n < 0$. Consider the map $\pi : \hat{M} \to M$ sending each sequence $(x_n)_{n\le 0}$ to its term x_0 of order zero. Observe that $\pi(\hat{M}) = M$. Finally, define $\hat{f} : \hat{M} \to \hat{M}$ to be the shift by one unit to the left:

$$\hat{f}(\ldots,x_n,\ldots,x_0) = (\ldots,x_n,\ldots,x_0,f(x_0)). \tag{2.4.2}$$

It is clear that $\hat{f}$ is well defined and satisfies $\pi \circ \hat{f} = f \circ \pi$. Moreover, $\hat{f}$ is invertible: the inverse is the shift to the right:

$$(\ldots,y_n,\ldots,y_{-1},y_0) \mapsto (\ldots,y_n,\ldots,y_{-2},y_{-1}).$$

If M is a measurable space then we may turn $\hat{M}$ into a measurable space by endowing it with the σ-algebra generated by the *measurable cylinders*

$$[A_k,\ldots,A_0] = \{(x_n)_{n\le 0} \in \hat{M} : x_i \in A_i \text{ for } i = k,\ldots,0\}, \tag{2.4.3}$$

where $k \le 0$ and A_k, $\ldots$, A_0 are measurable subsets of M. Then π is a measurable map, since

$$\pi^{-1}(A) = [A]. \tag{2.4.4}$$

Moreover, $\hat{f}$ is measurable if f is measurable:

$$\hat{f}^{-1}([A_k,\ldots,A_0]) = \big[A_k,\ldots,A_{-2},A_{-1} \cap f^{-1}(A_0)\big]. \tag{2.4.5}$$

The inverse of $\hat{f}$ is also measurable, since

$$\hat{f}([A_k,\ldots,A_0]) = [A_k,\ldots,A_0,M]. \tag{2.4.6}$$

Analogously, if M is a topological space then we may turn $\hat{M}$ into a topological space by endowing it with the topology generated by the *open cylinders* $[A_k,\ldots,A_0]$, where $k \le 0$ and A_k, $\ldots$, A_0 are open subsets of M. The relations (2.4.4) and (2.4.6) show that π and $\hat{f}^{-1}$ are continuous, whereas (2.4.5) shows that $\hat{f}$ is continuous if f is continuous. Observe that if M admits a countable basis $\mathcal{U}$ of open sets then the cylinders $[A_k,\ldots,A_0]$ with $k \ge 0$ and $A_0,\ldots,A_k \in \mathcal{U}$ constitute a countable basis of open sets for $\hat{M}$.

If M is a metric space, with distance d, then the following function is a distance on $\hat{M}$:

$$\hat{d}(\hat{x},\hat{y}) = \sum_{n=-\infty}^{0} 2^n \min\{d(x_n,y_n),1\}, \tag{2.4.7}$$

where $\hat{x} = (x_n)_{n\le 0}$ and $\hat{y} = (y_n)_{n\le 0}$. It follows immediately from the definition that if $\hat{x}$ and $\hat{y}$ belong to the same pre-image $\pi^{-1}(x)$ then

$$\hat{d}(\hat{f}^j(\hat{x}),\hat{f}^j(\hat{y})) \le 2^{-j}\hat{d}(\hat{x},\hat{y}) \quad \text{for every } j \ge 0.$$

So, every pre-image $\pi^{-1}(x)$ is a *stable set*, that is, a subset restricted to which the transformation $\hat{f}$ is uniformly contracting.

Example 2.4.2. Given any transformation $g : M \to M$, consider its maximal invariant set $M_g = \bigcap_{n=1}^{\infty} g^n(M)$. Clearly, $g(M_g) \subset M_g$. Suppose that

(i) M is compact and g is continuous or (ii) $\#g^{-1}(y) < \infty$ for every y.

Then (Exercise 2.4.3), the restriction $f = (g \mid M_g) : M_g \to M_g$ is surjective. This restriction contains all the interesting dynamics of g. For example, assuming that $f^n(M)$ is a measurable set for every n, every probability measure invariant under g is also invariant under f. Analogously, every point that is recurrent for g is also recurrent for f, at least in case (i). For this reason, we also refer to the natural extension of $f = (g \mid M_g)$ as the natural extension of g.

A set $\Lambda \subset M$ such that $f^{-1}(\Lambda) = \Lambda$ is called an *invariant set* of f. There is a corresponding notion for the transformation $\hat{f}$. The next proposition shows that every closed invariant set of f admits a unique lift to a closed invariant set of the transformation $\hat{f}$:

Proposition 2.4.3. *Assume that M is a topological space. If $\Lambda \subset M$ is a closed set invariant under f then $\hat{\Lambda} = \pi^{-1}(\Lambda)$ is the only closed set invariant under f and satisfying $\pi(\hat{\Lambda}) = \Lambda$.*

Proof. Since π is continuous, if Λ is closed then $\hat{\Lambda} = \pi^{-1}(\Lambda)$ is also closed. Moreover, if Λ is invariant under f then $\hat{\Lambda}$ is invariant under $\hat{f}$:

$$\hat{f}^{-1}(\hat{\Lambda}) = (\pi \circ \hat{f})^{-1}(\Lambda) = (f \circ \pi)^{-1}(\Lambda) = \pi^{-1}(\Lambda) = \hat{\Lambda}.$$

In the converse direction, let $\hat{\Lambda} \subset \hat{M}$ be a closed set invariant under $\hat{f}$ and such that $\pi(\hat{\Lambda}) = \Lambda$. It is clear that $\hat{\Lambda} \subset \pi^{-1}(\Lambda)$. To prove the other inclusion, we must show that, given any $x_0 \in \Lambda$, if $\hat{x} \in \pi^{-1}(x_0)$ then $\hat{x} \in \hat{\Lambda}$. Let us write $\hat{x} = (x_n)_{n \le 0}$. Consider $n \le 0$ and any neighborhood of $\hat{x}$ of the form

$$V = [A_n, \ldots, A_0], \quad A_n, \ldots, A_0 \text{ open subsets of } M.$$

By the definition of natural extension, $x_0 = f^{-n}(x_n)$ and, hence, $x_n \in f^n(\Lambda) = \Lambda$. Then, the hypothesis $\pi(\hat{\Lambda}) = \Lambda$ implies that $\pi(\hat{y}_n) = x_n$ for some $\hat{y}_n \in \hat{\Lambda}$. Since $\hat{\Lambda}$ is invariant under $\hat{f}$, we have that $\hat{f}^{-n}(\hat{y}_n) \in \hat{\Lambda}$. Moreover, the property $\pi(\hat{y}_n) = x_n$ implies that

$$\hat{f}^{-n}(\hat{y}_n) = (\ldots, y_{n,k}, \ldots, y_{n,-1}, y_{n,0} = x_n, x_{n-1}, \ldots, x_{-1}, x_0).$$

It follows that $\hat{f}^{-n}(\hat{y}_n) \in V$, since V contains $\hat{x}$ and its definition only depends on the coordinates indexed by $j \in \{n, \ldots, 0\}$. This proves that $\hat{x}$ is accumulated by elements of $\hat{\Lambda}$. Since $\hat{\Lambda}$ is closed, it follows that $\hat{x} \in \hat{\Lambda}$.

Now let $\hat{\mu}$ be an invariant measure of $\hat{f}$ and let $\mu = \pi_* \hat{\mu}$. The property $\pi \circ \hat{f} = f \circ \pi$ implies that μ is invariant under f:

$$f_* \mu = f_* \pi_* \hat{\mu} = \pi_* \hat{f}_* \hat{\mu} = \pi_* \hat{\mu} = \mu.$$

We say that $\hat{\mu}$ is a *lift* of μ. The next result, which is a kind of version of Proposition 2.4.3 for measures, is due to Rokhlin [Rok61]:

Proposition 2.4.4. *Assume that M is a complete separable metric space and $f : M \to M$ is continuous. Then every probability measure μ invariant under f*

admits a unique lift, that is, there is a unique measure $\hat{\mu}$ on $\hat{M}$ invariant under $\hat{f}$ and such that $\pi_*\hat{\mu} = \mu$.

Uniqueness is easy to establish and is independent of the hypotheses on the space M and the transformation f. Indeed, if $\hat{\mu}$ is a lift of μ then (2.4.4) and (2.4.5) imply that the measure of every cylinder is uniquely determined:

$$\hat{\mu}([A_k,\ldots,A_0]) = \hat{\mu}\big(\big[A_k \cap \cdots \cap f^{-k}(A_0)\big]\big) = \mu\big(A_k \cap \cdots \cap f^{-k}(A_0)\big). \quad (2.4.8)$$

The proof of existence will be proposed to the reader in Exercise 5.2.4, using ideas to be developed in Chapter 5. We will also see in Exercise 8.5.7 that those arguments remain valid in the somewhat more general setting of Lebesgue spaces. But existence of the lift is *not* true in general, for arbitrary probability spaces, as shown by the example in Exercise 1.15 in the book of Przytycki and Urbański [PU10]).

2.4.3 Exercises

2.4.1. Let M be a compact metric space and X be a set of continuous maps $f: M \to M$, endowed with a probability measure ν. Consider the skew-product $F: X^{\mathbb{N}} \times M \to X^{\mathbb{N}} \times M$ defined by $F((f_n)_n, x) = ((f_{n+1})_n, f_0(x))$. Show that F admits some invariant probability measure m of the form $m = \nu^{\mathbb{N}} \times \mu$. Moreover, a measure m of this form is invariant under F if and only if the measure μ is *stationary* for ν, that is, if and only if $\mu(E) = \int f_*\mu(E)\, d\nu(f)$ for every measurable set $E \subset M$.

2.4.2. Let $f: M \to M$ be a surjective transformation, $\hat{f}: \hat{M} \to \hat{M}$ be its natural extension and $\pi: \hat{M} \to M$ be the canonical projection. Show that if $g: N \to N$ is an invertible transformation such that $f \circ p = p \circ g$ for some map $p: N \to M$ then there exists a unique map $\hat{p}: N \to \hat{M}$ such that $\pi \circ \hat{p} = p$ and $\hat{p} \circ g = \hat{f} \circ \hat{p}$. Suppose that M and N are compact spaces and the maps p and g are continuous. Show that if p is surjective then $\hat{p}$ is surjective (and so $g: N \to N$ is an extension of $\hat{f}: \hat{M} \to \hat{M}$).

2.4.3. Check the claims in Example 2.4.2.

2.4.4. Show that if (M,d) is a complete separable metric space then the same holds for the space $(\hat{M},\hat{d})$ of the pre-orbits of any continuous surjective transformation $f: M \to M$.

2.4.5. The purpose of this exercise and the next is to generalize the notion of natural extension to finite families of commuting transformations. Let M be a compact space and $f_1,\ldots,f_q: M \to M$ be commuting surjective continuous transformations. Let $\hat{M}$ be the set of all sequences $(x_{n_1,\ldots,n_q})_{n_1,\ldots,n_q \le 0}$, indexed by the q-tuples of non-positive integer numbers, such that

$$f_i(x_{n_1,\ldots,n_i,\ldots,n_q}) = x_{n_1,\ldots,n_i+1,\ldots,n_q} \quad \text{for every } i \text{ and every } (n_1,\ldots,n_q).$$

Let $\pi: \hat{M} \to M$ be the map sending $(x_{n_1,\ldots,n_q})_{n_1,\ldots,n_q \le 0}$ to the point $x_{0,\ldots,0}$. For each i, let $\hat{f}_i: \hat{M} \to \hat{M}$ be the map sending $(x_{n_1,\ldots,n_i,\ldots n_q})_{n_1,\ldots,n_q \le 0}$ to $(x_{n_1,\ldots,n_i+1,\ldots n_q})_{n_1,\ldots,n_q \le 0}$.

(a) Prove that $\hat{M}$ is a compact space. Moreover, $\hat{M}$ is metrizable if M is metrizable.

(b) Show that every $\hat{f}_i : \hat{M} \to \hat{M}$ is a homeomorphism with $\pi \circ \hat{f}_i = f_i \circ \pi$. Moreover, these homeomorphisms commute.

(c) Prove that π is continuous and surjective. In particular, $\hat{M}$ is non-empty.

2.4.6. Let M be a compact space and $g_1, \dots, g_q : M \to M$ be commuting continuous transformations. Define $M_g = \bigcap_{n=1}^{\infty} g_1^n \cdots g_q^n(M)$.

(a) Check that $M_g = \bigcap_{n_1,\dots,n_q} g_1^{n_1} \cdots g_q^{n_q}(M)$, where the intersection is over all q-tuples $(n_1, \dots, n_q)$ with $n_i \geq 1$ for every i.

(b) Show that $g_i(M_g) \subset M_g$ and the restriction $f_i = g_i \mid M_g$ is surjective, for every i.

[Observation: It is clear that these restrictions f_i commute.]

2.4.7. Use the construction in Exercises 2.4.5 and 2.4.6 to extend the proof of Theorem 1.5.1 to the case when the transformations f_i are not necessarily invertible.

2.5 Arithmetic progressions

In this section we prove two fundamental results of combinatorial arithmetics, the theorem of van der Waerden and the theorem of Szemerédi, using the multiple recurrence theorems (Theorem 1.5.1 and Theorem 1.5.2) introduced in Section 1.5.

We call a *partition* of the set $\mathbb{Z}$ of integers numbers any finite family of pairwise disjoint sets $S_1, \dots, S_k \subset \mathbb{Z}$ whose union is the whole of $\mathbb{Z}$. Recall that a (finite) *arithmetic progression* is a sequence of the form

$$m+n, m+2n, \dots, m+qn, \quad \text{with } m \in \mathbb{Z} \text{ and } n, q \geq 1.$$

The number q is called the *length* of the progression.

The next theorem was originally proven by the Dutch mathematician Bartel van der Waerden [vdW27] in the 1920's:

Theorem 2.5.1 (van der Waerden). *Given any partition $\{S_1, \dots, S_l\}$ of $\mathbb{Z}$, there exists $j \in \{1, \dots, l\}$ such that S_j contains arithmetic progressions of every length. In other words, for every $q \geq 1$ there exist $m \in \mathbb{Z}$ and $n \geq 1$ such that $m + in \in S_j$ for every $1 \leq i \leq q$.*

Some time afterwards, the Hungarian mathematicians Pål Erdös and Pål Turan [ET36] conjectured the following statement, which is stronger than the theorem of van der Waerden: *any set $S \subset \mathbb{Z}$ whose upper density is positive contains arithmetic progressions of every length.* This was proven by another Hungarian mathematician, Endre Szemerédi [Sze75], almost four decades later. To state the theorem of Szemerédi precisely, we need to define the notion of upper density of a subset of $\mathbb{Z}$.

We call an *interval* of the set $\mathbb{Z}$ any subset I of the form $\{n \in \mathbb{Z} : a \leq n < b\}$ with $a \leq b$. The *cardinal* of an interval I is the number $\#I = b - a$. The *upper density* of a subset S of $\mathbb{Z}$ is the number

$$D_u(S) = \limsup_{\#I\to\infty} \frac{\#(S\cap I)}{\#I},$$

where I represents any interval of $\mathbb{Z}$. The *lower density* $D_l(S)$ of a subset S of $\mathbb{Z}$ is defined analogously, just replacing limit superior with limit inferior. In other words, $D_u(S)$ is the largest and $D_l(D)$ is the smallest number D such that

$$\frac{\#(S\cap I_j)}{\#I_j} \to D \quad \text{for some sequence of intervals } I_j \subset \mathbb{Z} \text{ with } \#I_j \to \infty.$$

In the next lemma we collect some simple properties of the upper and lower densities. The proof is left as an exercise (Exercise 2.5.1).

Lemma 2.5.2. *For any $S \subset \mathbb{Z}$,*

$$0 \le D_l(S) \le D_u(S) \le 1 \quad \text{and} \quad D_l(S) = 1 - D_u(\mathbb{Z}\setminus S).$$

Moreover, if $S_1,\dots,S_l$ is a partition of $\mathbb{Z}$ then

$$D_l(S_1)+\cdots+D_l(S_l) \le 1 \le D_u(S_1)+\cdots+D_u(S_l).$$

Example 2.5.3. Let S be the set of even numbers. For any interval $I \subset \mathbb{Z}$, we have $\#(S\cap I) = \#I/2$ if the cardinal of I is even and $\#(S\cap I) = (\#I \pm 1)/2$ if the cardinal of I is odd; the sign $\pm$ is positive if the smallest element of I is an even number and it is negative otherwise. It follows, immediately, that $D_u(S) = D_l(S) = 1/2$.

Example 2.5.4. Let S be the following subset of $\mathbb{Z}$:

$$\{1,3,4,7,8,9,13,14,15,16,21,22,23,24,25,31,32,33,34,35,36,43,\dots\}.$$

That is, for each $k \ge 1$ we include in S a block of k consecutive integers and then we omit the next k integer numbers. On the one hand, S contains intervals of every length. Consequently, $D_u(S) = 1$. On the other hand, the complement of S also contains intervals of every length. So, $D_l(S) = 1 - D_u(\mathbb{Z}\setminus S) = 0$.

Notice that, in both examples, the set S contains arithmetic progressions of every length. Actually, in Example 2.5.3 the set S even contains arithmetic progressions of infinite length. That is not true in Example 2.5.4, because in this case the complement of S contains arbitrarily long intervals.

Theorem 2.5.5 (Szemerédi). *If S is a subset of $\mathbb{Z}$ with positive upper density then it contains arithmetic progressions of every length.*

The theorem of van der Waerden is an easy consequence of the theorem of Szemerédi. Indeed, it follows from Lemma 2.5.2 that if $S_1,\dots,S_l$ is a partition of $\mathbb{Z}$ then there exists j such that $D_u(S_j) > 0$. By Theorem 2.5.5, such an S_j contains arithmetic progressions of every length.

The original proofs of these results were combinatorial. Then, Furstenberg (see [Fur81]) observed that the two theorems could also be deduced from ideas in ergodic theory: we will show in Section 2.5.1 how to obtain the theorem of van der Waerden from the multiple recurrence theorem of Birkhoff (Theorem 1.5.1); similar arguments yield the theorem of Szemerédi from the multiple recurrence theorem of Poincaré (Theorem 1.5.2), as we will see in Section 2.5.2.

The theory of Szemerédi remains a very active research area. In particular, alternative proofs of Theorem 2.5.5 have been given by other authors. Recently, this led to the following spectacular result of the British mathematician Ben Green and the Australian mathematician Terence Tao [GT08]: *the set of prime numbers contains arithmetic progressions of every length.* This is not a consequence of the theorem of Szemerédi, because the upper density of the set of prime numbers is zero, but the theorem of Szemerédi does have an important role in the proof. On the other hand, the Green–Tao theorem is a special case of yet another conjecture of Erdös: if $S \subset \mathbb{N}$ is such that the sum of the inverses diverges, that is, such that

$$\sum_{n\in S} \frac{1}{n} = \infty,$$

then S contains arithmetic progressions of every length. This more general statement remains open.

2.5.1 Theorem of van der Waerden

In this section we prove Theorem 2.5.1. The idea of the proof is to reduce the conclusion of the theorem to a claim about the shift map

$$\sigma : \Sigma \to \Sigma, \quad (\alpha_n)_{n\in\mathbb{Z}} \mapsto (\alpha_{n+1})_{n\in\mathbb{Z}}$$

in the space $\Sigma = \{1,2,\dots,l\}^{\mathbb{Z}}$ of two-sided sequences with values in the set $\{1,2,\dots,l\}$. This claim will then be proved using the multiple recurrence theorem of Birkhoff.

Observe that every partition $\{S_1,\dots,S_l\}$ of $\mathbb{Z}$ into $l \geq 2$ subsets determines an element $\underline{\alpha} = (\alpha_n)_{n\in\mathbb{Z}}$ of Σ, through $\alpha_n = i \Leftrightarrow n \in S_i$. Conversely, every $\underline{\alpha} \in \Sigma$ determines a partition of $\mathbb{Z}$ into subsets

$$S_i = \{n \in \mathbb{Z} : \alpha_n = i\}, \quad i = 1,\dots,l.$$

We are going to show that for every $\underline{\alpha} \in \Sigma$ and every $q \geq 1$, there exist $m \in \mathbb{Z}$ and $n \geq 1$ such that

$$\alpha_{m+n} = \cdots = \alpha_{m+qn}. \tag{2.5.1}$$

In view of what we have just observed, this means that for every partition $\{S_1,\dots,S_l\}$ and every $q \geq 1$ there exists $i \in \{1,\dots,l\}$ such that S_i contains some arithmetic progression of length q. Since there are finitely many S_i, that implies that some S_j contains arithmetic progressions of arbitrarily large lengths. This is the same as saying that S_i contains arithmetic progressions of every length because, clearly, every arithmetic progression of length q contains arithmetic progressions of every length smaller than q. This reduces the proof of the theorem to proving the claim in (2.5.1).

To that end, let us consider on Σ the distance defined by $d(\underline{\beta},\underline{\gamma}) = 2^{-N(\underline{\beta},\underline{\gamma})}$, where

$$N(\underline{\beta},\underline{\gamma}) = \max\{N \geq 0 : \beta_n = \gamma_n \text{ for every } n \in \mathbb{Z} \text{ with } |n| < N\}.$$

Note that

$$d(\underline{\beta},\underline{\gamma}) < 1 \quad \text{if and only if } \beta_0 = \gamma_0. \tag{2.5.2}$$

Since the metric space (Σ, d) is compact, the closure $Z = \overline{\{\sigma^n(\underline{\alpha}) : n \in \mathbb{Z}\}}$ of the trajectory of $\underline{\alpha}$ is also compact. Moreover, Z is invariant under the shift map. Let us consider the transformations $f_1 = \sigma, f_2 = \sigma^2, \ldots, f_q = \sigma^q$ defined from Z to Z. It is clear from the definition that these transformations commute with each other. So, we may use Theorem 1.5.1 to conclude that there exist $\underline{\theta} \in Z$ and a sequence $(n_k)_k \to \infty$ such that

$$\lim_k f_i^{n_k}(\underline{\theta}) = \underline{\theta} \quad \text{for every } i = 1, 2, \ldots, q.$$

Observe that $f_i^{n_j} = \sigma^{in_j}$. In particular, we may fix $n = n_j$ such that the iterates $\sigma^n(\underline{\theta}), \sigma^{2n}(\underline{\theta}), \ldots, \sigma^{qn}(\underline{\theta})$ are all within a distance of less than $1/2$ from the point $\underline{\theta}$. Consequently,

$$d\big(\sigma^{in}(\underline{\theta}), \sigma^{jn}(\underline{\theta})\big) < 1 \quad \text{for every } 1 \leq i, j \leq q.$$

Then, as $\underline{\theta}$ is in the closure Z of the orbit of $\underline{\alpha}$, we may find $m \in \mathbb{Z}$ such that $\sigma^m(\underline{\alpha})$ is so close to $\underline{\theta}$ that

$$d\big(\sigma^{m+in}(\underline{\alpha}), \sigma^{m+jn}(\underline{\alpha})\big) < 1 \quad \text{for every } 1 \leq i, j \leq q.$$

Taking into account the observation (2.5.2) and the definition of the shift map σ, this means that $\alpha_{m+n} = \cdots = \alpha_{m+qn}$, as we wanted to prove. This completes the proof of the theorem of van der Waerden.

2.5.2 Theorem of Szemerédi

Now let us prove Theorem 2.5.5. We use the same kind of dictionary between partitions of $\mathbb{Z}$ and sequences of integer numbers that was used in the previous section to prove the theorem of van der Waerden.

Let S be a subset of integer numbers with positive upper density, that is, such that there exist $c > 0$ and intervals $I_j = [a_j, b_j)$ of $\mathbb{Z}$ satisfying

$$\lim_j \#I_j = \infty \quad \text{and} \quad \lim_j \frac{\#(S \cap I_j)}{\#I_j} \geq c.$$

Let us associate with S the sequence $\underline{\alpha} = (\alpha_j)_{j \in \mathbb{Z}} \in \Sigma = \{0,1\}^{\mathbb{Z}}$ defined by

$$\alpha_j = 1 \Leftrightarrow j \in S.$$

Consider the shift map $\sigma : \Sigma \to \Sigma$ and the subset $A = \{\underline{\alpha} \in \Sigma : \alpha_0 = 1\}$ of Σ. Note that both A and its complement are open cylinders of Σ. Thus, A is both open and closed in Σ. Moreover, for every $j \in \mathbb{Z}$,

$$\sigma^j(\underline{\alpha}) \in A \Leftrightarrow \alpha_j = 1 \Leftrightarrow j \in S.$$

So, to prove the theorem of Szemerédi it suffices to show that for every $k \in \mathbb{N}$ there exist $m \in \mathbb{Z}$ and $n \geq 1$ such that

$$\sigma^{m+n}(\underline{\alpha}), \sigma^{m+2n}(\underline{\alpha}), \ldots, \sigma^{m+kn}(\underline{\alpha}) \in A. \tag{2.5.3}$$

For that, let us consider the sequence μ_j of probability measures defined on Σ by

$$\mu_j = \frac{1}{\#I_j} \sum_{i \in I_j} \delta_{\sigma^i(\underline{\alpha})}. \tag{2.5.4}$$

Since the space $\mathcal{M}_1(\Sigma)$ of all probability measures on Σ is compact (Theorem 2.1.5), up to replacing $(\mu_j)_j$ by some subsequence we may suppose that it converges in the weak* topology to some probability measure μ on Σ.

Observe that μ is a σ-invariant probability measure, for, given any continuous function $\varphi : \Sigma \to \mathbb{R}$,

$$\begin{aligned}\int (\varphi \circ \sigma)\, d\mu_j &= \frac{1}{\#I_j} \sum_{i \in I_j} \varphi(\sigma^i(\underline{\alpha})) + \frac{1}{\#I_j}\Big[\varphi(\sigma^{b_j}(\underline{\alpha})) - \varphi(\sigma^{a_j}(\underline{\alpha}))\Big] \\ &= \int \varphi\, d\mu_j + \frac{1}{\#I_j}\Big[\varphi(\sigma^{b_j}(\underline{\alpha})) - \varphi(\sigma^{a_j}(\underline{\alpha}))\Big]\end{aligned}$$

and, taking the limit when $j \to \infty$, it follows that $\int (\varphi \circ \sigma)\, d\mu = \int \varphi\, d\mu$. Observe also that $\mu(A) > 0$. Indeed, since A is closed, Theorem 2.1.2 ensures that

$$\mu(A) \geq \limsup_j \mu_j(A) = \limsup_j \frac{\#(S \cap I_j)}{\#I_j} \geq c.$$

Given any $k \geq 1$, consider $f_i = \sigma^i$ for $i = 1, \ldots, k$. It is clear that these transformations commute with each other. So, we are in a position to apply Theorem 1.5.2 to conclude that there exists some $n \geq 1$ such that

$$\mu\big(A \cap \sigma^{-n}(A) \cap \cdots \cap \sigma^{-kn}(A)\big) > 0.$$

Since A is open, this implies (Theorem 2.1.2) that

$$\mu_l\big(A \cap \sigma^{-n}(A) \cap \cdots \cap \sigma^{-kn}(A)\big) > 0$$

for every l sufficiently large. By the definition (2.5.4) of μ_l, this means that there exists some $m \in I_l$ such that

$$\sigma^m(\underline{\alpha}) \in A \cap \sigma^{-n}(A) \cap \cdots \cap \sigma^{-kn}(A).$$

In particular, $\sigma^{m+in}(\underline{\alpha}) \in A$ for every $i = 1, \ldots, k$, as we wanted to prove.

2.5.3 Exercises

2.5.1. Prove Lemma 2.5.2.

2.5.2. Show that the conclusion of Theorem 2.5.1 remains valid for partitions of finite subsets of $\mathbb{Z}$, as long as they are sufficiently large. More precisely: given $q, l \geq 1$ there exists $N \geq 1$ such that, for any partition of the set $\{1, 2, \ldots, N\}$ into l subsets, at least one of these subsets contains arithmetic progressions of length q.

2.5.3. A point $x \in M$ is said to be *super non-wandering* if, given any neighborhood U of x and any $k \geq 1$, there exists $n \geq 1$ such that $\bigcap_{j=0}^{k} f^{-jn}(U) \neq \emptyset$. Show that the theorem of van der Warden is equivalent to the following statement: every invertible transformation on a compact metric space has some super non-wandering point.

2.5.4. Prove the following generalization of the theorem of van der Waerden to arbitrary dimension, called the Grünwald theorem: given any partition $\mathbb{N}^k = S_1 \cup \cdots \cup S_l$ and any $q \geq 1$, there exist $j \in \{1, \ldots, l\}$, $d \in \mathbb{N}$ and $b \in \mathbb{N}^k$ such that

$$b + d(a_1, \ldots, a_k) \in S_j \quad \text{for any } 1 \leq a_i \leq q \text{ and any } 1 \leq i \leq k.$$

3

Ergodic theorems

In this chapter we present the fundamental results of ergodic theory. To motivate the kind of statements that we are going to discuss, let us consider a measurable set $E \subset M$ with positive measure and an arbitrary point $x \in M$. We want to analyze the set of iterates of x that visit E, that is,

$$\{j \geq 0 : f^j(x) \in E\}.$$

For example, the Poincaré recurrence theorem states that this set is infinite, for almost every $x \in E$. We would like to have more precise quantitative information. Let us call the *mean sojourn time* of x to E the value of

$$\tau(E,x) = \lim_{n\to\infty} \frac{1}{n} \#\{0 \leq j < n : f^j(x) \in E\}. \tag{3.0.1}$$

There is an analogous notion for flows, defined by

$$\tau(E,x) = \lim_{T\to\infty} \frac{1}{T} m\big(\{0 \leq t \leq T : f^t(x) \in E\}\big), \tag{3.0.2}$$

where m is the Lebesgue measure on the real line. It would be interesting to know, for example, under which conditions the mean sojourn time is positive. But before tackling this problem one must answer an even more basic question: when do the limits in (3.0.1)–(3.0.2) exist?

These questions go back to the work of the Austrian physicist Ludwig Boltzmann (1844–1906), who developed the kinetic theory of gases. Boltzmann was an emphatic supporter of the atomic theory, according to which gases are formed by a large number of small moving particles, constantly colliding with each other, at a time when this theory was still highly controversial. In principle, it should be possible to explain the behavior of a gas by applying the laws of classical mechanics to each one of these particles (molecules). In practice, this is not realistic because the number of molecules is huge.

The proposal of the kinetic theory was, then, to try and explain the behavior of gases at a macroscopic scale as the statistical combination of the motions of all its molecules. To formulate the theory in precise mathematical terms,

Boltzmann was forced to make an assumption that became known as the *ergodic hypothesis*. In modern language, the ergodic hypothesis claims that, for the kind of systems (Hamiltonian flows) that describe the motions of particles of a gas, *the mean sojourn time to any measurable set E exists and is equal to the measure of E, for almost every point x.*

Efforts to validate (or not) this hypothesis led to important developments, in mathematics (ergodic theory, dynamical systems) as well as in physics (statistical mechanics). In this chapter we concentrate on results concerning the *existence* of the mean sojourn time. The question of whether $\tau(E,x) = \mu(E)$ for almost every x is the subject of Chapter 4.

Denoting by φ the characteristic function of the set E, we may rewrite the expression on the right-hand side of (3.0.1) as

$$\lim_{n\to\infty} \frac{1}{n} \sum_{j=0}^{n-1} \varphi(f^j(x)). \tag{3.0.3}$$

This suggests a natural generalization of the original question: does the limit in (3.0.3) exist for more general functions φ, for example, for all integrable functions?

The ergodic theorem of von Neumann (Theorem 3.1.6) states that the limit in (3.0.3) does exist, in the space $L^2(\mu)$, for every function $\varphi \in L^2(\mu)$. The ergodic theorem of Birkhoff (Theorem 3.2.3) goes a lot further, by asserting that the convergence holds at μ-almost every point, for every $\varphi \in L^1(\mu)$. In particular, the limit in (3.0.1) is well defined for μ-almost every x (Theorem 3.2.1).

We give a direct proof of the theorem of von Neumann and we also show how it can be deduced from the theorem of Birkhoff. Concerning the latter, we are going to see that it can be obtained as a special case of an even stronger result, the subadditive ergodic theorem of Kingman (Theorem 3.3.3). This theorem asserts that ψ_n/n converges almost everywhere, for any sequence of functions ψ_n such that $\psi_{m+n} \le \psi_m + \psi_n \circ f^m$ for every m, n.

All these results remain valid for flows, as we comment upon in Section 3.4.

3.1 Ergodic theorem of von Neumann

In this section we state and prove the ergodic theorem of von Neumann. We begin by reviewing some general ideas concerning isometries in Hilbert spaces. See Appendices A.6 and A.7 for more information on this topic.

3.1.1 Isometries in Hilbert spaces

Let H be a Hilbert space and F be a closed subspace of H. Then,

$$H = F \oplus F^{\perp}, \tag{3.1.1}$$

where $F^\perp = \{w \in H : v \cdot w = 0 \text{ for every } v \in F\}$ is the orthogonal complement of F. The projection $P_F : H \to F$ associated with the decomposition (3.1.1) is called the *orthogonal projection* to F. It is uniquely characterized by

$$\|x - P_F(x)\| = \min\{\|x - v\| : v \in F\}.$$

Observe that $P_F(v) = v$ for every $v \in F$ and, consequently, $P_F^2 = P_F$.

Example 3.1.1. Consider the Hilbert space $L^2(\mu)$, with the inner product

$$\varphi \cdot \psi = \int \varphi \bar{\psi} \, d\mu.$$

Let F be the subspace of constant functions. Given any $\varphi \in L^2(\mu)$, we have that $(P_F(\varphi) - \varphi) \cdot 1 = 0$, that is,

$$P_F(\varphi) \cdot 1 = \varphi \cdot 1.$$

Since $P_F(\varphi)$ is a constant function, the expression on the left-hand side is equal to $P_F(\varphi)$. The expression on the right-hand side is equal to $\int \varphi \, d\mu$. Therefore, the orthogonal projection to the subspace F is given by

$$P_F(\varphi) = \int \varphi \, d\mu.$$

Recall that the adjoint operator $U^* : H \to H$ of a continuous linear operator $U : H \to H$ is defined by the relation

$$U^* u \cdot v = u \cdot Uv \quad \text{for every } u, v \in H. \tag{3.1.2}$$

The operator U is said to be an *isometry* if it preserves the inner product:

$$Uu \cdot Uv = u \cdot v \quad \text{for every } u, v \in H. \tag{3.1.3}$$

This is equivalent to saying that U preserves the norm of H (see Exercise A.6.9). Another equivalent condition is $U^* U = \mathrm{id}$. Indeed,

$$Uu \cdot Uv = u \cdot v \text{ for every } u, v \quad \Leftrightarrow \quad U^* Uu \cdot v = u \cdot v \text{ for every } u, v.$$

The property $U^* U = \mathrm{id}$ implies that U is injective. In general, an isometry need not be surjective. See Exercises 2.3.5 and 2.3.6. If an isometry is surjective then it is an isomorphism; such isometries are also called unitary operators.

Example 3.1.2. If $f : M \to M$ preserves a measure μ then, as we saw in Section 2.3.2, the Koopman operator $U_f : L^2(\mu) \to L^2(\mu)$ is an isometry. If f is invertible then U_f is a unitary operator.

We call the *set of invariant vectors* of a continuous linear operator $U : H \to H$ the subspace

$$I(U) = \{v \in H : Uv = v\}.$$

Observe that $I(U)$ is a closed vector subspace, since U is continuous and linear. When U is an isometry, we have that $I(U) = I(U^*)$:

Lemma 3.1.3. *If $U : H \to H$ is an isometry then $Uv = v$ if and only if $U^* v = v$.*

Proof. Since $U^*U = \mathrm{id}$, it is clear that $Uv = v$ implies $U^*v = v$. Now assume that $U^*v = v$. Then, $Uv \cdot v = v \cdot U^*v = v \cdot v = \|v\|^2$. So, using the fact that U preserves the norm of H,

$$\|Uv - v\|^2 = (Uv - v) \cdot (Uv - v) = \|Uv\|^2 - Uv \cdot v - v \cdot Uv + \|v\|^2 = 0.$$

This means that $Uv = v$.

To close this brief digression, let us quote a classical result from functional analysis, due to Marshall H. Stone, that permits the reduction of the study of Koopman operators of continuous time systems to the discrete case.

Let $U_t : H \to H$, $t \in \mathbb{R}$ be a 1-*parameter group* of linear operators on a Banach space: by this we mean that $U_0 = \mathrm{id}$ and $U_{t+s} = U_t U_s$ for every $t, s \in \mathbb{R}$. We say that the group is *strongly continuous* if

$$\lim_{t \to t_0} U_t v = U_{t_0} v, \quad \text{for every } t_0 \in \mathbb{R} \text{ and } v \in H.$$

Theorem 3.1.4 (Stone). *If $U_t : H \to H$, $t \in \mathbb{R}$ is a strongly continuous 1-parameter group of unitary operators on a complex Hilbert space then there exists a self-adjoint operator A, defined on a dense subspace $D(A)$ of H, such that $U_t \mid D(A) = e^{itA}$ for every $t \in \mathbb{R}$.*

A proof may be found in Yosida [Yos68, § IX.9] and a simple application is given in Exercise 3.1.5. The operator iA is called the *infinitesimal generator* of the group. It may be retrieved through

$$iAv = \lim_{t \to 0} \frac{1}{t}\big(U_t v - v\big). \tag{3.1.4}$$

See Yosida [Yos68, § IX.3] for a proof of the fact that the limit on the right-hand side exists for every v in a dense subspace of H.

Example 3.1.5. Let H be the Banach space of continuous functions $\varphi : S^1 \to \mathbb{C}$, with the norm of uniform convergence. Define $U_t(\varphi)(x) = \varphi(x+t)$ for every function $\varphi \in H$. Observe that $(U_t)_t$ is a strongly continuous 1-parameter group of isometries of H. The infinitesimal generator is given by

$$iA\phi(x) = \lim_{t \to 0} \frac{1}{t}\big(U_t\phi(x) - \phi(x)\big) = \lim_{t \to 0} \frac{1}{t}\big(\phi(x+t) - \phi(x)\big) = \phi'(x).$$

Its domain is the subset of functions of class C^1, which is well known to be dense in H.

3.1.2 Statement and proof of the theorem

Our first ergodic theorem is:

Theorem 3.1.6 (von Neumann). *Let $U : H \to H$ be an isometry in a Hilbert space H and P be the orthogonal projection to the subspace $I(U)$ of invariant* *vectors of U. Then,*

$$\lim_{n\to\infty}\frac{1}{n}\sum_{j=0}^{n-1}U^j v = Pv \quad \textit{for every } v\in H. \tag{3.1.5}$$

Proof. Let $L(U)$ be the set of vectors $v \in H$ of the form $v = Uu - u$ for some $u \in H$ and let $\bar{L}(U)$ be its closure. We claim that

$$I(U) = \bar{L}(U)^{\perp}. \tag{3.1.6}$$

This can be checked as follows. Consider any $v \in I(U)$ and $w \in \bar{L}(U)$. By Lemma 3.1.3, we have that $v \in I(U^*)$, that is, $U^*v = v$. Moreover, by definition of $\bar{L}(U)$, there are $u_n \in H$, $n \geq 1$ such that $(Uu_n - u_n)_n \to w$. Since

$$v\cdot(Uu_n - u_n) = v\cdot Uu_n - v\cdot u_n = U^*v\cdot u_n - v\cdot u_n = 0$$

for every n, we conclude that $v \cdot w = 0$. This proves that $I(U) \subset \bar{L}(U)^{\perp}$. Next, consider any $v \in \bar{L}(U)^{\perp}$. Then, in particular,

$$v\cdot(Uu-u) = 0 \quad \text{or, equivalently,} \quad U^*v\cdot u - v\cdot u = 0$$

for every $u \in H$. This means that $U^*v = v$. Using Lemma 3.1.3 once more, we deduce that $v \in I(U)$. This shows that $\bar{L}(U)^{\perp} \subset I(U)$, which completes the proof of (3.1.6). As a consequence, using (3.1.1),

$$H = I(U) \oplus \bar{L}(U). \tag{3.1.7}$$

Now we prove the identity (3.1.5), successively, for $v \in I(U)$, for $v \in \bar{L}(U)$ and for any $v \in H$. Begin by supposing that $v \in I(U)$. On the one hand, $Pv = v$. On the other hand,

$$\frac{1}{n}\sum_{j=0}^{n-1}U^j v = \frac{1}{n}\sum_{j=0}^{n-1}v = v$$

for every n, and so this sequence converges to v when $n \to \infty$. Combining these two observations we get (3.1.5) in this case.

Next, suppose that $v \in L(U)$. Then, by definition, there exists $u \in H$ such that $v = Uu - u$. It is clear that

$$\frac{1}{n}\sum_{j=0}^{n-1}U^j v = \frac{1}{n}\sum_{j=0}^{n-1}(U^{j+1}u - U^j u) = \frac{1}{n}(U^n u - u).$$

The norm of this last expression is bounded by $2\|u\|/n$ and, consequently, converges to zero when $n \to \infty$. This shows that

$$\lim_n \frac{1}{n}\sum_{j=0}^{n-1}U^j v = 0 \quad \text{for every } v \in L(U). \tag{3.1.8}$$

More generally, suppose that $v \in \bar{L}(U)$. Then, there exist vectors $v_k \in L(U)$ converging to v when $k \to \infty$. Observe that

$$\left\| \frac{1}{n}\sum_{j=0}^{n-1} U^j v - \frac{1}{n}\sum_{j=0}^{n-1} U^j v_k \right\| \leq \frac{1}{n}\sum_{j=0}^{n-1} \|U^j(v - v_k)\| \leq \|v - v_k\|$$

for every n and every k. Together with (3.1.8), this implies that

$$\lim_n \frac{1}{n}\sum_{j=0}^{n-1} U^j v = 0 \quad \text{for every } v \in \bar{L}(U). \tag{3.1.9}$$

Since (3.1.6) implies that $Pv = 0$ for every $v \in \bar{L}(U)$, this shows that (3.1.5) holds also when $v \in \bar{L}(U)$.

The general case of (3.1.5) follows immediately, as $H = I(U) \oplus \bar{L}(U)$.

3.1.3 Convergence in $L^2(\mu)$

Given a measurable transformation $f : M \to M$ and an invariant probability measure μ on M, we say that a measurable function $\psi : M \to \mathbb{R}$ is *invariant* if $\psi \circ f = \psi$ at μ-almost every point. The following result is a special case of Theorem 3.1.6:

Theorem 3.1.7. *Given any $\varphi \in L^2(\mu)$, let $\tilde{\varphi}$ be the orthogonal projection of φ to the subspace of invariant functions. Then the sequence*

$$\frac{1}{n}\sum_{j=0}^{n-1} \varphi \circ f^j \tag{3.1.10}$$

converges to $\tilde{\varphi}$ in the space $L^2(\mu)$. If f is invertible, then the sequence

$$\frac{1}{n}\sum_{j=0}^{n-1} \varphi \circ f^{-j} \tag{3.1.11}$$

also converges to $\tilde{\varphi}$ in $L^2(\mu)$.

Proof. Let $U = U_f : L^2(\mu) \to L^2(\mu)$ be the Koopman operator of (f, μ). Note that a function ψ is in the subspace $I(U)$ of invariant functions if and only if $\psi \circ f = \psi$ at μ-almost every point. By Theorem 3.1.6, the sequence in (3.1.10) converges in the space $L^2(\mu)$ to the orthogonal projection $\tilde{\varphi}$ of φ to the subspace $I(U)$. This proves the first claim.

The second one is analogous, taking instead $U = U_{f^{-1}}$, which is the inverse of U_f. We get that the sequence in (3.1.11) converges in $L^2(\mu)$ to the orthogonal projection of φ to the subspace $I(U_{f^{-1}})$. Observing that $I(U_{f^{-1}}) = I(U_f)$, we conclude that the limit of this sequence is just the same function $\tilde{\varphi}$ as before.

3.1.4 Exercises

3.1.1. Show that under the hypotheses of the von Neumann ergodic theorem one has the following stronger conclusion:

$$\lim_{n-m\to\infty} \frac{1}{n-m}\sum_{j=m}^{n-1} \varphi \circ f^j \to P(\varphi).$$

3.1.2. Use the previous exercise to show that, given any $A \subset M$ with $\mu(A) > 0$, the set of values of $n \in \mathbb{N}$ such that $\mu(A \cap f^{-n}(A)) > 0$ is syndetic. [Observation: We have seen a different proof of this fact in Exercise 1.2.5.]

3.1.3. Prove that the set $F = \{\varphi \in L^1(\mu) : \varphi \text{ is } f\text{-invariant}\}$ is a closed subspace of $L^1(\mu)$.

3.1.4. State and prove a version of the von Neumann ergodic theorem for flows.

3.1.5. Let $f_t : M \to M$, $t \in \mathbb{R}$ be a continuous flow on a compact metric space M and μ be an invariant probability measure. Check that the 1-parameter group $U_t : L^2(\mu) \to L^2(\mu)$, $t \in \mathbb{R}$ of Koopman operators $\varphi \mapsto U_t\varphi = \varphi \circ f_t$ is strongly continuous. Show that μ is ergodic if and only if 0 is a simple eigenvalue of the infinitesimal generator of the group.

3.2 Birkhoff ergodic theorem

The theorem that we present in this section was proven by George David Birkhoff,[1] the prominent American mathematician of his generation and author of many other fundamental contributions to dynamics. It is a substantial improvement of the von Neumann ergodic theorem, because its conclusion is stated in terms of convergence at μ-almost every point, which in this context is a stronger property than convergence in $L^2(\mu)$, as explained in Section 3.2.3.

3.2.1 Mean sojourn time

We start by stating the version of the theorem for mean sojourn times:

Theorem 3.2.1 (Birkhoff). *Let $f : M \to M$ be a measurable transformation and μ be a probability measure invariant under f. Given any measurable set $E \subset M$, the mean sojourn time*

$$\tau(E,x) = \lim_n \frac{1}{n}\#\{j = 0,1,\dots,n-1 : f^j(x) \in E\}$$

exists at μ-almost every point $x \in M$. Moreover, $\int \tau(E,x)\,d\mu(x) = \mu(E)$.

Observe that if $\tau(E,x)$ exists for some $x \in M$ then

$$\tau(E,f(x)) = \tau(E,x). \tag{3.2.1}$$

[1] His son Garret Birkhoff was also a mathematician, and is well known for his work in algebra. The notion of projective distance that we use in Section 12.3 was due to him.

Indeed, by definition,

$$\begin{aligned}\tau(E,f(x)) &= \lim_{n\to\infty}\frac{1}{n}\sum_{j=1}^{n}\mathcal{X}_E(f^j(x))\\ &= \lim_{n\to\infty}\frac{1}{n}\sum_{j=0}^{n-1}\mathcal{X}_E(f^j(x)) - \frac{1}{n}\big[\mathcal{X}_E(x)-\mathcal{X}_E(f^n(x))\big]\\ &= \tau(E,x) - \lim_{n\to\infty}\frac{1}{n}\big[\mathcal{X}_E(x)-\mathcal{X}_E(f^n(x))\big].\end{aligned}$$

Since the characteristic function is bounded, the last limit is equal to zero. This proves (3.2.1).

The next example shows that the mean sojourn time does not exist for *every* point, in general:

Example 3.2.2. Consider the number $x \in (0,1)$ defined by the decimal expansion $x = 0.a_1a_2a_3\ldots$, where $a_i = 1$ if $2^k \le i < 2^{k+1}$ with k even and $a_i = 0$ if $2^k \le i < 2^{k+1}$ with k odd. In other words,

$$x = 0.100111100000000111111111111111110\ldots,$$

where the lengths of the alternating blocks of 0s and 1s are given by successive powers of 2. Let $f : [0,1] \to [0,1]$ be the transformation defined in Section 1.3.1 and let $E = [0,1/10)$. That is, E is the set of all points whose decimal expansion starts with the digit 0. It is easy to check that if $n = 2^k - 1$ with k even then

$$\frac{1}{n}\sum_{j=0}^{n-1}\mathcal{X}_E(f^j(x)) = \frac{2^1+2^3+\cdots+2^{k-1}}{2^k-1} = \frac{2}{3}.$$

On the other hand, if one takes $n = 2^k - 1$ with k odd then

$$\frac{1}{n}\sum_{j=0}^{n-1}\mathcal{X}_E(f^j(x)) = \frac{2^1+2^3+\cdots+2^{k-2}}{2^k-1} = \frac{2^k-2}{3(2^k-1)} \to \frac{1}{3}$$

as $k \to \infty$. Thus, the mean sojourn time of x in the set E does not exist.

3.2.2 Time averages

As we observed previously,

$$\tau(E,x) = \lim_n \frac{1}{n}\sum_{j=0}^{n-1}\varphi(f^j(x)), \quad \text{where } \varphi = \mathcal{X}_E.$$

The next statement extends Theorem 3.2.1 to the case when φ is any integrable function:

Theorem 3.2.3 (Birkhoff). *Let $f : M \to M$ be a measurable transformation and μ be a probability measure invariant under f. Given any integrable function $\varphi : M \to \mathbb{R}$, the limit*

$$\tilde{\varphi}(x) = \lim_{n\to\infty} \frac{1}{n} \sum_{j=0}^{n-1} \varphi(f^j(x)) \tag{3.2.2}$$

exists at μ-almost every point $x \in M$. Moreover, the function $\tilde{\varphi}$ defined in this way is integrable and satisfies

$$\int \tilde{\varphi}(x)\, d\mu(x) = \int \varphi(x)\, d\mu(x).$$

In a little while, we will obtain this theorem as a special case of a more general result, the subadditive ergodic theorem. The limit $\tilde{\varphi}$ is called the *time average*, or *orbital average*, of φ. The next proposition shows that time averages are constant on the orbit of μ-almost every point, which generalizes (3.2.1):

Proposition 3.2.4. *Let $\varphi : M \to \mathbb{R}$ be an integrable function. Then,*

$$\tilde{\varphi}(f(x)) = \tilde{\varphi}(x) \quad \textit{for } \mu\textit{-almost every point } x \in M. \tag{3.2.3}$$

Proof. By definition,

$$\tilde{\varphi}(f(x)) = \lim_{n\to\infty} \frac{1}{n} \sum_{j=1}^{n} \varphi(f^j(x)) = \lim_{n\to\infty} \frac{1}{n} \sum_{j=0}^{n-1} \varphi(f^j(x)) + \frac{1}{n}\big[\varphi(f^n(x)) - \varphi(x)\big]$$
$$= \tilde{\varphi}(x) + \lim_{n\to\infty} \frac{1}{n}\big[\varphi(f^n(x)) - \varphi(x)\big].$$

We need the following lemma:

Lemma 3.2.5. *If ϕ is an integrable function then* $\lim_n (1/n)\phi(f^n(x)) = 0$ *for μ-almost every point $x \in M$.*

Proof. Fix any $\varepsilon > 0$. Since μ is invariant, we have that

$$\mu\big(\{x \in M : |\phi(f^n(x))| \geq n\varepsilon\}\big) = \mu\big(\{x \in M : |\phi(x)| \geq n\varepsilon\}\big)$$
$$= \sum_{k=n}^{\infty} \mu\Big(\{x \in M : k \leq \frac{|\phi(x)|}{\varepsilon} < k+1\}\Big).$$

Adding these expressions over $n \in \mathbb{N}$, we obtain

$$\sum_{n=1}^{\infty} \mu\big(\{x \in M : |\phi(f^n(x))| \geq n\varepsilon\}\big) = \sum_{k=1}^{\infty} k\mu\Big(\{x \in M : k \leq \frac{|\phi(x)|}{\varepsilon} < k+1\}\Big)$$
$$\leq \int \frac{|\phi|}{\varepsilon}\, d\mu.$$

Since ϕ is integrable, by assumption, all these expressions are finite. That implies that the set $B(\varepsilon)$ of all points x such that $|\phi(f^n(x))| \geq n\varepsilon$ for infinitely many values of n has zero measure (check Exercise A.1.6). Now, the definition of $B(\varepsilon)$ implies that for every $x \notin B(\varepsilon)$ there exists $p \geq 1$ such that $|\phi(f^n(x))| < n\varepsilon$ for every $n \geq p$. Consider the set $B = \bigcup_{i=1}^{\infty} B(1/i)$. Then B has zero measure and $\lim_n (1/n)\phi(f^n(x)) = 0$ for every $x \notin B$.

Applying Lemma 3.2.5 to the function $\phi = \varphi$ we obtain the identity in (3.2.3). This completes the proof of Proposition 3.2.4.

In general, the total measure subset of points for which the limit in (3.2.2) exists depends on the function φ under consideration. However, in some situations it is possible to choose such a set independent of the function. A useful example of such a situation is:

Theorem 3.2.6. *Let M be a compact metric space and $f : M \to M$ be a measurable map. Then there exists some measurable set $G \subset M$ with $\mu(G) = 1$ such that*

$$\frac{1}{n}\sum_{j=0}^{n-1} \varphi(f^j(x)) \to \tilde{\varphi}(x) \tag{3.2.4}$$

for every $x \in G$ and every continuous function $\varphi : M \to \mathbb{R}$.

Proof. By the Birkhoff ergodic theorem, for every continuous function φ there exists $G(\varphi) \subset M$ such that $\mu(G(\varphi)) = 1$ and (3.2.4) holds for every $x \in G(\varphi)$. By Theorem A.3.13, the space $C^0(M)$ of continuous functions admits some countable dense subset $\{\varphi_k : k \in \mathbb{N}\}$. Take

$$G = \bigcap_{k=1}^{\infty} G(\varphi_k).$$

Since the intersection is countable, it is clear that $\mu(G) = 1$. So, it suffices to prove that (3.2.4) holds for every continuous function φ whenever $x \in G$. This can be done as follows. Given $\varphi \in C^0(M)$ and any $\varepsilon > 0$, take $k \in \mathbb{N}$ such that

$$\|\varphi - \varphi_k\| = \sup\big\{|\varphi(x) - \varphi_k(x)| : x \in M\big\} \leq \varepsilon.$$

Then, given any point $x \in G$,

$$\limsup_n \frac{1}{n}\sum_{j=0}^{n-1} \varphi(f^j(x)) \leq \lim_n \frac{1}{n}\sum_{j=0}^{n-1} \varphi_k(f^j(x)) + \varepsilon = \tilde{\varphi}_k(x) + \varepsilon$$

$$\liminf_n \frac{1}{n}\sum_{j=0}^{n-1} \varphi(f^j(x)) \geq \lim_n \frac{1}{n}\sum_{j=0}^{n-1} \varphi_k(f^j(x)) - \varepsilon = \tilde{\varphi}_k(x) - \varepsilon.$$

This implies that

$$\limsup_n \frac{1}{n}\sum_{j=0}^{n-1}\varphi(f^j(x)) - \liminf_n \frac{1}{n}\sum_{j=0}^{n-1}\varphi(f^j(x)) \le 2\varepsilon.$$

Since ε is arbitrary, it follows that the limit $\tilde{\varphi}(x)$ exists, as stated.

In general, one can not say anything about the speed of convergence in Theorem 3.2.3. For example, it follows from a theorem of Kakutani and Petersen (check pages 94 to 99 of Petersen [Pet83]) that if the measure μ is ergodic[2] and non-atomic then, given any sequence $(a_n)_n$ of positive real numbers with $\lim_n a_n = 0$, there exists some bounded measurable function φ with

$$\limsup_n \frac{1}{a_n}\left|\frac{1}{n}\sum_{j=0}^{n-1}\varphi(f^j(x)) - \int \varphi\, d\mu\right| = +\infty.$$

Another interesting observation is that there is no analogue of the Birkhoff ergodic theorem for *infinite* invariant measures. Indeed, suppose that μ is a σ-finite, but infinite, invariant measure of a transformation $f: M \to M$. We say that a measurable set $W \subset M$ is *wandering* if the pre-images $f^{-i}(W)$, $i \ge 0$ are pairwise disjoint. Suppose that μ is ergodic and *conservative*, that is, such that every wandering set has zero measure. Then, given any sequence $(a_n)_n$ of positive real numbers,

1. either, for every $\varphi \in L^1(\mu)$,

$$\liminf_n \frac{1}{a_n}\sum_{j=0}^{n-1}\varphi\circ f^j = 0 \quad \text{at almost every point;}$$

2. or, there exists $(n_k)_k \to \infty$ such that, for every $\varphi \in L^1(\mu)$,

$$\lim_k \frac{1}{a_{n_k}}\sum_{j=0}^{n_k-1}\varphi\circ f^j = \infty \quad \text{at almost every point.}$$

This and other related facts about infinite measures are proved in Section 2.4 of Aaronson [Aar97].

3.2.3 Theorem of von Neumann and consequences

The theorem of von Neumann (Theorem 3.1.7) may also be deduced directly from the theorem of Birkhoff, as we are going to explain.

2 We say that an invariant measure μ is *ergodic* if $f^{-1}(A) = A$ up to measure zero implies that either $\mu(A) = 0$ or $\mu(A^c) = 0$. The study of ergodic measures will be the subject of the next chapter.

Consider any $\varphi \in L^2(\mu)$ and let $\tilde{\varphi}$ be the corresponding time average. We start by showing that $\tilde{\varphi} \in L^2(\mu)$ and its norm satisfies $\|\tilde{\varphi}\|_2 \le \|\varphi\|_2$. Indeed,

$$|\tilde{\varphi}| \le \lim_n \frac{1}{n} \sum_{j=0}^{n-1} |\varphi \circ f^j| \quad \text{and, hence,} \quad |\tilde{\varphi}|^2 \le \lim_n \left(\frac{1}{n} \sum_{j=0}^{n-1} |\varphi \circ f^j| \right)^2.$$

Then, by the Fatou lemma (Theorem A.2.10),

$$\left[\int |\tilde{\varphi}|^2 \, d\mu \right]^{1/2} \le \liminf_n \left[\int \left(\frac{1}{n} \sum_{j=0}^{n-1} |\varphi \circ f^j| \right)^2 d\mu \right]^{1/2}. \tag{3.2.5}$$

We can use the Minkowski inequality (Theorem A.5.3) to bound the sequence on the right-hand side from above:

$$\left[\int \left(\frac{1}{n} \sum_{j=0}^{n-1} |\varphi \circ f^j| \right)^2 d\mu \right]^{1/2} \le \frac{1}{n} \sum_{j=0}^{n-1} \left[\int |\varphi \circ f^j|^2 \, d\mu \right]^{1/2}. \tag{3.2.6}$$

Since μ is invariant under f, the expression on the right-hand side is equal to $\left[\int |\varphi|^2 \, d\mu\right]^{1/2}$. So, (3.2.5) and (3.2.6) imply that $\|\tilde{\varphi}\|_2 \le \|\varphi\|_2 < \infty$.

Now let us show that $(1/n)\sum_{j=0}^{n-1} \varphi \circ f^j$ converges to $\tilde{\varphi}$ in $L^2(\mu)$. Initially, suppose that the function φ is bounded, that is, there exists $C > 0$ such that $|\varphi| \le C$. Then,

$$\left| \frac{1}{n} \sum_{j=0}^{n-1} \varphi \circ f^j \right| \le C \quad \text{for every } n \quad \text{and} \quad |\tilde{\varphi}| \le C.$$

Then we may use the dominated convergence theorem (Theorem A.2.11) to conclude that

$$\lim_n \int \left(\frac{1}{n} \sum_{j=0}^{n-1} \varphi \circ f^j - \tilde{\varphi} \right)^2 d\mu = \int \left(\lim_n \frac{1}{n} \sum_{j=0}^{n-1} \varphi \circ f^j - \tilde{\varphi} \right)^2 d\mu = 0.$$

In other words, $(1/n)\sum_{j=0}^{n-1} \varphi \circ f^j$ converges to $\tilde{\varphi}$ in the space $L^2(\mu)$. We are left to extend this conclusion to arbitrary functions φ in $L^2(\mu)$. For that, let us consider some sequence (φ_k) of bounded functions such that $(\varphi_k)_k$ converges to φ. For example:

$$\varphi_k(x) = \begin{cases} \varphi(x) & \text{if } |\varphi(x)| \le k \\ 0 & \text{otherwise.} \end{cases}$$

Denote by $\tilde{\varphi}_k$ the corresponding time averages. Given any $\varepsilon > 0$, let k_0 be fixed such that $\|\varphi - \varphi_k\|_2 < \varepsilon/3$ for every $k \ge k_0$. Note that $\|(\varphi - \varphi_k) \circ f^j\|_2$ is equal to $\|\varphi - \varphi_k\|_2$ for every $j \ge 0$, because the measure μ is invariant. Thus,

$$\left\| \frac{1}{n} \sum_{j=0}^{n-1} (\varphi - \varphi_k) \circ f^j \right\|_2 \le \|\varphi - \varphi_k\|_2 < \varepsilon/3 \quad \text{for every } n \ge 1 \text{ and } k \ge k_0. \tag{3.2.7}$$

Observe also that $\tilde{\varphi} - \tilde{\varphi}_k$ is the time average of the function $\varphi - \varphi_k$. So, the argument in the previous paragraph gives that

$$\|\tilde{\varphi} - \tilde{\varphi}_k\|_2 \le \|\varphi - \varphi_k\|_2 < \varepsilon/3 \quad \text{for every } k \ge k_0. \tag{3.2.8}$$

By assumption, for every $k \ge 1$ there exists $n_0(k) \ge 1$ such that

$$\Big\| \frac{1}{n} \sum_{j=0}^{n-1} \varphi_k \circ f^j - \tilde{\varphi}_k \Big\|_2 < \varepsilon/3 \quad \text{for every } n \ge n_0(k). \tag{3.2.9}$$

Adding (3.2.7), (3.2.8), (3.2.9) we get that

$$\Big\| \frac{1}{n} \sum_{j=0}^{n-1} \varphi \circ f^j - \tilde{\varphi} \Big\|_2 < \varepsilon \quad \text{for every } n \ge n_0(k_0).$$

This completes the proof of the theorem of von Neumann from the theorem of Birkhoff.

Exercise 3.2.5 contains an extension of these conclusions to any $L^p(\mu)$ space.

Corollary 3.2.7. *The time average $\tilde{\varphi}$ of any function $\varphi \in L^2(\mu)$ coincides with the orthogonal projection $P(\varphi)$ of φ to the subspace of invariant functions.*

Proof. On the one hand, Theorem 3.1.7 gives that $(1/n)\sum_{j=0}^{n-1} \varphi \circ f^j$ converges to $P(\varphi)$ in $L^2(\mu)$. On the other hand, we have just shown that this sequence converges to $\tilde{\varphi}$ in the space $L^2(\mu)$. So, by uniqueness of the limit, $P(\varphi) = \tilde{\varphi}$.

Corollary 3.2.8. *If $f : M \to M$ is invertible then the time averages of any function $\varphi \in L^2(\mu)$ relative to f and to f^{-1} coincide at μ-almost every point:*

$$\lim_n \frac{1}{n} \sum_{j=0}^{n-1} \varphi \circ f^{-j} = \lim_n \frac{1}{n} \sum_{j=0}^{n-1} \varphi \circ f^j \quad \textit{at } \mu\textit{-almost every point.} \tag{3.2.10}$$

Proof. The limit on the left-hand side of (3.2.10) is the orthogonal projection of φ to the subspace of functions invariant under f^{-1}, whereas the limit on the right-hand side is the orthogonal projection of φ to the subspace of functions invariant under f. It is clear that these two subspaces are exactly the same. Thus, the two limits coincide in $L^2(\mu)$.

3.2.4 Exercises

3.2.1. Let $X = \{x_1, \dots, x_r\}$ be a finite set and $\sigma : X \to X$ be a permutation. We call σ a *cyclic permutation* if it admits a unique orbit (containing all r elements of X).

1. Prove that, for any cyclic permutation σ and any function $\varphi : X \to \mathbb{R}$,

$$\lim_{n\to\infty} \frac{1}{n} \sum_{i=0}^{n-1} \varphi(\sigma^i(x)) = \frac{\varphi(x_1) + \cdots + \varphi(x_r)}{r}.$$

2. More generally, prove that for any permutation σ and any function φ

$$\lim_{n\to\infty}\frac{1}{n}\sum_{i=0}^{n-1}\varphi(\sigma^i(x)) = \frac{\varphi(x)+\varphi(\sigma(x))+\cdots+\varphi(\sigma^{p-1}(x))}{p},$$

where $p \geq 1$ is the cardinality of the orbit of x.

3.2.2. Check that Lemma 3.2.5 can also be deduced from the Birkhoff ergodic theorem and then we may even weaken the hypothesis: it suffices to suppose that ϕ is measurable and $\psi = \phi\circ f - \phi$ is integrable.

3.2.3. A function $\varphi : \mathbb{Z} \to \mathbb{R}$ is said to be *uniformly quasi-periodic* if for every $\varepsilon > 0$ there exists $L(\varepsilon) \in \mathbb{N}$ such that every interval $\{n+1,\ldots,n+L(\varepsilon)\}$ in the set of integers contains some τ such that $|\varphi(k+\tau)-\varphi(k)| < \varepsilon$ for every $k \in \mathbb{Z}$. Any such τ is called an *ε-quasi-period* of f.

(a) Prove that if φ is uniformly quasi-periodic then φ is bounded.

(b) Show that for every $\varepsilon > 0$ there exists $\rho \geq 1$ such that

$$\left|\frac{1}{\rho}\sum_{j=n\rho+1}^{(n+1)\rho}\varphi(j) - \frac{1}{\rho}\sum_{j=1}^{\rho}\varphi(j)\right| < 2\varepsilon \quad \text{for every } n \geq 1.$$

(c) Show that the sequence $(1/n)\sum_{j=1}^{n}\varphi(j)$ converges to some real number when $n \to \infty$.

(d) More generally, prove that $\lim_n (1/n)\sum_{k=1}^{n}\varphi(x+k)$ exists for every $x \in \mathbb{Z}$ and is independent of x.

3.2.4. Prove that for Lebesgue-almost every $x \in [0,1]$, the geometric mean of the integer numbers $a_1,\ldots,a_n,\ldots$ in the continued fraction expansion of x converges to some real number: in other words, there exists $b \in \mathbb{R}$ such that $\lim_n (a_1a_2\cdots a_n)^{1/n} = b$. [Observation: Compare with Exercise 4.2.12.]

3.2.5. Let $\varphi : M \to \mathbb{R}$ be an integrable function and $\tilde{\varphi}$ be the corresponding time average, given by Theorem 3.2.3. Show that if $\varphi \in L^p(\mu)$ for some $p > 1$ then $\tilde{\varphi} \in L^p(\mu)$ and $\|\tilde{\varphi}\|_p \leq \|\varphi\|_p$. Moreover,

$$\frac{1}{n}\sum_{j=0}^{n-1}\varphi\circ f^j$$

converges to $\tilde{\varphi}$ in the space $L^p(\mu)$.

3.2.6. Prove the Birkhoff ergodic theorem for flows: if μ is a probability measure invariant under a flow f and $\varphi \in L^1(\mu)$ then the function

$$\tilde{\varphi}(x) = \lim_{T\to\infty}\frac{1}{T}\int_0^T \varphi(f^t(x))\,dt$$

is defined at μ-almost every point and $\int\tilde{\varphi}\,d\mu = \int\varphi\,d\mu$.

3.2.7. Prove that if a continuous transformation $f : M \to M$ of a compact metric space M admits exactly one invariant probability measure μ, and this measure is such that $\mu(A) > 0$ for every non-empty open set $A \subset M$, then every orbit of f is dense in M.

3.3 Subadditive ergodic theorem

A sequence of functions $\varphi_n : M \to \mathbb{R}$ is said to be *subadditive* for a transformation $f : M \to M$ if

$$\varphi_{m+n} \le \varphi_m + \varphi_n \circ f^m \quad \text{for every } m, n \ge 1. \tag{3.3.1}$$

Example 3.3.1. A sequence $\varphi_n : M \to \mathbb{R}$ is *additive* for the transformation f if $\varphi_{m+n} = \varphi_m + \varphi_n \circ f^m$ for every $m, n \ge 1$. For example, the time sums

$$\varphi_n(x) = \sum_{j=0}^{n-1} \varphi(f^j(x))$$

of any function $\varphi : M \to \mathbb{R}$ form an additive sequence. In fact, every additive sequence is of this form, with $\varphi = \varphi_1$. Of course, additive sequences are also subadditive.

For the next example we need the notion of the *norm* of a square matrix A of dimension d, which is defined as follows:

$$\|A\| = \sup\left\{\frac{\|Av\|}{\|v\|} : v \in \mathbb{R}^d \setminus \{0\}\right\}. \tag{3.3.2}$$

Compare with (2.3.1). It follows directly from the definition that the norm of the product of two matrices is less than or equal to the product of the norms of those matrices:

$$\|AB\| \le \|A\| \, \|B\|. \tag{3.3.3}$$

Example 3.3.2. Let $A : M \to \mathrm{GL}(d)$ be a measurable function with values in the *linear group*, that is, the set $\mathrm{GL}(d)$ of invertible square matrices of dimension d. Define

$$\phi^n(x) = A(f^{n-1}(x)) \cdots A(f(x))A(x)$$

for every $n \ge 1$ and $x \in M$. Then the sequence $\varphi_n(x) = \log \|\phi^n(x)\|$ is subadditive. Indeed,

$$\phi^{m+n}(x) = \phi^n(f^m(x))\phi^m(x).$$

and so, using (3.3.3),

$$\begin{aligned} \varphi_{m+n}(x) &= \log \|\phi^n(f^m(x))\phi^m(x)\| \\ &\le \log \|\phi^m(x)\| + \log \|\phi^n(f^m(x))\| = \varphi_m(x) + \varphi_n(f^m(x)) \end{aligned}$$

for every m, n and x.

Recall that, given any function $\varphi : M \to \mathbb{R}$, we denote by $\varphi^+ : M \to \mathbb{R}$ its positive part, which is defined by $\varphi^+(x) = \max\{\varphi(x), 0\}$.

Theorem 3.3.3 (Kingman). *Let μ be a probability measure invariant under a transformation $f : M \to M$ and let $\varphi_n : M \to \mathbb{R}$, $n \ge 1$ be a subadditive sequence of measurable functions such that $\varphi_1^+ \in L^1(\mu)$. Then $(\varphi_n/n)_n$ converges at*

μ-almost every point to some function $\varphi : M \to [-\infty, +\infty)$ that is invariant under f. Moreover, $\varphi^+ \in L^1(\mu)$ and

$$\int \varphi \, d\mu = \lim_n \frac{1}{n} \int \varphi_n \, d\mu = \inf_n \frac{1}{n} \int \varphi_n \, d\mu \in [-\infty, +\infty).$$

The proof of Theorem 3.3.3 that we are going to present is due to Avila and Bochi [AB], who started from a proof of the Birkhoff ergodic theorem (Theorem 3.2.3) by Katznelson and Weiss [KW82]. An important observation is that Theorem 3.2.3 is *not* used in the arguments. This allows us to obtain the theorem of Birkhoff as a particular case of Theorem 3.3.3.

3.3.1 Preparing the proof

A sequence $(a_n)_n$ in $[-\infty, +\infty)$ is said to be *subadditive* if $a_{m+n} \le a_m + a_n$ for every $m, n \ge 1$.

Lemma 3.3.4. *If $(a_n)_n$ is a subadditive sequence then*

$$\lim_n \frac{a_n}{n} = \inf_n \frac{a_n}{n} \in [-\infty, \infty). \tag{3.3.4}$$

Proof. If $a_m = -\infty$ for some m then, by subadditivity, $a_n = -\infty$ for every $n > m$. In that case, both sides of (3.3.4) are equal to $-\infty$ and so the lemma holds. From now on let us assume that $a_n \in \mathbb{R}$ for every n.

Let $L = \inf_n (a_n/n) \in [-\infty, +\infty)$ and B be any real number larger than L. Then we may find $k \ge 1$ such that

$$\frac{a_k}{k} < B.$$

For $n > k$, we may write $n = kp + q$, where p and q are integers such that $p \ge 1$ and $1 \le q \le k$. Then, by subadditivity,

$$a_n \le a_{kp} + a_q \le p a_k + a_q \le p a_k + \alpha,$$

where $\alpha = \max\{a_i : 1 \le i \le k\}$. Hence,

$$\frac{a_n}{n} \le \frac{pk}{n} \frac{a_k}{k} + \frac{\alpha}{n}.$$

Observe that pk/n converges to 1 and α/n converges to zero when $n \to \infty$. So, since $a_k/k < B$, we have that

$$L \le \frac{a_n}{n} < B$$

for every n sufficiently large. Making $B \to L$, we conclude that

$$\lim_n \frac{a_n}{n} = L = \inf_n \frac{a_n}{n}.$$

This completes the argument.

Now let $(\varphi_n)_n$ be as in Theorem 3.3.3. By subadditivity,

$$\varphi_n \le \varphi_1 + \varphi_1 \circ f + \cdots + \varphi_1 \circ f^{n-1}.$$

This relation remains valid if we replace φ_n and φ_1 by their positive parts φ_n^+ and φ_1^+. Hence, the hypothesis that $\varphi_1^+ \in L^1(\mu)$ implies that $\varphi_n^+ \in L^1(\mu)$ for every n. Moreover, the hypothesis that $(\varphi_n)_n$ is subadditive implies that

$$a_n = \int \varphi_n \, d\mu, \quad n \ge 1$$

is a subadditive sequence in $[-\infty, +\infty)$. Therefore, by Lemma 3.3.4, the limit

$$L = \lim_n \frac{a_n}{n} = \inf_n \frac{a_n}{n} \in [-\infty, \infty)$$

exists. Define $\varphi_- : M \to [-\infty, \infty]$ and $\varphi_+ : M \to [-\infty, \infty]$ through

$$\varphi_-(x) = \liminf_n \frac{\varphi_n}{n}(x) \quad \text{and} \quad \varphi_+(x) = \limsup_n \frac{\varphi_n}{n}(x).$$

Clearly, $\varphi_-(x) \le \varphi_+(x)$ for every $x \in M$. We are going to prove that

$$\int \varphi_- \, d\mu \ge L \ge \int \varphi_+ \, d\mu, \tag{3.3.5}$$

as long as each function φ_n is bounded from below. Consequently, the two functions φ_- and φ_+ coincide at μ-almost every point and their integral is equal to L. Thus, the theorem will be proven in this case, with $\varphi = \varphi_- = \varphi_+$ (the fact that φ is invariant under f is part of Exercise 3.3.2). At the end, we remove that boundedness assumption using a truncation trick.

3.3.2 Key lemma

In this section we assume that $\varphi_- > -\infty$ at every point. Fix $\varepsilon > 0$ and define, for each $k \in \mathbb{N}$,

$$E_k = \big\{x \in M : \varphi_j(x) \le j\big(\varphi_-(x) + \varepsilon\big) \text{ for some } j \in \{1, \ldots, k\}\big\}.$$

It is clear that $E_k \subset E_{k+1}$ for every k. Moreover, the definition of $\varphi_-(x)$ implies that $M = \bigcup_k E_k$. Define also

$$\psi_k(x) = \begin{cases} \varphi_-(x) + \varepsilon & \text{if } x \in E_k \\ \varphi_1(x) & \text{if } x \in E_k^c. \end{cases}$$

It follows from the definition of E_k that $\varphi_1(x) > \varphi_-(x) + \varepsilon$ for every $x \in E_k^c$. Combining this fact with the previous observations, we see that the sequence $(\psi_k(x))_k$ is non-increasing and converges to $\varphi_-(x) + \varepsilon$, for every $x \in M$. In particular, by the monotone convergence theorem (Theorem A.2.9),

$$\int \psi_k \, d\mu \to \int (\varphi_- + \varepsilon) \, d\mu \quad \text{as } k \to \infty.$$

The crucial step in the proof of the theorem is the following estimate:

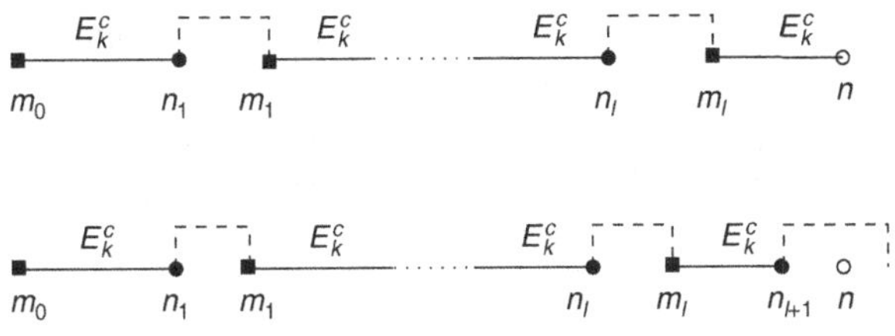

Figure 3.1. Decomposition of the trajectory of a point

Lemma 3.3.5. *For every $n > k \geq 1$ and μ-almost every $x \in M$,*

$$\varphi_n(x) \leq \sum_{i=0}^{n-k-1} \psi_k(f^i(x)) + \sum_{i=n-k}^{n-1} \max\{\psi_k, \varphi_1\}(f^i(x)).$$

Proof. Take $x \in M$ such that $\varphi_-(x) = \varphi_-(f^j(x))$ for every $j \geq 1$ (this holds at μ-almost every point, according to Exercise 3.3.2). Consider the sequence, possibly finite, of integer numbers

$$m_0 \leq n_1 < m_1 \leq n_2 < m_2 \leq \dots \tag{3.3.6}$$

defined inductively as follows (see also Figure 3.1).

Define $m_0 = 0$. Let n_j be the smallest integer greater than or equal to m_{j-1} satisfying $f^{n_j}(x) \in E_k$ (if it exists). Then, by the definition of E_k, there exists m_j such that $1 \leq m_j - n_j \leq k$ and

$$\varphi_{m_j - n_j}(f^{n_j}(x)) \leq (m_j - n_j)(\varphi_-(f^{n_j}(x)) + \varepsilon). \tag{3.3.7}$$

This completes the definition of the sequence (3.3.6). Now, given $n \geq k$, let $l \geq 0$ be the largest integer such that $m_l \leq n$. By subadditivity,

$$\varphi_{n_j - m_{j-1}}(f^{m_{j-1}}(x)) \leq \sum_{i=m_{j-1}}^{n_j - 1} \varphi_1(f^i(x))$$

for every $j = 1, \dots, l$ such that $m_{j-1} \neq n_j$, and analogously for $\varphi_{n-m_l}(f^{m_l}(x))$. Thus,

$$\varphi_n(x) \leq \sum_{i \in I} \varphi_1(f^i(x)) + \sum_{j=1}^{l} \varphi_{m_j - n_j}(f^{n_j}(x)) \tag{3.3.8}$$

where $I = \bigcup_{j=1}^{l} [m_{j-1}, n_j) \bigcup [m_l, n)$. Observe that

$$\varphi_1(f^i(x)) = \psi_k(f^i(x)) \quad \text{for every} \quad i \in \bigcup_{j=1}^{l} [m_{j-1}, n_j) \cup [m_l, \min\{n_{l+1}, n\}),$$

since $f^i(x) \in E_k^c$ in all these cases. Moreover, since φ_- is constant on orbits (see Exercise 3.3.2) and $\psi_k \geq \varphi_- + \varepsilon$, the relation (3.3.7) gives that

$$\varphi_{m_j - n_j}(f^{n_j}(x)) \leq \sum_{i=n_j}^{m_j - 1} (\varphi_-(f^i(x)) + \varepsilon) \leq \sum_{i=n_j}^{m_j - 1} \psi_k(f^i(x))$$

for every $j = 1, \dots, l$. In this way, using (3.3.8), we conclude that

$$\varphi_n(x) \le \sum_{i=0}^{\min\{n_{l+1}, n\}-1} \psi_k(f^i(x)) + \sum_{i=n_{l+1}}^{n-1} \varphi_1(f^i(x)).$$

Since $n_{l+1} > n - k$, the lemma is proven.

3.3.3 Estimating φ_-

Towards establishing (3.3.5), in this section we prove the following lemma:

Lemma 3.3.6. $\int \varphi_- \, d\mu = L$

Proof. Suppose for a while that φ_n/n is uniformly bounded from below, that is, that there exists $\kappa > 0$ such that $\varphi_n/n \ge -\kappa$ for every n. Applying the lemma of Fatou (Theorem A.2.10) to the sequence of non-negative functions $\varphi_n/n + \kappa$, we get that φ_- is integrable and

$$\int \varphi_- \, d\mu \le \lim_n \int \frac{\varphi_n}{n} \, d\mu = L.$$

To prove the opposite inequality, observe that Lemma 3.3.5 implies

$$\frac{1}{n} \int \varphi_n \, d\mu \le \frac{n-k}{n} \int \psi_k \, d\mu + \frac{k}{n} \int \max\{\psi_k, \varphi_1\} \, d\mu. \tag{3.3.9}$$

Note that $\max\{\psi_k, \varphi_1\} \le \max\{\varphi_- + \varepsilon, \varphi_1^+\}$, and this last function is integrable. So, the limit superior of the last term in (3.3.9) as $n \to \infty$ is less than or equal to zero. So, making $n \to \infty$ we get that $L \le \int \psi_k \, d\mu$ for every k. Then, making $k \to \infty$, we conclude that

$$L \le \int \varphi_- \, d\mu + \varepsilon.$$

Finally, making $\varepsilon \to 0$ we get that $L \le \int \varphi_- \, d\mu$. This proves the lemma when φ_n/n is uniformly bounded from below.

We are left to remove this hypothesis. Define, for each $\kappa > 0$,

$$\varphi_n^\kappa = \max\{\varphi_n, -\kappa n\} \quad \text{and} \quad \varphi_-^\kappa = \max\{\varphi_-, -\kappa\}.$$

The sequence $(\varphi_n^\kappa)_n$ satisfies all the conditions of Theorem 3.3.3: indeed, it is subadditive and the positive part of φ_1^κ is integrable. Moreover, it is clear that $\varphi_-^\kappa = \liminf_n (\varphi_n^\kappa / n)$. So, the argument in the previous paragraph shows that

$$\int \varphi_-^\kappa \, d\mu = \inf_n \frac{1}{n} \int \varphi_n^\kappa \, d\mu. \tag{3.3.10}$$

By the monotone convergence theorem (Theorem A.2.9), we also have that

$$\int \varphi_n \, d\mu = \inf_\kappa \int \varphi_n^\kappa \, d\mu \quad \text{and} \quad \int \varphi_- \, d\mu = \inf_\kappa \int \varphi_-^\kappa \, d\mu. \tag{3.3.11}$$

Combining the relations (3.3.10) and (3.3.11), we get that

$$\int \varphi_- \, d\mu = \inf_\kappa \int \varphi_-^\kappa = \inf_\kappa \inf_n \frac{1}{n} \int \varphi_n^\kappa \, d\mu = \inf_n \frac{1}{n} \int \varphi_n \, d\mu = L.$$

This completes the proof of the lemma.

3.3.4 Bounding φ_+

To complete the proof of (3.3.5), we are now going to show that $\int \varphi_+ \, d\mu \le L$ as long as $\inf_x \varphi_n(x)$ is finite for every n. Let us start by proving the following auxiliary result:

Lemma 3.3.7. *For any fixed k,*

$$\limsup_n \frac{\varphi_{kn}}{n} = k \limsup_n \frac{\varphi_n}{n}.$$

Proof. The inequality $\le$ is clear, since φ_{kn}/kn is a subsequence of φ_n/n. To prove the opposite inequality, let us write $n = kq_n + r_n$ with $r_n \in \{1, \ldots, k\}$. By subadditivity,

$$\varphi_n \le \varphi_{kq_n} + \varphi_{r_n} \circ f^{kq_n} \le \varphi_{kq_n} + \psi \circ f^{kq_n},$$

where $\psi = \max\{\varphi_1^+, \ldots, \varphi_k^+\}$. Observe that $n/q_n \to k$ when $n \to \infty$. Moreover, as $\psi \in L^1(\mu)$, we may use Lemma 3.2.5 to see that $\psi \circ f^n/n$ converges to zero at μ-almost every point. Hence, dividing all the terms in the previous relation by n and taking the limit superior as $n \to \infty$, we get that

$$\limsup_n \frac{1}{n}\varphi_n \le \limsup_n \frac{1}{n}\varphi_{kq_n} + \limsup_n \frac{1}{n}\psi \circ f^{kq_n} = \frac{1}{k} \limsup_q \frac{1}{q}\varphi_{kq},$$

as stated in the lemma.

Lemma 3.3.8. *Suppose that* $\inf_x \varphi_n(x) > -\infty$ *for every n. Then $\int \varphi_+ \, d\mu \le L$.*

Proof. For each k and $n \ge 1$, consider $\theta_n = -\sum_{j=0}^{n-1} \varphi_k \circ f^{jk}$. Observe that

$$\int \theta_n \, d\mu = -n \int \varphi_k \, d\mu \quad \text{for every } n, \tag{3.3.12}$$

since f^k preserves the measure μ. Since the sequence $(\varphi_n)_n$ is subadditive, $\theta_n \le -\varphi_{kn}$ for every n. Hence, using Lemma 3.3.7,

$$\theta_- = \liminf_n \frac{\theta_n}{n} \le -\limsup_n \frac{\varphi_{kn}}{n} = -k \limsup_n \frac{\varphi_n}{n} = -k\varphi_+,$$

and so

$$\int \theta_- \, d\mu \le -k \int \varphi_+ \, d\mu. \tag{3.3.13}$$

Observe also that the sequence $(\theta_n)_n$ is additive: $\theta_{m+n} = \theta_m + \theta_n \circ f^{km}$ for every $m, n \ge 1$. Since $\theta_1 = -\varphi_k$ is bounded from above by $-\inf \varphi_k$, we also have that the function θ_1^+ is bounded and, consequently, integrable. Thus, we may apply Lemma 3.3.6, together with the equality (3.3.12), to conclude that

$$\int \theta_- \, d\mu = \inf_n \int \frac{\theta_n}{n} \, d\mu = -\int \varphi_k \, d\mu. \tag{3.3.14}$$

Putting (3.3.13) and (3.3.14) together we get that

$$\int \varphi_+ \, d\mu \le \frac{1}{k} \int \varphi_k \, d\mu.$$

Finally, taking the infimum over k we get that $\int \varphi_+ \, d\mu \le L$.

Lemmas 3.3.6 and 3.3.8 imply the relation (3.3.5) and, thus, Theorem 3.3.3 is proven when $\inf \varphi_k > -\infty$ for every k. In the general case, consider

$$\varphi_n^\kappa = \max\{\varphi_n, -\kappa n\} \quad \text{and} \quad \varphi_-^\kappa = \max\{\varphi_-, -\kappa\} \quad \text{and} \quad \varphi_+^\kappa = \max\{\varphi_+, -\kappa\}$$

for every constant $\kappa > 0$. The previous arguments may be applied to the sequence $(\varphi_n^\kappa)_n$ for each fixed $\kappa > 0$. Hence, $\varphi_+^\kappa = \varphi_-^\kappa$ at μ-almost every point for every $\kappa > 0$. Since $\varphi_-^\kappa \to \varphi_-$ and $\varphi_+^\kappa \to \varphi_+$ when $\kappa \to \infty$, it follows that $\varphi_- = \varphi_+$ at μ-almost every point. The proof of Theorem 3.3.3 is complete.

3.3.5 Lyapunov exponents

We have observed previously that every sequence of time sums

$$\varphi_n = \sum_{j=0}^{n-1} \varphi \circ f^j, \quad n \ge 1$$

is additive and, in particular, subadditive. Therefore, the ergodic theorem of Birkhoff (Theorem 3.2.3) is a particular case of Theorem 3.3.3. Another important consequence of the subadditive ergodic theorem is the theorem of Furstenberg–Kesten that we state next.

Let $f : M \to M$ be a measurable transformation and μ be an invariant probability measure. Consider any measurable function $\theta : M \to \mathrm{GL}(d)$ with values in the group $\mathrm{GL}(d)$. The *cocycle* defined by θ over f is the sequence of functions defined by

$$\phi^n(x) = \theta(f^{n-1}(x)) \cdots \theta(f(x))\theta(x) \text{ for } n \ge 1 \quad \text{and} \quad \phi^0(x) = \mathrm{id}$$

for every $x \in M$. We leave it to the reader to check that

$$\phi^{m+n}(x) = \phi^n(f^m(x)) \cdot \phi^m(x) \quad \text{for every } m, n \in \mathbb{N} \text{ and } x \in M. \tag{3.3.15}$$

It is also easy to check that, conversely, any sequence $(\phi^n)_n$ with this property is the cocycle defined by $\theta = \phi^1$ over the transformation f.

Theorem 3.3.9 (Furstenberg–Kesten). *If* $\log^+ \|\theta\| \in L^1(\mu)$ *then*

$$\lambda_{\max}(x) = \lim_n \frac{1}{n} \log \|\phi^n(x)\|$$

exists at μ-almost every point. Moreover, $\lambda_{\max}^{+} \in L^1(\mu)$ and

$$\int \lambda_{\max} \, d\mu = \lim_n \frac{1}{n} \int \log \|\phi^n\| \, d\mu = \inf_n \frac{1}{n} \int \log \|\phi^n\| \, d\mu.$$

If $\log^+ \|\theta^{-1}\| \in L^1(\mu)$ then

$$\lambda_{\min}(x) = \lim_n -\frac{1}{n} \log \|\phi^n(x)^{-1}\|$$

exists at μ-almost every point. Moreover, $\lambda_{\min} \in L^1(\mu)$ and

$$\int \lambda_{\min} \, d\mu = \lim_n -\frac{1}{n} \int \log \|(\phi^n)^{-1}\| \, d\mu = \sup_n -\frac{1}{n} \int \log \|(\phi^n)^{-1}\| \, d\mu.$$

To deduce this result from Theorem 3.3.3 it suffices to note that the sequences

$$\varphi_n^{\max}(x) = \log \|\phi^n(x)\| \quad \text{and} \quad \varphi_n^{\min}(x) = \log \|\phi^n(x)^{-1}\|$$

are subadditive (recall Example 3.3.2).

The multiplicative ergodic theorem of Oseledets, which we are going to state in the following, provides a major refinement of the conclusion of the Furstenberg–Kesten theorem. It asserts that, under the same hypotheses as Theorem 3.3.9, for μ-almost every $x \in M$ there exist a positive integer $k = k(x)$ and real numbers $\lambda_1(x) > \cdots > \lambda_k(x)$ and a *filtration* (that is, a decreasing sequence of vector subspaces)

$$\mathbb{R}^d = V_x^1 > \cdots > V_x^k > V_x^{k+1} = \{0\} \tag{3.3.16}$$

such that, for every $i \in \{1, \ldots, k\}$ and μ-almost every $x \in M$,

(a1) $k(f(x)) = k(x)$ and $\lambda_i(f(x)) = \lambda_i(x)$ and $\theta(x) \cdot V_x^i = V_{f(x)}^i$;

(b1) $\lim_n \frac{1}{n} \log \|\phi^n(x)v\| = \lambda_i(x)$ for every $v \in V_x^i \setminus V_x^{i+1}$;

(c1) $\lim_n \frac{1}{n} \log |\det \phi^n(x)| = \sum_{i=1}^{k} d_i(x)\lambda_i(x)$, where $d_i(x) = \dim V_x^i - \dim V_x^{i+1}$.

Moreover, the numbers $k(x)$ and $\lambda_1(x), \ldots, \lambda_k(x)$ and the subspaces $V_x^1, \ldots, V_x^k$ depend measurably on the point x.

The numbers $\lambda_i(x)$ are called the *Lyapunov exponents* of θ at the point x. They satisfy $\lambda_1 = \lambda_{\max}$ and $\lambda_k = \lambda_{\min}$. For this reason, we also call $\lambda_{\max}(x)$ and $\lambda_{\min}(x)$ the *extremal Lyapunov exponents* of θ at the point x. Each $d_i(x)$ is called the *multiplicity* of the Lyapunov exponent $\lambda_i(x)$.

When f is invertible, we may extend the sequence ϕ^n to the whole of $\mathbb{Z}$, through

$$\phi^{-n}(x) = \phi^n(f^{-n}(x))^{-1} \quad \text{for every } n \geq 1 \text{ and } x \in M.$$

Assuming also that $\log^+ \|\theta^{-1}\| \in L^1(\mu)$, one obtains a stronger conclusion than before: more than a filtration, there is a decomposition

$$\mathbb{R}^d = E_x^1 \oplus \cdots \oplus E_x^k \tag{3.3.17}$$

such that, for every $i = 1, \dots, k$,

(a2) $\theta(x) \cdot E_x^i = E_{f(x)}^i$ and $V_x^i = V_x^{i+1} \oplus E_x^i$; so, $\dim E_x^i = \dim V_x^i - \dim V_x^{i+1}$;

(b2) $\lim_{n \to \pm\infty} \frac{1}{n} \log \|\phi^n(x)v\| = \lambda_i(x)$ for every $v \in E_x^i$ different from zero;

(c2) $\lim_{n \to +\infty} \frac{1}{n} \log |\det \phi^n(x)| = \sum_{i=1}^{k} d_i(x)\lambda_i(x)$, where $d_i(x) = \dim E_x^i$.

The reader will find a much more detailed discussion of these results, including proofs, in Chapter 4 of [Via14].

3.3.6 Exercises

3.3.1. Give a direct proof of the Birkhoff ergodic theorem (Theorem 3.2.3), using the approach in the proof of Theorem 3.3.3.

3.3.2. Given a subadditive sequence $(\varphi_n)_n$ with $\varphi_1^+ \in L^1(\mu)$, show that the functions

$$\varphi_- = \liminf_n \frac{\varphi_n}{n} \quad \text{and} \quad \varphi_+ = \limsup_n \frac{\varphi_n}{n}$$

are f-invariant, that is, they satisfy $\varphi_-(x) = \varphi_- \circ f(x)$ and $\varphi_+(x) = \varphi_+ \circ f(x)$ for μ-almost every $x \in M$.

3.3.3. State and prove the subadditive ergodic theorem for flows.

3.3.4. Let M be a compact manifold and $f : M \to M$ be a diffeomorphism of class C^1 that preserves the Lebesgue measure. Check that

$$\sum_{i=1}^{k(x)} d_i(x)\lambda_i(x) = 0 \quad \text{at } \mu\text{-almost every point } x \in M,$$

where $\lambda_i(x)$, $i = 1, \dots, k(x)$ are the Lyapunov exponents of Df at the point x and $d_i(x)$, $i = 1, \dots, k(x)$ are the corresponding multiplicities.

3.3.5. Let $(\varphi_n)_n$ be a subadditive sequence of functions for some transformation $f : M \to M$. We call the *time constant* of $(\varphi_n)_n$ the number

$$\lim_n \frac{1}{n} \int \varphi_n \, d\mu$$

when it exists. Assuming that the limit does exist and is finite, show that we may write $\varphi_n = \psi_n + \gamma_n$ for each n, in such a way that $(\psi_n)_n$ is an additive sequence and $(\gamma_n)_n$ is a subadditive sequence with time constant equal to zero.

3.3.6. Under the assumptions of the Furstenberg–Kesten theorem, show that the sequence $\psi_n = (1/n) \log \|\phi^n\|$ is *uniformly integrable*, in the following sense: for every $\varepsilon > 0$ there exists $\delta > 0$ such that

$$\mu(E) < \delta \quad \Rightarrow \quad \int_E \psi_n^+ \, d\mu < \varepsilon \text{ for every } n.$$

3.3.7. Under the assumptions of the Furstenberg–Kesten theorem, let Ψ_k denote the time average of the function $\psi_k = (1/k) \log \|\phi^k\|$ relative to the transformation f^k. Show that $\lambda_{\max}(x) \le \Psi_k(x)$ for every k and μ-almost every x. Using Exercise 3.3.6, show that for every $\rho > 0$ and μ-almost every x there exists k such that $\Psi_k(x) \le \lambda_{\max}(x) + \rho$.

3.4 Discrete time and continuous time

Most of the time we focus our presentation in the context of dynamical systems with discrete time. However, almost everything that was said so far extends, more or less straightforwardly, to systems with continuous time. One reason why the two theories are so similar is that one may relate systems of either kind to systems of the other kind, through certain constructions. That is the subject of the present section. For simplicity, we stick to the case of invertible systems. The general case may be handled using the notion of natural extension, which was described in Section 2.4.2.

3.4.1 Suspension flows

Our first construction associates with every invertible map $f: M \to M$ and every measurable function $\tau: M \to (0, \infty)$ a flow $g^t: N \to N$, $t \in \mathbb{R}$, that we call the *suspension of f with return time τ*, whose recurrence properties are directly related to the recurrence properties of f. In particular, we associate a measure ν invariant under this flow with every measure μ invariant under f. For this construction we assume that the function τ is such that

$$\sum_{j=1}^{\infty} \tau(f^j(x)) = \sum_{j=1}^{\infty} \tau(f^{-j}(x)) = +\infty \tag{3.4.1}$$

for every $x \in M$. That is the case, for instance, if τ is bounded away from zero.

The first step is to construct the domain N of the suspension flow. Let us consider the transformation $F: M \times \mathbb{R} \to M \times \mathbb{R}$ defined by

$$F(x,s) = (f(x), s - \tau(x)).$$

Note that F is invertible. Let $\sim$ be the equivalence relation defined in $M \times \mathbb{R}$ by

$$(x,s) \sim (\tilde{x}, \tilde{s}) \quad \Leftrightarrow \quad \text{there exists } n \in \mathbb{Z} \text{ such that } F^n(x,s) = (\tilde{x}, \tilde{s}).$$

We denote by N the set of equivalence classes and by $\pi: M \times \mathbb{R} \to N$ the canonical projection associating with every $(x,s) \in M \times \mathbb{R}$ the corresponding equivalence class.

Now consider the flow $G^t: M \times \mathbb{R} \to M \times \mathbb{R}$ given by $G^t(x,s) = (x, s+t)$. It is clear that $G^t \circ F = F \circ G^t$ for every $t \in \mathbb{R}$. This ensures that G^t, $t \in \mathbb{R}$ induces a flow g^t, $t \in \mathbb{R}$ in the quotient space N, given by

$$g^t(\pi(x,s)) = \pi(G^t(x,s)) \quad \text{for every } x \in M \text{ and } s, t \in \mathbb{R}. \tag{3.4.2}$$

Indeed, if $\pi(x,s) = \pi(\tilde{x}, \tilde{s})$ then there exists $n \in \mathbb{Z}$ such that $F^n(x,s) = (\tilde{x}, \tilde{s})$. Hence,

$$G^t(\tilde{x}, \tilde{s}) = G^t \circ F^n(x,s) = F^n \circ G^t(x,s)$$

and so $\pi(G^t(x,s)) = \pi(G^t(\tilde{x}, \tilde{s}))$. This shows that the flow g^t, $t \in \mathbb{R}$ is well defined.

To better understand how this flow is related to the transformation f, we need to revisit the construction from a more concrete point of view. Let us consider $D = \{(x,s) \in M \times \mathbb{R} : 0 \le s < \tau(x)\}$. We claim that D is a *fundamental domain* for the equivalence relation $\sim$, that is, it contains exactly one representative of each equivalence class. Uniqueness of the representative is immediate: just observe that if $(x,s) \in D$ then $F^n(x,s) = (x_n, s_n)$ with $s_n < 0$ for every $n > 0$ and $s_n > \tau(f^n(x))$ for every $n > 0$. To prove existence, we need the condition (3.4.1): it ensures that the iterates $(x_n, s_n) = F^n(x,s)$ of any (x,s) satisfy

$$\lim_{n\to+\infty} s_n = -\infty \quad \text{and} \quad \lim_{n\to-\infty} s_n = +\infty.$$

Then, taking n maximum such that $s_n \ge 0$, we find that $(x_n, s_n) \in D$. In this way, the claim is proved. Now observe that the claim means that the restriction of the projection π to domain D is a bijection over N. Thus, we may identify N with D and, in particular, we may consider g^t, $t \in \mathbb{R}$ as a flow in D.

In just the same way, we may identify M with the subset $\Sigma = \pi(M \times \{0\})$ of N. Observing that

$$g^{\tau(x)}(\pi(x,0)) = \pi(x, \tau(x)) = \pi(f(x), 0), \tag{3.4.3}$$

we see that, through this identification, the transformation $f : M \to M$ corresponds to the first-return map (or *Poincaré return map*) of the suspension flow to Σ. See Figure 3.2.

Now let μ be a measure on M invariant under f. Let us denote by ds the Lebesgue measure on the real line $\mathbb{R}$. It is clear that the (infinite) measure $\mu \times ds$ is invariant under the flow G^t, $t \in \mathbb{R}$. Moreover, it is invariant under the transformation F, since μ is invariant under f. We call the *suspension of μ with return time τ* the measure ν defined on N by

$$\nu = \pi_*(\mu \times ds \mid D). \tag{3.4.4}$$

In other words, ν is the measure such that

$$\int \psi \, d\nu = \int \int_0^{\tau(x)} \psi(\pi(x,s)) \, ds \, d\mu(x)$$

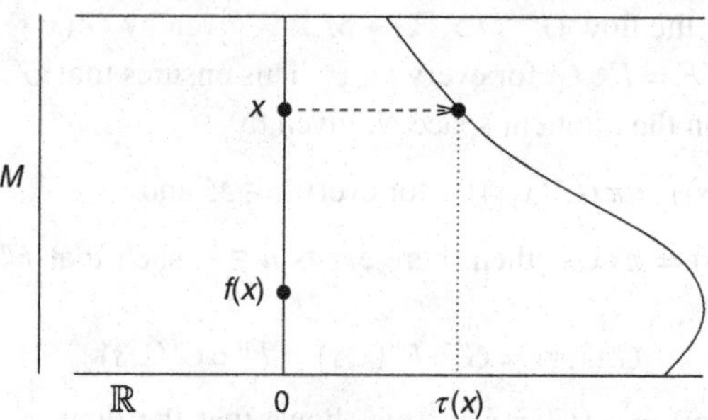

Figure 3.2. Suspension flow

for every bounded measurable function $\psi : N \to (0,\infty)$. In particular,

$$\nu(N) = \int 1\, d\nu = \int \tau(x)\, d\mu(x) \tag{3.4.5}$$

is finite if and only if the function τ is integrable with respect to μ.

Proposition 3.4.1. *The flow g^t, $t \in \mathbb{R}$ preserves the measure ν.*

Proof. Let us fix $t \in \mathbb{R}$. Given any measurable set $B \subset N$, let $\hat{B} = \pi^{-1}(B) \cap D$. By the definition of ν, we have that $\nu(B) = (\mu \times ds)(\hat{B})$. For each $n \in \mathbb{Z}$, let $\hat{B}_n$ be the set of all pairs $(x,s) \in \hat{B}$ such that $G^{-t}(x,s) \in F^n(D)$ and let $B_n = \pi(\hat{B}_n)$. Since D is a fundamental domain, $\{\hat{B}_n : n \in \mathbb{Z}\}$ is a partition of $\hat{B}$ and $\{B_n : n \in \mathbb{Z}\}$ is a partition of B. Moreover, $\hat{B}_n = \pi^{-1}(B_n) \cap D$ and, consequently, $\nu(B_n) = (\mu \times ds)(\hat{B}_n)$ for every n. The definition of the suspension flow gives that

$$\pi^{-1}\big(g^{-t}(B_n)\big) = G^{-t}\big(\pi^{-1}(B_n)\big) = G^{-t}\big(\bigcup_{k\in\mathbb{Z}} F^k(\hat{B}_n)\big) = \bigcup_{k\in\mathbb{Z}} F^k\big(G^{-t}(\hat{B}_n)\big).$$

Observing that $F^{-n}(G^{-t}(\hat{B}_n)) \subset D$, we conclude that

$$\nu\big(g^{-t}(B_n)\big) = (\mu \times ds)\big(\pi^{-1}(g^{-t}(B_n)) \cap D\big) = (\mu \times ds)\big(F^{-n}(G^{-t}(\hat{B}_n))\big).$$

As the measure $\mu \times ds$ is invariant under both F and G^t, the last expression is equal to $(\mu \times ds)(\hat{B}_n)$. Therefore,

$$\nu(g^{-t}(B)) = \sum_{n\in\mathbb{Z}} \nu(g^{-t}(B_n)) = \sum_{n\in\mathbb{Z}} (\mu \times ds)(\hat{B}_n) = (\mu \times ds)(\hat{B}) = \nu(B).$$

This proves that ν is invariant under the flow g^t, $t \in \mathbb{R}$.

In Exercise 3.4.2 we invite the reader to relate the recurrence properties of the systems (f,μ) and (g^t,ν).

3.4.2 Poincaré maps

Next, we present a kind of inverse for the construction described in the previous section. Let $g^t : N \to N$, $t \in \mathbb{R}$ be a measurable flow and ν be an invariant measure. Let $\Sigma \subset N$ be a *cross-section* of the flow, that is, a subset of N such that for every $x \in \Sigma$ there exists $\tau(x) \in (0,\infty]$ such that $g^t(x) \notin \Sigma$ for every $t \in (0,\tau(x))$ and $g^{\tau(x)}(x) \in \Sigma$ whenever $\tau(x)$ is finite. We call $\tau(x)$ the *first-return time* of x to Σ. Our goal is to construct, starting from ν, a measure μ that is invariant under the *first-return map* (or *Poincaré return map*)

$$f : \{x \in \Sigma : \tau(x) < \infty\} \to \Sigma, \quad f(x) = g^{\tau(x)}(x).$$

Observe that this map is injective.

For each $\rho > 0$, denote $\Sigma_\rho = \{x \in \Sigma : \tau(x) \geq \rho\}$. Given $A \subset \Sigma_\rho$ and $\delta \in (0,\rho]$, consider $A_\delta = \{g^t(x) : x \in A \text{ and } 0 \leq t < \delta\}$. Observe that the map $(x,t) \mapsto g^t(x)$

is a bijection from $A \times [0,\delta)$ to A_δ. Assume that Σ is endowed with a σ-algebra of measurable subsets such that

1. the function τ and the maps f and f^{-1} are measurable;
2. if $A \subset \Sigma_\rho$ is measurable then $A_\delta \subset N$ is measurable, for every $\delta \in (0,\rho]$.

Lemma 3.4.2. *Let A be a measurable subset of Σ_ρ for some $\rho > 0$. Then, the function $\delta \mapsto \nu(A_\delta)/\delta$ is constant in the interval $(0,\rho]$.*

Proof. Consider any $\delta \in (0,\rho]$ and $l \geq 1$. It is clear that

$$A_\delta = \bigcup_{i=0}^{l-1} g^{i\delta/l}(A_{\delta/l}),$$

and this is a disjoint union. Using that ν is invariant under the flow g^t, $t \in \mathbb{R}$, we conclude that $\nu(A_\delta) = l\nu(A_{\delta/l})$ for every $\delta \in (0,\rho]$ and every $l \geq 1$. Then, $\nu(A_{r\delta}) = r\nu(A_\delta)$ for every $\delta \in (0,\rho]$ and every rational number $r \in (0,1)$. Using, furthermore, the fact that both sides of this relation vary monotonically with r, we get that the equality remains true for every real number $r \in (0,1)$. This implies the conclusion of the lemma.

Given any measurable subset A of Σ_ρ, $\rho > 0$, let us define

$$\mu(A) = \frac{\nu(A_\delta)}{\delta} \quad \text{for any } \delta \in (0,\rho]. \tag{3.4.6}$$

Next, given any measurable subset A of Σ, let

$$\mu(A) = \sup_\rho \mu(A \cap \Sigma_\rho). \tag{3.4.7}$$

See Figure 3.3. We leave it to the reader to check that μ is a measure in Σ (Exercise 3.4.1). We call it the *flux* of ν through Σ under the flow.

Proposition 3.4.3. *Suppose that the measure ν is finite. Then the measure μ in Σ is invariant under the Poincaré map f.*

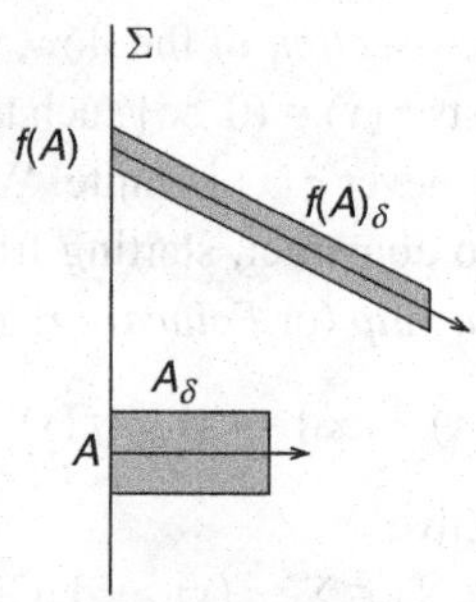

Figure 3.3. Flux of a measure through a cross-section

Proof. Start by observing that the map f is essentially surjective: the complement of the image $f(\Sigma)$ has measure zero. Indeed, suppose that there exists a set E with $\mu(E) > 0$ contained in $\Sigma \setminus f(\Sigma)$. It is no restriction to assume that $E \subset \Sigma_\rho$ for some $\rho > 0$. Then, $\nu(E_\rho) > 0$. Since ν is finite, by assumption, we may apply the Poincaré recurrence theorem to the flow g^{-t}, $t \in \mathbb{R}$. We get that there exists $z \in E_\rho$ such that $g^{-s}(z) \in E_\rho$ for arbitrarily large values of $s > 0$. By definition, $z = g^t(y)$ for some $y \in E$ and some $t \in (0, \rho]$. By construction, the backward trajectory of y intersects Σ. Hence, there exists $x \in \Sigma$ such that $f(x) = y$. This contradicts the choice of E. Thus, the claim is proved.

Given any measurable set $B \subset \Sigma$, let us denote $A = f^{-1}(B)$. Moreover, given $\varepsilon > 0$, let us consider a countable partition of B into measurable subsets B^i satisfying the following conditions: for every i there is $\rho_i > 0$ such that

1. B^i and $A^i = f^{-1}(B^i)$ are contained in Σ_{ρ_i};
2. $\sup(\tau \mid A^i) - \inf(\tau \mid A^i) < \varepsilon\rho_i$.

Next, choose $t_i < \inf(\tau \mid A^i) \le \sup(\tau \mid A^i) < s_i$ such that $s_i - t_i < \varepsilon\rho_i$. Fix $\delta_i = \rho_i/2$. Then, using the fact that f is essentially surjective,

$$g^{t_i}(A^i_{\delta_i}) \supset B^i_{\delta_i-(s_i-t_i)} \quad \text{and} \quad g^{s_i}(A^i_{\delta_i}) \subset B^i_{\delta_i+(s_i-t_i)}.$$

Hence, using the hypothesis that ν is invariant,

$$\nu(A^i_{\delta_i}) = \nu(g^{t_i}(A^i_{\delta_i})) \ge \nu(B^i_{\delta_i-(s_i-t_i)})$$
$$\nu(A^i_{\delta_i}) = \nu(g^{s_i}(A^i_{\delta_i})) \le \nu(B^i_{\delta_i+(s_i-t_i)}).$$

Dividing by δ_i we get that

$$\mu(A^i) \ge 1 - \frac{(s_i - t_i)}{\delta}\mu(B^i) > (1 - 2\varepsilon)\mu(B^i)$$
$$\mu(A^i) \le 1 + \frac{(s_i - t_i)}{\delta}\mu(B^i) < (1 + 2\varepsilon)\mu(B^i).$$

Finally, adding over all the values of i, we conclude that

$$(1 - 2\varepsilon)\mu(A) \le \mu(B) \le (1 + 2\varepsilon)\mu(A).$$

Since ε is arbitrary, this proves that the measure μ is invariant under f.

3.4.3 Exercises

3.4.1. Check that the function μ defined by (3.4.6)–(3.4.7) is a measure.

3.4.2. In the context of Section 3.4.1, suppose that M is a topological space and $f : M \to M$ and $\tau : M \to (0, \infty)$ are continuous. Let $g^t : N \to N$ be the suspension flow and ν be the suspension of some Borel measure μ invariant under f.

(a) Show that if $x \in M$ is recurrent for the transformation f then $\pi(x, s) \in N$ is recurrent for the flow g^t, for every $s \in \mathbb{R}$.

(b) Show that if $\pi(x, s) \in N$ is recurrent for the flow g^t, for some $s \in \mathbb{R}$, then $x \in M$ is recurrent for f.

(c) Conclude that the set of recurrent points of f has total measure for μ if and only if the set of recurrent points of g^t, $t \in \mathbb{R}$ has total measure for ν. In particular, this happens if at least one of the measures μ or ν is finite.

3.4.3. Let $g^t : N \to N$, $t \in \mathbb{R}$ be the flow defined by a vector field X of class C^1 on a compact Riemannian manifold N. Assume that this flow preserves the volume measure ν associated with the Riemannian metric. Let Σ be a hypersurface of N transverse to X and ν_Σ be the volume measure on Σ associated with the restriction of the Riemannian metric. Define $\phi : \Sigma \to (0,\infty)$ through $\phi(y) = |X(y) \cdot n(y)|$, where $n(\cdot)$ is a unit vector field orthogonal to Σ. Show that the measure $\eta = \phi\nu_\Sigma$ is invariant under the Poincaré map $f : \Sigma \to \Sigma$ of the flow. Indeed, η coincides with the flux of ν through Σ.

3.4.4. The following construction has a significant role in the theory of interval exchanges. Let $\hat{N} \subset \mathbb{R}^4_+$ be the set of all 4-tuples $(\lambda_1, \lambda_2, h_1, h_2)$ of positive real numbers, endowed with the standard volume measure $\hat{\nu} = d\lambda_1 d\lambda_2 dh_1 dh_2$. Define

$$F : \hat{N} \to \hat{N}, \quad F(\lambda_1, \lambda_2, h_1, h_2) = \begin{cases} (\lambda_1 - \lambda_2, \lambda_2, h_1, h_1 + h_2) & \text{if } \lambda_1 > \lambda_2 \\ (\lambda_1, \lambda_2 - \lambda_1, h_1 + h_2, h_2) & \text{if } \lambda_1 < \lambda_2. \end{cases}$$

(F is not defined when $\lambda_1 = \lambda_2$.) Let N be the quotient of $\hat{N}$ by the equivalence relation $z \sim \tilde{z} \Leftrightarrow F^n(z) = \tilde{z}$ for some $n \in \mathbb{Z}$ and let $\pi : \hat{N} \to N$ be the canonical projection. Define

$$G^t : \hat{N} \to \hat{N}, t \in \mathbb{R}, \quad G^t(\lambda_1, \lambda_2, h_1, h_2) = (e^t \lambda_1, e^t \lambda_2, e^{-t} h_1, e^{-t} h_2).$$

Let $\hat{a} : \hat{N} \to (0,\infty)$ be the functional given by $\hat{a}(\lambda_1, \lambda_2, h_1, h_2) = \lambda_1 h_1 + \lambda_2 h_2$. For each $c > 0$, let $\hat{N}_c$ be the subset of all $x \in \hat{N}$ such that $\hat{a}(x) = c$, let $\hat{\nu}_c$ be the volume measure defined on $\hat{N}_c$ by the restriction of the Riemannian metric of $\mathbb{R}^4_+$ and let $\hat{\eta}_c = \hat{\nu}_c / \|\operatorname{grad} \hat{a}\|$.

(a) Show that F preserves the functional $\hat{a}$ and so there exists a functional $a : N \to (0,\infty)$ such that $a \circ \pi = \hat{a}$. Show that G^t commutes with F and preserves $\hat{a}$. Hence, $(G^t)_t$ induces a flow $(g^t)_t$ in the quotient space N that preserves the functional a. Check that F and $(G^t)_t$ preserve $\hat{\nu}$ and $\hat{\eta}_c$ for every c.

(b) Check that $D = \{(\lambda_1, \lambda_2, h_1, h_2) : \lambda_1 + \lambda_2 \geq 1 > \max\{\lambda_1, \lambda_2\}\}$ is a fundamental domain for $\sim$. Consider the measure $\nu = \pi_*(\hat{\nu} \mid D)$ on N. Check that the definition does not depend on the choice of the fundamental domain and show that ν is invariant under the flow $(g^t)_t$. Is the measure ν finite?

(c) Check that $\Sigma = \pi(\{(\lambda_1, \lambda_2, h_1, h_2) : \lambda_1 + \lambda_2 = 1\})$ is a cross-section of the flow $(g^t)_t$. Calculate the Poincaré map $f : \Sigma \to \Sigma$ and the corresponding first-return time function τ. Calculate the flux μ of the measure ν through Σ. Is the measure μ finite?

(d) For every $c > 0$, let $N_c = \pi(\hat{N}_c)$ and $\eta_c = \pi_*(\hat{\eta}_c \cap D)$. Show that N_c and η_c are invariant under $(g^t)_t$, for every $c > 0$. Check that $\eta_c(N_c) < \infty$ for every c. Conclude that ν-almost every point is recurrent for the flow $(g^t)_t$.

4
Ergodicity

The theorems presented in the previous chapter fully establish the first part of Boltzmann's ergodic hypothesis: for any measurable set E, the mean sojourn time $\tau(E,x)$ is well defined for almost every point x. The second part of the ergodic hypothesis, that is, the claim that $\tau(E,x)$ should coincide with the measure of E for almost every x, is a statement of a different nature and is the subject of the present chapter.

In this chapter we always take μ to be a probability measure invariant under some measurable transformation $f : M \to M$. We say that the system (f,μ) is *ergodic* if, given any measurable set E, we have $\tau(E,x) = \mu(E)$ for μ-almost every point $x \in M$. We are going to see that this is equivalent to saying that the system is dynamically indivisible, in the sense that every invariant set has either full measure or zero measure. Other equivalent formulations of the ergodicity property are discussed in Section 4.1. One of them is that time averages coincide with space averages: for every integrable function φ,

$$\lim_n \frac{1}{n} \sum_{j=0}^{n-1} \varphi(f^j(x)) = \int \varphi \, d\mu \quad \text{at } \mu\text{-almost every point.}$$

In Section 4.2 we illustrate, by means of examples, several techniques to prove or disprove ergodicity. Most of them will be utilized again later in more complex situations. Next, we take the following viewpoint: we fix the dynamical systems and analyze the properties of ergodic measures within the space of all invariant measures of that dynamical system. As we are going to see in Section 4.3, the ergodic measures are precisely the extremal elements of that space.

In Section 4.4 we give a brief outline of the historical development of ergodic theory in the context of conservative systems. The main highlights are *KAM theory*, thus denominated in homage to Andrey Kolmogorov, Vladimir Arnold and Jürgen Moser, and *hyperbolic dynamics*, which was initiated by Steven Smale, Dmitry Anosov, Yakov Sinai and their collaborators. The two theories deal with distinct types of dynamical behavior, elliptic and hyperbolic, and they reach opposing conclusions: roughly speaking, hyperbolic systems are ergodic whereas elliptic systems are not.

4.1 Ergodic systems

We use the expressions "the measure μ is ergodic with respect to the transformation f" or "the transformation f is ergodic with respect to the measure μ" to mean the same thing, namely, that the system (f, μ) is ergodic. Recall that, by definition, this means that the mean sojourn time in any measurable set of μ-almost every point coincides with the measure of that set. This condition can be rephrased in several equivalent ways, as we are going to see next.

4.1.1 Invariant sets and functions

A measurable function $\varphi : M \to \mathbb{R}$ is said to be *invariant* if $\varphi = \varphi \circ f$ at μ-almost every point. In other words, φ is invariant if it is constant on every trajectory of f outside a zero measure subset. Moreover, we say that a measurable set $B \subset M$ is *invariant* if its characteristic function $\mathcal{X}_B$ is an invariant function. Equivalently, B is invariant if it differs from its pre-image $f^{-1}(B)$ by a zero measure set:

$$\mu(B \Delta f^{-1}(B)) = 0.$$

Exercise 1.1.4 collects some equivalent formulations of this property. It is easy to check that the family of all invariant sets is a σ-algebra, that is, it is closed under countable unions and intersections and under passage to the complement.

Example 4.1.1. Let $f : [0,1] \to [0,1]$ be the decimal expansion transformation introduced in Section 1.3.1, and μ be the Lebesgue measure. Clearly, the set $A = \mathbb{Q} \cap [0,1]$ of rational numbers is invariant. Other interesting examples are the sets of points $x = 0.a_1 a_2 \ldots$ in $[0,1]$ with prescribed proportions of digits a_i with each value $k \in \{0, \ldots, 9\}$. More precisely, given any vector $p = (p_0, \ldots, p_9)$ such that $p_i \geq 0$ for every i and $\sum_i p_i = 1$, define

$$A_p = \left\{ x : \lim_n \frac{1}{n} \#\{1 \leq i \leq n : a_i = k\} = p_k \text{ for } k = 0, \ldots, 9 \right\}.$$

Observe that if $x = 0 \cdot a_1 a_2 \ldots$ then every point $y \in f^{-1}(x)$ may be written as $y = 0 \cdot b a_1 a_2 \ldots$ with $b \in \{0, \ldots, 9\}$. It is clear that the extra digit b does not affect the proportion of digits with any of the values 0, ..., 9 in the decimal expansion. Thus, $y \in A_p$ if and only if $x \in A_p$. This implies that A_p is indeed invariant under f.

Example 4.1.2. Let $\varphi : [0,1] \to \mathbb{R}$ be a function in $L^1(\mu)$. According to the ergodic theorem of Birkhoff (Theorem 3.2.3), the time average $\tilde{\varphi}$ is an invariant function. So, every level set

$$B_c = \{x \in [0,1]; \tilde{\varphi}(x) = c\}$$

is an invariant set. Observe also that every invariant function is of this form: it is clear that if φ is invariant then it coincides with its time average $\tilde{\varphi}$ at μ-almost every point.

The next proposition collects a few equivalent ways to define ergodicity. We say that a function φ is constant at μ-almost every point if there exists $c \in \mathbb{R}$ such that $\varphi(x) = c$ for μ-almost every $x \in M$.

Proposition 4.1.3. *Let μ be an invariant probability measure of a measurable transformation $f : M \to M$. The following conditions are all equivalent:*

(i) For every measurable set $B \subset M$ one has $\tau(B,x) = \mu(B)$ for μ-almost every point.
(ii) For every measurable set $B \subset M$ the function $\tau(B,\cdot)$ is constant at μ-almost every point.
(iii) For every integrable function $\varphi : M \to \mathbb{R}$ one has $\tilde{\varphi}(x) = \int \varphi \, d\mu$ for μ-almost every point.
(iv) For every integrable function $\varphi : M \to \mathbb{R}$ the time average $\tilde{\varphi} : M \to \mathbb{R}$ is constant at μ-almost every point.
(v) For every invariant integrable function $\psi : M \to \mathbb{R}$ one has $\psi(x) = \int \psi \, d\mu$ for μ-almost every point.
(vi) Every invariant integrable function $\psi : M \to \mathbb{R}$ is constant at μ-almost every point.
(vii) For every invariant subset A we have either $\mu(A) = 0$ or $\mu(A) = 1$.

Proof. It is immediate that (i) implies (ii), that (iii) implies (iv) and that (v) implies (vi). It is also clear that (v) implies (iii) and (vi) implies (iv), because the time average is an invariant function (recall Proposition 3.2.4). Analogously, (iii) implies (i) and (iv) implies (ii), because the mean sojourn time is a time average (of the characteristic function of B). We are left to prove the following implications:

(ii) implies (vii): Let A be an invariant set. Then $\tau(A,x) = 1$ for μ-almost every $x \in A$ and $\tau(A,x) = 0$ for μ-almost every $x \in A^c$. Since $\tau(A,\cdot)$ is assumed to be constant at μ-almost every point, it follows that $\mu(A) = 0$ or $\mu(A) = 1$.

(vii) implies (v): Let ψ be an invariant integrable function. Then every level set

$$B_c = \{x \in M : \psi(x) \le c\}$$

is an invariant set. So, the hypothesis implies that $\mu(B_c) \in \{0,1\}$ for every $c \in \mathbb{R}$. Since $c \mapsto \mu(B_c)$ is non-decreasing, it follows that there exists $\bar{c} \in \mathbb{R}$

such that $\mu(B_c) = 0$ for every $c < \bar{c}$ and $\mu(B_c) = 1$ for every $c \geq \bar{c}$. Then $\psi = \bar{c}$ at μ-almost every point. Hence, $\int \psi \, d\mu = \bar{c}$ and so $\psi = \int \psi \, d\mu$ at μ-almost every point.

4.1.2 Spectral characterization

The next proposition characterizes the ergodicity property in terms of the Koopman operator $U_f(\varphi) = \varphi \circ f$:

Proposition 4.1.4. *Let μ be an invariant probability measure of a measurable transformation $f : M \to M$. The following conditions are equivalent:*

(i) (f, μ) is ergodic.
(ii) For any pair of measurable sets A and B one has

$$\lim_n \frac{1}{n} \sum_{j=0}^{n-1} \mu\big(f^{-j}(A) \cap B\big) = \mu(A)\mu(B). \tag{4.1.1}$$

(iii) For any functions $\varphi \in L^p(\mu)$ and $\psi \in L^q(\mu)$, with $1/p + 1/q = 1$, one has

$$\lim_n \frac{1}{n} \sum_{j=0}^{n-1} \int (U_f^j \varphi) \psi \, d\mu = \int \varphi \, d\mu \int \psi \, d\mu. \tag{4.1.2}$$

Proof. It is clear that (iii) implies (ii): just take $\varphi = \mathcal{X}_A$ and $\psi = \mathcal{X}_B$. To show that (ii) implies (i), let A be an invariant set. Taking $A = B$ in hypothesis (ii), we get that

$$\mu(A) = \lim_n \frac{1}{n} \sum_{j=0}^{n-1} \mu\big(f^{-j}(A) \cap A\big) = \mu(A)^2.$$

This implies that $\mu(A) = 0$ or $\mu(A) = 1$.

Now it suffices to prove that (i) implies (iii). Consider any $\varphi \in L^p(\mu)$ and $\psi \in L^q(\mu)$. By ergodicity and the ergodic theorem of Birkhoff (Theorem 3.2.3) we have that

$$\frac{1}{n} \sum_{j=0}^{n-1} U_f^j \varphi \to \int \varphi \, d\mu \tag{4.1.3}$$

at μ-almost every point. Initially, assume that $|\varphi| \leq k$ for some $k \geq 1$. Then, for every $n \in \mathbb{N}$,

$$\left| \left(\frac{1}{n} \sum_{j=0}^{n-1} U_f^j \varphi \right) \psi \right| \leq k|\psi|.$$

So, since $k|\psi| \in L^1(\mu)$, we may use the dominated convergence theorem (Theorem A.2.11) to conclude that

$$\int \left(\frac{1}{n} \sum_{j=0}^{n-1} U_f^j \varphi \right) \psi \, d\mu \to \int \varphi \, d\mu \int \psi \, d\mu.$$

This proves the claim (4.1.2) when φ is bounded. All that is left to do is remove this restriction. Given any $\varphi \in L^p(\mu)$ and $k \geq 1$, define

$$\varphi_k(x) = \begin{cases} k & \text{if } \varphi(x) > k \\ \varphi(x) & \text{if } \varphi(x) \in [-k,k] \\ -k & \text{if } \varphi(x) < -k. \end{cases}$$

Fix $\varepsilon > 0$. By the previous argument, for every $k \geq 1$ one has

$$\left| \int \left(\frac{1}{n} \sum_{j=0}^{n-1} U_f^j \varphi_k \right) \psi \, d\mu - \int \varphi_k \, d\mu \int \psi \, d\mu \right| < \varepsilon \tag{4.1.4}$$

if n is large enough (depending on k). Next, observe that $\|\varphi_k - \varphi\|_p \to 0$ when $k \to \infty$: this is clear when $p = \infty$, because $\varphi_k = \varphi$ for every $k > \|\varphi\|_\infty$; for $p < \infty$ use the monotone convergence theorem (Theorem A.2.9). Hence, using the Hölder inequality (Theorem A.5.5), we have that

$$\left| \int (\varphi_k - \varphi) \, d\mu \int \psi \, d\mu \right| \leq \|\varphi_k - \varphi\|_p \left| \int \psi \, d\mu \right| < \varepsilon, \tag{4.1.5}$$

for every k sufficiently large. Similarly,

$$\begin{aligned} \left| \int \frac{1}{n} \sum_{j=0}^{n-1} U_f^j (\varphi_k - \varphi) \, \psi \, d\mu \right| &\leq \frac{1}{n} \sum_{j=0}^{n-1} \left| \int U_f^j (\varphi_k - \varphi) \, \psi \, d\mu \right| \\ &\leq \frac{1}{n} \sum_{j=0}^{n-1} \|U_f^j (\varphi_k - \varphi)\|_p \, \|\psi\|_q \, d\mu \\ &= \|\varphi_k - \varphi\|_p \, \|\psi\|_q < \varepsilon, \end{aligned} \tag{4.1.6}$$

for every n and every k sufficiently large, *independent of n*. Fix k so that (4.1.5) and (4.1.6) hold and then take n sufficiently large such that (4.1.4) also holds. Summing the three relations (4.1.4) to (4.1.6), we get that

$$\left| \int \left(\frac{1}{n} \sum_{j=0}^{n-1} U_f^j \varphi \right) \psi \, d\mu - \int \varphi \, d\mu \int \psi \, d\mu \right| < 3\varepsilon$$

for every n sufficiently large. This gives condition (iii).

In the case $p = q = 2$, the condition (4.1.2) may be expressed in terms of the inner product $\cdot$ in the space $L^2(\mu)$. In this way we get that (f, μ) is ergodic if and only if:

$$\lim_n \frac{1}{n} \sum_{j=0}^{n-1} \left[(U_f^n \varphi) - (\varphi \cdot 1) \right] \cdot \psi = 0 \quad \text{for every } \varphi, \psi \in L^2(\mu). \tag{4.1.7}$$

We will use a few times the following elementary facts: given any measurable sets A and B,

$$\begin{aligned}|\mu(A)-\mu(B)| &= |\mu(A\setminus B)-\mu(B\setminus A)| \\ &\le \mu(A\setminus B)+\mu(B\setminus A)=\mu(A\Delta B),\end{aligned} \tag{4.1.8}$$

and given any sets A_1, A_2, B_1, B_2,

$$(A_1\cap A_2)\Delta(B_1\cap B_2)\subset(A_1\Delta B_1)\cup(A_2\Delta B_2). \tag{4.1.9}$$

Corollary 4.1.5. *Assume that the condition* (4.1.1) *in Proposition 4.1.4 holds for every A and B in some algebra $\mathcal{A}$ that generates the σ-algebra of measurable sets. Then (f,μ) is ergodic.*

Proof. Let A and B be arbitrary measurable sets. By the approximation theorem (Theorem A.1.19), given any $\varepsilon > 0$ there are A_0 and B_0 in $\mathcal{A}$ such that $\mu(A\Delta A_0) < \varepsilon$ and $\mu(B\Delta B_0) < \varepsilon$. Observe that

$$\begin{aligned}\left|\mu(f^{-j}(A)\cap B)-\mu(f^{-j}(A_0)\cap B_0)\right| &\le \mu(f^{-j}(A)\Delta f^{-j}(A_0))+\mu(B\Delta B_0) \\ &= \mu(A\Delta A_0)+\mu(B\Delta B_0) < 2\varepsilon\end{aligned}$$

(the equality uses the fact that μ is an invariant measure) for every j and

$$|\mu(A)\mu(B)-\mu(A_0)\mu(B_0)| \le \mu(A\Delta A_0)+\mu(B\Delta B_0) < 2\varepsilon.$$

Then, the hypothesis

$$\lim_n \frac{1}{n}\sum_{j=0}^{n-1}\mu(f^{-j}(A_0)\cap B_0)=\mu(A_0)\mu(B_0)$$

implies that

$$\begin{aligned}-4\varepsilon &\le \liminf_n \frac{1}{n}\sum_{j=0}^{n-1}\mu(f^{-j}(A)\cap B)-\mu(A)\mu(B) \\ &\le \limsup_n \frac{1}{n}\sum_{j=0}^{n-1}\mu(f^{-j}(A)\cap B)-\mu(A)\mu(B)\le 4\varepsilon.\end{aligned}$$

Since ε is arbitrary, this proves that the condition (4.1.1) holds for all pairs of measurable sets. According to Proposition 4.1.4, it follows that the system is ergodic.

In the same spirit, it suffices to check part (iii) of Proposition 4.1.4 on dense subsets:

Corollary 4.1.6. *Assume that the condition* (4.1.2) *in Proposition 4.1.4 for every φ and ψ in dense subsets of $L^p(\mu)$ and $L^q(\mu)$, respectively. Then (f,μ) is ergodic.*

We leave the proof of this fact to the reader (see Exercise 4.1.3).

4.1.3 Exercises

4.1.1. Let $(M,\mathcal{A})$ be a measurable space and $f: M \to M$ be a measurable transformation. Prove that if $p \in M$ is a periodic point of period k, then the measure $\mu_p = \frac{1}{k}(\delta_p + \delta_{f(p)} + \cdots + \delta_{f^{k-1}(p)})$ is ergodic.

4.1.2. Let μ be an invariant probability measure, not necessarily ergodic, of a measurable transformation $f: M \to M$. Show that the following limit exists for any pair of measurable sets A and B:

$$\lim_n \frac{1}{n} \sum_{i=0}^{n-1} \mu\big(f^{-i}(A) \cap B\big).$$

4.1.3. Show that an invariant probability measure μ is ergodic for a transformation $f: M \to M$ if and only if any one of the following conditions holds:

(a) $\mu(\bigcup_{n\geq 0} f^{-n}(A)) = 1$ for every measurable set A with $\mu(A) > 0$;

(b) given any measurable sets A, B with $\mu(A)\mu(B) > 0$, there is $n \geq 1$ such that $\mu\big(f^{-n}(A) \cap B\big) > 0$;

(c) the convergence in condition (iii) of Proposition 4.1.4 holds for some choice of p, q and some dense subset of functions $\varphi \in L^p(\mu)$ and $\psi \in L^q(\mu)$;

(d) there is $p \in [1,\infty]$ such that every invariant function $\varphi \in L^p(\mu)$ is constant at μ-almost every point;

(e) every integrable function φ satisfying $\varphi \circ f \geq \varphi$ at μ-almost every point (or $\varphi \circ f \leq \varphi$ at μ-almost every point) is constant at μ-almost every point.

4.1.4. Take M to be a metric space. Prove that an invariant probability measure μ is ergodic for $f: M \to M$ if and only if the time average of every bounded uniformly continuous function $\varphi: M \to \mathbb{R}$ is constant at μ-almost every point.

4.1.5. Take M to be a metric space. We call the *basin* of an invariant probability measure μ the set $B(\mu)$ of all points $x \in M$ such that

$$\lim_{n\to\infty} \frac{1}{n} \sum_{j=0}^{n-1} \varphi(f^j(x)) = \int \varphi \, d\mu$$

for every bounded continuous function $\varphi: M \to \mathbb{R}$. Check that the basin is an invariant set. Moreover, if μ is ergodic then $B(\mu)$ has full μ-measure.

4.1.6. Show that if μ and η are distinct ergodic probability measures of a transformation $f: M \to M$, then η and μ are mutually singular.

4.1.7. Let μ be a probability measure invariant under some transformation $f: M \to M$. Show that the product measure $\mu_2 = \mu \times \mu$ is invariant under the transformation $f_2: M \times M \to M \times M$ defined by $f_2(x,y) = (f(x), f(y))$. Moreover, if (f_2, μ_2) is ergodic then (f, μ) is ergodic. Is the converse true?

4.1.8. Let μ be a probability measure invariant under some transformation $f: M \to M$. Assume that (f^n, μ) is ergodic for every $n \geq 1$. Show that if φ is a non-constant eigenfunction of the Koopman operator U_f then the eigenvalue is *not* a root of unity and any set restricted to which φ is constant has zero measure.

4.2 Examples

In this section we use a number of examples to illustrate different methods for checking whether a system is ergodic or not.

4.2.1 Rotations on tori

Initially, let us consider the case of a rotation $R_\theta : S^1 \to S^1$ on the circle $S^1 = \mathbb{R}/\mathbb{Z}$. As observed in Section 1.3.3, the Lebesgue measure m is invariant under R_θ. We want to analyze the ergodic behavior of the system (R_θ, m) for different values of θ.

If θ is rational, say $\theta = p/q$ in irreducible form, $R_\theta^q(x) = x$ for every $x \in S^1$. Then, given any segment $I \subset S^1$ with length less than $1/q$, the set

$$A = I \cup R_\theta(I) \cup \cdots \cup R_\theta^{q-1}(I)$$

is invariant under R_θ and its Lebesgue measure satisfies $0 < m(A) < 1$. Thus, if θ is rational then the Lebesgue measure is *not* ergodic. The converse is much more interesting:

Proposition 4.2.1. *If θ is irrational then R_θ is ergodic relative to the Lebesgue measure.*

We are going to mention two different proofs of this fact. The first one, which we detail below, uses some simple facts from Fourier analysis. The second one, which we leave as an exercise (Exercise 4.2.6), is based on a density point argument similar to the one we will use in Section 4.2.2 to prove that the decimal expansion map is ergodic relative to the Lebesgue measure.

We denote by $L^2(m)$ the Hilbert space of measurable functions ψ whose square is integrable, that is, such that:

$$\int |\psi|^2 \, dm < \infty.$$

It is convenient to consider functions with values in $\mathbb{C}$, and we will do so. We use the well-known fact that the family of functions

$$\phi_k : S^1 \to \mathbb{C},\ x \mapsto e^{2\pi ikx}, \quad k \in \mathbb{Z}$$

is a Hilbert basis of this space: given any $\varphi \in L^2(m)$ there exists a unique sequence $(a_k)_{k\in\mathbb{Z}}$ of complex numbers such that

$$\varphi(x) = \sum_{k\in\mathbb{Z}} a_k e^{2\pi ikx} \quad \text{for almost every } x \in S^1. \tag{4.2.1}$$

This is called the *Fourier series* expansion of $\varphi \in L^2(m)$. Then

$$\varphi\big(R_\theta(x)\big) = \sum_{k\in\mathbb{Z}} a_k e^{2\pi ik\theta} e^{2\pi ikx}. \tag{4.2.2}$$

Assume that φ is invariant. Then (4.2.1) and (4.2.2) coincide. By uniqueness of the coefficients in the Fourier series expansion, this happens if and only if

$$a_k e^{2\pi ik\theta} = a_k \quad \text{for every } k \in \mathbb{Z}.$$

The hypothesis that θ is irrational means that $e^{2\pi ik\theta} \neq 1$ for every $k \neq 0$. Hence, the relation that we just obtained implies that $a_k = 0$ for every $k \neq 0$. In other words, $\varphi(z) = a_0$ for m-almost every $z \in S^1$. This proves that every invariant L^2 function is constant m-almost everywhere. In particular, the characteristic function $\varphi = \mathcal{X}_A$ of any invariant set $A \subset S^1$ is constant at m-almost every point. This is the same as saying that A has either zero measure or full measure. Hence, by Proposition 4.1.3, the measure m is ergodic.

These observations extend in a natural way to the rotation on the d-torus $\mathbb{T}^d$, for any $d \geq 1$:

Proposition 4.2.2. *If $\theta = (\theta_1, \dots, \theta_d)$ is rationally independent then the rotation $R_\theta : \mathbb{T}^d \to \mathbb{T}^d$ is ergodic relative to the Lebesgue measure.*

This may be proved by the same argument as in the case $d = 1$, using the fact (see Exercise 4.2.1) that the family of functions

$$\phi_{k_1,\dots,k_d} : \mathbb{T}^d \to \mathbb{C},\ (x_1,\dots,x_d) \mapsto e^{2\pi i(k_1x_1+\cdots+k_dx_d)}, \quad (k_1,\dots,k_d) \in \mathbb{Z}^d$$

is a Hilbert basis of the space $L^2(m)$.

Corollary 4.2.3. *If $\theta = (\theta_1, \dots, \theta_d)$ is rationally independent then the rotation $R_\theta : \mathbb{T}^d \to \mathbb{T}^d$ is minimal, that is, every orbit $\mathcal{O}(x) = \{R_\theta^n(x) : n \in \mathbb{N}\}$ is dense in $\mathbb{T}^d$.*

Proof. Let us consider in $\mathbb{T}^d$ the *flat distance*, defined by:

$$d([\xi],[\eta]) = \inf\{d(\xi',\eta') : \xi',\eta' \in \mathbb{R}^d, \xi' \sim \xi, \eta' \sim \eta\}.$$

Observe that this distance is preserved by every rotation. Let $\{U_k : k \in \mathbb{N}\}$ be a countable basis of open sets of $\mathbb{T}^d$ and m be the Lebesgue measure on $\mathbb{T}^d$. By ergodicity, there is $W \subset \mathbb{T}^d$, with total Lebesgue measure, such that $\tau(U_k, x) = m(U_k) > 0$ for every k and every $x \in W$. In particular, the orbit of x is dense in $\mathbb{T}^d$ for every $x \in W$. Now consider an arbitrary point $x \in M$ and consider any $y \in W$. Then, for every $\delta > 0$ there exists $k \geq 1$ such that $d(f^k(y), x) < \delta$. It follows that $d(f^{n+k}(y), f^n(x)) < \delta$ for every $n \geq 1$. Since the orbit of y is dense, this implies that the orbit of x is δ-dense, that is, it intersects the δ-neighborhood of every point. Since δ is arbitrary, this implies that the orbit of x is dense in the ambient torus.

In fact, the irrational rotations on the circle or, more generally, on any torus have a much stronger property than ergodicity: they are *uniquely ergodic*, meaning that they admit a unique invariant probability measure (which is the Lebesgue measure, of course). Uniquely ergodic systems are studied in Chapter 6.

4.2.2 Decimal expansion

Consider the transformation $f : [0,1] \to [0,1]$, $f(x) = 10x - [10x]$ introduced in Section 1.3.1. We have seen that f preserves the Lebesgue measure m.

Proposition 4.2.4. *The transformation f is ergodic relative to the Lebesgue measure m.*

Proof. By Proposition 4.1.3, it suffices to prove that every invariant set A has total measure. The main ingredient is the derivation theorem (Theorem A.2.15), according to which almost every point of A is a density point of A. More precisely (see also Exercise A.2.9), m-almost every point $a \in A$ satisfies

$$\liminf_{\varepsilon \to 0} \left\{ \frac{m(I \cap A)}{m(I)} : I \text{ an interval such that } a \in I \subset B(a,\varepsilon) \right\} = 1. \tag{4.2.3}$$

Let us fix a density point $a \in A$. Since the set of points of the form $m/10^k$, $k \in \mathbb{N}$, $0 \le m \le 10^k$ has zero measure, it is no restriction to suppose that a is *not* of that form. Let us consider the family of intervals

$$I(k,m) = \left(\frac{m-1}{10^k}, \frac{m}{10^k} \right), \quad k \in \mathbb{N}, \quad m = 1, \dots, 10^k.$$

It is clear that for each $k \in \mathbb{N}$ there exists a unique $m = m_k$ such that $I(k, m_k)$ contains the point a. Denote $I_k = I(k, m_k)$. The property (4.2.3) implies that

$$\frac{m(I_k \cap A)}{m(I_k)} \to 1 \quad \text{when } k \to \infty.$$

Observe also that each f^k is an *affine* bijection from I_k to the interval $(0,1)$. This has the following immediate consequence, which is crucial for our argument:

Lemma 4.2.5 (Distortion). *For every $k \in \mathbb{N}$, one has*

$$\frac{m(f^k(E_1))}{m(f^k(E_2))} = \frac{m(E_1)}{m(E_2)} \tag{4.2.4}$$

for any measurable subsets E_1 and E_2 of I_k.

Applying this fact to $E_1 = I_k \cap A$ and $E_2 = I_k$ we find that

$$\frac{m\big(f^k(I_k \cap A)\big)}{m\big((0,1)\big)} = \frac{m(I_k \cap A)}{m(I_k)}.$$

Clearly, $m\big((0,1)\big) = 1$. Moreover, as we take A to be invariant, $f^k(I_k \cap A)$ is contained in A. In this way we get that

$$m(A) \ge \frac{m(I_k \cap A)}{m(I_k)} \quad \text{for every } k.$$

Since the sequence on the right-hand side converges to 1 when $k \to \infty$, it follows that $m(A) = 1$, as we wanted to prove.

The proof of Lemma 4.2.5 relies on the fact that the transformation f is affine on each interval $\big((m-1)/10, m/10\big)$; that may give the impression that the method of proof that we just presented is restricted to a very special class of examples. In fact, this is not so—much to the contrary.

The reason is that there are many situations where one can obtain a slightly weaker version of the statement of Lemma 4.2.5 that is, nevertheless, still sufficient to conclude the proof of ergodicity. In a few words, instead of the claim that the two sides of (4.2.4) are equal, one can often show that the quotient between the two terms is bounded by some uniform constant. That is called the *bounded distortion property*. As an illustration of these ideas, in Section 4.2.4 we prove that the Gauss transformation is ergodic.

Next, we describe an application of Proposition 4.2.4 in the context of number theory. We say that a number $x \in \mathbb{R}$ is 10-*normal* if every block of digits $(b_1, \dots, b_l)$, $l \geq 1$ occurs with frequency 10^{-l} in the decimal expansion of x. Rational numbers are never 10-normal, of course, and it is also easy to give irrational examples, such as $x = 0.101001000100001000001\cdots$. Moreover, it is not difficult to construct 10-normal numbers, for example, the Champernowne constant $x = 0.12345678910111213141516171819202122\cdots$, which is obtained by concatenation of the successive natural numbers.

However, it is usually difficult to check whether a given number is 10-normal or not. For example, that remains unknown for the numbers π, e and even $\sqrt{2}$. On the other hand, using the previous proposition one can easily prove that almost every number is 10-normal:

Proposition 4.2.6. *The set of* 10-*normal numbers* $x \in \mathbb{R}$ *has full Lebesgue measure in the real line.*

Proof. Since the fact of being 10-normal or not is independent of the integer part of the number, we only need to show that almost every $x \in [0,1]$ is 10-normal. Consider $f : [0,1] \to [0,1]$ defined by $f(x) = 10x - [10x]$. For each block $(b_1, \dots, b_l) \in \{0, \dots, 9\}^l$, consider the interval

$$I_{b_1,\dots,b_l} = \Big[\frac{\kappa}{10^l}, \frac{\kappa+1}{10^l}\Big) \quad \text{with } \kappa = \sum_{i=1}^{l} b_i 10^{l-i}.$$

Recall that if $x = 0.a_0 a_1 \cdots a_k a_{k+1} \cdots$ then $f^k(x) = 0.a_k a_{k+1} \cdots$ for every $k \geq 1$. Hence, $f^k(x) \in I_{b_1,\dots,b_l}$ if and only if $(a_k, \dots, a_{k+l-1}) = (b_1, \dots, b_l)$. So, the mean sojourn time $\tau(I_{b_1,\dots,b_l}, x)$ is equal to the frequency of the block $(b_1, \dots, b_l)$ in the decimal expansion of x. Using the Birkhoff ergodic theorem and the fact that the transformation f is ergodic with respect to the Lebesgue measure m, we conclude that for every $(b_1, \dots, b_l)$ there exists a full Lebesgue measure subset $B(b_1, \dots, b_l)$ of the interval $[0,1]$ such that

$$\tau(I_{b_1,\dots,b_l}, x) = m(I_{b_1,\dots,b_l}) = \frac{1}{10^l} \quad \text{for every } x \in B(b_1, \dots, b_l).$$

Let B be the intersection of $B(b_1,\dots,b_l)$ over all values of $b_1,\dots,b_l$ in $\{0,\dots,9\}$ and every $l \ge 1$. Then $m(B) = 1$ and every $x \in B$ is 10-normal.

More generally, for any integer $d \ge 2$, we say that $x \in \mathbb{R}$ is *d-normal* if every block $(b_1,\dots,b_l) \in \{0,\dots,d-1\}^l$, $l \ge 1$ occurs with frequency d^{-l} in the expansion of x in base d. Finally, we say that x is a *normal number* if it is d-normal for every $d \ge 2$. Everything that was said before for $d = 10$ extends immediately to general d. In particular, the set of d-normal numbers has full Lebesgue measure for every $d \ge 2$. Taking the intersection over all the values of d, we conclude that *Lebesgue-almost every real number is normal* (Borel normal theorem).

4.2.3 Bernoulli shifts

Let $(X,\mathcal{C},\nu)$ be a probability space. In this section we consider the product space $\Sigma = X^{\mathbb{N}}$, endowed with the product σ-algebra $\mathcal{B} = \mathcal{C}^{\mathbb{N}}$ and the product measure $\mu = \nu^{\mathbb{N}}$. As explained in Appendix A.2.3, this means that: Σ is the set of all sequences $(x_n)_{n\in\mathbb{N}}$ with $x_n \in X$ for every n; $\mathcal{B}$ is the σ-algebra generated by the measurable cylinders

$$[m;A_m,\dots,A_n] = \{(x_i)_{i\in\mathbb{N}} : x_i \in A_i \text{ for } m \le i \le n\}$$

with $m \le n$ and $A_i \in \mathcal{C}$ for each i; and μ is the probability measure on Σ characterized by

$$\mu([m;A_m,\dots,A_n]) = \prod_{i=m}^{n} \nu(A_i). \tag{4.2.5}$$

We may think of the elements of Σ as representing the results of a sequence of random experiments with values in X and all subject to the same probability distribution ν: given any measurable set $A \subset X$, the probability of $x_i \in A$ is equal to $\nu(A)$ for every i. Moreover, in this model the results of the successive experiments are independent: indeed, the relation (4.2.5) means that the probability of $x_i \in A_i$ for every $m \le i \le n$ is the product of the probabilities of the individual events $x_i \in A_i$.

In this section we introduce a dynamical system $\sigma : \Sigma \to \Sigma$ on the space Σ, called the *shift map*, which preserves the measure μ. The pair (σ,μ) is called a *Bernoulli shift*. The main result is that every Bernoulli shift is an ergodic system.

It is worth pointing out that $\mathbb{N}$ may be replaced with $\mathbb{Z}$ throughout the construction. That is, we may take Σ to be the space of two-sided sequences $(\dots,x_{-n},\dots,x_0,\dots,x_n,\dots)$. Up to minor adjustments, which we leave to the reader, all that follows remains valid in that case. In addition, in the two-sided case the shift map is invertible.

The *shift map* $\sigma : \Sigma \to \Sigma$ is defined by

$$\sigma\big((x_n)_n\big) = (x_{n+1})_n.$$

That is, by definition, σ sends each sequence $(x_0, x_1, \ldots, x_n, \ldots)$ to the sequence $(x_1, \ldots, x_n, \ldots)$. Observe that the pre-image of any cylinder is still a cylinder:

$$\sigma^{-1}([m; A_m, \ldots, A_n]) = [m+1; A_m, \ldots, A_n]. \tag{4.2.6}$$

It follows that the map σ is measurable with respect to the σ-algebra $\mathcal{B}$. Moreover,

$$\mu\big(\sigma^{-1}([m; A_m, \ldots, A_n])\big) = \nu(A_m) \cdots \nu(A_n) = \mu\big([m; A_m, \ldots, A_n]\big),$$

and (using Lemma 1.3.1) that ensures that the measure μ is invariant under σ.

Proposition 4.2.7. *Every Bernoulli shift (σ, μ) is ergodic.*

Proof. Let A be an invariant measurable set. We want to prove that $\mu(A) = 0$ or $\mu(A) = 1$. We use the following fact:

Lemma 4.2.8. *If B and C are finite unions of pairwise disjoint cylinders, then*

$$\mu\big(B \cap \sigma^{-j}(C)\big) = \mu(B)\mu(\sigma^{-j}(C)) = \mu(B)\mu(C),$$

for every j sufficiently large.

Proof. First, suppose that B and C are both cylinders: $B = [k; B_k, \ldots, B_l]$ and $C = [m; C_m, \ldots, C_n]$. Then,

$$\sigma^{-j}(C) = [m+j; C_m, \ldots, C_n] \quad \text{for each } j.$$

Consider any j large enough that $m + j > l$. Then,

$$\begin{aligned} B \cap \sigma^{-j}(C) &= \{(x_n)_n : x_k \in B_k, \ldots, x_l \in B_l, x_{m+j} \in C_m, \ldots, x_{n+j} \in C_n\} \\ &= [k; B_k, \ldots, B_l, X, \ldots, X, C_m, \ldots, C_n], \end{aligned}$$

where X appears exactly $m+j-l-1$ times. By the definition (4.2.5), this gives that

$$\mu\big(B \cap \sigma^{-j}(C)\big) = \prod_{i=k}^{l} \nu(B_i)\, 1^{m+j-l-1} \prod_{i=m}^{n} \nu(C_i) = \mu(B)\mu(C).$$

This proves the conclusion of the lemma when both sets are cylinders. The general case follows easily, using the fact that μ is finitely additive.

Proceeding with the proof of Proposition 4.2.7, suppose for a while that the invariant set A belongs to the algebra $\mathcal{B}_0$ whose elements are the finite unions of pairwise disjoint cylinders. Then, on the one hand, we may apply the previous lemma with $B = C = A$ to conclude that $\mu(A \cap \sigma^{-j}(A)) = \mu(A)^2$ for every large j. On the other hand, since A is invariant, the left-hand side of this identity is equal to $\mu(A)$ for every j. It follows that $\mu(A) = \mu(A)^2$, which means that either $\mu(A) = 0$ or $\mu(A) = 1$.

Now let A be an arbitrary invariant set. By the approximation theorem (Theorem A.1.19), given any $\varepsilon > 0$ there exists $B \in \mathcal{B}_0$ such that $\mu(A\Delta B) < \varepsilon$. By Lemma 4.2.8 we may fix j such that

$$\mu\big(B \cap \sigma^{-j}(B)\big) = \mu(B)\mu(\sigma^{-j}(B)) = \mu(B)^2. \tag{4.2.7}$$

Using (4.1.8) and (4.1.9) and the fact that μ is invariant, we get that

$$\big|\mu\big(A \cap \sigma^{-j}(A)\big) - \mu\big(B \cap \sigma^{-j}(B)\big)\big| \le 2\mu(A\Delta B) < 2\varepsilon \tag{4.2.8}$$

(a similar fact was deduced during the proof of Corollary 4.1.5). Moreover,

$$\big|\mu(A)^2 - \mu(B)^2\big| \le 2\big|\mu(A) - \mu(B)\big| < 2\varepsilon. \tag{4.2.9}$$

Putting the relations (4.2.7), (4.2.8) and (4.2.9) together, we conclude that $|\mu(A) - \mu(A)^2| < 4\varepsilon$. Since ε is arbitrary, we deduce that $\mu(A) = \mu(A)^2$ and, hence, either $\mu(A) = 0$ or $\mu(A) = 1$.

When X is a topological space and $\mathcal{C}$ is the corresponding Borel σ-algebra, we may endow Σ with the *product topology* which, by definition, is the topology generated by the cylinders $[m; A_m, \dots, A_n]$ where $A_m, \dots, A_n$ are open subsets of X. The property (4.2.6) implies that the shift map $\sigma : \Sigma \to \Sigma$ is continuous with respect to this topology. The theorem of Tychonoff (see [Dug66]) asserts that Σ is compact if X is compact.

A relevant special case is when X is a finite set endowed with the discrete topology, that is, such that every subset of X is open. A map $f : M \to M$ in a topological space M is said to be *transitive* if there exists some $x \in M$ whose trajectory $f^n(x)$, $n \ge 1$ is dense in M. We leave it to the reader (Exercise 4.2.2) to prove the following result:

Proposition 4.2.9. *Let X be a finite set and Σ be either $X^{\mathbb{N}}$ or $X^{\mathbb{Z}}$. Then the shift map $\sigma : \Sigma \to \Sigma$ is transitive. Moreover, the set of all periodic points of σ is dense in Σ.*

The following informal statement, which is one of many versions of the *monkey paradox*, illustrates the meaning of the ergodicity of the Bernoulli measure μ: *A monkey hitting keys at random on a typewriter keyboard for an infinite amount of time will almost surely type the complete text of "Os Lusíadas".*[1]

To "prove" this statement we need to formulate it a bit more precisely. The possible texts typed by the monkey correspond to the sequences $(x_n)_{n\in\mathbb{N}}$ in the (finite) set X of all the characters on the keyboard: letters, digits, space, punctuation signs, and so on. Denote by $\sigma : \Sigma \to \Sigma$ the shift map in the space $\Sigma = X^{\mathbb{N}}$. It is assumed that each character $* \in X$ has a positive probability p_*

[1] Monumental epic poem by the 16th-century Portuguese poet Luis de Camões.

of being hit at each time. This corresponds to the probability measure

$$\nu = \sum_{*\in X} p_* \delta_*$$

on the set X. Furthermore, it is assumed that the character hit at each time is independent of all the previous ones. This means that the distribution of the sequences of characters $(x_n)_n$ is governed by the Bernoulli probability measure $\mu = \nu^{\mathbb{N}}$.

The text of "Os Lusíadas" corresponds to a certain finite (albeit very long) sequence of characters $(l_0, \dots, l_N)$. Consider the cylinder $L = [0; l_0, \dots, l_N]$. Then

$$\mu(L) = \prod_{j=1}^{N} p_{l_j}$$

is positive (although very small). A sequence $(x_n)_n$ contains a complete copy of "Os Lusíadas" precisely if $\sigma^k\big((x_n)_n\big) \in L$ for some $k \geq 0$. By the Birkhoff ergodic theorem and the fact that (σ, μ) is ergodic, the set K of values of k for which that happens satisfies

$$\lim_n \frac{1}{n} \#\big(K \cap [0, n-1]\big) = \mu(L) > 0, \tag{4.2.10}$$

with full probability. In particular, for almost all sequences $(x_n)_n$ the set K is infinite, which means that $(x_n)_n$ contains infinitely many copies of "Os Lusíadas". Actually, (4.2.10) yields an even stronger conclusion: still with full probability, the copies of our poem correspond to a positive (although small) fraction of all the typed characters. In other words, on average, the monkey types a new copy of "Os Lusíadas" every so many (a great many) years.

4.2.4 Gauss map

As we have seen in Section 1.3.2, the gauss map $G(x) = 1/x - [1/x]$ has an invariant probability measure μ equivalent to the Lebesgue measure, namely:

$$\mu(E) = \frac{1}{\log 2} \int_E \frac{dx}{1+x}. \tag{4.2.11}$$

Proposition 4.2.10. *The system (G, μ) is ergodic.*

This can be proved using a more elaborate version of the method introduced in Section 4.2.2. We are going to outline the arguments in the proof, referring to Section 4.2.2 for those parts that are common to both situations and addressing in more detail the main new difficulty.

Let A be an invariant set with positive measure. We want to show that $\mu(A) = 1$. On the one hand, it remains true that for almost every point $a \in [0,1]$ there exists a sequence of intervals I_k containing a and such that G^k maps I_k

bijectively and differentiably onto $(0,1)$. Indeed, such intervals can be found as follows. First, consider

$$I(1,m) = \left(\frac{1}{m+1}, \frac{1}{m}\right),$$

for each $m \geq 1$. Next, define, by recurrence,

$$I(k,m_1,\ldots,m_k) = I(1,m_1) \cap G^{-k+1}\big(I(k-1,m_2,\ldots,m_k)\big)$$

for $m_1,\ldots,m_k \geq 1$. Then, it suffices to take as I_k the interval $I(k,m_1,\ldots,m_k)$ that contains a. This is well defined for every $k \geq 1$ and every point a in the complement of a countable set, namely, the set $\bigcup_{k=0}^{\infty} G^{-k}(\{0,1\})$.

On the other hand, although the restriction of G^k to each I_k is a differentiable bijection, it is not affine. For that reason, the analogue of (4.2.4) cannot hold in the present case. This difficulty is by passed by the result that follows, which is an example of *distortion* control: it is important to note that the constant K is independent of I_k, E_1, E_2 and, most of all, k.

Proposition 4.2.11 (Bounded distortion). *There exists $K > 1$ such that, given any $k \geq 1$ and any interval I_k such that G^k restricted to I_k is a differentiable bijection,*

$$\frac{\mu(G^k(E_1))}{\mu(G^k(E_2))} \leq K\frac{\mu(E_1)}{\mu(E_2)}$$

for any measurable subsets E_1 and E_2 of the interval I_k.

For the proof of this proposition we need the following two auxiliary results:

Lemma 4.2.12. *For every $x \in (0,1]$ we have*

$$|G'(x)| \geq 1 \quad \text{and} \quad |(G^2)'(x)| \geq 2 \quad \text{and} \quad |G''(x)/G'(x)^2| \leq 2.$$

Proof. Recall that $G(x) = 1/x - m$ on each interval $(1/(m+1), 1/m]$. Therefore,

$$G'(x) = -\frac{1}{x^2} \quad \text{and} \quad G''(x) = \frac{2}{x^3}.$$

The first identity implies that $|G'(x)| \geq 1$ for every $x \in (0,1]$. Moreover, $|G'(x)| \geq 2$ whenever $x \leq 2/3$. On the other hand, $x \geq 2/3$ implies that $G(x) = 1/x - 1 < 2/3$ and, consequently, $G'(G(x)) \geq 2$. Combining these observations we find that $|(G^2)'(x)| = |G'(x)|\,|G'(G(x))| \geq 2$ for every $x \in (0,1]$. Finally, $|G''(x)/G'(x)^2| = 2|x| \leq 2$ also for every $x \in (0,1]$.

Lemma 4.2.13. *There exists $C > 1$ such that, given any $k \geq 1$ and any interval I_k such that G^k restricted to I_k is a differentiable bijection,*

$$\frac{|(G^k)'(x)|}{|(G^k)'(y)|} \leq C \quad \textit{for any } x \textit{ and } y \textit{ in } I_k.$$

Proof. Let g be a local inverse of G, that is, a differentiable function defined on some interval and such that $G(g(z)) = z$ for every z in the domain of definition. Note that

$$\left[\log|G' \circ g(z)|\right]' = \frac{G''(g(z))g'(z)}{G'(g(z))} = \frac{G''(g(z))}{G'(g(z))^2}.$$

Therefore, the last estimate in Lemma 4.2.12 implies that

$$\left|\left[\log|G' \circ g(z)|\right]'\right| \le 2 \quad \text{for every } g \text{ and every } z. \tag{4.2.12}$$

In other words, every function of the form $\log|G' \circ g|$ admits 2 as a Lipschitz constant. Observe also that if $x, y \in I_k$ then

$$\log\frac{|(G^k)'(x)|}{|(G^k)'(y)|} = \sum_{j=0}^{k-1} \log|G'(G^j(x))| - \log|G'(G^j(y))|$$

$$= \sum_{j=1}^{k} \log|G' \circ g_j(G^j(x))| - \log|G' \circ g_j(G^j(y))|,$$

where g_j denotes a local inverse of G defined on the interval $[G^j(x), G^j(y)]$. Using the estimate (4.2.12), we get that

$$\log\frac{|(G^k)'(x)|}{|(G^k)'(y)|} \le 2\sum_{j=1}^{k} |G^j(x) - G^j(y)| = 2\sum_{i=0}^{k-1} |G^{k-i}(x) - G^{k-i}(y)|. \tag{4.2.13}$$

Now, the first two estimates in Lemma 4.2.12 imply that

$$|G^k(x) - G^k(y)| \ge 2^{[i/2]}|G^{k-i}(x) - G^{k-i}(y)|$$

for every $i = 0, \ldots, k$. Replacing in (4.2.13), we conclude that

$$\log\frac{|(G^k)'(x)|}{|(G^k)'(y)|} \le 2\sum_{i=0}^{k-1} 2^{-[i/2]}|G^k(x) - G^k(y)| \le 8|G^k(x) - G^k(y)| \le 8.$$

Now it suffices to take $C = e^8$.

Proof of Proposition 4.2.11. Let m be the Lebesgue measure on $[0,1]$. It follows from Lemma 4.2.13 that

$$\frac{m(G^k(E_1))}{m(G^k(E_2))} = \frac{\int_{E_1} |(G^k)'|\,dm}{\int_{E_2} |(G^k)'|\,dm} \le C\frac{m(E_1)}{m(E_2)}.$$

On the other hand, the definition (4.2.11) implies that

$$\frac{1}{2\log 2}m(E) \le \mu(E) \le \frac{1}{\log 2}m(E)$$

for every measurable set $E \subset [0,1]$. Combining these two relations, we find that

$$\frac{\mu(G^k(E_1))}{\mu(G^k(E_2))} \le 2\frac{m(G^k(E_1))}{m(G^k(E_2))} \le 2C\frac{m(E_1)}{m(E_2)} \le 4C\frac{\mu(E_1)}{\mu(E_2)}.$$

Hence, it suffices to take $K = 4C$.

We are ready to conclude that (G,μ) is ergodic. Let A be an invariant set with $\mu(A)>0$. Then A also has positive Lebesgue measure, since μ is absolutely continuous with respect to the Lebesgue measure. Let a be a density point of A whose future trajectory is contained in the open interval $(0,1)$. Consider the sequence $(I_k)_k$ of the intervals $I(k,m_1,\dots,m_k)$ that contain a. It follows from Lemma 4.2.12 that

$$\operatorname{diam} I_k \le \sup\left\{\frac{1}{|(G^k)'(x)|} : x\in I_k\right\} \le 2^{-[k/2]}$$

for every $k\ge 1$. In particular, the diameter of I_k converges to zero and so

$$\frac{\mu(I_k\cap A)}{\mu(I_k)} \to 1 \quad \text{when } k\to\infty. \tag{4.2.14}$$

Let us take $E_1 = I_k\cap A^c$ and $E_2 = I_k$. By Proposition 4.2.11,

$$\frac{\mu(G^k(I_k\cap A^c))}{\mu(G^k(I_k))} \le K\frac{\mu(I_k\cap A^c)}{\mu(I_k)}.$$

Observe that $G^k(I_k\cap A^c) = A^c$ up to a zero measure set, because the set A is assumed to be invariant. Recall also that $G^k(I_k) = (0,1)$, which has full measure. Therefore, the previous inequality may be written as

$$\mu(A^c) \le K\frac{\mu(I_k\cap A^c)}{\mu(I_k)}.$$

According to (4.2.14), the expression on the right-hand side converges to zero when $k\to\infty$. It follows that $\mu(A^c)=0$, as we wanted to prove.

4.2.5 Linear endomorphisms of the torus

Recall that we call the torus of dimension d (or just d-torus) the quotient space $\mathbb{T}^d = \mathbb{R}^d/\mathbb{Z}^d$, that is, the space of all equivalence classes of the equivalence relation defined in $\mathbb{R}^d$ by $x\sim y \Leftrightarrow x-y\in\mathbb{Z}^d$. This quotient inherits from $\mathbb{R}^d$ the structure of a differentiable manifold of dimension d. In what follows we assume that $\mathbb{T}^d$ is also endowed with the flat Riemannian metric, which makes it locally isometric to the Euclidean space $\mathbb{R}^d$. Let m be the volume measure associated with this Riemannian metric (see Appendix A.4.5).

Let A be a d-by-d matrix with integer coefficients and determinant different from zero. Then $A(\mathbb{Z}^d)\subset\mathbb{Z}^d$ and, consequently, A induces a transformation

$$f_A : \mathbb{T}^d \to \mathbb{T}^d, \quad f_A([x]) = [A(x)],$$

where $[x]$ denotes the equivalence class that contains $x\in\mathbb{R}^d$. These transformations are called *linear endomorphisms* of the torus.

Note that f_A is differentiable and the derivative $Df_A([x])$ at each point is canonically identified with A. In particular, the Jacobian $\det Df_A([x])$ is constant equal to $\det A$. It follows (Exercise 4.2.9) that the degree of f is equal to $|\det A|$. In particular, f_A is invertible if and only if $|\det A| = 1$. In this case, the inverse is the transformation $f_{A^{-1}}$ induced by the inverse matrix A^{-1}; observe that A^{-1} is also a matrix with integer coefficients.

In any case, f_A preserves the Lebesgue measure on $\mathbb{T}^d$. This may be seen as follows. Since f_A is a local diffeomorphism, the pre-image of any measurable set D with sufficiently small diameter consists of $|\det A|$ (= degree of f_A) pairwise disjoint sets D_i, each of which is mapped diffeomorphically onto D. By the formula of change of variables, $m(D) = |\det A| m(D_i)$ for every i. This proves that $m(D) = m(f_A^{-1}(D))$ for every measurable set D with small diameter. Hence, f_A does preserve the Lebesgue measure m. Next we prove the following fact:

Theorem 4.2.14. *The system (f_A, m) is ergodic if and only if no eigenvalue of the matrix A is a root of unity.*

Proof. Suppose that no eigenvalue of A is a root of unity. Consider any function $\varphi \in L^2(m)$ and let

$$\varphi([x]) = \sum_{k \in \mathbb{Z}^d} c_k e^{2\pi i(k \cdot x)}$$

be its Fourier series expansion (with $k \cdot x = k_1 x_1 + \cdots + k_d x_d$). The coefficients $c_k \in \mathbb{C}$ satisfy

$$\sum_{k \in \mathbb{Z}^d} |c_k|^2 = \|\varphi\|_2^2 < \infty. \tag{4.2.15}$$

Then, the Fourier series expansion of $\varphi \circ f_A$ is:

$$\varphi(f_A([x])) = \sum_{k \in \mathbb{Z}^d} c_k e^{2\pi i(k \cdot A(x))} = \sum_{k \in \mathbb{Z}^d} c_k e^{2\pi i(A^*(k) \cdot x)},$$

where A^* denotes the adjoint of A. Suppose that φ is an invariant function, that is, $\varphi \circ f_A = \varphi$ at m-almost every point. Then, since the Fourier series expansion is unique, we must have

$$c_{A^*(k)} = c_k \quad \text{for every } k \in \mathbb{Z}. \tag{4.2.16}$$

We claim that the trajectory of every $k \neq 0$ under the transformation A^* is infinite. Indeed, if the trajectory of some $k \neq 0$ were finite then there would exist $l, r \in \mathbb{Z}$ with $r > 0$ such that $A^{(l+r)*}(k) = A^{l*}(k)$. This could only happen if A^* had some eigenvalue λ such that $\lambda^r = 1$. Since A and A^* have the same eigenvalues, that would mean that A has some eigenvalue which is a root of unity, which is excluded by the hypothesis. Hence, the trajectory of every $k \neq 0$ is infinite, as claimed. Then the identity (4.2.16), together with (4.2.15), implies that $c_k = 0$ for every $k \neq 0$. Thus, $\varphi = c_0$ at m-almost every point. This proves that the system (f_A, m) is ergodic.

To prove the converse, assume that A admits some eigenvalue which is a root of unity. Then the same holds for A^* and, hence, there exists $r > 0$ such that 1 is an eigenvalue of A^{r*}. Since A^{r*} has integer coefficients, it follows (see

Exercise 4.2.8) that there exists some $k \in \mathbb{Z}^d \setminus \{0\}$ such that $A^{r*}(k) = k$. Fix k and consider the function $\varphi \in L^2(m)$ defined by

$$\varphi([x]) = \sum_{i=0}^{r-1} e^{2\pi i(A^{i*}(k)\cdot x)} = \sum_{i=0}^{r-1} e^{2\pi i(k\cdot A^i(x))}.$$

Then φ is an invariant function for f_A and it is not constant at m-almost every point. Hence, (f_A, m) is not ergodic.

4.2.6 Hopf argument

In this section we present an alternative, more geometric, method to prove the ergodicity of certain linear endomorphisms of the torus. This is based on an argument introduced by Eberhard F. Hopf in his pioneering work [Hop39] on the ergodicity of geodesic flows on surfaces with negative Gaussian curvature.

In the present linear context, the Hopf argument may be used whenever $|\det A| = 1$ and the matrix A is hyperbolic, that is, A has no eigenvalues in the unit circle. But its strongest point is that it may be extended to much more general differentiable systems, not necessarily linear. Some of these extensions are mentioned in Section 4.4.

The hypothesis that the matrix A is hyperbolic means that the space $\mathbb{R}^d$ may be written as a direct sum $\mathbb{R}^d = E^s \oplus E^u$ such that:

1. $A(E^s) = E^s$ and all the eigenvalues of $A \mid E^s$ have absolute value smaller than 1;
2. $A(E^u) = E^u$ and all the eigenvalues of $A \mid E^u$ have absolute value bigger than 1.

Then there exist constants $C > 0$ and $\lambda < 1$ such that

$$\begin{aligned} \|A^n(v^s)\| &\le C\lambda^n \|v^s\| \quad \text{for every } v^s \in E^s \text{ and every } n \ge 0, \\ \|A^{-n}(v^u)\| &\le C\lambda^n \|v^u\| \quad \text{for every } v^u \in E^u \text{ and every } n \ge 0. \end{aligned} \tag{4.2.17}$$

Example 4.2.15. Consider $A = \begin{pmatrix} 2 & 1 \\ 1 & 1 \end{pmatrix}$. The eigenvalues of A are

$$\lambda_u = \frac{3+\sqrt{5}}{2} > 1 > \lambda_s = \frac{3-\sqrt{5}}{2} > 0$$

and the corresponding eigenspaces are:

$$E^u = \left\{(x,y) \in \mathbb{R}^2 : y = \frac{\sqrt{5}-1}{2}x\right\} \text{ and } E^s = \left\{(x,y) \in \mathbb{R}^2 : y = -\frac{\sqrt{5}+1}{2}x\right\}.$$

The family of all affine subspaces of $\mathbb{R}^d$ of the form $v + E^s$, with $v \in \mathbb{R}^d$, defines a partition $\mathcal{F}^s$ of $\mathbb{R}^d$ that we call *stable foliation* and whose elements we call *stable leaves* of A. This partition is invariant under A, meaning that

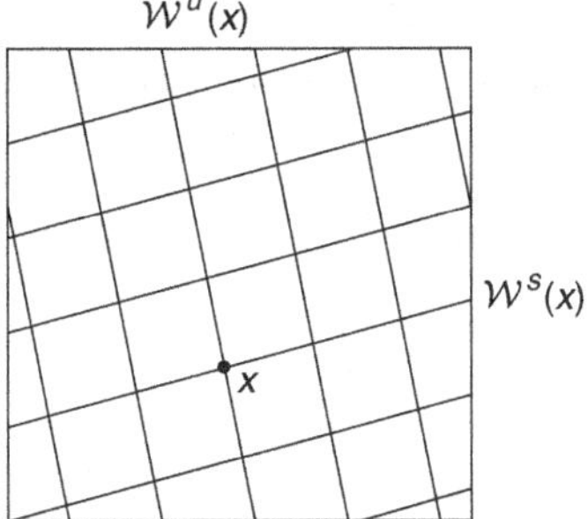

Figure 4.1. Stable foliation and unstable foliation in the torus

the image of any stable leaf is still a stable leaf. Moreover, by (4.2.17), the transformation A contracts distances uniformly inside each stable leaf. Analogously, the family of all affine subspaces of $\mathbb{R}^d$ of the form $v + E^u$ with $v \in \mathbb{R}^d$ defines the *unstable foliation* $\mathcal{F}^u$ of $\mathbb{R}^d$, whose elements are called *unstable leaves*. The unstable foliation is also invariant and the transformation A expands distances uniformly inside unstable leaves.

Mapping $\mathcal{F}^s$ and $\mathcal{F}^u$ by the canonical projection $\pi : \mathbb{R}^d \to \mathbb{T}^d$, we obtain foliations $\mathcal{W}^s$ and $\mathcal{W}^u$ of the torus that we call *stable foliation* and *unstable foliation* of the transformation f_A. See Figure 4.1. The previous observations show that these foliations are invariant under f_A. Moreover:

(i) $d(f_A^j(x), f_A^j(y)) \to 0$ when $j \to +\infty$, for any points x and y in the same stable leaf;
(ii) $d(f_A^j(y), f_A^j(z)) \to 0$ when $j \to -\infty$, for any points y and z in the same unstable leaf.

We are going to use this geometric information to prove that (f_A, m) is ergodic. To that end, let $\varphi : \mathbb{T}^d \to \mathbb{R}$ be any continuous function and consider the time averages

$$\varphi^+(x) = \lim_n \frac{1}{n} \sum_{j=0}^{n-1} \varphi(f_A^j(x)) \quad \text{and} \quad \varphi^-(x) = \lim_n \frac{1}{n} \sum_{j=0}^{n-1} \varphi(f_A^{-j}(x)),$$

which are defined for m-almost every $x \in \mathbb{T}^d$. By Corollary 3.2.8, there exists a full measure set $X \subset \mathbb{T}^d$ such that

$$\varphi^+(x) = \varphi^-(x) \quad \text{for every } x \in X. \tag{4.2.18}$$

Let us denote by $\mathcal{W}^s(x)$ and $\mathcal{W}^u(x)$, respectively, the stable leaf and the unstable leaf of f_A through each point $x \in \mathbb{T}^d$.

Lemma 4.2.16. *The function φ^+ is constant on each leaf of $\mathcal{W}^s$: if $\varphi^+(x)$ exists and $y \in \mathcal{W}^s(x)$ then $\varphi^+(y)$ exists and it is equal to $\varphi^+(x)$. Analogously, φ^- is constant on each leaf of $\mathcal{W}^u$.*

Proof. According to property (i) above, $d(f_A^j(x), f_A^j(y))$ converges to zero when $j \to \infty$. Noting that φ is uniformly continuous (because its domain is compact), it follows that

$$\varphi(f_A^j(x)) - \varphi(f_A^j(y)) \to 0 \quad \text{when } j \to \infty.$$

In particular, the Cesàro limit

$$\lim_n \frac{1}{n} \sum_{j=0}^{n-1} \varphi(f_A^j(x)) - \varphi(f_A^j(y))$$

is also zero. That implies that $\varphi^+(y)$ exists and is equal to $\varphi^+(x)$. The argument for φ^- is entirely analogous.

Given any open subset R of the torus and any $x \in R$, denote by $\mathcal{W}^s(x,R)$ the connected component of $\mathcal{W}^s(x) \cap R$ that contains x and by $\mathcal{W}^u(x,R)$ the connected component of $\mathcal{W}^u(x) \cap R$ that contains x. We call R a *rectangle* if $\mathcal{W}^s(x,R)$ intersects $\mathcal{W}^u(y,R)$ at a unique point, for every x and y in R. See Figure 4.2.

Lemma 4.2.17. *Given any rectangle $R \subset \mathbb{T}^d$, there exists a measurable set $Y_R \subset X \cap R$ such that $m(R \setminus Y_R) = 0$ and, given any x and y in Y_R, there exist points x' and y' in $X \cap R$ such that $x' \in \mathcal{W}^s(x,R)$ and $y' \in \mathcal{W}^s(y,R)$ and $y' \in \mathcal{W}^u(x')$.*

Proof. Let us denote by m_x^s the Lebesgue measure on the stable leaf $\mathcal{W}^s(x)$ of each point $x \in \mathbb{T}^d$. Note that $m(R \setminus X) = 0$, since X has full measure in $\mathbb{T}^d$. Then, by the theorem of Fubini,

$$m_x^s\big(\mathcal{W}^s(x,R) \setminus X\big) = 0 \quad \text{for } m\text{-almost every } x \in R.$$

Define $Y_R = \{x \in X \cap R : m_x^s\big(\mathcal{W}^s(x,R) \setminus X\big) = 0\}$. Then Y_R has full measure in R. Given $x, y \in R$, consider the map $\pi : \mathcal{W}^s(x,R) \to \mathcal{W}^s(y,R)$ defined by

$$\pi(x') = \text{intersection between } \mathcal{W}^u(x',R) \text{ and } \mathcal{W}^s(y,R).$$

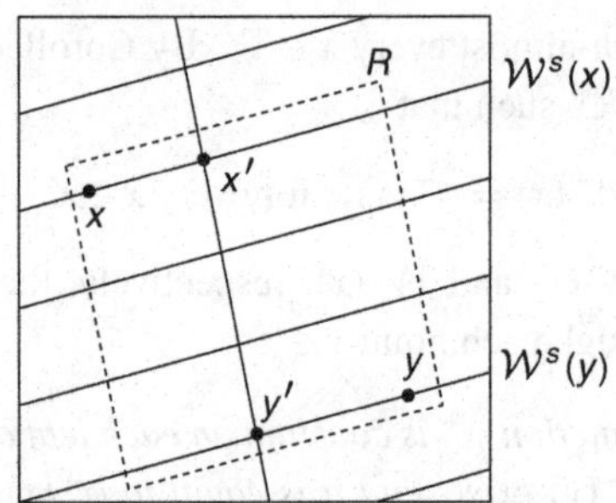

Figure 4.2. Rectangle in $\mathbb{T}^d$

This map is affine and, consequently, it has the following property, called *absolute continuity*:

$$m_x^s(E) = 0 \quad \Leftrightarrow \quad m_y^s(\pi(E)) = 0.$$

In particular, the image of $\mathcal{W}^s(x,R) \cap X$ has full measure in $\mathcal{W}^s(y,R)$ and, consequently, it intersects $\mathcal{W}^s(y,R) \cap X$. So, there exists $x' \in \mathcal{W}^s(x,R) \cap X$ whose image $y' = \pi(x')$ is in $\mathcal{W}^s(y,R) \cap X$. Observing that x' and y' are in the same unstable leaf, by the definition of π, we see that these points satisfy the conditions in the conclusion of the lemma.

Consider any rectangle R. Given any x, y in Y_R, consider the points x', y' in X given by Lemma 4.2.17. Using Lemma 4.2.16 as well, we obtain

$$\varphi^-(x) = \varphi^+(x) = \varphi^+(x') = \varphi^-(x') = \varphi^-(y') = \varphi^+(y') = \varphi^+(y) = \varphi^-(y).$$

This shows that the functions φ^+ and φ^- coincide with one another and are constant in Y_R.

Now let $R_1, \dots, R_N$ be a finite cover of the torus by rectangles. Consider the set

$$Y = \bigcup_{j=1}^{N} Y_j, \qquad \text{where } Y_j = Y_{R_j}.$$

Observe that $m(Y) = 1$, since $Y \cap R_j \supset Y_j$ has full measure in R_j for every j. We claim that $\varphi^+ = \varphi^-$ is constant on the whole Y. Indeed, given any $k, l \in \{1, \dots, N\}$ we may find $j_0 = k, j_1, \dots, j_{n-1}, j_n = l$ such that each R_{j_i} intersects $R_{j_{i-1}}$ (that is just because the torus is path-connected). Recalling that R_j is an open set and Y_j is a full measure subset, we get that each Y_{j_i} intersects $Y_{j_{i-1}}$. Then, $\varphi^+ = \varphi^-$ is constant on the union of all the Y_{j_i}. This proves our claim.

In this way, we have shown that the time averages $\varphi^\pm$ of any continuous function φ are constant at m-almost every point. Consequently (see Exercise 4.1.4), the system (f_A, m) is ergodic.

4.2.7 Exercises

4.2.1. Prove Proposition 4.2.2.

4.2.2. Prove Proposition 4.2.9.

4.2.3. Let $I = [0,1]$ and $f : I \to I$ be the function defined by

$$f(x) = \begin{cases} 2x & \text{if } 0 \le x < 1/3 \\ 2x - 2/3 & \text{if } 1/3 \le x < 1/2 \\ 2x - 1/3 & \text{if } 1/2 \le x < 2/3 \\ 2x - 1 & \text{if } 2/3 \le x \le 1. \end{cases}$$

Show that f is ergodic with respect to the Lebesgue measure m.

4.2.4. Let X be a finite set and $\Sigma = X^{\mathbb{N}}$. Prove that every infinite compact subset of Σ invariant under the shift map $\sigma : \Sigma \to \Sigma$ contains some non-periodic point.

4.2.5. Let X be a topological space, endowed with the corresponding Borel σ-algebra $\mathcal{C}$, and let $\Sigma = X^{\mathbb{N}}$. Show that if X has a countable basis of open sets then the Borel σ-algebra of Σ (for the product topology) coincides with the product σ-algebra $\mathcal{B} = \mathcal{C}^{\mathbb{N}}$. The same is true for $\Sigma = X^{\mathbb{Z}}$ and $\mathcal{B} = \mathcal{C}^{\mathbb{Z}}$.

4.2.6. In this exercise we propose an alternative proof of Proposition 4.2.1. Assume that θ is irrational. Let A be an invariant set with positive measure. Recalling that the orbit $\{R_\theta^n(a) : n \in \mathbb{Z}\}$ of every $a \in S^1$ is dense in S^1, show that no point of S^1 is a density point of A^c. Conclude that $\mu(A) = 1$.

4.2.7. Assume that θ is irrational. Let $\varphi : S^1 \to \mathbb{R}$ be any continuous function. Show that

$$\tilde{\varphi}(x) = \lim_{n\to\infty} \frac{1}{n} \sum_{j=0}^{n-1} \varphi(R_\theta^j(x)) \tag{4.2.19}$$

exists at *every* point and, in fact, the limit is uniform. Deduce that $\tilde{\varphi}$ is constant at every point. Conclude that R_θ has a unique invariant probability measure.

4.2.8. Let A be a square matrix of dimension d with rational coefficients and let λ be a rational eigenvalue of A. Show that there exists some eigenvector with integer coefficients, that is, some $k \in \mathbb{Z}^d \setminus \{0\}$ such that $Ak = \lambda k$.

4.2.9. Show that if $f : M \to M$ is a local diffeomorphism on a compact Riemannian manifold then

$$\operatorname{degree} f = \int |\det Df| \, dm,$$

where m denotes the volume measure induced by the Riemannian metric of M, normalized in such a way that $m(M) = 1$. In particular, for any square matrix A of dimension d with integer coefficients, the degree of the linear endomorphism $f_A : \mathbb{T}^d \to \mathbb{T}^d$ is equal to $|\det A|$.

4.2.10. A number $x \in (0,1)$ has continued fraction expansion *of bounded type* if the sequence $(a_n)_n$ constructed in Section 1.3.2 is bounded. Prove that the set $\mathcal{L} \subset (0,1)$ of points with continued fraction expansion of bounded type has Lebesgue measure zero.

4.2.11. Let $f : M \to M$ be a measurable transformation, μ be an ergodic invariant measure and $\varphi : M \to \mathbb{R}$ be a measurable function with $\int \varphi \, d\mu = +\infty$. Prove that $\lim_n (1/n) \sum_{j=0}^{n-1} \varphi(f^j(x)) = +\infty$ for μ-almost every $x \in M$.

4.2.12. Observe that the number b in Exercise 3.2.4 is independent of x in a set with full Lebesgue measure. Prove that the arithmetic mean of the numbers $a_1, \dots, a_n, \dots$ goes to infinity: $\lim_n (1/n)(a_1 + \cdots + a_n) = +\infty$.

4.3 Properties of ergodic measures

In this section we take the transformation $f : M \to M$ to be fixed and we analyze the set $\mathcal{M}_e(f)$ of probability measures that are ergodic with respect to f as a subset of the space $\mathcal{M}_1(f)$ of all probability measures invariant under f.

Recall that a measure ν is said to be *absolutely continuous* with respect to another measure μ if $\mu(E) = 0$ implies $\nu(E) = 0$. Then we write $\nu \ll \mu$. This relation is transitive: if $\nu \ll \mu$ and $\mu \ll \lambda$ then $\nu \ll \lambda$. The first result asserts that the ergodic probability measures are minimal for this order relation:

Lemma 4.3.1. *If μ and ν are invariant probability measures such that μ is ergodic and ν is absolutely continuous with respect to μ then $\mu = \nu$.*

Proof. Let $\varphi : M \to \mathbb{R}$ be any bounded measurable function. Since μ is invariant and ergodic, the time average

$$\tilde{\varphi}(x) = \lim_{n\to\infty} \frac{1}{n} \sum_{j=0}^{n-1} \varphi(f^j(x))$$

is constant: $\tilde{\varphi}(x) = \int \varphi \, d\mu$ at μ-almost every point. Since $\nu \ll \mu$, it follows that the equality also holds at ν-almost every point. In particular,

$$\int \varphi \, d\nu = \int \tilde{\varphi} \, d\nu = \int \varphi \, d\mu$$

(the first equality is part of the Birkhoff ergodic theorem). Therefore, the integrals of each bounded measurable function φ with respect to μ and with respect to ν coincide. In particular, considering characteristic functions, we conclude that $\mu = \nu$.

It is clear that if μ_1 and μ_2 are probability measures invariant under the transformation f then so is $(1-t)\mu_1 + t\mu_2$, for any $t \in (0,1)$. This means that the space $\mathcal{M}_1(f)$ of all probability measures invariant under f is *convex*. The next proposition asserts that the ergodic probability measures are the *extremal elements* of this convex set:

Proposition 4.3.2. *An invariant probability measure μ is ergodic if and only if it is not possible to write it as $\mu = (1-t)\mu_1 + t\mu_2$ with $t \in (0,1)$ and $\mu_1, \mu_2 \in \mathcal{M}_1(f)$ with $\mu_1 \neq \mu_2$.*

Proof. To prove the "if" claim, assume that μ is not ergodic. Then there exists some invariant set A with $0 < \mu(A) < 1$. Define μ_1 and μ_2 to be the normalized restriction of μ to the set A and to its complement A^c, respectively:

$$\mu_1(E) = \frac{\mu(E \cap A)}{\mu(A)} \quad \text{and} \quad \mu_2(E) = \frac{\mu(E \cap A^c)}{\mu(A^c)}.$$

Since A and A^c are invariant sets and μ is an invariant measure, both μ_1 and μ_2 are still invariant probability measures. Moreover,

$$\mu = \mu(A)\mu_1 + \mu(A^c)\mu_2$$

and, consequently, μ is not extremal.

To prove the converse, assume that μ is ergodic and $\mu = (1-t)\mu_1 + t\mu_2$ for some $t \in (0,1)$. It is clear that $\mu(E) = 0$ implies $\mu_1(E) = \mu_2(E) = 0$, that is, μ_1 and μ_2 are absolutely continuous with respect to μ. Hence, by Lemma 4.3.1, $\mu_1 = \mu = \mu_2$. This shows that μ is extremal.

Let us also point out that distinct ergodic measures "live" in disjoint subsets of the space M (see also Exercise 4.3.6):

Lemma 4.3.3. *Assume that the σ-algebra of M admits some countable generating subset Γ. Let $\{\mu_i : i \in \mathcal{I}\}$ be an arbitrary family of ergodic probability measures, all distinct. Then these measures μ_i are mutually singular: there exist pairwise disjoint measurable subsets $\{P_i : i \in \mathcal{I}\}$ invariant under f and such that $\mu_i(P_i) = 1$ for every $i \in \mathcal{I}$.*

Proof. Let $\mathcal{A}$ be the algebra generated by Γ. Note that $\mathcal{A}$ is countable, since it coincides with the union of the (finite) algebras generated by the finite subsets of Γ. For each $i \in \mathcal{I}$, define

$$P_i = \bigcap_{A \in \mathcal{A}} \{x \in M : \tau(A,x) = \mu_i(A)\}.$$

Since μ_i is ergodic, $\{x \in M : \tau(A,x) = \mu_i(A)\}$ has full measure for each A. Using that $\mathcal{A}$ is countable, it follows that $\mu_i(P_i) = 1$ for every $i \in \mathcal{I}$. Moreover, if there exists $x \in P_i \cap P_j$ then $\mu_i(A) = \tau(A,x) = \mu_j(A)$ for every $A \in \mathcal{A}$. In other words, $\mu_i = \mu_i$. This proves that the P_i are pairwise disjoint.

Now assume that $f : M \to M$ is a continuous transformation in a topological space M. We say that f is *transitive* if there exists some $x \in M$ such that $\{f^n(x) : n \in \mathbb{N}\}$ is dense in M. The next lemma provides a useful characterization of transitivity. Recall that a topological space M is called a *Baire space* if the intersection of any countable family of open dense subsets is dense in M. Every complete metric space is a Baire space and the same is true for every locally compact topological space (see [Dug66]).

Lemma 4.3.4. *Let M be a Baire space with a countable basis of open sets. Then $f : M \to M$ is transitive if and only if for every pair of open sets U and V there exists $k \geq 1$ such that $f^{-k}(U)$ intersects V.*

Proof. Assume that f is transitive and let $x \in M$ be a point whose orbit $\{f^n(x) : n \in \mathbb{N}\}$ is dense. Then there exists $m \geq 1$ such that $f^m(x) \in V$ and (using the fact that $\{f^n(x) : n > m\}$ is also dense) there exists $n > m$ such that $f^n(x) \in U$. Take $k = n - m$. Then $f^m(x) \in f^{-k}(U) \cap V$. This proves the "only if" part of the statement.

To prove the converse, let $\{U_j : j \in \mathbb{N}\}$ be a countable basis of open subsets of M. The hypothesis ensures that the open set $\bigcup_{k=1}^{\infty} f^{-k}(U_j)$ is dense in M for every $j \in \mathbb{N}$. Then the intersection

$$X = \bigcap_{j=1}^{\infty} \bigcup_{k=1}^{\infty} f^{-k}(U_j)$$

is a dense subset of M. In particular, it is non-empty. On the other hand, by definition, if $x \in X$ then for every $j \in \mathbb{N}$ there exists $k \geq 1$ such that $f^k(x) \in U_j$.

Since the U_j constitute a basis of open subsets of M, this means that $\{f^k(x) : k \in \mathbb{N}\}$ is dense in M.

Proposition 4.3.5. *Let M be a Baire space with a countable basis of open sets. If μ is an ergodic probability measure then the restriction of f to the support of μ is transitive.*

Proof. Start by noting that $\operatorname{supp}\mu$ has a countable basis of open sets, because it is a subspace of M, and it is a Baire space, since it is closed in M. Let U and V be open subsets of $\operatorname{supp}\mu$. By the definition of support, $\mu(U) > 0$ and $\mu(V) > 0$. Define $B = \bigcup_{k=1}^{\infty} f^{-k}(U)$. Then $\mu(B) > 0$, because $B \supset U$, and $f^{-1}(B) \subset B$. By ergodicity (see Exercise 1.1.4) it follows that $\mu(B) = 1$. Then B must intersect V. This proves that there exists $k \geq 1$ such that $f^{-k}(U)$ intersects V. By Lemma 4.3.4, it follows that the restriction $f : \operatorname{supp}\mu \to \operatorname{supp}\mu$ is transitive.

4.3.1 Exercises

4.3.1. Let M be a topological space M with a countable basis of open sets, $f : M \to M$ be a measurable transformation and μ be an ergodic probability measure. Show that the orbit $\{f^n(x) : n \geq 0\}$ of μ-almost every point $x \in M$ is dense in the support of μ.

4.3.2. Let $f : M \to M$ be a continuous transformation in a compact metric space. Given a function $\varphi : M \to \mathbb{R}$, prove that there exists an invariant probability measure μ_φ such that

$$\int \varphi \, d\mu_\varphi = \sup_{\eta \in \mathcal{M}_1(f)} \int \varphi \, d\eta.$$

4.3.3. Let $g : E \to E$ be a transformation induced by $f : M \to M$, that is, a transformation of the form $g(x) = f^{\rho(x)}(x)$ with $\rho : E \to \mathbb{N}$ (see Section 1.4.2). Let ν be an invariant probability measure of g and ν_g be the invariant measure of f defined by (1.4.5). Assume that $\nu_\rho(M) < \infty$ and denote $\mu = \nu_\rho/\nu_\rho(M)$. Show that (f, μ) is ergodic if and only if (g, ν) is ergodic.

4.3.4. Let $f : M \to M$ be a continuous transformation in a separable complete metric space. Given any invariant probability measure μ, let $\hat{\mu}$ be its lift to the natural extension $\hat{f} : \hat{M} \to \hat{M}$ (see Section 2.4.2). Show that $(\hat{f}, \hat{\mu})$ is ergodic if and only if (f, μ) is ergodic.

4.3.5. Let $f : M \to M$ be a measurable transformation and μ be an invariant measure. Let $g^t : N \to N$, $t \in \mathbb{R}$ be a suspension flow of f and ν be the corresponding suspension of the measure μ (see Section 3.4.1). Assume that $\nu(N) < \infty$ and denote $\hat{\nu} = \nu/\nu(N)$. Show that $\hat{\nu}$ is ergodic for the flow $(g^t)_t$ if and only if μ is ergodic for f.

4.3.6. Show that for finite or countable families of ergodic measures the conclusion of Lemma 4.3.3 holds even if the σ-algebra is not countably generated.

4.3.7. Give an example of a metric space M and a transformation $f : M \to M$ such that there exists a sequence of ergodic Borel measures μ_n converging, in the weak* topology, to a non-ergodic invariant measure μ.

4.3.8. Let M be a metric space, $f : M \to M$ be a continuous transformation and μ be an ergodic probability measure. Show that $\frac{1}{n}\sum_{j=0}^{n-1} f_*^j \nu$ converges to μ in the weak* topology for any probability measure ν on M absolutely continuous with respect to μ, but not necessarily invariant.

4.3.9. Let $X = \{1,\ldots,d\}$ and $\sigma : \Sigma \to \Sigma$ be the shift map in $\Sigma = X^{\mathbb{N}}$ or $\Sigma = X^{\mathbb{Z}}$.

(1) Show that for every $\delta > 0$ there exists $k \geq 1$ such that, given $x^1,\ldots,x^s \in \Sigma$ and $m_1,\ldots,m_s \geq 1$, there exists a periodic point $y \in \Sigma$ with period n_s and such that $d(f^{j+n_i}(y), f^j(x^i)) < \delta$ for every $0 \leq j < m_i$, where $n_1 = 0$ and $n_i = (m_1 + k) + \cdots + (m_{i-1} + k)$ for $1 < i \leq s$.

(2) Let $\varphi : \Sigma \to \mathbb{R}$ be a continuous function and $\tilde{\varphi}$ be its Birkhoff average. Show that, given $\varepsilon > 0$, points $x^1,\ldots,x^s \in \Sigma$ where the Birkhoff average of φ is well defined, and numbers $\alpha^1,\ldots,\alpha^s > 0$ such that $\sum_i \alpha^i = 1$, there exists a periodic point $y \in \Sigma$ satisfying $|\tilde{\varphi}(y) - \sum_i \alpha^i \tilde{\varphi}(x^i)| < \varepsilon$.

(3) Conclude that the set $\mathcal{M}_e(\sigma)$ of ergodic probability measures is dense in the space $\mathcal{M}_1(\sigma)$ of all invariant probability measures.

4.4 Comments in conservative dynamics

The ergodic theorem of Birkhoff, proven in the 1930's, provided a solid mathematical foundation to the statement of the Boltzmann ergodic hypothesis, but left entirely open the question of its *veracity*. In this section we briefly survey the main results obtained since then, in the context of *conservative* systems, that is, dynamical systems that preserve a volume measure on a manifold.

Let us start by mentioning that, in a certain abstract sense, the majority of conservative systems are ergodic. That is the sense of the theorem that we state next, which was proven in the early 1940's by John Oxtoby and Stanislav Ulam [OU41]. Recall that a subset of a Baire space is called *residual* if it may be written as a countable intersection of open and dense subsets. By the definition of Baire space, every residual subset is dense.

Theorem 4.4.1 (Oxtoby, Ulam). *For every compact Riemannian manifold M there exists a residual subset $\mathcal{R}$ of the space* $\mathrm{Homeo}_{\mathrm{vol}}(M)$ *of all conservative homeomorphisms of M such that every element of $\mathcal{R}$ is ergodic.*

The results presented below imply that the conclusion of this theorem is no longer true when one replaces $\mathrm{Homeo}_{\mathrm{vol}}(M)$ by the space $\mathrm{Diffeo}^k_{\mathrm{vol}}(M)$ of conservative diffeomorphisms of class C^k, at least for $k > 3$. Essentially nothing is known in this regard in the cases $k = 2$ and $k = 3$. On the other hand, Artur Avila, Sylvain Crovisier and Amie Wilkinson have recently announced a C^1 version of the previous theorem: *for every compact Riemannian manifold M, there exists a residual subset $\mathcal{R}$ of the space* $\mathrm{Diffeo}^1_{\mathrm{vol}}(M)$ *of conservative diffeomorphisms of class C^1 such that every $f \in \mathcal{R}$ with positive entropy $h_{\mathrm{vol}}(f)$ is ergodic.* The notion of entropy will be studied in Chapter 9.

4.4.1 Hamiltonian systems

The systems that interested Boltzmann, relative to the motion of gas molecules, may, in principle, be described by the laws of Newtonian classical mechanics. In the so-called Hamiltonian formalism of classical mechanics, the states of the system are represented by "generalized coordinates" $q_1,\dots,q_d$ and "generalized momenta" $p_1,\dots,p_d$, and the system's evolution is described by the solutions of the Hamilton–Jacobi equations:

$$\frac{dq_j}{dt} = \frac{\partial H}{\partial p_j} \quad \text{and} \quad \frac{dp_j}{dt} = -\frac{\partial H}{\partial q_j}, \quad j = 1,\dots,d, \tag{4.4.1}$$

where H (the total energy of the system) is a C^2 function of the variables $q = (q_1,\dots,q_d)$ and $p = (p_1,\dots,p_d)$; the integer $d \geq 1$ is the number of degrees of freedom.

Example 4.4.2 (Harmonic pendulum)**.** Let $d = 1$ and $H(q,p) = p^2/2 - g\cos q$, where g is a positive constant and $(q,p) \in \mathbb{R}^2$. The Hamilton–Jacobi equations

$$\frac{dq}{dt} = p \quad \text{and} \quad \frac{dp}{dt} = -g\sin q$$

describe the motion of a pendulum subject to a constant gravitational field: the coordinate q measures the angle with respect to the position of (stable) equilibrium and p measures the angular momentum. Then $p^2/2$ is the kinetic energy and $-g\cos q$ is the potential energy. Thus, the *Hamiltonian* H is the total energy.

Note that H is always a *first integral* of the system, that is, it is constant along the flow trajectories:

$$\frac{dH}{dt} = \sum_{j=1}^{d} \frac{\partial H}{\partial q_j}\frac{dq_j}{dt} + \frac{\partial H}{\partial p_j}\frac{dp_j}{dt} \equiv 0.$$

Thus, we may consider the restriction of the flow to each *energy hypersurface* $H_c = \{(q,p) : H(q,p) = c\}$. The volume measure $dq_1 \cdots dq_d dp_1 \cdots dp_d$ is called the *Liouville measure*. Observe that the divergence of the vector field

$$F = \left(-\frac{\partial H}{\partial p_1},\dots,-\frac{\partial H}{\partial p_d},\frac{\partial H}{\partial q_1},\dots,\frac{\partial H}{\partial q_d}\right)$$

is identically zero. Thus (recall Section 1.3.6) the Liouville measure is invariant under the Hamiltonian flow. It follows (see Exercise 1.3.12) that the restriction of the flow to each energy hypersurface H_c admits an invariant measure μ_c that is given by

$$\mu_c(E) = \int_E \frac{ds}{\|\operatorname{grad} H\|} \quad \text{for every measurable set } E \subset H_c,$$

where ds denotes the volume element on the hypersurface. Then, the ergodic hypothesis may be viewed as claiming that, in general, Hamiltonian systems are ergodic with respect to this invariant measure μ_c on (almost) every energy hypersurface.

The first important result in this context was announced by Andrey Kolmogorov at the International Congress of Mathematicians ICM 1954 and was substantiated, soon afterwards, by the works of Vladimir Arnold and Jürgen Moser. This led to the deep theory of so-called *almost integrable systems* that is known as KAM theory, in homage to its founders, and to which several other mathematicians contributed in a decisive manner, including Helmut Rüssmann, Michael Herman, Eduard Zehnder, Jean-Christophe Yoccoz and Jürgen Pöschel, among others. Let us explain what is meant by "almost integrable".

A Hamiltonian system with d degrees of freedom is said to be *integrable* (in the sense of Liouville) if it admits d first integrals $I_1, \dots, I_d$:

- *independent:* that is, such that the gradients

$$\operatorname{grad} I_j = \left(\frac{\partial I_j}{\partial q_1}, \frac{\partial I_j}{\partial p_1}, \dots, \frac{\partial I_j}{\partial q_d}, \frac{\partial I_j}{\partial p_d}\right), \quad 1 \le j \le d,$$

  are linearly independent at every point on an open and dense subset of the domain;
- *in involution:* that is, such that the Poisson brackets

$$\{I_j, I_k\} = \sum_{i=1}^{d} \left[\frac{\partial I_j}{\partial q_i}\frac{\partial I_k}{\partial p_i} - \frac{\partial I_j}{\partial p_i}\frac{\partial I_k}{\partial q_i}\right]$$

  are all identically zero.

It follows from the previous remarks that every system with $d = 1$ degree of freedom is integrable: the Hamiltonian H itself is a first integral. Another important example:

Example 4.4.3. For any number $d \ge 1$ of degrees of freedom, assume that the Hamiltonian H depends only on the variables $p = (p_1, \dots, p_d)$. Then the Hamilton–Jacobi equations (4.4.1) reduce to

$$\frac{dq_j}{dt} = \frac{\partial H}{\partial p_j}(p) \quad \text{and} \quad \frac{dp_j}{dt} = -\frac{\partial H}{\partial q_j}(p) = 0.$$

The second equation means that each p_j is a first integral; it is easy to see that the first integrals are independent and in involution. Then the expression on the right-hand side of the first equation is independent of time. Hence, the solution is given by

$$q_j(t) = q_j(0) + \frac{\partial H}{\partial p_i}(p(0))\,t.$$

As we are going to comment in the following, this example is totally typical of integrable systems.

A classical theorem of Liouville asserts that if the system is integrable then the Hamilton–Jacobi equations may be solved completely by quadratures. In the proof (see Arnold [Arn78]) one constructs certain functions $\varphi = (\varphi_1,\dots,\varphi_d)$ with values in $\mathbb{T}^d$ which, together with the first integrals $I = (I_1,\dots,I_d) \in \mathbb{R}^d$, constitute *canonical coordinates* of the system (they are called *action-angle coordinates*). What we mean by "canonical" is that the coordinate change

$$\Psi : (q,p) \mapsto (\varphi,I)$$

preserves the form of the Hamilton–Jacobi equations: (4.4.1) becomes

$$\frac{d\varphi_j}{dt} = \frac{\partial H'}{\partial I_j} \quad \text{and} \quad \frac{dI_j}{dt} = -\frac{\partial H'}{\partial \varphi_j}, \tag{4.4.2}$$

where $H' = H \circ \Psi^{-1}$ is the expression of the Hamiltonian in the new coordinates. Since the I_j are first integrals, the second equation yields

$$0 = \frac{dI_j}{dt} = -\frac{\partial H'}{\partial \varphi_j}.$$

This means that H does not depend on the variables φ_j and so we are in the type of situation described in Example 4.4.3. Each trajectory of the Hamiltonian flow is constrained inside a torus $\{I = \text{const}\}$ and, according to the first equation in (4.4.2), it is linear in the coordinate φ:

$$\varphi_j(t) = \varphi_j(0) + \omega_j(I)t, \quad \text{where } \omega_j(I) = \frac{\partial H'}{\partial I_j}(I).$$

In terms of the original coordinates (q,p), we conclude that the trajectories of the Hamiltonian flow are given by

$$t \mapsto \Psi^{-1}(\varphi(0) + \omega(I)t, I) = \Phi_{\varphi(0),I}(\omega(I)t), \tag{4.4.3}$$

where $\Phi_{\varphi(0),I} : \mathbb{R}^d \to M$ is a $\mathbb{Z}^d$-periodic function and $\omega(I) = (\omega_1(I),\dots,\omega_d(I))$ is called a *frequency vector*. We say that the trajectory is *quasi-periodic*.

4.4.2 Kolmogorov–Arnold–Moser theory

It is clear that integrable systems are never ergodic. However, since integrability is a very rare property, this alone would not be an obstruction to most Hamiltonian systems being ergodic. Nevertheless, the fundamental result that we state next asserts that generic integrable systems are *robustly* non-ergodic: every nearby Hamiltonian flow is also non-ergodic.

Let H_0 be an integrable Hamiltonian, written in action-angle coordinates (φ,I). More precisely, let B^d be a ball in $\mathbb{R}^d$ and assume that $H_0(\varphi,I)$ is defined for every $(\varphi,I) \in \mathbb{T}^d \times B^d$ but depends only on the coordinate I. We call H_0 *non-degenerate* if its Hessian matrix is invertible:

$$\det\left(\frac{\partial^2 H_0}{\partial I_i \partial I_j}\right)_{i,j} \neq 0 \quad \text{at every point.} \tag{4.4.4}$$

Observe that the Hessian matrix of H_0 coincides with the Jacobian matrix of the function $I \mapsto \omega(I)$. Therefore, the *twist condition* (4.4.4) means that the map assigning to each value of I the corresponding frequency vector $\omega(I)$ is a local diffeomorphism.

The next theorem means that, under this condition, most of the invariant tori of the Hamiltonian flow of H_0 persist for any nearby system:

Theorem 4.4.4. *Let H_0 be an integrable non-degenerate Hamiltonian of class C^∞. Then there exists a neighborhood $\mathcal{V}$ of H_0 in the space $C^\infty(\mathbb{T}^d \times B^d, \mathbb{R})$ such that for every $H \in \mathcal{V}$ there exists a compact set $K \subset \mathbb{T}^d \times B^d$ satisfying:*

(i) K is a union of differentiable tori of the form $\{(\varphi, u(\varphi)) : \varphi \in \mathbb{T}^d\}$ each of which is invariant under the Hamiltonian flow of H;
(ii) the restriction of the Hamiltonian flow of H to each of these tori is conjugate to a linear flow on $\mathbb{T}^d$;
(iii) the set K has positive volume and, in fact, the volume of its complement goes to zero when $H \to H_0$.

In particular, the Hamiltonian flow of H cannot be ergodic.

The latter is because the set K may be decomposed into positive volume subsets that are also unions of invariant tori and, thus, are invariant. The proof of the theorem shows that the persistence or not of a given invariant torus of H_0 is intimately related to the arithmetic properties of the corresponding frequency vector. Let us explain this.

Given $c > 0$ and $\tau > 0$, we say that a vector $\omega_0 \in \mathbb{R}^d$ is (c, τ)-*Diophantine* if

$$|k \cdot \omega_0| \geq \frac{c}{\|k\|^\tau} \quad \text{for every } k \in \mathbb{Z}^d, \tag{4.4.5}$$

where $\|k\| = |k_1| + \cdots + |k_d|$. Diophantine vectors are rationally independent; in fact, the condition (4.4.5) means that ω_0 is badly approximated by rationally dependent vectors. We say that ω_0 is τ-Diophantine if it is (c, τ)-Diophantine for some $c > 0$. The set of τ-Diophantine vectors is non-empty if and only if $\tau \geq d - 1$; moreover, it has full measure in $\mathbb{R}^d$ if τ is strictly larger than $d - 1$ (see Exercise 4.4.1).

While proving Theorem 4.4.4, it is shown that, given $c > 0$, $\tau \geq d - 1$ and any compact set $\Omega \subset \omega(B^d)$, one can find a neighborhood $\mathcal{V}$ of H_0 such that, for every $H \in \mathcal{V}$ and every (c, τ)-Diophantine vector $\omega_0 \in \Omega$, the Hamiltonian flow of H admits a differentiable invariant torus restricted to which the flow is conjugate to the linear flow $t \mapsto \varphi(t) = \varphi(0) + t\omega_0$.

Next, we discuss a version of Theorem 4.4.4 for discrete time systems or, more precisely, symplectic transformations. We call a *symplectic manifold* (see Arnold [Arn78, Chapter 8]) any differentiable manifold M endowed with a *symplectic form*, that is, a non-degenerate differential 2-form θ. By "non-degenerate" we mean that for every $x \in M$ and every $u \neq 0$ there exists

v such that $\theta_x(u,v) \neq 0$. Existence of a symplectic form on M implies that the dimension is even: write $\dim M = 2d$. Moreover, the d-th power $\theta^d = \theta \wedge \cdots \wedge \theta$ is a volume form on M.

A differentiable transformation $f : M \to M$ is said to be *symplectic* if it preserves the symplectic form, meaning that $\theta_x(u,v) = \theta_{f(x)}(Df(x)u, Df(x)v)$ for every $x \in M$ and any $u, v \in T_x M$. Then, in particular, f preserves the volume form θ^d.

Example 4.4.5. Let $M = \mathbb{R}^{2d}$, with coordinates $(q_1,\dots,q_d,p_1,\dots,p_d)$, and let θ be the differential 2-form defined by

$$\theta_x = dq_1 \wedge dp_1 + \cdots + dq_d \wedge dp_d \tag{4.4.6}$$

for every x. Then θ is a symplectic form on M. Actually, a classical theorem of Darboux states that for every symplectic form there exists some atlas of the manifold such that the expression of the symplectic form in any local chart is of the type (4.4.6). Consider any transformation of the form

$$f_0(q_1,\dots,q_d,p_1,\dots,p_d) = (q_1 + \omega_1(p),\dots,q_d + \omega_d(p),p_1,\dots,p_d).$$

Using

$$Df_0 \cdot \frac{\partial}{\partial q_j} = \frac{\partial}{\partial q_j} \quad \text{and} \quad Df_0 \cdot \frac{\partial}{\partial p_j} = \frac{\partial}{\partial p_j} + \sum_i \frac{\partial \omega_i}{\partial p_j}\frac{\partial}{\partial q_i},$$

we see that f_0 is symplectic with respect to the form θ.

Example 4.4.6 (Cotangent bundle). Let M be a manifold of class C^r with $r \geq 3$. By definition, the *cotangent space* T_q^*M at each point $q \in M$ is the dual of the tangent space T_qM, and the *cotangent bundle* of M is the disjoint union $T^*M = \bigcup_{q\in M} T_q^*M$ of all cotangent spaces. See Appendix A.4.3. The cotangent bundle is a manifold of class C^{r-1} and the canonical projection $\pi^* : T^*M \to M$ mapping each T_q^*M to the corresponding base point q is a map of class C^{r-1}.

A very important feature of the cotangent bundle is that it always admits a canonical symplectic form, that is, one that depends only on the manifold M. That can be seen as follows. Let α be the differential 1-form on T^*M defined by

$$\alpha_{(q,p)} : T_{(q,p)}(T^*M) \to \mathbb{R}, \qquad \alpha_{(q,p)} = p \circ D\pi^*(q,p)$$

for each $(q,p) \in T^*M$. It is clear that α is well defined and of class C^{r-2}. Consider the exterior derivative $\theta^* = d\alpha$. One can check (for instance, using local coordinates) that θ^* is non-degenerate at every point and, thus, is a symplectic form in T^*M.

There is no corresponding statement for the tangent bundle TM. However, once we fix a Riemannian metric on M it is possible to endow the tangent bundle with a (non-canonical) symplectic form:

Example 4.4.7 (Tangent bundle)**.** Let M be a Riemannian manifold of class C^r with $r \geq 3$. Then we may identify the tangent bundle TM with the cotangent bundle T^*M through the map $\Xi : TM \to T^*M$ that maps each point (q, v) with $v \in T_qM$ to the point (q, p) with $p \in T_q^*M$ defined by

$$p : T_qM \to \mathbb{R}, \quad p(w) = v \cdot_q w.$$

Indeed, Ξ is a diffeomorphism and it maps fibers of TM to fibers of T^*M, preserving the base point. In particular, we may use Ξ to transport the symplectic form θ^* in Example 4.4.6 to a symplectic form θ in TM:

$$\theta_{(q,v)}\big(w_1, w_2\big) = \theta^*_{\Xi(q,v)}\big(D\Xi(q,v)w_1, D\Xi(q,v)w_2\big)$$

for any $w_1, w_2 \in T_{(q,v)}(TM)$. It is clear from the construction that, unlike θ^*, this form θ depends on the Riemannian metric in M.

By analogy with the case of flows, we call a transformation f_0 *integrable* if there exist coordinates $(q, p) \in \mathbb{T}^d \times B^d$ such that $f_0(q, p) = (q + \omega(p), p)$ for every (q, p). Moreover, we say that f_0 is *non-degenerate* if

$$\text{the map } p \mapsto \omega(p) \text{ is a local diffeomorphism.} \tag{4.4.7}$$

Theorem 4.4.8. *Let f_0 be a non-degenerate integrable transformation of class C^∞. Then there exists a neighborhood $\mathcal{V}$ of f_0 in the space $C^\infty(\mathbb{T}^d \times B^d, \mathbb{R}^d)$ such that for every symplectic transformation[2] $f \in \mathcal{V}$ there exists a compact set $K \subset \mathbb{T}^d \times B^d$ satisfying:*

(i) K is a union of differentiable tori of the form $\{(q, u(q)) : q \in \mathbb{T}^d\}$, each of which is invariant under f;
(ii) the restriction of the transformation f to each of these tori is conjugate to a translation on $\mathbb{T}^d$;
(iii) the set K has positive volume and, in fact, the volume of the complement converges to zero when $f \to f_0$.

In particular, the transformation f cannot be ergodic.

Just as in the previous (continuous time) situation, the set K is formed by tori restricted to which the dynamics is conjugate to a Diophantine rotation.

Theorems 4.4.4 and 4.4.8 extend to systems of class C^r with r finite but sufficiently large, depending on the dimension. For example, the version of Theorem 4.4.8 for $d = 1$ is true for $r > 3$ and false for $r < 3$; in the boundary case $r = 3$, parts (i) and (ii) of the theorem remain valid but part (iii) does not.

The notion of Hamiltonian flow extends to any symplectic manifold (M, θ), as follows. Let $H : M \to \mathbb{R}$ be a function of class C^2 and $dH(z) : T_zM \to \mathbb{R}$ denote its derivative at each point $z \in M$. By the definition of symplectic form,

[2] Relative to the canonical symplectic form (4.4.6).

each $\theta_z : T_zM \times T_zM \to \mathbb{R}$ is a non-degenerate alternate 2-form. Hence, there exists exactly one vector $X_H(z) \in T_zM$ such that

$$\theta_z(X_H(z), v) = dH(z)v \quad \text{for every } v \in T_zM.$$

The map $z \mapsto X_H(z)$ is a vector field of class C^1 on the manifold M. This is called the *Hamiltonian vector field* associated with H. The corresponding flow, given by the differential equation

$$\frac{dz}{dt} = X_H(z), \tag{4.4.8}$$

is the *Hamiltonian flow* associated with H. We leave it to the reader to check that (4.4.8) corresponds precisely to the Hamilton–Jacobi equations (4.4.1) when $M = \mathbb{R}^{2d}$ and θ is the symplectic form in Example 4.4.5.

4.4.3 Elliptic periodic points

The ideas behind the results stated in the previous section may be used to describe the behavior of conservative systems in the neighborhood of elliptic periodic points. Let us explain this briefly, starting with the symplectic case in dimension 2.

When M is a surface, the notions of symplectic form and area form coincide. Thus, a differentiable transformation $f : M \to M$ is symplectic if and only if it preserves area. Let $\zeta \in M$ be an *elliptic* fixed point, that is, such that the eigenvalues of $Df(\zeta)$ are in the unit circle. Let λ and $\bar{\lambda}$ be the eigenvalues. We say that the fixed point ζ is *non-degenerate* if $\lambda^k \neq 1$ for every $1 \leq k \leq 4$. Then, by the Birkhoff normal form theorem (see Arnold [Arn78, Appendix 7]), there exist canonical coordinates $(x, y) \in \mathbb{R}^2$ in the neighborhood of the fixed point, with $\zeta = (0, 0)$, such that the transformation f has the form:

$$f(\theta, \rho) = (\theta + \omega_0 + \omega_1 \rho, \rho) + R(\theta, \rho) \quad \text{with } |R(\theta, \rho)| \leq C|\rho|^2, \tag{4.4.9}$$

where $(\theta, \rho) \in S^1 \times \mathbb{R}$ are the "polar" coordinates defined by

$$x = \sqrt{\rho}\cos 2\pi\theta \quad \text{and} \quad y = \sqrt{\rho}\sin 2\pi\theta.$$

Observe that the *normal form* $f_0 : (\theta, \rho) \mapsto (\theta + \omega_0 + \omega_1\rho, \rho)$ is integrable. Moreover, f_0 satisfies the twist condition (4.4.7) as long as $\omega_1 \neq 0$ (this condition does not depend on the choice of the canonical coordinates, just on the transformation f). Then one may apply the methods of Theorem 4.4.8 to conclude that there exists a set K with positive area that is formed by invariant circles with *Diophantine rotation numbers*, that is, such that the restriction of f to each of these circles is conjugate to a Diophantine rotation. Even more, the fixed point ζ is a density point of this set:

$$\lim_{r \to 0} \frac{m(B(\zeta, r) \setminus K)}{m(B(\zeta, r))} = 0,$$

where $B(\zeta, r)$ represents the ball of radius $r > 0$ around ζ.

We will refer to points ζ as in the previous paragraph as *generic elliptic fixed points*. An important consequence of what we just said is that *generic elliptic fixed points of area-preserving transformations are stable:* the trajectory of any point close to ζ remains close to ζ for all times, as it is "trapped" on the inside of some small invariant circle. This feature does not extend to higher dimensions, as we will explain shortly.

Still in dimension two, we want to mention other important dynamical phenomena that take place in the neighborhood of generic elliptic fixed points. Let us start by presenting a very useful tool, known as the *Poincaré–Birkhoff fixed point theorem* or *Poincaré last theorem*. The statement was proposed by Poincaré, who also presented some special cases, a few months before his death; the general case was proved by Birkhoff [Bir13] in the following year.

Let $A = S^1 \times [a,b]$, with $0 < a < b$, and let $f : A \to A$ be a homeomorphism that preserves each of the boundary components of the annulus A. We say that f is a *twist homeomorphism* if it rotates the two boundary components in opposite senses or, more precisely, if there exists some lift $F : \mathbb{R} \times [a,b] \to \mathbb{R} \times [a,b]$, $F(\theta,\rho) = (\Theta(\theta,\rho), R(\theta,\rho))$, of the map f to the universal cover of the annulus, such that

$$\big[\Theta(\theta,a) - \theta\big]\big[\Theta(\theta,b) - \theta\big] < 0 \quad \text{for every } \theta \in \mathbb{R}. \tag{4.4.10}$$

Theorem 4.4.9 (Poincaré–Birkhoff fixed point). *If $f : A \to A$ is a twist homeomorphism that preserves area then f admits at least two fixed points in the interior of A.*

As mentioned previously, every generic elliptic fixed point ζ is accumulated by invariant circles with Diophantine rotation numbers. Any two such disks bound an annulus around ζ. Applying Theorem 4.4.9 (or, more precisely, its corollary in Exercise 4.4.6) one gets that any such annulus contains, at least, a pair of periodic orbits with the same period.

In a sense, these pairs of periodic orbits are what is left of the invariant circles of the normal form f_0 with *rational* rotation numbers, which are usually destroyed by the addition of the term R in (4.4.9). Their periods go to infinity when one approaches ζ. Generically, one of these periodic orbits is hyperbolic (saddle points) and the other one is elliptic. An example is sketched in Figure 4.3: the elliptic fixed point ζ is surrounded by a hyperbolic periodic orbit and an elliptic periodic orbit, marked with the letters p and q, respectively, both with period 4. Two invariant circles around ζ are also represented.

The Swiss mathematician Eduard Zehnder proved that, generically, the hyperbolic periodic orbits exhibit transverse homoclinic points, that is, their stable manifolds and unstable manifolds intersect transversely, as depicted in Figure 4.3. This implies that the geometry of the stable manifolds and unstable manifolds is extremely complex. Moreover, the elliptic periodic orbits satisfy the genericity conditions mentioned previously. This means that all

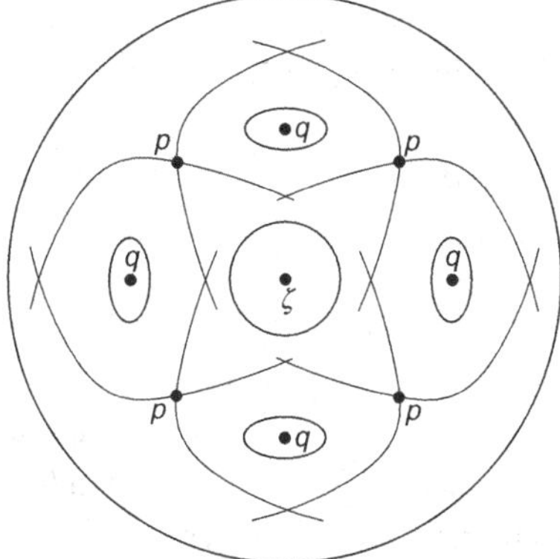

Figure 4.3. Invariant circles, periodic orbits and homoclinic intersections in the neighborhood of a generic elliptic fixed point

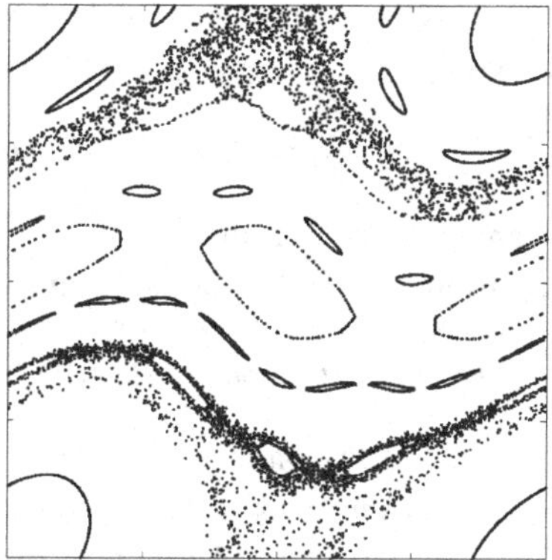

Figure 4.4. Computational evidence for the presence of invariant circles, elliptic islands and transverse homoclinic intersections

the dynamical complexity that we are describing in the neighborhood of ζ is reproduced in the neighborhood of each one of these "satellite" elliptic orbits (which have their own "satellites", etc.).

Moreover, a theory developed by the French physicist Serge Aubry and the American mathematician John Mather shows that ζ is also accumulated by certain infinite, totally disconnected invariant sets, restricted to which the transformation f is minimal (all the orbits are dense). In a sense, these *Aubry–Mather sets* are a souvenir of the invariant circles of the normal form f_0 with irrational *non-Diophantine* rotation numbers that are also typically destroyed by the addition of the perturbation term R in (4.4.9).

Figure 4.4 illustrates a good part of what we have been saying. It depicts several computer-calculated trajectories of an area-preserving transformation. The behavior of these trajectories suggests the presence of invariant circles, elliptic satellites with their own invariant circles, and even hyperbolic orbits with associated transverse homoclinic intersections. One can also observe the presence of certain trajectories with "chaotic" behavior, apparently related to those homoclinic intersections.

More generally, let $f : M \to M$ be a symplectic diffeomorphism on a symplectic manifold M of any (even) dimension $2d \geq 2$. We say that a fixed point $\zeta \in M$ is *elliptic* if all the eigenvalues of the derivative $Df(\zeta)$ are in the unit circle. Let $\lambda_1, \bar{\lambda}_1, \dots, \lambda_d, \bar{\lambda}_d$ be those eigenvalues. We say that ζ is *non-degenerate* if $\lambda_1^{k_1} \dots \lambda_d^{k_d} \neq 1$ for every $(k_1, \dots, k_d) \in \mathbb{Z}^d$ with $|k_1| + \cdots + |k_d| \leq 4$ (in particular, the eigenvalues are all distinct). Then, by the Birkhoff normal form theorem (see Arnold [Arn78, Appendix 7]), there exist canonical coordinates $(x_1, \dots, x_d, y_1, \dots, y_d) \in \mathbb{R}^{2d}$ in a neighborhood of ζ such that $\zeta = (0, \dots, 0, 0, \dots, 0)$ and the transformation f has the form

$$f(\theta, \rho) = (\theta + \omega_0 + \omega_1(\rho), \rho) + R(\theta, \rho) \quad \text{with } \|R(\theta, \rho)\| \leq \text{const} \, \|\rho\|^2,$$

where $\omega_0 \in \mathbb{R}^d$, $\omega_1 : \mathbb{R}^d \to \mathbb{R}^d$ is a linear map and $(\theta, \rho) \in \mathbb{T}^d \times \mathbb{R}^d$ are the "polar" coordinates defined by

$$x_j = \sqrt{\rho_j} \cos 2\pi \theta_j \quad \text{and} \quad y_j = \sqrt{\rho_j} \sin 2\pi \theta_j, \quad j = 1, \dots, d.$$

Assuming that ω_1 is an isomorphism (this is yet another generic condition on the transformation f), we have that the normal form

$$f_0 : (\theta, \rho) \mapsto (\theta + \omega_0 + \omega_1(\rho), \rho)$$

is integrable and satisfies the twist condition (4.4.7). Applying the ideas of Theorem 4.4.8, one concludes that ζ is a density point of a set K formed by invariant tori of dimension d, restricted to which the transformation f is conjugate to a Diophantine rotation.

In particular, symplectic transformations with generic elliptic fixed (or periodic) points are never ergodic. Observe, on the other hand, that for $d > 1$ a torus of dimension d does not separate the ambient space M into two connected components. Therefore, the argument we used before to conclude that generic elliptic fixed points on surfaces are stable does not extend to higher dimensions. In fact, it is known that when $d > 1$ elliptic fixed points are usually *unstable:* trajectories starting arbitrarily close to the fixed point may escape from a fixed neighborhood of it. This is related to the phenomenon known as *Arnold diffusion*, which is a very active research topic in this area.

Finally, let us mention that this theory also applies to continuous time conservative systems. We say that a stationary point ζ of a Hamiltonian flow is *elliptic* if all the eigenvalues of the derivative of the vector field at the point ζ are pure imaginary numbers. Arguments similar to those in the discrete time case show that, under generic hypothesis, ζ is a density point of a set formed by invariant tori of dimension d restricted to each of which the Hamiltonian flow is conjugate to a linear flow.

Moreover, there are corresponding results for periodic trajectories of Hamiltonian flows. One way to obtain such results is by considering a cross-section to the flow at some point of the periodic trajectory and applying the previous ideas to the corresponding Poincaré map. In this way one finds that, under generic conditions, elliptic periodic trajectories of Hamiltonian flows are accumulated by sets with positive volume consisting of invariant tori of the flow.

The theory of Kolmogorov–Arnold–Moser has many other applications, in a wide variety of situations in mathematics that go beyond the scope of this book. The reader may find more complete information in the following references: Arnold [Arn78], Bost [Bos86], Yoccoz [Yoc92], de la Llave [dlL93] and Arnold, Kozlov and Neishtadt [AKN06], among others.

4.4.4 Geodesic flows

Let M be a compact Riemannian manifold. Some of the notions that are used here are recalled in Appendix A.4.

It follows from the theory of ordinary differential equations that for each $(x,v) \in TM$ there exists a unique geodesic $\gamma_{x,v} : \mathbb{R} \to M$ of the manifold M such that $\gamma_{x,v}(0) = x$ and $\dot{\gamma}_{x,v}(0) = v$. Moreover, the family of transformations defined by

$$f^t : (x,v) \mapsto (\gamma_{x,v}(t), \dot{\gamma}_{x,v}(t))$$

is a flow on the tangent bundle TM, which is called the *geodesic flow* of M. We denote by T^1M the unit tangent bundle, formed by the pairs $(x,v) \in TM$ with $\|v\| = 1$. The unit tangent bundle is invariant under the geodesic flow.

Equivalently, the geodesic flow may be defined as the Hamiltonian flow in the tangent bundle TM (with the symplectic structure defined in Example 4.4.7) associated with the Hamiltonian function $H(x,v) = \|v\|^2$. So, $(f^t)_t$ preserves the Liouville measure of the tangent bundle.

In this context, the Liouville measure may be described as follows. Every inner product in a finite-dimensional vector space induces a *volume element*[3] in that space, relative to which the cube spanned by any orthonormal basis has volume 1. In particular, the Riemannian metric induces a volume element dv on each tangent space T_xM. Integrating this volume element along M, we get a volume measure dx on the manifold itself. The Liouville measure of TM is given, locally, by the product $dxdv$. Moreover, its restriction m to the unit tangent bundle is given, locally, by the product $dxd\alpha$, where $d\alpha$ is the measure of angle on the unit sphere of T_xM.

The fact that H is a first integral means that the norm $\|v\|$ is constant along trajectories of the flow. In particular, $(f^t)_t$ leaves the unit tangent bundle

[3] That is, a volume form defined up to sign: the sign is not determined because the inner product does not detect the orientation of the vector space.

invariant. Furthermore, the geodesic flow preserves the restriction m of the Liouville measure to T^1M. However, the behavior of geodesic flows is, usually, very different from the dynamics of the almost integrable systems that we described in Section 4.4.2.

For example, the Austrian mathematician Eberhard F. Hopf [Hop39] proved in 1939 that if M is a compact surface with *negative Gaussian curvature* at every point then its geodesic flow is ergodic. Almost three decades later, his theorem was extended to manifolds in any dimension, through the following remarkable result of the Russian mathematician Dmitry Anosov [Ano67]:

Theorem 4.4.10 (Anosov). *Let M be a compact manifold with negative sectional curvature. Then the geodesic flow on the unit tangent bundle is ergodic with respect to the Liouville measure on T^1M.*

Thus, the geodesic flows of manifolds with negative curvature were the first important class of Hamiltonian systems for which the ergodic hypothesis could be validated rigorously.

4.4.5 Anosov systems

There are two fundamental steps in the proof of Theorem 4.4.10. The first one is to show that every geodesic flow on a manifold with negative curvature is *uniformly hyperbolic*. This means that every trajectory γ of the flow is contained in invariant submanifolds $W^s(\gamma)$ and $W^u(\gamma)$ that intersect each other transversely along γ and satisfy:

- every trajectory in $W^s(\gamma)$ is exponentially asymptotic to γ in the future;
- every trajectory in $W^u(\gamma)$ is exponentially asymptotic to γ in the past;

(see Figure 4.5), with exponential convergence rates that are uniform, that is, independent of γ. Moreover, the geodesic flow is transitive. The second main step in the proof of Theorem 4.4.10 consists of showing that every transitive, uniformly hyperbolic flow (or transitive *Anosov flow*) of class C^2 that preserves volume is ergodic. We will comment on this last issue in a little while.

There exists a corresponding notion for discrete time systems: we say that a diffeomorphism $f : N \to N$ on a compact Riemannian manifold is *uniformly*

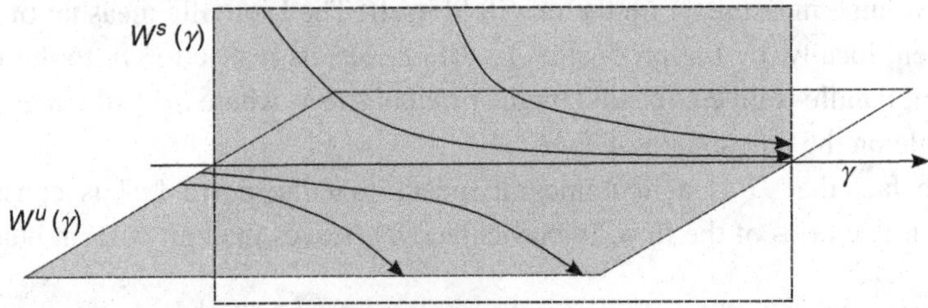

Figure 4.5. Hyperbolic behavior

hyperbolic (or an *Anosov diffeomorphism*) if the tangent space to the manifold at every point $z \in N$ admits a direct sum decomposition $T_zN = E^s_z \oplus E^u_z$ such that the decomposition is invariant under the derivative of f:

$$Df(z)E^s_z = E^s_{f(z)} \quad \text{and} \quad Df(z)E^u_z = E^u_{f(z)} \quad \text{for every } z \in N, \tag{4.4.11}$$

and the derivative contracts E^s_z and expands E^u_z, uniformly:

$$\sup_{z\in N} \|Df(z) \mid E^s_z\| < 1 \quad \text{and} \quad \sup_{z\in N} \|Df(z)^{-1} \mid E^u_z\| < 1 \tag{4.4.12}$$

(for some choice of a norm compatible with the Riemannian metric on M).

One can prove that for each $z \in N$ the set $W^s(z)$ of points whose forward trajectory is asymptotic to the trajectory of z is a differentiable (immersed) submanifold of N tangent to E^s_z at the point z; analogously, the set $W^s(z)$ of points whose backward trajectory is asymptotic to the trajectory of z is a differentiable submanifold tangent to E^u_z at the point z. These submanifolds form foliations (that is, decompositions of N into differentiable submanifolds) that are invariant under the diffeomorphism:

$$f(W^s(z)) = W^s(f(z)) \quad \text{and} \quad f(W^u(z)) = W^u(f(z)) \quad \text{for every } z \in N.$$

We call $W^s(z)$ the *stable manifold* (or *stable leaf*) and $W^u(z)$ the *unstable manifold* (or *unstable leaf*) of the point $z \in M$.

Concerning the second part of the proof of Theorem 4.4.10, the crucial technical tool to prove that every transitive, uniformly hyperbolic diffeomorphism of class C^2 that preserves volume is ergodic is the following theorem of Anosov and Sinai [AS67]:

Theorem 4.4.11 (Absolute continuity). *The stable and unstable foliations of any Anosov diffeomorphism (or flow) of class C^2 are absolutely continuous:*

1. *if $X \subset N$ has zero volume then $X \cap W^s(x)$ has volume zero inside $W^s(x)$ for almost every $x \in N$;*
2. *if $Y \subset \Sigma$ is a zero volume subset of some submanifold Σ transverse to the stable foliation, then the union of the stable manifolds through the points of Y has zero volume in N;*

and analogously for the unstable foliation.

Ergodicity of the system may then be deduced using the Hopf argument, which we introduced in a special case in Section 4.2.6. Let us explain this. Given any continuous function $\varphi : N \to \mathbb{R}$, let E_φ be the set of all points $z \in N$ for which the forward and backward time averages, $\varphi^+(z)$ and $\varphi^-(z)$, are well defined and coincide. This set E_φ has full volume, as we have seen in Corollary 3.2.8. Observe also that φ^+ is constant on each stable manifold and φ^- is constant on each unstable manifold. So, by the first part of Theorem 4.4.11, the intersection $Y_z = W^u(z) \cap E_\varphi$ has full volume in $W^u(z)$ for almost every $z \in N$. Moreover, $\varphi^- = \varphi^+$ is constant on each Y_z. Fix any such z. The transitivity hypothesis implies that the union of all stable manifolds through the points of $W^u(z)$ is the whole ambient manifold N. Hence, using the second part of Theorem 4.4.11, the union of the stable manifolds through the points of Y_z has full volume in N. Clearly, φ^+ is constant on this union. This shows that the time average of every continuous function φ is constant on a full measure set. Hence, f is ergodic.

We close this section by observing that all the known examples of Anosov diffeomorphisms are transitive. The corresponding statement for Anosov flows is false (see Verjovsky [Ver99]). Another open problem in this setting is whether ergodicity still holds when the Anosov system is only of class C^1. It is known (see [Bow75b, RY80]) that in this case the absolute continuity theorem (Theorem 4.4.11) is false, in general.

4.4.6 Billiards

As we have seen in Sections 4.4.2 and 4.4.3, non-ergodic systems are quite common in the realm of Hamiltonian flows and symplectic transformations. However, this fact alone is not sufficient to invalidate the ergodic hypothesis of Boltzmann in the context where it was formulated. Indeed, ideal gases are a special class of systems and it is conceivable that ergodicity could be typical in this more restricted setting, even it is not typical for general Hamiltonian systems.

In the 1960's, the Russian mathematician and theoretical physicist Yakov Sinai [Sin63] conjectured that Hamiltonian systems formed by spherical hard balls that hit each other elastically are ergodic. Hard ball systems (see Example 4.4.13 for a precise definition) had been proposed as a model for the behavior of ideal gases by the American scientist Josiah Willard Gibbs who, together with Boltzmann and Scottish mathematician and theoretical physicist James Clark Maxwell, created the area of statistical mechanics. The *ergodic hypothesis of Boltzmann–Sinai*, as Sinai's conjecture is often referred to, is the main topic in the present section.

In fact, we are going to discuss the problem of ergodicity for somewhat more general systems, called *billiards*, whose formal definition was first given by Birkhoff in the 1930's.

In its simplest form, a billiard is given by a bounded connected domain $\Omega \subset \mathbb{R}^2$, called the *billiard table*, whose boundary $\partial\Omega$ is formed by a finite number of differentiable curves. We call the *corners* those points of the boundary where it fails to be differentiable; by hypothesis, they constitute a finite set $\mathcal{C} \subset \partial\Omega$. One considers a point particle moving uniformly along straight lines inside Ω, with elastic reflections on the boundary. That is, whenever the particle hits $\partial\Omega \setminus \mathcal{C}$ it is reflected in such a way that the angle of incidence equals the angle of reflection. When the particle hits some corner it is absorbed: its trajectory is not defined from then on.

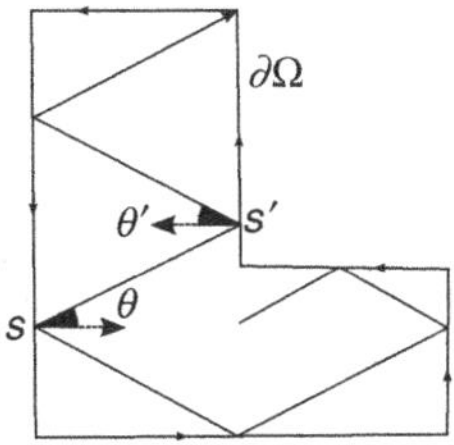

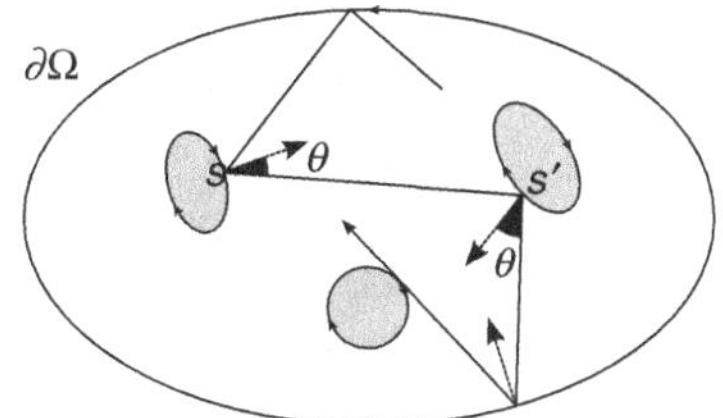

Figure 4.6. Dynamics of billiards

Let us denote by $\mathbf{n}$ the unit vector field orthogonal to the boundary $\partial\Omega$ and pointing to the inside of Ω. It defines an orientation in $\partial\Omega\setminus\mathcal{C}$: a vector t tangent to the boundary is *positive* if the basis $\{t,\mathbf{n}\}$ of $\mathbb{R}^2$ is positive. It is clear that the motion of the particle is characterized completely by the sequence of collisions with the boundary. Moreover, each such collision may be described by the position $s \in \partial\Omega$ and the angle of reflection $\theta \in (-\pi/2,\pi/2)$. Therefore, the evolution of the billiard is governed by the transformation

$$f : (\partial\Omega\setminus\mathcal{C}) \times (-\pi/2,\pi/2) \to \partial\Omega \times (-\pi/2,\pi/2), \tag{4.4.13}$$

that associates with each collision (s,θ) the subsequent one (s',θ'). See Figure 4.6.

In the example on the left-hand side of Figure 4.6 the billiard table is a polygon, that is, the boundary consists of a finite number of straight line segments. The one trajectory represented in the figure hits one of the corners. Nearby trajectories, to either side, collide with distinct boundary segments, with very different angles of incidence. In particular, it is clear that the billiard transformation (4.4.13) cannot be continuous. Discontinuities may occur even in the absence of corners. For example, on the right-hand side of Figure 4.6 the boundary has four connected components, all of which are differentiable curves. Consider the trajectory represented in the figure, tangent to one of the boundary components. Nearby trajectories, to either side, hit with different boundary components. Consequently, the billiard map is discontinuous in this case also.

Example 4.4.12 (Circular billiard table)**.** On the left-hand side of Figure 4.7 we represent a billiard in the unit ball $\Omega \subset \mathbb{R}^2$. The corresponding billiard transformation is given by

$$f : (s,\theta) \mapsto (s - (\pi - 2\theta),\theta).$$

The behavior of this transformation is described geometrically on the right-hand side of Figure 4.7. Observe that f preserves the area measure $ds\,d\theta$ and satisfies the twist condition (4.4.4). Note also that f is integrable (in the sense of Section 4.4.2) and, in particular, the area measure is not ergodic. We will see in a while (Theorem 4.4.14) that every planar billiard preserves a natural measure equivalent to the area measure on $\partial\Omega \times (-\pi/2,\pi/2)$.

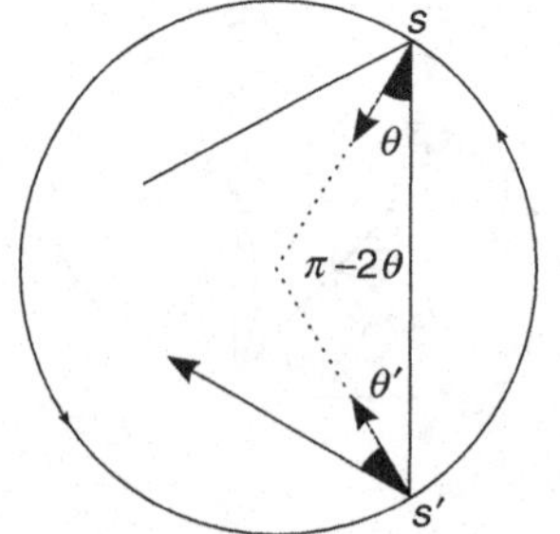

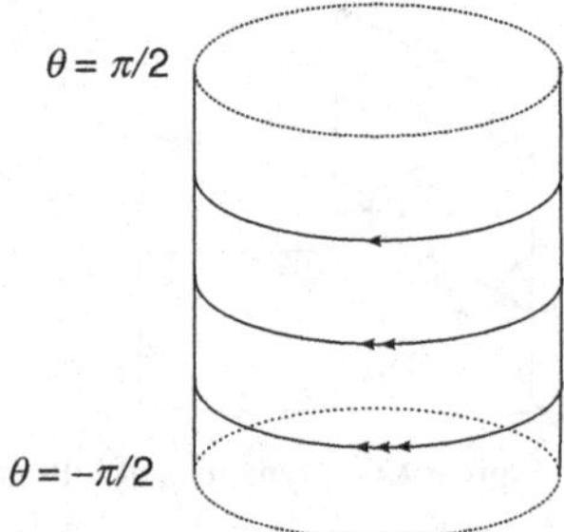

Figure 4.7. Billiard on a circular table

Then, using the previous observations, the KAM theory allows us to prove that billiards with almost circular tables are not ergodic with respect to that invariant measure.

The definition of billiard extends immediately to bounded connected domains Ω in any Euclidean space $\mathbb{R}^d$, $d \geq 1$, whose boundary consists of a finite number of differentiable hypersurfaces intersecting each other along submanifolds with codimension larger than 1. We denote by $\mathcal{C}$ the union of the submanifolds. As before, we endow $\partial\Omega$ with the orientation induced by the unit vector $\mathbf{n}$ orthogonal to the boundary and pointing to the "inside" of Ω. Elastic reflections on the boundary are defined by the following two conditions: (i) the incident trajectory segment, the reflected trajectory segment and the orthogonal vector $\mathbf{n}$ are co-planar and (ii) the angle of incidence equals the angle of reflection. The billiard transformation is defined as in (4.4.13), having as domain

$$\{(s, v) \in (\partial\Omega \setminus \mathcal{C}) \times S^{d-1} : v \cdot \mathbf{n}(s) > 0\}.$$

Even more generally, we may take as a billiard table any bounded connected domain in a Riemannian surface, whose boundary is formed by a finite number of differentiable hypersurfaces intersecting along higher codimension submanifolds. The definitions are analogous, except that the trajectories between consecutive reflections on the boundary are given by segments of geodesics and angles are measured according to the Riemannian metric on the manifolds.

Example 4.4.13 (Ideal gases and billiards)**.** Ideally, a gas is formed by a large number N of molecules ($N \approx 10^{27}$) that move uniformly along straight lines, between collisions, and collide with each other elastically. Check the right-hand side of Figure 4.8. For simplicity, let us assume that the molecules are identical spheres and that they are contained in the torus[4] of dimension

[4] One may replace the torus $\mathbb{T}^d$ by a more plausible container, such as the d-dimensional cube $[0, 1]^d$, for example. However, the analysis is a bit more complicated in that case, because we must take into account the collisions of the balls with the container's walls.

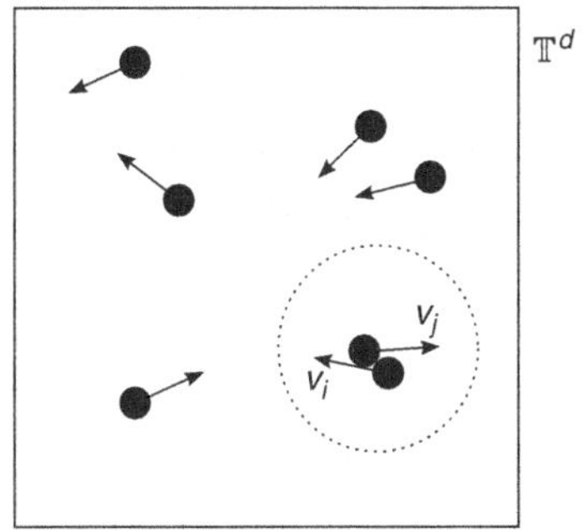

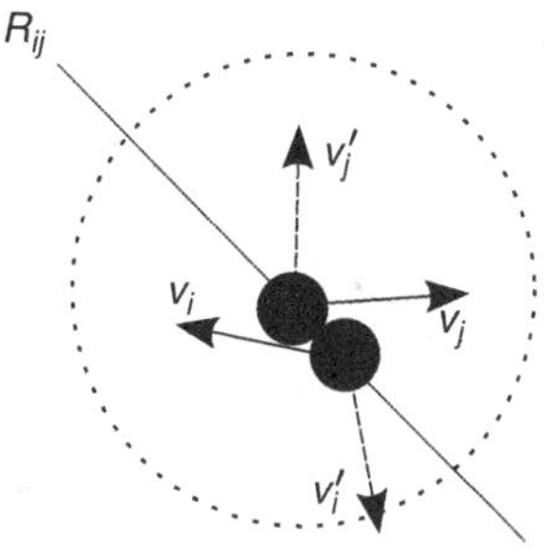

Figure 4.8. Model for an ideal gas

$d \geq 2$. Let us also assume that all the molecules move with constant unit speed. This system can be modelled by a billiard, as follows.

For $1 \leq i \leq N$, denote by $p_i \in \mathbb{T}^d$ the position of the center of the i-th molecule M_i. Let $\rho > 0$ be the radius of each molecule. Then, each state of the system is entirely described by a value of $p = (p_1, \ldots, p_N)$ in the set

$$\Omega = \{p = (p_1, \ldots, p_N) \in \mathbb{T}^{Nd} : \|p_i - p_j\| \geq 2\rho \text{ for every } i \neq j\}$$

(this set is connected, as long as the radius ρ is sufficiently small).

In the absence of collisions, the point p moves along a straight line inside Ω, with constant speed. When two molecules M_i and M_j collide, $\|p_i - p_j\| = 2\rho$ and the velocity vectors change in the following way. Let v_i and v_j be the velocity vectors of the two molecules immediately before the collision and let R_{ij} be the straight line through p_i and p_j. The elasticity hypothesis means that the velocity vectors v_i' and v_j' immediately after the collision are given by (check the right-hand side of Figure 4.8):

(i) the components of v_i and v_i' in the direction of R_{ij} are symmetric and the same is true for v_j and v_j';
(ii) the components of v_i and v_i' in the direction orthogonal to R_{ij} are equal and the same is true for v_j and v_j'.

This means, precisely, that the point p undergoes elastic reflection on the hypersurface $\{p \in \partial\Omega : \|p_i - p_j\| = 2\rho\}$ of the boundary of Ω (see Exercise 4.4.4). Therefore, the motion of the point p corresponds exactly to the evolution of the billiard in the table Ω.

The next result places billiards well inside the domain of interest of ergodic theory. Let ds be the volume measure induced on the boundary $\partial\Omega$ by the Riemannian metric of the ambient manifold; in the planar case (that is, when $\Omega \subset \mathbb{R}^2$), ds is just the arc-length. Denote by $d\theta$ the angle measure on each hemisphere $\{v \in S^{d-1} : v \cdot \mathbf{n}(s) > 0\}$.

Theorem 4.4.14. *The transformation f preserves the measure $\nu = \cos\theta\, ds\, d\theta$ on the domain $\{(s, v) \in \partial\Omega \times S^{d-1} : v \cdot \mathbf{n}(s) > 0\}$.*

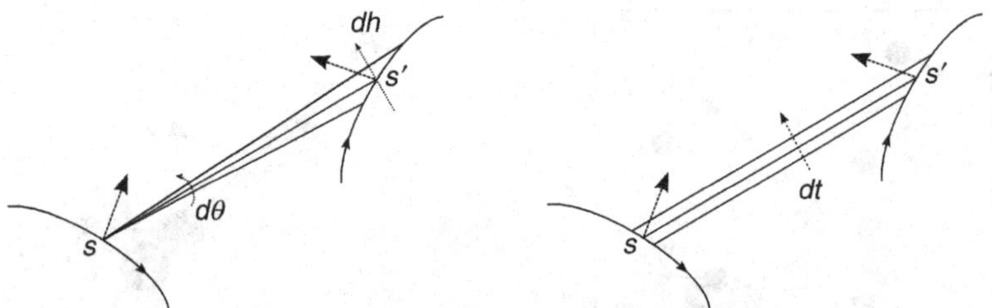

Figure 4.9. Calculating the derivative of the billiard map

In what follows we sketch the proof for planar billiards. The reader should have no trouble checking that all the arguments extend naturally to arbitrary dimension.

Consider any family of trajectories starting from a given boundary point (this means that s is fixed), as represented on the left-hand side of Figure 4.9. Let this family be parameterized by the angle of reflection θ. Denote by $\ell(s,s')$ the length of the line segment connecting s to s'. Then $\ell(s,s')d\theta = dh = \cos\theta' ds'$ and, thus,

$$\frac{\partial s'}{\partial \theta} = \frac{\ell(s,s')}{\cos\theta'}.$$

To calculate the derivative of θ' with respect to θ, observe that the variation of θ' is the sum of two components: the first one corresponds to the variation of θ, whereas the second one arises from the variation of the normal vector $\mathbf{n}(s')$ as the collision point s' varies. By the definition of curvature, this second component is equal to $\kappa(s')ds'$. It follows that $d\theta' = d\theta + \kappa(s')ds'$ and, consequently,

$$\frac{\partial \theta'}{\partial \theta} = 1 + \kappa(s')\frac{\partial s'}{\partial \theta} = 1 + \kappa(s')\frac{\ell(s,s')}{\cos\theta'}.$$

This can be summarized as follows:

$$Df(s,\theta)\frac{\partial}{\partial \theta} = \frac{\ell(s,s')}{\cos\theta'}\frac{\partial}{\partial s'} + \left(1 + \kappa(s')\frac{\ell(s,s')}{\cos\theta'}\right)\frac{\partial}{\partial \theta'}. \tag{4.4.14}$$

Next, consider any family of parallel trajectories, as represented on the right-hand side of Figure 4.9. Let this family be parameterized by the arc-length t in the direction orthogonal to the trajectories. The variations of s and s' along this family are given by $-\cos\theta ds = dt = \cos\theta' ds'$. Since the trajectories all have the same direction, the variations of the angles θ and θ' arise, solely, from the variations of the normal vectors $\mathbf{n}(s)$ and $\mathbf{n}(s')$ as s and s' vary. That is, $d\theta = \kappa(s)ds$ and $d\theta' = \kappa(s')ds'$. Therefore,

$$Df(s,\theta)\left(-\frac{1}{\cos\theta}\frac{\partial}{\partial s} - \frac{\kappa(s)}{\cos\theta}\frac{\partial}{\partial \theta}\right) = \frac{1}{\cos\theta'}\frac{\partial}{\partial s'} + \frac{\kappa(s')}{\cos\theta'}\frac{\partial}{\partial \theta'}. \tag{4.4.15}$$

Let $J(s,\theta)$ be the matrix of the derivative $Df(s,\theta)$ with respect to the bases $\{\partial/\partial s, \partial/\partial\theta\}$ and $\{\partial/\partial s', \partial/\partial\theta'\}$. The relations (4.4.14) and (4.4.15) imply that

$$\det J(s,\theta) = \frac{\begin{vmatrix} \frac{\ell(s,s')}{\cos\theta'} & \frac{1}{\cos\theta'} \\ 1+\kappa(s')\frac{\ell(s,s')}{\cos\theta'} & \frac{\kappa(s')}{\cos\theta'} \end{vmatrix}}{\begin{vmatrix} 0 & -\frac{1}{\cos\theta} \\ 1 & -\frac{\kappa(s)}{\cos\theta} \end{vmatrix}} = \frac{\cos\theta}{\cos\theta'}. \tag{4.4.16}$$

So, by change of variables,

$$\int \varphi\, d\nu = \int \varphi(s',\theta')\cos\theta'\, ds'\, d\theta' = \int \varphi(f(s,\theta))\cos\theta' \frac{\cos\theta}{\cos\theta'} ds\, d\theta$$
$$= \int \varphi(f(s,\theta))\cos\theta\, ds\, d\theta = \int (\varphi\circ f)\, d\nu$$

for every bounded measurable function φ. This proves that f preserves the measure $\nu = \cos\theta ds\, d\theta$, as we stated.

We call a billiard *dispersing* if the boundary of the billiard table is strictly convex at every point, when viewed from the inside. In the planar case, with the orientation conventions that we adopted, this means that the curvature κ is negative at every point. Figure 4.10 presents two examples. In the first one, $\Omega \subset \mathbb{R}^2$ and the boundary is a connected set formed by the union of five differentiable curves. In the second example, $\Omega \subset \mathbb{T}^2$ and the boundary has three connected components, all of which are differentiable and convex.

The class of dispersing billiards was introduced by Sinai in his 1970 article [Sin70]. The denomination "dispersing" refers to the fact that in such billiards any (thin) beam of parallel trajectories becomes divergent upon reflection on the boundary, as illustrated on the left-hand side of Figure 4.10. Sinai observed that dispersing billiards are hyperbolic systems, in a non-uniform sense: invariant sub-bundles E^s_z and E^u_z as in (4.4.11) exist at *almost every* point and, instead of (4.4.12), we have that the derivative is contracting along E^s_z and expanding along E^u_z *asymptotically*, that is, for sufficiently large iterates (depending on the point z).

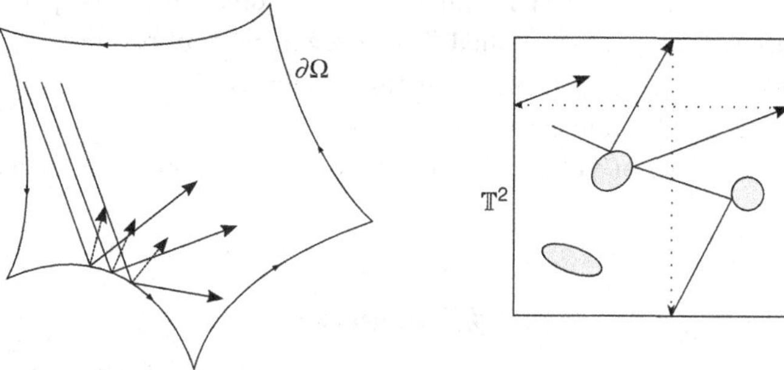

Figure 4.10. Dispersive billiards

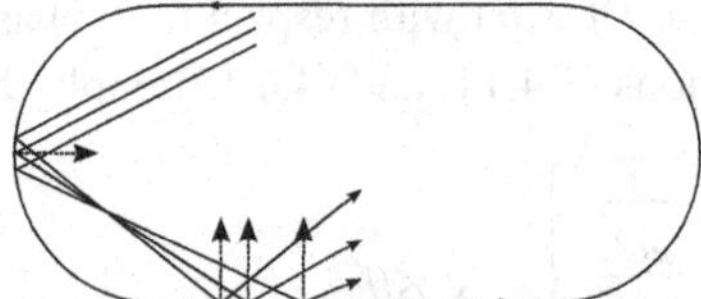

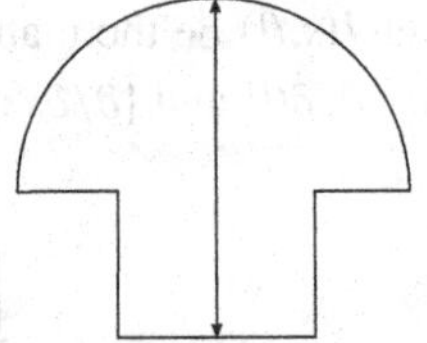

Figure 4.11. Bunimovich stadium and mushroom

The billiards associated with ideal gases (Example 4.4.13) with $N = 2$ molecules are dispersing: it is easy to see that $\{(p_1, p_2) \in \mathbb{R}^{2d} : \|p_1 - p_2\| = 2\rho\}$ is a convex hypersurface. Consequently, these billiards are hyperbolic, in the sense of the previous paragraph. Using a subtle version of the Hopf argument, Sinai proved in [Sin70] that such billiards are ergodic, at least when $d = 2$. This was later extended to arbitrary dimension $d \geq 2$ by Sinai and his student Nikolai Chernov [SC87], still in the case $N = 2$. Thus, dispersing billiards were the first class of billiards for which ergodicity was proven rigorously.

The case $N \geq 3$ of the Boltzmann–Sinai ergodic hypothesis is a lot more difficult because the corresponding billiards are *not* dispersing: the hypersurface

$$\{(p_1, p_2, \ldots, p_N) \in \mathbb{R}^{Nd} : \|p_1 - p_2\| = 2\rho\}$$

has cylinder geometry, with zero curvature along the direction of the variables p_i, $i > 2$. Such billiards are called *semi-dispersing*. Most results in this setting are due to the Hungarian mathematicians András Krámli, Nándor Simányi and Domoko Szász. In [KSS91, KSS92] they proved hyperbolicity and ergodicity for $N = 3$ and also for $N = 4$ assuming that $d \geq 3$. Later, Simányi [Sim02] proved hyperbolicity for the general case: any number of spheres, in any dimension. The problem of ergodicity remains open, in general, although there are many other partial results.

There are now several known examples of ergodic billiards that are not dispersing. This even includes some billiards whose boundary curvature is non-negative at every point. The best-known example is the *Bunimovich stadium*, whose boundary is formed by two semi-circles and two straight line segments. See Figure 4.11. This billiard is hyperbolic, but this property arises from a different mechanism, called *defocusing*: a beam of parallel trajectories reflecting on a concave segment of the billiard table wall starts by focusing, but then gets dispersed. Another interesting example is the *Bunimovich mushroom*: hyperbolic behavior and elliptic behavior coexist on disjoint invariant sets both with positive measure.

4.4.7 Exercises

4.4.1. We say that $\omega \in \mathbb{R}^d$ is τ-Diophantine if it is (c, τ)-Diophantine, that is, if it satisfies (4.4.5), for some $c > 0$. Prove that the set of τ-Diophantine vectors is non-empty if and only if $\tau \geq d-1$. Moreover, show that the set has full Lebesgue measure in $\mathbb{R}^d$ whenever τ is strictly larger than $d-1$.

4.4.2. Consider a billiard on a rectangular table. Check that every trajectory that does not hit any corner either is periodic or is dense in the billiard table.

4.4.3. Show that every billiard on an acute triangle exhibits some periodic trajectory. [Observation: the same is true for right triangles, but the problem is open for obtuse triangles.]

4.4.4. Consider the billiard model for ideal gases in Example 4.4.13. Check that elastic collisions between any two molecules correspond to the elastic reflections of the billiard point particle on the boundary of Ω.

4.4.5. Prove Theorem 4.4.9 under the additional hypothesis that the function $\rho \mapsto \Theta(\theta,\rho)$ is monotone (increasing or decreasing) for every $\theta \in \mathbb{R}$.

4.4.6. Consider the context of Theorem 4.4.9 but, instead of (4.4.10), assume that f rotates the two boundary components of A with different velocities: there exists some lift $F : \mathbb{R} \times [a,b] \to \mathbb{R} \times [a,b]$ and there exist $p,q \in \mathbb{Z}$ with $q \geq 1$, such that, denoting $F^q = (\Theta^q, R^q)$,

$$\big[\Theta^q(\theta,a) - p - \theta\big]\big[\Theta^q(\theta,b) - p - \theta\big] < 0 \quad \text{for every } \theta \in \mathbb{R}. \tag{4.4.17}$$

Show that f has two periodic orbits with period q in the interior of A, at least.

4.4.7. Let Ω be a convex domain in the plane whose boundary $\partial\Omega$ is a differentiable curve. Show that the billiard on Ω has infinitely many periodic orbits.

5
Ergodic decomposition

For convex subsets of vector spaces with finite dimension, it is clear that every element of the convex set may be written as a convex combination of the extremal elements. For example, every point in a triangle may be written as a convex combination of the vertices of the triangle. In view of the results in Section 4.3, it is natural to ask whether a similar property holds in the space of invariant probability measures, that is, whether every invariant measure is a convex combination of ergodic measures.

The ergodic decomposition theorem, which we prove in this chapter (Theorem 5.1.3), asserts that the answer is positive, except that the number of "terms" in this combination is not necessarily finite, not even countable. This theorem has several important applications; in particular, it permits the reduction of the proof of many results to the case when the system is ergodic.

We are going to deduce the ergodic decomposition theorem from another important result from measure theory, the Rokhlin disintegration theorem. The simplest instance of this theorem holds when we have a partition of a probability space (M,μ) into finitely many measurable subsets $P_1,\ldots,P_N$ with positive measure. Then, obviously, we may write μ as a linear combination

$$\mu = \mu(P_1)\mu_1 + \cdots + \mu(P_N)\mu_N$$

of its normalized restrictions $\mu_i(E) = \mu(E \cap P_i)/\mu(P_i)$ to each of the partition elements. The Rokhlin disintegration theorem (Theorem 5.1.11) states that this type of disintegration of the probability measure is possible for any partition $\mathcal{P}$ (possibly uncountable!) that can be obtained as the limit of an increasing sequence of finite partitions.

5.1 Ergodic decomposition theorem

Before stating the ergodic decomposition theorem, let us analyze a couple of examples that help motivate and clarify its content:

Example 5.1.1. Let $f : [0,1] \to [0,1]$ be given by $f(x) = x^2$. The Dirac measures δ_0 and δ_1 are invariant and ergodic for f. It is also clear that $x = 0$ and $x = 1$ are the unique recurrent points for f and so every invariant probability measure μ must satisfy $\mu(\{0,1\}) = 1$. Then, $\mu = \mu(\{0\})\delta_0 + \mu(\{1\})\delta_1$ is a (finite) convex combination of the ergodic measures.

Example 5.1.2. Let $f : \mathbb{T}^2 \to \mathbb{T}^2$ be given by $f(x,y) = (x+y,y)$. The Lebesgue measure m on the torus is preserved by f. Observe that every horizontal circle $H_y = S^1 \times \{y\}$ is invariant under f and the restriction $f : H_y \to H_y$ is the rotation R_y. Let m_y be the Lebesgue measure on H_y. Observe that m_y is also invariant under f. Moreover, m_y is ergodic whenever y is irrational. On the other hand, by the Fubini theorem,

$$m(E) = \int m_y(E)\,dy \quad \text{for every measurable set } E. \tag{5.1.1}$$

The identity is not affected if we consider the integral restricted to the subset of irrational values of y. Then (5.1.1) presents m as an (uncountable) convex combination of ergodic measures.

5.1.1 Statement of the theorem

Let us start by introducing some useful terminology. In what follows, $(M, \mathcal{B}, \mu)$ is a probability space and $\mathcal{P}$ is a partition of M into measurable subsets. We denote by $\pi : M \to \mathcal{P}$ the canonical projection that assigns to each point $x \in M$ the element $\mathcal{P}(x)$ of the partition that contains it. This projection map endows $\mathcal{P}$ with the structure of a probability space, as follows. Firstly, by definition, a subset $\mathcal{Q}$ of $\mathcal{P}$ is measurable if and only if its pre-image

$$\pi^{-1}(\mathcal{Q}) = \text{union of all } P \in \mathcal{P} \text{ that belong to } \mathcal{Q}$$

is a measurable subset of M. It is easy to check that this definition is consistent: the family $\hat{\mathcal{B}}$ of measurable subsets is a σ-algebra in $\mathcal{P}$. Then, we define the *quotient measure* $\hat{\mu}$ by

$$\hat{\mu}(\mathcal{Q}) = \mu(\pi^{-1}(\mathcal{Q})) \quad \text{for every } \mathcal{Q} \in \hat{\mathcal{B}}.$$

Theorem 5.1.3 (Ergodic decomposition). *Let M be a complete separable metric space, $f : M \to M$ be a measurable transformation and μ be an invariant probability measure. Then there exist a measurable set $M_0 \subset M$ with $\mu(M_0) = 1$, a partition $\mathcal{P}$ of M_0 into measurable subsets and a family $\{\mu_P : P \in \mathcal{P}\}$ of probability measures on M, satisfying*

(i) $\mu_P(P) = 1$ *for $\hat{\mu}$-almost every $P \in \mathcal{P}$;*
(ii) $P \mapsto \mu_P(E)$ *is measurable, for every measurable set $E \subset M$;*
(iii) μ_P *is invariant and ergodic for $\hat{\mu}$-almost every $P \in \mathcal{P}$;*
(iv) $\mu(E) = \int \mu_P(E)\,d\hat{\mu}(P)$, *for every measurable set $E \subset M$.*

Part (iv) of the theorem means that μ is a convex combination of the ergodic probability measures μ_P, where the "weight" of each μ_P is determined by the probability measure $\hat{\mu}$. Part (ii) ensures that the integral in (iv) is well defined. Moreover (see Exercise 5.1.3), it implies that the map $\mathcal{P} \to \mathcal{M}_1(M)$ given by $P \mapsto \mu_P$ is measurable.

5.1.2 Disintegration of a measure

We are going to deduce Theorem 5.1.3 from an important result in measure theory, the Rokhlin disintegration theorem, which has many other applications. To state this theorem we need the following notion.

Definition 5.1.4. A *disintegration* of μ with respect to a partition $\mathcal{P}$ is a family $\{\mu_P : P \in \mathcal{P}\}$ of probability measures on M such that, for every measurable set $E \subset M$:

(i) $\mu_P(P) = 1$ for $\hat{\mu}$-almost every $P \in \mathcal{P}$;
(ii) the map $\mathcal{P} \to \mathbb{R}$, defined by $P \mapsto \mu_P(E)$ is measurable;
(iii) $\mu(E) = \int \mu_P(E)\, d\hat{\mu}(P)$.

Recall that the partition $\mathcal{P}$ inherits from M a natural structure of probability space, with a σ-algebra $\hat{\mathcal{B}}$ and a probability measure $\hat{\mu}$. The measures μ_P are called *conditional probabilities* of μ with respect to $\mathcal{P}$.

Example 5.1.5. Let $\mathcal{P} = \{P_1, \dots, P_n\}$ be a finite partition of M into measurable subsets with $\mu(P_i) > 0$ for every i. The quotient measure $\hat{\mu}$ is given by $\hat{\mu}(\{P_i\}) = \mu(P_i)$. Consider the normalized restriction μ_i of μ to each P_i:

$$\mu_i(E) = \frac{\mu(E \cap P_i)}{\mu(P_i)} \quad \text{for every measurable set } E \subset M.$$

Then $\{\mu_1, \dots, \mu_n\}$ is a disintegration of μ with respect to $\mathcal{P}$: it is clear that $\mu(E) = \sum_{i=1}^{n} \hat{\mu}(\{P_i\})\mu_i(E)$ for every measurable set $E \subset M$.

This construction extends immediately to countable partitions. In the next example we treat an uncountable case:

Example 5.1.6. Let $M = \mathbb{T}^2$ and $\mathcal{P}$ be the partition of M into horizontal circles $S^1 \times \{y\}$, $y \in S^1$. Let m be the Lebesgue measure on $\mathbb{T}^2$ and $\hat{m}$ be the Lebesgue measure on S^1. Denote by m_y the Lebesgue measure (arc-length) on each horizontal circle $S^1 \times \{y\}$. By the Fubini theorem,

$$m(E) = \int m_y(E)\, d\hat{m}(y) \quad \text{for every measurable set } E \subset \mathbb{T}^2.$$

Hence, $\{m_y : y \in S^1\}$ is a disintegration of m with respect to $\mathcal{P}$.

The next proposition asserts that disintegrations are essentially unique, when they exist. The hypothesis is very general: it holds, for example, if M is a topological space with a countable basis of open sets and $\mathcal{B}$ is the Borel σ-algebra:

Proposition 5.1.7. *Assume that the σ-algebra $\mathcal{B}$ admits some countable generator. If $\{\mu_P : P \in \mathcal{P}\}$ and $\{\mu'_P : P \in \mathcal{P}\}$ are disintegrations of μ with respect to $\mathcal{P}$, then $\mu_P = \mu'_P$ for $\hat{\mu}$-almost every $P \in \mathcal{P}$.*

Proof. Let Γ be a countable generator of the σ-algebra $\mathcal{B}$ and $\mathcal{A}$ be the algebra generated by Γ. Note that $\mathcal{A}$ is countable, since it coincides with the union of the (finite) algebras generated by the finite subsets of Γ. For each $A \in \mathcal{A}$, consider the sets

$$\mathcal{Q}_A = \{P \in \mathcal{P} : \mu_P(A) > \mu'_P(A)\} \quad \text{and} \quad \mathcal{R}_A = \{P \in \mathcal{P} : \mu_P(A) < \mu'_P(A)\}.$$

If $P \in \mathcal{Q}_A$ then P is contained in $\pi^{-1}(\mathcal{Q}_A)$ and, using property (i) in the definition of disintegration, $\mu_P(A \cap \pi^{-1}(\mathcal{Q}_A)) = \mu_P(A)$. Otherwise, P is disjoint from $\pi^{-1}(\mathcal{Q}_A)$ and, hence, $\mu_P(A \cap \pi^{-1}(\mathcal{Q}_A)) = 0$. Moreover, these conclusions remain valid when one takes μ'_P in the place of μ_P. Hence, using property (iii) in the definition of disintegration,

$$\mu\big(A \cap \pi^{-1}(\mathcal{Q}_A)\big) = \begin{cases} \int_{\mathcal{P}} \mu_P\big(A \cap \pi^{-1}(\mathcal{Q}_A)\big)\, d\hat{\mu}(P) = \int_{\mathcal{Q}_A} \mu_P(A)\, d\hat{\mu}(P) \\[2ex] \int_{\mathcal{P}} \mu'_P\big(A \cap \pi^{-1}(\mathcal{Q}_A)\big)\, d\hat{\mu}(P) = \int_{\mathcal{Q}_A} \mu'_P(A)\, d\hat{\mu}(P). \end{cases}$$

Since $\mu_P(A) > \mu'_P(A)$ for every $P \in \mathcal{Q}_A$, this implies that $\hat{\mu}(\mathcal{Q}_A) = 0$ for every $A \in \mathcal{A}$. A similar argument shows that $\hat{\mu}(\mathcal{R}_A) = 0$ for every $A \in \mathcal{A}$. So,

$$\bigcup_{A \in \mathcal{A}} \mathcal{Q}_A \cup \mathcal{R}_A$$

is also a subset of $\mathcal{P}$ with measure zero. For every P in the complement of this subset, the measures μ_P and μ'_P coincide on the generating algebra $\mathcal{A}$ and, consequently, they coincide on the whole σ-algebra $\mathcal{B}$.

On the other hand, disintegrations may fail to exist:

Example 5.1.8. Let $f : S^1 \to S^1$ be an irrational rotation and $\mathcal{P}$ be the partition of S^1 whose elements are the orbits $\{f^n(x) : n \in \mathbb{Z}\}$ of f. Assume that there exists a disintegration $\{\mu_P : P \in \mathcal{P}\}$ of the Lebesgue measure μ with respect to $\mathcal{P}$. Consider the iterates $\{f_*\mu_P : P \in \mathcal{P}\}$ of the conditional probabilities. Since the partition elements are invariant sets, $f_*\mu_P(P) = \mu_P(P) = 1$ for $\hat{\mu}$-almost every P. It is clear that, given any measurable set $E \subset M$,

$$P \mapsto f_*\mu_P(E) = \mu_P(f^{-1}(E))$$

is a measurable function. Moreover, since μ is an invariant measure,

$$\mu(E) = \mu(f^{-1}(E)) = \int \mu_P(f^{-1}(E))\, d\hat{\mu}(P) = \int f_*\mu_P(E)\, d\hat{\mu}(P).$$

These observations show that $\{f_*\mu_P : P \in \mathcal{P}\}$ is a disintegration of μ with respect to the partition $\mathcal{P}$. By uniqueness (Proposition 5.1.7), it follows that $f_*\mu_P = \mu_P$ for $\hat{\mu}$-almost every P. That is, almost every conditional probability μ_P is invariant. That is a contradiction, because $P = \{f^n(x) : n \in \mathbb{Z}\}$ is an infinite countable set and so there can be no invariant probability measure giving P a positive weight.

The theorem of Rokhlin states that disintegrations always exist if the partition $\mathcal{P}$ is the limit of an increasing sequence of countable partitions and the space M is reasonably well behaved. The precise statement is given in the section that follows.

5.1.3 Measurable partitions

We say that $\mathcal{P}$ is a *measurable partition* if there exists some measurable set $M_0 \subset M$ with full measure such that, restricted to M_0,

$$\mathcal{P} = \bigvee_{n=1}^{\infty} \mathcal{P}_n$$

for some increasing sequence $\mathcal{P}_1 \prec \mathcal{P}_2 \prec \cdots \prec \mathcal{P}_n \prec \cdots$ of countable partitions (see also Exercise 5.1.1). By $\mathcal{P}_i \prec \mathcal{P}_{i+1}$ we mean that every element of $\mathcal{P}_{i+1}$ is contained in some element of $\mathcal{P}_i$ or, equivalently, every element of $\mathcal{P}_i$ coincides with a union of elements of $\mathcal{P}_{i+1}$. Then we say that $\mathcal{P}_i$ is *coarser* than $\mathcal{P}_{i+1}$ or, equivalently, that $\mathcal{P}_{i+1}$ is *finer* than $\mathcal{P}_i$.

Represent by $\bigvee_{n=1}^{\infty} \mathcal{P}_n$ the partition whose elements are the non-empty intersections of the form $\bigcap_{n=1}^{\infty} P_n$ with $P_n \in \mathcal{P}_n$ for every n. Equivalently, this is the coarser partition such that

$$\mathcal{P}_n \prec \bigvee_{n=1}^{\infty} \mathcal{P}_n \quad \text{for every } n.$$

It follows immediately from the definition that every countable partition is measurable. It is easy to find examples of uncountable measurable partitions:

Example 5.1.9. Let $M = \mathbb{T}^2$, endowed with the Lebesgue measure m, and let $\mathcal{P}$ be the partition of M into horizontal circles $S^1 \times \{y\}$. Then $\mathcal{P}$ is a measurable partition. To see that, consider

$$\mathcal{P}_n = \{S^1 \times I(i,n) : i = 1, \ldots, 2^n\},$$

where $I(i,n)$, $1 \le i \le 2^n$ is the segment of $S^1 = \mathbb{R}/\mathbb{Z}$ corresponding to the interval $[(i-1)/2^n, i/2^n) \subset \mathbb{R}$. The sequence $(\mathcal{P}_n)_n$ is increasing and $\mathcal{P} = \bigvee_{n=1}^{\infty} \mathcal{P}_n$.

On the other hand, not all partitions are measurable:

Example 5.1.10. Let $f : M \to M$ be a measurable transformation and μ be an ergodic probability measure. Let $\mathcal{P}$ be the partition of M whose elements are the orbits of f. Then $\mathcal{P}$ is *not* measurable, unless f exhibits an orbit with full measure. Indeed, suppose that there exists a non-decreasing sequence $\mathcal{P}_1 \prec \mathcal{P}_2 \prec \cdots \prec \mathcal{P}_n \prec \cdots$ of countable partitions such that $\mathcal{P} = \bigvee_{n=1}^{\infty} \mathcal{P}_n$ restricted to some full measure subset. This last condition implies that almost every orbit of f is contained in some element P_n of the partition $\mathcal{P}_n$. In other words, up to measure zero, every element of $\mathcal{P}_n$ is invariant under f. By ergodicity, it follows that for every n there exists exactly one $P_n \in \mathcal{P}_n$ such that $\mu(P_n) = 1$. Denote $P = \bigcap_{n=1}^{\infty} P_n$. Then P is an element of the partition $\bigvee_{n=1}^{\infty} \mathcal{P}_n = \mathcal{P}$, that is, P is an orbit of f, and it has $\mu(P) = 1$.

Theorem 5.1.11 (Rokhlin disintegration). *Assume that M is a complete separable metric space and $\mathcal{P}$ is a measurable partition. Then the probability measure μ admits some disintegration with respect to $\mathcal{P}$.*

Theorem 5.1.11 is proven in Section 5.2. The hypothesis that $\mathcal{P}$ is measurable is, actually, also necessary for the conclusion of the theorem (see Exercise 5.2.2).

5.1.4 Proof of the ergodic decomposition theorem

At this point we are going to use Theorem 5.1.11 to prove the ergodic decomposition theorem. Let $\mathcal{U}$ be a countable basis of open sets of M and $\mathcal{A}$ be the algebra generated by $\mathcal{U}$. Note that $\mathcal{A}$ is countable and that it generates the Borel σ-algebra of M. By the ergodic theorem of Birkhoff, for every $A \in \mathcal{A}$ there exists a set $M_A \subset M$ with $\mu(M_A) = 1$ such that the mean sojourn time $\tau(A,x)$ is well defined for every $x \in M_A$. Let $M_0 = \bigcap_{A \in \mathcal{A}} M_A$. Note that $\mu(M_0) = 1$, since the intersection is countable.

Now consider the partition $\mathcal{P}$ of M_0 defined as follows: two points $x, y \in M_0$ are in the same element of $\mathcal{P}$ if and only if $\tau(A,x) = \tau(A,y)$ for every $A \in \mathcal{A}$. We claim that this partition is measurable. To prove that it is so, consider any enumeration $\{A_k : k \in \mathbb{N}\}$ of the elements of the algebra $\mathcal{A}$ and let $\{q_k : k \in \mathbb{N}\}$ be an enumeration of the rational numbers. For each $n \in \mathbb{N}$, consider the partition $\mathcal{P}_n$ of M_0 defined as follows: two points $x, y \in M_0$ are in the same element of $\mathcal{P}_n$ if and only if, given any $i, j \in \{1, \dots, n\}$,

$$\text{either } \tau(A_i,x) \le q_j \text{ and } \tau(A_i,y) \le q_j$$
$$\text{or } \tau(A_i,x) > q_j \text{ and } \tau(A_i,y) > q_j.$$

It is clear that every $\mathcal{P}_n$ is a finite partition, with no more than 2^{n^2} elements. It follows immediately from the definition that x and y are in the same element of $\bigvee_{n=1}^{\infty} \mathcal{P}_n$ if and only if $\tau(A_i,x) = \tau(A_i,y)$ for every i. This means that

$$\mathcal{P} = \bigvee_{n=1}^{\infty} \mathcal{P}_n,$$

which implies our claim.

So, by Theorem 5.1.11, there exists some disintegration $\{\mu_P : P \in \mathcal{P}\}$ of μ with respect to $\mathcal{P}$. Parts (i), (ii) and (iv) of Theorem 5.1.3 are contained in the definition of disintegration. To prove part (iii) it suffices to show that μ_P is invariant and ergodic for $\hat{\mu}$-almost every P, which is what we do now.

Consider the family of probability measures $\{f_*\mu_P : P \in \mathcal{P}\}$. Observe that every $P \in \mathcal{P}$ is an invariant set, since mean sojourn times are constant on orbits. It follows that

$$f_*\mu_P(P) = \mu_P(f^{-1}(P)) = \mu_P(P) = 1.$$

Moreover, given any measurable set $E \subset M$, the function

$$P \mapsto f_*\mu_P(E) = \mu_P(f^{-1}(E))$$

is measurable and, using the fact that μ is invariant under f,

$$\mu(E) = \mu(f^{-1}(E)) = \int \mu_P(f^{-1}(E))\, d\hat{\mu}(P) = \int f_*\mu_P(E)\, d\hat{\mu}(P).$$

This shows that $\{f_*\mu_P : P \in \mathcal{P}\}$ is a disintegration of μ with respect to $\mathcal{P}$. By uniqueness (Proposition 5.1.7), it follows that $f_*\mu_P = \mu_P$ for almost every P.

We are left to prove that μ_P is ergodic for almost every P. Since $\mu(M_0) = 1$, we have that $\mu_P(M_0 \cap P) = 1$ for almost every P. Hence, it is enough to prove that, given any $P \in \mathcal{P}$ and any measurable set $E \subset M$, the mean sojourn time $\tau(E,x)$ is well defined for every $x \in M_0 \cap P$ and is constant on that set. Fix P and denote by $\mathcal{C}$ the class of all measurable sets E for which this holds. By construction, $\mathcal{C}$ contains the generating algebra $\mathcal{A}$. Observe that if $E_1, E_2 \in \mathcal{C}$ with $E_1 \supset E_2$ then $E_1 \setminus E_2 \in \mathcal{C}$:

$$\tau(E_1 \setminus E_2, x) = \tau(E_1, x) - \tau(E_2, x)$$

is well defined and it is constant on $M_0 \cap P$. In particular, if $E \in \mathcal{C}$ then E^c is also in $\mathcal{C}$. Analogously, $\mathcal{C}$ is closed under countable pairwise disjoint unions: if $E_j \in \mathcal{C}$ are pairwise disjoint then

$$\tau\Big(\bigcup_j E_j, x\Big) = \sum_j \tau(E_j, x)$$

is well defined and it is constant on $M_0 \cap P$. It is easy to deduce that $\mathcal{C}$ is a monotone class: given any sequences $A_n, B_n \in \mathcal{C}$, $n \geq 1$ with $A_n \subset A_{n+1}$ and $B_n \supset B_{n+1}$ for every n, the two previous observations yield

$$\bigcup_{n=1}^{\infty} A_n = A_1 \cup \bigcup_{n=1}^{\infty} (A_{n+1} \setminus A_n) \in \mathcal{C} \quad \text{and} \quad \bigcap_{n=1}^{\infty} B_n = \Big(\bigcup_{n=1}^{\infty} B_n^c\Big)^c \in \mathcal{C}.$$

By Theorem A.1.18, it follows that $\mathcal{C}$ contains the Borel σ-algebra of M.

This concludes the proof of Theorem 5.1.3 from Theorem 5.1.11.

5.1.5 Exercises

5.1.1. Show that a partition $\mathcal{P}$ is measurable if and only if there exist measurable subsets $M_0, E_1, E_2, \dots, E_n, \dots$ such that $\mu(M_0) = 1$ and, restricted to M_0,

$$\mathcal{P} = \bigvee_{n=1}^{\infty} \{E_n, M \setminus E_n\}.$$

5.1.2. Let μ be an ergodic probability measure for a transformation f. Then μ is also invariant under f^k for any $k \geq 2$. Describe the ergodic decomposition of μ for the iterate f^k.

5.1.3. Let M be a metric space and X be a measurable space. Prove that the following conditions are all equivalent:

(a) the map $\nu : X \to \mathcal{M}_1(M)$, $x \mapsto \nu_x$ is measurable;

(b) the map $X \to \mathbb{R}$, $x \mapsto \int \varphi \, d\nu_x$ is measurable, for every bounded continuous function $\varphi : M \to \mathbb{R}$;

(c) the map $X \to \mathbb{R}$, $x \mapsto \int \psi \, d\nu_x$ is measurable, for every bounded measurable function $\psi : M \to \mathbb{R}$;

(d) the map $X \to \mathbb{R}$, $x \mapsto \nu_x(E)$ is measurable, for every measurable set $E \subset M$.

5.1.4. Prove that if $\{\mu_P : P \in \mathcal{P}\}$ is a disintegration of μ then

$$\int \psi \, d\mu = \int \left(\int \psi \, d\mu_P \right) d\hat{\mu}(P)$$

for every bounded measurable function $\psi : M \to \mathbb{R}$.

5.1.5. Let μ be a probability measure invariant under a measurable transformation $f : M \to M$. Let $\hat{f} : \hat{M} \to \hat{M}$ be the natural extension of f and $\hat{\mu}$ be the lift of μ (Section 2.4.2). Relate the ergodic decomposition of μ to the ergodic decomposition of $\hat{\mu}$.

5.1.6. When M is a compact metric space, we may obtain the ergodic decomposition of an invariant probability measure μ by taking for M_0 the subset of points $x \in M$ such that

$$\mu_x = \lim_n \frac{1}{n} \sum_{j=0}^{n-1} \delta_{f^j(x)}$$

exists in the weak* topology and taking for $\mathcal{P}$ the partition of M_0 defined by $\mathcal{P}(x) = \mathcal{P}(y) \Leftrightarrow \mu_x = \mu_y$. Check the details of this alternative proof of Theorem 5.1.3 for compact metric spaces.

5.1.7. Let $\sigma : \Sigma \to \Sigma$ be the shift map in $\Sigma = \{1, \dots, d\}^{\mathbb{Z}}$. Consider the partition $\mathcal{W}^s$ of Σ into "stable sets"

$$\mathcal{W}^s((a_n)_n) = \{(x_n)_n : x_n = a_n \text{ for every } n \geq 0\}.$$

Given any probability measure μ invariant under σ, let $\{\mu_P : P \in \mathcal{P}\}$ be an ergodic decomposition of μ. Check that $\mathcal{P} \prec \mathcal{W}^s$, restricted to a full measure subset of M.

5.2 Rokhlin disintegration theorem

Now we prove Theorem 5.1.11. Fix any increasing $\mathcal{P}_1 \prec \mathcal{P}_2 \prec \cdots \prec \mathcal{P}_n \prec \cdots$ of countable partitions such that $\mathcal{P} = \bigvee_{n=1}^{\infty} \mathcal{P}_n$ restricted to some full measure set $M_0 \subset M$. As before, we use $\mathcal{P}_n(x)$ to denote the element of $\mathcal{P}_n$ that contains a given point $x \in M$.

5.2.1 Conditional expectations

Let $\psi : M \to \mathbb{R}$ be any bounded measurable function. For each $n \geq 1$, define $e_n(\psi) : M \to \mathbb{R}$ as follows:

$$e_n(\psi,x) = \begin{cases} \dfrac{1}{\mu(\mathcal{P}_n(x))} \displaystyle\int_{\mathcal{P}_n(x)} \psi \, d\mu & \text{if } \mu(\mathcal{P}_n(x)) > 0 \\ 0 & \text{otherwise.} \end{cases} \tag{5.2.1}$$

Since the partitions $\mathcal{P}_n$ are countable, the second case of the definition corresponds to a subset of points with total measure zero. Observe also that $e_n(\psi)$ is constant on each $P_n \in \mathcal{P}_n$; let us denote by $E_n(\psi, P_n)$ the value of that constant. Then,

$$\int \psi \, d\mu = \sum_{P_n} \int_{P_n} \psi \, d\mu = \sum_{P_n} \mu(P_n) E_n(\psi, P_n) = \int e_n(\psi) \, d\mu \tag{5.2.2}$$

for every $n \in \mathbb{N}$ (the sums involve only partition elements $P_n \in \mathcal{P}_n$ with positive measure).

Lemma 5.2.1. *Given any bounded measurable function $\psi : M \to \mathbb{R}$, there exists a subset M_ψ of M with $\mu(M_\psi) = 1$ such that*

(i) $e(\psi,x) = \lim_n e_n(\psi,x)$ exists for every $x \in M_\psi$;
(ii) $e(\psi) : M_\psi \to \mathbb{R}$ is measurable and constant on each $P \in \mathcal{P}$;
(iii) $\int \psi \, d\mu = \int e(\psi) \, d\mu$.

Proof. Initially, suppose that $\psi \geq 0$. For each $\alpha < \beta$, let $S(\alpha,\beta)$ be the set of points $x \in M$ such that

$$\liminf_n e_n(\psi,x) < \alpha < \beta < \limsup_n e_n(\psi,x).$$

It is clear that the sequence $e_n(\psi,x)$ diverges if and only if $x \in S(\alpha,\beta)$ for some pair of rational numbers $\alpha < \beta$. In other words, the limit $e(\psi,x)$ exists if and only if x belongs to the intersection M_ψ of all $S(\alpha,\beta)^c$ with rational $\alpha < \beta$. As this is a countable intersection, in order to prove that $\mu(M_\psi) = 1$ it suffices to show that $\mu(S(\alpha,\beta)) = 0$ for every $\alpha < \beta$. We do this next.

Let α and β be fixed and denote $S = S(\alpha,\beta)$. Given $x \in S$, fix any sequence of integers $1 \leq a_1^x < b_1^x < \cdots < a_i^x < b_i^x < \cdots$ such that

$$e_{a_i^x}(\psi,x) < \alpha \quad \text{and} \quad e_{b_i^x}(\psi,x) > \beta \quad \text{for every } i \geq 1.$$

Define A_i to be the union of the partition elements $A_i(x) = \mathcal{P}_{a_i^x}(x)$ and B_i to be the union of the partition elements $B_i(x) = \mathcal{P}_{b_i^x}(x)$ obtained in this way, for all points $x \in S$. By construction, $S \subset A_{i+1} \subset B_i \subset A_i$ for every $i \geq 1$. In particular, S is contained in the set

$$\tilde{S} = \bigcap_{i=1}^{\infty} B_i = \bigcap_{i=1}^{\infty} A_i.$$

Since the sequence $\mathcal{P}_n$, $n \geq 1$, is monotone increasing, given any two of the sets $A_i(x) = \mathcal{P}_{a_i^x}(x)$ that form A_i, either they are disjoint or one is contained in the other. It follows that the maximal sets $A_i(x)$ are pairwise disjoint and, hence, they constitute a partition of A_i. Hence, adding only over such maximal sets with positive measure,

$$\int_{A_i} \psi \, d\mu = \sum_{A_i(x)} \int_{A_i(x)} \psi \, d\mu \leq \sum_{A_i(x)} \alpha \, \mu(A_i(x)) = \alpha \, \mu(A_i),$$

for every $i \geq 1$. Analogously,

$$\int_{B_i} \psi \, d\mu = \sum_{B_i(x)} \int_{B_i(x)} \psi \, d\mu \geq \sum_{B_i(x)} \beta \, \mu(B_i(x)) = \beta \, \mu(B_i).$$

Since $A_i \supset B_i$ and we are assuming that $\psi \geq 0$, it follows that

$$\alpha \, \mu(A_i) \geq \int_{A_i} \psi \, d\mu \geq \int_{B_i} \psi \, d\mu \geq \beta \, \mu(B_i),$$

for every $i \geq 1$. Taking the limit as $i \to \infty$, we find that $\alpha \, \mu(\tilde{S}) \geq \beta \, \mu(\tilde{S})$. This implies that $\mu(\tilde{S}) = 0$ and, hence, $\mu(S) = 0$. This proves the claim when ψ is non-negative. The general case follows immediately, since we may always write $\psi = \psi^+ - \psi^-$, where $\psi^\pm$ are measurable, non-negative and bounded. Note that $e_n(\psi) = e_n(\psi^+) - e_n(\psi^-)$ for every $n \geq 1$ and, hence, the conclusion of the lemma holds for ψ if it holds for ψ^+ and ψ^-. This ends the proof of claim (i).

The other claims are simple consequences of the definition. The fact that $e(\psi)$ is measurable follows directly from Proposition A.1.31. Since $\mathcal{P}_n$ is coarser than $\mathcal{P}$, it is clear that $e_n(\psi)$ is constant on each $P \in \mathcal{P}$, restricted to a subset of M with full measure. Hence, the same is true for $e(\psi)$. This proves part (ii). Observe also that $|e_n(\psi)| \leq \sup |\psi|$ for every $n \geq 1$. Hence, we may use the dominated convergence theorem to pass to the limit in (5.2.2). In this way, we get part (iii).

We are especially interested in the case when ψ is a characteristic function: $\psi = \mathcal{X}_A$ for some measurable set $A \subset M$. In this case, the definition means that

$$e(\psi, x) = \lim_n \frac{\mu(\mathcal{P}_n(x) \cap A)}{\mu(\mathcal{P}_n(x))}. \tag{5.2.3}$$

We denote by $\mathcal{P}_A$ the subset of elements P of the partition $\mathcal{P}$ that intersect M_ψ. Observe that $\hat{\mu}(\mathcal{P}_A) = 1$. Moreover, we define $E(A) : \mathcal{P}_A \to \mathbb{R}$ by setting

$E(A,P) = e(\psi,x)$ for any $x \in M_\psi \cap P$. Note that $e(\psi) = E(A) \circ \pi$. Hence, the function $E(A)$ is measurable and satisfies:

$$\int \psi \, d\mu = \int e(\psi) \, d\mu = \int E(A) \, d\hat{\mu}. \tag{5.2.4}$$

5.2.2 Criterion for σ-additivity

The hypothesis that the ambient space M is complete separable metric space is used in the proof of the important criterion for σ-additivity that we now state and prove:

Proposition 5.2.2. *Let M be a complete separable metric space and $\mathcal{A}$ be an algebra generated by a countable basis $\mathcal{U} = \{U_k : k \in \mathbb{N}\}$ of open sets of M. Let $\mu : \mathcal{A} \to [0,1]$ be an additive function with $\mu(\emptyset) = 0$. Then μ extends to a probability measure on the Borel σ-algebra of M.*

First, let us outline the proof. We consider the product space $\Sigma = \{0,1\}^{\mathbb{N}}$, endowed with the topology generated by the cylinders

$$[0; a_0, \dots, a_s] = \{(i_k)_{k \in \mathbb{N}} : i_0 = a_0, \dots, i_s = a_s\}, \quad s \geq 0.$$

Note that Σ is compact (Exercise A.1.11). Using the fact that M is a complete metric space, we will show that the map

$$\gamma : M \to \Sigma, \quad \gamma(x) = \left(\mathcal{X}_{U_k}(x)\right)_{k \in \mathbb{N}}$$

is a measurable embedding of M inside Σ. Moreover, the function μ yields an additive function ν defined on the algebra $\mathcal{A}_\Sigma$ generated by the cylinders of Σ. This algebra is compact (Definition A.1.15), since every element is compact. Hence, ν extends to a probability measure on the Borel σ-algebra of Σ; we still represent this extension by ν. We will show that the image $\gamma(M)$ has full measure for ν. Then, the image $\gamma_*^{-1}\nu$ is a probability measure on the Borel σ-algebra of M. Finally, we will check that this probability measure is an extension of the function μ.

Now let us detail these arguments. In what follows, given any set $A \subset M$, we denote $A^1 = A$ and $A^0 = A^c$.

Lemma 5.2.3. *The image $\gamma(M)$ is a Borel subset of Σ.*

Proof. Let $x \in M$ and $(i_k)_k = \gamma(x)$. It is clear that

(A) $\bigcap_{j=0}^{k} U_j^{i_j} \neq \emptyset$ for every $k \in \mathbb{N}$,

since x is in the intersection. Moreover, since $\mathcal{U}$ is a basis of open sets of M,

(B) there exists $k \in \mathbb{N}$ such that $i_k = 1$ and $\operatorname{diam} U_k \leq 1$, and

(C) for every $k \in \mathbb{N}$ such that $i_k = 1$ there exists $l(k) > k$ such that $i_{l(k)} = 1$ and $\bar{U}_{l(k)} \subset U_k$ and $\operatorname{diam} U_{l(k)} \leq \operatorname{diam} U_k / 2$.

Conversely, suppose that $(i_k)_k \in \Sigma$ satisfies conditions (A), (B) and (C). We are going to show that there exists $x \in M$ such that $\gamma(x) = (i_k)_k$. For that, define

$$F_n = \bigcap_{k=0}^{n} V_k,$$

where $V_k = U_k^c$ if $i_k = 0$ and $V_k = \bar{U}_{l(k)}$ if $i_k = 1$. Then $(F_n)_n$ is a decreasing sequence of closed sets. Condition (A) assures that $F_n \neq \emptyset$ for every $n \geq 1$. Conditions (B) and (C) imply that the diameter of F_n converges to zero as $n \to \infty$. Then, since M is a complete metric space, the intersection $\bigcap_n F_n$ contains some point x. By construction, F_n is contained in $\bigcap_{k=0}^{n} U_k^{i_k}$ for every n. It follows that

$$x \in \bigcap_{k=0}^{\infty} U_k^{i_k},$$

that is, $\gamma(x) = (i_k)_k$. In this way, we have shown that the image of γ is perfectly characterized by the conditions (A), (B) and (C).

To conclude the proof it suffices to show that the subset described by each of these conditions may be constructed from cylinders through countable unions and intersections. Given $k \in \mathbb{N}$, let $N(k)$ be the set of $(k+1)$-tuples $(a_0, \dots, a_k)$ in $\{0,1\}$ such that $U_0^{a_0} \cap \cdots \cap U_k^{a_k} \neq \emptyset$. Condition (A) corresponds to the subset

$$\bigcap_{k=0}^{\infty} \bigcup_{(a_0,\dots,a_k)\in N(k)} [0; a_0, \dots, a_k].$$

Let $D = \{k \in \mathbb{N} : \operatorname{diam} U_k \leq 1\}$. Then, condition (B) corresponds to

$$\bigcup_{k\in D} \bigcup_{(a_0,\dots,a_{k-1})} [0; a_0, \dots, a_{k-1}, 1].$$

Finally, given any $k \in \mathbb{N}$, let $L(k)$ be the set of all $l > k$ such that $\bar{U}_l \subset U_k$ and $\operatorname{diam} U_l \leq \operatorname{diam} U_k/2$. Condition (C) corresponds to the subset

$$\bigcap_{k=0}^{\infty} \bigcup_{a_0,\dots,a_{k-1}} \Big([0; a_0, \dots, a_{k-1}, 0] \\ \cup \bigcup_{l\in L(k)} \bigcup_{a_{k+1},\dots,a_{l-1}} [0; a_0, \dots, a_{k-1}, 1, a_{k+1}, \dots, a_{l-1}, 1] \Big).$$

This completes the proof of the lemma.

Corollary 5.2.4. *The map $\gamma : M \to \gamma(M)$ is a measurable bijection whose inverse is also measurable.*

Proof. Given any points $x \neq y$ in M, there exists $k \in \mathbb{N}$ such that U_k contains one of the points but not the other. This ensures that γ is injective. For any $s \geq 0$ and $a_0, \dots, a_s \in \{0,1\}$,

$$\gamma^{-1}([0; a_0, \dots, a_s]) = U_0^{a_0} \cap \cdots \cap U_s^{a_s}. \tag{5.2.5}$$

This implies that γ is measurable, because the cylinders generate the Borel σ-algebra of Σ. Next, observe that

$$\gamma\,(U_0^{a_0}\cap\cdots\cap U_s^{a_s}) = [0;a_0,\ldots,a_s]\cap\gamma\,(M) \tag{5.2.6}$$

for every $s, a_0,\ldots,a_s$. Using Lemma 5.2.3, it follows that $\gamma\,(U_0^{a_0}\cap\cdots\cap U_s^{a_s})$ is a Borel subset of Σ for every $s, a_0,\ldots,a_s$. This proves that the inverse transformation γ^{-1} is measurable.

Now we are ready to prove that μ extends to a probability measure on the Borel σ-algebra of M, as claimed in Proposition 5.2.2. For that, let us consider the algebra $\mathcal{A}_\Sigma$ generated by the cylinders of Σ. Note that the elements of $\mathcal{A}$ are the finite pairwise disjoint unions of cylinders. In particular, every element of $\mathcal{A}_\Sigma$ is compact and, consequently, $\mathcal{A}_\Sigma$ is a compact algebra (Definition A.1.15). Define:

$$\nu([0;a_0,\ldots,a_s]) = \mu\big(U_0^{a_0}\cap\cdots\cap U_s^{a_s}\big), \tag{5.2.7}$$

for every $s\geq 0$ and $a_0,\ldots,a_s$ in $\{0,1\}$. Then ν is an additive function in the set of all cylinders, with values in $[0,1]$. It extends in a natural way to an additive function defined on the algebra $\mathcal{A}_\Sigma$, which we still denote as ν.

It is clear that $\nu(\Sigma)=1$. Moreover, since the algebra $\mathcal{A}_\Sigma$ is compact, we may use Theorem A.1.14 to conclude that the function $\nu:\mathcal{A}_\Sigma\to[0,1]$ is σ-additive. Hence, by Theorem A.1.13, the function ν extends to a probability measure defined on the Borel σ-algebra of Σ. Given any cover $\mathcal{C}$ of $\gamma\,(M)$ by cylinders, it follows from the definition (5.2.7) that

$$\nu\Big(\bigcup_{C\in\mathcal{C}} C\Big) = \mu\Big(\bigcup_{C\in\mathcal{C}}\gamma^{-1}(C)\Big) = \mu(M) = 1.$$

Taking the infimum over all covers, we conclude that $\nu(\gamma\,(M))=1$.

By Corollary 5.2.4, the image $\gamma_*^{-1}\nu$ is a Borel probability measure on M. By definition, and using the relation (5.2.6),

$$\begin{aligned}\gamma_*^{-1}\nu\big(U_0^{a_0}\cap\cdots\cap U_s^{a_s}\big) &= \nu\big(\gamma\,\big(U_0^{a_0}\cap\cdots\cap U_s^{a_s}\big)\big) = \nu\big([0;a_0,\ldots,a_s]\cap\gamma\,(M)\big)\\ &= \nu([0;a_0,\ldots,a_s]) = \mu\big(U_0^{a_0}\cap\cdots\cap U_s^{a_s}\big)\end{aligned}$$

for any $s, a_0,\ldots,a_s$. This implies that $\gamma_*^{-1}\nu$ is an extension of the function $\mu:\mathcal{A}\to[0,1]$. Therefore, the proof of Proposition 5.2.2 is complete.

5.2.3 Construction of conditional measures

Let $\mathcal{U}=\{U_k : k\in\mathbb{N}\}$ be a basis of open sets of M and $\mathcal{A}$ be the algebra generated by $\mathcal{U}$. It is clear that $\mathcal{A}$ generates the Borel σ-algebra of M. Observe also that $\mathcal{A}$ is countable: it coincides with the union of the (finite) algebras generated by the subsets $\{U_k : 0\leq k\leq n\}$, for every $n\geq 1$. Define

$$\mathcal{P}_* = \bigcap_{A\in\mathcal{A}}\mathcal{P}_A.$$

Then $\hat{\mu}(\mathcal{P}_*) = 1$, since the intersection is countable. For each $P \in \mathcal{P}_*$, define

$$\mu_P : \mathcal{A} \to [0,1], \quad \mu_P(A) = E(A,P). \tag{5.2.8}$$

In particular, $\mu_P(M) = E(M,P) = 1$. It is clear that μ_P is an additive function: the definition (5.2.3) gives that

$$A \cap B = \emptyset \quad \Rightarrow \quad E\big(A \cup B, P\big) = E(A,P) + E(B,P) \quad \text{for every } P \in \mathcal{P}_*.$$

By Proposition 5.2.2, it follows that this function extends to a probability measure defined on the Borel σ-algebra of M, which we still denote as μ_P. We are left to check that this family of measures $\{\mu_P : P \in \mathcal{P}_*\}$ satisfies all the conditions in the definition of disintegration (Definition 5.1.4).

Let us start with condition (i). Let $P \in \mathcal{P}_*$ and, for every $n \geq 1$, let P_n be the element of the partition $\mathcal{P}_n$ that contains P. Observe that if $A \in \mathcal{A}$ is such that $A \cap P_n = \emptyset$ for some n then

$$\mu_P(A) = E(A,P) = \lim_m \frac{\mu\big(A \cap P_m\big)}{\mu(P_m)} = 0,$$

since $P_m \subset P_n$ for every $m \geq n$. Fix n. For each $s \geq 0$, let P_n^s be the union of all sets of the form $U_0^{a_0} \cap \cdots \cap U_s^{a_s}$ that intersect P_n. By the previous observation, the cylinders of length $s+1$ that are not in P_n^s have measure zero for μ_P. Therefore, $\mu_P(P_n^s) = 1$ for every $s \geq 0$. Passing to the limit when $s \to \infty$, we conclude that $\mu_P(U) = 1$ for every open set U that contains P_n. Since the measure μ_P is regular (Proposition A.3.2), it follows that $\mu_P(P_n) = 1$. Passing to the limit when $n \to \infty$, we find that $\mu_P(P) = 1$ for every $P \in \mathcal{P}_*$.

Finally, let $\mathcal{C}$ denote the family of all measurable sets $E \subset M$ for which conditions (ii) and (iii) hold. By construction (recall Lemma 5.2.1), given any $A \in \mathcal{A}$, the function $P \mapsto \mu_P(A) = E(A,P)$ is measurable and satisfies

$$\mu(A) = \int E(A,P)\, d\hat{\mu}(P) = \int \mu_P(A)\, d\hat{\mu}(P).$$

This means that $\mathcal{A} \subset \mathcal{C}$. We claim that $\mathcal{C}$ is a monotone class. Indeed, suppose that B is the union of an increasing sequence $(B_j)_j$ of sets in $\mathcal{C}$. Then, by Proposition A.1.31,

$$P \mapsto \mu_P(B) = \sup_j \mu_P(B_j) \quad \text{is a measurable function}$$

and, using the monotone convergence theorem,

$$\mu(B) = \lim_n \mu(B_n) = \lim_n \int \mu_P(B_n)\, d\hat{\mu} = \int \lim_n \mu_P(B_n)\, d\hat{\mu} = \int \mu_P(B)\, d\hat{\mu}.$$

This means that $B \in \mathcal{C}$. Analogously, if B is the intersection of a decreasing sequence of sets in $\mathcal{C}$ then $P \mapsto \mu_P(B)$ is measurable and $\mu(B) = \int \mu_P(B)\, d\hat{\mu}(P)$. That is, $B \in \mathcal{C}$. This proves that $\mathcal{C}$ is a monotone class, as we claimed. By Theorem A.1.18 it follows that $\mathcal{C}$ coincides with the Borel σ-algebra of M.

The proof of Theorem 5.1.11 is complete.

5.2.4 Exercises

5.2.1. Let $\mathcal{P}$ and $\mathcal{Q}$ be measurable partitions of $(M,\mathcal{B},\mu)$ such that $\mathcal{P} \prec \mathcal{Q}$ up to measure zero. Let $\{\mu_P : P \in \mathcal{P}\}$ be a disintegration of μ with respect to $\mathcal{P}$ and, for every $P \in \mathcal{P}$, let $\{\mu_{P,Q} : Q \in \mathcal{Q}, Q \subset P\}$ be a disintegration of μ_P with respect to $\mathcal{Q}$. Let $\pi : \mathcal{Q} \to \mathcal{P}$ be the canonical projection, such that $Q \subset \pi(Q)$ for almost every $Q \in \mathcal{Q}$. Show that $\{\mu_{\pi(Q),Q} : Q \in \mathcal{Q}\}$ is a disintegration of μ with respect to $\mathcal{Q}$.

5.2.2. (Converse to the theorem of Rokhlin) Let M be a complete separable metric space. Show that if $\mathcal{P}$ satisfies the conclusion of Theorem 5.1.11, that is, if μ admits a disintegration with respect to $\mathcal{P}$, then the partition $\mathcal{P}$ is measurable.

5.2.3. Let $\mathcal{P}_1 \prec \cdots \prec \mathcal{P}_n \prec \cdots$ be an increasing sequence of countable partitions such that the union $\bigcup_n \mathcal{P}_n$ generates the σ-algebra $\mathcal{B}$ of measurable sets, up to measure zero. Show that the conditional expectation $e(\psi) = \lim_n e_n(\psi)$ coincides with ψ at almost every point, for every bounded measurable function.

5.2.4. Prove Proposition 2.4.4, using Proposition 5.2.2.

6
Unique ergodicity

This chapter is dedicated to a distinguished class of dynamical systems, characterized by the fact that they admit exactly one invariant probability measure. Initially, in Section 6.1, we give alternative formulations of this property and we analyze the properties of the unique invariant measure.

The relation between unique ergodicity and minimality is another important theme. A dynamical system is said to be *minimal* if every orbit is dense in the ambient space. As we observe in Section 6.2, every uniquely ergodic system is minimal, restricted to the support of the invariant measure, but the converse is not true, in general.

The main construction of uniquely ergodic transformations is algebraic in nature. In Section 6.3 we introduce the notion of the Haar measure of a topological group and we show that every transitive translation on a compact metrizable topological group is minimal and even uniquely ergodic: the Haar measure is the unique invariant probability measure.

In Section 6.4 we present a remarkable application of the idea of unique ergodicity in the realm of arithmetics: the theorem of Hermann Weyl on the equidistribution of polynomial sequences.

Throughout this chapter, unlcss stated otherwise, it is understood that M is a compact metric space and $f : M \to M$ is a continuous transformation.

6.1 Unique ergodicity

We say that a transformation $f : M \to M$ is *uniquely ergodic* if it admits exactly one invariant probability measure. The corresponding notion for flows is defined in precisely the same way. This denomination is justified by the observation that the invariant probability measure μ is necessarily ergodic. Indeed, suppose there existed some invariant set $A \subset M$ with $0 < \mu(A) < 1$. Then the normalized restriction of μ to A, defined by

$$\mu_A(E) = \frac{\mu(E \cap A)}{\mu(A)} \quad \text{for every measurable set } E \subset M,$$

would be an invariant probability measure, different from μ, which would contradict the assumption that f is uniquely ergodic.

Proposition 6.1.1. *The following conditions are equivalent:*

(i) f admits a unique invariant probability measure;
(ii) f admits a unique ergodic probability measure;
(iii) for every continuous function $\varphi : M \to \mathbb{R}$, the sequence of time averages $n^{-1}\sum_{j=0}^{n-1}\varphi(f^j(x))$ converges at every point to a constant;
(iv) for every continuous function $\varphi : M \to \mathbb{R}$, the sequence of time averages $n^{-1}\sum_{j=0}^{n-1}\varphi\circ f^j$ converges uniformly to a constant.

Proof. It is easy to see that (ii) implies (i). Indeed, since invariant measure is a convex combination of ergodic measures (Theorem 5.1.3), if there is a unique ergodic probability measure then the invariant probability measure is also unique. It is clear that (iv) implies (iii), since uniform convergence implies pointwise convergence. To see that (iii) implies (ii), suppose that μ and ν are ergodic probability measures of f. Then, given any continuous function φ : $M \to \mathbb{R}$,

$$\lim_n \frac{1}{n}\sum_{j=0}^{n-1}\varphi(f^j(x)) = \begin{cases} \int \varphi\, d\mu & \text{at } \mu\text{-almost every point} \\ \int \varphi\, d\nu & \text{at } \nu\text{-almost every point.} \end{cases}$$

Since, by assumption, the limit does not depend on the point x, it follows that

$$\int \varphi\, d\mu = \int \varphi\, d\nu$$

for every continuous function φ. Using Proposition A.3.3 we find that $\mu = \nu$.

We are left to prove that (i) implies (iv). Start by recalling that f admits some invariant probability measure μ (by Theorem 2.1). The idea is to show that if (iv) does not hold then there exists some probability measure $\nu \neq \mu$ and, hence, (i) does not hold either. Suppose then that (iv) does not hold, that is, that there exists some continuous function $\varphi : M \to \mathbb{R}$ such that $n^{-1}\sum_{j=0}^{n-1}\varphi\circ f^j$ does not converge uniformly to any constant; in particular, it does not converge uniformly to $\int \varphi\, d\mu$. By definition, this means that there exists $\varepsilon > 0$ such that for every $k \geq 1$ there exist $n_k \geq k$ and $x_k \in M$ such that

$$\left| \frac{1}{n_k}\sum_{j=0}^{n_k-1}\varphi(f^j(x_k)) - \int \varphi\, d\mu \right| \geq \varepsilon. \tag{6.1.1}$$

Let us consider the sequence of probability measures

$$\nu_k = \frac{1}{n_k}\sum_{j=0}^{n_k-1}\delta_{f^j(x_k)}.$$

Since the space $\mathcal{M}_1(M)$ of probability measures on M is compact for the weak* topology (Theorem 2.1.5), up to replacing this sequence by a subsequence, we may suppose that it converges to some probability measure ν on M. By Lemma 2.2.4 applied to the Dirac measure δ_x, the probability measure ν is invariant under f. On the other hand, the fact that $(\nu_k)_k$ converges to ν in the weak* topology implies that

$$\int \varphi \, d\nu = \lim_k \int \varphi \, d\nu_k = \lim_k \frac{1}{n_k} \sum_{j=0}^{n_k-1} \varphi(f^j(x_k)).$$

Then, recalling (6.1.1), we have that

$$\left| \int \varphi \, d\nu - \int \varphi \, d\mu \right| \geq \varepsilon.$$

In particular, $\nu \neq \mu$. This concludes the argument.

6.1.1 Exercises

6.1.1. Give an example of a transformation $f : M \to M$ in a compact metric space such that $(1/n)\sum_{j=0}^{n-1} \varphi \circ f^j$ converges uniformly, for every continuous function $\varphi : M \to \mathbb{R}$, but f is not uniquely ergodic.

6.1.2. Let $f : M \to M$ be a transitive continuous transformation in a compact metric space. Show that if $(1/n)\sum_{j=0}^{n-1} \varphi \circ f^j$ converges uniformly, for every continuous function $\varphi : M \to \mathbb{R}$, then f is uniquely ergodic.

6.1.3. Let $f : M \to M$ be an isometric homeomorphism in a compact metric space M. Show that if μ is an ergodic measure for f then, for every $n \in \mathbb{N}$, the function $\varphi(x) = d(x, f^n(x))$ is constant on the support of μ.

6.2 Minimality

Let $\Lambda \subset M$ be a closed invariant set of $f : M \to M$. We say that Λ is *minimal* if it coincides with the closure of the orbit $\{f^n(x) : n \geq 0\}$ of every point $x \in \Lambda$. We say that the transformation f is minimal if the ambient M is a minimal set.

Recall that the support of a measure μ is the set of all points $x \in M$ such that $\mu(V) > 0$ for every neighborhood V of x. It follows immediately from the definition that the complement of the support is an open set: if $x \notin \operatorname{supp} \mu$ then there exists an open neighborhood V such that $\mu(V) = 0$; then V is contained in the complement of the support. Therefore, $\operatorname{supp} \mu$ is a closed set.

It is also easy to see that the support of any invariant measure is an invariant set, in the following sense: $f(\operatorname{supp} \mu) \subset \operatorname{supp} \mu$. Indeed, let $x \in \operatorname{supp} \mu$ and let V be any neighborhood of $y = f(x)$. Since f is continuous, $f^{-1}(V)$ is a neighborhood of x. Then $\mu(f^{-1}(V)) > 0$, because $x \in \operatorname{supp} \mu$. Hence, using that μ is invariant, $\mu(V) > 0$. This proves that $y \in \operatorname{supp} \mu$.

Proposition 6.2.1. *If $f : M \to M$ is uniquely ergodic then the support of the unique invariant probability measure μ is a minimal set.*

Proof. Suppose that there exists $x \in \operatorname{supp}\mu$ whose orbit $\{f^j(x) : j \geq 0\}$ is not dense in the support of μ. This means that there exists some open subset U of M such that $U \cap \operatorname{supp}\mu$ is non-empty and

$$f^j(x) \notin U \cap \operatorname{supp}\mu \quad \text{for every } j \geq 0. \tag{6.2.1}$$

Let ν be any accumulation point of the sequence of probability measures

$$\nu_n = n^{-1} \sum_{j=0}^{n-1} \delta_{f^j(x)}, \quad n \geq 1$$

with respect to the weak* topology. Accumulation points do exist, by Theorem 2.1.5, and ν is an invariant probability measure, by Lemma 2.2.4. The condition (6.2.1) means that $\nu_n(U) = 0$ for every $n \geq 1$. Hence, using Theorem 2.1.2 (see also part 3 of Exercise 2.1.1) we have that $\nu(U) = 0$. This implies that no point of U is in the support of μ, which contradicts the fact that $U \cap \operatorname{supp}\mu$ is non-empty.

The converse to Proposition 6.2.1 is false in general:

Theorem 6.2.2 (Furstenberg). *There exists some real-analytic diffeomorphism $f : \mathbb{T}^2 \to \mathbb{T}^2$ that is minimal, preserves the Lebesgue measure m on the torus, but is not ergodic for m. In particular, f is not uniquely ergodic.*

In the remainder of this section we give a brief sketch of the proof of this result. A detailed presentation may be found in the original paper of Furstenberg [Fur61], as well as in Mañé [Mañ87]. In Section 7.3.1 we mention other examples of minimal transformations that are not uniquely ergodic.

To prove Theorem 6.2.2, we look for a transformation $f : \mathbb{T}^2 \to \mathbb{T}^2$ of the form $f(x,y) = (x + \alpha, y + \phi(x))$, where α is an irrational number and $\phi : S^1 \to \mathbb{R}$ is a real-analytic function with $\int \phi(x)\, dx = 0$. Note that f preserves the Lebesgue measure on $\mathbb{T}^2$. Let us also consider the map $f_0 : \mathbb{T}^2 \to \mathbb{T}^2$ given by $f_0(x,y) = (x+\alpha, y)$. Note that no orbit of f_0 is dense in $\mathbb{T}^2$ and that the system (f_0, m) is not ergodic.

Let us consider the *cohomological equation*

$$u(x+\alpha) - u(x) = \phi(x). \tag{6.2.2}$$

If ϕ and α are such that (6.2.2) admits some measurable solution $u : S^1 \to \mathbb{R}$ then (f_0, m) and (f, m) are ergodically equivalent (see Exercise 6.2.1) and, consequently, (f, m) is not ergodic. On the other hand, one can show that if (6.2.2) admits no continuous solution then f is minimal (the converse to this fact is Exercise 6.2.2). Therefore, it suffices to find ϕ and α such that the cohomological equation admits a measurable solution but not a continuous solution.

It is convenient to express these conditions in terms of the Fourier expansion $\phi(x) = \sum_{n\in\mathbb{Z}} a_n e^{2\pi i n x}$. To ensure that ϕ is real-analytic it is enough to require that:

$$\text{there exists } \rho < 1 \text{ such that } |a_n| \leq \rho^n \text{ for every } n \text{ sufficiently large.} \tag{6.2.3}$$

Indeed, in that case the series $\sum_{n\in\mathbb{Z}} a_n z^n$ converges uniformly on every corona $\{z \in \mathbb{C} : r \leq |z| \leq r^{-1}\}$ with $r > \rho$. In particular, its sum in the unit circle, which coincides with ϕ, is a real-analytic function. Since we want ϕ to take values in the real line and to have zero average, we must also require:

$$a_0 = 0 \quad \text{and} \quad a_{-n} = \bar{a}_n \text{ for every } n \geq 1. \tag{6.2.4}$$

According to Exercise 6.2.3, the cohomological equation admits a solution in the space $L^2(m)$ if and only if

$$\sum_{n=1}^{\infty} \left| \frac{a_n}{e^{2\pi n i \alpha} - 1} \right|^2 < \infty. \tag{6.2.5}$$

Moreover, the solution is uniquely determined: $u = \sum_{n\in\mathbb{Z}} b_n e^{2\pi i n x}$ with

$$b_n = \frac{a_n}{e^{2\pi i n \alpha} - 1} \quad \text{for every } n \in \mathbb{Z}. \tag{6.2.6}$$

Fejér's theorem (see [Zyg68]) states that if u is a continuous function then the sequence of partial sums of its Fourier expansion converges Cesàro uniformly to u:

$$\frac{1}{n}\sum_{k=1}^{n}\left(\sum_{j=-k}^{k} b_j e^{2\pi i j x}\right) \text{ converges uniformly to } u(x). \tag{6.2.7}$$

Hence, to ensure that u is not continuous it suffices to require:

$$\left(\sum_{j=-k}^{k} b_j\right)_k \text{ is not Cesàro convergent.} \tag{6.2.8}$$

In this way, the problem is reduced to finding α and $(a_n)_n$ that satisfy (6.2.3), (6.2.4), (6.2.5) and (6.2.8). Exercise 6.2.4 hints at the issues involved in the choice of such objects.

6.2.1 Exercises

6.2.1. Show that if u is a measurable solution of the cohomological equation (6.2.2) then $h : \mathbb{T}^2 \to \mathbb{T}^2$, $h(x,y) = (x, y + u(x))$ is an ergodic equivalence between (f_0, m) and (f, m), that is, h is an invertible measurable transformation that preserves the measure m and conjugates the two maps f and f_0. Deduce that (f, m) cannot be ergodic.

6.2.2. Show that if u is a continuous solution of the cohomological equation (6.2.2) then $h : \mathbb{T}^2 \to \mathbb{T}^2$, $h(x,y) = (x, y + u(x))$ is a topological conjugacy between f_0 and f. In particular, f cannot be transitive.

6.2.3. Check that if $u(x) = \sum_{n\in\mathbb{Z}} b_n e^{2\pi inx}$ is a solution of (6.2.2) then

$$b_n = \frac{a_n}{e^{2\pi in\alpha} - 1} \quad \text{for every } n \in \mathbb{Z}. \tag{6.2.9}$$

Moreover, $u \in L^2(m)$ if and only if $\sum_{n=1}^{\infty} |b_n|^2 < \infty$.

6.2.4. We say that an irrational number α is *Diophantine* if there exist $c > 0$ and $\tau > 0$ such that $|q\alpha - p| \geq c|q|^{-\tau}$ for any $p, q \in \mathbb{Z}$ with $q \neq 0$. Show that the condition (6.2.5) is satisfied whenever α is Diophantine and ϕ satisfies (6.2.3).

6.2.5. (Theorem of Gottschalk) Let $f : M \to M$ be a continuous map in a compact metric space M. Show that the closure of the orbit of a point $x \in M$ is a minimal set if and only if $R_\varepsilon = \{n \in \mathbb{Z} : d(x, f^n(x)) < \varepsilon\}$ is a syndetic set for every $\varepsilon > 0$.

6.2.6. Let $f : M \to M$ be a continuous map in a compact metric space M. We say that $x, y \in M$ are *close* if $\inf_n d(f^n(x), f^n(y)) = 0$. Show that if $x \in M$ is such that the closure of its orbit is a minimal set then, for every neighborhood U of x and every point y close to x, there exists an increasing sequence $(n_i)_i$ such that $f^{n_{i_1}+\cdots+n_{i_k}}(x)$ and $f^{n_{i_1}+\cdots+n_{i_k}}(y)$ are in U for any $i_1 < \cdots < i_k$ and $k \geq 1$.

6.2.7. (Theorem of Hindman) A theorem of Auslander and Ellis (see [Fur81, Theorem 8.7]) states that in the conditions of Exercise 6.2.6 the closure of the orbit of every $y \in M$ contains some point x that is close to y and such that the closure of its orbit is a minimal set. Deduce the following refinement of the theorem of van der Waerden: given any decomposition $\mathbb{N} = S_1 \cup \cdots \cup S_q$ of the set of natural numbers into pairwise disjoint sets, there exists j such that S_j contains a sequence $n_1 < \cdots < n_i < \cdots$ such that $n_{i_1} + \cdots + n_{i_k} \in S_j$ for every $k \geq 1$ and any $i_1 < \cdots < i_k$.

6.3 Haar measure

We are going to see that every compact topological group carries a remarkable probability measure, called the *Haar measure*, that is invariant under every translation and every surjective group endomorphism. Assuming that the group is metrizable, every transitive translation is uniquely ergodic, with the Haar measure as the unique invariant probability measure.

6.3.1 Rotations on tori

Fix $d \geq 1$ and a rationally independent vector $\theta = (\theta_1, \ldots, \theta_d)$. As we have seen in Section 4.2.1, the rotation $R_\theta : \mathbb{T}^d \to \mathbb{T}^d$ is ergodic with respect to the Lebesgue measure m on the torus. Our goal now is to show that, in fact, R_θ is uniquely ergodic.

According to Proposition 6.1.1, we only have to show that, given any continuous function $\varphi : \mathbb{T}^d \to \mathbb{R}$, there exists $c_\varphi \in \mathbb{R}$ such that

$$\varphi_n = \frac{1}{n}\sum_{j=0}^{n-1} \varphi \circ R_\theta^j \quad \text{converges to } c_\varphi \text{ at every point.} \tag{6.3.1}$$

Take $c_\varphi = \int \varphi\, dm$. By ergodicity, the sequence $(\varphi_n)_n$ of time averages converges to c_φ at m-almost every point. In particular, $\varphi_n(x) \to c_\varphi$ for a dense subset of values of $x \in \mathbb{T}^d$.

Let d be the distance induced in the torus $\mathbb{T}^d = \mathbb{R}^d/\mathbb{Z}^d$ by the usual norm in $\mathbb{R}^d$: the distance between any two points in the torus is the minimum of the distances between all their representatives in $\mathbb{R}^d$. It is clear that the rotation R_θ preserves that distance:

$$d(R_\theta(x), R_\theta(y)) = d(x,y) \quad \text{for every } x, y \in \mathbb{T}^d.$$

Then, using that φ is continuous, given any $\varepsilon > 0$ we may find $\delta > 0$ such that

$$d(x,y) < \delta \Rightarrow d(R_\theta^j(x), R_\theta^j(y)) < \delta \Rightarrow |\varphi(R_\theta^j(x)) - \varphi(R_\theta^j(y))| < \varepsilon$$

for every $j \geq 0$. Then,

$$d(x,y) < \delta \Rightarrow |\varphi_n(x) - \varphi_n(y)| < \varepsilon \quad \text{for every } n \geq 1.$$

Since ε does not depend on n, this proves that the sequence $(\varphi_n)_n$ is equicontinuous.

This allows us to use the theorem of Ascoli to prove the claim (6.3.1), as follows. Suppose that there exists $\bar{x} \in \mathbb{T}^d$ such that $(\varphi_n(\bar{x}))_n$ does not converge to c_φ. Then there exists $c \neq c_\varphi$ and some subsequence $(n_k)_k$ such that $\varphi_{n_k}(\bar{x})$ converges to c when $k \to \infty$. By the theorem of Ascoli, up to restricting to a subsequence we may suppose that $(\varphi_{n_k})_k$ is uniformly convergent. Let ψ be its limit. Then ψ is a continuous function such that $\psi(x) = c_\varphi$ for a dense subset of values of $x \in \mathbb{T}^d$ but $\psi(\bar{x}) = c$ is different from c_φ. It is clear that such a function does not exist. This contradiction proves our claim that R_θ is uniquely ergodic.

6.3.2 Topological groups and Lie groups

Recall that a *topological group* is a group $(G, \cdot)$ endowed with a topology with respect to which the two operations

$$G \times G \to G,\ (g,h) \mapsto gh \qquad \text{and} \qquad G \to G,\ g \mapsto g^{-1} \tag{6.3.2}$$

are continuous. In all that follows it is assumed that the topology of G is such that every set consisting of a single point is closed. When G is a manifold and the operations in (6.3.2) are differentiable, we say that $(G, \cdot)$ is a *Lie group*. See Exercise 6.3.1.

The Euclidean space $\mathbb{R}^d$ is a topological group, and even a Lie group, relative to addition $+$, and the same holds for the torus $\mathbb{T}^d$. Recall that $\mathbb{T}^d$ is the quotient of $\mathbb{R}^d$ by its subgroup $\mathbb{Z}^d$. This construction may be generalized as follows:

Example 6.3.1. Given any closed normal subgroup H of a topological group G, let G/H be the set of equivalence classes for the equivalence relation defined in G by $x \sim y \Leftrightarrow x^{-1}y \in H$. Denote by xH the equivalence class that contains each $x \in G$. Consider the following group operation in G/H:

$$xH \cdot yH = (x \cdot y)H.$$

The hypothesis that H is a normal subgroup ensures that this operation is well defined. Let $\pi : G \mapsto G/H$ be the canonical projection, given by $\pi(x) = xH$. Consider in G/H the quotient topology, defined in the following way: a function $\psi : G/H \to X$ is continuous if and only if $\psi \circ \pi : G \to X$ is continuous. The hypothesis that H is closed ensures that the points are closed subsets of G/H. It follows easily from the definitions that G/H is a topological group. Recall also that if the group G is abelian then all subgroups are normal.

Example 6.3.2 (Linear group)**.** The set $G = \mathrm{GL}(d,\mathbb{R})$ of invertible real matrices of dimension d is a Lie group for the multiplication of matrices, called *real linear group* of dimension d. Indeed, G may be identified with an open subset of the Euclidean space $\mathbb{R}^{(d^2)}$ and, thus, has a natural structure of a differentiable manifold. Moreover, it follows directly from the definitions that the multiplication of matrices and the inversion map $A \mapsto A^{-1}$ are differentiable with respect to this manifold structure. G has many important Lie subgroups, such as the *special linear group* $\mathrm{SL}(d,\mathbb{R})$, consisting of the matrices with determinant 1, and the *orthogonal group* $\mathrm{O}(d,\mathbb{R})$, formed by the orthogonal matrices.

We call *left-translation* and *right-translation* associated with an element g of the group G, respectively, the maps

$$L_g : G \to G,\ L_g(h) = gh \quad \text{and} \quad R_g : G \to G,\ R_g(h) = hg.$$

An *endomorphism* of G is a continuous map $\phi : G \to G$ that preserves the group operation, that is, such that $\phi(gh) = \phi(g)\phi(h)$ for every $g, h \in G$. When ϕ is an invertible endomorphism, that is, a bijection whose inverse is also an endomorphism, we call it an *automorphism*.

Example 6.3.3. Let $A \in \mathrm{GL}(d,\mathbb{Z})$; in other words, A is an invertible matrix of dimension d with integer coefficients. Then, as we have seen in Section 4.2.5, A induces an endomorphism $f_A : \mathbb{T}^d \to \mathbb{T}^d$. It can be shown that every endomorphism of the torus $\mathbb{T}^d$ is of this form.

A topological group is *locally compact* if every $g \in G$ has some compact neighborhood. For example, every Lie group is locally compact. On the other hand, the additive group of rational numbers, with the topology inherited from the real line, is not locally compact.

The following theorem is the starting point of the ergodic theory of locally compact groups:

Theorem 6.3.4 (Haar). *Let G be a locally compact topological group. Then:*

(i) There exists some Borel measure μ_G on G that is invariant under all left-translations, finite on compact sets and positive on open sets;
(ii) If η is a measure invariant under all left-translations and finite on compact sets then $\eta = c\mu_G$ for some $c > 0$.
(iii) $\mu_G(G) < \infty$ if and only if G is compact.

We are going to sketch the proof of parts (i) and (ii) in the special case when G is a Lie group. It will be apparent that in this case μ_G is a volume measure on G. The proof of part (iii), for any topological group, is proposed in Exercise 6.3.4.

Starting with part (i), let e be the unit element and $d \geq 1$ be the dimension of the Lie group. Consider any inner product $\cdot$ in the tangent space T_eG. For each $g \in G$, represent by $\mathcal{L}_g : T_eG \to T_gG$ the derivative of the left-translation L_g at the point e. Next, consider the inner product defined in T_gG in the following way:

$$u \cdot v = \mathcal{L}_g^{-1}(u) \cdot \mathcal{L}_g^{-1}(v) \quad \text{for every } u, v \in T_gG.$$

It is clear that this inner product depends differentiably on g. Therefore, it defines a Riemannian metric in G. It is also clear from the construction that this metric is invariant under left-translations: noting that $\mathcal{L}_{hg} = DL_h(g)\mathcal{L}_g$, we see that

$$\begin{aligned} DL_h(g)(u) \cdot DL_h(g)(v) &= \mathcal{L}_{hg}^{-1} DL_h(g)(u) \cdot \mathcal{L}_{hg}^{-1} DL_h(g)(v) \\ &= \mathcal{L}_g^{-1}(u) \cdot \mathcal{L}_g^{-1}(v) = u \cdot v \end{aligned}$$

for any $g, h \in G$ and $u, v \in T_gG$. Let μ_G be the volume measure induced by this Riemannian metric. This measure may be characterized in the following way. Given any $x = (x_1, \ldots, x_d)$ in G, consider

$$\rho(x) = \det \begin{pmatrix} g_{1,1}(x) & \cdots & g_{1,d}(x) \\ \ddots & \ddots & \ddots \\ g_{d,1}(x) & \cdots & g_{d,d}(x) \end{pmatrix} \quad \text{where} \quad g_{i,j} = \frac{\partial}{\partial x_i} \cdot \frac{\partial}{\partial x_j}.$$

Then $\mu_G(B) = \int_B |\rho(x)|\, dx_1 \cdots dx_d$, for any measurable set B contained in the domain of the local coordinates. Noting that the function ρ is continuous and non-zero, for every local chart, it follows that μ_G is positive on open sets and finite on compact sets. Moreover, since the Riemannian metric is invariant under left-translations, the measure μ_G is also invariant under left-translations.

Now we move on to discussing part (ii) of Theorem 6.3.4. Let ν any measure as in the statement. Denote by $B(g, r)$ the open ball of center g and radius r, relative to the distance associated with the Riemannian metric. In other words, $B(g, r)$ is the set of all points in G that may be connected to g by some curve of length less than r. Fix $\rho > 0$ such that $\nu(B(e, \rho))$ is finite (such a ρ does exist because G is locally compact and ν is finite on compact sets). We claim that

$$\limsup_{r\to 0} \frac{\nu(B(g,r))}{\mu_G(B(g,r))} \leq \frac{\nu(B(e,\rho))}{\mu_G(B(e,\rho))} \tag{6.3.3}$$

for every $g \in G$. This may be seen as follows.

First, the limit on the left-hand side of the inequality does not depend on g, because both measures are assumed to be invariant under left-translations. Therefore, it is enough to consider the case $g = e$. Let $(r_n)_n$ be any sequence converging to zero and such that:

$$\lim_n \frac{\nu(B(e,r_n))}{\mu_G(B(e,r_n))} = \limsup_{r\to 0} \frac{\nu(B(e,r))}{\mu_G(B(e,r))}. \tag{6.3.4}$$

By the Vitali lemma (Theorem A.2.16), we may find $(g_j)_j$ in $B(e,\rho)$ and $(n_j)_j$ in $\mathbb{N}$ such that

1. the balls $B(g_j, r_{n_j})$ are contained in $B(e,\rho)$ and they are pairwise disjoint;
2. the union of these balls has full μ_G-measure in $B(e,\rho)$.

Moreover, given any $a \in \mathbb{R}$ smaller than the limit in (6.3.4), we may suppose that the integers n_j are sufficiently large that $\nu(B(g_j, r_{n_j})) \geq a\mu_G(B(g_j, r_{n_j}))$ for every j. It follows that

$$\nu(B(e,\rho)) \geq \sum_j \nu(B(g_j, r_{n_j})) \geq \sum_j a\mu_G(B(g_j, r_{n_j})) = a\mu_G(B(e,\rho)).$$

Since a may be taken arbitrarily close to (6.3.4), this proves the claim (6.3.3).

Next, we claim that ν is absolutely continuous with respect to μ_G. Indeed, let b be any number larger than the quotient on the right-hand side of (6.3.3). Given any measurable set $B \subset G$ with $\mu_G(B) = 0$, and given any $\varepsilon > 0$, let $\{B(g_j, r_j) : j\}$ be a cover of B by balls of small radii, such that $\nu(B(g_j, r_j)) \leq b\mu(B(g_j, r_j))$ and $\sum_j \mu_G(B(g_j, r_j)) \leq \varepsilon$. Then,

$$\nu(B) \leq \sum_j \nu(B(g_j, r_j)) \leq b\sum_j \mu(B(g_j, r_j)) \leq b\varepsilon.$$

Since $\varepsilon > 0$ is arbitrary, it follows that $\nu(B) = 0$. Therefore, $\nu \ll \mu_G$, as claimed. Now, by the Lebesgue derivation theorem (Theorem A.2.15),

$$\frac{d\nu}{\mu_G}(g) = \lim_{r\to 0} \frac{1}{\mu(B(g,r))} \int_{B(g,r)} \frac{d\nu}{\mu_G} d\mu_G = \lim_{r\to 0} \frac{\nu(B(g,r))}{\mu(B(g,r))}$$

for μ-almost every $g \in G$. The limit on the left-hand side does not depend on g and, by (6.3.3), it is finite. Let $c \in \mathbb{R}$ be that limit. Then $\nu = c\mu_G$, as stated in part (ii) of Theorem 6.3.4.

In the case when the group G is compact, it follows from Theorem 6.3.4 that there exists a unique probability measure that is invariant under left-translations, positive on open sets and finite on compact sets. This probability measure μ_G is called the *Haar measure* of the group. For example, the normalized Lebesgue measure is the Haar measure on the torus $\mathbb{T}^d$. See also Exercises 6.3.5 and 6.3.6. The Haar measure features some additional properties:

Corollary 6.3.5. *Assume that G is compact. Then the Haar measure μ_G is invariant under right-translations and under every surjective endomorphism of G.*

Proof. Given any $g \in G$, consider the probability measure $(R_g)_*\mu_G$. Observe that $L_h \circ R_g = R_g \circ L_h$ for every $h \in G$. Hence,

$$(L_h)_*(R_g)_*\mu_G = (R_g)_*(L_h)_*\mu_G = (R_g)_*\mu_G.$$

In other words, $(R_g)_*\mu_G$ is invariant under every left-translation. By uniqueness, it follows that $(R_g)_*\mu_G = \mu_G$ for every $g \in G$, as claimed.

Given any surjective endomorphism $\phi : G \to G$, consider the probability $\phi_*\mu_G$. Given any $h \in G$, choose some $g \in \phi^{-1}(h)$. Observe that $L_h \circ \phi = \phi \circ L_g$. Hence,

$$(L_h)_*\phi_*\mu_G = \phi_*(L_g)_*\mu_G = \phi_*\mu_G.$$

In other words, $\phi_*\mu_G$ is invariant under every left-translation. By uniqueness, it follows that $\phi_*\mu_G = \mu_G$, as claimed.

More generally, when we do not assume G to be compact, the argument in Corollary 6.3.5 shows that for every $g \in G$ there exists $\lambda(g) > 0$ such that

$$(L_g)_*\mu_G = \lambda(g)\mu_G.$$

The map $G \to (0,\infty)$, $g \mapsto \lambda(g)$ is a group homomorphism.

6.3.3 Translations on compact metrizable groups

We call a distance d in a topological group G *left-invariant* if it is invariant under every left-translation: $d(L_h(g_1), L_h(g_2)) = d(g_1, g_2)$ for every g_1, g_2, $h \in G$. Analogously, we call a distance *right-invariant* if it is invariant under every right-translation. In this section we always take the group G to be compact and metrizable. We start by observing that it is always possible to choose the distance in G in such a way that it is invariant under all the translations:

Lemma 6.3.6. *If G is a compact metrizable topological group then there exists some distance compatible with the topology of G that is both left-invariant and right-invariant.*

Proof. Let $(U_n)_n$ be a countable basis of neighborhoods of the unit element e of G. By Lemma A.3.4, for every n there exists a continuous function $\varphi_n : G \to [0,1]$ such that $\varphi_n(e) = 0$ and $\varphi_n(z) = 1$ for every $z \in G \setminus U_n$. Define

$$\varphi : G \to [0,1], \quad \varphi(z) = \sum_{n=1}^{\infty} 2^{-n}\varphi_n(z).$$

Then, φ is continuous and $\varphi(e) = 0 < \varphi(z)$ for every $z \neq e$. Now define

$$d(x,y) = \sup\{|\varphi(gxh) - \varphi(gyh)| : (g,h) \in G^2\} \tag{6.3.5}$$

for every $x, y \in G$. The supremum is finite, since we take G to be compact. It is easy to see that d is a distance in G. Indeed, note that $d(x,y) = 0$ means that $\varphi(gxh) = \varphi(gyh)$ for every $g, h \in G$. In particular, taking $g = e$ and $h = y^{-1}$, we get that $\varphi(xy^{-1}) = \varphi(e)$. By the construction of φ, this implies that $x = y$. All the other axioms of the notion of distance follow directly from the definition of d. It is also clear from the definition that d is invariant under both left-translations and right-translations.

We are left to prove that the distance d is compatible with the topology of the group G. It is easy to check that, given any neighborhood V of a point $x \in G$, there exists $\delta > 0$ such that $B(x,\delta) \subset V$. Indeed, since $U = x^{-1}V$ is a neighborhood of $e \in G$, the properties of φ ensure that there exists $\delta > 0$ such that $\varphi(z) \leq 1 - \delta$ for every $z \notin U$. Then, $y \notin V$ implies that $\varphi(x^{-1}y) \leq 1 - \delta$ or, in other words, that $|\varphi(e) - \varphi(x^{-1}y)| \geq \delta$. Taking $g = x^{-1}$ and $h = e$ in the definition (6.3.5), we see that this last inequality implies that $d(x,y) \geq \delta$, that is, $y \notin B(x,\delta)$. Now let us check the converse: given $x \in G$ and $\delta > 0$, there exists some neighborhood V of x contained in $B(x,\delta)$. By continuity, for every pair $(g,h) \in G^2$ there exists an open neighborhood $U \times V \times W$ of (g,x,h) in G^3 such that

$$|\varphi(gxh) - \varphi(g'x'h')| \leq \delta/2 \quad \text{for every } (g',x',h') \in U \times V \times W. \tag{6.3.6}$$

The sets $U \times W$ obtained in this way, with x fixed and g, h variable, form an open cover of G^2. Let $U_i \times W_i$, $i = 1, \ldots, k$ be a finite subcover and V_i, $i = 1, \ldots, k$ be the corresponding neighborhoods of x. Take $V = \bigcap_{i=1}^{k} V_i$ and consider any $y \in V$. Given any $(g,h) \in G^2$, the condition (6.3.6) implies that $|\varphi(gxh) - \varphi(gyh)| \leq \delta/2$. It follows that $d(x,y) \leq \delta/2$ and, consequently, $y \in B(x,\delta)$.

Example 6.3.7. Given a matrix $A \in \mathrm{GL}(d,\mathbb{R})$, denote by $\|A\|$ its operator norm, that is, $\|A\| = \sup\{\|Av\| : \|v\| = 1\}$. Observe that $\|OA\| = \|A\| = \|AO\|$ for every O in the orthogonal group $\mathrm{O}(d,\mathbb{R})$. Define

$$d(A,B) = \log(1 + \|A^{-1}B - \mathrm{id}\| + \|B^{-1}A - \mathrm{id}\|).$$

Then d is a distance in $\mathrm{GL}(d,\mathbb{R})$, invariant under left-translations:

$$d(CA,CB) = \log(1 + \|A^{-1}C^{-1}CB - \mathrm{id}\| + \|B^{-1}C^{-1}CA - \mathrm{id}\|) = d(A,B)$$

for every $C \in \mathrm{GL}(d,\mathbb{R})$. This distance is not invariant under right-translations in $\mathrm{GL}(d,\mathbb{R})$ (Exercise 6.3.3). However, it is right-invariant (and left-invariant) restricted to the orthogonal group $\mathrm{O}(d,\mathbb{R})$: for every $O \in \mathrm{O}(d,\mathbb{R})$,

$$\begin{aligned} d(AO,CO) &= \log(1 + \|O^{-1}A^{-1}BO - \mathrm{id}\| + \|O^{-1}B^{-1}AO - \mathrm{id}\|) \\ &= \log(1 + \|O^{-1}(A^{-1}B - \mathrm{id})O\| + \|O^{-1}(B^{-1}A - \mathrm{id})O\|) \\ &= d(A,B). \end{aligned}$$

Theorem 6.3.8. *Let G be a compact metrizable topological group and let $g \in G$. The following conditions are equivalent:*

(i) L_g is uniquely ergodic;
(ii) L_g is ergodic with respect to μ_G;
(iii) the subgroup $\{g^n : n \in \mathbb{Z}\}$ generated by g is dense in G.

Proof. It is clear that (i) implies (ii). To prove that (ii) implies (iii), consider the invariant distance d given by Lemma 6.3.6. Let H be the closure of $\{g^n : n \in \mathbb{Z}\}$ and consider the continuous function

$$\varphi(x) = \min\{d(x,y); y \in H\}.$$

Observe that this function is invariant under L_g: using that $gH = H$, we get that

$$\begin{aligned} \varphi(x) &= \min\{d(x,y) : y \in H\} = \min\{d(gx,gy) : y \in H\} \\ &= \min\{d(gx,z) : z \in H\} = \varphi(gx) \quad \text{for every } x \in G. \end{aligned}$$

Since H is closed, $\varphi(x) = 0$ if and only if $x \in H$. If $H \neq G$ then $\mu_G(G \setminus H) > 0$, as the Haar measure is positive on open sets. In that case, the function φ is not constant at μ_G-almost every point, which implies that L_g cannot be ergodic with respect to μ_G.

Finally, to prove that (iii) implies (i), let us show that if μ is a probability measure invariant under L_g then $\mu = \mu_G$. For that, it suffices to check that μ is invariant under every left-translation in G. Fix $h \in G$. Since μ is invariant under L_g,

$$\int \varphi(x)\,d\mu(x) = \int \varphi(g^n x)\,d\mu(x)$$

for every $n \in \mathbb{N}$ and every continuous function $\varphi : G \to \mathbb{R}$. On the other hand, the hypothesis ensures that there exists a sequence of natural numbers $n_j \to \infty$ such that $g^{n_j} \to h$. Given any (uniformly) continuous function $\varphi : G \to \mathbb{R}$ and any $\varepsilon > 0$, fix $\delta > 0$ such that $|\varphi(x) - \varphi(y)| < \varepsilon$ whenever $d(x,y) < \delta$. If j is sufficiently large,

$$d(g^{n_j}x, hx) = d(g^{n_j}, h) < \delta \quad \text{for every } x \in G.$$

Hence, $|\varphi(g^{n_j}x) - \varphi(hx)| < \varepsilon$ for every x and, consequently,

$$\left|\int \big(\varphi(x) - \varphi(hx)\big)\,d\mu\right| = \left|\int \big(\varphi(g^{n_j}x) - \varphi(hx)\big)\,d\mu\right| < \varepsilon.$$

Since ε is arbitrary, it follows that $\int \varphi\, d\mu = \int \varphi \circ L_h\, d\mu$ for every continuous function φ and every $h \in G$. This implies that μ is invariant under L_h for every $h \in G$, as claimed.

6.3.4 Odometers

Odometers, or *adding machines*, are mathematical models for the mechanisms that register the distance (number of kilometers) travelled by a car or the amount of electricity (number of energy units) consumed in a house. They come with a dynamic, which consists in advancing the counter by one unit each time. The main difference with respect to real-life odometers is that our idealized counters allow for an infinite number of digits.

Fix any number basis $d \geq 2$, for example $d = 10$, and consider the set $X = \{0, 1, \ldots, d-1\}$, endowed with the discrete topology. Let $M = X^{\mathbb{N}}$ be the set of all sequences $\alpha = (\alpha_n)_n$ with values in X, endowed with the product topology. This topology is metrizable: it is compatible, for instance, with the distance defined in M by

$$d(\alpha, \alpha') = 2^{-N(\alpha,\alpha')} \quad \text{where} \quad N(\alpha, \alpha') = \min\{j \geq 0 : \alpha_j \neq \alpha'_j\}. \qquad (6.3.7)$$

Observe also that M is compact, being the product of compact spaces (theorem of Tychonoff).

Let us introduce in M the following operation of "sum with transport": given $\alpha = (\alpha_n)_n$ and $\beta = (\beta_n)_n$ in M, define $\alpha + \beta = (\gamma_n)_n$ as follows. First,

- if $\alpha_0 + \beta_0 < d$ then $\gamma_0 = \alpha_0 + \beta_0$ and $\delta_1 = 0$;
- if $\alpha_0 + \beta_0 \geq d$ then $\gamma_0 = \alpha_0 + \beta_0 - d$ and $\delta_1 = 1$.

Next, for every $n \geq 1$,

- if $\alpha_n + \beta_n + \delta_n < d$ then $\gamma_n = \alpha_n + \beta_n + \delta_n$ and $\delta_{n+1} = 0$;
- if $\alpha_n + \beta_n + \delta_n \geq d$ then $\gamma_n = \alpha_n + \beta_n + \delta_n - d$ and $\delta_{n+1} = 1$.

The auxiliary sequence $(\delta_n)_n$ corresponds precisely to the transports. The map $+ : M \times M \to M$ defined in this way turns M into an abelian topological group and the distance (6.3.7) is invariant under all the translations (Exercise 6.3.8).

Now consider the "translation by 1" $f : M \to M$ defined by

$$f\big((\alpha_n)_n\big) = (\alpha_n)_n + (1, 0, \ldots, 0, \ldots) = (0, \ldots, 0, \alpha_k + 1, \alpha_{k+1}, \ldots, \alpha_n, \ldots)$$

where $k \geq 0$ is the smallest value of n such that $\alpha_n < d-1$; if there exists no such k, that is, if $(\alpha_n)_n$ is the constant sequence equal to $d-1$, then the image $f((\alpha_n)_n)$ is the constant sequence equal to 0. We leave it to the reader to check that this transformation $f : M \to M$ is uniquely ergodic (Exercise 6.3.9).

It is possible to genralize this construction somewhat, in the following way. Take $M = \prod_{n=0}^{\infty}\{0, 1, \ldots, d_n - 1\}$, where $(d_n)_n$ is any sequence of integer numbers larger than 1. Just as in the previous particular case, this set has the

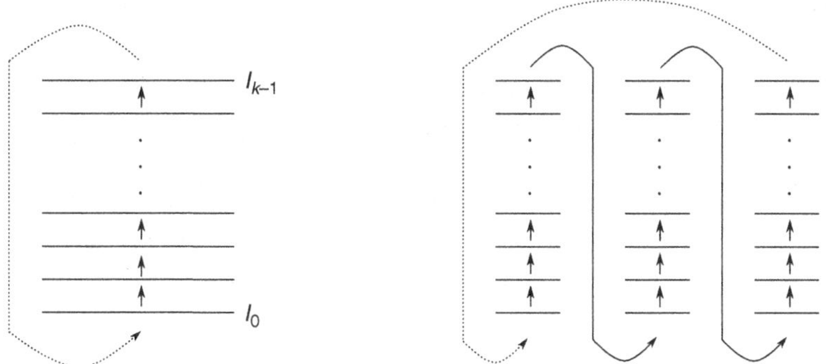

Figure 6.1. Example of the piling method

structure of a metrizable compact abelian group and the "translation by 1" is uniquely ergodic.

Example 6.3.9. A *(simple) pile* in an interval[1] I is an ordered family $\mathcal{S}$ of pairwise disjoint subintervals $I_0, \ldots, I_{k-1}$ with the same length and whose union is I. Write $I_k = I_0$. We associate with $\mathcal{S}$ the transformation $f : I \to I$ whose restriction to each I_j is the translation mapping I_j to I_{j+1}. Graphically, we represent the subintervals "piled up" on top of each other in order: from the *bottom* I_0 to the *top* I_{k-1}. Then f is nothing but the translation "upwards", except at the top of the pile. See the left-hand side of Figure 6.1.

Let us consider a sequence $(\mathcal{S}_n)_n$ of piles in the same interval I, constructed as follows. Fix any integer number $d \geq 2$. Take $\mathcal{S}_0 = \{I\}$. For each $n \geq 1$, take as $\mathcal{S}_n$ the pile obtained by dividing $\mathcal{S}_{n-1}$ into d columns, all with the same width, and piling them up on top of each other. This procedure is described on the right-hand side of Figure 6.1 for $d = 3$. Let $f_n : I \to I$ be the transformation associated with each $\mathcal{S}_n$. We leave it to the reader to show (Exercise 6.3.10) that the sequence $(f_n)_n$ converges at every point to a transformation $f : I \to I$ that preserves the Lebesgue measure. Moreover, this transformation f is uniquely ergodic.

This is only one of the simplest applications of the so-called *piling method*, which is a very effective tool to produce examples with interesting properties. The reader may find a detailed discussion of this method in Section 6 of Friedman [Fri69]. Another application, a bit more elaborate, will be given in Example 8.2.3.

Example 6.3.10 (Substitutions). We are going to mention briefly a construction of a combinatorial nature that generalizes the definition of odometer and provides several other interesting examples of minimal and even uniquely ergodic systems. For more information, including about the relations between

[1] For definiteness, take all intervals to be closed on the left and open on the right.

such systems and the odometer, we recommend the book of Queffélec [Que87] and the paper of Ferenczi, Fisher and Talet [FFT09].

We call a *substitution* in a finite alphabet $\mathcal{A}$ any map associating with each letter $\alpha \in \mathcal{A}$ a word $s(\alpha)$ formed by a finite number of letters of $\mathcal{A}$. A few examples, for $\mathcal{A} = \{0,1\}$: *Thue–Morse substitution* $s(0) = 01$ and $s(1) = 10$; *Fibonacci substitution* $s(0) = 01$ and $s(1) = 0$; *Feigenbaum substitution* $s(0) = 11$ and $s(1) = 10$; *Cantor substitution* $s(0) = 010$ and $s(1) = 111$; and *Chacon substitution* $s(0) = 0010$ and $s(1) = 1$. We may iterate a substitution by defining $s^1(\alpha) = s(\alpha)$ and

$$s^{k+1}(\alpha) = s(\alpha_1)\cdots s(\alpha_n) \quad \text{if} \quad s^k(\alpha) = \alpha_1 \cdots \alpha_n.$$

We call a substitution s *primitive* (or *aperiodic*) if there exists $k \geq 1$ such that for any $\alpha, \beta \in \mathcal{A}$ the word $s^k(\alpha)$ contains the letter β.

Let $\mathcal{A}$ be endowed with the discrete topology and $\Sigma = \mathcal{A}^{\mathbb{N}}$ be the space of all sequences in $\mathcal{A}$, endowed with the product topology. Denote by $S : \Sigma \to \Sigma$ the map induced in that space by a given substitution s: the image of each $(a_0, \ldots, a_n, \ldots) \in \Sigma$ is the sequence of the letters that constitute the word obtained when one concatenates the finite words $s(a_0), \ldots, s(a_n), \ldots$ Suppose that there exists some letter $\alpha_0 \in \mathcal{A}$ such that the word $s(\alpha_0)$ has length larger than 1 and starts with the letter α_0. That is the case for all the examples listed above. Then (Exercise 6.3.11), S admits a unique fixed point $x = (x_n)_n$ with $x_0 = \alpha_0$.

Consider the restriction $\sigma : \mathcal{X} \to \mathcal{X}$ of the shift map $\sigma : \Sigma \to \Sigma$ to the closure $\mathcal{X} \subset \Sigma$ of the orbit $\{\sigma^n(x) : n \geq 0\}$ of the point x. If the substitution s is primitive then $\sigma : \mathcal{X} \to \mathcal{X}$ is minimal and uniquely ergodic (see Section 5 in [Que87]). That holds, for instance, for the Thue–Morse, Fibonacci and Feigenbaum substitutions.

6.3.5 Exercises

6.3.1. Let G be a manifold and $\cdot$ be a group operation in G such that the map $(g,h) \mapsto g \cdot h$ is of class C^1. Show that $g \mapsto g^{-1}$ is also of class C^1.

6.3.2. Let G be a compact topological space such that every point admits a countable basis of neighborhoods and let $\cdot$ be a group operation in G such that the map $(g,h) \mapsto g \cdot h$ is continuous. Show that $g \mapsto g^{-1}$ is also continuous.

6.3.3. Show that the distance d in Example 6.3.7 is not right-invariant.

6.3.4. Prove part (iii) of Theorem 6.3.4: a locally compact group G is compact if and only if its Haar measure is finite.

6.3.5. Identify $\mathrm{GL}(1,\mathbb{R})$ with the multiplicative group $\mathbb{R} \setminus \{0\}$. Check that the measure μ defined on $\mathrm{GL}(1,\mathbb{R})$ by

$$\int_{\mathrm{GL}(1,\mathbb{R})} \varphi \, d\mu = \int_{\mathbb{R}\setminus\{0\}} \frac{\varphi(x)}{|x|} \, dx$$

is both left-invariant and right-invariant. Find a measure invariant under all the translations of $\mathrm{GL}(1,\mathbb{C})$.

6.3.6. Identify $\mathrm{GL}(2,\mathbb{R})$ with $\{(a_{11},a_{12},a_{21},a_{22}) \in \mathbb{R}^4 : a_{11}a_{22} - a_{12}a_{21} \neq 0\}$, in such a way that $\det(a_{11},a_{12},a_{21},a_{22}) = a_{11}a_{22} - a_{12}a_{21}$. Show that the measure μ defined by

$$\int_{\mathrm{GL}(2,\mathbb{R})} \varphi \, d\mu = \int \frac{\varphi(x_{11},x_{12},x_{21},x_{22})}{|\det(x_{11},x_{12},x_{21},x_{22})|^2} dx_{11}dx_{12}dx_{21}dx_{22}$$

is both left-invariant and right-invariant. Find a measure invariant under all the translations of $\mathrm{GL}(2,\mathbb{C})$.

6.3.7. Let G be a compact metrizable group and let $g \in G$. Check that the following conditions are equivalent:

(1) L_g is uniquely ergodic;

(2) L_g is transitive: there is $x \in G$ such that $\{g^n x : n \in \mathbb{Z}\}$ is dense in G;

(3) L_g is minimal: $\{g^n y : n \in \mathbb{Z}\}$ is dense in G for every $y \in G$.

6.3.8. Show that the operation $+ : M \times M \to M$ defined in Section 6.3.4 is continuous and endows M with the structure of an abelian group. Moreover, every translation in this group preserves the distance defined in (6.3.7).

6.3.9. Let $f : M \to M$ be an odometer, as defined in Section 6.3.4, with $d = 10$. Given $b_0, \dots, b_{k-1}$ in $\{0, \dots, 9\}$, denote by $[b_0, \dots, b_{k-1}]$ the set of all sequences $\beta \in M$ with $\beta_0 = b_0, \dots, \beta_{k-1} = b_{k-1}$. Show that

$$\lim_n \frac{1}{n} \#\big\{0 \le j < n : f^j(x) \in [b_0, \dots, b_{k-1}]\big\} = \frac{1}{10^k}$$

for every $x \in M$. Moreover, this limit is uniform. Conclude that f admits a unique invariant probability measure and calculate that measure explicitly.

6.3.10. Check the claims in Example 6.3.9.

6.3.11. Prove that if s is a substitution in a finite alphabet $\mathcal{A}$ and $\alpha \in \mathcal{A}$ is such that $s(\alpha)$ has length larger than 1 and starts with the letter α, then the transformation $S : \Sigma \to \Sigma$ defined in Example 6.3.10 admits a unique fixed point that starts with the letter $\alpha \in \mathcal{A}$.

6.4 Theorem of Weyl

In this section we use ideas that were discussed previously to prove a beautiful theorem of Hermann Weyl [Wey16] about the distribution of polynomial sequences.

Consider any polynomial function $P : \mathbb{R} \to \mathbb{R}$ with real coefficients and degree $d \ge 1$:

$$P(x) = a_0 + a_1 x + a_2 x^2 + \cdots + a_d x^d.$$

Composing P with the canonical projection $\mathbb{R} \to S^1$, we obtain a polynomial function $P_* : \mathbb{R} \to S^1$ with values on the circle $S^1 = \mathbb{R}/\mathbb{Z}$. Define

$$z_n = P_*(n), \quad \text{for every } n \ge 1.$$

We may think of z_n as the fractional part of the real number $P(n)$. We want to understand how the sequence $(z_n)_n$ is distributed on the circle.

Definition 6.4.1. We say that a sequence $(x_n)_n$ in S^1 is *equidistributed* if, for any continuous function $\varphi : S^1 \to \mathbb{R}$,

$$\lim_{n\to\infty} \frac{1}{n} \sum_{j=1}^{n} \varphi(x_j) = \int \varphi(x)\, dx.$$

According to Exercise 6.4.1, this is equivalent to saying that, for every segment $I \subset S^1$, the fraction of terms of the sequence that are in I is equal to the corresponding length $m(I)$.

Theorem 6.4.2 (Weyl). *If at least one of the coefficients $a_1, a_2, \ldots, a_d$ is irrational then the sequence $z_n = P_*(n)$, $n \in \mathbb{N}$ is equidistributed.*

In order to develop some intuition about this theorem, let us start by considering the special case $d = 1$. In this case the polynomial function reduces to $P(x) = a_0 + a_1 x$. Let us consider the transformation

$$f : S^1 \to S^1, \quad f(\theta) = \theta + a_1.$$

By assumption, the coefficient a_1 is irrational. Therefore, as we have seen in Section 6.3.1, this transformation admits a unique invariant probability measure, which is the Lebesgue measure m. Consequently, given any continuous function $\varphi : S^1 \to \mathbb{R}$ and any point $\theta \in S^1$,

$$\lim_{n\to\infty} \frac{1}{n} \sum_{j=1}^{n} \varphi(f^j(\theta)) = \int \varphi\, dm.$$

Take $\theta = a_0$. Then $f^j(\theta) = a_0 + a_1 j = z_j$. Hence, the previous relation yields

$$\lim_{n\to\infty} \frac{1}{n} \sum_{j=1}^{n} \varphi(z_j) = \int \varphi\, dm.$$

This is precisely what it means to say that z_j is equidistributed.

6.4.1 Ergodicity

Now we extend the previous arguments to any degree $d \geq 1$. Consider the transformation $f : \mathbb{T}^d \to \mathbb{T}^d$ defined on the d-dimensional torus $\mathbb{T}^d$ by the following expression:

$$f(\theta_1, \theta_2, \ldots, \theta_d) = (\theta_1 + \alpha, \theta_2 + \theta_1, \ldots, \theta_d + \theta_{d-1}), \qquad (6.4.1)$$

where α is an irrational number to be chosen later. Note that f is invertible: the inverse is given by

$$f^{-1}(\theta_1, \theta_2, \ldots, \theta_d) = (\theta_1 - \alpha, \theta_2 - \theta_1 + \alpha, \ldots, \theta_d - \theta_{d-1} + \cdots + (-1)^{d-1}\theta_1 + (-1)^d \alpha).$$

Note also that the derivative of f at each point is given by the matrix

$$\begin{pmatrix} 1 & 0 & 0 & \cdots & 0 & 0 \\ 1 & 1 & 0 & \cdots & 0 & 0 \\ 0 & 1 & 1 & \cdots & 0 & 0 \\ \vdots & \vdots & \vdots & \vdots & \vdots & \vdots \\ 0 & 0 & 0 & \cdots & 1 & 1 \end{pmatrix},$$

whose determinant is 1. This ensures that f preserves the Lebesgue measure on the torus (recall Lemma 1.3.5).

Proposition 6.4.3. *The Lebesgue measure on $\mathbb{T}^d$ is ergodic for f.*

Proof. We are going to use a variation of the Fourier series expansion argument in Proposition 4.2.1. Let $\varphi : \mathbb{T}^d \to \mathbb{R}$ be any function in $L^2(m)$. Write

$$\varphi(\theta) = \sum_{n\in\mathbb{Z}^d} a_n e^{2\pi i n\cdot\theta}$$

with $\theta = (\theta_1, \dots, \theta_d)$, $n = (n_1, \dots, n_d)$ and $n\cdot\theta = n_1\theta_1 + \cdots + n_d\theta_d$. The L^2-norm of φ is given by

$$\sum_{n\in\mathbb{Z}^d} |a_n|^2 = \int |\varphi(\theta)|^2 \, d\theta_1 \cdots d\theta_d < \infty. \tag{6.4.2}$$

Observe that

$$\begin{aligned} \varphi(f(\theta)) &= \sum_{n\in\mathbb{Z}^d} a_n e^{2\pi i(n_1(\theta_1+\alpha)+n_2(\theta_2+\theta_1)+\cdots+n_d(\theta_d+\theta_{d-1}))} \\ &= \sum_{n\in\mathbb{Z}^d} a_n e^{2\pi i n_1\alpha} e^{2\pi i L(n)\cdot\theta}, \end{aligned}$$

where $L(n) = (n_1+n_2, n_2+n_3, \dots, n_{d-1}+n_d, n_d)$. Suppose that the function φ is invariant, that is, $\varphi\circ f = \varphi$ at almost every point. Then

$$a_n e^{2\pi i n_1\alpha} = a_{L(n)} \quad \text{for every } n\in\mathbb{Z}^d. \tag{6.4.3}$$

This implies that a_n and $a_{L(n)}$ have the same absolute value. On the other hand, the integrability relation (6.4.2) implies that there exists at most a finite number of terms with any given absolute value different from zero. It follows that $a_n = 0$ for every $n \in \mathbb{Z}^d$ whose orbit $L^j(n)$, $j\in\mathbb{Z}$ is infinite. Observing the expression of L, we deduce that $a_n = 0$ except, possibly, if $n_2 = \cdots = n_d = 0$. For the remaining values of n, that is, for every $n = (n_1, 0, \dots, 0)$, one has that $L(n) = n$ and, thus, the relation (6.4.3) becomes

$$a_n = a_n e^{2\pi i n_1\alpha}.$$

Since α is irrational, the last factor is different from 1 whenever n_1 is non-zero. Therefore, this relation implies that $a_n = 0$ also for $n = (n_1, 0, \dots, 0)$ with $n_1 \neq 0$. In this way, we have shown that if φ is an invariant function then all the terms in its Fourier series vanish except, possibly, the constant term. This means that φ is constant at almost every point, which proves that the Lebesgue measure is ergodic for f.

6.4.2 Unique ergodicity

The last step in the proof of Theorem 6.4.2 is the following result:

Proposition 6.4.4. *The transformation f is uniquely ergodic: the Lebesgue measure on the torus is the unique invariant probability measure.*

Proof. The proof is by induction on the degree d of the polynomial P. The case of degree 1 was treated previously. Therefore, we only need to explain how the case of degree d may be deduced from the case of degree $d-1$. For that, we write $\mathbb{T}^d = \mathbb{T}^{d-1} \times S^1$ and

$$f : \mathbb{T}^{d-1} \times S^1 \to \mathbb{T}^{d-1} \times S^1, \quad f(\theta_0, \eta) = (f_0(\theta_0), \eta + \theta_{d-1}), \tag{6.4.4}$$

where $\theta_0 = (\theta_1, \ldots, \theta_{d-1})$ and $f_0(\theta_0) = (\theta_1 + \alpha, \theta_2 + \theta_1, \ldots, \theta_{d-1} + \theta_{d-2})$. By induction, the transformation

$$f_0 : \mathbb{T}^{d-1} \to \mathbb{T}^{d-1}$$

is uniquely ergodic. Let us denote by $\pi : \mathbb{T}^d \to \mathbb{T}^{d-1}$ the projection $\pi(\theta) = \theta_0$.

Lemma 6.4.5. *For any probability measure μ invariant under f, the projection $\pi_*\mu$ coincides with the Lebesgue measure m_0 on $\mathbb{T}^{d-1}$.*

Proof. Given any measurable set $E \subset \mathbb{T}^{d-1}$,

$$(\pi_*\mu)(f_0^{-1}(E)) = \mu(\pi^{-1} f_0^{-1}(E)).$$

Using that $\pi \circ f = f_0 \circ \pi$ and the fact that μ is f-invariant, we get that the expression on the right-hand side is equal to

$$\mu(f^{-1}\pi^{-1}(E)) = \mu(\pi^{-1}(E)) = (\pi_*\mu)(E).$$

Therefore, $(\pi_*\mu)(f_0^{-1}(E)) = (\pi_*\mu)(E)$ for every measurable subset E, that is, $\pi_*\mu$ is invariant under f_0. It is clear that $\pi_*\mu$ is a probability measure. Since f_0 is uniquely ergodic, it follows that $\pi_*\mu$ coincides with the Lebesgue measure m_0 on $\mathbb{T}^{d-1}$.

Now suppose that μ, besides being invariant, is also ergodic for f. By Theorem 3.2.6, and by ergodicity, the set $G(\mu) \subset M$ of all points $\theta \in \mathbb{T}^d$ such that

$$\lim_n \frac{1}{n} \sum_{j=0}^{n-1} \varphi(f^j(\theta)) = \int \varphi \, d\mu \quad \text{for any continuous function } \varphi : \mathbb{T}^d \to \mathbb{R} \tag{6.4.5}$$

has full measure. Let $G_0(\mu)$ be the set of all $\theta_0 \in \mathbb{T}^{d-1}$ such that $G(\mu)$ intersects $\{\theta_0\} \times S^1$. In other words, $G_0(\mu) = \pi(G(\mu))$. It is clear that $\pi^{-1}(G_0(\mu))$ contains $G(\mu)$ and, thus, has full measure. Hence, using Lemma 6.4.5,

$$m_0(G_0(\mu)) = \mu(\pi^{-1}(G_0(\mu))) = 1. \tag{6.4.6}$$

For the same reasons, this relation remains valid for the Lebesgue measure:

$$m_0(G_0(m)) = m(\pi^{-1}(G_0(m))) = 1. \tag{6.4.7}$$

The identities (6.4.6) and (6.4.7) imply that the intersection between $G_0(\mu)$ and $G_0(m)$ has full measure for m_0. So, in particular, these two sets cannot be disjoint. Let θ_0 be any point in the intersection. By definition, $G(\mu)$ intersects $\{\theta_0\} \times S^1$. But the next result asserts that $G(m)$ *contains* $\{\theta_0\} \times S^1$:

Lemma 6.4.6. *If $\theta_0 \in G_0(m)$ then $\{\theta_0\} \times S^1$ is contained in $G(m)$.*

Proof. The crucial observation is that the measure m is invariant under every transformation of the form

$$R_\beta : \mathbb{T}^{d-1} \times S^1 \to \mathbb{T}^{d-1} \times S^1, \quad (\zeta, \eta) \mapsto (\zeta, \eta + \beta).$$

The hypothesis $\theta_0 \in G_0(m)$ means that there exists some $\eta \in S^1$ such that $(\theta_0, \eta) \in G(m)$, that is,

$$\lim_n \frac{1}{n} \sum_{j=0}^{n-1} \varphi(f^j(\theta_0, \eta)) = \int \varphi \, dm$$

for every continuous function $\varphi : \mathbb{T}^d \to \mathbb{R}$. Any other point of $\{\theta_0\} \times S^1$ may be written as $(\theta_0, \eta + \beta) = R_\beta(\theta_0, \eta)$ for some $\beta \in S^1$. Recalling (6.4.1), we see that

$$f\big(R_\beta(\tau_0, \zeta)\big) = (\tau_1 + \alpha, \tau_2 + \tau_1, \ldots, \tau_{d-1} + \tau_{d-2}, \zeta + \beta + \tau_{d-1}) = R_\beta\big(f(\tau_0, \zeta)\big)$$

for every $(\tau_0, \zeta) \in \mathbb{T}^{d-1} \times S^1$. Hence, by induction,

$$f^j(\theta_0, \eta + \beta) = f^j\big(R_\beta(\theta_0, \eta)\big) = R_\beta\big(f^j(\theta_0, \eta)\big)$$

for every $j \geq 1$. Therefore, given any continuous function $\varphi : \mathbb{T}^d \to \mathbb{R}$,

$$\begin{aligned}
\lim_n \frac{1}{n} \sum_{j=0}^{n-1} \varphi(f^j(\theta_0, \eta + \beta)) &= \lim \frac{1}{n} \sum_{j=0}^{n-1} (\varphi \circ R_\beta)(f^j(\theta_0, \eta)) \\
&= \int (\varphi \circ R_\beta) \, dm = \int \varphi \, dm.
\end{aligned}$$

This proves that $(\theta_0, \eta + \beta)$ is in $G(m)$ for every $\beta \in S^1$, as stated.

It follows from what we said so far that $G(\mu)$ and $G(m)$ intersect each other at some point of $\{\theta_0\} \times S^1$. In view of the definition (6.4.5), this implies that the two measures have the same integral for every continuous function. According to Proposition A.3.3, this implies that $\mu = m$, as we wanted to prove.

Corollary 6.4.7. *The orbit of every point $\theta \in \mathbb{T}^d$ is equidistributed on the torus $\mathbb{T}^d$, in the sense that*

$$\lim_n \frac{1}{n} \sum_{j=0}^{n-1} \psi(f^j(\theta)) = \int \psi \, dm$$

for every continuous function $\psi : \mathbb{T}^d \to \mathbb{R}$.

Proof. This follows immediately from Propositions 6.1.1 and 6.4.4.

6.4.3 Proof of the theorem of Weyl

To complete the proof of Theorem 6.4.2, we introduce the polynomial functions $p_1, \dots, p_d$ defined by

$$\begin{aligned} p_d(x) &= P(x) \quad \text{and} \\ p_{j-1}(x) &= p_j(x+1) - p_j(x) \quad \text{for } j = 2, \dots, d. \end{aligned} \tag{6.4.8}$$

Lemma 6.4.8. *The polynomial $p_j(x)$ has degree j, for every $1 \le j \le d$. Moreover, $p_1(x) = \alpha x + \beta$ with $\alpha = d! a_d$.*

Proof. By definition, $p_d(x) = P(x)$ has degree d. Hence, to prove the first claim it suffices to show that if $p_j(x)$ has degree j then $p_{j-1}(x)$ has degree $j-1$. In order to do that, let

$$p_j(x) = b_j x^j + b_{j-1} x^{j-1} + \cdots + b_0,$$

where $b_j \neq 0$. Then

$$\begin{aligned} p_j(x+1) &= b_j(x+1)^j + b_{j-1}(x+1)^{j-1} + \cdots + b_0 \\ &= b_j x^j + (jb_j + b_{j-1})x^{j-1} + \cdots + b_0. \end{aligned}$$

Subtracting one expression from the other, we get that

$$p_{j-1}(x) = (jb_j)x^{j-1} + b'_{j-2}x^{j-2} + \cdots + b'_0$$

has degree $j-1$. This proves the first claim in the lemma. This calculation also shows that the main coefficient of $p_{j-1}(x)$ (the coefficient of the term with highest degree) can be obtained multiplying by j the main coefficient of $p_j(x)$. Consequently, the main coefficient of p_1 must be equal to $d! a_q$, as claimed in the last part of the lemma.

Lemma 6.4.9. *For every $n \ge 0$,*

$$f^n\big(p_1(0), p_2(0), \dots, p_d(0)\big) = \big(p_1(n), p_2(n), \dots, p_d(n)\big).$$

Proof. The proof is by induction on n. Since the case $n = 0$ is obvious, we only need to treat the inductive step. Recall that f was defined in (6.4.1). If

$$f^{n-1}(p_1(0), p_2(0), \dots, p_d(0)) = (p_1(n-1), p_2(n-1), \dots, p_d(n-1))$$

then $f^n(p_1(0), p_2(0), \ldots, p_d(0))$ is equal to

$$(p_1(n-1)+\alpha, p_2(n-1)+p_1(n-1), \ldots, p_d(n-1)+p_{d-1}(n-1)).$$

Using the definition (6.4.8) and Lemma 6.4.8, we find that this expression is equal to

$$(p_1(n), p_2(n), \ldots, p_d(n)),$$

and that proves the lemma.

Finally, we are ready to prove that the sequence $z_n = P_*(n)$, $n \in \mathbb{N}$ is equidistributed. We treat two cases separately.

First, suppose that the main coefficient a_d of $P(x)$ is irrational. Then the number α in Lemma 6.4.8 is irrational and, thus, the results in Section 6.4.2 are valid for the transformation $f : \mathbb{T}^d \to \mathbb{T}^d$. Let $\varphi : S^1 \to \mathbb{R}$ be any continuous function. Consider $\psi : \mathbb{T}^d \to \mathbb{R}$ defined by

$$\psi(\theta_1, \theta_2, \ldots, \theta_d) = \varphi(\theta_d).$$

Fix $\theta = (p_1(0), p_2(0), \ldots, p_d(0))$. Using Lemma 6.4.9 and Corollary 6.4.7, we get that

$$\lim_n \frac{1}{n}\sum_{j=0}^{n-1} \varphi(z_n) = \lim_n \frac{1}{n}\sum_{j=0}^{n-1} \psi(f^n(\theta)) = \int \psi\, dm = \int \varphi\, dx.$$

This ends the proof of Theorem 6.4.2 in the case when a_d is irrational.

Now suppose that a_d is rational. Write $a_d = p/q$ with $p \in \mathbb{Z}$ and $q \in \mathbb{N}$. It is clear that we may write z_n as a sum

$$z_n = x_n + y_n, \quad x_n = a_d n^d \quad \text{and} \quad y_n = Q_*(n)$$

where $Q(x) = a_0 + a_1 x + \cdots + a_{d-1}x^{d-1}$ and $Q_* : \mathbb{R} \to S^1$ is given by $Q_* = \pi \circ Q$. To begin with, observe that

$$x_{n+q} - x_n = \frac{p}{q}(n+q)^d - \frac{p}{q}n^d$$

is an integer, for every $n \in \mathbb{N}$. This means that the sequence x_n, $n \in \mathbb{N}$ is periodic (with period q) in the circle $\mathbb{R}/\mathbb{Z}$. In particular, it takes no more than q distinct values. Observe also that, since a_d is rational, the hypothesis of the theorem implies that some of the coefficients $a_1, \ldots, a_{d-1}$ of Q are irrational. Hence, by induction on the degree, the sequence y_n, $n \in \mathbb{N}$ is equidistributed. More than that, the subsequences

$$y_{qn+r} = Q_*(qn+r), \quad n \in \mathbb{Z}$$

are equidistributed for every $r \in \{0, 1, \ldots, q-1\}$. In fact, as the reader may readily check, these sequences may be written as $y_{nq+r} = Q_*^{(r)}(n)$ for some polynomial $Q^{(r)}$ that also has degree $d-1$ and, thus, the induction hypothesis applies to each one of them as well. From these two observations it follows that every subsequence z_{qn+r}, $n \in \mathbb{Z}$ is equidistributed. Consequently, z_n, $n \in \mathbb{N}$ is also equidistributed. This completes the proof of Theorem 6.4.2.

6.4.4 Exercises

6.4.1. Show that a sequence $(z_j)_j$ is equidistributed on the circle if and only if

$$\lim_n \frac{1}{n}\#\{1 \le j \le n : z_j \in I\} = m(I)$$

for every segment $I \subset S^1$, where $m(I)$ denotes the length of I.

6.4.2. Show that the sequence $(\sqrt{n} \mod \mathbb{Z})_n$ is equidistributed on the circle. Does the same hold for the sequence $(\log n \mod \mathbb{Z})_n$?

6.4.3. Koksma [Kok35] proved that the sequence $(a^n \mod \mathbb{Z})_n$ is equidistributed on the circle for Lebesgue-almost every $a > 1$. That is not true for *every* $a > 1$. Indeed, consider the *golden ratio* $a = (1+\sqrt{5})/2$. Check that the sequence $(a^n \mod \mathbb{Z})_n$ converges to $0 \in S^1$ when $n \to \infty$; in particular, it is not equidistributed on the circle.

7
Correlations

The models of dynamical systems that interest us the most, transformations and flows, are deterministic: the state of the system at any time determines the whole future trajectory; when the system is invertible, the past trajectory is equally determined. However, these systems may also present stochastic (that is, "partly random") behavior: at some level coarser than that of individual trajectories, information about the past is gradually lost as the system is iterated. That is the subject of the present chapter.

In probability theory one calls the *correlation* between two random variables X and Y the number

$$C(X,Y) = \mathbb{E}\big[(X - \mathbb{E}[X])(Y - \mathbb{E}[Y])\big] = \mathbb{E}[XY] - \mathbb{E}[X]\mathbb{E}[Y].$$

Note that the expression $(X - \mathbb{E}[X])(Y - \mathbb{E}[Y])$ is positive if X and Y are on the same side (either larger or smaller) of their respective means, $\mathbb{E}[X]$ and $\mathbb{E}[Y]$, and it is negative otherwise. Therefore, the sign of $C(X,Y)$ indicates whether the two variables exhibit, predominantly, the same behavior or opposite behaviors, relative to their means. Furthermore, correlation close to zero indicates that the two behaviors are little, if at all, related to each other.

Given an invariant probability measure μ of a dynamical system $f : M \to M$ and given measurable functions $\varphi, \psi : M \to \mathbb{R}$, we want to analyze the evolution of the correlations

$$C_n(\varphi, \psi) = C(\varphi \circ f^n, \psi)$$

when time n goes to infinity. We may think of φ and ψ as quantities that are measured in the system, such as temperature, acidity (pH), kinetic energy, and so forth. Then $C_n(\varphi, \psi)$ measures how much the value of φ at time n is correlated with the value of ψ at time zero; to what extent one value "influences" the other.

For example, if $\varphi = \mathcal{X}_A$ and $\psi = \mathcal{X}_B$ are characteristic functions, then $\psi(x)$ provides information on the position of the initial point x, whereas $\varphi(f^n(x))$ informs on the position of its n-th iterate $f^n(x)$. If the correlation $C_n(\varphi, \psi)$ is small, then the first information is of little use to make predictions about the second one. That kind of behavior, where correlations approach zero as time n increases, is quite common in important models, as we are going to see.

We start by introducing the notions of (strong) mixing and weak mixing systems, and by studying their basic properties (Section 7.1). In Sections 7.2 and 7.3 we discuss these notions in the context of Markov shifts, which generalize Bernoulli shifts, and of interval exchanges, which are an extension of the class of circle rotations. In Section 7.4 we analyze, in quantitative terms, the speed of decay of correlations for certain classes of functions.

7.1 Mixing systems

Let $f : M \to M$ be a measurable transformation and μ be an invariant probability measure. The *correlations sequence* of two measurable functions $\varphi, \psi : M \to \mathbb{R}$ is defined by

$$C_n(\varphi,\psi) = \int (\varphi \circ f^n)\psi \, d\mu - \int \varphi \, d\mu \int \psi \, d\mu, \qquad n \in \mathbb{N}. \tag{7.1.1}$$

We say that the system (f,μ) is *mixing* if

$$\lim_n C_n(\mathcal{X}_A, \mathcal{X}_B) = \lim_n \mu(f^{-n}(A) \cap B) - \mu(A)\mu(B) = 0, \tag{7.1.2}$$

for any measurable sets $A, B \subset M$. In other words, when n grows the probability of the event $\{x \in B \text{ and } f^n(x) \in A\}$ converges to the product of the probabilities of the events $\{x \in B\}$ and $\{f^n(x) \in A\}$.

Analogously, given a flow $f^t : M \to M$, $t \in \mathbb{R}$ and an invariant probability measure μ, we define

$$C_t(\varphi,\psi) = \int (\varphi \circ f^t)\psi \, d\mu - \int \varphi \, d\mu \int \psi \, d\mu, \qquad t \in \mathbb{R} \tag{7.1.3}$$

and we say that the system (f^t,μ) is mixing if

$$\lim_{t\to+\infty} C_t(\mathcal{X}_A, \mathcal{X}_B) = \lim_{t\to+\infty} \mu(f^{-t}(A) \cap B) - \mu(A)\mu(B) = 0, \tag{7.1.4}$$

for any measurable sets $A, B \subset M$.

7.1.1 Properties

A mixing system is necessarily ergodic. Indeed, suppose that there exists some invariant set $A \subset M$ with $0 < \mu(A) < 1$. Taking $B = A^c$, we get $f^{-n}(A) \cap B = \emptyset$ for every n. Then, $\mu(f^{-n}(A) \cap B) = 0$ for every n, whereas $\mu(A)\mu(B) \neq 0$. In particular, (f,μ) is not mixing. The example that follows shows that ergodicity is strictly weaker than mixing:

Example 7.1.1. Let $\theta \in \mathbb{R}$ be an irrational number. As we have seen in Section 4.2.1, the rotation $R_\theta : S^1 \to S^1$ is ergodic with respect to the Lebesgue measure m. However, (R_θ, m) is not mixing. Indeed, if $A, B \subset S^1$ are two small intervals then $R_\theta^{-n}(A) \cap B$ is empty and, thus, $m(R_\theta^{-n}(A) \cap B) = 0$ for infinitely many values of n. Since $m(A)m(B) \neq 0$, it follows that the condition in (7.1.2) does not hold.

It is clear from the definition (7.1.2) that if (f,μ) is mixing then (f^k,μ) is mixing for every $k \in \mathbb{N}$. The corresponding statement for ergodicity is false: the map $f(x) = 1-x$ on the set $\{0,1\}$ is ergodic with respect to the measure $(\delta_0+\delta_1)/2$ but the second iterate f^2 is not.

Lemma 7.1.2. *Assume that* $\lim_n \mu(f^{-n}(A)\cap B) = \mu(A)\mu(B)$ *for every pair of sets* A *and* B *in an algebra* $\mathcal{A}$ *that generates the* σ*-algebra of measurable sets. Then* (f,μ) *is mixing.*

Proof. Let $\mathcal{C}$ be the family of all measurable sets A such that $\mu(f^{-n}(A)\cap B)$ converges to $\mu(A)\mu(B)$ for every $B \in \mathcal{A}$. By assumption, $\mathcal{C}$ contains $\mathcal{A}$. We claim that $\mathcal{C}$ is a monotone class. Indeed, let $A = \bigcup_k A_k$ be the union of an increasing sequence $A_1 \subset \cdots \subset A_k \subset \cdots$ of elements of $\mathcal{C}$. Given $\varepsilon > 0$, there exists $k_0 \geq 1$ such that

$$\mu(A) - \mu(A_k) = \mu(A \setminus A_k) < \varepsilon$$

for every $k \geq k_0$. Moreover, for every $n \geq 1$,

$$\begin{aligned}\mu\big(f^{-n}(A)\cap B\big) - \mu\big(f^{-n}(A_k)\cap B\big) &= \mu\big(f^{-n}(A\setminus A_k)\cap B\big)\\ &\leq \mu(f^{-n}(A\setminus A_k)) = \mu(A\setminus A_k) < \varepsilon.\end{aligned}$$

For each fixed $k \geq k_0$, the fact that $A_k \in \mathcal{C}$ ensures that there exists $n(k) \geq 1$ such that

$$|\mu\big(f^{-n}(A_k)\cap B\big) - \mu(A_k)\mu(B)| < \varepsilon \quad \text{for every } n \geq n(k).$$

Adding these three inequalities we conclude that

$$|\mu\big(f^{-n}(A)\cap B\big) - \mu(A)\mu(B)| < 3\varepsilon \quad \text{for every } n \geq n(k_0).$$

Since $\varepsilon > 0$ is arbitrary, this shows that $A \in \mathcal{C}$. In the same way, one proves that the intersection of any decreasing sequence of elements of $\mathcal{C}$ is still an element of $\mathcal{C}$. So, $\mathcal{C}$ is indeed a monotone class. By the monotone class theorem (Theorem A.1.18), it follows that $\mathcal{C}$ contains every measurable set: for every measurable set A one has

$$\lim_n \mu\big(f^{-n}(A)\cap B\big) = \mu(A)\mu(B) \quad \text{for every } B \in \mathcal{A}.$$

All that is left to do is to deduce that this property holds for every measurable set B. This follows from precisely the same kind of arguments as we have just detailed, as the reader may readily check.

Example 7.1.3. Every Bernoulli shift (recall Section 4.2.3) is mixing. Indeed, given any two cylinders $A = [p;A_p,\dots,A_q]$ and $B = [r;B_r,\dots,B_s]$,

$$\begin{aligned}\mu\big(f^{-n}(A)\cap B\big) &= \mu([r;B_r,\dots,B_s,X,\dots,X,A_p,\dots,A_q])\\ &= \mu([r;B_r,\dots,B_s])\mu([p;A_p,\dots,A_q]) = \mu(A)\mu(B)\end{aligned}$$

for every $n > s - p$. Let $\mathcal{A}$ be the algebra generated by the cylinders: its elements are the finite pairwise disjoint unions of cylinders. It follows from what we have just said that $\mu(f^{-n}(A) \cap B) = \mu(A)\mu(B)$ for every pair of sets $A, B \in \mathcal{A}$ and every n sufficiently large. Since $\mathcal{A}$ generates the σ-algebra of measurable sets, we may use Lemma 7.1.2 to conclude that the system is mixing, as stated.

Example 7.1.4. Let $g : S^1 \to S^1$ be defined by $g(x) = kx$, where $k \geq 2$ is an integer number, and let m be the Lebesgue measure on the circle. The system (g, m) is equivalent to a Bernoulli shift, in the following sense. Let $X = \{0, 1, \ldots, k-1\}$ and let $f : M \to M$ be the shift map in $M = X^{\mathbb{N}}$. Consider the product measure $\mu = \nu^{\mathbb{N}}$ in M, where ν is the probability measure defined by $\nu(A) = \#A/k$ for every $A \subset X$. The map

$$h : M \to S^1, \quad h\big((a_n)_n\big) = \sum_{n=1}^{\infty} \frac{a_{n-1}}{k^n}$$

is a bijection, restricted to a full measure subset, and both h and its inverse are measurable. Moreover, $h_*\mu = m$ and $h \circ f = g \circ h$ at almost every point. We say that h is an *ergodic equivalence* between (g, m) and (f, μ). Through it, properties of one system may be translated to corresponding properties for the other system. In particular, recalling Example 7.1.3, we get that (g, m) is mixing: given any measurable sets $A, B \subset S^1$,

$$m\big(g^{-n}(A) \cap B\big) = \mu\Big(h^{-1}\big(g^{-n}(A) \cap B\big)\Big) = \mu\big(f^{-n}(h^{-1}(A)) \cap h^{-1}(B)\big)$$
$$\to \mu(h^{-1}(A))\mu(h^{-1}(B)) = m(A)m(B) \quad \text{when } n \to \infty.$$

Example 7.1.5. For surjective endomorphisms of the torus (Section 4.2.5) mixing and ergodicity are equivalent properties: the system (f_A, m) is mixing if and only if no eigenvalue of the matrix A is a root of unity (compare Theorem 4.2.14). In Exercise 7.1.4 we invite the reader to prove this fact; a stronger statement will appear in Exercise 8.4.2. More generally, relative to the Haar measure, a surjective endomorphism of a compact group is mixing if and only if it is ergodic. In fact, even stronger statements are true, as we will comment upon in Section 9.5.3.

Let us also discuss the topological version of the notion of a mixing system. For that, take the ambient M to be a topological space. A transformation $f : M \to M$ is said to be *topologically mixing* if, given any non-empty open sets $U, V \subset M$, there exists $n_0 \in \mathbb{N}$ such that $f^{-n}(U) \cap V$ is non-empty for every $n \geq n_0$. This is similar to but strictly stronger than the hypothesis of Lemma 4.3.4: in the lemma we asked $f^{-n}(U)$ to intersect V for *some* value of n, whereas now we request that to happen for every n sufficiently large.

Proposition 7.1.6. *If (f, μ) is mixing then the restriction of f to the support of μ is topologically mixing.*

Proof. Denote $X = \operatorname{supp}(\mu)$. Let $A, B \subset X$ be open sets. By the definition of support of a measure, $\mu(A) > 0$ and $\mu(B) > 0$. Hence, since μ is mixing, there exists n_0 such that $\mu(f^{-n}(A) \cap B) > \mu(A)\mu(B)/2 > 0$ for every $n \geq n_0$. In particular, $\mu(f^{-n}(A) \cap B) \neq \emptyset$ for every $n \geq n_0$.

It follows directly from this proposition that if f admits some invariant probability measure μ that is mixing and positive on open sets, then f is topologically mixing. For example, given any finite set $X = \{1, \dots, d\}$, the shift map

$$f : X^{\mathbb{Z}} \to X^{\mathbb{Z}} \qquad (\text{or } f : X^{\mathbb{N}} \to X^{\mathbb{N}})$$

is topologically mixing. Indeed, for any probability measure ν supported on the whole of X, the Bernoulli measure $\mu = \nu^{\mathbb{Z}}$ (or $\mu = \nu^{\mathbb{N}}$) is mixing and positive on open sets, as we have seen in Example 7.1.3. Analogously, by Example 7.1.4, every transformation $f : S^1 \to S^1$ of the form $f(x) = kx$ with $k \geq 2$ is topologically mixing.

Example 7.1.7. Translations in a metrizable group G are never topologically mixing. Indeed, consider any left-translation L_g (the case of right-translations is analogous). We may suppose that g is not the unit element e since otherwise it is obvious that L_g is not topologically mixing. Fix some distance d invariant under all the translations of the group G (recall Lemma 6.3.6) and let $\alpha = d(e, g^{-1})$. Consider $U = V =$ ball of radius $\alpha/4$ around e. Every $L_g^{-n}(U)$ is a ball of radius $\alpha/4$. Assume that $L_g^{-n}(U)$ intersects V. Then $L_g^{-n}(U)$ is contained in the ball of radius $3\alpha/4$ and, thus, $L_g^{-n-1}(U)$ is contained in the ball of radius $3\alpha/4$ around g^{-1}. Consequently, $L_g^{-n-1}(U)$ does not intersect V. Since n is arbitrary, this shows that L_g is not topologically mixing.

7.1.2 Weak mixing

A system (f, μ) is *weak mixing* if, given any measurable sets $A, B \subset M$,

$$\lim_n \frac{1}{n} \sum_{j=0}^{n-1} |C_j(\mathcal{X}_A, \mathcal{X}_B)| = \lim_{n\to\infty} \frac{1}{n} \sum_{j=0}^{n-1} \left| \mu(f^{-j}(A) \cap B) - \mu(A)\mu(B) \right| = 0. \quad (7.1.5)$$

It is clear from the definition that every mixing system is also weak mixing. On the other hand, every weak mixing system is ergodic. Indeed, if $A \subset M$ is an invariant set then

$$\lim_n \frac{1}{n} \sum_{j=0}^{n-1} |C_j(\mathcal{X}_A, \mathcal{X}_{A^c})| = \mu(A)\mu(A^c)$$

and, hence, the hypothesis implies that $\mu(A) = 0$ or $\mu(A^c) = 0$.

Example 7.1.8. Translations in metrizable compact groups are never weak mixing with respect to the Haar measure μ (or any other invariant measure positive on open sets). Indeed, as observed in Example 7.1.7, it is always possible to choose open sets U and V such that $f^{-n}(U) \cap V$ is empty for at least one in every two consecutive values n. Then,

$$\liminf_n \frac{1}{n} \sum_{j=0}^{n-1} \left| \mu(f^{-j}(U) \cap V) - \mu(U)\mu(V) \right| \geq \frac{1}{2}\mu(U)\mu(V) > 0.$$

In this way we get several examples of ergodic systems, even uniquely ergodic ones, that are not weak mixing.

We are going to see in Section 7.3.2 that the family of interval exchanges contains many systems that are weak mixing (and uniquely ergodic) but are not mixing.

The proof of the next result is analogous to the proof of Lemma 7.1.2 and is left to the reader:

Lemma 7.1.9. *Assume that* $\lim_n (1/n) \sum_{j=0}^{n-1} |\mu(f^{-j}(A) \cap B) - \mu(A)\mu(B)| = 0$ *for every pair of sets A and B in some algebra $\mathcal{A}$ that generates the σ-algebra of measurable sets. Then (f, μ) is weak mixing.*

Example 7.1.10. Given a system (f, μ), let us consider the product transformation $f_2 : M \times M \to M \times M$ given by $f_2(x,y) = (f(x), f(y))$. It is easy to see that f_2 preserves the product measure $\mu_2 = \mu \times \mu$. If (f_2, μ_2) is ergodic then (f, μ) is ergodic: just note that if $A \subset M$ is invariant under f and $\mu(A) \in (0,1)$ then $A \times A$ is invariant under f_2 and $\mu_2(A \times A) \in (0,1)$.

The converse is not true in general, that is, (f_2, μ_2) may not be ergodic even if (f, μ) is ergodic. For example, if $f : S^1 \to S^1$ is an irrational rotation and d is a distance invariant under rotations, then any neighborhood $\{(x,y) : d(x,y) < r\}$ of the diagonal is invariant under f_2.

The next result shows that this type of phenomenon cannot occur in the category of weak mixing systems:

Proposition 7.1.11. *The following conditions are equivalent:*

(i) (f, μ) is weak mixing;
(ii) (f_2, μ_2) is weak mixing;
(iii) (f_2, μ_2) is ergodic.

Proof. To prove that (i) implies (ii), consider any measurable sets A, B, C, D in M. Then

$$\begin{aligned}
&\left| \mu_2(f_2^{-j}(A \times B) \cap (C \times D)) - \mu_2(A \times B)\mu_2(C \times D) \right| \\
&\qquad = \left| \mu(f^{-j}(A) \cap C)\mu(f^{-j}(B) \cap D) - \mu(A)\mu(B)\mu(C)\mu(D) \right| \\
&\qquad \leq \left| \mu(f^{-j}(A) \cap C) - \mu(A)\mu(C) \right| + \left| \mu(f^{-j}(B) \cap D) - \mu(B)\mu(D) \right|.
\end{aligned}$$

Therefore, the hypothesis (i) implies that

$$\lim_n \frac{1}{n}\sum_{j=0}^{n-1}\left|\mu_2(f_2^{-j}(A\times B)\cap(C\times D)) - \mu_2(A\times B)\mu_2(C\times D)\right| = 0.$$

It follows that

$$\lim_n \frac{1}{n}\sum_{j=0}^{n-1}\left|\mu_2(f_2^{-j}(X)\cap Y) - \mu_2(X)\mu_2(Y)\right| = 0$$

for any X, Y in the algebra generated by the products $E \times F$ of measurable subsets of M, that is, the algebra formed by the finite pairwise disjoint unions of such products. Since this algebra generates the σ-algebra of measurable subsets of $M \times M$, we may use Lemma 7.1.9 to conclude that (f_2, μ_2) is weak mixing.

It is immediate that (i) implies (iii). To prove that (iii) implies (i), observe that

$$\frac{1}{n}\sum_{j=0}^{n-1}\left[\mu\left(f^{-j}(A)\cap B\right) - \mu(A)\mu(B)\right]^2$$

$$= \frac{1}{n}\sum_{j=0}^{n-1}\left[\mu\left(f^{-j}(A)\cap B\right)^2 - 2\mu(A)\mu(B)\mu\left(f^{-j}(A)\cap B\right) + \mu(A)^2\mu(B)^2\right].$$

The right-hand side may be rewritten as

$$\frac{1}{n}\sum_{j=0}^{n-1}\left[\mu_2\left(f_2^{-j}(A\times A)\cap(B\times B)\right) - \mu_2(A\times A)\mu_2(B\times B)\right]$$

$$- 2\mu(A)\mu(B)\frac{1}{n}\sum_{j=0}^{n-1}\left[\mu\left(f^{-j}(A)\cap B\right) - \mu(A)\mu(B)\right].$$

Since (f_2, μ_2) is ergodic and, consequently, (f, μ) is also ergodic, part (ii) of Proposition 4.1.4 gives that both terms in this expression converge to zero. In this way, we conclude that

$$\lim_n \frac{1}{n}\sum_{j=0}^{n-1}\left[\mu\left(f^{-j}(A)\cap B\right) - \mu(A)\mu(B)\right]^2 = 0$$

for any measurable sets $A, B \subset M$. Using Exercise 7.1.2, we deduce that (f, μ) is weak mixing.

7.1.3 Spectral characterization

In this section we discuss equivalent formulations of the notions of mixing and weak mixing systems, in terms of the Koopman operator.

Proposition 7.1.12. *The following conditions are equivalent:*

(i) (f,μ) *is mixing.*

(ii) There exist $p,q \in [1,\infty]$ *with* $1/p+1/q=1$ *such that* $C_n(\varphi,\psi) \to 0$ *for any* $\varphi \in L^p(\mu)$ *and* $\psi \in L^q(\mu)$.

(iii) The condition in part (ii) holds for φ *in some dense subset of* $L^p(\mu)$ *and* ψ *in some dense subset of* $L^q(\mu)$.

Proof. Condition (i) is the special case of (ii) for characteristic functions. Since the correlations $(\varphi,\psi) \mapsto C_n(\varphi,\psi)$ are bilinear functions, condition (i) implies that $C_n(\varphi,\psi) \to 0$ for any simple functions φ and ψ. This implies (iii), since the simple functions form a dense subset of $L^r(\mu)$ for any $r \geq 1$.

To show that (iii) implies (ii), let us begin by observing that as correlations $C_n(\varphi,\psi)$ are equicontinuous functions of φ and ψ. Indeed, given $\varphi_1,\varphi_2 \in L^p(\mu)$ and $\psi_1,\psi_2 \in L^q(\mu)$, the Hölder inequality (Theorem A.5.5) gives that

$$\left|\int(\varphi_1 \circ f^n)\psi_1\,d\mu - \int(\varphi_2 \circ f^n)\psi_2\,d\mu\right| \leq \|\varphi_1-\varphi_2\|_p\,\|\psi_1\|_q + \|\varphi_2\|_p\,\|\psi_1-\psi_2\|_q.$$

Moreover,

$$\left|\int\varphi_1\,d\mu\int\psi_1\,d\mu - \int\varphi_2\,d\mu\int\psi_2\,d\mu\right| \leq \|\varphi_1-\varphi_2\|_1\,\|\psi_1\|_1 + \|\varphi_2\|_1\,\|\psi_1-\psi_2\|_1.$$

Adding these inequalities, and noting that $\|\cdot\|_1 \leq \|\cdot\|_r$ for every $r \geq 1$, we get that

$$\left|C_n(\varphi_1,\psi_1) - C_n(\varphi_2,\psi_2)\right| \leq 2\|\varphi_1-\varphi_2\|_p\,\|\psi_1\|_q + 2\|\varphi_2\|_p\,\|\psi_1-\psi_2\|_q \quad (7.1.6)$$

for every $n \geq 1$. Then, given $\varepsilon > 0$ and any $\varphi \in L^p(\mu)$ and $\psi \in L^q(\mu)$, we may take φ' and ψ' in the dense subsets mentioned in the hypothesis such that

$$\|\varphi-\varphi'\|_p < \varepsilon \quad \text{and} \quad \|\psi-\psi'\|_q < \varepsilon.$$

In particular, $\|\varphi'\|_p < \|\varphi\|_p + \varepsilon$ and $\|\psi'\|_q < \|\psi\|_q + \varepsilon$. Then, (7.1.6) gives that

$$|C_n(\varphi,\psi)| \leq |C_n(\varphi',\psi')| + 2\varepsilon(\|\varphi\|_p + \|\psi\|_q + 2\varepsilon) \quad \text{for every } n.$$

Moreover, by hypothesis, $|C_n(\varphi',\psi')| < \varepsilon$ for every n sufficiently large. Since ε is arbitrary, these two inequalities imply that $C_n(\varphi,\psi)$ converges to zero when $n \to \infty$. This proves property (ii).

The same argument proves the following version of Proposition 7.1.12 for the weak mixing property:

Proposition 7.1.13. *The following conditions are equivalent:*

(i) (f,μ) *is weak mixing.*

(ii) There exist $p,q \in [1,\infty]$ *with* $1/p+1/q=1$ *such that* $(1/n)\sum_{j=1}^{n}|C_j(\varphi,\psi)|$ *converges to* 0 *for any* $\varphi \in L^p(\mu)$ *and* $\psi \in L^q(\mu)$.

(iii) The condition in part (ii) holds for φ *in some dense subset of* $L^p(\mu)$ *and* ψ *in some dense subset of* $L^q(\mu)$.

In the case $p = q = 2$, we may express the correlations in terms of the inner product $\cdot$ in the Hilbert space $L^2(\mu)$:

$$C_n(\varphi, \psi) = \left[U_f^n \varphi - (\varphi \cdot 1)\right] \cdot \psi \quad \text{for every } \varphi, \psi \in L^2(\mu).$$

Therefore, Proposition 7.1.12 gives that (f, μ) is mixing if and only if

$$\lim_n \left[U_f^n \varphi - (\varphi \cdot 1)\right] \cdot \psi = 0 \quad \text{for every } \varphi, \psi \in L^2(\mu), \tag{7.1.7}$$

and Proposition 7.1.13 gives that (f, μ) is weak mixing if and only if

$$\lim_n \frac{1}{n} \sum_{j=1}^{n} \left|\left[U_f^j \varphi - (\varphi \cdot 1)\right] \cdot \psi\right| = 0 \quad \text{for every } \varphi, \psi \in L^2(\mu). \tag{7.1.8}$$

The condition (7.1.7) means that $U_f^n \varphi$ converges *weakly* to $\varphi \cdot 1 = \int \varphi \, d\mu$, while (7.1.8) is a Cesàro version of that assertion. Compare both conditions with the characterization of ergodicity in (4.1.7).

Corollary 7.1.14. *Let $f : M \to M$ be a mixing transformation relative to some invariant probability measure μ. Let ν be any probability measure on M, absolutely continuous with respect to μ. Then $f_*^n \nu$ converges pointwise to μ, that is, $\nu(f^{-n}(B)) \to \mu(B)$ for every measurable set $B \subset M$.*

Proof. Let $\varphi = \mathcal{X}_B$ and $\psi = d\nu/d\mu$. Note that $\varphi \in L^\infty(\mu)$ and $\psi \in L^1(\mu)$. Hence, by Proposition 7.1.12,

$$\int (\mathcal{X}_B \circ f^n) \frac{d\nu}{d\mu} \, d\mu = \int (U_f^n \varphi) \psi \, d\mu \to \int \varphi \, d\mu \int \psi \, d\mu = \int \mathcal{X}_B d\mu \int \frac{d\nu}{d\mu} \, d\mu.$$

The sequence on the left-hand side coincides with $\int (\mathcal{X}_B \circ f^n) \, d\nu = \nu(f^{-n}(B))$. The right-hand side is equal to $\mu(B) \int 1 \, d\nu = \mu(B)$.

7.1.4 Exercises

7.1.1. Show that (f, μ) is mixing if and only if $\mu(f^{-n}(A) \cap A) \to \mu(A)^2$ for every measurable set A.

7.1.2. Let $(a_n)_n$ be a bounded sequence of real numbers. Prove that

$$\lim_n \frac{1}{n} \sum_{j=1}^{n} |a_j| = 0$$

if and only if there exists $E \subset \mathbb{N}$ with density zero at infinity (that is, with $\lim_n (1/n) \#(E \cap \{0, \ldots, n-1\}) = 0$) such that the restriction of $(a_n)_n$ to the complement of E converges to zero when $n \to \infty$. Deduce that

$$\lim_n \frac{1}{n} \sum_{j=1}^{n} |a_j| = 0 \quad \Leftrightarrow \quad \lim_n \frac{1}{n} \sum_{j=1}^{n} (a_j)^2 = 0.$$

7.1.3. Prove that if μ is weak mixing for f then μ is weak mixing for every iterate f^k, $k \geq 1$.

7.1.4. Show that if no eigenvalue of $A \in \mathrm{SL}(d,\mathbb{R})$ is a root of unity then the linear endomorphism $f_A : \mathbb{T}^d \to \mathbb{T}^d$ induced by A is mixing, with respect to the Haar measure.

7.1.5. Let $f : M \to M$ be a measurable transformation in a metric space. Check that an invariant probability measure μ is mixing if and only if $(f_*^n \eta)_n$ converges to μ in the weak* topology for every probability measure η absolutely continuous with respect to μ.

7.1.6. (Multiple von Neumann ergodic theorem). Show that if (f,μ) is weak mixing then

$$\frac{1}{N}\sum_{n=0}^{N-1}(\varphi_1 \circ f^n)\cdots(\varphi_k \circ f^{kn}) \to \int \varphi_1\,d\mu \cdots \int \varphi_k\,d\mu \quad \text{in } L^2(\mu),$$

for any bounded measurable functions $\varphi_1, \ldots, \varphi_k$.

7.2 Markov shifts

In this section we introduce an important class of systems that generalizes the notion of Bernoulli shift. As explained previously, Bernoulli shifts model sequences of identical experiments such that the outcome of each experiment is *independent* of all the others. In the definition of Markov shifts we weaken this independence condition: we allow each outcome to depend on the preceding one, but not the others. More generally, Markov shifts may be used to model the so-called *finite memory* processes, that is, sequences of experiments for which there exists $k \geq 1$ such that the outcome of each experiment depends only on the outcomes of the k previous experiments. In this regard, see Exercise 7.2.4.

To define a Markov shift, let $(X,\mathcal{A})$ be a measurable space and $\Sigma = X^{\mathbb{N}}$ (or $\Sigma = X^{\mathbb{Z}}$) be the space of all sequences in X, endowed with the product σ-algebra. Let us consider the shift map

$$\sigma : \Sigma \to \Sigma, \quad \sigma\big((x_n)_n\big) = (x_{n+1})_n.$$

Let us be given a family $\{P(x,\cdot) : x \in X\}$ of probability measures on X that depend measurably on the point x. They will be called *transition probabilities*: for each measurable set $E \subset X$, the number $P(x,E)$ is meant to represent the probability that $x_{n+1} \in E$, given that $x_n = x$. A probability measure p in X is called a *stationary measure*, relative to the family of transition probabilities, if it satisfies

$$\int P(x,E)\,dp(x) = p(E), \quad \text{for every measurable set } E \subset X. \tag{7.2.1}$$

Heuristically, this means that, relative to p, a probability of the event $x_{n+1} \in E$ is equal to the probability of the event $x_n \in E$.

Fix any stationary measure p (assuming it exists) and then define

$$\mu([m;A_m,\dots,A_n]) \\ = \int_{A_m} dp(x_m) \int_{A_{m+1}} dP(x_m,x_{m+1}) \cdots \int_{A_n} dP(x_{n-1},x_n) \tag{7.2.2}$$

for every cylinder $[m;A_m,\dots,A_n]$ of Σ. One can show (check Exercise 7.2.1) that this function extends to a probability measure in the σ-algebra generated by the cylinders. This probability measure is invariant under the shift map σ, because the right-hand side of (7.2.2) does not depend on m. Every probability measure μ obtained in this way is called a *Markov measure*; moreover, the system (σ,μ) is called a *Markov shift*.

Example 7.2.1 (Bernoulli measure)**.** Suppose that $P(x,\cdot)$ does not depend on x, that is, that there exists a probability measure ν on X such that $P(x,\cdot)=\nu$ for every $x\in X$. Then

$$\int P(x,E)\,dp(x) = \int \nu(E)\,dp(x) = \nu(E)$$

for every probability measure p and every measurable set $E\subset X$. Therefore, there exists exactly one stationary measure, namely $p=\nu$. The definition in (7.2.2) gives

$$\begin{aligned}\mu([m;A_m,\dots,A_n]) &= \int_{A_m} d\nu(x_m) \int_{A_{m+1}} d\nu(x_{m+1}) \cdots \int_{A_n} d\nu(x_n) \\ &= \nu(A_m)\nu(A_{m+1})\cdots\nu(A_n).\end{aligned}$$

Example 7.2.2. Suppose that the set X is finite, say $X=\{1,\dots,d\}$ for some $d\geq 2$. Any family of transition probabilities $P(x,\cdot)$ on X is completely characterized by the values

$$P_{i,j} = P(i,\{j\}), \quad 1\leq i,j\leq d. \tag{7.2.3}$$

Moreover, a measure p on the set X is completely characterized by the values $p_i=p(\{i\})$, $1\leq i\leq d$. With these notations, the definition (7.2.1) translates to

$$\sum_{i=1}^{d} p_i P_{i,j} = p_j, \quad \text{for every } 1\leq j\leq d. \tag{7.2.4}$$

Moreover, a Markov measure μ is determined by

$$\mu([m;a_m,\dots,a_n]) = p_{a_m} P_{a_m,a_{m+1}} \cdots P_{a_{n-1},a_n}. \tag{7.2.5}$$

In the remainder of this book we always restrict ourselves to *finite* Markov shifts, that is, to the context of Example 7.2.2. We take the set X endowed with the discrete topology and the corresponding Borel σ-algebra. Observe that the matrix

$$P = (P_{i,j})_{1\leq i,j\leq d}$$

defined by (7.2.3) satisfies the following conditions:

(i) $P_{i,j} \geq 0$ for every $1 \leq i,j \leq d$;
(ii) $\sum_{j=1}^{d} P_{i,j} = 1$ for every $1 \leq i \leq d$.

We say that P is a *stochastic matrix*. Conversely, any matrix that satisfies (i) and (ii) defines a family of transition probabilities on the set X. Observe also that, denoting $p = (p_1, \ldots, p_d)$, the relation (7.2.4) corresponds to

$$P^* p = p, \tag{7.2.6}$$

where P^* denotes the transpose of the matrix P. In other words, the stationary measures correspond precisely to the eigenvectors of the transposed matrix for the eigenvalue 1. Using the following classical result, one can show that such eigenvalues always exist:

Theorem 7.2.3 (Perron–Frobenius). *Let A be a $d \times d$ matrix with non-negative coefficients. Then there exists $\lambda \geq 0$ and some vector $v \neq 0$ with non-negative coefficients such that $Av = \lambda v$ and $\lambda \geq |\gamma|$ for every eigenvalue γ of A.*

If A has some power whose coefficients are all positive then $\lambda > 0$ and it has some eigenvector v whose coefficients are all positive. Indeed, $\lambda > |\gamma|$ for any other eigenvalue γ of A. Moreover, the eigenvalue λ has multiplicity 1 *and it is the only eigenvalue of A having some eigenvector with non-negative coefficients.*

A proof of the Perron–Frobenius theorem may be found in Meyers [Mey00], for example. Applying this theorem to the matrix $A = P^*$, we conclude that there exist $\lambda \geq 0$ and $p \neq 0$ with $p_i \geq 0$ for every i, such that

$$\sum_{i=1}^{d} p_i P_{i,j} = \lambda p_j, \quad \text{for every } 1 \leq j \leq d.$$

Adding over $j = 1, \ldots, d$ we get that

$$\sum_{j=1}^{d} \sum_{i=1}^{d} p_i P_{i,j} = \lambda \sum_{j=1}^{d} p_j.$$

Using property (ii) of the stochastic matrix, the left-hand side of this equality may be written as

$$\sum_{i=1}^{d} p_i \sum_{j=1}^{d} P_{i,j} = \sum_{i=1}^{d} p_i.$$

Comparing the last two equalities and recalling that the sum of the coefficients of p is a positive number, we conclude that $\lambda = 1$. This proves our claim that there always exist vectors $p \neq 0$ satisfying (7.2.6).

When P^n has positive coefficients for some $n \geq 1$, it follows from Theorem 7.2.3 that the eigenvector is unique up to scaling, and it may be chosen with positive coefficients.

Example 7.2.4. In general, p is not unique and it may also happen that there is no eigenvalue with positive coefficients. For example, consider:

$$P = \begin{pmatrix} 1-a & a & 0 & 0 & 0 \\ b & 1-b & 0 & 0 & 0 \\ 0 & 0 & 1-c & c & 0 \\ 0 & 0 & d & 1-d & 0 \\ e & 0 & 0 & 0 & 1-e \end{pmatrix}$$

where $a,b,c,d,e \in (0,1)$. A vector $p = (p_1,p_2,p_3,p_4,p_5)$ satisfies $P^*p = p$ if and only if $ap_1 = bp_2$ and $cp_3 = dp_4$ and $p_5 = 0$. Therefore, the eigenspace has dimension 2 and no eigenvector has positive coefficients.

On the other hand, suppose that p is such that $p_i = 0$ for some i. Let μ be the corresponding Markov measure and let $\Sigma_i = (X \setminus \{i\})^{\mathbb{N}}$ (or $\Sigma_i = (X \setminus \{i\})^{\mathbb{Z}}$). Then $\mu(\Sigma_i) = 1$, since $\mu([n;i]) = p_i = 0$ for every n. This means that we may eliminate the symbol i, and still have a system that is equivalent to the original one. Therefore, up to removing from the set X a certain number of superfluous symbols, we may always take the eigenvector p to have positive coefficients.

Denote by Σ_P the set of all sequences $(x_n)_n \in \Sigma$ satisfying

$$P_{x_n,x_{n+1}} > 0 \quad \text{for every } n, \tag{7.2.7}$$

that is, such that all the transitions are "allowed" by P. It is clear from the definition that Σ_P is invariant under the shift map σ. The transformations $\sigma : \Sigma_P \to \Sigma_P$ constructed in this way are called *shifts of finite type* and will be studied in more detail in Section 10.2.2.

Lemma 7.2.5. *The set Σ_P is closed in Σ and, given any solution of $P^*p = p$ with positive coefficients, the support of the corresponding Markov measure μ coincides with Σ_P.*

Proof. Let $x^k = (x_n^k)_n$, $k \in \mathbb{N}$ be any sequence in Σ_P and suppose that it converges in Σ to some $x = (x_n)_n$. By the definition of the topology in Σ, this means that for every n there exists $k_n \geq 1$ such that $x_n^k = x_n$ for every $k \geq k_n$. So, given any n, taking $k \geq \max\{k_n, k_{n+1}\}$ we conclude that $P_{x_n,x_{n+1}} = P_{x_n^k,x_{n+1}^k} > 0$. This shows that $x \in \Sigma_P$ and that proves the first part of the lemma.

To prove the second part, recall that the cylinders $[m;x_m,\ldots,x_n]$ form a basis of neighborhoods of any $x = (x_n)_n$ in Σ. If $x \in \Sigma_P$ then

$$\mu([m;x_m,\ldots,x_n]) = p_{x_m} P_{x_m,x_{m+1}} \cdots P_{x_{n-1},x_n} > 0$$

for every cylinder and, thus, $x \in \operatorname{supp}\mu$. If $x \notin \Sigma_P$ then there exists n such that $P_{x_n,x_{n+1}} = 0$. In that case, $\mu([n;x_n,x_{n+1}]) = 0$ and so $x \notin \operatorname{supp}\mu$.

Example 7.2.6. There are three possibilities for the support of a Markov measure in Example 7.2.4. If $p = (p_1,p_2,0,0,0)$ with $p_1,p_2 > 0$ then we may eliminate the symbols $3,4,5$. All the sequences of symbols $1,2$ are admissible.

Hence $\operatorname{supp}\mu = \{1,2\}^{\mathbb{N}}$. Analogously, if $p = (0,0,p_3,p_4,0)$ with $p_3,p_4 > 0$ then $\operatorname{supp}\mu = \{3,4\}^{\mathbb{N}}$. In all the other cases, $p = (p_1,p_2,p_3,p_4,0)$ with $p_1,p_2,p_3,p_4 > 0$. Eliminating the symbol 5, we get that the set of admissible sequences is

$$\Sigma_P = \{1,2\}^{\mathbb{N}} \cup \{3,4\}^{\mathbb{N}}.$$

Both sets in this union are invariant and have positive measure. So, in this case the Markov shift (σ,μ) is not ergodic. But it follows from the theory presented in the next section that in the previous two cases the system (σ,μ) is indeed ergodic.

In the next lemma we collect some simple properties of stochastic matrices that will be useful in what follows:

Lemma 7.2.7. *Let P be a stochastic matrix and $p = (p_1,\dots,p_d)$ be a solution of $P^*p = p$. For every $n \ge 0$, denote by $P^n_{i,j}$, $1 \le i,j \le d$ the coefficients of the matrix P^n. Then:*

(i) $\sum_{j=1}^d P^n_{i,j} = 1$ for every $1 \le i \le d$ and every $n \ge 1$;
(ii) $\sum_{i=1}^d p_i P^n_{i,j} = p_j$ for every $1 \le j \le d$ and every $n \ge 1$;
(iii) the hyperplane $H = \{(h_1,\dots,h_d) : h_1 + \dots + h_d = 0\}$ is invariant under the matrix P^*.

Proof. Condition (ii) in the definition of stochastic matrix may be written as $Pu = u$, with $u = (1,\dots,1)$. Then $P^n u = u$ for every $n \ge 1$. This is just another way of writing claim (i). Analogously, $P^*p = p$ implies that $(P^*)^n p = p$ for every $n \ge 1$, which is another way of writing claim (ii). Observe that H is the orthogonal complement of vector u. Since u is invariant under P, it follows that H is invariant under the transposed matrix P^*, as claimed in (iii).

7.2.1 Ergodicity

In this section we always take $p = (p_1,\dots,p_d)$ to be a solution of $P^*p = p$ with $p_i > 0$ for every i, normalized in such a way that $\sum_i p_i = 1$. Let μ be the corresponding Markov measure. We want to understand which conditions the stochastic matrix P must satisfy for the system (σ,μ) to be ergodic.

We say that a stochastic matrix P is *irreducible* if for every $1 \le i,j \le d$ there exists $n \ge 0$ such that $P^n_{i,j} > 0$. In other words, P is irreducible if any outcome i may be followed by any outcome j, after a certain number n of steps which may depend on i and j.

Theorem 7.2.8. *The Markov shift (σ,μ) is ergodic if and only if the matrix P is irreducible.*

The remainder of the present section is dedicated to the proof of this theorem. We start by proving the following useful estimate:

Lemma 7.2.9. *Let $A = [m; a_m, \ldots, a_q]$ and $B = [r; b_r, \ldots, b_s]$ be cylinders of Σ with $r > q$. Then*

$$\mu(A \cap B) = \mu(A)\mu(B)\frac{P^{r-q}_{a_q,b_r}}{p_{b_r}}.$$

Proof. We may write $A \cap B$ as a disjoint union

$$A \cap B = \bigcup_x [m; a_m, \ldots, a_q, x_{q+1}, \ldots, x_{r-1}, b_r, \ldots, b_s],$$

over all $x = (x_{q+1}, \ldots, x_{r-1}) \in X^{r-q-1}$. Then,

$$\mu(A \cap B) = \sum_x p_{a_m} P_{a_m,a_{m+1}} \cdots P_{a_{q-1},a_q} P_{a_q,x_{q+1}} \cdots P_{x_{r-1},b_r} P_{b_r,b_{r+1}} \cdots P_{b_{s-1},b_s}$$

$$= \mu(A) \sum_x P_{a_q,x_{q+1}} \cdots P_{x_{r-1},b_r} \frac{1}{p_{b_r}} \mu(B).$$

The sum in this last expression is equal to $P^{r-q}_{a_q,b_r}$. Therefore,

$$\mu(A \cap B) = \mu(A)\mu(B)P^{r-q}_{a_q,b_r}/p_{b_r},$$

as stated.

Lemma 7.2.10. *A stochastic matrix P is irreducible if and only if*

$$\lim_n \frac{1}{n}\sum_{l=0}^{n-1} P^l_{i,j} = p_j \quad \textit{for every } 1 \le i,j \le d. \tag{7.2.8}$$

Proof. Assume that (7.2.8) holds. Recall that $p_j > 0$ for every j. Then, given any $1 \le i,j \le d$, we have $P^l_{i,j} > 0$ for infinitely many values of l. In particular, P is irreducible.

To prove the converse, consider $A = [0; i]$ and $B = [0; j]$. By Lemma 7.2.9,

$$\frac{1}{n}\sum_{l=0}^{n-1} \mu\big(A \cap \sigma^{-l}(B)\big) = \frac{1}{p_j}\mu(A)\mu(B)\frac{1}{n}\sum_{l=0}^{n-1} P^l_{i,j}.$$

According to Exercise 4.1.2, the left-hand side converges when $n \to \infty$. Therefore,

$$Q_{i,j} = \lim_n \frac{1}{n}\sum_{l=0}^{n-1} P^l_{i,j}$$

exists for every $1 \le i,j \le d$. Consider the matrix $Q = (Q_{i,j})_{i,j}$, that is,

$$Q = \lim_n \frac{1}{n}\sum_{l=0}^{n-1} P^l. \tag{7.2.9}$$

Using Lemma 7.2.7(ii) and taking the limit when $n \to \infty$, we get that

$$\sum_{i=1}^{d} p_i Q_{i,j} = p_j \quad \text{for every } 1 \le j \le d. \tag{7.2.10}$$

Observe also that, given any $k \geq 1$,

$$P^k Q = \lim_n \frac{1}{n} \sum_{l=0}^{n-1} P^{k+l} = \lim_n \frac{1}{n} \sum_{l=0}^{n-1} P^l = Q. \qquad (7.2.11)$$

It follows that $Q_{i,j}$ does not depend on i. Indeed, suppose that there exist r and s such that $Q_{r,j} < Q_{s,j}$. Of course, we may choose s in such a way that the right-hand side of this inequality is larger. Since P is irreducible, there exists k such that $P^k_{s,r} > 0$. Hence, using (7.2.11) followed by Lemma 7.2.7(i),

$$Q_{s,j} = \sum_{i=1}^{d} P^k_{s,i} Q_{i,j} < \left(\sum_{i=1}^{d} P^k_{s,i} \right) Q_{s,j} = Q_{s,j},$$

which is a contradiction. This contradiction proves that $Q_{i,j}$ does not depend on i, as claimed. Write $Q_j = Q_{i,j}$ for any i. The property (7.2.10) gives that

$$p_j = \sum_{i=1}^{d} Q_{i,j} p_i = Q_j \left(\sum_{i=1}^{d} p_i \right) = Q_j,$$

for every j. This finishes the proof of the lemma.

Proof of Theorem 7.2.8. Suppose that μ is ergodic. Let $A = [0;i]$ and $B = [0;j]$. By Proposition 4.1.4,

$$\lim_n \frac{1}{n} \sum_{l=0}^{n-1} \mu\big(A \cap \sigma^{-l}(B)\big) = \mu(A)\mu(B) = p_i p_j. \qquad (7.2.12)$$

On the other hand, by Lemma 7.2.9, we have that $\mu(A \cap \sigma^{-l}(B)) = p_i P^l_{i,j}$. Using this identity in (7.2.12) and dividing both sides by p_i we find that

$$\lim_n \frac{1}{n} \sum_{l=0}^{n-1} P^l_{i,j} = p_j.$$

Note that j is arbitrary. So, by Lemma 7.2.10, this proves that P is irreducible.

Now suppose that the matrix P is irreducible. We want to conclude that μ is ergodic. According to Corollary 4.1.5, it is enough to prove that

$$\lim_n \frac{1}{n} \sum_{l=0}^{n-1} \mu\big(A \cap \sigma^{-l}(B)\big) = \mu(A)\mu(B) \qquad (7.2.13)$$

for any A and B in the algebra generated by the cylinders. Since the elements of this algebra are the finite pairwise disjoint unions of cylinders, it suffices to consider the case when A and B are cylinders, say $A = [m; a_m, \ldots, a_q]$ and $B = [r; b_r, \ldots, b_s]$. Observe also that the validity of (7.2.13) is not affected if one replaces B by some pre-image $\sigma^{-j}(B)$. So, it is no restriction to suppose that $r > q$. Then, by Lemma 7.2.9,

$$\frac{1}{n} \sum_{l=0}^{n-1} \mu\big(A \cap \sigma^{-l}(B)\big) = \mu(A)\mu(B) \frac{1}{p_{b_r}} \frac{1}{n} \sum_{l=0}^{n-1} P^{r-q+l}_{a_q, b_r}$$

for every n. By Lemma 7.2.10,

$$\lim_n \frac{1}{n}\sum_{l=0}^{n-1} P^{r-q+l}_{a_q,b_r} = \lim_n \frac{1}{n}\sum_{l=0}^{n-1} P^{l}_{a_q,b_r} = p_{b_r}.$$

This proves the property (7.2.13) for the cylinders A and B.

7.2.2 Mixing

In this section we characterize the Markov shifts that are mixing, in terms of the corresponding stochastic matrix P. As before, we take p to be a normalized solution of $P^*p = p$ with positive coefficients and μ to be the corresponding Markov measure.

We say that a stochastic matrix P is *aperiodic* if there exists $n \geq 1$ such that $P^n_{i,j} > 0$ for every $1 \leq i,j \leq d$. In other words, P is aperiodic if some power P^n has only positive coefficients. The relation between the notions of aperiodicity and irreducibility is analyzed in Exercise 7.2.6.

Theorem 7.2.11. *The Markov shift (σ, μ) is mixing if and only if the matrix P is aperiodic.*

For the proof of Theorem 7.2.11 we need the following fact:

Lemma 7.2.12. *A stochastic matrix P is aperiodic if and only if*

$$\lim_l P^l_{i,j} = p_j \quad \textit{for every } 1 \leq i,j \leq d. \tag{7.2.14}$$

Proof. Since we assume that $p_j > 0$ for every j, it is clear that (7.2.14) implies that $P^l_{i,j} > 0$ for every i,j and every l sufficiently large.

Now suppose that P is aperiodic. Then we may apply the theorem of Perron–Frobenius (Theorem 7.2.3) to the matrix $A = P^*$. Since p is an eigenvector of A with positive coefficients, we get that $\lambda = 1$ and all the other eigenvalues of A are smaller than 1 in absolute value. By Lemma 7.2.7(iii), the hyperplane H formed by the vectors $(h_1, \ldots, h_d)$ with $h_1 + \cdots + h_d = 0$ is invariant under A. It is clear that H is transverse to the direction of p. Then the decomposition

$$\mathbb{R}^d = \mathbb{R}p \oplus H \tag{7.2.15}$$

is invariant under A and the restriction of A to the hyperplane H is a contraction, meaning that its spectral radius is smaller than 1. It follows that the sequence $(A^l)_l$ converges to the projection on the first coordinate of (7.2.15), that is, to the matrix B characterized by $Bp = p$ and $Bh = 0$ for every $h \in H$. In other words, $(P^l)_l$ converges to B^*. Observe that

$$B_{i,j} = p_i \quad \text{for every } 1 \leq i,j \leq d.$$

Therefore, $\lim_n P^l_{i,j} = B_{j,i} = p_j$ for every i,j.

Proof of Theorem 7.2.11. Suppose that the measure μ is mixing. Let $A = [0;i]$ and $B = [0;j]$. By Lemma 7.2.9, we have that $\mu(A \cap \sigma^{-l}(B)) = p_i P^l_{i,j}$ for every l. Therefore,

$$p_i \lim_l P^l_{i,j} = \lim_l \mu\big(A \cap \sigma^{-l}(B)\big) = \mu(A)\mu(B) = p_i p_j.$$

Dividing both sides by p_i we get that $\lim_l P^l_{i,j} = p_j$. According to Lemma 7.2.12, this proves that P is aperiodic.

Now suppose that the matrix P is aperiodic. We want to conclude that μ is mixing. According to Lemma 7.1.2, it is enough to prove that

$$\lim_l \mu\big(A \cap \sigma^{-l}(B)\big) = \mu(A)\mu(B) \tag{7.2.16}$$

for any A and B in the algebra generated by the cylinders. Since the elements of this algebra are the finite pairwise disjoint unions of cylinders, it suffices to treat the case when A and B are cylinders, say $A = [m; a_m, \ldots, a_q]$ and $B = [r; b_r, \ldots, b_s]$. By Lemma 7.2.9,

$$\mu\big(A \cap \sigma^{-l}(B)\big) = \mu(A)\mu(B)\frac{1}{p_{b_r}} P^{r-q+l}_{a_q,b_r}$$

for every $l > q - r$. Then, using Lemma 7.2.12,

$$\begin{aligned}\lim_l \mu\big(A \cap \sigma^{-l}(B)\big) &= \mu(A)\mu(B)\frac{1}{p_{b_r}} \lim_l P^{r-q+l}_{a_q,b_r} \\ &= \mu(A)\mu(B)\frac{1}{p_{b_r}} \lim_l P^{l}_{a_q,b_r} = \mu(A)\mu(B)\end{aligned}$$

This proves the property (7.2.16) for cylinders A and B.

Example 7.2.13. In Example 7.2.4 we found different types of Markov measures, depending on the choice of the probability eigenvector p. In the first case, $p = (p_1, p_2, 0, 0, 0)$ and the measure μ is supported on $\{1,2\}^{\mathbb{N}}$. Once the superfluous symbols 3, 4, 5 have been removed, the stochastic matrix reduces to

$$P = \begin{pmatrix} 1-a & a \\ b & 1-b \end{pmatrix}.$$

Since this matrix is aperiodic, the Markov measure μ is mixing. The second case is entirely analogous. In the third case, $p = (p_1, p_2, p_3, p_4, 0)$ and, after removing the superfluous symbol 5, the stochastic matrix reduces to

$$P = \begin{pmatrix} 1-a & a & 0 & 0 \\ b & 1-b & 0 & 0 \\ 0 & 0 & 1-c & c \\ 0 & 0 & d & 1-d \end{pmatrix}.$$

This matrix is not irreducible and, hence, the Markov measures that one finds in this case are not ergodic (recall also Example 7.2.6).

Example 7.2.14. It is not difficult to find examples of irreducible matrices that are not aperiodic:

$$P = \begin{pmatrix} 0 & 1/2 & 0 & 1/2 \\ 1/2 & 0 & 1/2 & 0 \\ 0 & 1/2 & 0 & 1/2 \\ 1/2 & 0 & 1/2 & 0 \end{pmatrix}.$$

Indeed, we see that $P^n_{i,j} > 0$ if and only if n has the same parity as $i-j$. Note that

$$P^2 = \begin{pmatrix} 1/2 & 0 & 1/2 & 0 \\ 0 & 1/2 & 0 & 1/2 \\ 1/2 & 0 & 1/2 & 0 \\ 0 & 1/2 & 0 & 1/2 \end{pmatrix}.$$

Exercise 7.2.6 shows that every irreducible matrix has a form of this type.

7.2.3 Exercises

7.2.1. Let $X = \{1,\dots,d\}$ and $P = (P_{i,j})_{i,j}$ be a stochastic matrix and $p = (p_i)_i$ be a probability vector such that $P^*p = p$. Show that the function defined on the set of all cylinders by

$$\mu([m;a_m,\dots,a_n]) = p_{a_m} P_{a_m,a_{m+1}} \cdots P_{a_{n-1},a_n}$$

extends to a measure on the Borel σ-algebra of $\Sigma = X^{\mathbb{N}}$ (or $\Sigma = X^{\mathbb{Z}}$), invariant under the shift map $\sigma : \Sigma \to \Sigma$.

7.2.2. Prove that every weak mixing Markov shift is actually mixing.

7.2.3. Let $X = \{1,\dots,d\}$ and let μ be a Markov measure for the shift map $\sigma : X^{\mathbb{Z}} \to X^{\mathbb{Z}}$. Does it follow that μ is also a Markov measure for the inverse $\sigma^{-1} : \Sigma \to \Sigma$?

7.2.4. Let X be a finite set and $\Sigma = X^{\mathbb{Z}}$ (or $\Sigma = X^{\mathbb{N}}$). Let μ be a probability measure on Σ, invariant under the shift map $\sigma : \Sigma \to \Sigma$. Given $k \geq 0$, we say that μ has *memory k* if

$$\frac{\mu([m-l;a_{m-l},\dots,a_{m-1},a_m])}{\mu([m-l;a_{m-l},\dots,a_{m-1}])} = \frac{\mu([m-k;a_{m-k},\dots,a_{m-1},a_m])}{\mu([m-k;a_{m-k},\dots,a_{m-1}])}$$

for every $l \geq k$, every m and every $(a_n)_n \in \Sigma$ (by convention, the equality holds whenever at least one of the denominators is zero). Check that the measures with memory 0 are the Bernoulli measures and the measures with memory 1 are the Markov measures. Show that every measure with memory $k \geq 2$ is equivalent to a Markov measure in the space $\tilde{\Sigma} = \tilde{X}^{\mathbb{Z}}$ (or $\tilde{\Sigma} = \tilde{X}^{\mathbb{N}}$), where $\tilde{X} = X^k$.

7.2.5. The goal is to show that the set of all measures with finite memory is dense in the space $\mathcal{M}_1(\sigma)$ of all probability measures invariant under the shift map $\sigma : \Sigma \to \Sigma$. Given any invariant probability measure μ and any $k \geq 1$, consider the function μ_k defined on the set of all cylinders by

- $\mu_k = \mu$ for cylinders with length less than or equal to k;
- for every $l \geq k$, every m and every $(a_n)_n \in \Sigma$,

$$\frac{\mu_k([m-l;a_{m-l},\dots,a_{m-1},a_m])}{\mu_k([m-l;a_{m-l},\dots,a_{m-1}])} = \frac{\mu([m-k;a_{m-k},\dots,a_{m-1},a_m])}{\mu([m-k;a_{m-k},\dots,a_{m-1}])}.$$

Show that μ_k extends to a probability measure on the Borel σ-algebra of Σ, invariant under the shift map and with memory k. Check that $\lim_k \mu_k = \mu$ in the weak* topology.

7.2.6. Let P be an irreducible stochastic matrix. The goal is to show that there exist $\kappa \geq 1$ and a partition of X into κ subsets such that the restriction of P^κ to each of these subsets is aperiodic. To do so:
(1) For every $i \in X$, define $R(i) = \{n \geq 1 : P^n_{i,i} > 0\}$. Show that $R(i)$ is closed under addition: if $n_1, n_2 \in R(i)$ then $n_1 + n_2 \in R(i)$.
(2) Let $R \subset \mathbb{N}$ be closed under addition and let $\kappa \geq 1$ be the greatest common divisor of its elements. Show that there exists $m \geq 1$ such that $R \cap [m, \infty) = \kappa\mathbb{N} \cap [m, \infty)$.
(3) Show that the greatest common divisor κ of the elements of $R(i)$ does not depend on $i \in X$ and that P is aperiodic if and only if $\kappa = 1$.
(4) Assume that $\kappa \geq 2$. Find a partition $\{X_r : 0 \leq r < \kappa\}$ of X such that the restriction of P^κ to each X_r is aperiodic.

7.3 Interval exchanges

By definition, an *interval exchange* is a bijection of the interval $[0,1)$ with a finite number of discontinuities and whose restriction to every continuity subinterval is a translation. Figure 7.1 describes an example with four continuity subintervals. To fix ideas, we always take the transformation to be continuous on the right, that is, we take all continuity subintervals to be closed on the left and open on the right.

As a direct consequence of the definition, every interval exchange preserves the Lebesgue measure on $[0,1)$. These transformations exhibit a very rich dynamical behavior and they also have important connections with many other systems, such as polygonal billiards, conservative flows on surfaces and Teichmüller flows. For example, the construction that we sketch next shows that interval exchanges arise naturally as Poincaré return maps of conservative vector fields on surfaces.

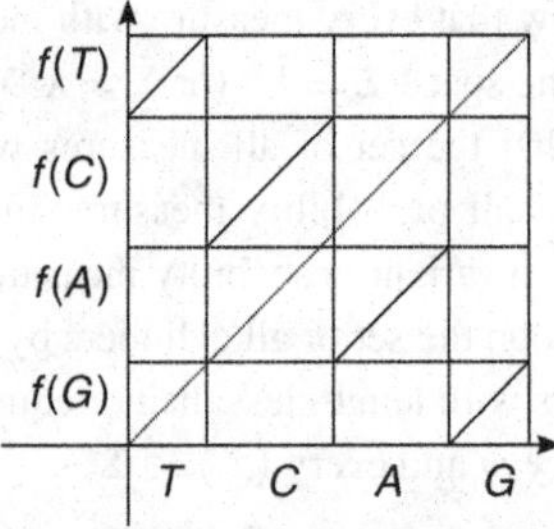

Figure 7.1. An interval exchange

Example 7.3.1. Let S be an orientable surface and ω be an area form in S, that is, a differential 2-form that is non-degenerate at every point. We may associate with every vector field X a differential 1-form β, defined by

$$\beta_x(v) = \omega_x(X(x), v) \quad \text{for every vector } v \in T_x S.$$

Observe that X and β have the same zeros. Moreover, at all other points the kernel of β coincides with the direction of the vector field. The 1-form β permits the definition of the notion of "transverse arc-length" of curves $c : [a,b] \to S$, as follows:

$$\ell(c) = \int_c \beta = \int_a^b \beta_{c(t)}(\dot{c}(t))\, dt.$$

Note that the flow trajectories have transverse arc-length zero. However, for curves transverse to the flow, the measure ℓ is equivalent to the usual arc-length measure, in the sense that they have the same zero measure sets. We can show (see Exercise 7.3.1) that the 1-form β is closed if and only if X preserves area. Then, using the theorem of Green, the Poincaré maps of the flow preserve the transverse length. With an additional hypothesis on the zeros of X, the first-return map $f : \Sigma \to \Sigma$ to any cross-section Σ is well defined and is continuous outside a finite subset of Σ. Then, parameterizing Σ by transverse arc length, f is an interval exchange.

An interval exchange is determined by two ingredients. The first one, of a combinatorial nature, concerns the number of continuity subintervals, the order of these subintervals and the order of their images inside the interval $[0,1)$. This may be informed by assigning a label (for example, a letter) to each continuity subinterval and to its image, and by listing these labels in their corresponding orders, in two horizontal rows. For example, in the case described in Figure 7.1, we obtain

$$\pi = \begin{pmatrix} T & C & A & G \\ G & A & C & T \end{pmatrix}.$$

Note that the choice of the labels is arbitrary. We denote by $\mathcal{A}$, and call *alphabet*, the set of all labels.

The second ingredient, of a metric nature, concerns the lengths of the subintervals. This may be expressed through a vector with positive coefficients, indexed by the alphabet: each coefficient determines the length of the corresponding continuity subinterval (and of its image). In the case of Figure 7.1 this *length vector* has the form

$$\lambda = (\lambda_T, \lambda_C, \lambda_A, \lambda_G).$$

The sum of the coefficients of a length vector is always equal to 1.

Then, the interval exchange $f : [0,1) \to [0,1)$ associated with each pair (π, λ) is defined as follows. For every label $\alpha \in \mathcal{A}$, denote by I_α the corresponding continuity subinterval and define $w_\alpha = v_1 - v_0$, where v_0 is the sum of the

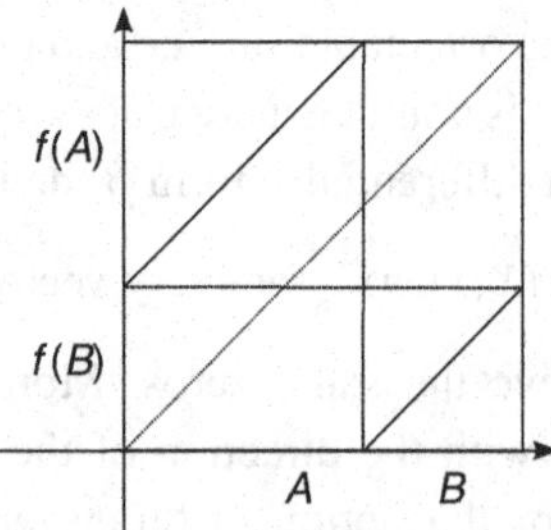

Figure 7.2. Rotation viewed as an exchange of two intervals

lengths λ_β corresponding to all labels β to the left of α on the top row of π and v_1 is the sum of the lengths λ_γ corresponding to all the labels γ to the left of α on the bottom row of π. Then

$$f(x) = x + w_\alpha \quad \text{for every } x \in I_\alpha.$$

The vector $w = (w_\alpha)_{\alpha \in \mathcal{A}}$ is called the *translation vector*. Clearly, for each fixed π, the translation vector is a linear function of the length vector $\lambda = (\lambda_\alpha)_{\alpha \in \mathcal{A}}$.

Example 7.3.2. The simplest interval exchanges have only two continuity subintervals. See Figure 7.2. Choosing the alphabet $\mathcal{A} = \{A, B\}$, we get

$$\pi = \begin{pmatrix} A & B \\ B & A \end{pmatrix} \quad \text{and} \quad f(x) = \begin{cases} x + \lambda_B & \text{for } x \in I_A \\ x - \lambda_A = x + \lambda_B - 1 & \text{for } x \in I_B. \end{cases}$$

This transformation corresponds precisely to the rotation R_{λ_B} if we identify $[0,1)$ with the circle S^1 in the natural way. In this sense, the class of interval exchanges are a generalization of the family of circle rotations.

7.3.1 Minimality and ergodicity

As we saw previously, a circle rotation R_θ is minimal if and only if θ is irrational. Moreover, in that case R_θ is also uniquely ergodic. Given that almost every number is irrational, this means that minimality and unique ergodicity are typical in the family of circle rotations. In this section we discuss the two properties in the broader context of interval exchanges.

Let us start with an observation that has no analogue for rotations. We say that the combinatorics π of an interval exchange *reducible* if there exists some position such that the labels to the left of that position in the two rows of π are exactly the same. For example,

$$\pi = \begin{pmatrix} B & X & O & L & F & D \\ X & O & B & F & D & L \end{pmatrix}$$

is reducible, as the labels to the left of the fourth position are the same in both rows: B, O and X. As a consequence, for any length vector λ, the interval exchange f defined by (π,λ) leaves the subinterval $I_B \cup I_O \cup I_X$ invariant. In particular, f cannot be minimal, not even transitive. In what follows we always assume the combinatorics π to be irreducible.

It is natural to ask whether the interval exchange is minimal whenever the length vector $\lambda = (\lambda_\alpha)_{\alpha\in\mathcal{A}}$ is rationally independent, that is, whenever

$$\sum_{\alpha\in\mathcal{A}} n_\alpha \lambda_\alpha \neq 0$$

for every non-zero vector $(n_\alpha)_{\alpha\in\mathcal{A}}$ with integer coefficients. This turns out to be true but, in fact, the hypothesis of rational independence is a bit too strong: we are going to present a somewhat more general condition that still implies minimality.

We denote by ∂I_α the left endpoint of each subinterval I_α. We say that a pair (π,λ) satisfies the *Keane condition* if the trajectories of these points are disjoint:

$$f^m(\partial I_\alpha) \neq \partial I_\beta \quad \text{for every } m \geq 1 \text{ and any } \alpha,\beta \in \mathcal{A} \text{ with } \partial I_\beta \neq 0 \qquad (7.3.1)$$

(note that there always exist $\bar{\alpha}$ and $\bar{\beta}$ such that $f(\partial I_{\bar{\alpha}}) = 0 = \partial I_{\bar{\beta}}$). We leave the proof of the next lemma as an exercise (Exercise 7.3.2):

Lemma 7.3.3. *(1) If the pair (π,λ) satisfies the Keane condition then the combinatorics matrix π is irreducible.*

(2) If π is irreducible and λ is rationally independent then the pair (π,λ) satisfies the Keane condition.

Since the subset of rationally independent vectors has full Lebesgue measure, it follows that the Keane condition is satisfied for almost every length vector λ, if π is irreducible.

Example 7.3.4. In the case of two subintervals (recall Example 7.3.2), the interval exchange has the form $f^m(x) = x + m\lambda_B \mod \mathbb{Z}$. Then, the Keane condition means that

$$m\lambda_B \neq \lambda_A + n \quad \text{and} \quad \lambda_A + m\lambda_B \neq \lambda_A + n$$

for every $m \in \mathbb{N}$ and $n \in \mathbb{Z}$. It is clear that this holds if and only if the vector (λ_A, λ_B) is rationally independent.

Example 7.3.5. For exchanges of three or more intervals, the Keane condition is strictly weaker than the rational independence of the length vector. Consider, for example,

$$\pi = \begin{pmatrix} A & B & C \\ C & A & B \end{pmatrix}.$$

Then $f^m(x) = x + m\lambda_C \mod \mathbb{Z}$ and, thus, the Keane condition means that

$$\{m\lambda_C, \lambda_A + m\lambda_C, \lambda_A + \lambda_B + m\lambda_C\} \quad \text{and} \quad \{\lambda_A + n, \lambda_A + \lambda_B + n\}$$

are disjoint for every $m \in \mathbb{N}$ and $n \in \mathbb{Z}$. Equivalently,

$$p\lambda_C \notin \{q, \lambda_A + q\} \quad \text{for every } p \in \mathbb{Z} \text{ and } q \in \mathbb{Z}.$$

This may hold even when $(\lambda_A, \lambda_B, \lambda_C)$ is rationally dependent.

The following result was proved by Michael Keane [Kea75]:

Theorem 7.3.6 (Keane). *If (π, λ) satisfies the Keane condition then the interval exchange f is minimal.*

Example 7.3.7. The Keane condition is not necessary for minimality. For example, consider the interval exchange defined by (π, λ), where

$$\pi = \begin{pmatrix} A & B & C & D \\ D & C & B & A \end{pmatrix},$$

$\lambda_A = \lambda_C$, $\lambda_B = \lambda_D$ and $\lambda_A/\lambda_B = \lambda_C/\lambda_D$ is irrational. Then (π, λ) does not satisfy the Keane condition and yet f is minimal.

As observed previously, every minimal circle rotation is also uniquely ergodic. This is still true for exchanges of three intervals, but not in general. Indeed, Keane gave an example of an exchange of four intervals exhibiting two ergodic probability measures, notwithstanding the fact that the combinatorics matrix π is irreducible and the length vector λ is rationally independent.

Keane conjectured that, nevertheless, it should be true that almost every interval exchange is uniquely ergodic. The following remarkable result, obtained independently by Howard Masur [Mas82] and William Veech [Vee82], established this conjecture:

Theorem 7.3.8 (Masur, Veech). *Assume that π is irreducible. Then, for Lebesgue-almost every length vector λ, the interval exchange defined by (π, λ) is uniquely ergodic.*

Earlier, Michael Keane and Gérard Rauzy [KR80] had shown that unique ergodicity holds for a residual (Baire second category) subset of length vectors whenever the combinatorics is irreducible.

7.3.2 Mixing

The interval exchanges provide many examples of systems that are uniquely ergodic and weak mixing but not (strongly) mixing.

Indeed, the theorem of Masur–Veech (Theorem 7.3.8) asserts that almost every interval exchange is uniquely ergodic. Another deep theorem, due to Artur Avila and Giovanni Forni [AF07], states that, circle rotations (more precisely, interval exchanges with a unique discontinuity point) excluded, almost every interval exchange is weak mixing. The topological version of this fact had been proved by Arnaldo Nogueira and Donald Rudolph [NR97].

On the other hand, a result of Anatole Katok [Kat80] that we are going to discuss below asserts that interval exchanges are never mixing:

Theorem 7.3.9. *Let $f : [0,1) \to [0,1)$ be an interval exchange and μ be an invariant probability measure. Then (f,μ) is not mixing.*

Proof. We may take μ to be ergodic, for otherwise the conclusion is obvious. If μ has some atom then its support is a periodic orbit and, thus, μ cannot be mixing. Hence, we may also take μ to be non-atomic. Denote by m the Lebesgue measure on the interval and consider the map

$$h : [0,1) \to [0,1), \quad h(x) = \mu([0,x]).$$

Then h is a homeomorphism and satisfies $h_*\mu = m$. Consequently, the map $g = h \circ f \circ h^{-1} : [0,1) \to [0,1)$ has finitely many discontinuity points and preserves the Lebesgue measure. In particular, the restriction of g to each continuity subinterval is a translation. Therefore, g is also an interval exchange. It is clear that (f,μ) is mixing if and only if (g,m) is mixing. Therefore, to prove Theorem 7.3.9 it is no restriction to suppose that μ is the Lebesgue measure m. We do that from now on.

Our goal is to find a measurable set X such that $m(X \cap f^{-n}(X))$ does not converge to $m(X)^2$ when $n \to \infty$. Let $d = \#\mathcal{A}$.

Lemma 7.3.10. *Every interval $J = [a,b)$ contained in some I_β admits a partition $\{J_1,\dots,J_s\}$ into no more than $d+2$ subintervals of the form $J_i = [a_i,b_i)$ and admits natural numbers $t_1,\dots,t_s \geq 1$ such that*

(i) $f^n(J_i) \cap J = \emptyset$ for every $0 < n < t_i$ and $1 \leq i \leq s$;
(ii) $f^{t_i} \mid J_i$ is a translation for every $1 \leq i \leq s$;
(iii) $\{f^{t_1}(J_1),\dots,f^{t_s}(J_s)\}$ is a partition of J;
(iv) the intervals $f^n(J_i)$, $1 \leq i \leq s$, $0 \leq n < t_i$ are pairwise disjoint;
(v) $\bigcup_{n=0}^{\infty} f^n(J) = \bigcup_{i=1}^{s} \bigcup_{n=0}^{t_i-1} f^n(J_i)$.

Proof. Let B be the set formed by the endpoints a and b of J together with the endpoints ∂I_α, $\alpha \in \mathcal{A}$ minus the origin. Then $\#B \leq d+1$. Let $B_J \subset J$ be the set of points $x \in J$ for which there exists $m \geq 1$ such that $f^m(x) \in B$ and $f^n(x) \notin J$ for every $0 < n < m$. The fact that f is injective, together with the definition of m, implies that the map

$$B_J \to B, \quad x \mapsto f^m(x)$$

is injective. In particular, $\#B_J \leq \#B \leq d+1$. Consider the partition of J into subintervals $J_i = [a_i,b_i)$ with endpoints a_i, b_i in the set $B_J \cup \{a,b\}$. This partition has at most $d+2$ elements. By the Poincaré recurrence theorem, for each i there exists $t_i \geq 1$ such that $f^{t_i}(J_i)$ intersects J. Take t_i minimum with this property. Part (i) of the lemma is an immediate consequence. By the definition of B_J, the restriction of f^{t_i} to the interval J_i is a translation, as stated in part (ii), and its image is contained in J. Moreover, the images $f^{t_i}(J_i)$, $1 \le i \le s$ are pairwise disjoint, since f is injective and the t_i are the first-return times to J. In particular,

$$\sum_{i=1}^{s} m(f^{t_i}(J_i)) = \sum_{i=1}^{s} m(J_i) = m(J)$$

and so $\bigcup_{i=1}^{s} f^{t_i}(J_i) = J$. This proves part (iii). Part (iv) also follows directly from the fact that f is injective and the t_i are the first-return times to J. Finally, part (v) is a direct consequence of part (iii).

Consider any interval J contained in some I_β. By ergodicity, the invariant set $\bigcup_{n=0}^{\infty} f^n(J)$ has full measure. By part (v) of Lemma 7.3.10, this set is a finite union of intervals closed on the left and open on the right. Therefore,

$$\bigcup_{n=0}^{\infty} f^n(J) = \bigcup_{i=1}^{s} \bigcup_{n=0}^{t_i-1} f^n(J_i) = I.$$

So, by Lemma 7.3.10(iv), the family $\mathcal{P}_J = \{f^n(J_i) : 1 \le i \le s \text{ and } 0 \le n < t_i\}$ is a partition of I.

Lemma 7.3.11. *Given $\delta > 0$ and $N \ge 1$, we may choose the interval J in such a way that* $\operatorname{diam} \mathcal{P}_J < \delta$ *and $t_i \ge N$ for every i.*

Proof. It is clear that $\operatorname{diam} f^n(J_i) = \operatorname{diam} J_i \le \operatorname{diam} J$ for every i and every n. Hence, $\operatorname{diam} \mathcal{P}_J < \delta$ as long as we pick J with diameter smaller than δ. To get the second property in the statement, take any point $x \in I$ such that $f^n(x) \ne \partial I_\alpha$ for every $0 \le n < N$ and $\alpha \in \mathcal{A}$. We claim that $f^n(x) \ne x$ for every $0 < n < N$. Otherwise, since f^n is a translation in the neighborhood of x, we would have $f^n(y) = y$ for every point y in that neighborhood, which would contradict the hypothesis that (f, m) is ergodic. This proves our claim. Now it suffices to take $J = [x, x+\varepsilon)$ with $\varepsilon < \min_{0<n<N} d(x, f^n(x))$ to ensure that $t_i \ge N$ for every i.

Lemma 7.3.12. *For every $1 \le i \le s$ there exist $s_i \le d+2$ and natural numbers $\{t_{i,1}, \dots, t_{i,s_i}\}$ such that $t_{i,j} \ge t_i$ and, given any set A in the algebra $\mathcal{A}_J$ generated by $\mathcal{P}_J$, there exists $t_{i,j}$ such that*

$$m\big(A \cap f^{-t_{i,j}}(A)\big) \ge \frac{1}{(d+2)^2} m(A). \tag{7.3.2}$$

Proof. Applying Lemma 7.3.10 to each of the intervals J_i, $1 \le i \le s$ we find $s_i \le d+2$, a partition $\{J_{i,j} : 1 \le j \le s_i\}$ of the interval J_i and natural numbers $t_{i,j}$ such that each $t_{i,j}$ is the first-return time of the points of $J_{i,j}$ to J_i. It is clear that $t_{i,j} \ge t_i$, since t_i is the first-return time of any point of J_i to the interval J. The fact that $J_{i,j} \subset f^{-t_{i,j}}(J_i)$ implies that

$$f^n(J_i) = \bigcup_{j=1}^{s_i} f^n(J_{i,j}) \subset \bigcup_{j=1}^{s_i} f^{-t_{i,j}}(f^n(J_i)) \quad \text{for every } n \ge 0.$$

Since the algebra $\mathcal{A}_J$ is formed by the finite pairwise disjoint unions of intervals $f^n(J_i)$, $0 \le n < t_i$, it follows that

$$A \subset \bigcup_{i=1}^{s} \bigcup_{j=1}^{s_i} f^{-t_{i,j}}(A) \quad \text{for every } A \in \mathcal{A}_J.$$

In particular, $m(A) \le \sum_{i=1}^{s} \sum_{j=1}^{s_i} m(A \cap f^{-t_{i,j}}(A))$. Recalling that $s \le d+2$ and $s_i \le d+2$ for every i, this implies (7.3.2).

We are ready to conclude the proof of Theorem 7.3.9. For that, let us fix a measurable set $X \subset [0,1)$ with

$$0 < m(X) < \frac{1}{4(d+2)^2}.$$

By Lemma 7.3.11, given any $N \ge 1$ we may find an interval $J \subset [0,1)$ such that all the first-return times $t_i \ge N$ and there exists $A \in \mathcal{A}_J$ such that

$$m(X \Delta A) < \frac{1}{4} m(X)^2. \tag{7.3.3}$$

Applying Lemma 7.3.12, we get that there exists $t_{i,j} \ge t_i \ge N$ such that:

$$\begin{aligned} m\big(X \cap f^{-t_{ij}}(X)\big) &\ge m\big(A \cap f^{-t_{ij}}(A)\big) - 2m(X \Delta A) \\ &\ge \frac{1}{(d+2)^2} m(A) - \frac{1}{2} m(X)^2. \end{aligned}$$

The relation (7.3.3) implies that $m(A) \ge (3/4)m(X)$. Therefore,

$$\begin{aligned} m\big(X \cap f^{-t_{ij}}(X)\big) &\ge \frac{3}{4} \frac{1}{(d+2)^2} m(X) - \frac{1}{2} m(X)^2 \\ &\ge 3m(X)^2 - \frac{1}{2} m(X)^2 > 2m(X)^2. \end{aligned}$$

This proves that $\limsup_n m(X \cap f^{-n}(X)) \ge 2m(X)^2$, and so the system (f, m) cannot be mixing.

7.3.3 Exercises

7.3.1. Let ω be an area form on a surface. Let X be a differentiable vector field on S and β be the differential 1-form defined on S by $\beta_x = \omega_x(X(x), \cdot)$. Show that β is closed if and only if X preserves the area measure.

7.3.2. Prove Lemma 7.3.3.

7.3.3. Show that if (π, λ) satisfies the Keane condition then f has no periodic points. [Observation: This is a step in the proof of Theorem 7.3.6.]

7.3.4. Let $f : [0,1) \to [0,1)$ be an irreducible interval exchange and let $a \in (0,1)$ be the largest of all the discontinuity points of f and f^{-1}. The *Rauzy–Veech renormalization* $R(f) : [0,1) \to [0,1)$ is defined by $R(f)(x) = g(ax)/a$, where g is the first-return map of f to the interval $[0,a)$. Check that $R(f)$ is an interval exchange with the same number of continuity subintervals as f, or less. If f is described by the data (π, λ), how can we describe $R(f)$?

7.3.5. Given $d \geq 2$ and a bijection $\sigma : \mathbb{N} \to \mathbb{N}$ without periodic points, consider the transformation $f : [0,1] \to [0,1]$ where each $f(x)$ is obtained by permuting the digits of the base d expansion of x as prescribed by σ. More precisely, if $x = \sum_{n=1}^{\infty} a_n d^{-n}$ with $a_n \in \{0,\dots,d-1\}$ and infinitely many values of n such that $a_n < d-1$, then $f(x) = \sum_{n=1}^{\infty} a_{\sigma(n)} d^{-n}$. Show that f preserves the Lebesgue measure m in the interval and that (f,m) is mixing.

7.4 Decay of correlations

In this section we discuss how quickly the correlations sequence $C_n(\varphi,\psi)$ decays to zero in a mixing system. Since we are dealing with deterministic systems, we cannot expect interesting estimates to hold for arbitrary functions. However, as we are going to see, such estimates do exist in many important cases, if we restrict φ and ψ to suitable subsets of functions. Given that the correlations $(\varphi,\psi) \mapsto C_n(\varphi,\psi)$ are bilinear functions, it is natural to consider subsets that are vector subspaces.

We say that (f,μ) has *exponential decay of correlations* on a given vector space $\mathcal{V}$ if there exists $\lambda < 1$ and for every $\varphi,\psi \in \mathcal{V}$ there exists $A(\varphi,\psi) > 0$ such that

$$|C_n(\varphi,\psi)| \leq A(\varphi,\psi)\lambda^n \quad \text{for every } n \geq 1. \tag{7.4.1}$$

There are similar notions (polynomial decay, for instance) where the exponential λ^n is replaced by some other sequence converging to zero.

To illustrate the theory, let us analyze the issue of decay of correlations in the context of one-sided Markov shifts. That will also allow us to introduce several ideas that will be useful later in more general situations. Let $f : M \to M$ be the shift map in $M = X^{\mathbb{N}}$, where $X = \{1,\dots,d\}$ is a finite set. Let $P = (P_{i,j})_{i,j}$ be an aperiodic stochastic matrix and $p = (p_i)_i$ be the positive eigenvector of P^*, normalized by $p_1 + \cdots + p_d = 1$. Let μ be the Markov measure defined in M by (7.2.2).

Consider $L = G^{-1}P^*G$, where G is the diagonal matrix whose coefficients are $p_1,\dots,p_d$. The coefficients of L are given by

$$L_{i,j} = \frac{p_j}{p_i} P_{j,i} \quad \text{for each } 1 \leq i,j \leq d.$$

Recall that we denote $u = (1,\dots,1)$ and $H = \{(h_1,\dots,h_d) : h_1 + \cdots + h_d = 0\}$. Let

$$V = \{(v_1,\dots,v_d) : p_1 v_1 + \cdots + p_d v_d = 0\}.$$

Then $G(u) = p$ and $G(V) = H$. Recalling (7.2.15), it follows that the decomposition

$$\mathbb{R}^d = \mathbb{R}u \oplus V \tag{7.4.2}$$

is invariant under L and all the eigenvalues of the restriction of L to V are smaller than 1 in absolute value. We say that L has the *spectral gap property* if the largest eigenvalue is simple and all the rest of the spectrum is contained in a closed disk with strictly smaller radius.

The *transfer operator* of the shift map f is the linear operator $\mathcal{L}_f$ mapping each function $\psi : M \to \mathbb{R}$ to the function $\mathcal{L}_f \psi : M \to \mathbb{R}$ defined by

$$\mathcal{L}_f \psi(x_1,\ldots,x_n,\ldots,) = \sum_{x_0=1}^{d} L_{x_1,x_0} \psi(x_0,x_1,\ldots,x_n,\ldots). \tag{7.4.3}$$

The transfer operator is dual to the Koopman operator U_f, in the following sense:

$$\int \varphi(\mathcal{L}_f \psi)\, d\mu = \int (U_f \varphi)\psi\, d\mu \tag{7.4.4}$$

for any bounded measurable functions φ, ψ. Let us prove this fact.

We call a function $\varphi : M \to \mathbb{R}$ *locally constant* if there is $n \geq 0$ such that every $\varphi(x)$ depends only on the first n coordinates $x_0,\ldots,x_{n-1}$ of the point x. For example, characteristic functions of cylinders are locally constant functions. Since every bounded measurable function is a uniform limit of linear combinations of characteristic functions of cylinders, it follows that every bounded measurable function is the uniform limit of some sequence of locally constant functions. Thus, to prove (7.4.4) it suffices to consider the case when φ and ψ are both locally constant.

Then, consider functions φ and ψ that depend only on the first n coordinates. By the definition of Markov measure,

$$\int \varphi(\mathcal{L}_f \psi)\, d\mu = \sum_{a_1,\ldots,a_n} p_{a_1} P_{a_1,a_2} \cdots P_{a_{n-1},a_n} \varphi(a_1,\ldots,a_n) \mathcal{L}_f \psi(a_1,\ldots,a_n).$$

Using the definition of the transfer operator, the right-hand side of this expression is equal to

$$\sum_{a_0,a_1,\ldots,a_n} p_{a_0} P_{a_0,a_1} P_{a_1,a_2} \cdots P_{a_{n-1},a_n} \varphi(a_1,\ldots,a_n) \psi(a_0,a_1,\ldots,a_n).$$

Observe that $\varphi(a_1,\ldots,a_n) = U_f \varphi(a_0,a_1,\ldots,a_n)$. So, using once more the definition of the Markov measure, this last expression is equal to $\int (U_f \varphi)\psi\, d\mu$. This proves the duality property (7.4.4).

As a consequence, we may write the correlations sequence in terms of the iterates of the transfer operator:

$$C_n(\varphi,\psi) = \int (U_f^n \varphi)\psi - \int \varphi\, d\mu \int \psi\, d\mu = \int \varphi(\mathcal{L}_f^n \psi - \int \psi\, d\mu) d\mu. \tag{7.4.5}$$

The property $Lu = u$ means that $\sum_j L_{i,j} = 1$ for every j. This has the following useful consequence:

$$\sup |\mathcal{L}_f \psi| \leq \sup |\psi| \quad \text{for every } \psi. \tag{7.4.6}$$

Taking $\varphi \equiv 1$ in (7.4.4) we get the following special case, which will also be useful later:

$$\int \mathcal{L}_f \psi \, d\mu = \int \psi \, d\mu \quad \text{for every } \psi. \tag{7.4.7}$$

Now let us denote by $\mathcal{E}_0$ the subset of functions ψ that depend only on the first coordinate. The map $\psi \mapsto (\psi(1), \dots, \psi(d))$ is an isomorphism between $\mathcal{E}_0$ and the Euclidean space $\mathbb{R}^d$. Moreover, the definition

$$\mathcal{L}_f \psi(x_1) = \sum_{x_0=1}^{d} L_{x_1,x_0} \psi(x_0)$$

shows that the restriction of the transfer operator to $\mathcal{E}_0$ corresponds precisely to the operator $L : \mathbb{R}^d \to \mathbb{R}^d$. Note also that the hyperplane $V \subset \mathbb{R}^d$ corresponds to the subset of $\psi \in \mathcal{E}_0$ such that $\int \psi \, d\mu = 0$. Consider in $\mathcal{E}_0$ the norm defined by $\|\psi\|_0 = \sup |\psi|$.

Fix any number λ between 1 and the spectral radius of L restricted to V. Every function $\psi \in \mathcal{E}_0$ may be written:

$$\psi = c + v \quad \text{with} \quad c = \int \psi \, d\mu \in \mathbb{R}u \quad \text{and} \quad v = \psi - \int \psi \, d\mu \in V.$$

Then the spectral gap property implies that there exists $B > 1$ such that

$$\sup \left| \mathcal{L}_f^n \psi - \int \psi \, d\mu \right| \leq B \|\psi\|_0 \lambda^n \quad \text{for every } n \geq 1. \tag{7.4.8}$$

Using (7.4.5), it follows that

$$|C_n(\varphi, \psi)| \leq B \|\varphi\|_0 \|\psi\|_0 \lambda^n \quad \text{for every } n \geq 1.$$

In this way, we have shown that *every aperiodic Markov shift has exponential decay of correlations in* $\mathcal{E}_0$.

With a little more effort, one can improve this result, by extending the conclusion to a much larger space of functions. Consider in M the distance defined by

$$d\big((x_n)_n, (y_n)_n\big) = 2^{-N(x,y)} \quad \text{where } N(x,y) = \min\{n \geq 0 : x_n \neq y_n\}.$$

Fix any $\theta > 0$ and denote by $\mathcal{E}$ the set of functions φ that are θ-Hölder, that is, such that

$$K_\theta(\varphi) = \sup \left\{ \frac{|\varphi(x) - \varphi(y)|}{d(x,y)^\theta} : x \neq y \right\} \quad \text{is finite.}$$

It is clear that $\mathcal{E}$ contains all the locally constant functions. We claim:

Theorem 7.4.1. *Every aperiodic Markov shift (f, μ) has exponential decay of correlations in the space $\mathcal{E}$ of θ-Hölder functions, for any $\theta > 0$.*

Observe that $\mathcal{L}_f(\mathcal{E}) \subset \mathcal{E}$. The function $\|\psi\| = \sup |\psi| + K_\theta(\psi)$ is a complete norm in $\mathcal{E}$ and the linear operator $\mathcal{L}_f : \mathcal{E} \to \mathcal{E}$ is continuous relative to this norm (Exercise 7.4.1). One way to prove the theorem is by showing that this operator has the spectral gap property, with invariant decomposition

$$\mathcal{E} = \mathbb{R}u \oplus \left\{ \psi \in \mathcal{E} : \int \psi \, d\mu = 0 \right\}.$$

Once that is done, exactly the same argument that we used before for $\mathcal{E}_0$ proves the exponential decay of correlations in $\mathcal{E}$. We do not present the details here (but we will come back to this theme, in a much more general context, near the end of Section 12.3). Instead, we give a direct proof that (7.4.8) may be extended to the space $\mathcal{E}$.

Given $\psi \in \mathcal{E}$ and $x = (x_1, \dots, x_n, \dots) \in M$, we have

$$\mathcal{L}_f^k \psi(x) = \sum_{a_1,\dots,a_k=1}^{d} L_{x_1,a_k} \cdots L_{a_2,a_1} \psi(a_1, \dots, a_k, x_1, \dots, x_n, \dots)$$

for every $k \geq 1$. Then, given $y = (y_1, \dots, y_n, \dots)$ with $x_1 = y_1 = j$,

$$|\mathcal{L}_f^k \psi(x) - \mathcal{L}_f^k \psi(y)| \leq \sum_{a_1,\dots,a_k=1}^{d} L_{j,a_k} \cdots L_{a_2,a_1} K_\theta(\psi) 2^{-k\theta} d(x,y)^\theta .$$

Using the property $\sum_{i=1}^d L_{j,i} = 1$, we conclude that

$$|\mathcal{L}_f^k \psi(x) - \mathcal{L}_f^k \psi(y)| \leq K_\theta(\psi) 2^{-k\theta} d(x,y)^\theta \leq K_\theta(\psi) 2^{-k\theta} . \qquad (7.4.9)$$

Given any function φ, denote by $\pi \varphi$ the function that depends only on the first coordinate and coincides with the mean of φ on each cylinder $[0;i]$:

$$\pi \varphi(i) = \frac{1}{p_i} \int_{[0;i]} \varphi \, d\mu .$$

It is clear that $\sup |\pi \varphi| \leq \sup |\varphi|$ and $\int \pi \varphi \, d\mu = \int \varphi \, d\mu$. The inequality (7.4.9) implies that

$$\sup |\mathcal{L}_f^k \psi - \pi(\mathcal{L}_f^k \psi)| \leq K_\theta(\psi) 2^{-k\theta} \quad \text{for every } k \geq 1.$$

Then, using the property (7.4.6),

$$\sup |\mathcal{L}_f^{k+l} \psi - \mathcal{L}_f^l \pi(\mathcal{L}_f^k \psi)| \leq K_\theta(\psi) 2^{-k\theta} \quad \text{for every } k, l \geq 1. \qquad (7.4.10)$$

Moreover, the properties (7.4.6) and (7.4.7) imply that

$$\sup |\pi(\mathcal{L}_f^k \psi)| \leq \sup |\psi| \quad \text{and} \quad \int \pi(\mathcal{L}_f^k \psi) \, d\mu = \int \psi \, d\mu .$$

Therefore, the property (7.4.8) gives that

$$\sup \left| \mathcal{L}_f^l \pi(\mathcal{L}_f^k \psi) - \int \psi \, d\mu \right| \leq B \sup |\psi| \lambda^l \quad \text{for every } l \geq 1. \qquad (7.4.11)$$

Adding (7.4.10) and (7.4.11), we get that

$$\sup \left| \mathcal{L}_f^{k+l} \psi - \int \psi \, d\mu \right| \leq K_\theta(\psi) 2^{-k\theta} + B \sup |\psi| \lambda^l \quad \text{for every } k, l \geq 1.$$

Fix $\sigma < 1$ such that $\sigma^2 \geq \max\{2^{-\theta}, \lambda\}$. Then the previous inequality (with $l \approx n/2 \approx k$) gives

$$\sup \left| \mathcal{L}_f^n \psi - \int \psi \, d\mu \right| \leq B \| \psi \| \sigma^{n-1} \quad \text{for every } n. \tag{7.4.12}$$

Now Theorem 7.4.1 follows from the same argument that we used before for $\mathcal{E}_0$, with (7.4.12) in the place of (7.4.8).

7.4.1 Exercises

7.4.1. Show that $\|\varphi\| = \sup|\varphi| + K_\theta(\varphi)$ defines a complete norm in the space $\mathcal{E}$ of θ-Hölder functions and the transfer operator $\mathcal{L}_f$ is continuous relative to this norm.

7.4.2. Let $f : M \to M$ be a local diffeomorphism on a compact manifold M and $d \geq 2$ be the degree of f. Assume that there exists $\sigma > 1$ such that $\|Df(x)v\| \geq \sigma \|v\|$ for every $x \in M$ and every vector v tangent to M at the point x. Fix $\theta > 0$ and let $\mathcal{E}$ be the space of θ-Hölder functions $\varphi : M \to \mathbb{R}$. For every $\varphi \in \mathcal{E}$, define

$$\mathcal{L}_f \varphi : M \to \mathbb{R}, \quad \mathcal{L}_f \varphi(y) = \frac{1}{d} \sum_{x \in f^{-1}(y)} \varphi(x).$$

(a) Show that $\inf \varphi \leq \inf \mathcal{L}_f \varphi \leq \sup \mathcal{L}_f \varphi \leq \sup \varphi$ and $K_\theta(\mathcal{L}_f \varphi) \leq \sigma^{-\theta} K_\theta(\varphi)$ for every $\varphi \in \mathcal{E}$.

(b) Conclude that $\mathcal{L}_f : E \to E$ is a continuous linear operator (relative to the norm defined in Exercise 7.4.1) with $\|\mathcal{L}_f\| = 1$.

(c) Show that, for every $\varphi \in E$, the sequence $(\mathcal{L}_f^n \varphi)_n$ converges to a constant $\nu_\varphi \in \mathbb{R}$ when $n \to \infty$. Moreover, there exists $C > 0$ such that

$$\|\mathcal{L}_f^n \varphi - \nu_\varphi\| \leq C \sigma^{-n\theta} \|\varphi\| \quad \text{for every } n \text{ and every } \varphi \in \mathcal{E}.$$

(d) Conclude that the operator $\mathcal{L}_f : \mathcal{E} \to \mathcal{E}$ has the spectral gap property.

(e) Show that the map $\varphi \mapsto \nu_\varphi$ extends to a Borel probability measure on M (recall Theorem A.3.12).

8
Equivalent systems

This chapter is devoted to the *isomorphism problem* in ergodic theory: under what conditions should two systems (f, μ) and (g, ν) be considered "the same" and how does one decide, for given systems, whether they are in those conditions?

The fundamental notion is called *ergodic equivalence*: two systems are said to be ergodically equivalent if, restricted to subsets with full measure, the corresponding transformations are conjugated by some invertible map that preserves the invariant measures. Through such a map, properties of either system may be translated to corresponding properties of the other system.

Although this is a natural notion of isomorphism in the context of ergodic theory, it is not an easy one to handle. In general, the only way to prove that two given systems are equivalent is by exhibiting the equivalence map more or less explicitly. On the other hand, the most usual way to show that two systems are not equivalent is by finding some property that holds for one but not the other.

Thus, it is useful to consider a weaker notion, called *spectral equivalence*: two systems are spectrally equivalent if their Koopman operators are conjugated by some unitary operator. Two ergodically equivalent systems are always spectrally equivalent, but the converse is not true.

The idea of spectral equivalence leads to a rich family of invariants, related to the spectrum of the Koopman operator, that must have the same value for any two systems that are equivalent and, thus, may be used to exclude that possibility. Other invariants, of non-spectral nature, have an equally crucial role. The most important of all, the entropy, will be treated in Chapter 9.

The notions of ergodic equivalence and spectral equivalence, and the relations between them, are studied in Sections 8.1 and 8.2, respectively. In Sections 8.3 and 8.4 we study two classes of systems with opposite dynamical features: transformations with *discrete spectrum*, that include the ergodic translations on compact abelian groups, and transformations with a *Lebesgue spectrum*, which have the Bernoulli shifts as the main example.

These two classes of systems, as well as others that we introduced previously (ergodicity, strong mixing, weak mixing) are invariants of spectral equivalence and, hence, also of ergodic equivalence. Finally, in Section 8.5 we discuss a third notion of equivalence, that we call *ergodic isomorphism*, especially in the context of Lebesgue spaces.

8.1 Ergodic equivalence

Let μ and ν be probability measures invariant under measurable transformations $f : M \to M$ and $g : N \to N$, respectively. We say that the systems (f,μ) and (g,ν) are *ergodically equivalent* if one can find measurable sets $X \subset M$ and $Y \subset N$ with $\mu(M \setminus X) = 0$ and $\nu(N \setminus Y) = 0$, and a measurable bijection $\phi : X \to Y$ with measurable inverse, such that

$$\phi_*\mu = \nu \quad \text{and} \quad \phi \circ f = g \circ \phi.$$

We leave it to the reader to check that this is indeed an equivalence relation, that is, reflexive, symmetric and transitive.

Observe also that the sets X and Y in the definition may be chosen to be invariant under f and g, respectively. Indeed, consider $X_0 = \bigcap_{n=0}^{\infty} f^{-n}(X)$. It is clear from the definition that $X_0 \subset X$ and $f(X_0) \subset X_0$. Since $\mu(X) = 1$ and the intersection is countable, we have that $\mu(X_0) = 1$. Analogously, $Y_0 = \bigcap_{n=0}^{\infty} g^{-n}(Y)$ is a measurable subset of Y such that $\nu(Y_0) = 1$ and $g(Y_0) \subset Y_0$. Moreover, by construction, $Y_0 = \phi(X_0)$. Therefore, the restriction of ϕ to X_0 is still a bijection onto Y_0.

Example 8.1.1. Let $f : [0,1] \to [0,1]$ be defined by $f(x) = 10x - [10x]$. As we saw in Section 1.3.1, this transformation preserves the Lebesgue measure m on $[0,1]$. If one represents each number $x \in [0,1]$ by its decimal expansion $x = 0.a_0a_1a_2\ldots$, the transformation f corresponds, simply, to shifting all the digits of x one unit to the left. That motivates us to consider:

$$\phi : \{0,1,\ldots,9\}^{\mathbb{N}} \to [0,1], \quad \phi\big((a_n)_n\big) = \sum_{n=0}^{\infty} \frac{a_n}{10^{n+1}} = 0.a_0a_1a_2\ldots$$

It is clear that ϕ is surjective. On the other hand, it is not injective, since certain real numbers have more than one decimal expansion: for example, $0.1000000\ldots = 0.099999\ldots$ Actually, this happens if and only the number admits a finite decimal expansion, that is, such that all but finitely many digits are zero. The set of such numbers is countable and, hence, is irrelevant from the point of view of the Lebesgue measure. More precisely, let us consider the set $X \subset \{0,1,\ldots,9\}^{\mathbb{N}}$ of all sequences with an infinite number of symbols different from zero and the set $Y \subset [0,1]$ of all numbers whose decimal expansion is infinite (hence, unique). Then the restriction of ϕ to X is a bijection onto Y.

It is easy to check that both $\phi \mid X$ and its inverse are measurable: use the fact that the image of the intersection of X with each cylinder $[0;a_0,\ldots,a_{m-1}]$ is the intersection of Y with some interval of length 10^{-m}. This observation also shows that $\phi_*\nu = m$, where ν denotes the Bernoulli measure on $\{0,1,\dots,9\}^{\mathbb{N}}$ that assigns equal weights to all the digits. Moreover, denoting by σ the shift map in $\{0,1,\dots,9\}^{\mathbb{N}}$, we have that

$$\phi \circ \sigma\big((a_n)_n\big) = 0, a_1 a_2 \dots a_n \dots = f \circ \phi\big((a_n)_n\big)$$

for every $(a_n)_n \in X$. This proves that (f,m) is ergodically equivalent to the Bernoulli shift (σ,ν).

Suppose that (f,μ) and (g,ν) are ergodically equivalent. A measurable set $A \subset M$ is invariant under $f : M \to M$ if and only if $\phi(A)$ is invariant under $g : N \to N$. Moreover, $\nu(\phi(A)) = \mu(A)$. Therefore, (f,μ) is ergodic if and only if (g,ν) is ergodic. It is just as easy to obtain similar conclusions for the mixing and the weak mixing properties. Indeed, essentially all the properties that we study in this book are *invariants of ergodic equivalence*, that is, if they hold for a given system then they also hold for any system that is ergodically equivalent to it. An exception is unique ergodicity, which is a property of a different nature, since it concerns solely the transformation.

This also means that these properties may be used to try to distinguish systems that are not ergodically equivalent. Still, that is usually a difficult task. For example, nothing of what was said so far is of much help towards answering the following question: are the shift maps

$$\sigma : \{1,2\}^{\mathbb{Z}} \to \{1,2\}^{\mathbb{Z}} \quad \text{and} \quad \zeta : \{1,2,3\}^{\mathbb{Z}} \to \{1,2,3\}^{\mathbb{Z}}, \tag{8.1.1}$$

endowed with the corresponding Bernoulli measures giving the same weights to all the symbols, ergodically equivalent? It is easy to see that σ and ζ are not *topologically* conjugate (for example: ζ has three fixed points, whereas σ has only two), but the existence of an ergodic equivalence is a much more delicate issue. In fact, this type of question motivates most of the content of the present chapter and also leads to the notion of entropy, which is the subject of Chapter 9.

Example 8.1.2. Let $\sigma : M \to M$ be the shift map in $M = X^{\mathbb{N}}$ and let $\mu = \nu^{\mathbb{N}}$ be a Bernoulli measure. Let $\hat{\sigma} : \hat{M} \to \hat{M}$ be the natural extension of σ and $\hat{\mu}$ be the lift of μ (Section 2.4.2). Moreover, let $\tilde{\sigma} : \tilde{M} \to \tilde{M}$ be the shift map in $\tilde{M} = X^{\mathbb{Z}}$ and $\tilde{\mu} = \nu^{\mathbb{Z}}$ be the corresponding Bernoulli measure. Then, $(\hat{\sigma},\hat{\mu})$ is ergodically equivalent to $(\tilde{\sigma},\tilde{\mu})$. An equivalence may be constructed as follows.

By definition, $\hat{M}$ is the space of pre-orbits of $\sigma : M \to M$, that is, of all the sequences $\hat{x} = (\dots,x_{-n},\dots,x_0)$ in M such that $\sigma(x_{-j}) = x_{-j+1}$ for every $j \geq 1$. Moreover, each x_{-j} is a sequence $(x_{-j,i})_{i\in\mathbb{N}}$ in X. So, the previous relation means that

$$x_{-j,i+1} = x_{-j+1,i} \quad \text{for every } i,j \in \mathbb{N}. \tag{8.1.2}$$

Consider the map $\phi : \hat{M} \to \tilde{M}, \hat{x} \mapsto \tilde{x}$ given by

$$\tilde{x}_n = x_{0,n} = x_{-1,n+1} = \cdots \quad \text{and} \quad \tilde{x}_{-n} = x_{-n,0} = x_{-n-1,1} = \cdots .$$

We leave it to the reader to check that ϕ is indeed an ergodic equivalence between the natural extension $(\hat{\sigma}, \hat{\mu})$ and the two-sided shift map $(\tilde{\sigma}, \tilde{\mu})$.

8.1.1 Exercises

8.1.1. Let $f : [0,1] \to [0,1]$ be the transformation defined by $f(x) = 2x - [2x]$ and m be the Lebesgue measure on $[0,1]$. Exhibit a map $g : [0,1] \to [0,1]$ and a probability measure ν invariant under g such that (g, ν) is ergodically equivalent to (f, μ) and the support of ν has empty interior.

8.1.2. Let $f : \{1,\dots,k\}^{\mathbb{N}} \to \{1,\dots,k\}^{\mathbb{N}}$ and $g : \{1,\dots,l\}^{\mathbb{N}} \to \{1,\dots,l\}^{\mathbb{N}}$ be one-sided shift maps, endowed with Bernoulli measures μ and ν, respectively. Show that, for every set $X \subset \{1,\dots,k\}^{\mathbb{N}}$ with $f^{-1}(X) = X$ and $\mu(X) = 1$, there exists $x \in X$ such that $\#(X \cap f^{-1}(x)) = k$. Conclude that if $k \neq l$ then (f, μ) and (g, ν) cannot be ergodically equivalent.

8.1.3. Let $X = \{1,\dots,d\}$ and consider the shift map $\sigma : X^{\mathbb{N}} \to X^{\mathbb{N}}$ endowed with a Markov measure μ. Given any cylinder $C = [0; c_0, \dots, c_l]$ in $X^{\mathbb{N}}$, let μ_C be the normalized restriction of μ to C. Show that there exists an induced transformation $\sigma_C : C \to C$ (see Section 1.4.2) preserving μ_C and such that (σ_C, μ_C) is ergodically equivalent to a Bernoulli shift $(\sigma_{\mathbb{N}}, \nu)$ in $\mathbb{N}^{\mathbb{N}}$.

8.2 Spectral equivalence

Let $f : M \to M$ and $g : N \to N$ be transformations preserving probability measures μ and ν, respectively. Let $U_f : L^2(\mu) \to L^2(\mu)$ and $U_g : L^2(\nu) \to L^2(\nu)$ be the corresponding Koopman operators. We say that (f, μ) and (g, ν) are *spectrally equivalent* if there exists some unitary operator $L : L^2(\mu) \to L^2(\nu)$ such that

$$U_g \circ L = L \circ U_f. \tag{8.2.1}$$

We leave it to the reader to check that the relation defined in this way is, indeed, an equivalence relation.

It is easy to see that if two systems are ergodically equivalent then they are spectrally equivalent. Indeed, suppose that there exists an invertible map $h : M \to N$ such that $\phi_* \mu = \nu$ and $\phi \circ f = g \circ \phi$. Then, the Koopman operator

$$U_\phi : L^2(\nu) \to L^2(\mu), \quad U_\phi(\psi) = \psi \circ \phi$$

is an isometry and is invertible: the inverse is the Koopman operator associated with ϕ^{-1}. In other words, U_ϕ is a unitary operator. Moreover,

$$U_f \circ U_\phi = U_{\phi \circ f} = U_{g \circ \phi} = U_\phi \circ U_g.$$

Therefore, $L = U_\phi$ is a spectral equivalence between the two systems.

The converse is false, as will be clear from the sequel. For example, all countably generated two-sided Bernoulli shifts are spectrally equivalent (Corollary 8.4.12); however, not all have the same entropy (Example 9.1.10) and so not all are ergodically equivalent.

8.2.1 Invariants of spectral equivalence

Recall that the spectrum $\mathrm{spec}(A)$ of a bounded linear operator $A: E \to E$ in a complex Banach space E consists of the complex numbers λ such that $A - \lambda \mathrm{id}$ is not invertible. We say that $\lambda \in \mathrm{spec}(A)$ is an *eigenvalue* if $A - \lambda \mathrm{id}$ is not injective, that is, if there exists $v \neq 0$ such that $Av = \lambda v$. Then, the dimension of the kernel of $A - \lambda \mathrm{id}$ is called the *multiplicity* of the eigenvalue.

By definition, the *spectrum* of a system (f, μ) is the spectrum of the corresponding Koopman operator $U_f : L^2(\mu) \to L^2(\mu)$. If (f, μ) is spectrally equivalent to (g, ν) then the two systems have the same spectrum: the relation (8.2.1) implies that

$$(U_g - \lambda \,\mathrm{id}) = L \circ (U_f - \lambda \,\mathrm{id}) \circ L^{-1} \tag{8.2.2}$$

and, consequently, $U_g - \lambda \,\mathrm{id}$ is invertible if and only if $U_f - \lambda \,\mathrm{id}$ is invertible.

In fact, the spectrum itself is a poor invariant: in particular, all the invertible ergodic systems with no atoms have the same spectrum (Exercise 8.2.1). However, the associated spectral measure does provide very useful invariants, as we are going to see. The simplest one is the set of atoms of the spectral measure, that is, the set of eigenvalues of the Koopman operator. Note that (8.2.2) also shows that a given λ is an eigenvalue of U_f if and only if it is an eigenvalue of U_g; besides, in that case the two multiplicities are equal.

Observe that 1 is always an eigenvalue of the Koopman operator, since $U_f \varphi = \varphi$ for every constant function φ. By Proposition 4.1.3(v), the system (f, μ) is ergodic if and only if this eigenvalue has multiplicity 1 for U_f. Thus, it follows from what we have just said that (f, μ) is ergodic if and only if any system (g, ν) spectrally equivalent to it is ergodic. In other words, ergodicity is an *invariant of spectral equivalence*.

Analogously, suppose that the system (f, μ) is mixing. Then, by Proposition 7.1.12,

$$\lim_n U_f^n \varphi \cdot \psi = \int \varphi \, d\mu \int \psi \, d\mu$$

for every $\varphi, \psi \in L^2(\mu)$. Now suppose that (g, ν) is spectrally equivalent to (f, μ). Let L be the unitary operator in (8.2.1). The inverse L^{-1} maps eigenvectors of U_g associated with the eigenvalue 1 to eigenvectors of U_f associated with the same eigenvalue 1. Since the two systems are ergodic (use the previous paragraph), this means that L^{-1} maps constant functions to constant functions. Since L^{-1} is unitary,

$$U_g^n \varphi \cdot \psi = L^{-1}(U_g^n \varphi) \cdot L^{-1} \psi = U_f^n(L^{-1}\varphi) \cdot L^{-1} \psi$$

and, hence, $\lim_n U_g^n\varphi\cdot\psi = \int L^{-1}\varphi\,d\mu \int L^{-1}\psi\,d\mu$ for every $\varphi,\psi\in L^2(\nu)$. Also,

$$\int L^{-1}\varphi\,d\mu = L^{-1}\varphi\cdot 1 = L^{-1}\varphi\cdot L^{-1}1 = \varphi\cdot 1 = \int \varphi\,d\nu$$

and, analogously, $\int L^{-1}\psi\,d\mu = \int \psi\,d\mu$. In this way, we have shown that

$$\lim_n U_g^n\varphi\cdot\psi = \int \varphi\,d\mu \int \psi\,d\mu,$$

for every $\varphi,\psi\in L^2(\nu)$, that is, (g,ν) is also mixing. This shows that the mixing property is an invariant of spectral equivalence.

The same argument may be used for the weak mixing property, though the theorem that we prove in Section 8.2.2 below gives us a more interesting proof of the fact that weak mixing is an invariant of spectral equivalence.

8.2.2 Eigenvalues and weak mixing

As we have seen, the Koopman operator $U_f: L^2(\mu)\to L^2(\mu)$ of a system (f,μ) is an isometry, that is, it satisfies $U_f^*U_f = \mathrm{id}$. If f is invertible then the Koopman operator is unitary, that is, it satisfies $U_f^*U_f = U_fU_f^* = \mathrm{id}$. In particular, in that case U_f is a normal operator. Then the property of weak mixing admits the following interesting characterization:

Theorem 8.2.1. *An invertible system (f,μ) is weak mixing if and only if the constant functions are the only eigenvectors of the Koopman operator.*

In particular, a system (f,μ) is weak mixing if and only if it is ergodic and 1 is the unique eigenvalue of U_f.

Proof. Suppose that (f,μ) is weak mixing. Let $\varphi\in L^2(\mu)$ be any (non-zero) eigenfunction of U_f and λ be the corresponding eigenvalue. Then

$$\int \varphi\,d\mu = \int U_f\varphi\,d\mu = \lambda\int\varphi\,d\mu,$$

and this implies that $\int\varphi\,d\mu = 0$ or $\lambda = 1$. In the first case,

$$C_j(\varphi,\bar\varphi) = \left|\int (U_f^j\varphi)\bar\varphi\,d\mu\right| = \left|\lambda^j\int\varphi\bar\varphi\,d\mu\right| = \int|\varphi|^2\,d\mu$$

for every $j\ge 1$ (recall that $|\lambda| = 1$). But then

$$\lim_n \frac{1}{n}\sum_{j=0}^{n-1} C_j(\varphi,\bar\varphi) = \int|\varphi|^2\,d\mu > 0,$$

contradicting the hypothesis that the system is weak mixing. In the second case, using that the system is ergodic, we find that φ is constant at μ-almost every point. This shows that if the system is weak mixing then the constant functions are the only eigenvectors.

Now suppose that the only eigenvectors of U_f are the constant functions. To conclude that (f, μ) is weak mixing, we must show that

$$\frac{1}{n}\sum_{j=0}^{n-1} C_j(\varphi, \psi)^2 \to 0 \quad \text{for any } \varphi, \psi \in L^2(\mu)$$

(recall Exercise 7.1.2). It follows immediately from the definition that

$$C_j(\varphi, \psi) = C_j(\varphi', \psi) \quad \text{where } \varphi' = \varphi - \int \varphi \, d\mu$$

and the integral of φ' vanishes. Hence, it is no restriction to suppose that $\int \varphi \, d\mu = 0$. Then, using the relation (A.7.6) for the unitary operator $L = U_f$, we get:

$$C_j(\varphi, \psi)^2 = \left| \int (U_f^j \varphi) \psi \, d\mu \right|^2 = \left| \int_{\mathbb{C}} z^j \, d\theta(z) \right|^2,$$

where $\theta = E\varphi \cdot \psi$. The expression on the right-hand side may be rewritten as follows:

$$\int_{\mathbb{C}} z^j \, d\theta(z) \int_{\mathbb{C}} \bar{z}^j \, d\bar{\theta}(z) = \int_{\mathbb{C}} \int_{\mathbb{C}} z^j \bar{w}^j \, d\theta(z) \, d\bar{\theta}(w).$$

Therefore, given any $n \geq 1$,

$$\frac{1}{n}\sum_{j=0}^{n-1} C_j(\varphi, \psi)^2 = \int_{\mathbb{C}} \int_{\mathbb{C}} \frac{1}{n}\sum_{j=0}^{n-1} (z\bar{w})^j \, d\theta(z) \, d\bar{\theta}(w). \tag{8.2.3}$$

We claim that the measure $\theta = E\varphi \cdot \psi$ is non-atomic. In fact, suppose that there exists $\lambda \in \mathbb{C}$ such that $\theta(\{\lambda\}) \neq 0$. Then, $E(\{\lambda\}) \neq 0$ and then we may use Proposition A.7.8 to conclude that the function $E(\{\lambda\})\varphi$ is an eigenvector of U_f. By the hypothesis about the operator U_f, this implies that $E(\{\lambda\})\varphi$ is constant at μ-almost every point. Hence,

$$E(\{\lambda\})\varphi \cdot \varphi = E(\{\lambda\})\varphi \int \bar{\varphi} \, d\mu = 0.$$

Lemma A.7.3 also gives that

$$E(\{\lambda\})\varphi \cdot \varphi = E(\{\lambda\})^2 \varphi \cdot \varphi = E(\{\lambda\})\varphi \cdot E(\{\lambda\})\varphi.$$

Putting these two identities together, we conclude that $E(\{\lambda\})\varphi = 0$, which contradicts the hypothesis. Thus, our claim is proved.

The sequence $n^{-1}\sum_{j=0}^{n-1}(z\bar{w})^j$ in (8.2.3) is bounded and (see Exercise 8.2.6) converges to zero on the complement of the diagonal $\Delta = \{(z, w) : z = w\}$. Moreover, the diagonal has measure zero:

$$(\theta \times \bar{\theta})(\Delta) = \int \theta(\{y\}) \, d\bar{\theta}(y) = 0,$$

because θ is non-atomic. Then we may use the monotone convergence theorem to conclude that (8.2.3) converges to zero when $n \to \infty$. This proves that (f, μ) is weak mixing if U_f has no non-constant eigenvectors.

Suppose that M is a topological space. We say that a continuous map $f : M \to M$ is *topologically weak mixing* if the Koopman operator U_f has no non-constant *continuous* eigenfunctions. The following fact is an easy consequence of Theorem 8.2.1:

Corollary 8.2.2. *If (f,μ) is weak mixing then the restriction of f to the support of μ is topologically weak mixing.*

Proof. Let φ be a continuous eigenfunction of U_f. By Theorem 8.2.1, the function φ is constant at μ-almost every point. Hence, by continuity, φ is constant (at every point) on the support of μ.

We mentioned in Section 7.3 that almost every interval exchange is weak mixing but not mixing. In the following we describe an explicit construction, based on an extension of ideas that were hinted at in Example 6.3.9. The reader may find this and other variations of those ideas in Section 7.4 of Kalikow and McCutcheon [KM10].

Example 8.2.3 (Chacon). Consider the sequence $(\mathcal{S}_n)_n$ of piles defined as follows. First, $\mathcal{S}_1 = \{[0,2/3)\}$. Next, for each $n \geq 1$, let $\mathcal{S}_n$ be the pile obtained by dividing $\mathcal{S}_{n-1}$ into three columns, with the same width, and piling those columns up on top of each other, with an additional interval inserted between the second pile and the third one, as illustrated in Figure 8.1.

For example, $\mathcal{S}_2 = \{[0,2/9),[2/9,4/9),[6/9,8/9),[4/9,6/9)\}$ and

$$\begin{aligned}\mathcal{S}_3 = \{&[0,2/27),[6/27,8/27),[18/27,20/27),[12/27,14/27),[2/27,4/27),\\ &[8/27,10/27),[20/27,22/27),[14/27,16/27),[24/27,26/27),\\ &[4/27,6/27),[10/27,12/27),[22/27,24/27),[16/27,18/27)\}.\end{aligned}$$

Note that each $\mathcal{S}_n$ is a pile in the interval $J_n = [0,1-3^{-n})$. The sequence $(f_n)_n$ of transformations associated with such piles converges at every point to a transformation $f : [0,1) \to [0,1)$ that preserves the Lebesgue measure m. This system (f,m) is weak mixing but not mixing (Exercise 8.2.7).

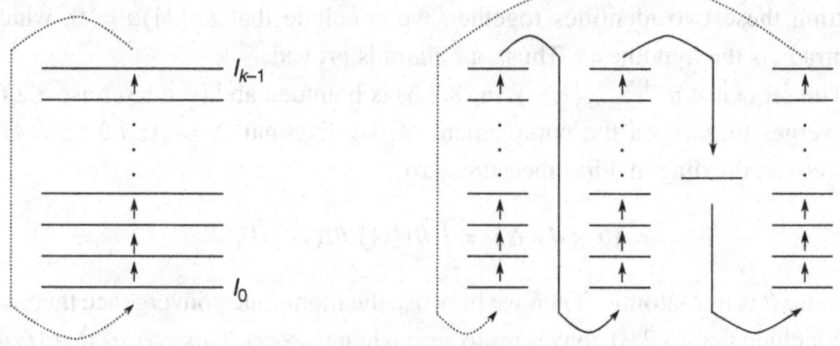

Figure 8.1. Constructing a weak mixing system that is not mixing

8.2.3 Exercises

8.2.1. Let (f,μ) be an invertible ergodic system with no atoms. Show that every λ in the unit circle $\{z\in\mathbb{C}:|z|=1\}$ is an *approximate eigenvalue* of the Koopman operator $U_f : L^2(\mu)\to L^2(\mu)$: there exists some sequence $(\varphi_n)_n$ such that $\|\varphi_n\|\to 1$ and $\|U_f\varphi_n-\lambda\varphi_n\|\to 0$. In particular, the spectrum of U_f coincides with the unit circle.

8.2.2. Let m be the Lebesgue measure on the circle and $U_\alpha : L^2(m)\to L^2(m)$ be the Koopman operator of the irrational rotation $R_\alpha : S^1\to S^1$. Calculate the eigenvalues of U_α and deduce that (R_α,m) and (R_β,m) are spectrally equivalent if and only if $\alpha=\pm\beta$. [Observation: Corollary 8.3.6 provides a more complete statement.]

8.2.3. Let m be the Lebesgue measure on the circle and, for each integer number $k\geq 2$, let $U_k : L^2(m)\to L^2(m)$ be the Koopman operator of the transformation $f_k : S^1\to S^1$ given by $f_k(x)=kx \mod \mathbb{Z}$. Check that if $p\neq q$ then (f_p,m) and (f_q,m) are not ergodically equivalent. Show that, for any $k\geq 2$,

$$L^2(m)=\{\text{constants}\}\oplus\bigoplus_{j=0}^{\infty}U_k^j(H_k),$$

where $H_k=\{\sum_{n\in\mathbb{Z}}a_n e^{2\pi inx} : a_n=0 \text{ if } k\mid n\}$ and the terms in the direct sum are pairwise orthogonal. Conclude that (f_p,m) and (f_q,m) are spectrally equivalent for any p and q.

8.2.4. Let $f : S^1\to S^1$ be the transformation given by $f(x)=kx \mod \mathbb{Z}$ and μ be the Lebesgue measure. Show that (f,μ) is weak mixing if and only if $|k|\geq 2$.

8.2.5. Prove that, for any invertible transformation f, if μ is ergodic for every iterate f^n and there exists $C>0$ such that

$$\limsup_n \mu\big(f^{-n}(A)\cap B\big)\leq C\mu(A)\mu(B),$$

for any measurable sets A and B, then μ is weak mixing. [Observation: This statement is due to Ornstein [Orn72]. In fact, he proved more: under these hypotheses the system is (strongly) mixing.]

8.2.6. Let z and w be two complex numbers with absolute value 1. Check that

(a) $\lim_n \frac{1}{n}\sum_{j=0}^{n-1}|z^j-1|=0$ if and only if $z=1$;

(b) $\lim_n \frac{1}{n}\sum_{j=0}^{n-1}(z\bar{w})^j=0$ if $z\neq w$.

8.2.7. Consider the system (f,m) in Example 8.2.3. Show that

(a) the system (f,m) is ergodic;

(b) the only eigenvalues of the Koopman operator $U_f : L^1(m)\to L^1(m)$ are the constant functions, and hence (f,m) is weak mixing;

(c) $\limsup_n m(f^n(A)\cap A)\geq 2/27$ if we take $A=[0,2/9)$; in particular, (f,m) is not mixing.

8.3 Discrete spectrum

In this section and the next we study two extreme cases, in terms of the type of spectral measure of the Koopman operator: systems with *discrete spectrum*, whose spectral measure is purely atomic, and systems with *Lebesgue spectrum*, whose spectral measure is equivalent to the Lebesgue measure on the unit circle.

We begin by describing some properties of the eigenvalues and eigenvectors of the Koopman operator. It is clear that all the eigenvalues are in the unit circle, since U_f is an isometry.

Proposition 8.3.1. *If $\varphi_1, \varphi_2 \in L^2(\mu)$ satisfy $U_f\varphi_1 = \lambda_1\varphi_1$ and $U_f\varphi_2 = \lambda_2\varphi_2$ with $\lambda_1 \neq \lambda_2$, then $\varphi_1 \cdot \varphi_2 = 0$. Moreover, the eigenvalues of U_f form a subgroup of the unit circle.*

If the system (f, μ) is ergodic then every eigenvalue of U_f is simple and the absolute value of every eigenfunction is constant at μ-almost every point.

Proof. The first claim follows from the identity

$$\varphi_1 \cdot \varphi_2 = U_f\varphi_1 \cdot U_f\varphi_2 = \lambda_1\varphi_1 \cdot \lambda_2\varphi_2 = \lambda_1\bar{\lambda}_2(\varphi_1 \cdot \varphi_2) = \lambda_1\lambda_2^{-1}(\varphi_1 \cdot \varphi_2),$$

since $\lambda_1\lambda_2^{-1} \neq 1$. This identity also shows that the set of all eigenvalues is closed under the operation $(\lambda_1, \lambda_2) \mapsto \lambda_1\lambda_2^{-1}$. Recalling that 1 is always an eigenvalue, it follows that this set is a multiplicative group.

Now assume that (f, μ) is ergodic and suppose that $U_f\varphi = \lambda\varphi$. Then $U_f(|\varphi|) = |U_f\varphi| = |\lambda\varphi| = |\varphi|$ at μ-almost every point. By ergodicity, this implies that $|\varphi|$ is constant at μ-almost every point. Next, suppose that $U_f\varphi_1 = \lambda\varphi_1$, $U_f\varphi_2 = \lambda\varphi_2$ and the functions φ_1 and φ_2 are not identically zero. Since $|\varphi_2|$ is constant at μ-almost every point, $\varphi_2(x) \neq 0$ for μ-almost every x. Then φ_1/φ_2 is well defined. Moreover,

$$U_f\left(\frac{\varphi_1}{\varphi_2}\right) = \frac{U_f(\varphi_1)}{U_f(\varphi_2)} = \frac{\lambda\varphi_1}{\lambda\varphi_2} = \frac{\varphi_1}{\varphi_2}.$$

By ergodicity, it follows that the quotient is constant at μ-almost every point. That is, $\varphi_1 = c\varphi_2$ for some $c \in \mathbb{C}$.

We say that a system (f, μ) has *discrete spectrum* if the eigenvectors of the Koopman operator $U_f : L^2(\mu) \to L^2(\mu)$ generate the Hilbert space $L^2(\mu)$. Observe that this implies that U_f is invertible and, hence, is a unitary operator. This terminology is justified by the following observation (recall also Theorem A.7.9):

Proposition 8.3.2. *A system (f, μ) has discrete spectrum if and only if its Koopman operator U_f has a spectral representation of the form*

$$T : \bigoplus_j L^2(\sigma_j)^{\chi_j} \to \bigoplus_j L^2(\sigma_j)^{\chi_j}, \quad (\varphi_{j,l})_{j,l} \mapsto \big(z \mapsto z\varphi_{j,l}(z)\big)_{j,l}, \tag{8.3.1}$$

where each σ_j is a Dirac measure at a point in the unit circle.

Proof. Suppose that U_f admits a spectral representation of the form (8.3.1) with $\sigma_j = \delta_{\lambda_j}$ for some λ_j in the unit circle. Each $L^2(\sigma_j)^{\chi_j}$ may be canonically identified with a subspace in the direct sum. The restriction of T to that subspace coincides with $\lambda_j \,\mathrm{id}$, since

$$z\varphi_{j,l}(z) = \lambda_j \varphi_{j,l}(z) \quad \text{at } \sigma_j\text{-almost every point.} \tag{8.3.2}$$

Let $(v_{j,l})_l$ be a Hilbert basis of $L^2(\sigma_j)^{\chi_j}$. Then $(v_{j,l})_{j,l}$ is a Hilbert basis of the direct sum formed by eigenvectors of T. Since T is unitarily conjugate to U_f, it follows that $L^2(\mu)$ admits a Hilbert basis formed by eigenvectors of the Koopman operator.

Now suppose that (f,μ) has discrete spectrum. Let $(\lambda_j)_j$ be the eigenvalues of U_f and, for each j, let $\sigma_j = \delta_{\lambda_j}$ and χ_j be the Hilbert dimension of the eigenspace $\ker(U_f - \lambda_j \,\mathrm{id})$. Note that the space $L^2(\sigma_j)$ is 1-dimensional, since every function is constant at σ_j-almost every point. Therefore, the Hilbert dimension of $L^2(\delta_{\lambda_j})^{\chi_j}$ is also equal to χ_j. Hence, there exists some unitary isomorphism

$$L_j : \ker(U_f - \lambda_j \,\mathrm{id}) \to L^2(\delta_{\lambda_j})^{\chi_j}.$$

It is clear that $L_j \circ U_f \circ L_j^{-1} = \lambda_j \,\mathrm{id}$. In other words, recalling the observation (8.3.2),

$$L_j \circ U_f \circ L_j^{-1} : (\varphi_{j,l})_l \mapsto \big(z \mapsto \lambda_j \varphi_{j,l}(z)\big)_l = \big(z \mapsto z\varphi_{j,l}(z)\big)_l. \tag{8.3.3}$$

The eigenspaces $\ker(U_f - \lambda_j \,\mathrm{id})$ generate $L^2(\mu)$, by hypothesis, and they are pairwise orthogonal, by Proposition 8.3.1. Hence, we may combine the operators L_j to obtain a unitary isomorphism $L : L^2(\mu) \to \bigoplus_j L^2(\sigma_j)^{\chi_j}$. The relation (8.3.3) gives that

$$L \circ U_f \circ L^{-1} : (\varphi_{j,l})_{j,l} \mapsto \big(z \mapsto z\varphi_{j,l}(z)\big)_l$$

is a spectral representation of U_f of the form we are looking for.

Example 8.3.3. Let m be the Lebesgue measure on the torus $\mathbb{T}^d$. Consider the Fourier basis $\{\phi_k(x) = e^{2\pi i k\cdot x} : k \in \mathbb{Z}^d\}$ of the Hilbert space $L^2(m)$. Let f be the rotation $R_\theta : \mathbb{T}^d \to \mathbb{T}^d$ corresponding to a given vector $\theta = (\theta_1, \dots, \theta_d)$. Then,

$$U_f \phi_k(x) = \phi_k(x+\theta) = e^{2\pi i k\cdot\theta} \phi_k(x) \quad \text{for every } x \in \mathbb{T}^d.$$

This shows that every ϕ_k is an eigenvector of U_f and, hence, (f,m) has discrete spectrum. Note that the group of eigenvalues is

$$G_\theta = \{e^{2\pi i k\cdot\theta} : k \in \mathbb{Z}^d\}, \tag{8.3.4}$$

that is, the subgroup of the unit circle generated by $\{e^{2\pi i\theta_j} : j = 1, \dots, d\}$.

More generally, every ergodic translation in a compact abelian group has discrete spectrum. Conversely, every ergodic system with discrete spectrum is ergodically isomorphic to a translation in a compact abelian group (the notion of ergodic isomorphism is discussed in Section 8.5). Another interesting result is that every subgroup of the unit circle is the group of eigenvalues of some ergodic system with discrete spectrum. These facts are proved in Section 3.3 of the book of Peter Walters [Wal82].

Proposition 8.3.4. *Suppose that (f,μ) and (g,ν) are ergodic and have discrete spectrum. Then (f,μ) and (g,ν) are spectrally equivalent if and only if their Koopman operators $U_f : L^2(\mu) \to L^2(\mu)$ and $U_g : L^2(\nu) \to L^2(\nu)$ have the same eigenvalues.*

Proof. It is clear that if the Koopman operators are conjugate then they have the same eigenvalues. To prove the converse, let $(\lambda_j)_j$ be the eigenvalues of the two operators. By Proposition 8.3.2, the eigenvalues are simple. For each j, let u_j and v_j be unit vectors in $\ker(U_f - \lambda_j \,\mathrm{id})$ and $\ker(U_g - \lambda_j \,\mathrm{id})$, respectively. Then $(u_j)_j$ and $(v_j)_j$ are Hilbert bases of $L^2(\mu)$ and $L^2(\nu)$, respectively. Consider the isomorphism $L : L^2(\mu) \to L^2(\nu)$ defined by $L(u_j) = v_j$. This operator is unitary, since it maps a Hilbert basis to a Hilbert basis, and it satisfies

$$L \circ U_f(u_j) = L(\lambda_j u_j) = \lambda_j v_j = U_g(v_j) = U_g \circ L(u_j)$$

for every j. By linearity, it follows that $L \circ U_f = U_g \circ L$. Therefore, (f,μ) and (g,ν) are spectrally equivalent.

Corollary 8.3.5. *If (f,μ) is ergodic, invertible and has discrete spectrum then (f,μ) is spectrally equivalent to (f^{-1},μ).*

Proof. It is clear that λ is an eigenvalue of U_f if and only if λ^{-1} is an eigenvalue of $U_{f^{-1}}$; moreover, in that case the eigenvectors are the same. Since the sets of eigenvalues are groups, it follows that the two operators have the same eigenvalues and the same eigenvectors. Apply Proposition 8.3.4.

Let m be the Lebesgue measure on the torus $\mathbb{T}^d$. Proposition 8.3.4 also allows us to classify the irrational rotations on the torus up to equivalence, ergodic and spectral:

Corollary 8.3.6. *Let $\theta = (\theta_1,\dots,\theta_d)$ and $\tau = (\tau_1,\dots,\tau_d)$ be rationally independent vectors and R_θ and R_τ be the corresponding rotations on the torus $\mathbb{T}^d$. The following conditions are equivalent:*

(i) (R_θ, m) and (R_τ, m) are ergodically equivalent;
(ii) (R_θ, m) and (R_τ, m) are spectrally equivalent;
(iii) there exists $L \in \mathrm{SL}(d,\mathbb{Z})$ such that $\theta = L\tau \mod \mathbb{Z}^d$.

We leave the proof to the reader (Exercise 8.3.2). In the special case of the circle, we get that two irrational rotations R_θ and R_τ are equivalent if and only if either $R_\theta = R_\tau$ or $R_\theta = R_\tau^{-1}$. See also Exercise 8.3.3.

8.3.1 Exercises

8.3.1. Suppose that (f,μ) has discrete spectrum and the Hilbert space $L^2(\mu)$ is separable (this is the case, for instance, if the σ-algebra of measurable sets is countably generated). Show that there exists a sequence $(n_k)_k$ converging to infinity such that $\|U_f^{n_k}\varphi - \varphi\|_2$ converges to zero when $k \to \infty$, for every $\varphi \in L^2(\mu)$.

8.3.2. Prove Corollary 8.3.6.

8.3.3. Let m be the Lebesgue measure on S^1 and $\theta = p/q$ and $\tau = r/s$ be two rational numbers, with $\gcd(p,q) = 1 = \gcd(r,s)$. Show that the rotations (R_θ, m) and (R_τ, m) are ergodically equivalent if and only if the denominators q and s are equal.

8.4 Lebesgue spectrum

This section is devoted to the class of systems whose Koopman operator has the following property (the reason for the terminology will become clear in Proposition 8.4.10):

Definition 8.4.1. Let $U : H \to H$ be an isometry in a Hilbert space. We say that U has *Lebesgue spectrum* if there exists some closed subspace $E \subset H$ such that

(i) $U(E) \subset E$;
(ii) $\bigcap_{n\in\mathbb{N}} U^n(E) = \{0\}$;
(iii) $\sum_{n\in\mathbb{N}} U^{-n}(E) = H$.

Given a probability measure μ, we denote by $L_0^2(\mu) = L_0^2(M,\mathcal{B},\mu)$ the orthogonal complement, inside the space $L^2(\mu) = L^2(M,\mathcal{B},\mu)$, of the subspace of constant functions. In other words,

$$L_0^2(\mu) = \{\varphi \in L^2(\mu) : \int \varphi\, d\mu = 0\}.$$

Note that $L_0^2(\mu)$ is invariant under the Koopman operator: $\varphi \in L_0^2(\mu)$ if and only if $U_f\varphi \in L_0^2(\mu)$. We say that the system (f,μ) has *Lebesgue spectrum* if the restriction of the Koopman operator to $L_0^2(\mu)$ has Lebesgue spectrum.

8.4.1 Examples and properties

We start by observing that all Bernoulli shifts have Lebesgue spectrum. It is convenient to treat one-sided shifts and the two-sided shifts separately.

Example 8.4.2. Consider a one-sided shift map $\sigma : X^{\mathbb{N}} \to X^{\mathbb{N}}$ and a Bernoulli measure $\mu = \nu^{\mathbb{N}}$ on $X^{\mathbb{N}}$. Let $E = L_0^2(\mu)$. Conditions (i) and (iii) in Definition 8.4.1 are obvious. To prove condition (ii), consider any function $\varphi \in L_0^2(\mu)$ in the intersection, that is, such that for every $n \in \mathbb{N}$ there exists a function $\psi_n \in L_0^2(\mu)$ satisfying $\varphi = \psi_n \circ \sigma^n$. We want to show that φ is constant at μ-almost every point. For each $c \in \mathbb{R}$, consider

$$A_c = \{x \in X^{\mathbb{N}} : \varphi(x) > c\}.$$

For each $n \in \mathbb{N}$, we may write $A_c = \sigma^{-n}(\{x \in X^{\mathbb{N}} : \psi_n(x) > c\})$. Then A_c belongs to the σ-algebra generated by the cylinders of the form $[n; C_n, \ldots, C_m]$ with $m \geq n$. Consequently, $\mu(A_c \cap C) = \mu(A_c)\mu(C)$ for every cylinder C of the form $C = [0; C_0, \ldots, C_{n-1}]$. Since n is arbitrary and the cylinders are a generating family, it follows that $\mu(A_c \cap B) = \mu(A_c)\mu(B)$ for every measurable set $B \subset X^{\mathbb{N}}$. Taking $B = A_c$ we conclude that $\mu(A_c) = \mu(A_c)^2$; in other words, $\mu(A_c) \in \{0, 1\}$ for every $c \in \mathbb{R}$. This proves that φ is constant at μ-almost every point, as stated.

Example 8.4.3. Now consider a two-sided shift map $\sigma : X^{\mathbb{Z}} \to X^{\mathbb{Z}}$ and a Bernoulli measure $\mu = \nu^{\mathbb{Z}}$. Let $\mathcal{A}$ be the σ-algebra generated by the cylinders of the form $[0; C_0, \ldots, C_m]$ with $m \geq 0$. Denote by $L_0^2(X^{\mathbb{Z}}, \mathcal{A}, \mu)$ the space of all functions $\varphi \in L_0^2(\mu)$ that are measurable with respect to the σ-algebra $\mathcal{A}$ (in other words, $\varphi(x)$ depends only on the coordinates x_n, $n \geq 0$ of the point). Take $E = L_0^2(X^{\mathbb{Z}}, \mathcal{A}, \mu)$. Condition (i) in Definition 8.4.1 is obvious. Condition (ii) follows from the same arguments that we used in Example 8.4.2. To prove condition (iii), note that $\bigcup_n U_\sigma^{-n}(E)$ contains the characteristic functions of all the cylinders. Therefore, it contains all the linear combinations of characteristic functions of sets in the algebra generated by the cylinders. This implies that the union is dense in $L_0^2(\mu)$, as we wanted to prove.

Lemma 8.4.4. *If (f, μ) has Lebesgue spectrum then* $\lim_n U_f^n \varphi \cdot \psi = 0$ *for every $\varphi \in L_0^2(\mu)$ and every $\psi \in L^2(\mu)$.*

Proof. Observe that the sequence $U_f^n \varphi \cdot \psi$ is bounded. Indeed, by the Cauchy–Schwarz inequality (Theorem A.5.4):

$$|U_f^n \varphi \cdot \psi| \leq \|U_f^n \varphi\|_2 \|\psi\|_2 = \|\varphi\|_2 \|\psi\|_2 \quad \text{for every } n.$$

So, it is enough to prove that every convergent subsequence $U_f^{n_j} \varphi \cdot \psi$ converges to zero. Furthermore, the set $\{U_f^n \varphi : n \in \mathbb{N}\}$ is bounded in $L^2(\mu)$, because U_f is an isometry. By the theorem of Banach–Alaoglu (Theorems A.6.1 and 2.3.1), every sequence in that set admits some weakly convergent subsequence. Hence, it is no restriction to suppose that $U_f^{n_j} \varphi$ converges weakly to some $\hat{\varphi} \in L^2(\mu)$.

Let E be a subspace satisfying the conditions in Definition 8.4.1. Initially, suppose that $\varphi \in U_f^{-k}(E)$ for some k. Then $U_f^{n_j} \varphi \in U_f^{n_j - k}(E)$. Hence, given any $l \in \mathbb{N}$, we have that $U_f^{n_j} \varphi \in U_f^l(E)$ for every j sufficiently large. It follows (see Exercise A.6.8) that $\hat{\varphi} \in U_f^l(E)$ for every $l \in \mathbb{N}$. By condition (ii) in the definition, this implies that $\hat{\varphi} = 0$ at μ-almost every point. In particular, $\lim_j U_f^{n_j} \varphi \cdot \psi = \hat{\varphi} \cdot \psi = 0$.

Now consider any $\varphi \in L^2_0(\mu)$. By condition (iii) in the definition, for every $\varepsilon > 0$ there exist $k \in \mathbb{N}$ and $\varphi_k \in U_f^{-k}(E)$ such that $\|\varphi - \varphi_k\|_2 \leq \varepsilon$. Using the Cauchy–Schwarz once more inequality:

$$|U_f^n\varphi \cdot \psi - U_f^n\varphi_k \cdot \psi| \leq \|\varphi - \varphi_k\|_2 \, \|\psi\|_2 \leq \varepsilon \|\psi\|_2$$

for every n. Recalling that $\lim_n U_f^n\varphi_k \cdot \psi = 0$ (by the previous paragraph), we find that

$$-\varepsilon \|\psi\|_2 \leq \liminf_n U_f^n\varphi \cdot \psi \leq \limsup_n U_f^n\varphi \cdot \psi \leq \varepsilon \|\psi\|_2.$$

Making $\varepsilon \to 0$, it follows that $\lim_n U_f^n\varphi \cdot \psi = 0$, as we wanted to prove.

Corollary 8.4.5. *If (f,μ) has Lebesgue spectrum then (f,μ) is mixing.*

Proof. It suffices to observe that

$$C_n(\varphi,\psi) = |U_f^n\varphi \cdot \psi - \left(\int \varphi \, d\mu\right) \cdot \psi| = |U_f^n\left(\varphi - \int \varphi \, d\mu\right) \cdot \psi|$$

and the function $\varphi' = \varphi - \int \varphi \, d\mu$ is in $L^2_0(\mu)$.

The converse to Corollary 8.4.5 is false, in general: in Example 8.4.13 we present certain mixing systems that do not have Lebesgue spectrum.

The class of systems with Lebesgue spectrum is invariant under spectral equivalence. Indeed, suppose that (f,μ) has Lebesgue spectrum and (g,ν) is spectrally equivalent to (f,μ). Let $L : L^2(\mu) \to L^2(\nu)$ be a unitary operator conjugating the Koopman operators U_f and U_g. It follows from the hypothesis and Corollary 8.4.5 that (f,μ) is weak mixing. Hence, by Theorem 8.2.1, the constant functions are the only eigenvectors of U_f. Then the same holds for U_g and so the conjugacy L maps constant functions to constant functions. Then, as L is unitary, its restriction to the orthogonal complement $L^2_0(\mu)$ is a unitary operator onto $L^2_0(\nu)$. Now, given any subspace $E \subset L^2_0(\mu)$ satisfying the conditions (i), (ii), (iii) in Definition 8.4.1 for U_f, it is clear that the subspace $L(E) \subset L^2_0(\nu)$ satisfies the corresponding conditions for U_g. Hence, (g,ν) has Lebesgue spectrum.

Given closed subspaces $V \subset W$ of a Hilbert space H, we denote by $W \ominus V$ the orthogonal complement of V inside W, that is,

$$W \ominus V = W \cap V^\perp = \{w \in W : v \cdot w = 0 \text{ for every } v \in V\}.$$

The proof of the following fact is discussed in the next section:

Proposition 8.4.6. *If $U : H \to H$ is an isometry and E_1 and E_2 are subspaces satisfying the conditions in Definition 8.4.1, then the orthogonal complements $E_1 \ominus U(E_1)$ and $E_2 \ominus U(E_2)$ have the same Hilbert dimension.*

This leads to the following definition: the *rank* of an operator $U : H \to H$ with Lebesgue spectrum is the Hilbert dimension of $E \ominus U(E)$ for any subspace E satisfying the conditions in Definition 8.4.1.

Then we define the *rank* of a system (f,μ) with Lebesgue spectrum to be the rank of the associated Koopman operator restricted to $L^2_0(\mu)$. It is clear that the rank is less than or equal to the Hilbert dimension of $L^2_0(\mu)$. In particular, if $L^2(\mu)$ is separable then the rank is countable, possibly finite. The majority of interesting examples fall into this category:

Example 8.4.7. Suppose that the probability space $(M,\mathcal{B},\mu)$ is *countably generated*, that is, there exists a countable family G of measurable subsets such that every element of $\mathcal{B}$ coincides, up to measure zero, with some element of the σ-algebra generated by G. Then $L^2(\mu)$ is separable: the algebra $\mathcal{A}$ generated by G is countable and the linear combinations with rational coefficients of characteristic functions of elements of $\mathcal{A}$ form a countable dense subset of $L^2(\mu)$.

It is interesting to point out that no examples are known of systems with Lebesgue spectrum of finite rank. For Bernoulli shifts, the rank coincides with the dimension of the corresponding $L^2(\mu)$:

Example 8.4.8. Let (σ,μ) be a one-sided Bernoulli shift (similar considerations apply in the two-sided case). As we have seen in Example 8.4.2, we may take $E = L^2_0(\mu)$. Then, denoting $x = (x_1,\dots,x_n,\dots)$ and recalling that $\mu = \nu^{\mathbb{N}}$,

$$\varphi \in E \ominus U_\sigma(E) \Leftrightarrow \int \varphi(x_0,x)\psi(x)\,d\mu(x_0,x) = 0 \qquad \forall \psi \in L^2_0(\mu)$$

$$\Leftrightarrow \int \left(\int \varphi(x_0,x)\,d\nu(x_0) \right) \psi(x)\,d\mu(x) = 0 \quad \forall \psi \in L^2_0(\mu).$$

Hence, $E \ominus U_\sigma(E) = \{\varphi \in L^2(\mu) : \int \varphi(x_0,x)\,d\nu(x_0) = 0 \text{ for } \mu\text{-almost every } x\}$. We claim that $\dim(E \ominus U_\sigma(E)) = \dim L^2(\mu)$. The inequality $\le$ is obvious. To prove the other inequality, fix any measurable function $\phi : X \to \mathbb{R}$ with $\int \phi\,d\nu = 0$ and $\int \phi^2\,d\nu = 1$. Consider the linear map $I : L^2(\mu) \to L^2(\mu)$ associating with each $\psi \in L^2(\mu)$ the function $I\psi(x_0,x) = \phi(x_0)\psi(x)$. The assumptions on ϕ imply that

$$I\psi \in E \ominus U_\sigma(E) \quad \text{and} \quad \|I\psi\|_2 = \|\psi\|_2 \quad \text{for every } \psi \in L^2(\mu).$$

This shows that $E \ominus U_\sigma(E)$ contains a subspace isometric to $L^2(\mu)$ and, hence, $\dim E \ominus U_\sigma(E) \ge \dim L^2(\mu)$. This concludes the argument.

We say that the shift is *of countable type* if the probability space X is countably generated. This is automatic, for example, if X is finite, or even countable. In that case, the space $\Sigma = X^{\mathbb{N}}$ (or $\Sigma = X^{\mathbb{Z}}$) is also countably generated: if G is a countable generator of X then the cylinders $[m; C_m,\dots,C_n]$ with $C_j \in G$ form a countable generator of Σ. Then, as observed in Example 8.4.7, the space $L^2(\mu)$ is separable. Therefore, it follows from Example 8.4.8 that *every Bernoulli shift of countable type has Lebesgue spectrum with countable rank.*

8.4.2 The invertible case

In this section we take the system (f,μ) to be invertible. In this context, the notion of Lebesgue spectrum may be formulated in a more transparent way:

Proposition 8.4.9. *Let $U : H \to H$ be a unitary operator in a Hilbert space H. Then U has Lebesgue spectrum if and only if there exists a closed subspace $F \subset H$ such that the iterates $U^k(F)$, $k \in \mathbb{Z}$ are pairwise orthogonal and satisfy*

$$H = \bigoplus_{k\in\mathbb{Z}} U^k(F).$$

Proof. Suppose that there exists some subspace F as in the statement. Take $E = \bigoplus_{k=0}^{\infty} U^k(F)$. Condition (i) in Definition 8.4.1 is immediate:

$$U(E) = \bigoplus_{k=1}^{\infty} U^k(F) \subset E.$$

As for condition (ii), note that $\varphi \in \bigcap_{n=0}^{\infty} U^n(E)$ means that $\varphi \in \bigoplus_{k=n}^{\infty} U^k(F)$ for every $n \geq 0$. This implies that φ is orthogonal to $U^k(F)$ for every $k \in \mathbb{Z}$. Hence, $\varphi = 0$. Finally, by hypothesis, we may write any $\varphi \in H$ as an orthogonal sum $\varphi = \sum_{k\in\mathbb{Z}} \varphi_k$ with $\varphi_k \in U^k(F)$ for every k. Then

$$\sum_{k=-n}^{\infty} \varphi_k \in \bigoplus_{k=-n}^{\infty} U^k(F) = U^{-n}(E)$$

for every n and the sequence on the left-hand side converges to φ when $n \to \infty$. This gives condition (iii) in the definition.

Now we prove the converse. Given E satisfying the conditions (i), (ii) and (iii) in the definition, take $F = E \ominus U(E)$. It is easy to see that the iterates of F are pairwise orthogonal. We claim that

$$\bigoplus_{k=0}^{\infty} U^k(F) = E. \tag{8.4.1}$$

Indeed, consider any $v \in E$. It follows immediately from the definition of F that there exist sequences $v_n \in U^n(F)$ and $w_n \in U^n(E)$ such that $v = v_0 + \cdots + v_{n-1} + w_n$ for each $n \geq 1$. We want to show that $(w_n)_n$ converges to zero, to conclude that $v = \sum_{j=0}^{\infty} v_n$. For that, observe that

$$\|v\|^2 = \sum_{j=0}^{n-1} \|v_j\|^2 + \|w_n\|^2 \quad \text{for every } n$$

and, thus, the series $\sum_{j=0}^{\infty} \|v_j\|^2$ is summable. Given $\varepsilon > 0$, fix $m \geq 1$ such that the sum of the terms with $j \geq m$ is less than ε. For every $n \geq m$,

$$\|w_m - w_n\|^2 = \|v_m + \cdots + v_{n-1}\|^2 = \|v_m\|^2 + \cdots + \|v_{n-1}\|^2 < \varepsilon.$$

This proves that $(w_n)_n$ is a Cauchy sequence in H. Let w be its limit. Since $w_n \in U^n(E) \subset U^m(E)$ for every $m \leq n$, taking the limit we get that $w \in U^m(E)$

for every m. By condition (ii) in the hypothesis, this implies that $w = 0$. Therefore, the proof of the claim (8.4.1) is complete. To conclude the proof of the proposition it suffices to observe that

$$\bigoplus_{k\in\mathbb{Z}}^{\infty} U^k(F) = \sum_{n=0}^{\infty} \bigoplus_{k=-n}^{\infty} U^k(F) = \sum_{n=0}^{\infty} U^{-n}(E).$$

Condition (iii) in the hypothesis implies that this subspace coincides with H.

In particular, an invertible system (f,μ) has Lebesgue spectrum if and only if there exists a closed subspace $F \subset L^2_0(\mu)$ such that

$$L^2_0(\mu) = \bigoplus_{k\in\mathbb{Z}} U_f^k(F). \tag{8.4.2}$$

The next result is the reason why systems with Lebesgue spectrum are denominated in this way, and it also leads naturally to the notion of rank:

Proposition 8.4.10. *Let $U: H \to H$ be a unitary operator in a Hilbert space. Let λ denote the Lebesgue measure on the unit circle. Then U has Lebesgue spectrum if and only if it admits a spectral representation*

$$T: L^2(\lambda)^\chi \to L^2(\lambda)^\chi \quad (\varphi_a)_a \mapsto (z \mapsto z\varphi_a(z))_a$$

for some cardinal χ. Moreover, χ is uniquely determined by U.

Proof. Let us start by proving the "if" claim. As we know, the Fourier family $\{z^n : n \in \mathbb{Z}\}$ is a Hilbert basis of the space $L^2(\lambda)$. Let V_n be the one-dimensional subspace generated by $\varphi(z) = z^n$. Then, $L^2(\lambda) = \bigoplus_{n\in\mathbb{Z}} V_n$ and, consequently,

$$L^2(\lambda)^\chi = \left(\bigoplus_{n\in\mathbb{Z}} V_n\right)^\chi = \bigoplus_{n\in\mathbb{Z}} V_n^\chi \tag{8.4.3}$$

(W^χ denotes the orthogonal direct sum of χ copies of a space W). Moreover, the restriction of T to each V_n^χ is a unitary operator onto V_{n+1}^χ. Take $F' = V_0^\chi$. The relation (8.4.3) means that the iterates $T^n(F') = V_n^\chi$ are pairwise orthogonal and their orthogonal direct sum is the space $L^2(\lambda)^\chi$. Using the conjugacy of T to the Koopman operator in $L^2_0(\mu)$, we conclude that there exists a subspace F in the conditions of Proposition 8.4.9.

Conversely, suppose that there exists F in the conditions of Proposition 8.4.9. Let $\{v_q : q \in Q\}$ be a Hilbert basis of F. Then $\{U^n(v_q) : n \in \mathbb{Z}, q \in Q\}$ is a Hilbert basis of H. Given $q \in Q$, denote by δ_q the element of the space $L^2(\lambda)^Q$ that is equal to 1 in the coordinate q and identically zero in all the other coordinates. Define

$$L: H \to L^2(\lambda)^Q, \quad L(U^n(v_q)) = z^n\delta_q \quad \text{for each } n \in \mathbb{Z} \text{ and } q \in Q.$$

Observe that L is a unitary operator, since $\{z^n\delta_q\}$ is a Hilbert basis of $L^2(\lambda)^Q$. Observe also that $LU = TL$. This provides the spectral representation in the statement of the proposition, with χ equal to the cardinal of the set Q, that is, equal to the Hilbert dimension of the subspace F.

Let $E \subset H$ be any subspace satisfying the conditions in Definition 8.4.1. Then the orthogonal difference $F = E \ominus U(E)$ satisfies the conclusion of Proposition 8.4.9, as we saw during the proof of that proposition. Moreover, according to the proof of Proposition 8.4.10, we may take the cardinal χ equal to the Hilbert dimension of F. Since χ is uniquely determined, the same holds for the Hilbert dimension of $E \ominus U(E)$. This proves Proposition 8.4.6 in the invertible case. In Exercise 8.4.3 we invite the reader to prove the general case.

We have just shown that the rank of a system with Lebesgue spectrum is well defined. Next, we are going to see that for invertible systems the rank is a complete invariant of spectral equivalence:

Corollary 8.4.11. *Two invertible systems with Lebesgue spectrum are spectrally equivalent if and only if they have the same rank.*

Proof. It is clear that two invertible systems are spectrally equivalent if and only if they admit the same spectral representation. By Proposition 8.4.10, this happens if and only if the value of the cardinal χ is the same, that is, if the rank is the same.

Corollary 8.4.12. *All two-sided Bernoulli shifts of countable type are spectrally equivalent.*

Proof. As we saw in the previous section, all Bernoulli shifts of countable type have countable rank.

Proofs of the facts that are quoted in the following may be found in Mañé [Mañ87, Section II.10]:

Example 8.4.13 (Gaussian shifts). Let $A = (a_{i,j})_{i,j\in\mathbb{Z}}$ be an infinite real matrix. We say that A is *positive definite* if every finite restriction $A_{m,n} = (a_{i,j})_{m\le i,j<n}$ is positive definite, for any $m < n$. We say that A is *symmetric* if $a_{i,j} = a_{j,i}$ for any $i,j \in \mathbb{Z}$. Let μ be a Borel probability measure on $\Sigma = \mathbb{R}^{\mathbb{Z}}$ (similar considerations hold for $\Sigma = \mathbb{R}^{\mathbb{N}}$). We say that μ is a *Gaussian measure* if there exists some symmetric positive definite matrix A such that $\mu([m; B_m, \dots, B_{n-1}])$ is equal to

$$\frac{1}{(\det A_{m,n})^{1/2}} \frac{1}{(2\pi)^{(n-m)/2}} \int_{B_m\times\cdots\times B_{n-1}} \exp\left(-\frac{1}{2}(A_{m,n}^{-1}z\cdot z)\right) dz$$

for any $m < n$ and any measurable sets $B_m, \dots, B_{n-1} \subset \mathbb{R}$. The reason for the factor on the left-hand side is explained in Exercise 8.4.4. A is called the *covariance matrix* of μ. It is uniquely determined by

$$a_{i,j} = \int x_i x_j \, d\mu(x) \quad \text{for each } i,j \in \mathbb{Z}.$$

For each symmetric positive definite matrix A there exists a unique Gaussian probability measure μ that has A as its covariance matrix. Moreover, μ is invariant under the shift map $\sigma : \Sigma \to \Sigma$ if and only if $a_{i,j} = a_{i+1,j+1}$ for any $i,j \in \mathbb{Z}$. In that case, the properties of the system (σ, μ) are directly related to the behavior of the *covariance sequence*

$$\alpha_n = a_{n,0} = U_\sigma^n x_0 \cdot x_0 \quad \text{for each } n \geq 0.$$

In particular, (f, μ) is mixing if and only if the covariance sequence converges to zero.

Now, Exercise 8.4.5 shows that if (f, μ) has Lebesgue spectrum then the covariance sequence is generated by some absolutely continuous probability measure ν on the unit circle, in the following sense:

$$\alpha_n = \int z^n d\nu(z) \quad \text{for each } n \geq 0.$$

(The Riemann–Lebesgue lemma asserts that if ν is a probability measure absolutely continuous with respect to the Lebesgue measure λ on the unit circle then the sequence $\int z^n d\nu(z)$ converges to zero when $n \to \infty$.) But Exercise 8.4.6 shows that not every sequence that converges to zero is of this form. Therefore, there exist Gaussian shifts (σ, μ) that are mixing but do not have Lebesgue spectrum.

8.4.3 Exercises

8.4.1. Show that every mixing Markov shift has Lebesgue spectrum with countable rank. [Observation: In Section 9.5.3 we mention stronger results.]

8.4.2. Let μ be the Haar measure on $\mathbb{T}^d$ and $f_A : \mathbb{T}^d \to \mathbb{T}^d$ be a surjective endomorphism. Assume that no eigenvalue of the matrix A is a root of unity. Check that every orbit of A^t in the set $\mathbb{Z}^d \setminus \{0\}$ is infinite and use this fact to conclude that (f_A, μ) has Lebesgue spectrum. Conversely, if (f_A, μ) has Lebesgue spectrum then no eigenvalue of A is a root of unity.

8.4.3. Complete the proof of Proposition 8.4.6, using Exercise 2.3.6 to reduce the general case to the invertible one.

8.4.4. Check that $\int_{\mathbb{R}} e^{-x^2/2}\, dx = \sqrt{2\pi}$. Use this fact to show that if A is a symmetric positive definite matrix of dimension $d \geq 1$ then

$$\int_{\mathbb{R}^d} \exp\big(-(A^{-1}z \cdot z)/2\big)\, dz = (\det A)^{1/2}(2\pi)^{d/2}.$$

8.4.5. Let (f, μ) be an invertible system with Lebesgue spectrum. Show that for every $\varphi \in L_0^2(\mu)$ there exists a probability measure ν absolutely continuous with respect to the Lebesgue measure λ on the unit circle $\{z \in \mathbb{C} : |z| = 1\}$ and such that $U_f^n \varphi \cdot \varphi = \int z^n d\nu(z)$ for every $n \in \mathbb{Z}$.

8.4.6. Let λ be the Lebesgue measure in the unit circle. Consider the linear operator $F : L^1(\lambda) \to c_0$ defined by

$$F(\varphi) = \left(\int z^n \varphi(z)\, d\lambda(z) \right)_n.$$

Show that F is continuous and injective but not surjective. Therefore, not every sequence of complex numbers $(\alpha_n)_n$ converging to zero may be written as $\alpha_n = \int z^n \, d\nu(z)$ for $n \geq 0$, for some probability measure ν absolutely continuous with respect to λ.

8.5 Lebesgue spaces and ergodic isomorphism

The main subject this section are the *Lebesgue spaces* (also called *standard probability spaces*), a class of probability spaces introduced by the Russian mathematician Vladimir A. Rokhlin [Rok62]. These spaces have a distinguished role in measure theory, for two reasons: on the one hand, they exhibit much better properties than a general probability space; on the other hand, they include most interesting examples. In particular, every complete separable metric space endowed with a Borel probability measure is a Lebesgue space.

Initially, we discuss yet another notion of equivalence, intermediate to ergodic equivalence and spectral equivalence, that we call *ergodic isomorphism.* One of the highlights is that for transformations in Lebesgue spaces the notions of ergodic equivalence and ergodic isomorphism turn out to coincide.

8.5.1 Ergodic isomorphism

Let $(M,\mathcal{B},\mu)$ be a probability space. We denote by $\tilde{\mathcal{B}}$ the quotient of the σ-algebra by the equivalence relation $A \sim B \Leftrightarrow \mu(A\Delta B) = 0$. Observe that if $A_k \sim B_k$ for every $k \in \mathbb{N}$ then $\bigcup_k A_k \sim \bigcup_k B_k$, $\bigcap_k A_k \sim \bigcap_k B_k$ and $A_k^c \sim B_k^c$ for every $k \in \mathbb{N}$. Therefore, the basic operations of set theory are well defined in the quotient $\tilde{\mathcal{B}}$. Moreover, the measure μ induces a measure $\tilde{\mu}$ on $\tilde{\mathcal{B}}$. The pair $(\tilde{\mathcal{B}},\tilde{\mu})$ is called the *measure algebra* of the probability space.

Now let $(M,\mathcal{B},\mu)$ and $(N,\mathcal{C},\nu)$ be two probability spaces, and $(\tilde{\mathcal{B}},\tilde{\mu})$ and $(\tilde{\mathcal{C}},\tilde{\nu})$ be their measure algebras. A *homomorphism* of measure algebras is a map $H : \tilde{\mathcal{B}} \to \tilde{\mathcal{C}}$ that preserves the operations of union, intersection and complement and also preserves the measures: $\mu(B) = \nu(H(B))$ for every $B \in \tilde{\mathcal{B}}$. If H is a bijection, we call it an *isomorphism* of measure algebras. In that case the inverse H^{-1} is also an isomorphism of measure algebras.

Every measurable map $h : M \to N$ satisfying $h_*\mu = \nu$ defines a homomorphism $\tilde{h} : \tilde{\mathcal{C}} \to \tilde{\mathcal{B}}$, through $B \mapsto h^{-1}(B)$. Moreover, if h is invertible then $\tilde{h}$ is an isomorphism. In the same way, transformations $f : M \to M$ and $g : N \to N$ preserving the measures in the corresponding probability spaces define homomorphisms $\tilde{f} : \tilde{\mathcal{B}} \to \tilde{\mathcal{B}}$ and $\tilde{g} : \tilde{\mathcal{C}} \to \tilde{\mathcal{C}}$, respectively. We say that the systems (f,μ) and (g,ν) are *ergodically isomorphic* if these homomorphisms are conjugate, that is, if $\tilde{f} \circ H = H \circ \tilde{g}$ for some isomorphism $H : \tilde{\mathcal{C}} \to \tilde{\mathcal{B}}$.

Ergodically equivalent systems are always ergodically isomorphic: given any ergodic equivalence h, it suffices to take $H = \tilde{h}$. We also have the following relation between ergodic isomorphism and spectral equivalence:

Proposition 8.5.1. *If two systems (f, μ) and (g, ν) are ergodically isomorphic then they are spectrally equivalent.*

Proof. Let $H : \tilde{\mathcal{C}} \to \tilde{\mathcal{B}}$ be an isomorphism such that $\tilde{f} \circ H = H \circ \tilde{g}$. Consider the linear operator $L : L^2(\nu) \to L^2(\mu)$ constructed as follows. Initially, $L(\mathcal{X}_C) = \mathcal{X}_{H(C)}$ for every $B \in \tilde{\mathcal{C}}$. Note that $\|L(\mathcal{X}_C)\| = \|\mathcal{X}_C\|$. Extend the definition to the set of simple functions, preserving linearity:

$$L\left(\sum_{j=1}^{k} c_j \mathcal{X}_{C_j}\right) = \sum_{j=1}^{k} c_j \mathcal{X}_{H(C_j)} \quad \text{for any } k \geq 1,\ c_j \in \mathbb{R} \text{ and } C_j \in \tilde{\mathcal{C}}.$$

The definition does not depend on the representation of the simple function as a linear combination of characteristic functions (Exercise 8.5.1). Moreover, $\|L(\varphi)\| = \|\varphi\|$ for every simple function. Recall that the set of simple functions is dense in $L^2(\nu)$. Then, by continuity, L extends uniquely to a linear isometry defined on the whole of $L^2(\nu)$. Observe that this isometry is invertible: the inverse is constructed in the same way, starting from the inverse of H. Finally,

$$U_f \circ L(\mathcal{X}_C) = U_f(\mathcal{X}_{H(C)}) = \mathcal{X}_{\tilde{f}(H(C))} = \mathcal{X}_{H(\tilde{g}(C))} = L(\mathcal{X}_{\tilde{g}(C)}) = L \circ U_g(\mathcal{X}_C)$$

for every $C \in \tilde{\mathcal{C}}$. By linearity, it follows that $U_f \circ L(\varphi) = L \circ U_g(\varphi)$ for every simple function; then, by continuity, the same holds for every $\varphi \in L^2(\nu)$.

Summarizing these observations, we have the following relation between the three equivalence relations:

ergodic equivalence $\Rightarrow$ *ergodic isomorphism* $\Rightarrow$ *spectral equivalence.*

In what follows we discuss some partial converses, starting with the relation between ergodic isomorphism and spectral equivalence.

The following result of Paul Halmos and John von Neumann [HvN42] broadens Proposition 8.3.4 and shows that for systems with discrete spectrum the notions of ergodic isomorphism and spectral equivalence coincide. The reader may find a proof in Section 3.2 of Walters [Wal75].

Theorem 8.5.2 (Discrete spectrum). *If (f, μ) and (g, ν) are ergodic systems with discrete spectrum then the following conditions are equivalent:*

1. *(f, μ) and (g, ν) are spectrally equivalent;*
2. *the Koopman operators of (f, μ) and (g, ν) have the same eigenvalues;*
3. *(f, μ) and (g, ν) are ergodically isomorphic.*

In particular, every invertible ergodic system with discrete spectrum is ergodically isomorphic to its inverse.

8.5.2 Lebesgue spaces

Let $(M,\mathcal{B},\mu)$ be any probability space. Initially, suppose that the measure μ is non-atomic, that is, that $\mu(\{x\})=0$ for every $x\in M$. Let $\mathcal{P}_1 \prec \cdots \prec \mathcal{P}_n \prec \cdots$ be an increasing sequence of finite partitions of M into measurable sets. We call the sequence *separating* if, given any two different points $x,y\in M$, there exists $n\geq 1$ such that $\mathcal{P}_n(x)\neq\mathcal{P}_n(y)$. In other words, the non-empty elements of the partition $\bigvee_{n=1}^{\infty}\mathcal{P}_n$ contain a unique point.

Let $M_{\mathcal{P}}$ be the subset one obtains by removing from M all the $P\in\bigcup_n\mathcal{P}_n$ with measure zero. Observe that $M_{\mathcal{P}}$ has full measure. We denote by $\mathcal{B}_{\mathcal{P}}$ and $\mu_{\mathcal{P}}$ the restrictions of $\mathcal{B}$ and μ, respectively, to $M_{\mathcal{P}}$. Let m be the Lebesgue measure on $\mathbb{R}$. The next proposition means that the separating sequence allows one to represent the probability space $(M_{\mathcal{P}},\mathcal{B}_{\mathcal{P}},\mu_{\mathcal{P}})$ as a kind of subspace of the real line. We say "kind of" because, in general, the image $\iota(M_{\mathcal{P}})$ *is not* a measurable subset of $\mathbb{R}$.

Proposition 8.5.3. *Given any separating sequence $(\mathcal{P}_n)_n$, there exists a compact totally disconnected set $K\subset\mathbb{R}$ and there exists a measurable injective map $\iota: M_{\mathcal{P}}\to K$ such that the closure of the image $\overline{\iota(P)}$ of every $P\in\bigcup_n\mathcal{P}_n$ is an open and closed subset of K with $m(\overline{\iota(P)})=\mu(P)$. In particular, $\iota_*\mu$ coincides with the restriction of the Lebesgue measure m to the set K.*

Proof. Let $\alpha_n=1+1/n$ for $n\geq 1$. We are going to construct a sequence of bijective maps $\psi_n:\mathcal{P}_n\to\mathcal{I}_n$, $n\geq 1$ satisfying:

(i) each $\mathcal{I}_n$ is a finite family of compact pairwise disjoint intervals;
(ii) each element of $\mathcal{I}_n$, $n>1$ is contained in some element of $\mathcal{I}_{n-1}$;
(iii) $m(\psi_n(P))=\alpha_n\mu(P)$ for every $P\in\mathcal{P}_n$ and every $n\geq 1$.

To do this, we start by writing $\mathcal{P}_1=\{P_1,\dots,P_N\}$. Consider any family $\mathcal{I}_1=\{I_1,\dots,I_N\}$ of compact pairwise disjoint intervals such that $m(I_j)=\alpha_1\mu(P_j)$ for every j. Let $\psi_1:\mathcal{P}_1\to\mathcal{I}_1$ be the map associating with each P_j the corresponding I_j. Now suppose that, for a given $n\geq 1$, we have already constructed maps ψ_1, $\dots$, ψ_n satisfying (i), (ii), (iii). For each $P\in\mathcal{P}_n$, let $I=\psi_n(P)$ and let $P_1,\dots,P_N$ be the elements of $\mathcal{P}_{n+1}$ contained in P. Take compact pairwise disjoint intervals $I_1,\dots,I_N\subset I$ satisfying $m(I_j)=\alpha_{n+1}\mu(P_j)$ for each $j=1,\dots,N$. This is possible because, by the induction hypothesis,

$$m(I)=\alpha_n\mu(P)=\alpha_n\sum_{j=1}^{N}\mu(P_j)>\alpha_{n+1}\sum_{j=1}^{N}\mu(P_j).$$

Then, define $\psi_{n+1}(P_j)=I_j$ for each $j=1,\dots,N$. Repeating this procedure for each $P\in\mathcal{P}_n$, we complete the definition of ψ_{n+1} and $\mathcal{I}_{n+1}$. It is clear that the conditions (i), (ii), (iii) are preserved. This finishes the construction.

Now, let $K = \bigcap_n \bigcup_{I \in \mathcal{I}_n} I$. It is clear that K is compact and its intersection with any $I \in \mathcal{I}_n$ is an open and closed subset of K. Moreover,

$$\max\{m(I) : I \in \mathcal{I}_n\} = \alpha_n \max\{\mu(P) : P \in \mathcal{P}_n\} \to 0 \quad \text{when } n \to \infty \tag{8.5.1}$$

because the sequence $(\mathcal{P}_n)_n$ is separating and the measure μ is non-atomic. Hence, K is totally disconnected. For each $x \in M_{\mathcal{P}}$, the intervals $\psi_n(\mathcal{P}_n(x))$ form a decreasing sequence of compact sets whose lengths decrease to zero. Define $\imath(x)$ to be the unique point in $\bigcap_n \psi_n(\mathcal{P}_n(x))$. The hypothesis that the sequence is separating ensures that $\imath$ is injective: if $x \neq y$ then there exists $n \geq 1$ such that $\mathcal{P}_n(x) \cap \mathcal{P}_n(y) = \emptyset$ and, thus, $\imath(x) \neq \imath(y)$. By construction, the pre-image of $K \cap I$ is in $\bigcup_n \mathcal{P}_n$ for every $I \in \bigcup_n \mathcal{I}_n$. Consider the algebra $\mathcal{A}$ formed by the finite disjoint unions of sets $K \cap I$ of this form. This algebra is generating and we have just checked that $\imath^{-1}(A)$ is a measurable set for every $A \in \mathcal{A}$. Therefore, the transformation $\imath$ is measurable.

To check the other properties in the statement of the proposition, begin by noting that, for every $n \geq 1$ and $P \in \mathcal{P}_n$,

$$\overline{\imath(P)} = \bigcap_{k=n}^{\infty} \bigcup_{Q} \psi_k(Q), \tag{8.5.2}$$

where the union is over all the $Q \in \mathcal{P}_k$ that are contained in P. To get the inclusion $\subset$ it suffices to note that $\imath(P) = \bigcup_Q \imath(Q)$ and $\imath(Q) \subset \psi(Q)$ for every $Q \in \mathcal{P}_k$ and every k. The converse follows from the fact that $\imath(P)$ intersects every $\psi_k(Q)$ (the intersection contains $\imath(Q)$) and the length of the $\psi_k(Q)$ converges to zero when $k \to \infty$. In this way, (8.5.2) is proven. It follows that

$$m(\overline{\imath(P)}) = \lim_k m\Big(\bigcup_Q \psi_k(Q)\Big) = \lim_k \sum_Q \alpha_k \mu(Q) = \lim_k \alpha_k \mu(P) = \mu(P).$$

Moreover, (8.5.2) means that $\overline{\imath(P)} = \bigcap_{k=n}^{\infty} \bigcup_I I$, where the union is over all the $I \in \mathcal{I}_k$ that are contained in $\psi_n(P)$. The right-hand side of this equality coincides with $K \cap \psi_n(P)$ and, hence, is an open and closed subset of K. It also follows from the construction that $\imath^{-1}(\overline{\imath(P)}) = P$. Consequently, $\imath_*\mu(\overline{\imath(P)}) = \mu(P) = m(\overline{\imath(P)})$ for every $P \in \bigcup_n \mathcal{P}_n$. Since the algebra of finite pairwise disjoint unions of sets $\overline{\imath(P)}$ generates the measurable structure of K, we conclude that $\imath_*\mu = m \mid K$.

We say that a probability space without atoms $(M, \mathcal{B}, \mu)$ is a *Lebesgue space* if, for some separating sequence, the image $\imath(M_{\mathcal{P}})$ is a Lebesgue measurable set. Actually, this property does not depend on the choice of the generating sequence (nor on the families $\mathcal{I}_n$ in the proof of Proposition 8.5.3), but we do not prove this fact here: the reader may find a proof in [Rok62, §2.2]. Exercise 8.5.6 shows that it is possible to define Lebesgue space in a more direct way, without using Proposition 8.5.3.

Note that if $\iota(M_{\mathcal{P}})$ is measurable then $\iota(P) = \iota(M_{\mathcal{P}}) \cap \psi_n(P)$ is measurable for every $P \in \mathcal{P}_n$ and every n. Hence, the inverse ι^{-1} is also a measurable transformation. Moreover, $m(\iota(M_{\mathcal{P}})) = \mu(M_{\mathcal{P}}) = 1 = m(K)$. Therefore, every Lebesgue space $(M, \mathcal{B}, \mu)$ is isomorphic, as a measure space, to a measurable subset of a compact totally disconnected subset of the real line.

Observe that if the cardinal of M is strictly larger than the cardinal of the continuum then $(M, \mathcal{B}, \mu)$ admits no separating sequence and, thus, cannot be a Lebesgue space. In Exercise 8.5.8 we propose another construction of probability spaces that are not Lebesgue spaces. Despite examples such as these, practically all the probability spaces we deal with are Lebesgue spaces:

Theorem 8.5.4. *If M is a complete separable metric space and μ is a Borel probability measure with no atoms then $(M, \mathcal{B}, \mu)$ is a Lebesgue space.*

Proof. Let $X \subset M$ be a countable dense subset and $\{B_n : n \in \mathbb{N}\}$ be an enumeration of the set of balls $B(x, 1/k)$ with $x \in X$ and $k \geq 1$. We are going to construct an increasing sequence $(\mathcal{P}_n)_n$ of finite partitions such that

(i) $\mathcal{P}_n$ is finer than $\{B_1, B_1^c\} \vee \cdots \vee \{B_n, B_n^c\}$, and
(ii) $E_n = \{x \in M : \mathcal{P}_n(x) \text{ is not compact}\}$ satisfies $\mu(E_n) \leq 2^{-n}$.

We start by considering $\mathcal{Q}_1 = \{B_1, B_1^c\}$. By Proposition A.3.7, there exist compact sets $K_1 \subset B_1$ and $K_2 \subset B_1^c$ such that $\mu(B_1 \setminus K_1) \leq 2^{-1}\mu(B_1)$ and $\mu(B_1^c \setminus K_2) \leq 2^{-1}\mu(B_1^c)$. Then take $\mathcal{P}_1 = \{K_1, B_1 \setminus K_1, K_2, B_1^c \setminus K_2\}$. Now, for each $n \geq 1$, assume that one has already constructed partitions $\mathcal{P}_1 \prec \cdots \prec \mathcal{P}_n$ satisfying (i) and (ii). Consider the partition $\mathcal{Q}_{n+1} = \mathcal{P}_n \vee \{B_{n+1}, B_{n+1}^c\}$ and let $Q_1, \ldots, Q_m$ be its elements. By Proposition A.3.7, there exist compact sets $K_j \subset Q_j$ such that $\mu(Q_j \setminus K_j) \leq 2^{-(n+1)}\mu(Q_j)$ for every $j = 1, \ldots, m$. Take

$$\mathcal{P}_{n+1} = \{K_1, Q_1 \setminus K_1, \ldots, K_m, Q_m \setminus K_m\}.$$

It is clear that $\mathcal{P}_{n+1}$ satisfies (i) and (ii). Therefore, our construction is complete.

All that is left is to show that the existence of such a sequence $(\mathcal{P}_n)_n$ implies the conclusion of the theorem. Property (i) ensures that the sequence is separating. Let $\iota : M_{\mathcal{P}} \to K$ be a map as in Proposition 8.5.3. Fix any $N \geq 1$ and consider any point $y \in K \setminus \iota(M_{\mathcal{P}})$. For each $n > N$, let I_n be the interval in the family $\mathcal{I}_n$ that contains y and let P_n be the element of $\mathcal{P}_n$ such that $\psi_n(P_n) = I_n$. Note that $(P_n)_n$ is a decreasing sequence. If they were all compact, there would be $x \in \bigcap_{n>N} P_n$ and, by definition, $\iota(x)$ would be equal to y. Since we are assuming that y is not in the image of ι, this proves that there exists $l > N$ such that P_l is not compact. Take $l > N$ minimum and let $I_l = \psi_l(P_l)$. Recall that $m(I_l) = \alpha_l \mu(P_l) \leq 2\mu(P_l)$. Let $\tilde{I}_N$ and $\tilde{P}_N$ be the unions of all these I_l and P_l, respectively, when we vary y on the whole $K \setminus \iota(M_{\mathcal{P}})$. On the one hand, $\tilde{I}_N$ contains $K \setminus \iota(M_{\mathcal{P}})$; on the other hand, $\tilde{P}_N$ is contained in $\bigcup_{l>N} E_l$. Moreover, the I_l are pairwise disjoint (because we took l minimum) and the same holds for the P_l. Hence,

$$m(\tilde{I}_N) \le 2\mu(\tilde{P}_N) \le 2\mu\Big(\bigcup_{l>N} E_l\Big) \le 2^{-N+1}.$$

Then the intersection $\bigcap_N \tilde{I}_N$ has Lebesgue measure zero and contains $K \setminus \iota(M_{\mathcal{P}})$. Since K is a Borel set, this shows that $\iota(M_{\mathcal{P}})$ is a Lebesgue measurable set.

The next result implies that all the Lebesgue spaces with no atoms are isomorphic:

Proposition 8.5.5. *If $(M,\mathcal{B},\mu)$ is a Lebesgue space with no atoms, there exists an invertible measurable map $h: M \to [0,1]$ (defined between subsets of full measure) such that $h_*\mu$ coincides with the Lebesgue measure on $[0,1]$.*

Proof. Let $\iota: M_{\mathcal{P}} \to K$ be a map as in Proposition 8.5.3. Consider the map $g: K \to [0,1]$ defined by $g(x) = m([a,x]\cap K)$, where $a = \min K$. It follows immediately from the definition that g is non-decreasing and Lipschitz:

$$g(x_2) - g(x_1) = m\big([x_1,x_2]\cap K\big) \le x_2 - x_1,$$

for any $x_1 < x_2$ in K. In particular, g is measurable. By monotonicity, the pre-image of any interval $[y_1,y_2] \subset [0,1]$ is a set of the form $[x_1,x_2]\cap K$ with $x_1,x_2 \in K$ and $g(x_1) = y_1$ and $g(x_2) = y_2$. In particular,

$$m\big([x_1,x_2]\cap K\big) = g(x_2) - g(x_1) = y_2 - y_1 = m([y_1,y_2]).$$

This shows that $g_*(m \mid K) = m \mid [0,1]$. Let Y be the set of points $y \in [0,1]$ such that $g^{-1}(\{y\}) = [x_1,x_2]\cap K$ with $x_1,x_2 \in K$ and $x_1 < x_2$. Let $X = g^{-1}(Y)$. Then $m(X) = m(Y) = 0$ because Y is countable. Moreover, the restriction $g: K \setminus X \to [0,1]\setminus Y$ is bijective. Its inverse is non-decreasing and, consequently, measurable. Now, take $h = g \circ \iota$. It follows from the previous observations that

$$h: M_{\mathcal{P}} \setminus \iota^{-1}(X) \to g(\iota(M_{\mathcal{P}})) \setminus Y$$

is a measurable bijection with measurable inverse such that $h_*\mu = m \mid [0,1]$.

Now we extend this discussion to general probability spaces $(M,\mathcal{B},\mu)$, possibly with atoms. Let $A \subset M$ be the set of all the atoms; note that A is at most countable, possibly finite. If the space is purely atomic, that is, if $\mu(A) = 1$, then, by definition, it is a Lebesgue space. More generally, let $M' = M \setminus A$, let $\mathcal{B}'$ be the restriction of $\mathcal{B}$ to M' and let μ' be the normalized restriction of μ to $\mathcal{B}'$. By definition, $(M,\mathcal{B},\mu)$ is a Lebesgue space if $(M',\mathcal{B}',\mu')$ is a Lebesgue space.

It is clear that Theorem 8.5.4 remains valid in the general case: every complete separable metric space endowed with a Borel probability measure, possibly with atoms, is a Lebesgue space. Moreover, Proposition 8.5.5 has the following extension to the atomic case: if $(M,\mathcal{B},\mu)$ is a Lebesgue space and

$A \subset M$ denotes the set of atoms of the measure μ, then there exists an invertible measurable map $h : M \to [0, 1 - \mu(A)] \cup A$ such that $h_*\mu$ coincides with m on the interval $[0, 1 - \mu(A)]$ and coincides with μ on A.

Proposition 8.5.6. *Let $(M, \mathcal{B}, \mu)$ and $(N, \mathcal{C}, \nu)$ be two Lebesgue spaces and $H : \tilde{\mathcal{C}} \to \tilde{\mathcal{B}}$ be an isomorphism between the corresponding measure algebras. Then there exists an invertible measurable map $h : M \to N$ such that $h_*\mu = \nu$ and $H = \tilde{h}$ for every $C \in \tilde{\mathcal{C}}$. Moreover, h is essentially unique: any two maps satisfying these conditions coincide at μ-almost every point.*

We are going to sketch the proof of this proposition in the non-atomic case. The arguments are based on the ideas and use the notations in the proof of Proposition 8.5.3.

Let us start with the uniqueness claim. Let $h_1, h_2 : M \to N$ be any two maps such that $(h_1)_*\mu = (h_2)_*\mu = \nu$. Suppose that $h_1(x) \neq h_2(x)$ for every x in a set $E \subset M$ with $\mu(E) > 0$. Let $(\mathcal{Q}_n)_n$ be a separating sequence in $(N, \mathcal{C}, \nu)$. Then $\mathcal{Q}_n(h_1(x)) \neq \mathcal{Q}_n(h_2(x))$ for every $x \in E$ and every n sufficiently large. Hence, we may fix n (large) and $E' \subset E$ with $\mu(E') > 0$ such that $\mathcal{Q}_n(h_1(x)) \neq \mathcal{Q}_n(h_2(x))$ for every $x \in E'$. Consequently, there exist $Q \in \mathcal{Q}_n$ and $E'' \subset E'$ with $\mu(E'') > 0$ such that $\mathcal{Q}_n(h_1(x)) = Q$ and $\mathcal{Q}_n(h_2(x)) \neq Q$ for every $x \in E''$. Therefore, $E'' \subset h_1^{-1}(Q) \setminus h_2^{-1}(Q)$. This implies that $\tilde{h}_1(Q) \neq \tilde{h}_2(Q)$ and, hence, $\tilde{h}_1 \neq \tilde{h}_2$.

Next we comment on the existence claim. Let $(\mathcal{P}'_n)_n$ and $(\mathcal{Q}'_n)_n$ be separating sequences in $(M, \mathcal{B}, \mu)$ and $(N, \mathcal{C}, \nu)$, respectively. Define $\mathcal{P}_n = \mathcal{P}'_n \vee H(\mathcal{Q}'_n)$ and $\mathcal{Q}_n = \mathcal{Q}'_n \vee H^{-1}(\mathcal{P}_n)$. Then $(\mathcal{P}_n)_n$ and $(\mathcal{Q}_n)_n$ are also separating sequences and $\mathcal{P}_n = H(\mathcal{Q}_n)$ for each n. Let $\imath : M_{\mathcal{P}} \to K$ be a map as in Proposition 8.5.3 and $\psi_n : \mathcal{P}_n \to \mathcal{I}_n$, $n \geq 1$ be the family of bijections used in its construction. Let $\jmath : N_{\mathcal{Q}} \to L$ and $\varphi_n : \mathcal{Q}_n \to \mathcal{J}_n$ be corresponding objects for $(N, \mathcal{C}, \nu)$. Since we are assuming that $(M, \mathcal{B}, \mu)$ and $(N, \mathcal{C}, \nu)$ are Lebesgue spaces, $\imath$ and $\jmath$ are invertible maps over subsets with full measure. Recall also that $m(\psi_n(P)) = \alpha_n \mu(P)$ for each $P \in \mathcal{P}_n$ and, analogously, $m(\varphi_n(Q)) = \alpha_n \nu(Q)$ for each $Q \in \mathcal{Q}_n$. Hence, $m(\psi_n(P)) = m(\varphi_n(Q))$ if $P = H(Q)$. Then, for each n,

$$\psi_n \circ H \circ \varphi_n^{-1} : \mathcal{J}_n \to \mathcal{I}_n \tag{8.5.3}$$

is a bijection that preserves length. Given $z \in K$ and $n \geq 1$, let I_n be the element of $\mathcal{I}_n$ that contains z and let J_n be the corresponding element of $\mathcal{J}_n$, via (8.5.3). By construction, $(J_n)_n$ is a nested sequence of compact intervals whose length converges to zero. Let $\phi(z)$ be the unique point in the intersection. In this way, one has defined a measurable map $\phi : K \to L$ that preserves the Lebesgue measure. It is clear from the construction that ϕ is invertible and the inverse is also measurable. Now it suffices to take $h = \jmath^{-1} \circ \phi \circ \imath$.

All that is left is to check that h is invertible. Applying the construction in the previous paragraph to the inverse H^{-1} we find $h' : N \to M$ such that $h'_*\nu = \mu$

and $H^{-1}=\tilde{h}'$. Then, $\widetilde{h'\circ h}=\tilde{h}\circ\tilde{h}'=\mathrm{id}$ and $\widetilde{h\circ h'}=\tilde{h}'\circ\tilde{h}=\mathrm{id}$. By uniqueness, it follows that $h'\circ h=\mathrm{id}$ and $h\circ h'=\mathrm{id}$ at almost every point.

Corollary 8.5.7. *Let $(M,\mathcal{B},\mu)$ and $(N,\mathcal{C},\nu)$ be two Lebesgue spaces and let $f:M\to M$ and $g:N\to N$ be measurable transformations preserving the measures in their corresponding domains. Then (f,μ) and (g,ν) are ergodically equivalent if and only if they are ergodically isomorphic.*

Proof. We only need to show that if the systems are ergodically isomorphic then they are ergodically equivalent. Let $H:\tilde{\mathcal{C}}\to\tilde{\mathcal{B}}$ be an ergodic isomorphism. By Proposition 8.5.6, there exists an invertible measurable map $h:M\to N$ such that $h_*\mu=\nu$ and $H=\tilde{h}$. Then,

$$\widetilde{h\circ f}=\tilde{f}\circ\tilde{h}=\tilde{f}\circ H=H\circ\tilde{g}=\tilde{h}\circ\tilde{g}=\widetilde{g\circ h}.$$

By the uniqueness part of Proposition 8.5.6, it follows that $h\circ f=g\circ h$ at μ-almost every point. This shows that h is an ergodic equivalence.

8.5.3 Exercises

8.5.1. Let $H:\tilde{\mathcal{C}}\to\tilde{\mathcal{B}}$ be a homomorphism of measure algebras. Show that

$$\sum_{i=1}^{l} b_i\mathcal{X}_{B_i}=\sum_{j=1}^{k} c_j\mathcal{X}_{C_j}\Rightarrow\sum_{i=1}^{l} b_i\mathcal{X}_{H(B_i)}=\sum_{j=1}^{k} c_j\mathcal{X}_{H(C_j)}.$$

8.5.2. Check that the homomorphism of measure algebras $\tilde{g}:\mathcal{C}\to\mathcal{B}$ induced by a measure-preserving map $g:M\to N$ is injective. Suppose that N is a Lebesgue space. Show that, given another measure-preserving map $h:M\to N$, the corresponding homomorphisms $\tilde{g}$ and $\tilde{h}$ coincide if and only if $g=h$ at almost every point.

8.5.3. Let $f:M\to M$ be a measurable transformation in a Lebesgue space $(M,\mathcal{B},\mu)$, preserving the measure μ. Show that (f,μ) is invertible at almost every point (that is, there exists an invariant full measure subset restricted to which f is a measurable bijection with measurable inverse) if and only if the corresponding homomorphism of measure algebras $\tilde{f}:\tilde{\mathcal{B}}\to\tilde{\mathcal{B}}$ is surjective.

8.5.4. Show that the Koopman operator of a system (f,μ) is surjective if and only if the corresponding homomorphism of measure algebras $\tilde{f}:\tilde{\mathcal{B}}\to\tilde{\mathcal{B}}$ is surjective. In Lebesgue spaces this happens if and only if the system is invertible at almost every point.

8.5.5. Show that every system (f,μ) with discrete spectrum in a Lebesgue space is invertible at almost every point.

8.5.6. Given a separating sequence $\mathcal{P}_1\prec\cdots\prec\mathcal{P}_n\prec\cdots$, we call a *chain* any sequence $(P_n)_n$ with $P_n\in\mathcal{P}_n$ and $P_{n+1}\subset P_n$ for every n. We say that a chain is *empty* if $\bigcap_n P_n=\emptyset$. Consider the map $\iota:M_{\mathcal{P}}\to K$ constructed in Proposition 8.5.3. Show that the image $\iota(M_{\mathcal{P}})$ is a Lebesgue measurable set and $m(K\setminus\iota(M_{\mathcal{P}}))=0$ if and only if the empty chains have zero measure in the following sense: for every $\delta>0$ there exists $B\subset M$ such that B is a union of elements of $\bigcup_n\mathcal{P}_n$ with $\mu(B)<\delta$ and every empty chain $(P_n)_n$ has $P_n\subset B$ for every n sufficiently large.

8.5.7. Prove the following extension of Proposition 2.4.4: If $f : M \to M$ preserves a probability measure μ and (M, μ) is a Lebesgue space then μ admits a (unique) lift $\hat{\mu}$ to the natural extension $\hat{f} : \hat{M} \to \hat{M}$.

8.5.8. Let M be a subset of $[0,1]$ with exterior measure $m^*(M) = 1$ but which is not a Lebesgue measurable set. Consider the σ-algebra $\mathcal{M}$ of all sets of the form $M \cap B$, where B is a Lebesgue measurable subset of $\mathbb{R}$. Check that $\mu(M \cap B) = m(B)$ defines a probability measure on $(M, \mathcal{M})$ such that $(M, \mathcal{M}, \mu)$ is not a Lebesgue space.

9
Entropy

The word *entropy* was invented in 1865 by the German physicist and mathematician Rudolf Clausius, one of the founding pioneers of thermodynamics. In the theory of systems in thermodynamical equilibrium, the entropy quantifies the degree of "disorder" in the system. The second law of thermodynamics states that, when an isolated system passes from an equilibrium state to another, the entropy of the final state is necessarily bigger than the entropy of the initial state. For example, when we join two containers with different gases (oxygen and nitrogen, say), the two gases mix with one another until reaching a new macroscopic equilibrium, where they are both uniformly distributed in the two containers. The entropy of the new state is larger than the entropy of the initial equilibrium, where the two gases were separate.

The notion of entropy plays a crucial role in different fields of science. An important example, which we explore in our presentation, is the field of information theory, initiated by the work of the American electrical engineer Claude Shannon in the mid 20th century. At roughly the same time, the Russian mathematicians Andrey Kolmogorov and Yakov Sinai were proposing a definition of the entropy of a system in ergodic theory. The main purpose was to provide an invariant of ergodic equivalence that, in particular, could distinguish two Bernoulli shifts. This Kolmogorov–Sinai entropy is the subject of this chapter.

In Section 9.1 we define the entropy of a transformation with respect to an invariant probability measure, by analogy with a similar notion in information theory. The theorem of Kolmogorov–Sinai, which we discuss in Section 9.2, is a fundamental tool for the actual calculation of the entropy in specific systems. In Section 9.3 we analyze the concept of entropy from a more local viewpoint, which is more closely related to Shannon's formulation of this concept. Next, in Section 9.4, we illustrate a few methods for calculating the entropy, by means of concrete examples.

In Section 9.5 we discuss the role of the entropy as an invariant of ergodic equivalence. The highlight is the theorem of Ornstein (Theorem 9.5.2), according to which any two-sided Bernoulli shifts are ergodically equivalent if and only if they have the same entropy. In that section we also introduce the class of Kolmogorov systems, which contains the Bernoulli shifts and is contained in the class of systems with Lebesgue spectrum. In both cases the inclusion is strict.

In the last couple of sections we present two complementary topics that will be useful later. The first one (Section 9.6) is the theorem of Jacobs, according to which the entropy behaves in an affine way with respect to the ergodic decomposition. The other (Section 9.7) concerns the notion of the Jacobian and its relations with the entropy.

9.1 Definition of entropy

To motivate the definition of Kolmogorov–Sinai entropy, let us look at the following basic situation in information theory. Consider some communication channel transmitting symbols from a certain alphabet $\mathcal{A}$, one after the other. This could be a telegraph transmitting group of dots and dashes, according to the old Morse code, an optical fiber, transmitting packets of zeros and ones, according to the ASCII binary code, or any other process of sequential transmission of information, such as our reader's going through the text of this book, one letter after the other. The objective is to measure the *entropy* of the channel, that is, the mean *quantity of information* it carries, per unit of time.

9.1.1 Entropy in information theory

It is assumed that each symbol has a given frequency, that is, a given probability of being used at any time in the communication. For example, if the channel is transmitting a text in English then the letter E is more likely to be used than the letter Z, say. The occurrence of rarer symbols, such as Z, restricts the kind of word or sentence in which they appear and, hence, is more informative than the presence of commoner symbols, such as E.

This suggests that *information* should be a function of *probability*: the more unlikely a symbol (or a *word*, defined as a finite sequence of symbols) is, the more information it carries.

The situation is actually more complicated, because for most communication codes the probability of using a given symbol also depends on the context. For example, still assuming that the channel transmits in English, any sequence of symbols S, Y, S, T, E must be followed by an M: in this case, in view of the symbols transmitted previously, this letter M is unavoidable, which also means that it carries no additional information.[1]

[1] We once participated in a "treasure hunt" that consisted in searching the woods for hidden letters that would form the name of a mathematical object. It just so happened that the first three letters that were found were Z, Z and Z. That unfortunate circumstance ended the game

On the other hand, in those situations where symbols are transmitted at random, independently of each other, the information carried by each symbol simply adds to the information conveyed by the previous ones. For example, if the transmission reflects the outcomes of the successive flipping of a fair coin, then the amount of information associated with the outcome (*Head*, *Tail*, *Tail*) must be equal to the *sum* of the amounts of information associated with each of the symbols *Head*, *Tail* and *Tail*. Now, by independence, the probability of the event (*Head*, *Tail*, *Tail*) is the *product* of the probabilities of the events *Head*, *Tail* and *Tail*.

This suggests that information should be defined in terms of the *logarithm* of the probability. In information theory it is usual to consider base 2 logarithms, because essentially all the communication channels one finds in practice are binary. However, there is no reason to stick to that custom in our setting: we will consider natural (base e) logarithms instead.

By definition, the *quantity of information* associated with a symbol $a \in \mathcal{A}$ is given by

$$I(a) = -\log p_a, \tag{9.1.1}$$

where p_a is the probability (frequency) of the symbol a. The *mean information* associated with the alphabet $\mathcal{A}$ is given by

$$I(\mathcal{A}) = \sum_{a\in\mathcal{A}} p_a I(a) = \sum_{a\in\mathcal{A}} -p_a \log p_a. \tag{9.1.2}$$

More generally, the quantity of information associated with a word $a_1 \dots a_n$ is

$$I(a_1 \dots a_n) = -\log p_{a_1 \dots a_n}, \tag{9.1.3}$$

where $p_{a_1 \dots a_n}$ denotes the probability of the word. In the independent case this coincides with the product $p_{a_1} \dots p_{a_n}$ of the probabilities of the symbols, but not in general. Denoting by $\mathcal{A}^n$ the set of all the words of length n, we define

$$I(\mathcal{A}^n) = \sum_{a_1,\dots,a_n} p_{a_1 \dots a_n} I(a_1,\dots,a_n) = \sum_{a_1,\dots,a_n} -p_{a_1 \dots a_n} \log p_{a_1 \dots a_n}. \tag{9.1.4}$$

Finally, the *entropy* of the communication channel is defined by:

$$I = \lim_n \frac{1}{n} I(\mathcal{A}^n). \tag{9.1.5}$$

We invite the reader to check that the sequence $I(\mathcal{A}^n)$ is subadditive and, thus, the limit in (9.1.5) does exist. This is also contained in the much more general theory that we are about to present.

prematurely, since at that point the remaining letters could add no information: there is only one mathematical object whose name includes the letter Z three times (the *Yoccoz puzzle*).

9.1.2 Entropy of a partition

We want to adapt these ideas to our context in ergodic theory. The main difference is that, while in information theory the alphabet $\mathcal{A}$ is usually discrete (finite or, at most, countable), that is not the case for the domain (space of states) of most interesting dynamical systems. That issue is dealt with by using partitions of the domain.

Let $(M, \mathcal{B}, \mu)$ be a probability space. In this chapter, by *partition* we always mean a **countable** (finite or infinite) family $\mathcal{P}$ of pairwise disjoint measurable subsets of M whose union has full measure. We denote by $\mathcal{P}(x)$ the element of the partition that contains a given point x. The *sum* $\mathcal{P} \vee \mathcal{Q}$ of two partitions $\mathcal{P}$ and $\mathcal{Q}$ is the partition whose elements are the intersections $P \cap Q$ with $P \in \mathcal{P}$ and $Q \in \mathcal{Q}$. More generally, given any countable family of partitions $\mathcal{P}_n$, we define

$$\bigvee_n \mathcal{P}_n = \left\{ \bigcap_n P_n : P_n \in \mathcal{P}_n \text{ for each } n \right\}.$$

With each partition $\mathcal{P}$ we associate the corresponding *information function*

$$I_{\mathcal{P}} : M \to \mathbb{R}, \quad I_{\mathcal{P}}(x) = -\log \mu(\mathcal{P}(x)). \tag{9.1.6}$$

It is clear that the function $I_{\mathcal{P}}$ is measurable. By definition, the *entropy* of the partition $\mathcal{P}$ is the mean of its information function, that is,

$$H_\mu(\mathcal{P}) = \int I_{\mathcal{P}}\, d\mu = \sum_{P \in \mathcal{P}} -\mu(P) \log \mu(P). \tag{9.1.7}$$

We always abide to the usual (in the theory of Lebesgue integration) convention that $0 \log 0 = \lim_{x \to 0} x \log x = 0$. See Figure 9.1.

Consider the function $\phi : (0, \infty) \to \mathbb{R}$ given by $\phi(x) = -x \log x$. One can readily check that $\phi'' < 0$. Therefore, ϕ is strictly concave:

$$t_1 \phi(x_1) + \cdots + t_k \phi(x_k) \le \phi(t_1 x_1 + \cdots + t_k x_k) \tag{9.1.8}$$

for every $x_1, \ldots, x_k > 0$ and $t_1, \ldots, t_k > 0$ with $t_1 + \cdots + t_k = 1$; moreover, the identity holds if and only if $x_1 = \cdots = x_k$. This observation will be useful on several occasions.

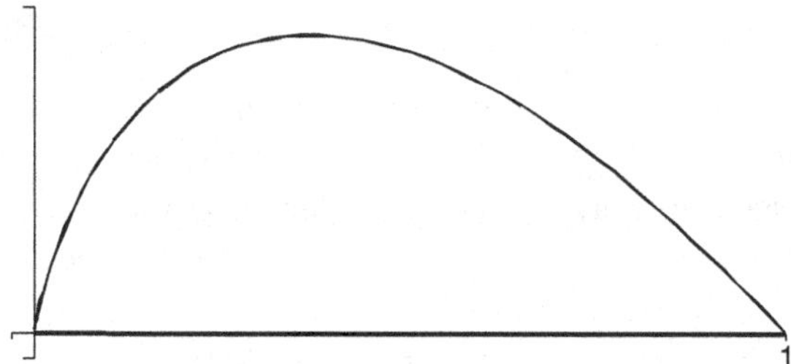

Figure 9.1. Graph of the function $\phi(x) = -x \log x$

We say that two partitions $\mathcal{P}$ and $\mathcal{Q}$ are *independent* if $\mu(P\cap Q)=\mu(P)\mu(Q)$ for every $P\in\mathcal{P}$ and every $Q\in\mathcal{Q}$. Then, $I_{\mathcal{P}\vee\mathcal{Q}}=I_{\mathcal{P}}+I_{\mathcal{Q}}$ and, therefore, $H_\mu(\mathcal{P}\vee\mathcal{Q})=H_\mu(\mathcal{P})+H_\mu(\mathcal{Q})$. In general, one has the inequality $\le$, as we are going to see in a while.

Example 9.1.1. Let $M=[0,1]$ be endowed with the Lebesgue measure. For each $n\ge 1$, consider the partition $\mathcal{P}^n$ of the interval M into the subintervals $\big((i-1)/10^n, i/10^n\big]$ with $1\le i\le 10^n$. Then,

$$H_\mu(\mathcal{P}^n)=\sum_{i=1}^{10^n}-10^{-n}\log 10^{-n}=n\log 10.$$

Example 9.1.2. Let $M=\{1,\dots,d\}^{\mathbb{N}}$ be endowed with a product measure $\mu=\nu^{\mathbb{N}}$. Denote $p_i=\nu(\{i\})$ for each $i\in\{1,\dots,d\}$. For each $n\ge 1$, let $\mathcal{P}^n$ be the partition of M into the cylinders $[0;a_1,\dots,a_n]$ of length n. The entropy of $\mathcal{P}^n$ is

$$\begin{aligned}
H_\mu(\mathcal{P}^n)&=\sum_{a_1,\dots,a_n}-p_{a_1}\cdots p_{a_n}\log\big(p_{a_1}\cdots p_{a_n}\big)\\
&=\sum_j\sum_{a_1,\dots,a_n}-p_{a_1}\cdots p_{a_j}\cdots p_{a_n}\log p_{a_j}\\
&=\sum_j\sum_{a_j}-p_{a_j}\log p_{a_j}\sum_{a_i,i\neq j}p_{a_1}\cdots p_{a_{j-1}}p_{a_{j+1}}\cdots p_{a_n}.
\end{aligned}$$

The last sum is equal to 1, since $\sum_i p_i=1$. Therefore,

$$H_\mu(\mathcal{P}^n)=\sum_{j=1}^{n}\sum_{a_j=1}^{d}-p_{a_j}\log p_{a_j}=\sum_{j=1}^{n}\sum_{i=1}^{d}-p_i\log p_i=-n\sum_{i=1}^{d}p_i\log p_i.$$

Lemma 9.1.3. *Every finite partition $\mathcal{P}$ has finite entropy: $H_\mu(\mathcal{P})\le\log\#\mathcal{P}$ and the identity holds if and only if $\mu(P)=1/\#\mathcal{P}$ for every $P\in\mathcal{P}$.*

Proof. Let $\mathcal{P}=\{P_1,P_2,\dots,P_n\}$ and consider $t_i=1/n$ and $x_i=\mu(P_i)$. By the concavity property (9.1.8):

$$\frac{1}{n}H_\mu(\mathcal{P})=\sum_{i=1}^{n}t_i\phi(x_i)\le\phi\Big(\sum_{i=1}^{n}t_ix_i\Big)=\phi\Big(\frac{1}{n}\Big)=\frac{\log n}{n}.$$

Therefore, $H_\mu(\mathcal{P})\le\log n$. Moreover, the identity holds if and only if $\mu(P_i)=1/n$ for every $i=1,\dots,n$.

Example 9.1.4. Let $M=[0,1]$ be endowed with the Lebesgue measure μ. Observe that the series $\sum_{k=1}^{\infty}1/(k(\log k)^2)$ is convergent. Let c be the value of the sum. Then, we may partition $[0,1]$ into intervals P_k with $\mu(P_k)=1/(ck(\log k)^2)$ for every k. Let $\mathcal{P}$ be the partition formed by these subintervals. Then,

$$H_\mu(\mathcal{P})=\sum_{k=1}^{\infty}\frac{\log c+\log k+2\log\log k}{ck(\log k)^2}.$$

By the ratio convergence criterion, the series on the right-hand side has the same behavior as the series $\sum_{k=1}^{\infty} 1/(k \log k)$ which, as we know (use the integral convergence criterion), is divergent. Therefore, $H_\mu(\mathcal{P}) = \infty$.

This shows that infinite partitions may have infinite entropy. From now, for the rest of the chapter, we always consider (countable) partitions with finite entropy.

The *conditional entropy* of a partition $\mathcal{P}$ with respect to another partition $\mathcal{Q}$ is the number

$$H_\mu(\mathcal{P}/\mathcal{Q}) = \sum_{P \in \mathcal{P}} \sum_{Q \in \mathcal{Q}} -\mu(P \cap Q) \log \frac{\mu(P \cap Q)}{\mu(Q)}. \tag{9.1.9}$$

Intuitively, it measures the amount of information provided by the partition $\mathcal{P}$ in addition to the information provided by the partition $\mathcal{Q}$. It is clear that $H_\mu(\mathcal{P}/\mathcal{M}) = H_\mu(\mathcal{P})$ for every $\mathcal{P}$, where $\mathcal{M}$ denotes the trivial partition $\mathcal{M} = \{M\}$. Moreover, if $\mathcal{P}$ and $\mathcal{Q}$ are independent then $H_\mu(\mathcal{P}/\mathcal{Q}) = H_\mu(\mathcal{P})$. In general, one has the inequality $\leq$, as we are going to see later.

Given two partitions $\mathcal{P}$ and $\mathcal{Q}$, we say that $\mathcal{P}$ is *coarser* than $\mathcal{Q}$ (or, equivalently, $\mathcal{Q}$ is finer than $\mathcal{P}$) and we write $\mathcal{P} \prec \mathcal{Q}$, if every element of $\mathcal{Q}$ is contained in some element of $\mathcal{P}$, up to measure zero. The sum $\mathcal{P} \vee \mathcal{Q}$ may also be defined as the coarsest of all the partitions $\mathcal{R}$ such that $\mathcal{P} \prec \mathcal{R}$ and $\mathcal{Q} \prec \mathcal{R}$.

Lemma 9.1.5. *Let $\mathcal{P}$, $\mathcal{Q}$ and $\mathcal{R}$ be partitions with finite entropy. Then,*

(i) $H_\mu(\mathcal{P} \vee \mathcal{Q}/\mathcal{R}) = H_\mu(\mathcal{P}/\mathcal{R}) + H_\mu(\mathcal{Q}/\mathcal{P} \vee \mathcal{R})$;

(ii) if $\mathcal{P} \prec \mathcal{Q}$ *then* $H_\mu(\mathcal{P}/\mathcal{R}) \leq H_\mu(\mathcal{Q}/\mathcal{R})$ *and* $H_\mu(\mathcal{R}/\mathcal{P}) \geq H_\mu(\mathcal{R}/\mathcal{Q})$;

(iii) $\mathcal{P} \prec \mathcal{Q}$ *if and only if* $H_\mu(\mathcal{P}/\mathcal{Q}) = 0$.

Proof. By definition,

$$\begin{aligned} H_\mu(\mathcal{P} \vee \mathcal{Q}/\mathcal{R}) &= \sum_{P,Q,R} -\mu(P \cap Q \cap R) \log \frac{\mu(P \cap Q \cap R)}{\mu(R)} \\ &= \sum_{P,Q,R} -\mu(P \cap Q \cap R) \log \frac{\mu(P \cap Q \cap R)}{\mu(P \cap R)} \\ &\quad + \sum_{P,Q,R} -\mu(P \cap Q \cap R) \log \frac{\mu(P \cap R)}{\mu(R)}. \end{aligned}$$

The sum on the right-hand side may be rewritten as

$$\begin{aligned} &\sum_{S \in \mathcal{P} \vee \mathcal{R}, Q \in \mathcal{Q}} -\mu(S \cap Q) \log \frac{\mu(S \cap Q)}{\mu(S)} + \sum_{P \in \mathcal{P}, R \in \mathcal{R}} -\mu(P \cap R) \log \frac{\mu(P \cap R)}{\mu(R)} \\ &= H_\mu(\mathcal{Q}/\mathcal{P} \vee \mathcal{R}) + H_\mu(\mathcal{P}/\mathcal{R}). \end{aligned}$$

This proves part (i). Next, observe that if $\mathcal{P} \prec \mathcal{Q}$ then

$$H_\mu(\mathcal{P}/\mathcal{R}) = \sum_P \sum_R \sum_{Q \subset P} -\mu(Q \cap R) \log \frac{\mu(P \cap R)}{\mu(R)}$$
$$\leq \sum_P \sum_R \sum_{Q \subset P} -\mu(Q \cap R) \log \frac{\mu(Q \cap R)}{\mu(R)} = H_\mu(\mathcal{Q}/\mathcal{R}).$$

This proves the first half of claim (ii). To prove the second half, note that for any $P \in \mathcal{P}$ and $R \in \mathcal{R}$,

$$\frac{\mu(R \cap P)}{\mu(P)} = \sum_{Q \subset P} \frac{\mu(Q)}{\mu(P)} \frac{\mu(R \cap Q)}{\mu(Q)}.$$

It is clear that $\sum_{Q \subset P} \mu(Q)/\mu(P) = 1$. Therefore, by (9.1.8),

$$\phi\left(\frac{\mu(R \cap P)}{\mu(P)}\right) \geq \sum_{Q \subset P} \frac{\mu(Q)}{\mu(P)} \phi\left(\frac{\mu(R \cap Q)}{\mu(Q)}\right)$$

for every $P \in \mathcal{P}$ and $R \in \mathcal{R}$. Consequently,

$$H_\mu(\mathcal{R}/\mathcal{P}) = \sum_{P,R} \mu(P) \phi\left(\frac{\mu(R \cap P)}{\mu(P)}\right) \geq \sum_{P,R} \mu(P) \sum_{Q \subset P} \frac{\mu(Q)}{\mu(P)} \phi\left(\frac{\mu(R \cap Q)}{\mu(Q)}\right)$$
$$= \sum_{Q,R} \mu(Q) \phi\left(\frac{\mu(R \cap Q)}{\mu(Q)}\right) = H_\mu(\mathcal{R}/\mathcal{Q}).$$

Finally, it follows from the definition in (9.1.9) that $H_\mu(\mathcal{P}/\mathcal{Q}) = 0$ if and only if

$$\mu(P \cap Q) = 0 \quad \text{or else} \quad \frac{\mu(P \cap Q)}{\mu(Q)} = 1$$

for every $P \in \mathcal{P}$ and every $Q \in \mathcal{Q}$. In other words, either Q is disjoint from P (up to measure zero) or else Q is contained in P (up to measure zero). This means that $H_\mu(\mathcal{P}/\mathcal{Q}) = 0$ if and only if $\mathcal{P} \prec \mathcal{Q}$.

In particular, taking $\mathcal{Q} = \mathcal{M}$ in part (ii) of the lemma we get that

$$H_\mu(\mathcal{R}/\mathcal{P}) \leq H_\mu(\mathcal{R}) \quad \text{for any partitions } \mathcal{R} \text{ and } \mathcal{P}. \tag{9.1.10}$$

Moreover, taking $\mathcal{R} = \mathcal{M}$ in part (i) we find that

$$H_\mu(\mathcal{P} \vee \mathcal{Q}) = H_\mu(\mathcal{P}) + H_\mu(\mathcal{Q}/\mathcal{P}) \leq H_\mu(\mathcal{P}) + H_\mu(\mathcal{Q}). \tag{9.1.11}$$

Let $f : M \to N$ be a measurable transformation and μ be a probability measure on M. Then, $f_*\mu$ is a probability measure on N. Moreover, if $\mathcal{P}$ is a partition of N then $f^{-1}(\mathcal{P}) = \{f^{-1}(P) : P \in \mathcal{P}\}$ is a partition of M. By definition,

$$\begin{aligned} H_\mu(f^{-1}(\mathcal{P})) &= \sum_{P \in \mathcal{P}} -\mu(f^{-1}(P)) \log \mu(f^{-1}(P)) \\ &= \sum_{P \in \mathcal{P}} -f_*\mu(P) \log f_*\mu(P) = H_{f*\mu}(\mathcal{P}). \end{aligned} \tag{9.1.12}$$

In particular, if $M = N$ and the measure μ is invariant under f then

$$H_\mu(f^{-1}(\mathcal{P})) = H_\mu(\mathcal{P}) \quad \text{for every partition } \mathcal{P}. \tag{9.1.13}$$

We also need the following continuity property:

Lemma 9.1.6. *Given $k \geq 1$ and $\varepsilon > 0$ there exists $\delta > 0$ such that, for any finite partitions $\mathcal{P} = \{P_1, \dots, P_k\}$ and $\mathcal{Q} = \{Q_1, \dots, Q_k\}$,*

$$\mu(P_i \Delta Q_i) < \delta \text{ for every } i = 1, \dots, k \quad \Rightarrow \quad H_\mu(\mathcal{Q}/\mathcal{P}) < \varepsilon.$$

Proof. Fix $\varepsilon > 0$ and $k \geq 1$. Since $\phi : [0,1] \to \mathbb{R}$, $\phi(x) = -x \log x$ is a continuous function, there exists $\rho > 0$ such that $\phi(x) < \varepsilon/k^2$ for every $x \in [0,\rho) \cup (1-\rho, 1]$. Let $\delta = \rho/k$. Given partitions $\mathcal{P}$ and $\mathcal{Q}$ as in the statement, denote by $\mathcal{R}$ the partition whose elements are the intersections $P_i \cap Q_j$ with $i \neq j$ and also the set $\bigcup_{i=1}^k P_i \cap Q_i$. Note that $\mu(P_i \cap Q_j) \leq \mu(P_i \Delta Q_i) < \delta$ for every $i \neq j$ and

$$\mu\Big(\bigcup_{i=1}^k P_i \cap Q_i\Big) \geq \sum_{i=1}^k \big(\mu(P_i) - \mu(P_i \Delta Q_i)\big) > \sum_{i=1}^k \big(\mu(P_i) - \delta\big) = 1 - \rho.$$

Therefore,

$$H_\mu(\mathcal{R}) = \sum_{R \in \mathcal{R}} \phi(\mu(R)) < \#\mathcal{R} \frac{\varepsilon}{k^2} \leq \varepsilon.$$

It is clear from the definition that $\mathcal{P} \vee \mathcal{Q} = \mathcal{P} \vee \mathcal{R}$. Then, using (9.1.11) and (9.1.10),

$$\begin{aligned} H_\mu(\mathcal{Q}/\mathcal{P}) &= H_\mu(\mathcal{P} \vee \mathcal{Q}) - H_\mu(\mathcal{P}) = H_\mu(\mathcal{P} \vee \mathcal{R}) - H_\mu(\mathcal{P}) \\ &= H_\mu(\mathcal{R}/\mathcal{P}) \leq H_\mu(\mathcal{R}) < \varepsilon. \end{aligned}$$

This proves the lemma.

9.1.3 Entropy of a dynamical system

Let $f : M \to M$ be a measurable transformation preserving a probability measure μ. The notion of the entropy of the system (f, μ) that we introduce in what follows is inspired by (9.1.5).

Given a partition $\mathcal{P}$ of M with finite entropy, denote

$$\mathcal{P}^n = \bigvee_{i=0}^{n-1} f^{-i}(\mathcal{P}) \quad \text{for each } n \geq 1.$$

Observe that the element $\mathcal{P}^n(x)$ that contains $x \in M$ is given by:

$$\mathcal{P}^n(x) = \mathcal{P}(x) \cap f^{-1}(\mathcal{P}(f(x))) \cap \cdots \cap f^{-n+1}(\mathcal{P}(f^{n-1}(x))).$$

It is clear that the sequence $\mathcal{P}^n$ is non-decreasing, that is, $\mathcal{P}^n \prec \mathcal{P}^{n+1}$ for every n. Therefore, the sequence of entropies $H_\mu(\mathcal{P}^n)$ is also non-decreasing. Another important fact is that this sequence is subadditive:

Lemma 9.1.7. *$H_\mu(\mathcal{P}^{m+n}) \le H_\mu(\mathcal{P}^m) + H_\mu(\mathcal{P}^n)$ for every $m, n \ge 1$.*

Proof. By definition, $\mathcal{P}^{m+n} = \bigvee_{i=0}^{m+n-1} f^{-i}(\mathcal{P}) = \mathcal{P}^m \vee f^{-m}(\mathcal{P}^n)$. Therefore, using (9.1.11),

$$H_\mu(\mathcal{P}^{m+n}) \le H_\mu(\mathcal{P}^m) + H_\mu(f^{-m}(\mathcal{P}^n)). \tag{9.1.14}$$

On the other hand, since the measure μ is invariant under f, the property (9.1.13) implies that $H_\mu(f^{-m}(\mathcal{P}^n)) = H_\mu(\mathcal{P}^n)$ for every m, n. Substituting this fact in (9.1.14) we get the conclusion of the lemma.

In view of Lemma 3.3.4, it follows from Lemma 9.1.7 that the limit

$$h_\mu(f, \mathcal{P}) = \lim_n \frac{1}{n} H_\mu(\mathcal{P}^n) \tag{9.1.15}$$

exists and coincides with the infinitum of the sequence on the left-hand side. We call $h_\mu(f, \mathcal{P})$ the *entropy of f with respect to the partition* $\mathcal{P}$. Observe that this entropy is a non-decreasing function of the partition:

$$\mathcal{P} \prec \mathcal{Q} \quad \Rightarrow \quad h_\mu(f, \mathcal{P}) \le h_\mu(f, \mathcal{Q}). \tag{9.1.16}$$

Indeed, if $\mathcal{P} \prec \mathcal{Q}$ then $\mathcal{P}^n \prec \mathcal{Q}^n$ for every n. Using Lemma 9.1.5, it follows that $H_\mu(\mathcal{P}^n) \le H_\mu(\mathcal{Q}^n)$ for every n, and this implies (9.1.16).

Finally, the *entropy* of the system (f, μ) is defined by

$$h_\mu(f) = \sup_{\mathcal{P}} h_\mu(f, \mathcal{P}), \tag{9.1.17}$$

where the supremum is taken over all the partitions with finite entropy. A useful observation is that the definition is not affected if we take the supremum only over the finite partitions (see Exercise 9.1.2).

Example 9.1.8. Suppose that the invariant measure μ is supported on a periodic orbit. In other words, there exist $x \in M$ and $k \ge 1$ such that $f^k(x) = x$ and the measure μ is given by

$$\mu = \frac{1}{k}\left(\delta_x + \delta_{f(x)} + \cdots + \delta_{f^{k-1}(x)}\right).$$

Note that this measure takes only a finite number of values (because the Dirac measure takes only the values 0 and 1). Hence, the entropy function $\mathcal{P} \mapsto H_\mu(\mathcal{P})$ also takes only finitely many values. In particular, $\lim_n n^{-1} H_\mu(\mathcal{P}^n) = 0$ for every partition $\mathcal{P}$. This proves that $h_\mu(f) = 0$.

Example 9.1.9. Consider the decimal expansion map $f : [0,1] \to [0,1]$, given by $f(x) = 10x - [10x]$. As observed previously, f preserves the Lebesgue measure μ on the interval. Let $\mathcal{P}$ be the partition of $[0,1]$ into the intervals of the form $\big((i-1)/10, i/10\big]$ with $i = 1, \dots, 10$. Then, $\mathcal{P}^n$ is the partition into the intervals of the form $\big((i-1)/10^n, i/10^n\big]$ with $i = 1, \dots, 10^n$. Using the calculation in Example 9.1.1, we get that

$$h_\mu(f,\mathcal{P}) = \lim_n \frac{1}{n} H_\mu(\mathcal{P}^n) = \log 10.$$

Using the theory in Section 9.2 (the theorem of Kolmogorov–Sinai and its corollaries), one can easily check that this is also the value of the entropy $h_\mu(f)$, that is, $\mathcal{P}$ realizes the supremum in the definition (9.1.17).

Example 9.1.10. Consider the shift map $\sigma : \Sigma \to \Sigma$ in $\Sigma = \{1,\dots,d\}^{\mathbb{N}}$ (or $\Sigma = \{1,\dots,d\}^{\mathbb{Z}}$), with a Bernoulli measure $\mu = \nu^{\mathbb{N}}$ (respectively, $\mu = \nu^{\mathbb{Z}}$). Let $\mathcal{P}$ be the partition of Σ into the cylinders $[0;a]$ with $a = 1,\dots,d$. Then, $\mathcal{P}^n$ is the partition into cylinders $[0;a_1,\dots,a_n]$ of length n. Using the calculation in Example 9.1.2 we conclude that

$$h_\mu(\sigma,\mathcal{P}) = \lim_n \frac{1}{n} H_\mu(\mathcal{P}^n) = \sum_{i=1}^{d} -p_i \log p_i. \tag{9.1.18}$$

The theory presented in Section 9.2 permits us to prove that this is also the value of the entropy $h_\mu(\sigma)$.

It follows from expression (9.1.18) that for every $x > 0$ there exists some Bernoulli shift (σ,μ) such that $h_\mu(\sigma) = x$. We use this observation a few times in what follows.

Lemma 9.1.11. *$h_\mu(f,\mathcal{Q}) \le h_\mu(f,\mathcal{P}) + H_\mu(\mathcal{Q}/\mathcal{P})$ for any partitions $\mathcal{P}$ and $\mathcal{Q}$ with finite entropy.*

Proof. By Lemma 9.1.5, for every $n \ge 1$,

$$\begin{aligned} H_\mu\big(\mathcal{Q}^{n+1}/\mathcal{P}^{n+1}\big) &= H_\mu\big(\mathcal{Q}^n \vee f^{-n}(\mathcal{Q})/\mathcal{P}^n \vee f^{-n}(\mathcal{P})\big) \\ &\le H_\mu\big(\mathcal{Q}^n/\mathcal{P}^n\big) + H_\mu\big(f^{-n}(\mathcal{Q})/f^{-n}(\mathcal{P})\big). \end{aligned}$$

The last term is equal to $H_\mu(\mathcal{Q}/\mathcal{P})$, because the measure μ is invariant under f. Therefore, the previous relation proves that

$$H_\mu\big(\mathcal{Q}^n/\mathcal{P}^n\big) \le n H_\mu\big(\mathcal{Q}/\mathcal{P}\big) \quad \text{for every } n \ge 1. \tag{9.1.19}$$

Using Lemma 9.1.5 once more, it follows that

$$H_\mu(\mathcal{Q}^n) \le H_\mu\big(\mathcal{P}^n \vee \mathcal{Q}^n\big) = H_\mu(\mathcal{P}^n) + H_\mu(\mathcal{Q}^n/\mathcal{P}^n) \le H_\mu(\mathcal{P}^n) + n H_\mu(\mathcal{Q}/\mathcal{P}).$$

Dividing by n and taking the limit when $n \to \infty$, we get the conclusion of the lemma.

Lemma 9.1.12. *$h_\mu(f,\mathcal{P}) = \lim_n H_\mu(\mathcal{P}/\bigvee_{j=1}^{n} f^{-j}(\mathcal{P}))$ for any partition $\mathcal{P}$ with finite entropy.*

Proof. Using Lemma 9.1.5(i) and the fact that the measure μ is invariant under f, we get that

$$H_\mu\Big(\bigvee_{j=0}^{n-1} f^{-j}(\mathcal{P})\Big) = H_\mu\Big(\bigvee_{j=1}^{n-1} f^{-j}(\mathcal{P})\Big) + H_\mu\Big(\mathcal{P}/\bigvee_{j=1}^{n-1} f^{-j}(\mathcal{P})\Big)$$
$$= H_\mu\Big(\bigvee_{j=0}^{n-2} f^{-j}(\mathcal{P})\Big) + H_\mu\Big(\mathcal{P}/\bigvee_{j=1}^{n-1} f^{-j}(\mathcal{P})\Big)$$

for every n. By recurrence, it follows that

$$H_\mu\Big(\bigvee_{j=0}^{n-1} f^{-j}(\mathcal{P})\Big) = H_\mu(\mathcal{P}) + \sum_{k=1}^{n-1} H_\mu\Big(\mathcal{P}/\bigvee_{j=1}^{k} f^{-j}(\mathcal{P})\Big).$$

Therefore, $h_\mu(f,\mathcal{P})$ is given by the Cesàro limit

$$h_\mu(f,\mathcal{P}) = \lim_n \frac{1}{n} H_\mu\Big(\bigvee_{j=0}^{n-1} f^{-j}(\mathcal{P})\Big) = \lim_n \frac{1}{n}\sum_{k=1}^{n-1} H_\mu\Big(\mathcal{P}/\bigvee_{j=1}^{k} f^{-j}(\mathcal{P})\Big).$$

On the other hand, Lemma 9.1.5(ii) ensures that $H_\mu(\mathcal{P}/\bigvee_{j=1}^{n} f^{-j}(\mathcal{P}))$ is a non-increasing sequence. In particular, $\lim_n H_\mu(\mathcal{P}/\bigvee_{j=1}^{n} f^{-j}(\mathcal{P}))$ exists and, consequently, coincides with the Cesàro limit in the previous identity.

Recall that $\mathcal{P}^n = \bigvee_{j=0}^{n-1} f^{-j}(\mathcal{P})$. When $f : M \to M$ is invertible, we also consider $\mathcal{P}^{\pm n} = \bigvee_{j=-n}^{n-1} f^{-j}(\mathcal{P})$.

Lemma 9.1.13. *Let $\mathcal{P}$ be a partition with finite entropy. For every $k \geq 1$, we have $h_\mu(f,\mathcal{P}) = h_\mu(f,\mathcal{P}^k)$ and, if f is invertible, $h_\mu(f,\mathcal{P}) = h_\mu(f,\mathcal{P}^{\pm k})$.*

Proof. Observe that, given any $n \geq 1$,

$$\bigvee_{j=0}^{n-1} f^{-j}(\mathcal{P}^k) = \bigvee_{j=0}^{n-1} f^{-j}\Big(\bigvee_{i=0}^{k-1} f^{-i}(\mathcal{P})\Big) = \bigvee_{l=0}^{n+k-2} f^{-l}(\mathcal{P}) = \mathcal{P}^{n+k-1}.$$

Therefore,

$$h_\mu(f,\mathcal{P}^k) = \lim_n \frac{1}{n} H_\mu(\mathcal{P}^{n+k-1}) = \lim_n \frac{1}{n} H_\mu(\mathcal{P}^n) = h_\mu(f,\mathcal{P}).$$

This proves the first part of the lemma. To prove the second part, note that

$$\bigvee_{j=0}^{n-1} f^{-j}(\mathcal{P}^{\pm k}) = \bigvee_{j=0}^{n-1} f^{-j}\Big(\bigvee_{i=-k}^{k-1} f^{-i}(\mathcal{P})\Big) = \bigvee_{l=-k}^{n+k-2} f^{-l}(\mathcal{P}) = f^{-k}(\mathcal{P}^{n+2k-1})$$

for every n and every k. Therefore,

$$h_\mu(f,\mathcal{P}^{\pm k}) = \lim_n \frac{1}{n} H_\mu(f^{-k}(\mathcal{P}^{n+2k-1})) = \lim_n \frac{1}{n} H_\mu(\mathcal{P}^{n+2k-1}) = h_\mu(f,\mathcal{P})$$

(the second equality uses the fact that μ is invariant under f).

Proposition 9.1.14. *One has $h_\mu(f^k) = kh_\mu(f)$ for every $k \in \mathbb{N}$. If f is invertible then $h_\mu(f^k) = |k|h_\mu(f)$ for every $k \in \mathbb{Z}$.*

Proof. It is clear that the identity holds for $k=0$, since $f^0 = \mathrm{id}$ and $h_\mu(\mathrm{id}) = 0$. Take k to be non-zero from now on. Let $g = f^k$ and $\mathcal{P}$ be any partition of M with finite entropy. Recalling that $\mathcal{P}^k = \mathcal{P} \vee f^{-1}(\mathcal{P}) \vee \cdots \vee f^{-k+1}(\mathcal{P})$, we see that

$$\mathcal{P}^{km} = \bigvee_{j=0}^{km-1} f^{-j}(\mathcal{P}) = \bigvee_{i=0}^{m-1} f^{-ki}\Big(\bigvee_{j=0}^{k-1} f^{-j}(\mathcal{P})\Big) = \bigvee_{i=0}^{m-1} g^{-i}(\mathcal{P}^k).$$

Therefore,

$$kh_\mu\big(f,\mathcal{P}\big) = \lim_m \frac{1}{m} H_\mu\big(\mathcal{P}^{km}\big) = h_\mu\big(g,\mathcal{P}^k\big). \tag{9.1.20}$$

Since $\mathcal{P} \prec \mathcal{P}^k$, this implies that $h_\mu(g,\mathcal{P}) \le kh_\mu(f,\mathcal{P}) \le h_\mu(g)$ for any $\mathcal{P}$. Taking the supremum over these partitions $\mathcal{P}$, it follows that $h_\mu(g) \le kh_\mu(f) \le h_\mu(g)$. This proves that $kh_\mu(f) = h_\mu(g)$, as stated.

Now suppose that f is invertible. Let $\mathcal{P}$ be any partition of M with finite entropy. For any $n \ge 1$,

$$H_\mu\Big(\bigvee_{j=0}^{n-1} f^{-j}(\mathcal{P})\Big) = H_\mu\Big(f^{-n+1}\Big(\bigvee_{i=0}^{n-1} f^{i}(\mathcal{P})\Big)\Big) = H_\mu\Big(\bigvee_{i=0}^{n-1} f^{i}(\mathcal{P})\Big),$$

because the measure μ is invariant under f. Dividing by n and taking the limit when $n \to \infty$, we get that

$$h_\mu(f,\mathcal{P}) = h_\mu(f^{-1},\mathcal{P}). \tag{9.1.21}$$

Taking the supremum over these partitions $\mathcal{P}$, it follows that $h_\mu(f) = h_\mu(f^{-1})$. Replacing f with f^k and using the first half of the proposition, we get that $h_\mu(f^{-k}) = h_\mu(f^k) = kh_\mu(f)$ for every $k \in \mathbb{N}$.

9.1.4 Exercises

9.1.1. Prove that $H_\mu(\mathcal{P}/\mathcal{R}) \le H_\mu(\mathcal{P}/\mathcal{Q}) + H_\mu(\mathcal{Q}/\mathcal{R})$ for any partitions $\mathcal{P}$, $\mathcal{Q}$ and $\mathcal{R}$.

9.1.2. Show that the supremum of $h_\mu(f,\mathcal{P})$ over the finite partitions coincides with the supremum over all the partitions with finite entropy.

9.1.3. Check that $\lim_n H_\mu(\bigvee_{i=0}^{k-1} f^{-i}(\mathcal{P}) / \bigvee_{j=k}^{n} f^{-j}(\mathcal{P})) = kh(f,\mathcal{P})$ for every partition $\mathcal{P}$ with finite entropy and every $k \ge 1$.

9.1.4. Let $f : M \to M$ be a measurable transformation preserving a probability measure μ.

(a) Assume that there exists an invariant set $A \subset M$ with $\mu(A) \in (0,1)$. Let μ_A and μ_B be the normalized restrictions of μ to the sets A and $B = A^c$, respectively. Show that $h_\mu(f) = \mu(A)h_{\mu_A}(f) + \mu(B)h_{\mu_B}(f)$.

(b) Suppose that μ is a convex combination $\mu = \sum_{i=1}^n a_i\mu_i$ of ergodic measures $\mu_1, \ldots, \mu_n$. Show that $h_\mu(f) = \sum_{i=1}^n a_i h_{\mu_i}(f)$.

[Observation: In Section 9.6 we prove much stronger results.]

9.1.5. Let $(M,\mathcal{B},\mu)$ and $(N,\mathcal{C},\nu)$ be probability spaces and $f : M \to M$ and $g : N \to N$ be measurable transformations preserving the measures μ and ν, respectively. We say that (g,ν) is a *factor* of (f,μ) if there exists a measurable map, not necessarily invertible, $\phi : (M,\mathcal{B}) \to (N,\mathcal{C})$ such that $\phi_*\mu = \nu$ and $\phi \circ f = g \circ \phi$ at almost every point. Show that in that case $h_\nu(g) \le h_\mu(f)$.

9.2 Theorem of Kolmogorov–Sinai

In general, the main difficulty in calculating the entropy lies in the calculation of the supremum in (9.1.17). The methods that we develop in this section permit the simplification of that task in many cases, by identifying certain partitions $\mathcal{P}$ that realize the supremum, that is, such that $h_\mu(f,\mathcal{P}) = h_\mu(f)$. The main result is:

Theorem 9.2.1 (Kolmogorov–Sinai). *Let $\mathcal{P}_1 \prec \cdots \prec \mathcal{P}_n \prec \cdots$ be a non-decreasing sequence of partitions with finite entropy such that $\bigcup_{n=1}^{\infty} \mathcal{P}_n$ generates the σ-algebra of measurable sets, up to measure zero. Then,*

$$h_\mu(f) = \lim_n h_\mu(f,\mathcal{P}_n).$$

Proof. The limit always exists, for property (9.1.16) implies that the sequence $h_\mu(f,\mathcal{P}_n)$ is non-decreasing. The inequality $\ge$ in the statement is a direct consequence of the definition of entropy. Therefore, we only need to show that $h_\mu(f,\mathcal{Q}) \le \lim_n h_\mu(f,\mathcal{P}_n)$ for every partition $\mathcal{Q}$ with finite entropy. We use the following fact, which is interesting in itself:

Proposition 9.2.2. *Let $\mathcal{A}$ be an algebra that generates the σ-algebra of measurable sets, up to measure zero. For every partition $\mathcal{Q}$ with finite entropy and every $\varepsilon > 0$ there exists some finite partition $\mathcal{P} \subset \mathcal{A}$ such that $H_\mu(\mathcal{Q}/\mathcal{P}) < \varepsilon$.*

Proof. The first step is to reduce the statement to the case when $\mathcal{Q}$ is finite. Denote by $Q_j, j = 1,2,\dots$ the elements of $\mathcal{Q}$. For each $k \ge 1$, consider the finite partition

$$\mathcal{Q}_k = \left\{ Q_1,\dots,Q_k, M \setminus \bigcup_{j=1}^{k} Q_j \right\}.$$

Lemma 9.2.3. *If $\mathcal{Q}$ is a partition with finite entropy then* $\lim_k H_\mu(\mathcal{Q}/\mathcal{Q}_k) = 0$.

Proof. Denote $Q_0 = M \setminus \bigcup_{j=1}^{k} Q_j$. By definition,

$$H_\mu(\mathcal{Q}/\mathcal{Q}_k) = \sum_{i=0}^{k} \sum_{j\ge 1} -\mu(Q_i \cap Q_j) \log \frac{\mu(Q_i \cap Q_j)}{\mu(Q_i)}.$$

All the terms with $i \ge 1$ vanish, since in that case $\mu(Q_i \cap Q_j)$ is equal to zero if $i \ne j$ and is equal to $\mu(Q_i)$ if $i = j$. For $i = 0$ we have that $\mu(Q_i \cap Q_j)$ is equal to zero if $j \le k$ and is equal to $\mu(Q_j)$ if $j > k$. Therefore,

$$H_\mu(\mathcal{Q}/\mathcal{Q}_k) = \sum_{j>k} -\mu(Q_j) \log \frac{\mu(Q_j)}{\mu(Q_0)} \le \sum_{j>k} -\mu(Q_j) \log \mu(Q_j).$$

The hypothesis that $\mathcal{Q}$ has finite entropy means that the expression on the right-hand side converges to zero when $k \to \infty$.

Given $\varepsilon > 0$, fix $k \geq 1$ such that $H_\mu(\mathcal{Q}/\mathcal{Q}_k) < \varepsilon/2$. Consider any $\delta > 0$. By the approximation theorem (Theorem A.1.19), for each $i = 1, \dots, k$ there exists $A_i \in \mathcal{A}$ such that

$$\mu(Q_i \Delta A_i) < \delta/(2k^2). \tag{9.2.1}$$

Define $P_1 = A_1$ and $P_i = A_i \setminus \bigcup_{j=1}^{i-1} A_j$ for $i = 2, \dots, k$ and $P_0 = M \setminus \bigcup_{j=1}^{k-1} A_j$. It is clear that $\mathcal{P} = \{P_1, \dots, P_k, P_0\}$ is a partition of M and also that $P_i \in \mathcal{A}$ for every i. For $i = 1, \dots, k$, we have $P_i \Delta A_i = P_i \setminus A_i = A_i \cap \left(\bigcup_{j=1}^{i-1} A_j\right)$. For any x in this set, there is $j < i$ such that $x \in A_i \cap A_j$. Since $Q_i \cap Q_j = \emptyset$, it follows that $x \in (A_i \setminus Q_i) \cup (A_j \setminus Q_j)$. This proves that

$$P_i \Delta A_i \subset \bigcup_{j=1}^{i} (A_j \setminus Q_j) \subset \bigcup_{j=1}^{i} (A_j \Delta Q_j),$$

and so $\mu(P_i \Delta A_i) < i\delta/(2k^2) \leq \delta/(2k)$. Together with (9.2.1), this implies that

$$\mu(P_i \Delta Q_i) < \delta/(2k^2) + \delta/(2k) \leq \delta/k \quad \text{for } i = 1, \dots, k. \tag{9.2.2}$$

Moreover, $P_0 \Delta Q_0 \subset \bigcup_{i=1}^{k} P_i \Delta Q_i$ since $\mathcal{P}$ and $\mathcal{Q}_k$ are partitions of M. Hence, (9.2.2) implies that

$$\mu(P_0 \Delta Q_0) < \delta. \tag{9.2.3}$$

By Lemma 9.1.6, the relations (9.2.2) and (9.2.3) imply that $H_\mu(\mathcal{Q}_k/\mathcal{P}) < \varepsilon/2$, as long as we take $\delta > 0$ sufficiently small. Then, by the inequality in Exercise 9.1.1,

$$H_\mu(\mathcal{Q}/\mathcal{P}) \leq H_\mu(\mathcal{Q}/\mathcal{Q}_k) + H_\mu(\mathcal{Q}_k/\mathcal{P}) < \varepsilon,$$

as stated.

Corollary 9.2.4. *If $(\mathcal{P}_n)_n$ is a sequence of partitions as in Theorem 9.2.1 then* $\lim_n H_\mu(\mathcal{Q}/\mathcal{P}_n) = 0$ *for every partition $\mathcal{Q}$ with finite entropy.*

Proof. For each n, let $\mathcal{A}_n$ be the algebra generated by $\bigcup_{j=1}^{n} \mathcal{P}_j$. Then $(\mathcal{A}_n)_n$ is a non-decreasing sequence and the union $\mathcal{A} = \bigcup_n \mathcal{A}_n$ is the algebra generated by $\bigcup_n \mathcal{P}_n$. Consider any $\varepsilon > 0$. By Proposition 9.2.2, there exists a finite partition $\mathcal{P} \subset \mathcal{A}$ such that $H_\mu(\mathcal{Q}/\mathcal{P}) < \varepsilon$. Hence, since $\mathcal{P}$ is finite, there exists $m \geq 1$ such that $\mathcal{P} \subset \mathcal{A}_m$ and, thus, $\mathcal{P}$ is coarser than $\mathcal{P}_m$. Then, using Lemma 9.1.5,

$$H_\mu(\mathcal{Q}/\mathcal{P}_n) \leq H_\mu(\mathcal{Q}/\mathcal{P}_m) \leq H_\mu(\mathcal{Q}/\mathcal{P}) < \varepsilon \quad \text{for every } n \geq m.$$

This completes the proof of the corollary.

We are ready to conclude the proof of Theorem 9.2.1. By Lemma 9.1.11,

$$h_\mu(f, \mathcal{Q}) \leq h_\mu(f, \mathcal{P}_n) + H_\mu(\mathcal{Q}/\mathcal{P}_n) \quad \text{for every } n.$$

Taking the limit as $n \to \infty$, we get that $h_\mu(f, \mathcal{Q}) \leq \lim_n h_\mu(f, \mathcal{P}_n)$ for every partition $\mathcal{Q}$ with finite entropy.

9.2.1 Generating partitions

In this section and the ones that follow, we deduce several useful consequences of Theorem 9.2.1.

Corollary 9.2.5. *Let $\mathcal{P}$ be a partition with finite entropy such that the union of the iterates $\mathcal{P}^n = \bigvee_{j=0}^{n-1} f^{-j}(\mathcal{P})$, $n \geq 1$ generates the σ-algebra of measurable sets, up to measure zero. Then, $h_\mu(f) = h_\mu(f,\mathcal{P})$.*

Proof. It suffices to apply Theorem 9.2.1 to the sequence $\mathcal{P}^n$ and to recall that $h_\mu(f,\mathcal{P}^n) = h_\mu(f,\mathcal{P})$ for every n, by Lemma 9.1.13.

Corollary 9.2.6. *Assume that the system (f,μ) is invertible. Let $\mathcal{P}$ be a partition with finite entropy such that the union of the iterates $\mathcal{P}^{\pm n} = \bigvee_{j=-n}^{n-1} f^{-j}(\mathcal{P})$, $n \geq 1$ generates the σ-algebra of measurable sets, up to measure zero. Then, $h_\mu(f) = h_\mu(f,\mathcal{P})$.*

Proof. It suffices to apply Theorem 9.2.1 to the sequence $\mathcal{P}^{\pm n}$ and to recall that $h_\mu(f,\mathcal{P}^{\pm n}) = h_\mu(f,\mathcal{P})$ for every n, by Lemma 9.1.13.

In particular, Corollaries 9.2.5 and 9.2.6 complete the calculation of the entropy of the decimal expansion and the Bernoulli shifts that we started in Examples 9.1.9 and 9.1.10, respectively.

In both situations, Corollary 9.2.5 and Corollary 9.2.6, we say that $\mathcal{P}$ is a *generating partition* or, simply, a *generator* of the system. Note, however, that this contains a certain abuse of language, since the conditions in the two corollaries are not equivalent. For example, for the shift map in $M = \{1,\dots,d\}^{\mathbb{Z}}$, the partition $\mathcal{P}$ into cylinders $\{[0;a] : a = 1,\dots,d\}$ is such that the union of the two-sided iterates $\mathcal{P}^{\pm n}$ generates the σ-algebra of measurable sets but the union of the one-sided iterates $\mathcal{P}^n$ does not. Whenever it is necessary to distinguish between these two concepts, we talk of a *one-sided* generator and a *two-sided* generator, respectively.

In this regard, let us point out that certain *invertible* systems admit one-sided generators. For example, if $f : S^1 \to S^1$ is an irrational rotation and $\mathcal{P} = \{I, S^1 \setminus I\}$ is a partition of the circle into two complementary intervals, then $\mathcal{P}$ is a one-sided generator (and also a two-sided generator, of course). However, this kind of situation is possible only for systems with entropy zero:

Corollary 9.2.7. *Assume that the system (f,μ) is invertible and there exists a partition $\mathcal{P}$ with finite entropy such that $\bigcup_{n=1}^{\infty} \mathcal{P}^n$ generates the σ-algebra of measurable sets, up to measure zero. Then, $h_\mu(f) = 0$.*

Proof. Combining Lemma 9.1.12 and Corollary 9.2.5, we get that

$$h_\mu(f) = h_\mu(f,\mathcal{P}) = \lim_n H_\mu(\mathcal{P}/f^{-1}(\mathcal{P}^n)).$$

Since $\bigcup_n \mathcal{P}^n$ generates the σ-algebra $\mathcal{B}$ of measurable sets, $\bigcup_n f^{-1}(\mathcal{P}^n)$ generates the σ-algebra $f^{-1}(\mathcal{B})$. Now notice that $f^{-1}(\mathcal{B}) = \mathcal{B}$, since f is invertible. Hence, Corollary 9.2.4 implies that $H_\mu(\mathcal{P}/f^{-1}(\mathcal{P}^n))$ converges to zero when $n \to \infty$. It follows that $h_\mu(f) = h_\mu(f,\mathcal{P}) = 0$.

Now take M to be a metric space and μ to be a Borel probability measure.

Corollary 9.2.8. *Let $\mathcal{P}_1 \prec \cdots \prec \mathcal{P}_n \prec \cdots$ be a non-decreasing sequence of partitions with finite entropy such that* $\operatorname{diam}\mathcal{P}_n(x) \to 0$ *for μ-almost every $x \in M$. Then,*

$$h_\mu(f) = \lim_n h_\mu(f,\mathcal{P}_n).$$

Proof. Let U be any non-empty open subset of M. The hypothesis ensures that for each $x \in U$ there exists $n(x)$ such that the set $P_x = \mathcal{P}_{n(x)}(x)$ is contained in U. It is clear that P_x belongs to the algebra $\mathcal{A}$ generated by $\bigcup_n \mathcal{P}_n$. Observe also that $\mathcal{A}$ is countable, since it consists of the finite unions of elements of the partitions $\mathcal{P}_n$ together with the complements of such unions. In particular, the map $x \mapsto P_x$ takes only countably many values. It follows that $U = \bigcup_{x\in U} P_x$ is in the σ-algebra generated by $\mathcal{A}$. This proves that the σ-algebra generated by $\mathcal{A}$ contains all the open sets and, thus, all the Borel sets. Now, the conclusion follows directly from Theorem 9.2.1.

Example 9.2.9. Let $f: S^1 \to S^1$ be a homeomorphism and μ be any invariant probability measure. Given a finite partition $\mathcal{P}$ of S^1 into subintervals, denote by $x_1,\dots,x_m$ their endpoints. For any $j \geq 1$, the partition $f^{-j}(\mathcal{P})$ consists of the subintervals of S^1 determined by the points $f^{-j}(x_i)$. This implies that, for each $n \geq 1$, the elements of $\mathcal{P}^n$ have their endpoints in the set

$$\{f^{-j}(x_i) : j = 0,\dots,n-1 \text{ and } i = 1,\dots,m\}.$$

In particular, $\#\mathcal{P}^n \leq mn$. Then, using Lemma 9.1.3,

$$h_\mu(f,\mathcal{P}) = \lim_n \frac{1}{n} H_\mu(\mathcal{P}^n) \leq \lim_n \frac{1}{n} \log \#\mathcal{P}^n = \lim_n \frac{1}{n} \log mn = 0.$$

It follows that $h_\mu(f) = 0$: to see that, it suffices to consider any sequence of finite partitions into intervals with diameter going to zero and to apply Corollary 9.2.8.

Corollary 9.2.10. *Let $\mathcal{P}$ be a partition with finite entropy such that we have* $\operatorname{diam}\mathcal{P}^n(x) \to 0$ *for μ-almost every $x \in M$. Then, $h_\mu(f) = h_\mu(f,\mathcal{P})$.*

Proof. It suffices to apply Corollary 9.2.8 to the sequence $\mathcal{P}^n$, recalling that $h_\mu(f,\mathcal{P}^n) = h_\mu(f,\mathcal{P})$ for every n.

Analogously, if f is invertible and $\mathcal{P}$ is a partition with finite entropy and such that $\operatorname{diam}\mathcal{P}^{\pm n}(x) \to 0$ for μ-almost every $x \in M$, then $h_\mu(f) = h_\mu(f,\mathcal{P})$.

It is known that generators do exist in most interesting cases, although it may be difficult to exhibit a generator explicitly. Indeed, suppose that the ambient M is a Lebesgue space. Rokhlin [Rok67a, §10] proved that if a system is aperiodic (that is, the set of periodic points has measure zero) and almost every point has a countable (finite or infinite) set of pre-images, then there exists some generator. In particular, every invertible aperiodic system admits some countable generator. In general, this generator may have infinite entropy. But Rokhlin also showed that every invertible aperiodic system with finite entropy admits some two-sided generator with finite entropy. Moreover (Krieger [Kri70]), this generator may be chosen to be finite if the system is ergodic.

9.2.2 Semi-continuity of the entropy

Next, we examine some properties of the *entropy function* that associates with each invariant measure μ of a given transformation f the value of the corresponding entropy $h_\mu(f)$. We are going to see that this function is usually not continuous. However, under quite broad assumptions, it is *upper semi-continuous*: given any $\varepsilon > 0$, one has $h_\nu(f) \le h_\mu(f) + \varepsilon$ for every ν sufficiently close to μ. That holds, in particular, for the class of transformations that we call *expansive*. These facts have important consequences, some of which are explored in Sections 9.2.3 and 9.6. Moreover, we return to this subject in Section 10.5.

Let us start by showing, through an example, that the entropy function may be discontinuous:

Example 9.2.11. Let $f : [0,1] \to [0,1]$ be the decimal expansion map. As we saw in Example 9.1.9, the entropy of f with respect to the Lebesgue measure m is $h_m(f) = \log 10$. For each $k \ge 1$, denote by F_k the set of fixed points of the iterate f^k. Observe that F_k is an invariant set with $\#F_k = 10^k$. Observe also that these sets are equidistributed, in the following sense: each interval $[(i-1)/10^k, i/10^k]$ contains exactly one point of F_k. Consider the sequence of measures

$$\mu_k = \frac{1}{10^k} \sum_{x \in F_k} \delta_x.$$

The previous observations imply (check!) that each μ_k is an invariant probability measure and the sequence $(\mu_k)_k$ converges to the Lebesgue measure m in the weak* topology. Since μ_k is supported on a finite set, the same argument as in Example 9.1.8 proves that $h_{\mu_k}(f) = 0$ for every k. In particular, we have that $\lim_k h_{\mu_k}(f) = 0 < h_m(f)$.

On the other hand, in general, consider any finite partition $\mathcal{P}$ of M whose *boundary*

$$\partial \mathcal{P} = \bigcup_{P \in \mathcal{P}} \partial P$$

satisfies $\mu(\partial \mathcal{P}) = 0$. By Theorem 2.1.2 or, more precisely, by the fact that the topology (2.1.5) is equivalent to the weak* topology, the function $\nu \mapsto \nu(P)$ is continuous at the point μ, for every $P \in \mathcal{P}$. Consequently, the function

$$\nu \mapsto H_\nu(\mathcal{P}) = \sum_{P \in \mathcal{P}} -\nu(P) \log \nu(P)$$

is also continuous at μ. The hypothesis on $\mathcal{P}$ also implies that $\mu(\partial \mathcal{P}^n) = 0$ for every $n \geq 1$, since

$$\partial \mathcal{P}^n \subset \partial \mathcal{P} \cup f^{-1}(\partial \mathcal{P}) \cup \cdots \cup f^{-n+1}(\partial \mathcal{P}).$$

Thus, the same argument shows that $\nu \mapsto H_\nu(\mathcal{P}^n)$ is continuous for every n.

Proposition 9.2.12. *Let $\mathcal{P}$ be a finite partition such that $\mu(\partial \mathcal{P}) = 0$. Then the function $\nu \mapsto h_\nu(f, \mathcal{P})$ is upper semi-continuous at μ.*

Proof. Recall that, by definition,

$$h_\nu(f, \mathcal{P}) = \inf_n \frac{1}{n} H_\nu(f, \mathcal{P}).$$

It is a well-known easy fact that the infimum of any family of continuous functions is an upper semi-continuous function.

Corollary 9.2.13. *Assume that there exists a finite partition $\mathcal{P}$ such that $\mu(\partial \mathcal{P}) = 0$ and $\bigcup_n \mathcal{P}^n$ generates the σ-algebra of measurable sets of M, up to measure zero. Then the function $\eta \mapsto h_\eta(f)$ is upper semi-continuous at μ.*

Proof. By Proposition 9.2.12, given $\varepsilon > 0$ there exists a neighborhood U of μ such that $h_\nu(f, \mathcal{P}) \leq h_\mu(f, \mathcal{P}) + \varepsilon$ for every $\nu \in V$. By definition, $h_\mu(f, \mathcal{P}) \leq h_\mu(f)$. By Corollary 9.2.5, the hypothesis implies that $h_\nu(f, \mathcal{P}) = h_\nu(f)$ for every ν. Therefore, $h_\nu(f) \leq h_\mu(f) + \varepsilon$ for every $\nu \in V$.

Now let us suppose that M is a compact metric space. As before, take μ to be a Borel probability measure. By definition, the diameter $\operatorname{diam} \mathcal{P}$ of a partition $\mathcal{P}$ is the supremum of the diameters of its elements. Then we have the following more specialized version of the previous corollary:

Corollary 9.2.14. *Assume that there exists $\varepsilon_0 > 0$ such that every finite partition $\mathcal{P}$ with* $\operatorname{diam} \mathcal{P} < \varepsilon_0$ *satisfies* $\lim_n \operatorname{diam} \mathcal{P}^n = 0$. *Then, the function $\mu \mapsto$*

$h_\mu(f)$ is upper semi-continuous. Consequently, that function is bounded and its supremum is attained for some measure μ.

Proof. As we saw in Corollary 9.2.10, the property $\lim_n \operatorname{diam} \mathcal{P}^n = 0$ implies that $\bigcup_n \mathcal{P}^n$ generates the σ-algebra of measurable sets. On the other hand, given any invariant probability measure μ, it is easy to choose[2] a partition $\mathcal{P}$ with diameter smaller than ε_0 and such that $\mu(\partial\mathcal{P}) = 0$. It follows from the previous corollary that the entropy function is upper semi-continuous at μ and, since μ is arbitrary, this gives the first claim in the statement.

The other claims are general consequences of semi-continuity and the fact that the domain of the entropy function, that is, the space $\mathcal{M}_1(f)$ of all invariant probability measures, is compact.

When f is invertible we may replace $\mathcal{P}^n$ by $\mathcal{P}^{\pm n} = \bigvee_{j=-n}^{n-1} f^{-j}(\mathcal{P})$ in the statement of Corollaries 9.2.13 and 9.2.14. The proof is analogous, using Corollary 9.2.5 and the version of Corollary 9.2.10 for invertible transformations.

9.2.3 Expansive transformations

Next, we discuss a rather broad class of transformations that satisfy the conditions in Corollary 9.2.14.

A continuous transformation $f: M \to M$ in a metric space is said to be *expansive* if there exists $\varepsilon_0 > 0$ (called a *constant of expansivity*) such that, given any $x, y \in M$ with $x \neq y$, there exists $n \in \mathbb{N}$ such that $d(f^n(x), f^n(y)) \geq \varepsilon_0$. That is, any two distinct orbits of f may be distinguished, at a macroscopic scale, at some stage of the iteration.

When f is invertible, there is also a two-sided version of the notion of expansivity, defined as follows: there exists $\varepsilon_0 > 0$ such that, given any $x, y \in M$ with $x \neq y$ there exists $n \in \mathbb{Z}$ such that $d(f^n(x), f^n(y)) \geq \varepsilon_0$. It is clear that (one-sided) expansive homeomorphisms are also two-sided expansive.

Example 9.2.15. Let $\sigma: \Sigma \to \Sigma$ be the shift map in $\Sigma = \{1, \dots, d\}^{\mathbb{N}}$. Consider in Σ the distance defined by $d((x_n)_n, (y_n)_n) = 2^{-N}$, where N is the smallest value of n such that $x_n \neq y_n$. Note that $d(\sigma^N(x), \sigma^N(y)) = 2^0 = 1$ if $x = (x_n)_n$ and $y = (y_n)_n$ are distinct. This shows that σ is an expansive transformation, with $\varepsilon_0 = 1$ as a constant of expansivity.

Analogously, the two-sided shift map $\sigma: \Sigma \to \Sigma$ in $\Sigma = \{1, \dots, d\}^{\mathbb{Z}}$ is two-sided expansive (but not one-sided expansive).

[2] For example: for each x choose $r_x \in (0, \varepsilon_0)$ such that the boundary of the ball of center x and radius r_x has measure zero. Then let $\mathcal{U}$ be a finite cover of M by such balls and take for $\mathcal{P}$ the partition defined by $\mathcal{U}$, that is, the partition whose elements are the maximal sets that, for each $U \in \mathcal{U}$, are either contained in U or disjoint from U; see Figure 2.1.

We leave it to the reader to check (Exercise 9.2.1) that the decimal expansion transformation $f(x) = 10x - [10x]$ is also expansive. On the other hand, torus rotations are never expansive.

Proposition 9.2.16. *Let $f : M \to M$ be an expansive transformation in a compact metric space and let $\varepsilon_0 > 0$ be a constant of expansivity. Then* $\lim_n \operatorname{diam} \mathcal{P}^n = 0$ *for every finite partition $\mathcal{P}$ with* $\operatorname{diam} \mathcal{P} < \varepsilon_0$.

Proof. It is clear that the sequence $\operatorname{diam} \mathcal{P}^n$ is non-increasing. Suppose that its infimum δ is positive. Then, for every $n \geq 1$ there exist points x_n and y_n such that $d(x_n, y_n) > \delta/2$ but x_n and y_n belong to the same element of $\mathcal{P}^n$ and, thus, satisfy

$$d(f^j(x_n), f^j(y_n)) \leq \operatorname{diam} \mathcal{P} < \varepsilon_0 \quad \text{for every } 0 \leq j < n.$$

By compactness, there exists a sequence $(n_j)_j \to \infty$ such that $(x_{n_j})_j$ and $(y_{n_j})_j$ converge to points x and y, respectively. Then, $d(x, y) \geq \delta/2 > 0$ but $d(f^j(x), f^j(y)) \leq \operatorname{diam} \mathcal{P} < \varepsilon_0$ for every $j \geq 0$. This contradicts the hypothesis that ε_0 is a constant of expansivity.

Corollary 9.2.17. *If $f : M \to M$ is an expansive transformation in a compact metric space then the entropy function is upper semi-continuous. Moreover, there exist invariant probability measures μ whose entropy $h_\mu(f)$ is maximum among all the invariant probability measures of f.*

Proof. Combine Proposition 9.2.16 with Corollary 9.2.14.

If the transformation f is invertible and two-sided expansive, we may replace $\mathcal{P}^n$ by $\mathcal{P}^{\pm n}$ in Proposition 9.2.16 and the conclusion of Corollary 9.2.17 also remains valid as stated.

9.2.4 Exercises

9.2.1. Show that the decimal expansion $f : [0,1] \to [0,1]$, $f(x) = 10x - [10x]$ is expansive and exhibit a constant of expansivity.

9.2.2. Check that for every $s > 0$ there exists some Bernoulli shift (σ, μ) whose entropy is equal to s.

9.2.3. Let $X = \{0\} \cup \{1/n : n \geq 1\}$ and consider the space $\Sigma = X^{\mathbb{N}}$ endowed with the distance $d((x_n)_n, (y_n)_n) = 2^{-N}|x_N - y_N|$, where $N = \min\{n \in \mathbb{N} : x_n \neq y_n\}$.
   (a) Verify that the shift map $\sigma : \Sigma \to \Sigma$ is not expansive.
   (b) For each $k \geq 1$, let ν_k be the probability measure on X that assigns weight $1/2$ to each of the points $1/k$ and $1/(k+1)$. Use the Bernoulli measures $\mu_k = \nu_k^{\mathbb{N}}$ to conclude that the entropy function of the shift is not upper semi-continuous.
   (c) Let μ be the Bernoulli measure associated with any probability vector $(p_x)_{x \in X}$ such that $\sum_{x \in X} -p_x \log p_x = \infty$. Show that $h_\mu(\sigma)$ is infinite.

9.2.4. Let $f : S^1 \to S^1$ be a covering map of degree $d \geq 2$ and μ be a probability measure invariant under f. Show that $h_\mu(f) \leq \log d$.

9.2.5. Let $\mathcal{P}$ and $\mathcal{Q}$ be two partitions with finite entropy. Show that if $\mathcal{P}$ is coarser than $\bigvee_{j=0}^{\infty} f^{-j}(\mathcal{Q})$ then $h_\mu(f,\mathcal{P}) \le h_\mu(f,\mathcal{Q})$.

9.2.6. Show that if $\mathcal{A}$ is an algebra that generates the σ-algebra of measurable sets, up to measure zero, then the supremum of $h_\mu(f,\mathcal{P})$ over the partitions with finite entropy (or even the finite partitions) $\mathcal{P} \subset \mathcal{A}$ coincides with $h_\mu(f)$.

9.2.7. Consider transformations $f : M \to M$ and $g : N \to N$ preserving probability measures μ and ν, respectively. Consider $f \times g : M \times N \to M \times N$ defined by $(f \times g)(x,y) = (f(x),g(y))$. Show that $f \times g$ preserves the product measure $\mu \times \nu$ and that $h_{\mu\times\nu}(f \times g) = h_\mu(f) + h_\nu(g)$.

9.3 Local entropy

The theorem of Shannon–McMillan–Breiman, which we discuss in this section, provides a complementary view of the concept of entropy, more detailed and more local in nature. We also mention a topological version of that idea, which is due to Brin–Katok.

Theorem 9.3.1 (Shannon–McMillan–Breiman). *Given any partition $\mathcal{P}$ with finite entropy, the limit*

$$h_\mu(f,\mathcal{P},x) = \lim_n -\frac{1}{n}\log \mu(\mathcal{P}^n(x)) \quad \textit{exists at } \mu\textit{-almost every point.} \tag{9.3.1}$$

The function $x \mapsto h_\mu(f,\mathcal{P},x)$ is μ-integrable, and the limit in (9.3.1) *also holds in $L^1(\mu)$. Moreover,*

$$\int h_\mu(f,\mathcal{P},x)\,d\mu(x) = h_\mu(f,\mathcal{P}).$$

If (f,μ) is ergodic then $h_\mu(f,\mathcal{P},x) = h_\mu(f,\mathcal{P})$ at μ-almost every point.

Recall that $\mathcal{P}^n(x) = \mathcal{P}(x) \cap f^{-1}(\mathcal{P}(f(x))) \cap \cdots \cap f^{-n+1}(\mathcal{P}(f^{n-1}(x)))$, that is, $\mathcal{P}^n(x)$ is formed by the points whose trajectories remain "close" to the trajectory of x during n iterates, in the sense that they visit the same elements of $\mathcal{P}$. Theorem 9.3.1 states that the measure of this set has a well-defined exponential rate of decay: at μ-almost every point,

$$\mu(\mathcal{P}^n(x)) \approx e^{-nh_\mu(f,\mathcal{P},x)} \quad \text{for every large } n.$$

The proof of the theorem is presented in Section 9.3.1.

The theorem of Brin–Katok that we state in the sequel belongs to the same family of results, but uses a different notion of proximity.

Definition 9.3.2. Let $f : M \to M$ be a continuous map in a compact metric space. Given $x \in M$, $n \ge 1$ and $\varepsilon > 0$, we call the *dynamical ball* of length n and radius ε around x the set

$$B(x,n,\varepsilon) = \{y \in M : d(f^j(x),f^j(y)) < \varepsilon \text{ for every } j = 0,1,\ldots,n-1\}.$$

In other words,

$$B(x,n,\varepsilon) = B(x,\varepsilon) \cap f^{-1}(B(f(x),\varepsilon)) \cap \cdots \cap f^{-n+1}(B(f^{n-1}(x),\varepsilon)).$$

Define

$$h^+_\mu(f,\varepsilon,x) = \limsup_n -\frac{1}{n}\log \mu(B(x,n,\varepsilon)) \quad \text{and}$$

$$h^-_\mu(f,\varepsilon,x) = \liminf_n -\frac{1}{n}\log \mu(B(x,n,\varepsilon)).$$

Theorem 9.3.3 (Brin–Katok). *Let μ be a probability measure invariant under f. The limits*

$$\lim_{\varepsilon\to 0} h^+_\mu(f,\varepsilon,x) \quad \textit{and} \quad \lim_{\varepsilon\to 0} h^-_\mu(f,\varepsilon,x)$$

exist and are equal, for μ-almost every point. Denoting by $h_\mu(f,x)$ their common value, the function $h_\mu(f,\cdot)$ is integrable and

$$h_\mu(f) = \int h_\mu(f,x)\,d\mu(x).$$

The proof of this result may be found in the original paper of Brin and Katok [BK83], and is not presented here.

Example 9.3.4 (Translations in compact groups). Let G be a compact metrizable group and μ be its Haar measure. Every translation of G has zero entropy with respect to μ. Indeed, consider in G any distance d that is invariant under all the translations (recall Lemma 6.3.6). Relative to such a distance,

$$L^j_g(B(x,\varepsilon)) = B(L^j_g(x),\varepsilon)$$

for every $g \in G$ and $j \in \mathbb{Z}$. Consequently, $B(x,n,\varepsilon) = B(x,\varepsilon)$ for every $n \geq 1$. Then,

$$h^\pm_\mu(L_g,\varepsilon,x) = \lim_n -\frac{1}{n}\log \mu(B(x,\varepsilon)) = 0$$

for every $\varepsilon > 0$ and $x \in G$. By the theorem of Brin–Katok, it follows that $h_\mu(L_g) = 0$ for every $g \in G$. The same argument applies to every right-translation R_g.

9.3.1 Proof of the Shannon–McMillan–Breiman theorem

Consider the sequence of functions $\varphi_n : M \to \mathbb{R}$ defined by

$$\varphi_n(x) = -\log \frac{\mu(\mathcal{P}^n(x))}{\mu(\mathcal{P}^{n-1}(f(x)))}.$$

By telescopic cancellation,

$$-\frac{1}{n}\log \mu(\mathcal{P}^n(x)) = -\frac{1}{n}\log \mu(\mathcal{P}(f^{n-1}(x))) + \frac{1}{n}\sum_{j=0}^{n-2}\varphi_{n-j}(f^j(x)) \tag{9.3.2}$$

for every n and every x.

Lemma 9.3.5. *The sequence* $-n^{-1}\log\mu(\mathcal{P}(f^{n-1}(x)))$ *converges to zero* μ*-almost everywhere and in* $L^1(\mu)$.

Proof. Start by noting that the function $x \mapsto -\log\mu(\mathcal{P}(x))$ is integrable:

$$\int |\log\mu(\mathcal{P}(x))|\,d\mu(x) = \int -\log\mu(\mathcal{P}(x))\,d\mu(x) = H_\mu(\mathcal{P}) < \infty.$$

Using Lemma 3.2.5, it follows that $-(n-1)^{-1}\log\mu(\mathcal{P}(f^{n-1}(x)))$ converges to zero at μ-almost every point. Moreover, it is clear that this conclusion is not affected if one replaces $n-1$ by n in the denominator. This proves the claim of μ-almost everywhere convergence. Next, using the fact that the measure μ is invariant and $H_\mu(\mathcal{P}) < \infty$,

$$\left\| -\frac{1}{n}\log\mu(\mathcal{P}(f^{n-1}(x)))\right\|_1 = \frac{1}{n}\int -\log\mu(\mathcal{P}(f^{n-1}(x)))\,d\mu(x) = \frac{1}{n}H_\mu(\mathcal{P})$$

converges to zero when $n \to \infty$. This proves the convergence in $L^1(\mu)$.

Next, we show that the last term in (9.3.2) also converges μ-almost everywhere and in $L^1(\mu)$.

Lemma 9.3.6. *The limit* $\varphi(x) = \lim_n \varphi_n(x)$ *exists at* μ*-almost every point.*

Proof. For each $n > 1$, denote by $\mathcal{Q}_n$ the partition of M defined by

$$\mathcal{Q}_n(x) = f^{-1}(\mathcal{P}^{n-1}(f(x))) = f^{-1}(\mathcal{P}(f(x))) \cap \cdots \cap f^{-n+1}(\mathcal{P}(f^{n-1}(x))).$$

Note that $\mu(\mathcal{P}^{n-1}(f(x)) = \mu(\mathcal{Q}_n(x))$ and $\mathcal{P}^n(x) = \mathcal{P}(x) \cap \mathcal{Q}_n(x)$. Therefore,

$$\frac{\mu(\mathcal{P}^n(x))}{\mu(\mathcal{P}^{n-1}(f(x)))} = \frac{\mu(\mathcal{P}(x) \cap \mathcal{Q}_n(x))}{\mu(\mathcal{Q}_n(x))}. \tag{9.3.3}$$

For each $P \in \mathcal{P}$ and $n > 1$, consider the conditional expectation (recall Section 5.2.1)

$$e_n(\mathcal{X}_P, x) = \frac{1}{\mu(\mathcal{Q}_n(x))}\int_{\mathcal{Q}_n(x)} \mathcal{X}_P\,d\mu = \frac{\mu(P \cap \mathcal{Q}_n(x))}{\mu(\mathcal{Q}_n(x))}.$$

Comparing with (9.3.3) we see that

$$e_n(\mathcal{X}_P, x) = \frac{\mu(\mathcal{P}^n(x))}{\mu(\mathcal{P}^{n-1}(f(x)))} \quad \text{for every } x \in P.$$

By Lemma 5.2.1, the limit $e(\mathcal{X}_P, x) = \lim_n e_n(\mathcal{X}_P, x)$ exists for μ-almost every $x \in M$ and, in particular, for μ-almost every $x \in P$. Since $P \in \mathcal{P}$ is arbitrary, this proves that

$$\lim_n \frac{\mu(\mathcal{P}^n(x))}{\mu(\mathcal{P}^{n-1}(f(x)))}$$

exists for μ-almost every point. Taking logarithms, we get that $\lim_n \varphi_n(x)$ exists for μ-almost every point, as stated.

Lemma 9.3.7. *The function $\Phi = \sup_n \varphi_n$ is integrable.*

Proof. As in the previous lemma, let us consider the partitions $\mathcal{Q}_n$ defined by $\mathcal{Q}_n(x) = f^{-1}(\mathcal{P}^{n-1}(f(x)))$. Fix any $P \in \mathcal{P}$. Given $x \in P$ and $t > 0$, it is clear that $\Phi(x) > t$ if and only if $\varphi_n(x) > t$ for some n. Moreover,

$$\varphi_n(x) > t \quad \Leftrightarrow \quad \mu\big(P \cap \mathcal{Q}_n(x)\big) < e^{-t}\mu(\mathcal{Q}_n(x))$$

and, in that case, $\varphi_n(y) > t$ for every $y \in P \cap \mathcal{Q}_n(x)$. Therefore, we may write the set $\{x \in P : \Phi(x) > t\}$ as a disjoint union $\bigcup_j (P \cap Q_j)$, where each Q_j belongs to some partition $\mathcal{Q}_{n(j)}$ and

$$\mu\big(P \cap Q_j\big) < e^{-t}\mu(Q_j) \quad \text{for every } j.$$

Consequently, for every $t > 0$ and $P \in \mathcal{P}$,

$$\mu(\{x \in P : \Phi(x) > t\}) = \sum_j \mu\big(P \cap Q_j\big) < e^{-t} \sum_j \mu(Q_j) \le e^{-t}. \tag{9.3.4}$$

Then (see Exercise 9.3.1),

$$\begin{aligned}\int \Phi\, d\mu = \sum_{P \in \mathcal{P}} \int_P \Phi\, d\mu &= \sum_{P \in \mathcal{P}} \int_0^\infty \mu(\{x \in P : \Phi(x) > t\})\, dt \\ &\le \sum_{P \in \mathcal{P}} \int_0^\infty \min\{e^{-t}, \mu(P)\}\, dt.\end{aligned}$$

The last integral may be rewritten as follows:

$$\int_0^{-\log\mu(P)} \mu(P)\, dt + \int_{-\log\mu(P)}^\infty e^{-t}\, dt = -\mu(P)\log\mu(P) + \mu(P).$$

Combining these two relations:

$$\int \Phi\, d\mu \le \sum_{P \in \mathcal{P}} -\mu(P)\log\mu(P) + \mu(P) = H_\mu(\mathcal{P}) + 1 < \infty.$$

This proves the lemma, since Φ is non-negative.

Lemma 9.3.8. *The function φ is integrable and $(\varphi_n)_n$ converges to φ in $L^1(\mu)$.*

Proof. We saw in Lemma 9.3.6 that $(\varphi_n)_n$ converges to φ at μ-almost every point. Since $0 \le \varphi_n \le \Phi$ for every n, we also have that $0 \le \varphi \le \Phi$. In particular, φ is integrable. Moreover, $|\varphi - \varphi_n| \le \Phi$ for every n and, thus, we may use the dominated convergence theorem (Theorem A.2.11) to conclude that

$$\lim_n \int |\varphi - \varphi_n|\, d\mu = \int \lim_n |\varphi - \varphi_n|\, d\mu = 0.$$

This proves the convergence in $L^1(\mu)$.

Lemma 9.3.9. *At μ-almost every point and in $L^1(\mu)$,*

$$\lim_n \frac{1}{n}\sum_{j=0}^{n-2} \varphi_{n-j}(f^j(x)) = \lim_n \frac{1}{n}\sum_{j=0}^{n-2} \varphi(f^j(x)).$$

Proof. By the Birkhoff ergodic theorem (Theorem 3.2.3), the limit on the right-hand side exists at μ-almost every point and in $L^1(\mu)$, indeed, it is equal to the time average of the function φ. Therefore, it is enough to show that the difference

$$\frac{1}{n}\sum_{j=0}^{n-2} (\varphi_{n-j} - \varphi) \circ f^j \tag{9.3.5}$$

converges to zero at μ-almost every point and in $L^1(\mu)$. Since the measure μ is invariant, $\|(\varphi_{n-j} - \varphi) \circ f^j\|_1 = \|\varphi_{n-j} - \varphi\|_1$ for every j. Hence,

$$\left\| \frac{1}{n}\sum_{j=0}^{n-2} (\varphi_{n-j} - \varphi) \circ f^j \right\|_1 \le \frac{1}{n}\sum_{j=0}^{n-2} \|\varphi_{n-j} - \varphi\|_1.$$

By Lemma 9.3.8 the sequence on the right-hand side converges to zero. This implies that (9.3.5) converges to zero in $L^1(\mu)$. We are left to prove that the sequence converges at μ-almost every point.

For each fixed $k \ge 2$, consider $\Phi_k = \sup_{i>k} |\varphi_i - \varphi|$. Note that $\Phi_k \le \Phi$ and, thus, $\Phi_k \in L^1(\mu)$. Moreover,

$$\frac{1}{n}\sum_{j=0}^{n-2} |\varphi_{n-j} - \varphi| \circ f^j = \frac{1}{n}\sum_{j=0}^{n-k-1} |\varphi_{n-j} - \varphi| \circ f^j + \frac{1}{n}\sum_{j=n-k}^{n-2} |\varphi_{n-j} - \varphi| \circ f^j$$

$$\le \frac{1}{n}\sum_{j=0}^{n-k-1} \Phi_k \circ f^j + \frac{1}{n}\sum_{j=n-k}^{n-2} \Phi \circ f^j.$$

By the Birkhoff ergodic theorem, the first term on the right-hand side converges to the time average $\tilde{\Phi}_k$ at μ-almost every point. By Lemma 3.2.5, the last term converges to zero at μ-almost every point: the lemma implies that $n^{-1}\Phi \circ f^{n-i}$ converges to zero for any fixed i. Hence,

$$\limsup_n \frac{1}{n}\sum_{j=0}^{n-2} |\varphi_{n-j} - \varphi|(f^j(x)) \le \tilde{\Phi}_k(x) \quad \text{at } \mu\text{-almost every point.} \tag{9.3.6}$$

We claim that $\lim_k \tilde{\Phi}_k(x) = 0$ at μ-almost every point. Indeed, the sequence $(\Phi_k)_k$ is non-increasing and, by Lemma 9.3.6, it converges to zero at μ-almost every point. By the monotone convergence theorem (Theorem A.2.9), it follows that $\int \Phi_k \, d\mu \to 0$ when $k \to \infty$. Another consequence is that $(\tilde{\Phi}_k)_k$ is non-increasing. Hence, using the monotone convergence theorem together with the Birkhoff ergodic theorem,

$$\int \lim_k \tilde{\Phi}_k \, d\mu = \lim_k \int \tilde{\Phi}_k \, d\mu = \lim_k \int \Phi_k \, d\mu = 0.$$

Since $\tilde{\Phi}_k$ is non-negative, it follows that $\lim_k \tilde{\Phi}_k = 0$ at μ-almost every point, as we claimed. Therefore, (9.3.6) implies that

$$\lim_n \frac{1}{n} \sum_{j=0}^{n-2} |\varphi_{n-j} - \varphi| \circ f^j = 0$$

at μ-almost every point. This completes the proof of the lemma.

It follows from (9.3.2) and Lemmas 9.3.5 and 9.3.9 that

$$h_\mu(f,\mathcal{P},x) = \lim_n -\frac{1}{n} \log \mu(\mathcal{P}^n(x))$$

exists at μ-almost every point and in $L^1(\mu)$: in fact, it coincides with the time average $\tilde{\varphi}(x)$ of the function φ. Then, in particular,

$$\begin{aligned}\int h_\mu(f,\mathcal{P},x)\, d\mu(x) &= \lim_n \frac{1}{n} \int -\log \mu(\mathcal{P}^n(x))\, d\mu(x) \\ &= \lim_n \frac{1}{n} H_\mu(\mathcal{P}^n) = h_\mu(f,\mathcal{P}).\end{aligned}$$

Moreover, if (f,μ) is ergodic then $h(f,\mathcal{P},x) = \tilde{\varphi}(x)$ is constant at μ-almost every point. That is, in that case $h_\mu(f,\mathcal{P},x) = h_\mu(f,\mathcal{P})$ for μ-almost every point. This closes the proof of Theorem 9.3.1.

9.3.2 Exercises

9.3.1. Check that, for any measurable function $\varphi : M \to (0,\infty)$,

$$\int \varphi \, d\mu = \int_0^\infty \mu(\{x \in M : \varphi(x) > t\})\, dt.$$

9.3.2. Use Theorem 9.3.1 to calculate the entropy of a Bernoulli shift in $\Sigma = \{1,\dots,d\}^{\mathbb{N}}$.

9.3.3. Show that the function $h_\mu(f,x)$ in Theorem 9.3.3 is f-invariant. Conclude that if (f,μ) is ergodic, then $h_\mu(f) = h_\mu(f,x)$ for μ-almost every x.

9.3.4. Suppose that (f,μ) is ergodic and let $\mathcal{P}$ be a partition with finite entropy. Show that given $\varepsilon > 0$ there exists $k \geq 1$ such that for every $n \geq k$ there exists $\mathcal{B}_n \subset \mathcal{P}^n$ such that

$$e^{-n(h_\mu(f,\mathcal{P})+\varepsilon)} < \mu(B) < e^{-n(h_\mu(f,\mathcal{P})-\varepsilon)} \quad \text{for every } B \in \mathcal{B}_n,$$

and the measure of the union of the elements of $\mathcal{B}_n$ is larger than $1-\varepsilon$.

9.4 Examples

In this section we illustrate the previous results through a few examples.

9.4.1 Markov shifts

Let $\Sigma = \{1,\dots,d\}^{\mathbb{N}}$ and $\sigma : \Sigma \to \Sigma$ be the shift map. Let μ be the Markov measure associated with a stochastic matrix $P = (P_{i,j})_{i,j}$ and a probability vector $p = (p_i)_i$. We are going to prove:

Proposition 9.4.1. $h_\mu(\sigma) = \sum_{a=1}^{d} p_a \sum_{b=1}^{d} -P_{a,b} \log P_{a,b}$.

Proof. Consider the partition $\mathcal{P}$ of Σ into cylinders $[0;a]$, $a = 1,\dots,d$. For each n, the iterate $\mathcal{P}^n$ is the partition into cylinders $[0;a_1,\dots,a_n]$ of length n. Recalling that $\mu([0;a_1,\dots,a_n]) = p_{a_1} P_{a_1,a_2} \cdots P_{a_{n-1},a_n}$, we see that

$$
\begin{aligned}
H_\mu(\mathcal{P}^n) &= \sum_{a_1,\dots,a_n} -p_{a_1} P_{a_1,a_2} \cdots P_{a_{n-1},a_n} \log \left(p_{a_1} P_{a_1,a_2} \cdots P_{a_{n-1},a_n}\right) \\
&= \sum_{a_1} -p_{a_1} \log p_{a_1} \sum_{a_2,\dots,a_n} P_{a_1,a_2} \cdots P_{a_{n-1},a_n} \\
&\quad + \sum_{j=1}^{n-1} \sum_{a_j,a_{j+1}} -\log P_{a_j,a_{j+1}} \sum p_{a_1} P_{a_1,a_2} \cdots P_{a_{n-1},a_n},
\end{aligned}
\tag{9.4.1}
$$

where the last sum is over all the values of $a_1,\dots,a_{j-1},a_{j+2},\dots,a_n$. On the one hand,

$$
\sum_{a_2,\dots,a_n} P_{a_1,a_2} \cdots P_{a_{n-1},a_n} = \sum_{a_n} P^n_{a_1,a_n} = 1,
$$

because P^n is a stochastic matrix. On the other hand,

$$
\begin{aligned}
\sum p_{a_1} P_{a_1,a_2} \cdots P_{a_{n-1},a_n} &= \sum_{a_1,a_n} p_{a_1} P^j_{a_1,a_j} P_{a_j,a_{j+1}} P^{n-j-1}_{a_{j+1},a_n} \\
&= \sum_{a_1} p_{a_1} P^j_{a_1,a_j} P_{a_j,a_{j+1}} = p_{a_j} P_{a_j,a_{j+1}},
\end{aligned}
$$

because P^{n-j-1} is a stochastic matrix and $pP^j = P^{*j} p = p$. Replacing these observations in (9.4.1), we get that

$$
\begin{aligned}
H_\mu(\mathcal{P}^n) &= \sum_{a_1} -p_{a_1} \log p_{a_1} + \sum_{j=1}^{n-1} \sum_{a_j,a_{j+1}} -p_{a_j} P_{a_j,a_{j+1}} \log P_{a_j,a_{j+1}} \\
&= \sum_{a} -p_a \log p_a + (n-1) \sum_{a,b} -p_a P_{a,b} \log P_{a,b}.
\end{aligned}
$$

It follows that $h_\mu(\sigma,\mathcal{P}) = \sum_{a,b} -p_a P_{a,b} \log P_{a,b}$. Since the family of all cylinders $[0;a_1,\dots,a_n]$ generates the σ-algebra of Σ, it follows from Corollary 9.2.5 that $h_\mu(\sigma) = h_\mu(\sigma,\mathcal{P})$. This completes the proof of theorem.

This conclusion remains valid for two-sided Markov shifts as well, that is, when $\Sigma = \{1,\dots,d\}^{\mathbb{Z}}$. The argument is analogous, using Corollary 9.2.6.

9.4.2 Gauss map

Now we calculate the entropy of the Gauss map $G(x) = (1/x) - [1/x]$ relative to the invariant probability measure

$$\mu(E) = \frac{1}{\log 2}\int_E \frac{dx}{1+x}, \tag{9.4.2}$$

which was already studied in Sections 1.3.2 and 4.2.4. The method that we are going to present extends to a much broader class of systems, including the expanding maps of the interval that are defined and discussed in Example 11.1.16.

Let $\mathcal{P}$ be the partition of $(0,1)$ into the subintervals $(1/(m+1), 1/m)$ with $m \ge 1$. As before, we denote $\mathcal{P}^n = \bigvee_{j=0}^{n-1} G^{-j}(\mathcal{P})$. The following facts are used in what follows:

(A) G^n maps each $P_n \in \mathcal{P}^n$ diffeomorphically onto $(0,1)$, for each $n \ge 1$.
(B) $\operatorname{diam} \mathcal{P}^n \to 0$ when $n \to \infty$.
(C) There exists $C > 1$ such that $|(G^n)'(y)|/|(G^n)'(x)| \le C$ for every $n \ge 1$ and any x and y in the same element of the partition $\mathcal{P}^n$.
(D) There exist $c_1, c_2 > 0$ such that $c_1 m(P_n) \le \mu(P_n) \le c_2 m(P_n)$ for every $n \ge 1$ and every $P_n \in \mathcal{P}_n$, where m denotes the Lebesgue measure.

It is immediate from the definition that each $P \in \mathcal{P}$ is mapped by G diffeomorphically onto $(0,1)$. Property (A) is a consequence, by induction on n. Using (A) and Lemma 4.2.12, we get that

$$\operatorname{diam} P_n \le \sup_{x \in P_n} \frac{1}{|(G^n)'(x)|} \le 2^{-[n/2]}$$

for every $n \ge 1$ and every $P_n \in \mathcal{P}^n$. This implies (B). Property (C) is given by Lemma 4.2.13. Finally, (D) follows directly from (9.4.2).

Proposition 9.4.2. $h_\mu(G) = \int \log |G'| \, d\mu$.

Proof. Consider the function $\psi_n(x) = -\log \mu(\mathcal{P}^n(x))$, for each $n \ge 1$. Observe that

$$H_\mu(\mathcal{P}^n) = \sum_{P_n \in \mathcal{P}^n} -\mu(P_n) \log \mu(P_n) = \int \psi_n(x) \, d\mu(x).$$

Property (D) gives that

$$-\log c_1 \ge \psi_n(x) + \log m(\mathcal{P}^n(x)) \ge -\log c_2.$$

By property (A), we have $\log m(\mathcal{P}^n(x)) = -\log |(G^n)'(y)|$ for some $y \in \mathcal{P}^n(x)$. Using property (C), it follows that

$$-\log c_1 + \log C \ge \psi_n(x) - \log |(G^n)'(x)| \ge -\log c_2 - \log C$$

for every x and every n. Consequently,

$$-\log(c_1/C) \geq H_\mu(\mathcal{P}^n) - \int \log|(G^n)'|\,d\mu \geq \log(c_2 C) \tag{9.4.3}$$

for every n. Since the measure μ is invariant under G,

$$\int \log|(G^n)'|\,d\mu = \sum_{j=0}^{n-1} \int \log|G'| \circ G^j\,d\mu = n \int |G'|\,d\mu.$$

Then, dividing (9.4.3) by n and taking the limit when $n \to \infty$,

$$h_\mu(f,\mathcal{P}) = \lim_n \frac{1}{n} H_\mu(\mathcal{P}^n) = \int \log|G'|\,d\mu.$$

Now, property (B) ensures that we may apply Corollary 9.2.10 to conclude that

$$h_\mu(G) = h_\mu(G,\mathcal{P}) = \int \log|G'|\,d\mu.$$

This completes the proof of the proposition.

The integral in the statement of the proposition may be calculated explicitly: we leave it to the reader to check (using integration by parts and the fact that $\sum_{j=1}^{\infty} 1/j^2 = \pi^2/6$) that

$$h_\mu(G) = \int \log|G'|\,d\mu = \int_0^1 \frac{-2\log x\,dx}{(1+x)\log 2} = \frac{\pi^2}{6\log 2} \approx 5.46\ldots$$

Then, recalling that (G,μ) is ergodic (Section 4.2.4), it follows from the theorem of Shannon–McMillan–Breiman (Theorem 9.3.1) that

$$\lim_n -\frac{1}{n} \log \mu(\mathcal{P}^n(x)) = \frac{\pi^2}{6\log 2} \quad \text{for } \mu\text{-almost every } x.$$

As the measure μ is comparable to the Lebesgue measure, up to a constant factor, this means that

$$\operatorname{diam} \mathcal{P}^n(x) \approx e^{-\frac{\pi^2 n}{6\log 2}} \qquad (\text{up to a factor } e^{\pm\varepsilon n})$$

for μ-almost every x and every n sufficiently large. Observe that $\mathcal{P}^n(x)$ is formed by the points y whose continued fraction expansion coincide with the continued fraction expansion of x up to order n.

9.4.3 Linear endomorphisms of the torus

Given a real number $x > 0$, we denote $\log^+ x = \max\{\log x, 0\}$. In this section we prove the following result:

Proposition 9.4.3. *Let $f_A : \mathbb{T}^d \to \mathbb{T}^d$ be the endomorphism induced on the torus $\mathbb{T}^d$ by some invertible matrix A with integer coefficients. Let μ be the*

Haar measure of $\mathbb{T}^d$. Then

$$h_\mu(f_A) = \sum_{i=1}^{d} \log^+ |\lambda_i|,$$

where $\lambda_1, \dots, \lambda_d$ are the eigenvalues of A, counted with multiplicity.

Initially, let us suppose that the matrix A is diagonalizable. Let $v_1, \dots, v_d$ be a normed basis of $\mathbb{R}^d$ such that $Av_i = \lambda_i v_i$ for each i. Let u be the number of eigenvalues of A with absolute value strictly larger than 1. We may take the eigenvalues to be numbered in such a way that $|\lambda_i| > 1$ if and only if $i \le u$. Given $x \in \mathbb{T}^d$, every point y in a neighborhood of x may be written in the form

$$y = x + \sum_{i=1}^{d} t_i v_i$$

with $t_1, \dots, t_d$ close to zero. Given $\varepsilon > 0$, denote by $D(x, \varepsilon)$ the set of points y of this form with $|t_i| < \varepsilon$ for every $i = 1, \dots, d$. Moreover, for each $n \ge 1$, consider

$$D(x,n,\varepsilon) = \{y \in \mathbb{T}^d : f_A^j(y) \in D(f_A^j(x), \varepsilon) \text{ for every } j = 0, \dots, n-1\}.$$

Observe that $f_A^j(y) = f_A^j(x) + \sum_{i=1}^{d} t_i \lambda_i^j v_i$ for every $n \ge 1$. Therefore,

$$D(x,n,\varepsilon) = \left\{ x + \sum_{i=1}^{d} t_i v_i : |\lambda_i^n t_i| < \varepsilon \text{ for } i \le u \text{ and } |t_i| < \varepsilon \text{ for } i > u \right\}.$$

Hence, there exists a constant $C_1 > 1$ that depends only on A, such that

$$C_1^{-1} \varepsilon^d \prod_{i=1}^{u} |\lambda_i|^{-n} \le \mu(D(x,n,\varepsilon)) \le C_1 \varepsilon^d \prod_{i=1}^{u} |\lambda_i|^{-n}$$

for every $x \in \mathbb{T}^d$, $n \ge 1$ and $\varepsilon > 0$. It is also clear that there exists a constant $C_2 > 1$ that depends only on A, such that

$$B(x, C_2^{-1}\varepsilon) \subset D(x, \varepsilon) \subset B(x, C_2 \varepsilon)$$

for $x \in \mathbb{T}^d$ and $\varepsilon > 0$ small. Then, $B(x,n,C_2^{-1}\varepsilon) \subset D(x,n,\varepsilon) \subset B(x,n,C_2\varepsilon)$ for every $n \ge 1$. Combining these two observations and taking $C = C_1 C_2^d$, we get that

$$C^{-1} \varepsilon^d \prod_{i=1}^{u} |\lambda_i^{-n}| \le \mu(B(x,n,\varepsilon)) \le C \varepsilon^d \prod_{i=1}^{u} |\lambda_i^{-n}|$$

for every $x \in \mathbb{T}^d$, $n \ge 1$ and $\varepsilon > 0$. Then,

$$h_\mu^+(f,\varepsilon,x) = h_\mu^-(f,\varepsilon,x) = \lim_n \frac{1}{n} \log \mu\big(B(x,n,\varepsilon)\big) = \sum_{i=1}^{u} \log |\lambda_i|$$

for $x \in \mathbb{T}$ and $\varepsilon > 0$ small. Hence, using the theorem of Brin–Katok (Theorem 9.3.3),

$$h_\mu(f) = h_\mu(f,x) = \sum_{i=1}^{u} \log |\lambda_i|$$

for μ-almost every point x. This proves Proposition 9.4.3 in the diagonalizable case.

The general case may be treated analogously, through writing the matrix A in canonical Jordan form. We leave this task to the reader (Exercise 9.4.2).

9.4.4 Differentiable maps

Here we take M to be a Riemannian manifold (check Appendix A.4.5) and $f: M \to M$ to be a *local diffeomorphism*, that is, a C^1 map whose derivative $Df(x): T_xM \to T_{f(x)}M$ at each point x is an isomorphism. We are going to state and discuss two important theorems, the *Margulis–Ruelle inequality* and the *Pesin entropy formula*, that relate the entropy $h_\mu(f)$ of an invariant measure to the Lyapunov exponents of the derivative Df.

Let μ be any probability measure invariant under f. According to the multiplicative ergodic theorem of Oseledets (see Section 3.3.5), for μ-almost every point $x \in M$ there exist $k(x) \geq 1$, real numbers $\lambda_1(x) > \cdots > \lambda_k(x)$ and a filtration $\mathbb{R}^d = V_x^1 > \cdots > V_x^{k(x)} > V_x^{k(x)+1} = \{0\}$ such that

$$Df(x)V_x^i = V_{f(x)}^i \quad \text{and} \quad \lim_n \frac{1}{n} \log \|Df^n(x)v\| = \lambda_i(x)$$

for every $v \in V_x^i \setminus V_x^{i+1}$, every $i \in \{1, \ldots, k(x)\}$ and μ-almost every $x \in M$. Moreover, all these objects depend measurably on the point $x \in M$. When the measure μ is ergodic, the number $k(x)$, the Lyapunov exponents $\lambda_i(x)$ and their multiplicities $m_i(x) = \dim V_x^i - \dim V_x^{i+1}$ are all constant on a full measure set.

Define $\rho^+(x)$ to be the sum of all positive Lyapunov exponents, counted with multiplicity:

$$\rho^+(x) = \sum_{i=1}^{k(x)} m_i(x)\lambda_i^+(x) \qquad \text{with } \lambda_i^+(x) = \max\{\lambda_i(x), 0\}.$$

The Margulis–Ruelle inequality asserts that the average of ρ^+ is always an upper bound for the entropy of (f, μ). Proofs can be found in Ruelle [Rue78] and Mañé [Mañ87, Section 4.12].

Theorem 9.4.4 (Margulis–Ruelle inequality).

$$h_\mu(f) \leq \int \rho^+ \, d\mu.$$

It may happen that all the Lyapunov exponents are positive: that is the case, for instance, for the expanding differentiable maps in Section 11.1. Then, $\rho^+(x)$ is simply the sum of all Lyapunov exponents, counted with multiplicity. Now, it is also part of the theorem of Oseledets (property (c1) in Section 3.3.5) that

$$\sum_{i=1}^{k(x)} m_i(x)\lambda_i(x) = \lim_n \frac{1}{n} \log|\det Df^n(x)|$$

at μ-almost every point. Observe that the right-hand side of this identity is a Birkhoff time average:

$$\lim_n \frac{1}{n} \log|\det Df^n(x)| = \lim_n \frac{1}{n} \sum_{j=0}^{n-1} \log|\det Df|(f^j(x)).$$

So, by the Birkhoff ergodic theorem, the integral of ρ^+ coincides with the integral of the function $\log|\det Df|$. Thus, in this case the Margulis–Ruelle inequality becomes:

$$h_\mu(f) \le \int \log|\det Df|\, d\mu.$$

Another interesting particular case is when f is a diffeomorphism. It follows from the version of the Oseledets theorem for invertible maps (also stated in Section 3.3.5) that the Lyapunov exponents of Df^{-1} are the numbers $-\lambda_i(x)$, with multiplicities $m_i(x)$. Then, applying Theorem 9.4.4 to the inverse and recalling (Proposition 9.1.14) that $h_\mu(f) = h_\mu(f^{-1})$, we get that

$$h_\mu(f) \le \int \rho^- \, d\mu, \tag{9.4.4}$$

where $\rho^-(x) = \sum_{i=1}^{k(x)} m_i(x) \max\{-\lambda_i(x), 0\}$.

Now let us suppose that the invariant measure μ is absolutely continuous with respect to the volume measure associated with the Riemannian structure of M (check Appendix A.4.5). In this case, assuming just a little bit more regularity, we have a much stronger result:

Theorem 9.4.5 (Pesin entropy formula). *Assume that the derivative Df is Hölder and the invariant measure μ is absolutely continuous. Then*

$$h_\mu(f) = \int \rho^+ \, d\mu.$$

This fundamental result was originally proven by Pesin [Pes77]. See also Mañé [Mañ87, Section 4.13] for an alternative proof.

The expression for the entropy of the Haar measure of a linear torus endomorphism, given in Proposition 9.4.3, is a special case of Theorem 9.4.5. Indeed, one can check that the Lyapunov exponents of a linear endomorphism f_A at every point coincide with the logarithms $\log|\lambda_i|$ of the absolute values of the eigenvalues of the matrix A, with the same multiplicities. Thus, in this context

$$\rho^+(x) \equiv \sum_{i=1}^{d} \log^+ |\lambda_i|.$$

Of course, the Haar measure is absolutely continuous. Another special case of the Pesin entropy formula will appear in Section 12.1.8: see (12.1.31)–(12.1.36).

Finally, let us point out that the assumption of absolute continuity is too strong: the conclusion of Theorem 9.4.5 still holds if the invariant measure is just "absolutely continuous along unstable manifolds". Roughly speaking, this technical condition means that the conditional probabilities of μ with respect to a certain measurable partition, whose elements are unstable disks,[3] are absolutely continuous with respect to the volume measures induced on each of the disks by the Riemannian metric of M. Moreover, and most striking, this sufficient condition is also necessary for the Pesin entropy formula to hold. For precise statements, related results and proofs, see [LS82, Led84, LY85a, LY85b].

9.4.5 Exercises

9.4.1. Show that every rotation $R_\theta : \mathbb{T}^d \to \mathbb{T}^d$ has entropy zero with respect to the Haar measure of the torus $\mathbb{T}^d$. [Observation: This is a special case of Example 9.3.4 but for the present statement we do not need the theorem of Brin–Katok.]

9.4.2. Complete the proof of Proposition 9.4.3.

9.4.3. Let $f : M \to M$ be a measurable transformation and μ be an ergodic probability measure. Let $B \subset M$ be a measurable set with $\mu(B) > 0$, $g : B \to B$ be the first-return map of f to B and ν be the normalized restriction of μ to the set B (recall Section 1.4.1). Show that $h_\mu(f) = \nu(B)h_\nu(g)$.

9.4.4. Let $f : M \to M$ be a measure-preserving transformation in a Lebesgue space (M,μ). Let $\hat{f} : \hat{M} \to \hat{M}$ be the natural extension of f and $\hat{\mu}$ be the lift of μ (Exercise 8.5.7). Show that $h_\mu(f) = h_{\hat{\mu}}(\hat{f})$.

9.4.5. Prove that if f is the time-1 of a smooth flow on a surface M then $h_\mu(f) = 0$ for every invariant ergodic measure μ. [Observation: Using Theorem 9.6.2 below, it follows that the entropy is zero for every invariant measure.]

9.5 Entropy and equivalence

The notion of entropy was originally introduced in ergodic theory as a means to distinguish systems that are not ergodically equivalent, especially in the case of systems that are spectrally equivalent and, thus, cannot be distinguished by spectral invariants. It is easy to see that the entropy is, indeed, an invariant of ergodic equivalence:

Proposition 9.5.1. *Let $f : M \to M$ and $g : N \to N$ be transformations preserving probability measures μ in M and ν in N, respectively. If (f,μ) is ergodically equivalent to (g,ν), then $h_\mu(f) = h_\nu(g)$.*

[3] Unstable disks are differentiably embedded disks that are contracted exponentially under negative iteration; in the non-invertible case, the definition is formulated in terms of the natural extension of the transformation.

Proof. Let $\phi : M \to N$ be an ergodic equivalence between the two systems. This means that $\phi_*\mu = \nu$ and there exist full measure subsets $X \subset M$ and $Y \subset N$ such that ϕ is a measurable bijection from X to Y, with measurable inverse. Moreover, as observed in Section 8.1, the sets X and Y may be chosen to be invariant. Given any partition $\mathcal{P}$ of M with finite entropy for μ, let $\mathcal{P}_X$ be its restriction to X and $\mathcal{Q}_Y = \phi(\mathcal{P}_X)$ be the image of $\mathcal{P}_X$ under ϕ. Then $\mathcal{Q} = \mathcal{Q}_Y \cup \{Y^c\}$ is a partition of N and, since X and Y are full measure subsets,

$$H_\nu(\mathcal{Q}) = \sum_{Q\in\mathcal{Q}_Y} -\nu(Q)\log\nu(Q) = \sum_{P\in\mathcal{P}_X} -\mu(P)\log\mu(P) = H_\mu(\mathcal{P}).$$

Since X and Y are both invariant, $\mathcal{P}_X^n$ is the restriction of $\mathcal{P}^n$ to the subset X and $\mathcal{Q}^n = \mathcal{Q}_Y^n \cup \{Y^c\}$ for every n. Moreover,

$$\mathcal{Q}_Y^n = \bigvee_{j=0}^{n-1} g^{-j}(\mathcal{Q}_Y) = \phi\Big(\bigvee_{j=0}^{n-1} f^{-j}(\mathcal{P}_X)\Big) = \phi(\mathcal{P}_X^n)$$

for every n. Thus, the previous argument proves that $H_\nu(\mathcal{Q}^n) = H_\mu(\mathcal{P}^n)$ for every n and so

$$h_\nu(g,\mathcal{Q}) = \lim_n \frac{1}{n} H_\nu(\mathcal{Q}^n) = \lim_n \frac{1}{n} H_\mu(\mathcal{P}^n) = h_\mu(f,\mathcal{P}).$$

Taking the supremum over $\mathcal{P}$, we conclude that $h_\nu(g) \ge h_\mu(f)$. The converse inequality is entirely analogous.

Using this observation, Kolmogorov and Sinai concluded that not all two-sided Bernoulli shifts are ergodically equivalent despite their being spectrally equivalent, as we saw in Corollary 8.4.12. This also shows that spectral equivalence is strictly weaker than ergodic equivalence. In fact, as observed in Exercise 9.2.2, for every $s > 0$ there exists some two-sided Bernoulli shift (σ,μ) such that $h_\mu(\sigma) = s$. Therefore, a sole class of spectral equivalence may contain a whole continuum of ergodic equivalence classes.

9.5.1 Bernoulli automorphisms

The converse to Proposition 9.5.1 is false, in general. Indeed, we saw in Example 9.2.9 (and Corollary 9.2.7) that all the circle rotations have entropy zero. But an irrational rotation is never ergodically equivalent to a rational rotation, since the former is ergodic and the latter is not. Besides, Corollary 8.3.6 shows that irrational rotations are also not ergodically equivalent to each other, in general. The case of rational rotations is treated in Exercise 8.3.3.

However, a remarkable result due to Donald Ornstein [Orn70] states that the entropy is a complete invariant for two-sided Bernoulli shifts:

Theorem 9.5.2 (Ornstein). *Two-sided Bernoulli shifts in Lebesgue spaces are ergodically equivalent if and only if they have the same entropy.*

We call *Bernoulli automorphism* any system that is ergodically equivalent to a two-sided Bernoulli shift. In the sequel we find several examples of such systems. The theorem of Ornstein may be reformulated as follows: two Bernoulli automorphisms in Lebesgue spaces are ergodically equivalent if and only if they have the same entropy.

Let us point out that the theorem of Ornstein does not extend to one-sided Bernoulli shifts. Indeed, Exercise 8.1.2 shows that in the non-invertible case there are other invariants of ergodic equivalence, including the degree of the transformation (the number of pre-images).

William Parry and Peter Walters [PW72b, PW72a, Wal73] proved, among other results, that one-sided Bernoulli shifts corresponding to probability vectors $p = (p_1, \dots, p_k)$ and $q = (q_1, \dots, q_l)$ are ergodically equivalent if and only if $k = l$ and the vector p is a permutation of the vector q. In Exercise 9.7.7, after introducing the notion of the Jacobian, we invite the reader to prove this fact.

9.5.2 Systems with entropy zero

In this section we study some properties of systems whose entropy is equal to zero. The main result (Proposition 9.5.5) is that such systems are invertible at almost every point, if the ambient space is a Lebesgue space. It is worthwhile comparing this statement with Corollary 9.2.7. At the end of the section we briefly discuss the spectral types of systems with entropy zero.

In what follows, $(M, \mathcal{B}, \mu)$ is a probability space and $f : M \to M$ is a measure-preserving transformation. In the second half of the section we take $(M \mathcal{B}, \mu)$ to be a Lebesgue space.

Lemma 9.5.3. *For every $\varepsilon > 0$ there exists $\delta > 0$ such that if $\mathcal{P}$ and $\mathcal{Q}$ are partitions with finite entropy and $H_\mu(\mathcal{P}/\mathcal{Q}) < \delta$ then for every $P \in \mathcal{P}$ there exists a union P' of elements of $\mathcal{Q}$ satisfying $\mu(P \Delta P') < \varepsilon$.*

Proof. Let $s = 1 - \varepsilon/2$ and $\delta = -(\varepsilon/2)\log s$. For each $P \in \mathcal{P}$ consider

$$\mathcal{S} = \{Q \in \mathcal{Q} : \mu(P \cap Q) \geq s\mu(Q)\}.$$

Let P' be the union of all the elements of $\mathcal{S}$. On the one hand,

$$\mu(P' \setminus P) = \sum_{Q \in \mathcal{S}} \mu(Q \setminus P) \leq \sum_{Q \in \mathcal{S}} (1 - s)\mu(Q) \leq \frac{\varepsilon}{2}. \tag{9.5.1}$$

On the other hand,

$$\begin{aligned} H_\mu(\mathcal{P}/\mathcal{Q}) &= \sum_{R \in \mathcal{P}} \sum_{Q \in \mathcal{Q}} -\mu(R \cap Q) \log \frac{\mu(R \cap Q)}{\mu(Q)} \\ &\geq \sum_{Q \notin \mathcal{S}} -\mu(P \cap Q) \log s = -\mu(P \setminus P') \log s. \end{aligned}$$

This implies that

$$\mu(P \setminus P') \le \frac{H_\mu(\mathcal{P}/\mathcal{Q})}{-\log s} < \frac{\delta}{-\log s} = \frac{\varepsilon}{2}. \tag{9.5.2}$$

Putting (9.5.1) and (9.5.2) together, we get the conclusion of the lemma.

The next lemma means that the rate $h_\mu(f,\mathcal{P})$ of information (relative to the partition $\mathcal{P}$) generated by the system at each iteration is zero if and only if *the future determines the present,* in the sense that information relative to the 0-th iterate may be deduced from the ensemble of information relative to the future iterates.

Lemma 9.5.4. *Let $\mathcal{P}$ be a partition with finite entropy. Then, $h_\mu(f,\mathcal{P}) = 0$ if and only if $\mathcal{P} \prec \bigvee_{j=1}^\infty f^{-j}(\mathcal{P})$.*

Proof. Suppose that $h_\mu(f,\mathcal{P})$ is zero. Using Lemma 9.1.12, we obtain that $H_\mu(\mathcal{P}/\bigvee_{j=1}^n f^{-j}(\mathcal{P}))$ converges to zero when $n \to \infty$. Then, by Lemma 9.5.3, for each $l \ge 1$ and each $P \in \mathcal{P}$ there exist $n_l \ge 1$ and a union P_l of elements of the partition $\bigvee_{j=1}^{n_l} f^{-j}(\mathcal{P})$ such that $\mu(P\Delta P_l) < 2^{-l}$. It is clear that every P_l is a union of elements of $\bigvee_{j=1}^\infty f^{-j}(\mathcal{P})$ and, thus, the same is true for every $\bigcup_{l=n}^\infty P_l$ and also for $P_* = \bigcap_{n=1}^\infty \bigcup_{l=n}^\infty P_l$. Moreover,

$$\mu\Big(P \setminus \bigcup_{l=n}^\infty P_l\Big) = 0 \quad \text{and} \quad \mu\Big(\bigcup_{l=n}^\infty P_l \setminus P\Big) \le 2^{-n}$$

for every n and, consequently, $\mu(P\Delta P_*) = 0$. This shows that every element of $\mathcal{P}$ coincides, up to measure zero, with a union of elements of $\bigvee_{j=1}^\infty f^{-j}(\mathcal{P})$, as claimed in the "only if" half of the statement.

The argument to prove the converse is similar to the one in Proposition 9.2.2. Suppose that $\mathcal{P} \prec \bigvee_{j=1}^\infty f^{-j}(\mathcal{P})$. Write $\mathcal{P} = \{P_j : j = 1,2,\dots\}$ and, for each $k \ge 1$, consider the finite partition $\mathcal{P}_k = \{P_1,\dots,P_k, M \setminus \bigcup_{j=1}^k P_j\}$. Given any $\varepsilon > 0$, Lemma 9.2.3 ensures that $H_\mu(\mathcal{P}/\mathcal{P}_k) < \varepsilon/2$ for every k sufficiently large. Fix k in these conditions and write $P_0 = M \setminus \bigcup_{j=1}^k P_j$. For each $n \ge 1$ and each $j = 1,\dots,k$, let Q_i^n be the union of the elements of $\bigvee_{j=1}^n f^{-j}(\mathcal{P})$ that intersect P_i. The hypothesis ensures that each $(Q_i^n)_n$ is a decreasing sequence whose intersection coincides with P_i up to measure zero. Then, given $\delta > 0$ there exists n_0 such that

$$\sum_{i=1}^k \mu(Q_i^n \setminus P_i) < \delta \quad \text{for every } n \ge n_0. \tag{9.5.3}$$

Define $R_i^n = Q_i^n \setminus \bigcup_{j=1}^{i-1} Q_j^n$ for $i = 1,\dots,k$ and also $R_0^n = M \setminus \bigcup_{j=1}^k Q_j^n$. It is clear from the construction that $\mathcal{R}_n = \{R_1^n,\dots,R_k^n,R_0^n\}$ is a partition of M coarser than $\bigvee_{j=1}^n f^{-j}(\mathcal{P})$. Since $R_i^n \subset Q_i^n$ and $P_i \subset Q_i^n$, and the elements of $\mathcal{P}$ are pairwise disjoint,

$$R_i^n \setminus P_i \subset Q_i^n \setminus P_i \quad \text{and} \quad P_i \setminus R_i^n = P_i \cap \Big(\bigcup_{j=1}^{i-1} Q_j^n\Big) \subset \bigcup_{j=1}^{i-1} (Q_j^n \setminus P_j)$$

for $i = 1, \dots, k$. Similarly, $R_0^n \subset P_0$ and $P_0 \setminus R_0^n = P_0 \cap \bigcup_{j=1}^k Q_i^n = \bigcup_{j=1}^k (Q_j^n \setminus P_j)$. Therefore, the relation (9.5.3) implies that

$$\mu(P_i \Delta R_i^n) < \delta \quad \text{for every } i = 0, 1, \dots, k \text{ and every } n \geq n_0.$$

Then, assuming that $\delta > 0$ is small, it follows from Lemmas 9.1.5 and 9.1.6 that

$$H_\mu\left(\mathcal{P}_k / \bigvee_{j=1}^n f^{-j}(\mathcal{P})\right) \leq H_\mu(\mathcal{P}_k / \mathcal{R}_n) < \varepsilon/2$$

for every $n \geq n_0$. Using Exercise 9.1.1, we get that $H_\mu\big(\mathcal{P} / \bigvee_{j=1}^n f^{-j}(\mathcal{P})\big) < \varepsilon$ for every $n \geq n_0$. In this way, it is shown that $H_\mu\big(\mathcal{P} / \bigvee_{j=1}^n f^{-j}(\mathcal{P})\big) \to 0$. By Lemma 9.1.12, it follows that $h_\mu(f, \mathcal{P}) = 0$.

As a consequence, we get that every system with entropy zero is invertible at almost every point:

Proposition 9.5.5. *Let $(M, \mathcal{B}, \mu)$ be a Lebesgue space and $f : M \to M$ be a measure-preserving transformation. If $h_\mu(f) = 0$ then (f, μ) is invertible: there exists a measurable transformation $g : M \to M$ that preserves the measure μ and satisfies $f \circ g = g \circ f = \mathrm{id}$ at μ-almost every point.*

Proof. Consider the homomorphism $\tilde{f} : \tilde{\mathcal{B}} \to \tilde{\mathcal{B}}$ induced by f in the measure algebra of $\mathcal{B}$ (these notions were introduced in Section 8.5). Recall that $\tilde{f}$ is always injective (Exercise 8.5.2). Given any $B \in \mathcal{B}$, consider the partition $\mathcal{P} = \{B, B^c\}$. The hypothesis $h_\mu(f) = 0$ implies that $h_\mu(f, \mathcal{P}) = 0$. By Lemma 9.5.4, it follows that $\mathcal{P} \prec \bigvee_{j=1}^\infty f^{-j}(\mathcal{P})$. This implies that $\mathcal{P} \subset f^{-1}(\mathcal{B})$, because $f^{-j}(\mathcal{P}) \subset f^{-1}(\mathcal{B})$ for every $j \geq 1$. Varying B, we conclude that $\mathcal{B} \subset f^{-1}(\mathcal{B})$. In other words, the homomorphism $\tilde{f}$ is surjective. Hence, $\tilde{f}$ is an isomorphism of measure algebras. Then, by Proposition 8.5.6, there exists a measurable map $g : M \to M$ preserving the measure μ and such that the corresponding homomorphism of measure algebra $\tilde{g} : \tilde{\mathcal{B}} \to \tilde{\mathcal{B}}$ is the inverse of $\tilde{f}$. In other words, $\tilde{f} \circ \tilde{g} = \tilde{g} \circ \tilde{f} = \mathrm{id}$. Then, (Exercise 8.5.2) $f \circ g = g \circ f = \mathrm{id}$, as claimed.

These arguments also prove the following fact, which will be useful in a while:

Corollary 9.5.6. *In the conditions of Proposition 9.5.5, every σ-algebra $\mathcal{A} \subset \mathcal{B}$ that satisfies $f^{-1}(\mathcal{A}) \subset \mathcal{A}$ up to measure zero also satisfies $f^{-1}(\mathcal{A}) = \mathcal{A}$ up to measure zero.*

Exercise 9.1.5 implies that if (f, μ) has entropy zero then the same is true for any factor. Therefore, the following fact is also an immediate consequence of the proposition:

Corollary 9.5.7. *In the conditions of Proposition 9.5.5, every factor of (f, μ) is invertible at almost every point.*

It is not completely understood how entropy relates to the spectrum type of a system, but there are several partial results, especially for systems with entropy zero.

Rokhlin [Rok67a, § 14] proved that every ergodic system with discrete spectrum defined in a Lebesgue space has entropy zero. This may also be deduced from the fact that, as we mentioned in Section 8.3, every ergodic system with discrete spectrum is ergodically isomorphic to a translation in a compact abelian group. As we saw in Corollary 8.5.7, in Lebesgue spaces ergodic isomorphism implies ergodic equivalence. Recall also that systems with discrete spectrum in Lebesgue spaces are always invertible (Exercise 8.5.5).

In that same work of Rokhlin it is shown that invertible systems with singular spectrum defined in Lebesgue spaces have entropy zero and the same holds for systems with Lebesgue spectrum of finite rank (if they exist). The case of infinite rank is the focus of the next section. We are going to see that there are systems with Lebesgue spectrum of infinite rank and entropy zero. On the other hand, we introduce the important class of so-called Kolmogorov systems, for which the entropy is necessarily positive, in a strong sense.

9.5.3 Kolmogorov systems

Let $(M,\mathcal{B},\mu)$ be a non-trivial probability space, that is, one such that not all measurable sets have measure 0 or 1. We use $\bigvee_\alpha \mathcal{U}_\alpha$ to denote the σ-algebra generated by any family of subsets $\mathcal{U}_\alpha$ of $\mathcal{B}$. Let $f : M \to M$ be a transformation preserving the measure μ.

Definition 9.5.8. We say that (f,μ) is a *Kolmogorov system* if there exists some σ-algebra $\mathcal{A} \subset \mathcal{B}$ such that

(i) $f^{-1}(\mathcal{A}) \subset \mathcal{A}$ up to measure zero;
(ii) $\bigcap_{n=0}^{\infty} f^{-n}(\mathcal{A}) = \{\emptyset, M\}$ up to measure zero;
(iii) $\bigvee_{n=0}^{\infty}\{B \in \mathcal{B} : f^{-n}(B) \in \mathcal{A}\} = \mathcal{B}$ up to measure zero.

We leave it to the reader to check that this property is an invariant of ergodic equivalence (it is *not* an invariant of spectral equivalence, as will be explained shortly).

If (f,μ) is a Kolmogorov system then (f^k,μ) is a Kolmogorov system, for every $k \geq 1$. Indeed, if $\mathcal{A}$ satisfies condition (i) then the sequence $f^{-j}(\mathcal{A})$ is non-increasing and, in particular, $f^{-k}(\mathcal{A}) \subset \mathcal{A}$. Then, the conditions (ii) and (iii) imply that

$$\bigcap_{n=0}^{\infty} f^{-kn}(\mathcal{A}) = \bigcap_{n=0}^{\infty} f^{-n}(\mathcal{A}) = \{\emptyset, M\} \quad \text{and}$$

$$\bigvee_{n=0}^{\infty}\{B\in\mathcal{B}:f^{-kn}(B)\in\mathcal{A}\}=\bigvee_{n=0}^{\infty}\{B\in\mathcal{B}:f^{-n}(B)\in\mathcal{A}\}=\mathcal{B}$$

up to measure zero. We say that (f,μ) is a *Kolmogorov automorphism* if it is invertible and a Kolmogorov system. Then the inverse (f^{-1},μ) is also a Kolmogorov system, as we will see.

Proposition 9.5.9. *Every Kolmogorov system has Lebesgue spectrum of infinite rank. If the σ-algebra $\mathcal{B}$ is countably generated then the rank is countable.*

Proof. Let $\mathcal{A}\subset\mathcal{B}$ be a σ-algebra satisfying the conditions in Definition 9.5.8. Let $E=L_0^2(M,\mathcal{A},\mu)$ be the subspace of functions $\varphi\in L_0^2(M,\mathcal{B},\mu)$ that are $\mathcal{A}$-measurable, that is, such that the pre-image $\varphi^{-1}(B)$ of every measurable set $B\subset\mathbb{R}$ is in $\mathcal{A}$ up to measure zero. We are going to show that E satisfies all the conditions in Definition 8.4.1.

Start by observing that $U_f(L_0^2(M,\mathcal{A},\mu))=L_0^2(M,f^{-1}(\mathcal{A}),\mu)$. Indeed, it is clear that if φ is $\mathcal{A}$-measurable then $U_f\varphi=\varphi\circ f$ is $f^{-1}(\mathcal{A})$-measurable. The inclusion $\subset$ follows immediately. Conversely, given any $B\in f^{-1}(\mathcal{A})$, take $A\in\mathcal{A}$ such that $B=f^{-1}(A)$ and let $c=\mu(A)=\mu(B)$. Then, $\mathcal{X}_B-c=U_f(\mathcal{X}_A-c)$ is in $U_f(L_0^2(M,\mathcal{A},\mu))$. This gives the other inclusion. So, the hypothesis that $f^{-1}(\mathcal{A})\subset\mathcal{A}$ up to measure zero ensures that $U_f(E)\subset E$.

It also follows that $U_f^n(L_0^2(M,\mathcal{A},\mu))=L_0^2(M,f^{-n}(\mathcal{A}),\mu)$ for every $n\geq 0$. Then,

$$\bigcap_{n=0}^{\infty}U_f^n(L_0^2(M,\mathcal{A},\mu))=L_0^2\Big(M,\bigcap_{n=0}^{\infty}f^{-n}(\mathcal{A}),\mu\Big).$$

Hence, the hypothesis that $\bigcap_{n=0}^{\infty}f^{-n}(\mathcal{A})=\{0,M\}$ up to measure zero implies that $\bigcap_{n=0}^{\infty}U_f^n(E)=\{0\}$.

Now consider $\mathcal{A}_n=\{B\in\mathcal{B}:f^{-n}(B)\in\mathcal{A})\}$. The sequence $(\mathcal{A}_n)_n$ is non-decreasing, because $f^{-1}(\mathcal{A})\subset\mathcal{A}$. Moreover, each φ is $\mathcal{A}_n$-measurable if and only if $U_f^n\varphi=\varphi\circ f^n$ is $\mathcal{A}$-measurable. This shows that $U_f^{-n}(L_0^2(M,\mathcal{A},\mu))=L_0^2(M,\mathcal{A}_n,\mu)$ for every $n\geq 0$. Observe also that

$$\sum_{n=0}^{\infty}L_0^2(M,\mathcal{A}_n,\mu)=L_0^2\Big(M,\bigvee_{n=0}^{\infty}\mathcal{A}_n,\mu\Big).\tag{9.5.4}$$

Indeed, it is clear that $L_0^2(M,\mathcal{A}_k,\mu)\subset L_0^2(M,\bigvee_{n=0}^{\infty}\mathcal{A}_n,\mu)$ for every k, since $\mathcal{A}_k$ is contained in $\bigvee_{n=0}^{\infty}\mathcal{A}_n$. The inclusion $\subset$ in (9.5.4) is an immediate consequence of this observation, since $L_0^2(M,\bigvee_{n=0}^{\infty}\mathcal{A}_n,\mu)$ is a Banach space. Now consider any $A\in\bigvee_{n=0}^{\infty}\mathcal{A}_n$. The approximation theorem (Theorem A.1.19) implies that for each $\varepsilon>0$ there exist n and $A_n\in\mathcal{A}_n$ such that $\mu(A\Delta A_n)<\varepsilon$. Then $(\mathcal{X}_{A_n})_n$ converges to $\mathcal{X}_A$ in the L^2-norm, and so $\mathcal{X}_A\in\sum_{n=0}^{\infty}L_0^2(M,\mathcal{A}_n,\mu)$.

This proves the inclusion $\supset$ in (9.5.4). In this way, we have shown that

$$\sum_{n=0}^{\infty} U_f^{-n}(L_0^2(M,\mathcal{A},\mu)) = L_0^2\Big(M, \bigvee_{n=0}^{\infty} \mathcal{A}_n, \mu\Big).$$

Therefore, the hypothesis that $\bigvee_{n=0}^{\infty} \mathcal{A}_n = \mathcal{B}$ up to measure zero implies that $\sum_{n=0}^{\infty} U_f^{-n}(E) = L_0^2(M,\mathcal{B},\mu)$.

This concludes the proof that (f,μ) has Lebesgue spectrum. To prove that the rank is infinite, we need the following lemma:

Lemma 9.5.10. *Let $\mathcal{A}$ be any σ-algebra satisfying the conditions in Definition 9.5.8. Then for every $A \in \mathcal{A}$ with $\mu(A) > 0$ there exists $B \subset A$ such that $0 < \mu(B) < \mu(A)$.*

Proof. Suppose that $\mathcal{A}$ has any element A with positive measure that does not satisfy the conclusion of the lemma. We claim that $A \cap f^{-k}(A)$ has measure zero for every $k \geq 1$. Then,

$$\mu\Big(f^{-i}(A) \cap f^{-j}(A)\Big) = \mu\Big(A \cap f^{-j+i}(A)\Big) = 0 \quad \text{for every } 0 \leq i < j.$$

Since $\mu(f^{-j}(A)) = \mu(A)$ for every $j \geq 0$, this implies that the measure μ is infinite, which is a contradiction. This contradiction reduces the proof of the lemma to proving our claim.

To do that, note that condition (i) implies that $f^{-k}(A) \in f^{-k}(\mathcal{A}) \subset \mathcal{A}$. Then it follows from the choice of A that $A \cap f^{-k}(A)$ must have either zero measure or full measure in A:

$$\mu\Big(A \cap f^{-k}(A)\Big) = 0 \quad \text{or else} \quad \mu(A \setminus f^{-k}(A)) = 0.$$

So, to prove the claim it suffices to exclude the second possibility. Suppose that $\mu(A \setminus f^{-k}(A)) = 0$. Then (Exercise 1.1.4), there exists $B \in \mathcal{A}$ such that $\mu(A \Delta B) = 0$ and $f^{-k}(B) = B$. It follows that $B = f^{-nk}(B)$ for every $n \geq 1$ and, thus,

$$B \in \bigcap_{n \in \mathbb{N}} f^{-nk}(\mathcal{A}) = \bigcap_{n \in \mathbb{N}} f^{-n}(\mathcal{A}).$$

By condition (ii), this means that the measure of B is either 0 or 1. Since $\mu(B) = \mu(A)$ is positive, it follows that $\mu(A) = \mu(B) = 1$. Then, the hypothesis about A implies that the σ-algebra $\mathcal{A}$ contains only sets with measure 0 or 1. By condition (iii), it follows that the same is true for the σ-algebra $\mathcal{B}$, which contradicts the assumption that the probability space is non-trivial.

On the way toward proving that the orthogonal complement $F = E \ominus U_f(E)$ has infinite dimension, let us start by checking that $F \neq \{0\}$. Indeed, otherwise we would have $U_f(E) = E$ and, thus, $U_f^n(E) = E$ for every $n \geq 1$. By condition (ii), that would imply that $E = \bigcap_n U_f^n(E) = \{0\}$. Then, by condition (iii), we would have $L_0^2(M,\mathcal{B},\mu) = \{0\}$ and that would contradict the hypothesis that the probability space is non-trivial.

Let φ be any non-zero element of F, fixed once and for all, and let N be the set of all $x \in M$ such that $\varphi(x) \neq 0$. Then, $N \in \mathcal{A}$ and $\mu(N) > 0$. It is convenient to consider the space $E' = L^2(M, \mathcal{A}, \mu) = E \oplus \{\text{constants}\}$. Observe that F coincides with $E' \ominus U_f(E')$, because the Koopman operator preserves the line of constant functions. Let E'_N be the subspace of functions $\psi \in E'$ that vanish outside N, that is, such that $\psi(x) = 0$ for every $x \in N^c$. By Lemma 9.5.10, we may find sets $A_j \in \mathcal{A}$, $j \geq 1$ with positive measure, contained in N and pairwise disjoint. Then, $\mathcal{X}_{A_j}$ is in E'_N for every j. Moreover, $A_i \cap A_j = \emptyset$ yields $\mathcal{X}_{A_i} \cdot \mathcal{X}_{A_j} = 0$ for every $i \neq j$. This implies that E'_N has infinite dimension.

Now denote by $U_f(E')_N$ the subspace of functions $\psi \in U_f(E')$ that vanish outside N. Denote $F_N = E'_N \ominus U_f(E')_N$. The fact that $\dim E'_N = \infty$ ensures that $\dim F_N = \infty$ or $\dim U_f(E')_N = \infty$ (or both). We are going to show that any of these alternatives implies that $\dim F = \infty$.

To treat the first alternative, it suffices to show that $F_N \subset F$. For that, since it is clear that $F_N \subset E'$, it suffices to check that F_N is orthogonal to $U_f(E')$. Consider any $\xi \in F_N$ and $\eta \in E'$. The function $(U_f\eta)\mathcal{X}_N = U_f(\eta\mathcal{X}_{f^{-1}(N)})$ is in $U_f(E')$ and vanishes outside N. In other words, it is in $U_f(E')_N$. Then $\xi \cdot U_f\eta = \xi \cdot (U_f\eta)\mathcal{X}_N = 0$, because ξ vanishes outside N and is orthogonal to $U_f(E')_N$. This completes the argument in this case.

Now we treat the second alternative. Given any $U_f\eta \in U_f(E')_N$ and any $n \in \mathbb{N}$, let $\eta_n = \eta\mathcal{X}_{R_n}$ with $R_n = \{x \in M : |\eta(x)| \leq n\}$. Then $(\eta_n)_n$ is a sequence of bounded functions converging to η in E'. Moreover, every η_n vanishes outside $f^{-1}(N)$, because η does. Then, $(U_f\eta_n)_n$ is a sequence of bounded functions that vanish outside N and, recalling that U_f is an isometry, this sequence converges to $U_f\eta$ in E'. This proves that the subspace of bounded functions is dense in $U_f(E')_N$. Then, since we are assuming that $\dim U_f(E')_N = \infty$, this subspace must also have infinite dimension. Choose $\{\xi_k : k \geq 1\} \subset E'$ such that $\{U_f\xi_k : k \geq 1\}$ is a linearly independent subset of $U_f(E')_N$ consisting of bounded functions. Observe that the products $\varphi(U_f\xi_k)$, $k \geq 1$ form a linearly independent subset of E'. Moreover, given any $\eta \in E'$,

$$\varphi(U_f\xi_k) \cdot (U_f\eta) = \int \varphi\,(\xi_k \circ f)\,(\bar{\eta} \circ f)\,d\mu = \int \varphi\,(\xi_k\bar{\eta}) \circ f\,d\mu = \varphi \cdot U_f(\bar{\xi}_k\eta).$$

This last expression is equal to zero because $\bar{\xi}_k\eta \in E'$ and the function $\varphi \in F$ is orthogonal to $U_f(E')$. Varying $\eta \in E'$, we conclude that $\varphi(U_f\xi_k)$ is orthogonal to $U_f(E')$ for every k. This shows that $\{\varphi(U_f\xi_k) : k \geq 1\}$ is contained in F and, thus, $\dim F = \infty$ also in this case.

This completes the proof that (f, μ) has infinite rank. When $\mathcal{B}$ is countably generated, $L^2_0(M, \mathcal{B}, \mu)$ is separable (Example 8.4.7) and so the rank is necessarily countable.

We say that a partition of $(M, \mathcal{B}, \mu)$ is trivial if all its elements have measure 0 or 1. Keep in mind that in the present chapter all partitions are assumed to be countable.

Proposition 9.5.11. *A system (f, μ) in a Lebesgue space is a Kolmogorov system if and only if $h_\mu(f, \mathcal{P}) > 0$ for every non-trivial partition with finite entropy. In particular, every Kolmogorov system has positive entropy.*

This result is due to Pinsker [Pin60] and to Rokhlin and Sinai [RS61]. The proof may also be found in Rokhlin [Rok67a, §13]. Let us point out, however, that the last part of the statement is an immediate consequence of the ideas in Section 9.5.2. Indeed, suppose that (f, μ) is a Kolmogorov system with zero entropy. By Corollary 9.5.6, any σ-algebra $\mathcal{A}$ that satisfies condition (i) in Definition 9.5.8 also satisfies $f^{-1}(\mathcal{A}) = \mathcal{A}$ up to measure zero. Then, condition (ii) implies that $\mathcal{A}$ is trivial and, by condition (iii), the σ-algebra $\mathcal{B}$ itself is trivial (contradicting the assumption we made at the beginning of this section).

It follows from Proposition 9.5.11 and the relation (9.1.21) that the inverse of a Kolmogorov automorphism is also a Kolmogorov automorphism. Unlike what happens for Bernoulli automorphisms (Exercise 9.5.1), in the Kolmogorov case the two systems (f, μ) and (f^{-1}, μ) need not be ergodically equivalent.

Example 9.5.12. The first example of an invertible system with countable Lebesgue spectrum that is not a Kolmogorov system was found by Girsanov in 1959, but was never published. Another example, a factor of a certain Gaussian shift (recall Example 8.4.13) with countable Lebesgue spectrum but whose entropy vanishes, was exhibited a few years later by Newton and Parry [NP66]. Also, Gurevič [Gur61] proved that the horocyclic flow on surfaces with constant negative curvature has entropy zero; sometime before, Parasyuk [Par53] had shown that such flows have countable Lebesgue spectrum.

As we saw in Theorem 8.4.11, all systems with countable Lebesgue spectrum are spectrally equivalent. Therefore, the fact that systems as in Example 9.5.12 do exist has the interesting consequence that being a Kolmogorov system is *not* an invariant of spectral equivalence.

Example 9.5.13. We saw in Examples 8.4.2 and 8.4.3 that all the Bernoulli shifts have Lebesgue spectrum. In both cases, one-sided and two-sided, we exhibited subspaces of $L_0^2(M, \mathcal{B}, \mu)$ of the form $E = L_0^2(M, \mathcal{A}, \mu)$ for some σ-algebra $\mathcal{A} \subset \mathcal{B}$. Therefore, the same argument proves that every Bernoulli shift is a Kolmogorov system. In particular, every Bernoulli automorphism is a Kolmogorov automorphism.

There exist Kolmogorov automorphisms that are not Bernoulli automorphisms. The first example, found by Ornstein, is quite elaborate. The following simple construction is due to Kalikow [Kal82]:

Example 9.5.14. Let $\sigma : \Sigma \to \Sigma$ be the shift map in $\Sigma = \{1, 2\}^{\mathbb{Z}}$ and μ be the Bernoulli measure associated with the probability vector $p = (1/2, 1/2)$.

Consider the map $f : \Sigma \times \Sigma \to \Sigma \times \Sigma$ defined as follows:

$$f\big((x_n)_n, (y_n)_n\big) = \big(\sigma((x_n)_n), \sigma^{\pm 1}((y_n)_n)\big)$$

where the sign is $-$ if $x_0 = 1$ and is $+$ if $x_0 = 2$. This map preserves the product measure $\mu \times \mu$. Kalikow showed that (f, μ) is a Kolmogorov automorphism but not a Bernoulli automorphism.

Consider any Kolmogorov automorphism that is not a Bernoulli automorphism and let $s > 0$ be its entropy. Consider any Bernoulli automorphism whose entropy is equal to s (see Exercise 9.2.2). The two systems have the same entropy but they cannot be ergodically equivalent, since being a Bernoulli automorphism is an invariant of ergodic equivalence. Therefore, the entropy is not a complete invariant of ergodic equivalence for Kolmogorov automorphisms. Actually (see Ornstein and Shields [OS73]), there exists a non-countable family of Kolmogorov automorphisms that have the same entropy and, yet, are not ergodically equivalent.

The properties of Bernoulli automorphisms described in Exercise 9.5.1 do not extend to the Kolmogorov case: there exist Kolmogorov automorphisms that are not ergodically equivalent to their inverses (see Ornstein and Shields [OS73]), and there are also Kolmogorov automorphisms that admit no k-th root for any value of $k \geq 1$ (Clark [Cla72]).

Closing this section, let us discuss the Kolmogorov property for two specific classes of systems: Markov shifts and automorphisms of compact groups.

Concerning the first class, Friedman and Ornstein [FO70] proved that every two-sided mixing Markov shift is a Bernoulli automorphism. Recall (Theorem 7.2.11) that a Markov shift is mixing if and only if the corresponding stochastic matrix is aperiodic. It follows from the theorem of Friedman and Ornstein that the entropy is still a complete invariant of ergodic equivalence in the context of two-sided mixing Markov shifts. Another interesting consequence is that every two-sided mixing Markov shift is a Kolmogorov automorphism. Observe, however, that this consequence admits a relatively easy direct proof (see Exercise 9.5.4).

As for the second class, every ergodic automorphism of a compact group is a Kolmogorov automorphism. This was proven by Rokhlin [Rok67b] for abelian groups and by Yuzvinskii [Yuz68] in the general case. In fact, ergodic automorphisms of metrizable compact groups are Bernoulli automorphisms (Lind [Lin77], Miles and Thomas [MT78]). In particular, every ergodic linear automorphism of the torus $\mathbb{T}^d$ is a Bernoulli automorphism; this had been proved by Katznelson [Kat71]. Recall (Theorem 4.2.14) that a linear automorphism f_A is ergodic if and only if no eigenvalue of the matrix A is a root of unity.

9.5.4 Exact systems

We say that a Kolmogorov system is *exact* if one may take the σ-algebra $\mathcal{A}$ in Definition 9.5.8 to be the σ-algebra $\mathcal{B}$ of all measurable sets. Note that in this case the conditions (i) and (iii) are automatically satisfied. Therefore, a system (f,μ) is exact if and only if the σ-algebra $\mathcal{B}$ is such that $\bigcap_{n=0}^{\infty} f^{-n}(\mathcal{B})$ is trivial, meaning that it only contains sets with measure 0 or 1. Equivalently, (f,μ) is exact if and only if

$$\bigcap_{n=0}^{\infty} U_f^n\big(L_0^2(M,\mathcal{B},\mu)\big) = \{0\}.$$

This observation also implies that, unlike the Kolmogorov property, exactness is an invariant of spectral equivalence.

We saw in Example 8.4.2 that every one-sided Bernoulli shift has Lebesgue spectrum. In order to do that, we considered the subspace $E = L_0^2(M,\mathcal{B},\mu)$. Therefore, the same argument proves that every one-sided Bernoulli shift is an exact system. A much larger class of examples, expanding maps endowed with their equilibrium states, is studied in Chapter 12.

It is immediate that invertible systems are never exact: in the invertible case $f^{-n}(\mathcal{B}) = \mathcal{B}$ up to measure zero, for every n; therefore, the exactness condition corresponds to saying that the σ-algebra $\mathcal{B}$ is trivial (which is excluded, by hypothesis).

Figure 9.2 summarizes the relations between the different classes of systems studied in this book. It is organized in three columns: systems with zero entropy (which are necessarily invertible, as we saw in Proposition 9.5.5), invertible systems with positive entropy and non-invertible systems.

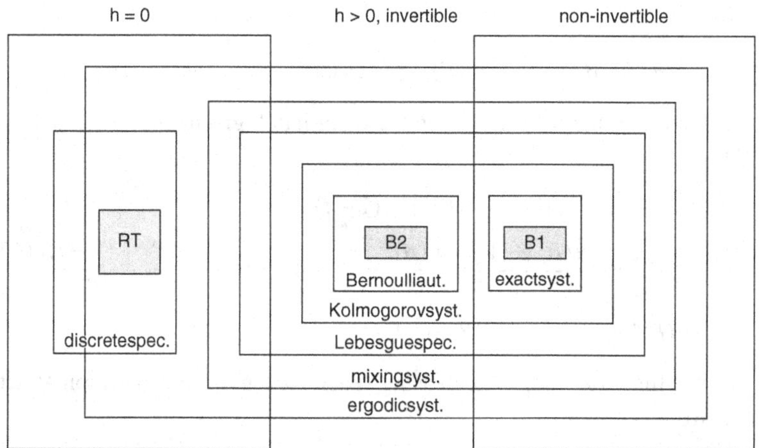

Figure 9.2. Relations between various of classes of systems (B1/B2 = one-sided/two-sided Bernoulli shifts, RT = irrational rotations on tori)

9.5.5 Exercises

9.5.1. Show that if (f,μ) is a Bernoulli automorphism then it is ergodically equivalent to its inverse (f^{-1},μ). Moreover, for every $k \geq 1$ there exists a Bernoulli automorphism (g,ν) that is a *k-th root* of (f,μ), that is, such that (g^k,ν) is ergodically equivalent to (f,μ). [Observation: Ornstein [Orn74] proved that, conversely, every k-th root of a Bernoulli automorphism is a Bernoulli automorphism.]

9.5.2. Use the notion of density point to show that the decimal expansion map $f(x) = 10x - [10x]$ is exact, relative to the Lebesgue measure.

9.5.3. Show that the Gauss map is exact, relative to its absolutely continuous invariant measure μ.

9.5.4. Show that the two-sided Markov shift associated with any aperiodic stochastic matrix P is a Kolmogorov automorphism.

9.5.5. Show that the one-sided Markov shift associated with any aperiodic stochastic matrix P is an exact system.

9.5.6. Prove that if (f,μ) is exact then $h_\mu(f,\mathcal{P}) > 0$ for every non-trivial partition $\mathcal{P}$ with finite entropy.

9.6 Entropy and ergodic decomposition

It is not difficult to show that the entropy $h_\mu(f)$ is always an *affine* function of the invariant measure μ:

Proposition 9.6.1. *Let μ and ν be probability measures invariant under a transformation $f: M \to M$. Then, $h_{t\mu+(1-t)\nu}(f) = th_\mu(f) + (1-t)h_\nu(f)$ for every $0 < t < 1$.*

Proof. Define $\phi(x) = -x\log x$ for $x > 0$. On the one hand, since the function ϕ is concave,

$$\phi(t\mu(B) + (1-t)\nu(B)) \geq t\phi(\mu(B)) + (1-t)\phi(\nu(B))$$

for every measurable set $B \subset M$. On the other hand, given any measurable set $B \subset M$,

$$\begin{aligned}
&\phi\big(t\mu(B) + (1-t)\nu(B)\big) - t\phi\big(\mu(B)\big) - (1-t)\phi\big(\nu(B)\big)\\
&\quad = -t\mu(B)\log\frac{t\mu(B) + (1-t)\nu(B)}{\mu(B)} - (1-t)\nu(B)\log\frac{t\mu(B) + (1-t)\nu(B)}{\nu(B)}\\
&\quad \leq -t\mu(B)\log t - (1-t)\nu(B)\log(1-t)
\end{aligned}$$

because the function $-\log$ is decreasing. Therefore, given any partition $\mathcal{P}$ with finite entropy,

$$\begin{aligned}
&H_{t\mu+(1-t)\nu}(\mathcal{P}) \geq tH_\mu(\mathcal{P}) + (1-t)H_\nu(\mathcal{P}) \quad \text{and}\\
&H_{t\mu+(1-t)\nu}(\mathcal{P}) \leq tH_\mu(\mathcal{P}) + (1-t)H_\nu(\mathcal{P}) - t\log t - (1-t)\log(1-t).
\end{aligned}$$

Consequently,

$$h_{t\mu+(1-t)\nu}(f,\mathcal{P}) = th_\mu(f,\mathcal{P}) + (1-t)h_\nu(f,\mathcal{P}). \qquad (9.6.1)$$

It follows, immediately, that $h_{t\mu+(1-t)\nu}(f) \le th_\mu(f) + (1-t)h_\nu(f)$. Moreover, the relations (9.1.16) and (9.6.1) imply that

$$h_{t\mu+(1-t)\nu}\Big(f,\mathcal{P}_1 \vee \mathcal{P}_2\Big) \ge th_\mu(f,\mathcal{P}_1) + (1-t)h_\nu(f,\mathcal{P}_2)$$

for any partitions $\mathcal{P}_1$ and $\mathcal{P}_2$. Taking the supremum on $\mathcal{P}_1$ and $\mathcal{P}_2$ we obtain that $h_{t\mu+(1-t)\nu}(f) \ge th_\mu(f) + (1-t)h_\nu(f)$.

In particular, given any invariant set $A \subset M$, we have that

$$h_\mu(f) = \mu(A)h_{\mu_A}(f) + \mu(A^c)h_{\mu_{A^c}}(f), \qquad (9.6.2)$$

where μ_A and μ_{A^c} denote the normalized restrictions of μ to the set A and its complement, respectively (this fact was obtained before, in Exercise 9.1.4). Another immediate consequence is the following version of Proposition 9.6.1 for finite convex combinations:

$$\mu = \sum_{i=1}^{n} t_i\mu_i \quad \Rightarrow \quad h_\mu(f) = \sum_{i=1}^{n} t_i h_{\mu_i}(f), \qquad (9.6.3)$$

for any invariant probability measures $\mu_1,\dots,\mu_n$ and any positive numbers $t_1,\dots,t_n$ with $\sum_{i=1}^n t_i = 1$.

A much deeper fact, due to Konrad Jacobs [Jac60, Jac63], is that the affinity property extends to the ergodic decomposition given by Theorem 5.1.3:

Theorem 9.6.2 (Jacobs). *Suppose that M is a complete separable metric space. Given any invariant probability measure μ, let $\{\mu_P : P \in \mathcal{P}\}$ be its ergodic decomposition. Then, $h_\mu(f) = \int h_{\mu_P}(f)\,d\hat{\mu}(P)$ (if one side is infinite, so is the other side).*

We are going to deduce this result from a general theorem about affine functionals in the space of probability measures, that we state in Section 9.6.1 and prove in Section 9.6.2.

9.6.1 Affine property

Let M be a complete separable metric space. We saw in Lemma 2.1.3 that the weak* topology in the space of probability measures $\mathcal{M}_1(M)$ is metrizable. Moreover (Exercise 2.1.3), the metric space $\mathcal{M}_1(M)$ is separable.

Let W be a probability measure on the Borel σ-algebra of $\mathcal{M}_1(M)$. The *barycenter* of W is the probability measure $\operatorname{bar}(W) \in \mathcal{M}_1(M)$ given by

$$\int \psi\, d\operatorname{bar}(W) = \int \Big(\int \psi\, d\eta \Big)\, dW(\eta) \qquad (9.6.4)$$

for every bounded measurable function $\psi : M \to \mathbb{R}$. We leave it to the reader to check that this relation determines $\operatorname{bar}(W)$ uniquely (Exercise 9.6.1) and that the barycenter is an affine function of the measure (Exercise 9.6.2).

Example 9.6.3. If W is a Dirac measure, that is, if $W = \delta_\nu$ for some $\nu \in \mathcal{M}_1(M)$, then $\operatorname{bar}(W) = \nu$. Using Exercise 9.6.2, we get the following generalization: if $W = \sum_{i=1}^\infty t_i \delta_{\nu_i}$ with $t_i \ge 0$ and $\sum_{i=1}^\infty t_i = 1$ and $\nu_i \in \mathcal{M}_1(M)$ for every i, then $\operatorname{bar}(W) = \sum_{i=1}^\infty t_i \nu_i$.

Example 9.6.4. Let $\{\mu_P : P \in \mathcal{P}\}$ be the ergodic decomposition of a probability measure μ invariant under a measurable transformation $f : M \to M$ and $\hat{\mu}$ be the associated quotient measure in $\mathcal{P}$ (recall Section 5.1.1). Let $W = \Xi_* \hat{\mu}$ be the image of the quotient measure $\hat{\mu}$ under the map $\Xi : \mathcal{P} \to \mathcal{M}$ that assigns to each $P \in \mathcal{P}$ the conditional probability μ_P. Then (Exercise 5.1.4),

$$\int \psi \, d\mu = \int \left(\int \psi \, d\mu_P \right) d\hat{\mu}(P) = \int \left(\int \psi \, d\eta \right) dW(\eta)$$

for every bounded measurable function $\psi : M \to \mathbb{R}$. This means that μ is the barycenter of W.

A set $\mathcal{M} \subset \mathcal{M}_1(M)$ is said to be *strongly convex* if $\sum_{i=1}^\infty t_i \nu_i \in \mathcal{M}$ for any $\nu_i \in \mathcal{M}$ and $t_i \ge 0$ with $\sum_{i=1}^\infty t_i = 1$.

Theorem 9.6.5. *Let $\mathcal{M}$ be a strongly convex subset of $\mathcal{M}_1(M)$ and $H : \mathcal{M} \to \mathbb{R}$ be a non-negative affine functional. If H is upper semi-continuous then*

$$H(\operatorname{bar}(W)) = \int H(\eta) \, dW(\eta)$$

for any probability measure W on $\mathcal{M}_1(M)$ with $W(\mathcal{M}) = 1$ and $\operatorname{bar}(W) \in \mathcal{M}$.

Before proving this result, let us explain how Theorem 9.6.2 may be obtained from it. The essential step is the following lemma:

Lemma 9.6.6. $h_\mu(f, \mathcal{Q}) = \int h_{\mu_P}(f, \mathcal{Q}) \, d\hat{\mu}(P)$ *for any finite partition $\mathcal{Q}$ of M.*

Proof. Let $\mathcal{M} = \mathcal{M}_1(f)$, the subspace of invariant probability measures, and $H : \mathcal{M} \to \mathbb{R}$ be the functional defined by $H(\eta) = h_\eta(f, \mathcal{Q})$. Let W be the image of the quotient measure $\hat{\mu}$ by the map $\Xi : \mathcal{P} \to \mathcal{M}$ that assigns to each $P \in \mathcal{P}$ the conditional probability μ_P. It is clear that $\mathcal{M}$ is strongly convex, $W(\mathcal{M}) = 1$ and (recall Example 9.6.4) the barycenter $\operatorname{bar}(W) = \mu$ is in $\mathcal{M}$. Proposition 9.6.1 shows that H is affine and it is clear that H is non-negative. In order to apply Theorem 9.6.5, we also need to check that H is upper semi-continuous.

Initially, suppose that f is the shift map in a space $\Sigma = X^{\mathbb{N}}$, where X is a finite set, and $\mathcal{Q}$ is the partition of Σ into cylinders $[0; a]$, $a \in X$. The point with this partition is that its elements are both open and closed subsets of Σ. In other words, $\partial Q = \emptyset$ for every $Q \in \mathcal{Q}$. By Proposition 9.2.12, it follows that the map $\eta \mapsto H(\eta) = h_\eta(f, \mathcal{Q})$ is upper semi-continuous at every point of $\mathcal{M}$. So, we may indeed apply Theorem 9.6.5 to the functional H. In this way we get that

$$h_\mu(f, \mathcal{Q}) = H(\mu) = H(\mathrm{bar}(W)) = \int H(\eta)\, dW(\eta)$$
$$= \int H(\mu_P)\, d\hat{\mu}(P) = \int h_{\mu_P}(f, \mathcal{Q})\, d\hat{\mu}(P).$$

Now we treat the general case of the lemma, by reduction to the previous paragraph. Given any finite partition $\mathcal{Q}$, let $\Sigma = \mathcal{Q}^{\mathbb{N}}$ and

$$h : M \to \Sigma, \qquad h(x) = \big(\mathcal{Q}(f^n(x))\big)_{n \in \mathbb{N}}.$$

Observe that $h \circ f = \sigma \circ h$, where $\sigma : \Sigma \to \Sigma$ denotes the shift map. To each measure η on M we may assign the measure $\eta' = h_* \eta$ on Σ. The previous relation ensures that if η is invariant under f then η' is invariant under σ. Moreover, if η is ergodic for f then η' is ergodic for σ. Indeed, if $B' \subset \Sigma$ is invariant under σ then $B = h^{-1}(B')$ is invariant under σ. Assuming that η is ergodic, it follows that $\eta'(B') = \eta(B)$ is either 0 or 1; hence, η' is ergodic.

By construction, $\mathcal{Q} = h^{-1}(\mathcal{Q}')$, where $\mathcal{Q}'$ denotes the partition of Σ into the cylinders $[0; Q]$, $Q \in \mathcal{Q}$. More generally, $\bigvee_{j=0}^{n-1} f^{-j}(\mathcal{Q}) = h^{-1}(\bigvee_{j=0}^{n-1} \sigma^{-j}(\mathcal{Q}'))$ and, thus,

$$H_\eta\Big(\bigvee_{j=0}^{n-1} f^{-j}(\mathcal{Q})\Big) = H_{\eta'}\Big(\bigvee_{j=0}^{n-1} \sigma^{-j}(\mathcal{Q}')\Big)$$

for every $n \in \mathbb{N}$. Dividing by n and taking the limit,

$$h_\eta(f, \mathcal{Q}) = h_{\eta'}(\sigma, \mathcal{Q}') \quad \text{for every } \eta \in \mathcal{M}. \tag{9.6.5}$$

Denote $\mu' = h_* \mu$ and $\mu'_P = h_*(\mu_P)$ for each P. For every bounded measurable function $\psi : \Sigma \to \mathbb{R}$,

$$\begin{aligned} \int \psi\, d\mu' = \int (\psi \circ h)\, d\mu &= \int \Big(\int (\psi \circ h)\, d\mu_P\Big) d\hat{\mu}(P) \\ &= \int \Big(\int \psi\, d\mu'_P\Big) d\hat{\mu}(P). \end{aligned} \tag{9.6.6}$$

As the measures μ'_P are ergodic, the relation (9.6.6) means that $\{\mu'_P : P \in \mathcal{P}\}$ is an ergodic decomposition of μ'. Then, according to the previous paragraph, $h_{\mu'}(\sigma, \mathcal{Q}') = \int h_{\mu'_P}(\sigma, \mathcal{Q}')\, d\hat{\mu}(P)$. By the relation (9.6.5) applied to $\eta = \mu$ and to $\eta = \mu_P$, this may be rewritten as

$$h_\mu(\sigma, \mathcal{Q}) = \int h_{\mu_P}(\sigma, \mathcal{Q})\, d\hat{\mu}(P),$$

which is precisely what we wanted to prove.

Proceeding with the proof of Theorem 9.6.2, consider any increasing sequence $\mathcal{Q}_1 \prec \cdots \prec \mathcal{Q}_n \prec \cdots$ of finite partitions of M such that $\operatorname{diam} \mathcal{Q}_n(x)$

converges to zero at every $x \in M$ (such a sequence may be constructed, for instance, from a family of balls centered at the points of a countable dense subset, with radii converging to zero). By Lemma 9.6.6,

$$h_\mu(f, \mathcal{Q}_n) = \int h_{\mu_P}(f, \mathcal{Q}_n)\, d\hat{\mu}(P) \tag{9.6.7}$$

for every n. According to (9.1.16), the sequence $h_\eta(f, \mathcal{Q}_n)$ is non-decreasing, for any invariant measure η. Moreover, by Corollary 9.2.5, its limit is equal to $h_\eta(f)$. Then, we may pass to the limit in (9.6.7) with the aid of the monotone convergence theorem. In this way we get that

$$h_\mu(f) = \int h_{\mu_P}(f)\, d\hat{\mu}(P),$$

as we wanted to prove. Note that the argument remains valid even when either of the two sides of this identity is infinite (then the other one is also infinite).

In this way, we reduced the proof of Theorem 9.6.2 to proving Theorem 9.6.5.

9.6.2 Proof of the Jacobs theorem

Now we prove Theorem 9.6.5. Let us start by proving that the barycenter function has the following continuity property: if W is concentrated in a neighborhood V of a given measure ν then the barycenter of W is close to ν. More precisely:

Lemma 9.6.7. *Let W be a probability measure on $\mathcal{M}_1(M)$ and $\nu \in \mathcal{M}_1(M)$. Given any finite set $\Phi = \{\phi_1, \dots, \phi_N\}$ of bounded continuous functions and any $\varepsilon > 0$, let $V = V(\nu, \Phi, \varepsilon)$ be as defined in* (2.1.1). *If $W(V) = 1$, then* $\operatorname{bar}(W) \in V$.

Proof. Consider any $i = 1, \dots, N$. By the definition of barycenter and the hypothesis that the complement of V has measure zero,

$$\left| \int \phi_i \, d\operatorname{bar}(W) - \int \phi_i \, d\nu \right| = \left| \int \left(\int \phi_i \, d\eta \right) dW(\eta) - \int \left(\int \phi_i \, d\nu \right) dW(\eta) \right|$$
$$\leq \int_V \left| \left(\int \phi_i \, d\eta - \int \phi_i \, d\nu \right) \right| dW(\eta).$$

By the definition of V, the last expression is smaller than ε. Therefore,

$$\left| \int \phi_i \, d\operatorname{bar}(W) - \int \phi_i \, d\nu \right| < \varepsilon$$

for every $i = 1, \dots, N$. In other words, $\operatorname{bar}(W) \in V$.

We also use the following simple property of non-negative affine functionals:

Lemma 9.6.8. *For any non-negative affine functional $H:\mathcal{M}\to\mathbb{R}$, probability measures $\nu_i\in\mathcal{M}$, $i\geq 1$ and non-negative numbers t_i, $i\geq 1$ with $\sum_{i=1}^{\infty}t_i=1$,*

$$H\left(\sum_{i=1}^{\infty}t_i\nu_i\right)\geq\sum_{i=1}^{\infty}t_iH(\nu_i).$$

Proof. Define $s_n=\sum_{i=1}^{n}t_i$ for every $n\geq 1$. Let $R_n=(1-s_n)^{-1}\sum_{i>n}t_i\nu_i$ if $s_n<1$; otherwise, pick R_n arbitrarily. Then,

$$\sum_{i=1}^{\infty}t_i\nu_i=\sum_{i=1}^{n}t_i\nu_i+(1-s_n)R_n.$$

Since H is affine and the expression on the right-hand side is a (finite) convex combination, it follows that

$$H\left(\sum_{i=1}^{\infty}t_i\nu_i\right)=\sum_{i=1}^{n}t_iH(\nu_i)+(1-s_n)H(R_n)\geq\sum_{i=1}^{n}t_iH(\nu_i)$$

for every n. Now just make n go to infinity.

Corollary 9.6.9. *If $H:\mathcal{M}\to\mathbb{R}$ is a non-negative affine functional then H is bounded.*

Proof. Suppose that H is not bounded: there exist $\nu_i\in\mathcal{M}$ such that $H(\nu_i)\geq 2^i$ for every $i\geq 1$. Consider $\nu=\sum_{i=1}^{\infty}2^{-i}\nu_i$. By Lemma 9.6.8,

$$H(\nu)\geq\sum_{i=1}^{\infty}2^{-i}H(\nu_i)=\infty.$$

This contradicts the fact that $H(\nu)$ is finite.

Now we are ready to prove the inequality $\geq$ in Theorem 9.6.5. Let us write $\mu=\operatorname{bar}(W)$. By the hypothesis of semi-continuity, given any $\varepsilon>0$ there exist $\delta>0$ and a finite family $\Phi=\{\phi_1,\dots,\phi_N\}$ of bounded continuous functions such that

$$H(\eta)<H(\mu)+\varepsilon\quad\text{for every }\eta\in\mathcal{M}\cap V(\mu,\Phi,\delta).\tag{9.6.8}$$

Since $\mathcal{M}_1(M)$ is a separable metric space, it admits a countable basis of open sets, and then so does any subspace. Let $\{V_1,\dots,V_n,\dots\}$ be a basis of open sets of $\mathcal{M}$, with the following properties:

(i) every V_n is contained in $\mathcal{M}\cap V(\nu_n,\Phi,\delta)$ for some $\nu_n\in\mathcal{M}$;
(ii) $H(\eta)<H(\nu_n)+\varepsilon$ for every $\eta\in V_n$.

Consider the partition $\{P_1,\dots,P_n,\dots\}$ of the space $\mathcal{M}$ defined by $P_1=V_1$ and $P_n=V_n\setminus(V_1\cup\cdots\cup V_{n-1})$ for every $n>1$. It is clear that the properties (i) and (ii) remain valid if we replace V_n by P_n. We claim that

$$\sum_n W(P_n)\nu_n\in V(\mu,\Phi,\delta).\tag{9.6.9}$$

Indeed, observe that

$$\left|\int \phi_i\,d\mu - \sum_n W(P_n)\int \phi_i\,d\nu_n\right| = \left|\sum_n \int_{P_n}\left(\int \phi_i\,d\eta - \int \phi_i\,d\nu_n\right)dW(\eta)\right|$$

for every i. Therefore, property (i) ensures that

$$\left|\int \phi_i\,d\mu - \sum_n W(P_n)\int \phi_i\,d\nu_n\right| < \sum_n \delta W(P_n) = \delta \quad \text{for every } i,$$

which is precisely what (9.6.9) means. Then, combining (9.6.8), (9.6.9) and Lemma 9.6.8,

$$\sum_n W(P_n)H(\nu_n) \le H\left(\sum_n W(P_n)\nu_n\right) < H(\mu)+\varepsilon.$$

On the other hand, property (ii) implies that

$$\begin{aligned}\int H(\eta)\,dW(\eta) - \sum_n W(P_n)H(\nu_n) &= \sum_n \int_{P_n} \left(H(\eta) - H(\nu_n)\right)dW\mu(\eta)\\ &< \sum_n \varepsilon W(P_n) = \varepsilon.\end{aligned}$$

Adding the last two inequalities, we get that $\int H(\eta)\,dW(\eta) < H(\mu)+2\varepsilon$. Since $\varepsilon > 0$ is arbitrary, this implies that $H(\mu) \ge \int H(\eta)\,dW(\eta)$.

Now we prove the inequality $\le$ in Theorem 9.6.5. Consider any sequence $(\mathcal{P}_n)_n$ of finite partitions of $\mathcal{M}$ such that, for every $\nu \in \mathcal{M}$, the diameter of $\mathcal{P}_n(\nu)$ converges to zero when n goes to infinity. For example, $\mathcal{P}_n = \bigvee_{i=1}^n \{V_i, V_i^c\}$, where $\{V_n : n \ge 1\}$ is any countable basis of open sets of $\mathcal{M}$. For each fixed n, consider the normalized restriction W_P of the measure W to each set $P \in \mathcal{P}_n$ (we consider only sets with positive measure: the union of all the elements of $\bigcup_n \mathcal{P}_n$ with $W(P) = 0$ has measure zero and so may be neglected):

$$W_P(A) = \frac{W(A \cap P)}{W(P)} \quad \text{for each measurable set } A \subset M.$$

It is clear that $W = \sum_{P\in\mathcal{P}_n} W(P)W_P$. Since the barycenter is an affine function (Exercise 9.6.2), it follows that

$$\operatorname{bar}(W) = \sum_{P\in\mathcal{P}_n} W(P)\operatorname{bar}(W_P)$$

and, therefore,

$$H(\operatorname{bar}(W)) = \sum_{P\in\mathcal{P}_n} W(P)H(\operatorname{bar}(W_P)).$$

Define $H_n(\eta) = H(\operatorname{bar}(W_{\mathcal{P}_n(\eta)}))$, for each $\eta \in \mathcal{M}$. Then the last identity above may be rewritten as follows:

$$H(\operatorname{bar}(W)) = \int H_n(\eta)\,dW(\eta) \quad \text{for every } n. \tag{9.6.10}$$

It follows directly from the definition of H_n that $0 \le H_n(\eta) \le \sup H$ for every n and every η. Recall that $\sup H < \infty$ (Corollary 9.6.9). We also claim that

$$\limsup_n H_n(\eta) \le H(\eta) \quad \text{for every } \eta \in \mathcal{M}. \tag{9.6.11}$$

This may be seen as follows. Given any neighborhood $V = V(\eta, \Phi, \varepsilon)$ of η, we have that $\mathcal{P}_n(\eta) \subset V$ for every large n, because the diameter of $\mathcal{P}_n(\eta)$ converges to zero. Then, assuming always that $W(\mathcal{P}_n(\eta))$ is positive,

$$W_{\mathcal{P}_n(\eta)}(V) \ge W_{\mathcal{P}_n(\eta)}(\mathcal{P}_n(\eta)) = 1.$$

By Lemma 9.6.7, it follows that $\operatorname{bar}(W_{\mathcal{P}_n(\eta)}) \in V$ for every large n. Now (9.6.11) is a direct consequence of the hypothesis that H is upper semi-continuous.

Applying the lemma of Fatou to the sequence $-H_n + \sup H$, we deduce that

$$\limsup_n \int H_n(\eta)\, dW(\eta) \le \int \limsup_n H_n(\eta)\, dW(\eta) \le \int H(\eta)\, dW(\eta). \tag{9.6.12}$$

Combining the relations (9.6.10) and (9.6.12), we get that

$$H(\operatorname{bar}(W)) \le \int H(\eta)\, dW(\eta),$$

as we wanted to prove.

Now the proof of Theorems 9.6.2 and 9.6.5 is complete.

9.6.3 Exercises

9.6.1. Check that, given any probability measure W on $\mathcal{M}_1(M)$, there exists a unique probability measure $\operatorname{bar}(W) \in \mathcal{M}_1(M)$ on M that satisfies (9.6.4).

9.6.2. Show that the barycenter function is strongly affine: if W_i, $i \ge 1$ are probability measures on $\mathcal{M}_1(M)$ and t_i, $i \ge 1$ are non-negative numbers with $\sum_{i=1}^{\infty} t_i = 1$, then

$$\operatorname{bar}\Big(\sum_{i=1}^{\infty} t_i W_i\Big) = \sum_{i=1}^{\infty} t_i \operatorname{bar}(W_i).$$

9.6.3. Show that if $\mathcal{M} \subset \mathcal{M}_1(M)$ is a closed convex set then $\mathcal{M}$ is strongly convex. Moreover, in that case $W(\mathcal{M}) = 1$ implies that $\operatorname{bar}(W) \in \mathcal{M}$.

9.6.4. Check that the inequality $\ge$ in Theorem 9.6.2 may also be obtained through the following, more direct, argument:

(1) Recalling that the function $\phi(x) = -x \log x$ is concave, show that $H_\mu(\mathcal{Q}) \ge \int H_{\mu_P}(\mathcal{Q})\, d\hat{\mu}(P)$ for every finite partition $\mathcal{Q}$.

(2) Deduce that $h_\mu(f, \mathcal{Q}) \ge \int h_{\mu_P}(f, \mathcal{Q})\, d\hat{\mu}(P)$ for every finite partition $\mathcal{Q}$.

(3) Conclude that $h_\mu(f) \ge \int h_{\mu_P}(f)\, d\hat{\mu}(P)$.

9.6.5. The inequality $\le$ in Theorem 9.6.2 is based on the fact that $h_\mu(f, \mathcal{Q}) \le \int h_{\mu_P}(f, \mathcal{Q})\, d\hat{\mu}(P)$ for every finite partition $\mathcal{Q}$, which is part of Lemma 9.6.6. Find what is wrong with the following "alternative proof":

Let $\mathcal{Q}$ be a finite partition. The theorem of Shannon–McMillan–Breiman ensures that $h_\mu(f, \mathcal{Q}) = \int h_\mu(f, \mathcal{Q}, x)\, d\mu(x)$, where

$$h_\mu(f, \mathcal{Q}, x) = \lim_n -\frac{1}{n} \log \mu(\mathcal{Q}^n(x)) = \lim_n -\frac{1}{n} \log \int \mu_P(\mathcal{Q}^n(x))\, d\hat{\mu}(P).$$

By the Jensen inequality applied to the convex function $\psi(x) = -\log x$,

$$\lim_n -\frac{1}{n}\log\int \mu_P(\mathcal{Q}^n(x))\,d\hat{\mu}(P) \leq \lim_n \int -\frac{1}{n}\log \mu_P(\mathcal{Q}^n(x))\,d\hat{\mu}(P).$$

Using the fact that $h_{\mu_P}(f,\mathcal{Q}) = h_{\mu_P}(f,\mathcal{Q},x)$ at almost every point (because the measure μ_P is ergodic),

$$\begin{aligned}\lim_n \int -\frac{1}{n}\log \mu_P(\mathcal{Q}^n(x))\,d\hat{\mu}(P) &= \int \lim_n -\frac{1}{n}\log \mu_P(\mathcal{Q}^n(x))\,d\hat{\mu}(P)\\ &= \int h_{\mu_P}(f,\mathcal{Q})\,d\hat{\mu}(P).\end{aligned}$$

This shows that $h_\mu(f,\mathcal{Q},x) \leq \int h_{\mu_P}(f,\mathcal{Q})\,d\hat{\mu}(P)$ for every finite partition $\mathcal{Q}$ and μ-almost every x. Consequently, $h_\mu(f,\mathcal{Q}) \leq \int h_{\mu_P}(f,\mathcal{Q})\,d\hat{\mu}(P)$ for every finite partition $\mathcal{Q}$.

9.7 Jacobians and the Rokhlin formula

Let U be an open subset and m be the Lebesgue measure of $\mathbb{R}^d$. Let $f : U \to U$ be a local diffeomorphism. By the formula of change of variables,

$$m(f(A)) = \int_A |\det Df(x)|\,dx \tag{9.7.1}$$

for any measurable subset A of a small ball restricted to which f is injective. The notion of a Jacobian that we present in this section extends this kind of relation to much more general transformations and measures. Besides introducing this concept, we are going to show that Jacobians do exist under quite general hypotheses. Most important, it is possible to express the system's entropy explicitly in terms of the Jacobian. Actually, we already encountered an interesting manifestation of this fact in Proposition 9.4.2.

Let $f : M \to M$ be a measurable transformation. We say that f is *locally invertible* if there exists some countable cover $\{U_k : U_k \geq 1\}$ of M by measurable sets such that $f(U_k)$ is a measurable set and the restriction $f \mid U_k : U_k \to f(U_k)$ is a bijection with measurable inverse, for every $k \geq 1$. Every measurable subset of some U_k is called an *invertibility domain*. Note that the image $f(A)$ of any invertibility domain A is a measurable set. It is also clear that if f is locally invertible then the pre-image $f^{-1}(y)$ of any $y \in M$ is countable: it contains at most one point in each U_k.

Let η be a probability measure on M, not necessarily invariant under f. A measurable function $\xi : M \to [0,\infty)$ is a *Jacobian* of f with respect to η if the restriction of ξ to any invertibility domain A is integrable with respect to η and satisfies

$$\eta(f(A)) = \int_A \xi\,d\eta. \tag{9.7.2}$$

It is important to note (see Exercise 9.7.1) that the definition does not depend on the choice of the family $\{U_k : k \geq 1\}$.

Example 9.7.1. Let $\sigma : \Sigma \to \Sigma$ be the shift map in $\Sigma = \{1,2,\dots,d\}^{\mathbb{N}}$ and μ be the Bernoulli measure associated with a probability vector $p = (p_1,\dots,p_d)$. The restriction of σ to each cylinder $[0;a]$ is an invertible map. Moreover, given any cylinder $[0;a,a_1,\dots,a_n] \subset [0;a]$,

$$\mu\big(\sigma([0;a,a_1,\dots,a_n])\big) = p_{a_1}\cdots p_{a_n} = \frac{1}{p_a}\mu\big([0;a,a_1,\dots,a_n]\big).$$

We invite the reader to deduce that $\mu\big(\sigma(A)\big) = (1/p_a)\mu(A)$ for every measurable set $A \subset [0;a]$. Therefore, the function $\xi((x_n)_n) = 1/p_{x_0}$ is a Jacobian of σ with respect to μ.

We say that a measure η is *non-singular* with respect to the transformation $f : M \to M$ if the image of any invertibility domain with measure zero also has measure zero: if $\eta(A) = 0$ then $\eta(f(A)) = 0$. For example, if $f : U \to U$ is a local diffeomorphism of an open subset of $\mathbb{R}^d$ and η is the Lebesgue measure, then η is non-singular. For any locally invertible transformation, every invariant probability measure is non-singular restricted to some full measure invariant set (Exercise 9.7.8).

It follows immediately from the definition (9.7.2) that if f admits a Jacobian with respect to a measure η then this measure is non-singular. The converse is also true:

Proposition 9.7.2. *Let $f : M \to M$ be a locally invertible transformation and η be a measure on M, non-singular with respect to f. Then there exists some Jacobian of f with respect to η and it is essentially unique: any two Jacobians coincide at η-almost every point.*

Proof. We start by proving existence. Given any countable cover $\{U_k : k \ge 1\}$ of M by invertibility domains of f, define $P_1 = U_1$ and $P_k = U_k \setminus (U_1 \cup \cdots \cup U_{k-1})$ for each $k > 1$. Then $\mathcal{P} = \{P_k : k \ge 1\}$ is a partition of M formed by invertibility domains. For each $P_k \in \mathcal{P}$, denote by η_k the measure defined on P_k by $\eta_k(A) = \eta(f(A))$. Equivalently, η_k is the image under $(f \mid P_k)^{-1}$ of the measure η restricted to $f(P_k)$. The hypothesis that η is non-singular implies that every η_k is absolutely continuous with respect to η restricted to P_k:

$$\eta(A) = 0 \quad \Rightarrow \quad \eta_k(A) = \eta(f(A)) = 0$$

for every measurable set $A \subset P_k$. Let $\xi_k = d\eta_k/d(\eta \mid P_k)$ be the Radon–Nikodym derivative (Theorem A.2.18). Then, ξ_k is a function defined on P_k, integrable with respect to η and satisfying

$$\eta(f(A)) = \eta_k(A) = \int_A \xi_k \, d\eta \tag{9.7.3}$$

for every measurable set $A \subset P_k$. Consider the function $\xi : M \to [0,\infty)$ whose restriction to each $P_k \in \mathcal{P}$ is given by ξ_k. Every subset of U_k may be written as a (disjoint) union of subsets of $P_1,\dots,P_k$. Applying (9.7.3) to each one of these subsets and summing the corresponding equalities, we get that

$$\eta(f(A)) = \int_A \xi \, d\eta \quad \text{for every measurable set } A \subset U_k \text{ and } k \geq 1.$$

This proves that ξ is a Jacobian of f with respect to η.

Now suppose that ξ and ζ are Jacobians of f with respect to η and there exists $B \subset M$ with $\eta(B) > 0$ such that $\xi(x) \neq \zeta(x)$ for every $x \in B$. Up to replacing B by a suitable subset, and exchanging the roles of ξ and ζ if necessary, we may suppose that $\xi(x) < \zeta(x)$ for every $x \in B$. Similarly, we may suppose that B is contained in some U_k. Then,

$$\eta(f(B)) = \int_B \xi \, d\eta < \int_B \zeta \, d\eta = \eta(f(B)).$$

This contradiction proves that the Jacobian is essentially unique.

From now on, we denote by $J_\eta f$ the (essentially unique) Jacobian of a locally invertible transformation $f : M \to M$ with respect to a measure η, when it exists.

By definition, $J_\eta f$ is integrable on each invertibility domain. If f is such that the number of pre-images of any $y \in M$ is bounded then the Jacobian is (globally) integrable: if $\ell \geq 1$ is the maximum number of pre-images then

$$\int J_\eta f \, d\eta = \sum_k \int_{P_k} J_\eta f \, d\eta = \sum_k \eta(f(P_k)) \leq \ell,$$

because every point $y \in M$ is in no more than ℓ images $f(P_k)$.

The following observation will be useful in the sequel. Let $Z \subset M$ be the set of points where the Jacobian $J_\eta f$ vanishes. Covering Z with a countable family of invertibility domains and using (9.7.2), we see that $f(Z)$ is a measurable set and $\eta(f(Z)) = 0$. In other words, the set of points $y \in M$ such that $J_\mu f(x) > 0$ for every $x \in f^{-1}(y)$ has total measure for η. When the probability measure η is invariant under f, it also follows that $\eta(f^{-1}(f(Z))) = \eta(f(Z)) = 0$ and so $\eta(Z) = 0$.

The main result in this section is the following formula for the entropy of an invariant measure:

Theorem 9.7.3 (Rokhlin formula)**.** *Let $f : M \to M$ be a locally invertible transformation and μ be a probability measure invariant under f. Assume that there is some partition $\mathcal{P}$ with finite entropy such that $\bigcup_n \mathcal{P}^n$ generates the σ-algebra of M, up to measure zero, and every $P \in \mathcal{P}$ is an invertibility domain of f. Then $h_\mu(f) = \int \log J_\mu f \, d\mu$.*

Proof. Let us consider the sequence of partitions $\mathcal{Q}_n = \bigvee_{j=1}^n f^{-j}(\mathcal{P})$. By Corollary 9.2.5 and Lemma 9.1.12,

$$h_\mu(f) = h_\mu(f,\mathcal{P}) = \lim_n H_\mu(\mathcal{P}/\mathcal{Q}_n). \tag{9.7.4}$$

By definition (as before, $\phi(x) = -x\log x$),

$$\begin{aligned} H_\mu(\mathcal{P}/\mathcal{Q}_n) &= \sum_{P\in\mathcal{P}}\sum_{Q_n\in\mathcal{Q}_n} -\mu(P\cap Q_n)\log\frac{\mu(P\cap Q_n)}{\mu(Q_n)} \\ &= \sum_{P\in\mathcal{P}}\sum_{Q_n\in\mathcal{Q}_n} \mu(Q_n)\phi\left(\frac{\mu(P\cap Q_n)}{\mu(Q_n)}\right). \end{aligned} \tag{9.7.5}$$

Let $e_n(\psi,x)$ be the conditional expectation of a function ψ with respect to the partition $\mathcal{Q}_n$ and $e(\psi,x)$ be its limit when n goes to infinity (these notions were introduced in Section 5.2.1: see (5.2.1) and Lemma 5.2.1). It is clear from the definition that

$$\frac{\mu(P\cap Q_n)}{\mu(Q_n)} = e_n(\mathcal{X}_P,x) \quad \text{for every } x\in Q_n \text{ and every } Q_n\in\mathcal{Q}_n.$$

Therefore,

$$\sum_{P\in\mathcal{P}}\sum_{Q_n\in\mathcal{Q}_n} \mu(Q_n)\phi\left(\frac{\mu(P\cap Q_n)}{\mu(Q_n)}\right) = \sum_{P\in\mathcal{P}}\int \phi(e_n(\mathcal{X}_P,x))\,d\mu(x). \tag{9.7.6}$$

By Lemma 5.2.1, the limit $e(\mathcal{X}_P,x) = \lim_n e_n(\mathcal{X}_P,x)$ exists at μ-almost every x. So, observing that the function ϕ is bounded, we may use the dominated convergence theorem to deduce from (9.7.4)–(9.7.6) that

$$h_\mu(f) = \sum_{P\in\mathcal{P}}\int \phi(e(\mathcal{X}_P,x))\,d\mu(x). \tag{9.7.7}$$

Now we need to relate the expression inside the integral to the Jacobian. This we do by means of Lemma 9.7.5 below. Beforehand, let us prove the following *change of variables formulas*:

Lemma 9.7.4. *For any probability measure η non-singular with respect to f, and any invertibility domain $A\subset M$ of f:*

(i) $\int_{f(A)}\varphi\,d\eta = \int_A(\varphi\circ f)J_\eta f\,d\eta$ for any measurable function $\varphi: f(A)\to\mathbb{R}$ such that the integrals are defined (possibly $\pm\infty$).
(ii) $\int_A\psi\,d\eta = \int_{f(A)}(\psi/J_\eta f)\circ(f\mid A)^{-1}\,d\eta$ for any measurable function $\psi: A\to\mathbb{R}$ such that the integrals are defined (possibly $\pm\infty$).

Proof. The definition (9.7.2) means that the formula in part (i) holds for the characteristic function $\varphi = \mathcal{X}_{f(A)}$ for any invertibility domain A. Thus, it holds for the characteristic function of any measurable subset of $f(A)$, since such a subset may be written as $f(B)$ for some invertibility domain $B\subset A$. Hence, by linearity, the identity extends to every simple function defined on $f(A)$. Using the monotone convergence theorem, we conclude that the identity holds for every non-negative measurable function. Using linearity once more, we get the general statement of part (i).

To deduce the claim in (ii), apply (i) to the function $\varphi = (\psi/J_\eta f) \circ (f \mid A)^{-1}$. Note that this function is well defined at η-almost every point for, as observed before, $J_\eta f(x) > 0$ for every x in the pre-image of η-almost every $y \in M$.

Lemma 9.7.5. *For every bounded measurable function $\psi : M \to \mathbb{R}$ and every probability measure η invariant under f,*

$$e(\psi,x) = \hat{\psi}(f(x)) \text{ for } \eta\text{-almost every } x, \quad \text{where } \hat{\psi}(y) = \sum_{z \in f^{-1}(y)} \frac{\psi}{J_\eta f}(z).$$

Proof. Recall that $\mathcal{Q}_n = \bigvee_{j=1}^{n} f^{-j}(\mathcal{P})$, that is, $\mathcal{Q}_n(x) = \bigcap_{j=1}^{n} f^{-1}(\mathcal{P}(f^i(x)))$ for each x. We also use the sequence of partitions $\mathcal{P}^n = \bigvee_{j=0}^{n-1} f^{-j}(\mathcal{P})$. Observe that $\mathcal{Q}_n(x) = f^{-1}(\mathcal{P}^{n-1}(f(x)))$ and $\mathcal{P}^n(x) = \mathcal{P}(x) \cap \mathcal{Q}_n(x)$ for every n and every x. Then,

$$\int_{\mathcal{P}^{n-1}(f(x))} \hat{\psi}\, d\eta = \sum_{P \in \mathcal{P}} \int_{f(P) \cap \mathcal{P}^{n-1}(f(x))} \frac{\psi}{J_\eta f} \circ (f \mid P)^{-1}\, d\eta.$$

Using the formula of change of variables in Lemma 9.7.4(ii), the expression on the right-hand side may be rewritten as

$$\sum_{P \in \mathcal{P}} \int_{P \cap \mathcal{Q}_n(x)} \psi(z)\, d\eta(z) = \int_{\mathcal{Q}_n(x)} \psi\, d\eta.$$

Therefore,

$$\int_{\mathcal{P}^{n-1}(f(x))} \hat{\psi}\, d\eta = \int_{\mathcal{Q}_n(x)} \psi\, d\eta. \tag{9.7.8}$$

Let $e'_{n-1}(\hat{\psi},x)$ be the conditional expectation of $\hat{\psi}$ with respect to the partition $\mathcal{P}^{n-1}$, as defined in Section 5.2.1, and let $e'(\hat{\psi},x)$ be its limit when n goes to infinity, given by Lemma 5.2.1. The hypothesis that η is invariant gives that $\eta\big(\mathcal{P}^{n-1}(f(x))\big) = \eta\big(\mathcal{Q}_n(x)\big)$. Dividing both sides of (9.7.8) by this number, we get that

$$e'_{n-1}(\hat{\psi}, f(x)) = e_n(\psi,x) \quad \text{for every } x \text{ and every } n > 1. \tag{9.7.9}$$

Then, taking the limit, $e'(\hat{\psi}, f(x)) = e(\psi,x)$ for η-almost every x. On the other hand, according to Exercise 5.2.3, the hypothesis implies that $e'(\hat{\psi},y) = \hat{\psi}(y)$ for η-almost every $y \in M$.

Let us apply this lemma to $\psi = \mathcal{X}_P$ and $\eta = \mu$. Since f is injective on every element of $\mathcal{P}$, each intersection $P \cap f^{-1}(y)$ either is empty or contains exactly one point. Therefore, it follows from Lemma 9.7.5 that $e(\mathcal{X}_P,x) = \hat{\mathcal{X}}_P(f(x))$, with

$$\hat{\mathcal{X}}_P(y) = \begin{cases} 1/J_\mu f\big((f \mid P)^{-1}(y)\big) & \text{if } y \in f(P) \\ 0 & \text{if } y \notin f(P). \end{cases}$$

Then, recalling that the measure μ is assumed to be invariant,

$$\int \phi(e(\mathcal{X}_P,x))\,d\mu(x) = \int \phi(\hat{\mathcal{X}}_P(y))\,d\mu(y)$$
$$= \int_{f(P)} \left(\frac{1}{J_\mu f}\log J_\mu f\right) \circ (f \mid P)^{-1}\,d\mu \quad = \int_P \log J_\mu f\,d\mu$$

(the last step uses the identity in part (ii) of Lemma 9.7.4). Replacing this expression in (9.7.7), we get that

$$h_\mu(f) = \sum_{P\in\mathcal{P}} \int_P \log J_\mu f\,d\mu = \int \log J_\mu f\,d\mu,$$

as stated in the theorem.

9.7.1 Exercises

9.7.1. Check that the definition of a Jacobian does not depend on the choice of the cover $\{U_k : k \geq 1\}$ by invertibility domains.

9.7.2. Let $\sigma : \Sigma \to \Sigma$ be the shift map in $\Sigma = \{1,2,\dots,d\}^{\mathbb{N}}$ and μ be the Markov measure associated with an aperiodic matrix P. Find the Jacobian of f with respect to μ.

9.7.3. Let $f : M \to M$ be a locally invertible transformation and η be a probability measure on M, non-singular with respect to f. Show that for every bounded measurable function $\psi : M \to \mathbb{R}$,

$$\int \psi\,d\eta = \int \sum_{z\in f^{-1}(x)} \frac{\psi}{J_\eta f}(z)d\eta(x).$$

9.7.4. Let $f : M \to M$ be a locally invertible transformation and η be a probability measure on M, non-singular with respect to f. Show that η is invariant under f if and only if

$$\sum_{z\in f^{-1}(x)} \frac{1}{J_\eta f(z)} = 1 \quad \text{for } \eta\text{-almost every } x \in M.$$

Moreover, if η is invariant under f then $J_\eta f \geq 1$ at μ-almost every point.

9.7.5. Let $f : M \to M$ be a locally invertible transformation and η be a probability measure on M, non-singular with respect to f. Show that, for every $k \geq 1$, there exists a Jacobian of f^k with respect to η and it is given by

$$J_\eta f^j(x) = \prod_{j=0}^{k-1} J_\eta f(f^j(x)) \quad \text{for } \eta\text{-almost every } x.$$

Assuming that f is invertible, what can be said about the Jacobian of f^{-1} with respect to η?

9.7.6. Let $f : M \to M$ and $g : N \to N$ be locally invertible transformations and let μ and ν be probability measures invariant under f and g, respectively. Assume that there exists an ergodic equivalence $\phi : M \to N$ between the systems (f,μ) and (g,ν). Show that $J_\mu f = J_\nu g \circ \phi$ at μ-almost every point.

9.7.7. Let $\sigma_k : \Sigma_k \to \Sigma_k$ and $\sigma_l : \Sigma_l \to \Sigma_l$ be the shift maps in $\Sigma_k = \{1,\dots,k\}^{\mathbb{N}}$ and $\Sigma_l = \{1,\dots,l\}^{\mathbb{N}}$. Let μ_k and μ_l be the Bernoulli measures on Σ_k and Σ_l, respectively, associated with probability vectors $p = (p_1,\dots,p_k)$ and $q = (q_1,\dots,q_l)$. Show that the systems (σ_k, μ_k) and (σ_l, μ_l) are ergodically equivalent if and only if $k = l$ and the vectors p and q are obtained from one another by permutation of the components.

9.7.8. Let μ be a probability measure invariant under a locally invertible transformation $f : M \to M$. Show that there exists a full measure set $N \subset M$ such that $N \subset f^{-1}(N)$ and μ restricted to N is non-singular with respect to the restriction $f : N \to N$. Conclude that f admits a Jacobian with respect to μ.

10
Variational principle

In 1965, the IBM researchers R. Adler, A. Konheim and M. McAndrew proposed [AKM65] a notion of *topological entropy*, inspired by the Kolmogorov–Sinai entropy that we studied in the previous chapter, but whose definition does not involve any invariant measure. This notion applies to any continuous transformation in a compact topological space.

Subsequently, Efim Dinaburg [Din70] and Rufus Bowen [Bow71, Bow75a] gave a different, yet equivalent, definition for continuous transformations in compact metric spaces. Despite being a bit more restrictive, the Bowen–Dinaburg definition has the advantage of making more transparent the meaning of this concept: the topological entropy is the rate of exponential growth of the number of orbits that can be distinguished within a certain precision, arbitrarily small. Moreover, Bowen extended the definition to non-compact spaces, which is also very useful for applications.

These definitions of topological entropy and their properties are studied in Section 10.1 where, in particular, we observe that the topological entropy is an invariant of topological equivalence (topological conjugacy). In Section 10.2 we analyze several concrete examples.

The main result is the following remarkable relation between the topological entropy and the entropies of the transformation with respect to its invariant measures:

Theorem 10.1 (Variational principle). *If $f : M \to M$ is a continuous transformation in a compact metric space then its topological entropy $h(f)$ coincides with the supremum of the entropies $h_\mu(f)$ of f with respect to all the invariant probability measures.*

This theorem was proved by Dinaburg [Din70, Din71], Goodman [Goo71a] and Goodwin [Goo71b]. Here, it arises as a special case of a more general statement, the variational principle for the pressure, which is due to Walters [Wal75].

The *pressure* $P(f, \phi)$ is a weighted version of the topological entropy $h(f)$, where the "weights" are determined by a continuous function $\phi : M \to \mathbb{R}$, which we call a *potential*. We study these notions and their properties in Section 10.3. The topological entropy corresponds to the special case when the potential is identically zero. The notion of pressure was brought from statistical mechanics to ergodic theory by the Belgium mathematician and theoretical physicist David Ruelle, one of the founders of differentiable ergodic theory, and was then extended by the British mathematician Peter Walters.

The variational principle (Theorem 10.1) extends to the setting of the pressure, as we are going to see in Section 10.4:

$$P(f,\phi) = \sup\left\{h_\mu(f) + \int \phi\, d\mu : \mu \text{ is invariant under } f\right\} \qquad (10.0.1)$$

for every continuous function $\phi : M \to \mathbb{R}$. An invariant probability measure μ is called an *equilibrium state* for the potential ϕ if it realizes the supremum in (10.0.1), that is, if $h_\mu(f) + \int \phi\, d\mu = P(f,\phi)$. The set of all equilibrium states is studied in Section 10.5.

10.1 Topological entropy

Initially, we present the definitions of Adler–Konheim–McAndrew and Bowen–Dinaburg and we prove that they are equivalent when the ambient is a compact metric space.

10.1.1 Definition via open covers

The original definition of the topological entropy is very similar to that of the Kolmogorov–Sinai entropy, with open covers in the place of partitions into measurable sets.

Let M be a compact topological space. An *open cover* of M is any family α of open sets whose union is the whole of M. By compactness, every open cover admits a *subcover* (that is, a subfamily that is still an open cover) with finitely many elements. We call the *entropy* of the open cover α the number

$$H(\alpha) = \log N(\alpha), \qquad (10.1.1)$$

where $N(\alpha)$ is the smallest number such that α admits some finite subcover with that number of elements.

Given two open covers α and β, we say that α is *coarser* than β (or β is finer than α), and we write $\alpha \prec \beta$, if every element of β is contained in some element of α. For example, if β is a subcover of α then $\alpha \prec \beta$. By Exercise 10.1.1,

$$\alpha \prec \beta \Rightarrow H(\alpha) \le H(\beta). \qquad (10.1.2)$$

Given open covers $\alpha_1, \dots, \alpha_n$, we denote by $\alpha_1 \vee \cdots \vee \alpha_n$ their *sum*, that is, the open cover whose elements are the intersections $A_1 \cap \cdots \cap A_n$ with $A_j \in \alpha_j$ for each j. Note that $\alpha_j \prec \alpha_1 \vee \cdots \vee \alpha_n$ for every j.

Let $f : M \to M$ be a continuous transformation. If α is an open cover of M then so is $f^{-j}(\alpha) = \{f^{-j}(A) : A \in \alpha\}$. For each $n \geq 1$, let us denote

$$\alpha^n = \alpha \vee f^{-1}(\alpha) \vee \cdots \vee f^{-n+1}(\alpha).$$

Using Exercise 10.1.2, we see that

$$H(\alpha^{m+n}) = H\big(\alpha^m \vee f^{-m}(\alpha^n)\big) \leq H(\alpha^m) + H(f^{-m}(\alpha^n)) \leq H(\alpha^m) + H(\alpha^n)$$

for every $m, n \geq 1$. In other words, the sequence $H(\alpha^n)$ is subadditive. Consequently (Lemma 3.3.4),

$$h(f,\alpha) = \lim_n \frac{1}{n} H(\alpha^n) = \inf_n \frac{1}{n} H(\alpha^n) \tag{10.1.3}$$

always exists and is finite. It is called the entropy of f with respect to the open cover α. The relation (10.1.2) implies that

$$\alpha \prec \beta \quad \Rightarrow \quad h(f,\alpha) \leq h(f,\beta). \tag{10.1.4}$$

Finally, we define the *topological entropy* of f to be

$$h(f) = \sup\{h(f,\alpha) : \alpha \text{ is an open cover of } M\}. \tag{10.1.5}$$

In particular, if β is a subcover of α then $h(f,\alpha) \leq h(f,\beta)$. Therefore, the definition (10.1.5) does not change when one restricts the supremum to the finite open covers.

Observe that the entropy $h(f)$ is a non-negative number, possibly infinite (see Exercise 10.1.6).

Example 10.1.1. Let $f : S^1 \to S^1$ be any homeomorphism (for example, a rotation R_θ) and let α be an open cover of the circle formed by a finite number of open intervals. Let $\partial\alpha$ be the set consisting of the endpoints of those intervals. For each $n \geq 1$, the open cover α^n is formed by intervals whose endpoints are in

$$\partial\alpha^n = \partial\alpha \cup f^{-1}(\partial\alpha) \cup \cdots \cup f^{-n+1}(\partial\alpha).$$

Note that $\#\alpha^n \leq \#\partial\alpha^n \leq n\#\partial\alpha$. Therefore,

$$h(f,\alpha) = \lim_n \frac{1}{n} H(\alpha^n) \leq \liminf_n \frac{1}{n} \log \#\alpha^n \leq \liminf_n \frac{1}{n} \log n = 0.$$

Proposition 10.1.12 below gives that $h(f) = \lim_k h(f,\alpha_k)$ for any sequence of open covers α_k with $\operatorname{diam} \alpha_k \to 0$. Then, considering open covers α_k by intervals of length less than $1/k$, we conclude from the previous calculation that $h(f) = 0$ for every homeomorphism of the circle.

Example 10.1.2. Let $\Sigma = \{1,\dots,d\}^{\mathbb{N}}$ and α be the cover of Σ by the cylinders $[0;a]$, $a = 1,\dots,d$. Consider the shift map $\sigma : \Sigma \to \Sigma$. For each n, the open cover α^n consists of the cylinders of length n:

$$\alpha^n = \{[0;a_0,\dots,a_{n-1}] : a_j = 1,\dots,d\}.$$

Therefore, $H(\alpha^n) = \log \#\alpha^n = \log d^n$ and, consequently, $h(f,\alpha) = \log d$. Observe also that $\operatorname{diam} \alpha^n$ converges to zero when $n \to \infty$, relative to the distance defined by (A.2.7). Then, it follows from Corollary 10.1.13 below that $h(f) = h(f,\alpha) = \log d$. The same holds for the two-sided shift $\sigma : \Sigma \to \Sigma$ in $\Sigma = \{1,\ldots,d\}^{\mathbb{Z}}$.

Now we show that the topological entropy is an invariant of topological equivalence. Let $f : M \to M$ and $g : N \to N$ be continuous transformations in compact topological spaces M and N. We say that g is a *topological factor* of f if there exists a surjective continuous map $\theta : M \to N$ such that $\theta \circ f = g \circ \theta$. When θ may be chosen to be invertible (a homeomorphism), we say that the two transformations are *topologically equivalent*, or *topologically conjugate*, and we call θ a *topological conjugacy* between f and g.

Proposition 10.1.3. *If g is a topological factor of f then $h(g) \le h(f)$. In particular, if f and g are topologically equivalent then $h(f) = h(g)$.*

Proof. Let $\theta : M \to N$ be a surjective continuous map such that $\theta \circ f = g \circ \theta$. Given any open cover α of N, the family

$$\theta^{-1}(\alpha) = \{\theta^{-1}(A) : A \in \alpha\}$$

is an open cover of M. Recall that, by definition, the iterated sum α^n is the open cover formed by the sets $\bigcap_{j=0}^{n-1} g^{-j}(A_j)$ with $A_0, A_1, \ldots, A_{n-1} \in \alpha$. Analogously, the iterated sum $\theta^{-1}(\alpha)^n$ consists of the sets $\bigcap_{j=0}^{n-1} f^{-j}(\theta^{-1}(A_j))$. Clearly,

$$\bigcap_{j=0}^{n-1} f^{-j}(\theta^{-1}(A_j)) = \bigcap_{j=0}^{n-1} \theta^{-1}(g^{-j}(A_j)) = \theta^{-1}\left(\bigcap_{j=0}^{n-1} g^{-j}(A_j)\right).$$

Noting that the sets of the form on the right-hand side of this identity constitute the pre-image $\theta^{-1}(\alpha^n)$ of α^n, we conclude that $\theta^{-1}(\alpha^n) = \theta^{-1}(\alpha)^n$. Since θ is surjective, a family $\gamma \subset \alpha^n$ covers N if and only if $\theta^{-1}(\gamma)$ covers M. Therefore,

$$H(\theta^{-1}(\alpha)^n) = H(\theta^{-1}(\alpha^n)) = H(\alpha^n).$$

Since n is arbitrary, it follows that $h(f,\theta^{-1}(\alpha)) = h(g,\alpha)$. Then, taking the supremum over all the open covers α of N:

$$h(g) = \sup_\alpha h(g,\alpha) = \sup_\alpha h(f,\theta^{-1}(\alpha)) \le h(f).$$

This proves the first part of the proposition. The second part is an immediate consequence, since in that case f is also a factor of g.

The converse to Proposition 10.1.3 is false, in general. For example, all the homeomorphisms of the circle have topological entropy equal to zero (recall Example 10.1.1) but they are not necessarily topologically equivalent (for example, the identity is not topologically equivalent to any other homeomorphism).

10.1.2 Generating sets and separated sets

Next, we present the definition of topological entropy of Bowen–Dinaburg. Let $f: M \to M$ be a continuous transformation in a metric space M, not necessarily compact, and let $K \subset M$ be any compact subset. When M is compact it suffices to consider $K = M$, as observed in (10.1.12) below.

Given $\varepsilon > 0$ and $n \in \mathbb{N}$, we say that a set $E \subset M$ *(n,ε)-generates* K if for every $x \in K$ there exists $a \in E$ such that $d(f^i(x), f^i(a)) < \varepsilon$ for every $i \in \{0, \dots, n-1\}$. In other words,

$$K \subset \bigcup_{a \in E} B(a, n, \varepsilon),$$

where $B(a,n,\varepsilon) = \{x \in M : d(f^i(x), f^i(a)) < \varepsilon \text{ for } i = 0, \dots, n-1\}$ is the *dynamical ball* of center a, length n and radius ε. Note that $\{B(x,n,\varepsilon) : x \in K\}$ is an open cover of K. Hence, by compactness, there always exist finite (n,ε)-generating sets.

Let us denote by $g_n(f,\varepsilon,K)$ the smallest cardinality of an (n,ε)-generating set of K. We define

$$g(f,\varepsilon,K) = \limsup_n \frac{1}{n} \log g_n(f,\varepsilon,K). \tag{10.1.6}$$

Observe that the function $\varepsilon \mapsto g(f,\varepsilon,K)$ is monotone non-increasing. Indeed, it is clear from the definition that if $\varepsilon_1 < \varepsilon_2$ then every (n,ε_1)-generating set is also (n,ε_2)-generating. Therefore, $g_n(f,\varepsilon_1,K) \geq g_n(f,\varepsilon_2,K)$ for every $n \geq 1$ and, taking the limit, $g(f,\varepsilon_1,K) \geq g(f,\varepsilon_2,K)$. This ensures, in particular, that

$$g(f,K) = \lim_{\varepsilon \to 0} g(f,\varepsilon,K) \tag{10.1.7}$$

exists. Finally, we define

$$g(f) = \sup\{g(f,K) : K \subset M \text{ compact}\}. \tag{10.1.8}$$

We also introduce the following dual notion. Given $\varepsilon > 0$ and $n \in \mathbb{N}$, we say that a set $E \subset K$ is *(n,ε)-separated* if, given $x, y \in E$, there exists $j \in \{0, \dots, n-1\}$ such that $d(f^j(x), f^j(y)) \geq \varepsilon$. In other words, if $x \in E$ then $B(x,n,\varepsilon)$ contains no other point of E. We denote by $s_n(f,\varepsilon,K)$ the largest cardinality of an (n,ε)-separated set. We define

$$s(f,\varepsilon,K) = \limsup_n \frac{1}{n} \log s_n(f,\varepsilon,K). \tag{10.1.9}$$

It is clear that if $0 < \varepsilon_1 < \varepsilon_2$, then every (n,ε_2)-separated set is also (n,ε_1)-separated. Therefore, $s_n(f,\varepsilon_1,K) \geq s_n(f,\varepsilon_2,K)$ for every $n \geq 1$ and, taking the limit, $s(f,\varepsilon_1,K) \geq s(f,\varepsilon_2,K)$. In particular,

$$s(f,K) = \lim_{\varepsilon \to 0} s(f,\varepsilon,K) \tag{10.1.10}$$

always exists. Finally, we define

$$s(f) = \sup\{s(f,K) : K \subset M \text{ compact}\}. \tag{10.1.11}$$

It is clear that $g(f,K_1) \le g(f,K_2)$ and $s(f,K_1) \le s(f,K_2)$ if $K_1 \subset K_2$. In particular,

$$g(f) = g(f,M) \quad \text{and} \quad s(f) = s(f,M) \quad \text{if } M \text{ is compact.} \tag{10.1.12}$$

Another interesting observation (Exercise 10.1.7) is that the definitions (10.1.8) and (10.1.11) are not affected when we restrict the supremum to compact sets with small diameter.

Proposition 10.1.4. *We have $g(f,K) = s(f,K)$ for every compact $K \subset M$. Consequently, $g(f) = s(f)$.*

Proof. For the proof we need the following lemma:

Lemma 10.1.5. $g_n(f,\varepsilon,K) \le s_n(f,\varepsilon,K) \le g_n(f,\varepsilon/2,K)$ *for every $n \ge 1$, every $\varepsilon > 0$ and every compact $K \subset M$.*

Proof. Let $E \subset K$ be an (n,ε)-separated set with maximal cardinality. Given any $y \in K \setminus E$, the set $E \cup \{y\}$ is not (n,ε)-separated, and so there exists $x \in E$ such that $d(f^i(x),f^i(y)) < \varepsilon$ for every $i \in \{0,\dots,n-1\}$. This shows that E is an (n,ε)-generating set of K. Consequently, $g_n(f,\varepsilon,K) \le \#E = s_n(f,\varepsilon,K)$.

To prove the other inequality, let $E \subset K$ be an (n,ε)-separated set and $F \subset M$ be an $(n,\varepsilon/2)$-generating set of K. The hypothesis ensures that, given any $x \in E$ there exists some $y \in F$ such that $d(f^i(x),f^i(y)) < \varepsilon/2$ for every $i \in \{0,\dots,n-1\}$. Let $\phi : E \to F$ be a map such that each $\phi(x)$ is a point y satisfying this condition. We claim that the map ϕ is injective. Indeed, suppose that $x,z \in E$ are such that $\phi(x) = y = \phi(z)$. Then

$$d(f^i(x),f^i(z)) \le d(f^i(x),f^i(y)) + d(f^i(y),f^i(z)) < \varepsilon/2 + \varepsilon/2$$

for every $i \in \{0,\dots,n-1\}$. Since E is (n,ε)-separated, this implies that $x = z$. Therefore, ϕ is injective, as we claimed. It follows that $\#E \le \#F$ and, since E and F are arbitrary, this proves that $s_n(f,\varepsilon,K) \le g_n(f,\varepsilon/2,K)$.

Then, given any $\varepsilon > 0$ and any compact $K \subset M$,

$$\begin{aligned} g(f,\varepsilon,K) &= \limsup_n \frac{1}{n} \log g_n(f,\varepsilon,K) \\ &\le \limsup_n \frac{1}{n} \log s_n(f,\varepsilon,K) = s(f,\varepsilon,K) \\ &\le \limsup_n \frac{1}{n} \log g_n(f,\frac{\varepsilon}{2},K) = g(f,\frac{\varepsilon}{2},K). \end{aligned} \tag{10.1.13}$$

Taking the limit when $\varepsilon \to 0$, we get that

$$\begin{aligned} g(f,K) = \lim_{\varepsilon\to 0} g(f,\varepsilon,K) &\le \lim_{\varepsilon\to 0} s(f,\varepsilon,K) = s(f,K) \\ &\le \lim_{\varepsilon\to 0} g(f,\frac{\varepsilon}{2},K) = g(f,K). \end{aligned}$$

This proves the first part of the proposition. The second part is an immediate consequence.

By definition, the *diameter* of an open cover α of a metric space M is the supremum of the diameters of all the sets $A \in \alpha$.

Proposition 10.1.6. *If M is a compact metric space then $h(f) = g(f) = s(f)$.*

Proof. By Proposition 10.1.4, it suffices to show that $s(f) \le h(f) \le g(f)$.

Start by fixing $\varepsilon > 0$ and $n \ge 1$. Let $E \subset M$ be an (n,ε)-separated set and α be any open cover of M with diameter less than ε. If x and y are in the same element of α^n then

$$d(f^i(x), f^i(y)) \le \operatorname{diam} \alpha < \varepsilon \quad \text{for every } i = 0, \dots, n-1.$$

In particular, each element of α^n contains at most one element of E. Consequently, $\#E \le N(\alpha^n)$. Taking E with maximal cardinality, we conclude that $s_n(f,\varepsilon,M) \le N(\alpha^n)$ for every $n \ge 1$. So,

$$\begin{aligned} s(f,\varepsilon,M) &= \limsup_n \frac{1}{n} \log s_n(f,\varepsilon,M) \\ &\le \lim_n \frac{1}{n} \log N(\alpha^n) = h(f,\alpha) \le h(f). \end{aligned} \tag{10.1.14}$$

Making $\varepsilon \to 0$, we find that $s(f) = s(f,M) \le h(f)$.

Next, given any open cover α of M, let $\varepsilon > 0$ be a Lebesgue number for α, that is, a positive number such that every ball of radius ε is contained in some element of α. Let $E \subset M$ be an (n,ε)-generating set of M with minimal cardinality. For each $x \in E$ and $i = 0, \dots, n-1$, there exists $A_{x,i} \in \alpha$ such that $B(f^i(x),\varepsilon)$ is contained in $A_{x,i}$. Then,

$$B(x,n,\varepsilon) \subset \bigcap_{i=0}^{n-1} f^{-i}(A_{x,i}).$$

Therefore, the hypothesis that E is a generating set implies that the family $\gamma = \{\bigcap_{i=0}^{n-1} f^{-i}(A_{x,i}) : x \in E\}$ is an open cover of M. Since $\gamma \subset \alpha^n$, it follows that $N(\alpha^n) \le \#E = g_n(f,\varepsilon,M)$ for every n. Therefore,

$$\begin{aligned} h(f,\alpha) = \lim_n \frac{1}{n} \log N(\alpha^n) &\le \liminf_n \frac{1}{n} \log g_n(f,\varepsilon,M) \\ &\le \limsup_n \frac{1}{n} \log g_n(f,\varepsilon,M) = g(f,\varepsilon,M). \end{aligned} \tag{10.1.15}$$

Making $\varepsilon \to 0$, we get that $h(f,\alpha) \le g(f,M) = g(f)$. Since the open cover α is arbitrary, it follows that $h(f) \le g(f)$.

We define the *topological entropy* of a continuous transformation $f : M \to M$ in a metric space M to be $g(f) = s(f)$. Proposition 10.1.6 shows that this definition is compatible with the one we gave in Section 10.1.1

for transformations in compact topological spaces. A relevant difference is that, while for compact spaces the topological entropy depends only on the topology (because $h(f)$ is defined solely in terms of the open sets), in the non-compact case the topological entropy may also depend on the distance function in M. In this regard, see Exercises 10.1.4 and 10.1.5. They also show that in the non-compact case the topological entropy is no longer an invariant of topological conjugacy, although it remains an invariant of *uniformly* continuous conjugacy.

Example 10.1.7. Assume that $f: M \to M$ does not expand distances, that is, that $d(f(x), f(y)) \le d(x,y)$ for every $x, y \in M$. Then the topological entropy of f is equal to zero. Indeed, the hypothesis implies that $B(x,n,\varepsilon) = B(x,\varepsilon)$ for every $n \ge 1$. Hence, a set E is (n,ε)-generating if and only if it is $(1,\varepsilon)$-generating. In particular, the sequence $g_n(f,\varepsilon,K)$ does not depend on n and, hence, $g(f,\varepsilon,K) = 0$ for every $\varepsilon > 0$ and every compact set K. Making $\varepsilon \to 0$ and taking the supremum over K we get that $g(f) = 0$ (analogously, $s(f) = 0$).

There are two important special cases: *contractions*, such that there exists $\lambda < 1$ satisfying $d(f(x), f(y)) \le \lambda d(x,y)$ for every $x, y \in M$; and *isometries*, such that $d(f(x), f(y)) = d(x,y)$ for every $x, y \in M$. We saw in Lemma 6.3.6 that every compact metrizable group admits a distance relative to which every translation is an isometry. Therefore, it also follows from the previous observations that the topological entropy of every translation in a compact metrizable group is zero.

Recalling that $g(f) = g(f,M)$ and $s(f) = s(f,M)$ when M is compact, we see that the conclusion of Proposition 10.1.6 may be rewritten as follows:

$$h(f) = \lim_{\varepsilon \to 0} \limsup_n \frac{1}{n} \log g_n(f,\varepsilon,M) = \lim_{\varepsilon \to 0} \limsup_n \frac{1}{n} \log s_n(f,\varepsilon,M).$$

From the proof of the proposition we may also obtain the following related identity:

Corollary 10.1.8. *If $f: M \to M$ is a continuous transformation in a compact metric space then*

$$h(f) = \lim_{\varepsilon \to 0} \liminf_n \frac{1}{n} \log g_n(f,\varepsilon,M) = \lim_{\varepsilon \to 0} \liminf_n \frac{1}{n} \log s_n(f,\varepsilon,M).$$

Proof. The relation (10.1.15) gives that

$$h(f,\alpha) \le \liminf_n \frac{1}{n} \log g_n(f,\varepsilon,M)$$

whenever $\varepsilon > 0$ is a Lebesgue number for the open cover α. Making $\varepsilon \to 0$, we conclude that

$$h(f) \le \lim_{\varepsilon \to 0} \liminf_n \frac{1}{n} \log g_n(f,\varepsilon,M). \tag{10.1.16}$$

The first inequality in Lemma 10.1.5 implies that

$$\lim_{\varepsilon\to 0}\liminf_n \frac{1}{n}\log g_n(f,\varepsilon,M) \le \lim_{\varepsilon\to 0}\liminf_n \frac{1}{n}\log s_n(f,\varepsilon,M). \tag{10.1.17}$$

Also, it is clear that

$$\lim_{\varepsilon\to 0}\liminf_n \frac{1}{n}\log s_n(f,\varepsilon,M) \le \lim_{\varepsilon\to 0}\limsup_n \frac{1}{n}\log s_n(f,\varepsilon,M). \tag{10.1.18}$$

As we have just observed, the expression on the right-hand side is equal to $h(f)$. Therefore, the inequalities (10.1.16)–(10.1.18) imply the conclusion.

10.1.3 Calculation and properties

We start by proving a version of Lemma 9.1.13 for the topological entropy. The proof is a bit more elaborate because, unlike what happens for partitions, given an open cover α the covers $(\alpha^k)^n$ and α^{n+k-1} need not coincide if the elements of α are not pairwise disjoint.

Example 10.1.9. Let $f : M \to M$ be the shift map in $M = \{1,2,3\}^{\mathbb{N}}$ (or $M = \{1,2,3\}^{\mathbb{Z}}$) and α be the open cover of M consisting of the cylinders $[0;\{1,2\}]$ and $[0;\{1,3\}]$. For each $n \ge 1$, the cover α^n consists of the 2^n cylinders of the form $[0;A_0,\dots,A_{n-1}]$ with $A_j = [0;\{1,2\}]$ or $A_j = [0;\{1,3\}]$. In particular, $\#\alpha^3 = 8$. On the other hand, $(\alpha^2)^2$ contains 12 elements: the 8 elements of α^3 together with the 4 cylinders of the form $[0;A_0,\{1\},A_2]$ with $A_j = [0;\{1,2\}]$ or $A_j = [0;\{1,3\}]$ for $j = 0$ and $j = 2$. Hence, $\alpha^{n+k-1} \ne (\alpha^k)^n$ for $n = k = 2$.

Proposition 10.1.10. *Let M be a compact topological space, $f : M \to M$ be a continuous transformation and α be an open cover of M. Then $h(f,\alpha) = h(f,\alpha^k)$ for every $k \ge 1$. Moreover, if $f : M \to M$ is a homeomorphism then $h(f,\alpha) = h(f,\alpha^{\pm k})$ for every $k \ge 1$, where $\alpha^{\pm k} = \bigvee_{j=-k}^{k-1} f^{-j}(\alpha)$.*

Proof. The main point is to show that the open covers $(\alpha^k)^n$ and α^{n+k-1} have the same entropy, for every $n \ge 1$. We use the following simple fact, which will be useful again later:

Lemma 10.1.11. *Given any open cover α and any $n,k \ge 1$,*

1. *α^{n+k-1} is a subcover of $(\alpha^k)^n$ and, in particular, $(\alpha^k)^n \prec \alpha^{n+k-1}$;*
2. *for any subcover β of $(\alpha^k)^n$ there exists a subcover γ of α^{n+k-1} such that $\#\gamma \le \#\beta$ and $\gamma \prec \beta$.*

Proof. By definition, every element α^{n+k-1} has the form $B = \bigcap_{l=0}^{n+k-2} f^{-l}(B_l)$ with $B_l \in \alpha$ for every l. It is clear that this may be written in the form $B = \bigcap_{i=0}^{n-1} f^{-i}\big(\bigcap_{j=0}^{k-1} f^{-j}(B_{i+j})\big)$ and, thus, $B \in (\alpha^k)^n$. This proves the first claim. Next, let β be a subcover of $(\alpha^k)^n$. Every element of β has the form

$$A = \bigcap_{i=0}^{n-1} f^{-i}\Big(\bigcap_{j=0}^{k-1} f^{-j}(A_{i,j})\Big) = \bigcap_{l=0}^{n+k-2} f^{-l}\Big(\bigcap_{i+j=l} A_{i,j}\Big),$$

with $A_{i,j} \in \alpha$. Consider $B = \bigcap_{l=0}^{n+k-2} f^{-l}(B_l)$, where $B_l = A_{i,j}$ for some pair (i,j) such that $i+j=l$. Observe that $A \subset B$ and $B \in \alpha^{n+k-1}$. Therefore, the family γ formed by all the sets B obtained in this way satisfies all the conditions in the second claim.

According to the relation (10.1.2), the first part of Lemma 10.1.11 implies that $H((\alpha^k)^n) \le H(\alpha^{n+k-1})$. Clearly, the second part of the lemma implies the opposite inequality. Hence,

$$H(\alpha^{n+k-1}) = H((\alpha^k)^n) \quad \text{for any } n,k \ge 1, \tag{10.1.19}$$

as we claimed. Therefore,

$$h(f,\alpha^k) = \lim_n \frac{1}{n} H((\alpha^k)^n) = \lim_n \frac{1}{n} H(\alpha^{n+k-1}) = h(f,\alpha) \quad \text{for every } k.$$

When f is invertible, it follows from the definitions that $\alpha^{\pm k} = f^k(\alpha^{2k})$. Using Exercise 10.1.3, we get that $h(f,\alpha^{\pm k}) = h(f,f^k(\alpha^{2k})) = h(f,\alpha^{2k}) = h(f,\alpha)$.

The next proposition and its corollary simplify the calculation of the topological entropy significantly in concrete examples. Recall that, when M is a metric space, the *diameter* of an open cover is defined to be the supremum of the diameters of its elements.

Proposition 10.1.12. *Assume that M is a compact metric space. Let $(\beta_k)_k$ be any sequence of open covers of M such that* $\operatorname{diam}\beta_k$ *converges to zero. Then*

$$h(f) = \sup_k h(f,\beta_k) = \lim_k h(f,\beta_k).$$

Proof. Given any open cover α, let $\varepsilon > 0$ be a Lebesgue number of α. Take $n \ge 1$ such that $\operatorname{diam}\beta_k < \varepsilon$ for every $k \ge n$. By the definition of Lebesgue number, it follows that every element of β_k is contained in some element of α. In other words, $\alpha \prec \beta_k$ and, hence, $h(f,\beta_k) \ge h(f,\alpha)$. In view of the definition (10.1.5), this proves that

$$\liminf_k h(f,\beta_k) \ge h(f).$$

It is also clear from the definitions that $h(f) \ge \sup_k h(f,\beta_k) \ge \limsup_k h(f,\beta_k)$. Combining these observations, we obtain the conclusion of the proposition.

Corollary 10.1.13. *Assume that M is a compact metric space. If β is an open cover such that*

(1) *the diameter of the one-sided iterated sum $\beta^k = \bigvee_{j=0}^{k-1} f^{-j}(\beta)$ converges to zero when $k \to \infty$, or*
(2) *$f : M \to M$ is a homeomorphism and the diameter of the two-sided iterated sum $\beta^{\pm k} = \bigvee_{j=-k}^{k-1} f^{-j}(\beta)$ converges to zero when $k \to \infty$,*

then $h(f) = h(f,\beta)$.

Proof. In case (1), Propositions 10.1.10 and 10.1.12 yield

$$h(f) = \lim_k h(f, \beta^k) = h(f, \beta).$$

The proof in case (2) is analogous.

Next, we check that the topological entropy behaves as one could expect with respect to positive iterates, at least when the transformation is uniformly continuous:

Proposition 10.1.14. *If $f : M \to M$ is a uniformly continuous transformation in a metric space then $h(f^k) = kh(f)$ for every $k \in \mathbb{N}$.*

Proof. Fix $k \geq 1$ and let $K \subset M$ be any compact set. Consider any $n \geq 1$ and $\varepsilon > 0$. It is clear that if $E \subset M$ is an (nk, ε)-generating set of K for the transformation f then it is also an (n, ε)-generating set of K for the iterate f^k. Therefore, $g_n(f^k, \varepsilon, K) \leq g_{nk}(f, \varepsilon, K)$. Hence,

$$g(f^k, \varepsilon, K) = \lim_n \frac{1}{n} g_n(f^k, \varepsilon, K) \leq \lim_n \frac{1}{n} g_{nk}(f, \varepsilon, K) = kg(f, \varepsilon, K).$$

Making $\varepsilon \to 0$ and taking the supremum over K, we see that $h(f^k) \leq kh(f)$.

The proof of the other inequality uses the assumption that f is uniformly continuous. Take $\delta > 0$ such that $d(x, y) < \delta$ implies $d(f^j(x), f^j(y)) < \varepsilon$ for every $j \in \{0, \ldots, k-1\}$. If $E \subset M$ is an (n, δ)-generating set of K for f^k then E is an (nk, ε)-generating set of K for f. Therefore, $g_{nk}(f, \varepsilon, K) \leq g_n(f^k, \delta, K)$. This shows that $kg(f, \varepsilon, K) \leq g(f^k, \delta, K)$. Making ε and δ go to zero, we get that $kg(f, K) \leq g(f^k, K)$ for every compact set K. Hence, $kh(f) \leq h(f^k)$.

In particular, Proposition 10.1.14 holds for every continuous transformation in a compact metric space. On the other hand, in the case of homeomorphisms in compact spaces the conclusion extends to negative iterates:

Proposition 10.1.15. *If $f : M \to M$ is a homeomorphism of a compact metric space then $h(f^{-1}) = h(f)$. Consequently, $h(f^n) = |n| h(f)$ for every $n \in \mathbb{Z}$.*

Proof. Let α be an open cover of M. For every $n \geq 1$, denote

$$\alpha_+^n = \alpha \vee f^{-1}(\alpha) \vee \cdots \vee f^{-n+1}(\alpha) \quad \text{and} \quad \alpha_-^n = \alpha \vee f(\alpha) \vee \cdots \vee f^{n-1}(\alpha).$$

Observe that $\alpha_-^n = f^{n-1}(\alpha_+^n)$. Moreover, γ is a finite subcover of α_+^n if and only if $f^{n-1}(\gamma)$ is a finite subcover of α_-^n. Since the two subcovers have the same number of elements, it follows that $H(\alpha_+^n) = H(\alpha_-^n)$. Therefore,

$$h(f, \alpha) = \lim_n \frac{1}{n} H(\alpha_+^n) = \lim_n \frac{1}{n} H(\alpha_-^n) = h(f^{-1}, \alpha).$$

Since α is arbitrary, this proves that $h(f) = h(f^{-1})$. The second part of the statement follows from combining the first part with Proposition 10.1.14.

The claim in Proposition 10.1.15 is generally false when the space M is not compact:

Example 10.1.16. Let $M = \mathbb{R}$ with the distance $d(x,y) = |x - y|$ and take $f : \mathbb{R} \to \mathbb{R}$ to be given by $f(x) = 2x$. We are going to check that $h(f) \neq h(f^{-1})$. Let $K = [0,1]$ and, given $n \geq 1$ and $\varepsilon > 0$, take $E \subset \mathbb{R}$ to be any (n,ε)-generating set of K. In particular, every point of $f^{n-1}(K) = [0,2^{n-1}]$ is within less than ε from some point of $f^{n-1}(E)$. Hence,

$$2\varepsilon\#E = 2\varepsilon\#f^{n-1}(E) \geq 2^{n-1}.$$

This proves that $g_n(f,\varepsilon,K) \geq 2^{n-2}/\varepsilon$ for every n and, thus, $g(f,\varepsilon,K) \geq \log 2$. It follows that $h(f) \geq g(f,K) \geq \log 2$. On the other hand, f^{-1} is a contraction and so it follows from Example 10.1.7 that its topological entropy $h(f^{-1})$ is zero.

10.1.4 Exercises

10.1.1. Let M be a compact topological space. Show that if α and β are open covers of M such that $\alpha \prec \beta$ then $H(\alpha) \leq H(\beta)$.

10.1.2. Let $f : M \to M$ be a continuous transformation and α, β be open covers of a compact topological space M. Show that $H(\alpha \vee \beta) \leq H(\alpha) + H(\beta)$ and $H(f^{-1}(\beta)) \leq H(\beta)$. Check that if f is surjective then $H(f^{-1}(\beta)) = H(\beta)$.

10.1.3. Let M be a compact topological space. Show that if $f : M \to M$ is a surjective continuous transformation and β is an open cover of M then $h(f,\beta) = h(f,f^{-1}(\beta))$. Moreover, if f is a homeomorphism then $h(f,\beta) = h(f,f(\beta))$.

10.1.4. Let $M = (0,\infty)$ and $f : M \to M$ be given by $f(x) = 2x$. Calculate the topological entropy of f when one considers in M:
(a) the usual distance $d(x,y) = |x - y|$;
(b) the distance $d(x,y) = |\log x - \log y|$.
[Observation: Hence, in non-compact spaces the topological entropy may depend on the distance function, not just the topology.]

10.1.5. Consider in M two distances d_1 and d_2 that are uniformly equivalent: for every $\varepsilon > 0$ there exists $\delta > 0$ such that

$$d_1(x,y) < \delta \Rightarrow d_2(x,y) < \varepsilon \quad \text{and} \quad d_2(x,y) < \delta \Rightarrow d_1(x,y) < \varepsilon.$$

Show that if $f : M \to M$ is continuous with respect to either of the two distances then the value of the topological entropy is the same relative to both distances.

10.1.6. Let $f : M \to M$ and $g : N \to N$ be continuous transformations in compact metric spaces. Show that if there exists a continuous injective map $\psi : M \to N$ such that $\psi \circ f = g \circ \psi$ then $h(f) \leq h(g)$. Use this fact to show that the topological entropy of the shift map $\sigma : [0,1]^{\mathbb{Z}} \to [0,1]^{\mathbb{Z}}$ is infinite (thus, the topological entropy of a homeomorphism of a compact space need not be finite). [Observation: The first claim remains valid for non-compact spaces, as long as we require the inverse $\psi^{-1} : \psi(M) \to M$ to be uniformly continuous.]

10.1.7. Show that if $K, K_1, \dots, K_l$ are compact sets such that K is contained in $K_1 \cup \cdots \cup K_l$ then $g(f,K) \le \max_j g(f,K_j)$. Conclude that, given any $\delta > 0$,

$$g(f) = \sup\{g(f,K) : K \text{ compact with } \operatorname{diam} K < \delta\}$$

and analogously for $s(f)$.

10.1.8. Prove that the *logistic* map $f : [0,1] \to [0,1]$, $f(x) = 4x(1-x)$ is topologically conjugate to the map $g : [0,1] \to [0,1]$ defined by $g(x) = 1 - |2x-1|$. Use this fact to calculate $h(f)$.

10.1.9. Let $\mathcal{A}$ be a finite alphabet and $\sigma : \Sigma \to \Sigma$ be the shift map in $\Sigma = \mathcal{A}^{\mathbb{N}}$. The *complexity* of a sequence $x \in \Sigma$ is defined by $c(x) = \lim_n n^{-1} \log c_n(x)$, where $c_n(x)$ is the number of distinct words of length n that appear in x. Show that this limit exists and coincides with the topological entropy of the restriction $\sigma : \mathcal{X} \to \mathcal{X}$ of the shift map to the closure $\mathcal{X}$ of the orbit of x. [Observation: One interesting application we have in mind is in the context of Example 6.3.10, where x is the fixed point of a substitution.]

10.1.10. Check that if θ is the fixed point of the Fibonacci substitution in $\mathcal{A} = \{0,1\}$ (see Example 6.3.10) then $c_n(\theta) = n+1$ for every n and so the complexity $c(\theta)$ is equal to zero. Hence, the topological entropy of the shift map $\sigma : \mathcal{X} \to \mathcal{X}$ associated with the Fibonacci substitution is equal to zero.

10.2 Examples

Let us use a few concrete situations to illustrate the ideas introduced in the previous section.

10.2.1 Expansive maps

Recall (Section 9.2.3) that a continuous transformation $f : M \to M$ in a compact metric space is said to be *expansive* if there exists $\varepsilon_0 > 0$ such that $d(f^j(x), f^j(y)) < \varepsilon_0$ for every $j \in \mathbb{N}$ implies that $x = y$. When $f : M \to M$ is invertible, we say that it is *two-sided expansive* if there exists $\varepsilon_0 > 0$ such that $d(f^j(x), f^j(y)) < \varepsilon_0$ for every $j \in \mathbb{Z}$ implies that $x = y$. In both cases, ε_0 is called a *constant of expansivity* for f.

Proposition 10.2.1. *If $\varepsilon_0 > 0$ is a constant of expansivity for f then*

(i) $h(f) = h(f,\alpha)$ for every open cover α with diameter less than ε_0;
(ii) $h(f) = g(f,\varepsilon,M) = s(f,\varepsilon,M)$ for every $\varepsilon < \varepsilon_0/2$.

In particular, $h(f) < \infty$.

Proof. Let α be any open cover of M with diameter less than ε_0. We claim that $\lim_k \operatorname{diam} \alpha^k = 0$. Indeed, suppose that this is not so. It is clear that the sequence of diameters is non-increasing. Then, there exists $\delta > 0$ and for each $k \ge 1$ there exist points x_k and y_k in the same element of α^k such that $d(x_k, y_k) \ge \delta$.

By compactness, we may find a subsequence $(k_j)_j$ such that both $x = \lim_j x_{k_j}$ and $y = \lim_j y_{k_j}$ exist. On the one hand, $d(x,y) \geq \delta$ and so $x \neq y$. On the other hand, the fact that x_k and y_k are in the same element of α^k implies that

$$d(f^i(x_k), f^i(y_k)) \leq \operatorname{diam}\alpha \quad \text{for every } 0 \leq i < k.$$

Passing to the limit, we get that $d(f^i(x), f^i(y)) \leq \operatorname{diam}\alpha < \varepsilon_0$ for every $i \geq 0$. This contradicts the hypothesis that ε_0 is a constant of expansivity for f. This contradiction proves our claim. Using Corollary 10.1.13, it follows that $h(f) = h(f,\alpha)$, as claimed in part (i).

To prove part (ii), let α be the open cover of M formed by the balls of radius ε. Note that α^n contains every dynamical ball $B(x,n,\varepsilon)$:

$$B(x,n,\varepsilon) = \bigcap_{j=0}^{n-1} f^{-j}\big(B(f^j(x),\varepsilon)\big) \quad \text{and each} \quad B(f^j(x),\varepsilon) \in \alpha.$$

If E is an (n,ε)-generating set of M then $\{B(a,n,\varepsilon) : a \in E\}$ is an open cover of M; in view of what we have just said, it is a subcover of α^n. Therefore (recall also Lemma 10.1.5),

$$N(\alpha^n) \leq g_n(f,\varepsilon,M) \leq s_n(f,\varepsilon,M) \quad \text{for every } n.$$

Passing to the limit, we get that $h(f,\alpha) \leq g(f,\varepsilon,M) \leq s(f,\varepsilon,M)$. Recall that $s(f,\varepsilon,M) \leq s(f,M) = h(f)$. Since $\operatorname{diam}\alpha < \varepsilon_0$, the first part of the proposition yields that $h(f) = h(f,\alpha)$. These relations imply part (ii).

The last claim in the proposition is a direct consequence, since $g(f,\varepsilon,M)$, $s(f,\varepsilon,M)$ and $h(f,\alpha)$ are always finite. Indeed, that $h(f,\alpha) < \infty$ for every open cover was observed right after the definition (10.1.3). Then (10.1.14) implies that $s(f,\varepsilon,M) < \infty$ and (10.1.13) implies that $g(f,\varepsilon,M) < \infty$ for every $\varepsilon > 0$.

Exercise 10.2.8 contains an extension of Proposition 10.2.1 to *h-expansive* transformations, due to Rufus Bowen [Bow72]. Exercise 10.1.6 shows that the topological entropy of a continuous transformation, or even a homeomorphism, in a compact metric space may be infinite, if one omits the expansivity assumption.

Next, we prove that for expansive maps the topological entropy is an upper bound on the rate of growth of the number of periodic points. Let $\operatorname{Fix}(f^n)$ denote the set of all points $x \in M$ such that $f^n(x) = x$.

Proposition 10.2.2. *If M is a compact metric space and $f: M \to M$ is expansive then*

$$\limsup_n \frac{1}{n} \log \#\operatorname{Fix}(f^n) \leq h(f).$$

Proof. Let ε_0 be a constant of expansivity for f and α be any open cover of M with $\operatorname{diam}\alpha < \varepsilon_0$. We claim that every element of α^n contains at most one point of $\operatorname{Fix}(f^n)$. Indeed, if $x, y \in \operatorname{Fix}(f^n)$ are in the same element of α^n then

$d(f^i(x), f^i(y)) < \operatorname{diam} \alpha < \varepsilon_0$ for every $i = 0, \dots, n-1$. Since $f^n(x) = x$ and $f^n(y) = y$, it follows that $d(f^i(x), f^i(y)) < \varepsilon_0$ for every $i \geq 0$. By expansivity, this implies that $x = y$, which proves our claim. It follows that

$$\limsup_n \frac{1}{n} \log \# \operatorname{Fix}(f^n) \leq \limsup_n \frac{1}{n} \log N(\alpha^n) = h(f, \alpha).$$

Taking the limit when the diameter of α goes to zero, we get the conclusion of the proposition.

In some interesting situations, one can show that the topological entropy actually coincides with the rate of growth of the number of periodic points:

$$\lim_n \frac{1}{n} \log \# \operatorname{Fix}(f^n) = h(f). \tag{10.2.1}$$

That is the case, for example, for the shifts of finite type, which we are going to study in Section 10.2.2 (check Proposition 10.2.5 below). More generally, (10.2.1) holds whenever $f : M \to M$ is an expanding transformation in a compact metric space, as we are going to see in Section 11.3.

10.2.2 Shifts of finite type

Let $X = \{1, \dots, d\}$ be a finite set and $A = (A_{i,j})_{i,j}$ be a *transition matrix*, that is, a square matrix of dimension $d \geq 2$ with coefficients in the set $\{0, 1\}$ and such that no row is identically zero: for every i there exists j such that $A_{i,j} = 1$. Consider the subset Σ_A of $\Sigma = X^{\mathbb{N}}$ consisting of all the sequences $(x_n)_n \in \Sigma$ that are *A-admissible*, meaning that

$$A_{x_n, x_{n+1}} = 1 \quad \text{for every } n \in \mathbb{N}. \tag{10.2.2}$$

It is clear that Σ_A is invariant under the shift map $\sigma : \Sigma \to \Sigma$, in the sense that $\sigma(\Sigma_A) \subset \Sigma_A$. Note also that Σ_A is closed in Σ and, hence, it is a compact metric space (this is similar to Lemma 7.2.5).

The restriction $\sigma_A : \Sigma_A \to \Sigma_A$ of the shift map $\sigma : \Sigma \to \Sigma$ to this invariant compact set is called the *one-sided shift of finite type* associated with A. The two-sided shift of finite type associated with a transition matrix A is defined analogously, considering $\Sigma = X^{\mathbb{Z}}$ and requiring (10.2.2) for every $n \in \mathbb{Z}$. In this case, as part of the definition of a transition matrix, we also require the columns (not just the rows) of A to be non-zero.

The restriction of the shift map $\sigma : \Sigma \to \Sigma$ to the support of any Markov measure is a shift of finite type:

Example 10.2.3. Given a stochastic matrix $P = (P_{i,j})_{i,j}$, define $A = (A_{i,j})_{i,j}$ by

$$A_{i,j} = \begin{cases} 1 & \text{if } P_{i,j} > 0 \\ 0 & \text{if } P_{i,j} = 0. \end{cases}$$

Note that A is a transition matrix: the definition of a stochastic matrix implies that no row P is identically zero (in the two-sided situation we must assume

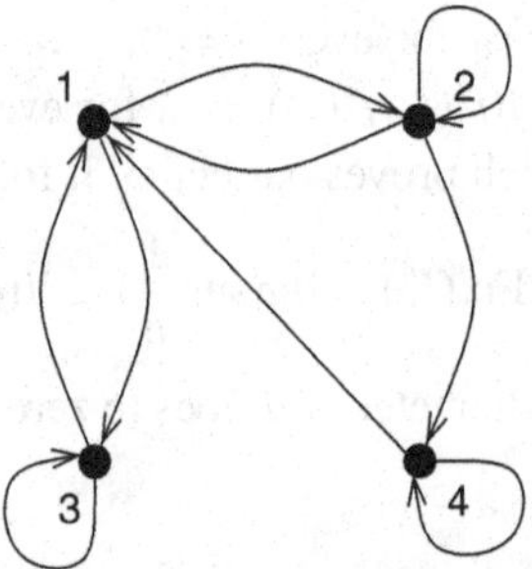

Figure 10.1. Graph associated with a transition matrix

that the columns of P are also not zero; this is automatic, for example, if the matrix P is aperiodic). Comparing (7.2.7) and (10.2.2), we see that a sequence is A-admissible if and only if it is P-admissible. Let μ be the Markov measure determined by a probability vector $p = (p_j)_j$ with positive coefficients and such that $P^*p = p$ (recall Example 7.2.2). By Lemma 7.2.5, the support of μ coincides with the set $\Sigma_A = \Sigma_P$ of all admissible sequences.

It is useful to associate with any transition matrix A the oriented graph whose vertices are the points of $X = \{1, \dots, d\}$ and such that there exists an edge from vertex a to vertex b if and only if $A_{a,b} = 1$. In other words,

$$G_A = \{(a,b) \in X \times X : A_{a,b} = 1\}.$$

For example, Figure 10.1 describes the graph associated with the matrix

$$A = \begin{pmatrix} 0 & 1 & 1 & 0 \\ 1 & 1 & 0 & 1 \\ 1 & 0 & 1 & 0 \\ 1 & 0 & 0 & 1 \end{pmatrix}.$$

A *path of length* $l \geq 1$ in the graph G_A is a sequence $a_0, \dots, a_l$ in X such that $A_{a_{i-1},a_i} = 1$ for every i, that is, such that there always exists an edge connecting a_{i-1} to a_i. Given $a, b \in X$ and $l \geq 1$, denote by $A^l_{a,b}$ the number of paths of length l starting at a and ending at b, that is, with $a_0 = a$ and $a_l = b$. Observe that:

1. $A^1_{a,b} = 1$ if there exists an edge connecting a to b and $A^1_{a,b} = 0$ otherwise. In other words, $A^1_{a,b} = A_{a,b}$ for every a, b.
2. The paths of length $l+m$ starting at a and ending at b are the concatenations of the paths of length l starting at a and ending at some point $z \in X$ with the paths of length m starting at that point z and ending at b. Therefore,

$$A^{l+m}_{a,b} = \sum_{z=1}^{d} A^l_{a,z} A^m_{z,b} \quad \text{for every } a, b \in X \text{ and every } l, m \geq 1.$$

It follows, by induction on l, that $A^l_{a,b}$ coincides with the coefficient in row a and column b of the matrix A^l.

The basic topological properties of shifts of finite type are analyzed in Exercise 10.2.2. In the proposition that follows we calculate the topological entropy of these transformations. We need a few prior observations about transition matrices.

Recall that the *spectral radius* $\rho(B)$ of a linear map $B:\mathbb{R}^d \to \mathbb{R}^d$ (that is, the largest absolute value of an eigenvalue of B) is given by

$$\rho(B) = \lim_n \|B^n\|^{1/n} = \lim_n |\operatorname{trc} B^n|^{1/n}, \tag{10.2.3}$$

where trc denotes the *trace* of the matrix and $\|\cdot\|$ denotes any norm in the vector space of linear maps (all norms are equivalent, as we are in finite dimension). Most of the time, one uses the *operator norm* $\|B\| = \sup\{\|Bv\|/\|v\| : v \neq 0\}$, but it will also be useful to consider the norm $\|\cdot\|_s$ defined by

$$\|B\|_s = \sum_{i,j=1}^{d} |B_{i,j}|.$$

Now take A to be a transition matrix. Since the coefficients of A are non-negative, we may use the Perron–Frobenius theorem (Theorem 7.2.3) to conclude that A admits a non-negative eigenvalue λ_A that is equal to the spectral radius. By our definition of the transition matrix, we also have that all the rows of A are non-zero. Then the same is true about A^n, for any $n \geq 1$ (Exercise 10.2.5). This implies that all the coefficients of the vector $A^n(1,\dots,1)$ are positive (and integer) and, thus,

$$\|A^n\| \geq \frac{\|A^n(1,\dots,1)\|}{\|(1,\dots,1)\|} \geq 1 \quad \text{for every } n \geq 1.$$

Using (10.2.3), we get that $\lambda_A = \rho(A) \geq 1$ for every transition matrix A.

Proposition 10.2.4. *The topological entropy $h(\sigma_A)$ of a shift of finite type $\sigma_A : \Sigma_A \to \Sigma_A$ is given by $h(\sigma_A) = \log \lambda_A$, where λ_A is the largest eigenvalue of the transition matrix A.*

Proof. We treat the case of one-sided shifts; the two-sided case is analogous, as the reader may readily check. Consider the open cover α of Σ_A formed by the restrictions

$$[0;a]_A = \{(x_j)_j \in \Sigma_A : x_0 = a\}$$

of the cylinders $[0;a]$ of Σ. For each $n \geq 1$, the open cover α^n is formed by the restrictions

$$[0;a_0,\dots,a_{n-1}]_A = \{(x_j)_j \in \Sigma_A : x_j = a_j \text{ for } j = 0,\dots,n-1\}$$

of the cylinders of length n. Observe that $[0;a_0,\dots,a_{n-1}]_A$ is non-empty if and only if $a_0,\dots,a_{n-1}$ is a path (of length $n-1$) in the graph G_A: it is evident that this condition is necessary; to see that it is also sufficient, use the assumption that for every i there exists j such that $A_{i,j} = 1$. Since the cylinders are pairwise disjoint, this observation shows that $N(\alpha^n)$ is equal to the total number of paths of length $n-1$ in the graph G_A. In other words,

$$N(\alpha^n) = \sum_{i,j=1}^{d} A_{i,j}^{n-1} = \|A^{n-1}\|_s.$$

By the spectral radius formula (10.2.3), it follows that

$$h(\sigma_A, \alpha) = \lim_n \frac{1}{n} \log N(\alpha^n) = \lim_n \frac{1}{n} \log \|A^{n-1}\|_s = \log \rho(A) = \log \lambda_A.$$

Finally, since $\operatorname{diam} \alpha^n \to 0$, Corollary 10.1.13 yields that $h(\sigma_A) = h(\sigma_A, \alpha)$.

Proposition 10.2.5. *If $\sigma_A : \Sigma_A \to \Sigma_A$ is a shift of finite type then*

$$h(\sigma_A) = \lim_n \frac{1}{n} \log \#\operatorname{Fix}(\sigma_A^n).$$

Proof. We treat the case of one-sided shifts, leaving the two-sided case for the reader. Note that $(x_k)_k \in \Sigma_A$ is a fixed point of σ_A^n if and only if $x_k = x_{k-n}$ for every $k \ge n$. In particular, every cylinder $[0; a_0, \dots, a_{n-1}]_A$ contains at most one element of $\operatorname{Fix}(\sigma_A^n)$. Moreover, the cylinder does contain a fixed point if and only if $a_0, \dots, a_{n-1}, a_0$ is a path (of length n) in the graph G_A. This proves that

$$\#\operatorname{Fix}(\sigma_A^n) = \sum_{i=1}^{d} A_{i,i}^n = \operatorname{trc} A^n$$

for every n. Consequently,

$$\lim_n \frac{1}{n} \log \#\operatorname{Fix}(\sigma_A^n) = \lim_n \frac{1}{n} \log \operatorname{trc} A^n = \log \rho(A).$$

Now the conclusion is a direct consequence of the previous proposition.

10.2.3 Topological entropy of flows

The definition of topological entropy extends easily to the context of continuous flows $\phi = \{\phi^t : M \to M : t \in \mathbb{R}\}$ in a metric space M, as we now explain.

Given $x \in M$ and $T > 0$ and $\varepsilon > 0$, the *dynamical ball* of center x, length T and radius $\varepsilon > 0$ is the set

$$B(x, T, \varepsilon) = \{y \in M : d(\phi^t(x), \phi^t(y)) < \varepsilon \text{ for every } 0 \le t \le T\}.$$

Let K be any compact subset of M. We say that $E \subset M$ is a *(T, ε)-generating* set for K if

$$K \subset \bigcup_{x \in E} B(x, T, \varepsilon),$$

and we say that $E \subset K$ is a *(T, ε)-separated* set if the dynamical ball $B(x, T, \varepsilon)$ of each $x \in E$ contains no other element of E.

Denote by $g_T(\phi,\varepsilon,K)$ the smallest cardinality of a (T,ε)-generating set of K and by $s_T(\phi,\varepsilon,K)$ the largest cardinality of a (T,ε)-separated set $E \subset K$. Then, take

$$g(\phi,K) = \lim_{\varepsilon\to 0}\limsup_{T\to\infty}\frac{1}{T}\log g_T(\phi,\varepsilon,K) \quad \text{and}$$

$$s(\phi,K) = \lim_{\varepsilon\to 0}\limsup_{T\to\infty}\frac{1}{T}\log s_T(\phi,\varepsilon,K)$$

and define

$$g(\phi) = \sup_K g(\phi,K) \quad \text{and} \quad s(\phi) = \sup_K s(\phi,K),$$

where both suprema are taken over all the compact sets $K \subset M$.

The next result, a continuous-time analogue of Proposition 10.1.4, ensures that these two last numbers coincide. We leave the proof up to the reader (Exercise 10.2.3). By definition, the *topological entropy* of the flow ϕ is the number $h(\phi) = g(\phi) = s(\phi)$.

Proposition 10.2.6. *We have* $g(\phi,K) = s(\phi,K)$ *for every compact* $K \subset M$. *Consequently,* $g(\phi) = s(\phi)$.

In the statement that follows we take the flow to be *uniformly continuous*, that is, such that for every $T > 0$ and $\varepsilon > 0$ there exists $\delta > 0$ such that

$$d(x,y) < \delta \quad \Rightarrow \quad d(\phi^t(x),\phi^t(y)) < \varepsilon \quad \text{for every } t \in [-T,T].$$

Observe that this is automatic for continuous flows when M is compact.

Proposition 10.2.7. *If the flow* ϕ *is uniformly continuous then its topological entropy* $h(\phi)$ *coincides with the topological entropy* $h(\phi^1)$ *of its time-*1 *map.*

Proof. It suffices to prove that $g(\phi,K) = g(\phi^1,K)$ for every compact $K \subset M$.

It is clear that if $E \subset M$ is (T,ε)-generating for K relative to the flow ϕ then E is also (n,ε)-generating for K relative to the time-1 map, for any $n \le T+1$. In particular, $g_n(\phi^1,\varepsilon,K) \le g_T(\phi,\varepsilon,K)$. It follows that

$$\limsup_n \frac{1}{n}\log g_n(\phi^1,\varepsilon,K) \le \limsup_{T\to\infty}\frac{1}{T}\log g_T(\phi,\varepsilon,K),$$

and so $g(\phi^1,K) \le g(\phi,K)$.

The hypothesis of uniform continuity is used for the opposite inequality. Given $\varepsilon > 0$, fix $\delta \in (0,\varepsilon)$ such that if $d(x,y) < \delta$ then $d(\phi^t(x),\phi^t(y)) < \varepsilon$ for every $t \in [0,1]$. If $E \subset M$ is an (n,δ)-generating set of K relative to ϕ^1 then E is a (T,ε)-generating set of K relative to the flow ϕ, for any $T \le n$. In particular, $g_T(\phi,\varepsilon,K) \le g_n(\phi^1,\delta,K)$. It follows that

$$\limsup_{T\to\infty}\frac{1}{T}\log g_T(\phi,\varepsilon,K) \le \limsup_n \frac{1}{n}\log g_n(\phi^1,\delta,K)$$

(given a sequence $(T_j)_j$ that realizes the supremum on the left-hand side, consider the sequence $(n_j)_j$ given by $n_j = [T_j]+1$). Making $\varepsilon \to 0$ (then $\delta \to 0$), we get that $g(\phi,K) \le g(\phi^1,K)$.

We have seen previously that for transformations the topological entropy is an invariant of topological (uniformly continuous) conjugacy. The same is true for flows: this follows from Proposition 10.2.7 and the obvious observation that any flow conjugacy also conjugates the corresponding time-1 maps. However, in the continuous-time context, one more often uses the concept of *topological equivalence*, which allows for rescaling of time. Clearly, topological equivalence need not preserve the topological entropy.

10.2.4 Differentiable maps

In this section we take M to be a Riemannian manifold (Appendix A.4.5). Let $f : M \to M$ be a differentiable map and $Df(x) : T_xM \to T_{f(x)}M$ denote the derivative of f at each point $x \in M$. Our goal is to prove that the norm of the derivative, defined by

$$\|Df(x)\| = \sup\left\{\frac{\|Df(x)v\|}{\|v\|} : v \in T_xM \text{ and } v \neq 0\right\},$$

determines an upper bound for the topological entropy $h(f)$ of f. For $x > 0$, we denote $\log^+ x = \max\{\log x, 0\}$.

Proposition 10.2.8. *Let $f : M \to M$ be a differentiable map in a Riemannian manifold of dimension d such that $\|Df\|$ is bounded. Then*

$$h(f) \le d \log^+ \sup \|Df\| < \infty.$$

Proof. Let $L = \sup\{\|Df(x)\| : x \in M\}$. By the mean value theorem,

$$d(f(x), f(y)) \le L d(x,y) \quad \text{for every } x, y \in M.$$

If $L \le 1$ then, as we have seen in Example 10.1.7, the entropy of f is zero. Thus, from now on we may suppose that $L > 1$.

Let $\mathcal{A}$ be an atlas of the manifold M consisting of charts $\varphi_\alpha : U_\alpha \to X_\alpha$ with $X_\alpha = (-2,2)^d$. Given any compact set $K \subset M$, we may find a finite family $\mathcal{A}_K \subset \mathcal{A}$ such that

$$\{\varphi_\alpha^{-1}((-1,1)^d) : \varphi_\alpha \in \mathcal{A}_K\}$$

covers K. Fix $B > 0$ such that $d(u,v) \le B d(\varphi_\alpha(u), \varphi_\alpha(v))$ for all $u, v \in [-1,1]^d$ and $\varphi_\alpha \in \mathcal{A}_K$. Given $n \ge 1$ and $\varepsilon > 0$, fix $\delta = (\varepsilon / B\sqrt{d}) L^{-n}$. Denote by $\delta\mathbb{Z}^d$ the set of all points of the form $(\delta k_1, \dots, \delta k_d)$ with $k_j \in \mathbb{Z}$ for every $j = 1, \dots, d$. Let $E \subset M$ be the union of the pre-images $\varphi_\alpha^{-1}(\delta\mathbb{Z}^d \cap (-1,1)^d)$, with $\varphi_\alpha \in \mathcal{A}_K$.

Note that every point of $(-1,1)^d$ is at a distance less than $\delta\sqrt{d}$ from some point of $\delta\mathbb{Z}^d \cap (-1,1)^d$. Therefore, for any $\varphi_\alpha \in \mathcal{A}_K$, every $x \in \varphi_\alpha^{-1}((-1,1)^d)$

is at a distance less than $B\delta\sqrt{d}$ from some point $a \in \varphi(\delta\mathbb{Z}^d \cap (-1,1)^d)$. Then, by the choice of δ,

$$d(f^j(x), f^j(a)) \leq L^j B\delta\sqrt{d} < L^n B\delta\sqrt{d} = \varepsilon$$

for every $j = 0, \ldots, n-1$. This proves that E is an (n,ε)-generating set for K. On the other hand, by construction,

$$\#E \leq \#\mathcal{A}_K \#\big(\delta\mathbb{Z}^d \cap (-1,1)^d\big) \leq \#\mathcal{A}_K (2/\delta)^d \leq \#\mathcal{A}_K (2B\sqrt{d}L^n/\varepsilon)^d,$$

so the expression on the right-hand side is an upper bound for $g_n(f,\varepsilon,K)$. Consequently,

$$g(f,\varepsilon,K) \leq \limsup_n \frac{1}{n} \log(2B\sqrt{d}L^n/\varepsilon)^d = d\log L.$$

Making $\varepsilon \to 0$ and taking the supremum over K, we get that $h(f) \leq d\log L$.

Combining Propositions 10.1.14 and 10.2.8, we find that

$$h(f) \leq \frac{1}{n} \log^+ \sup \|Df^n\| \quad \text{for every } n \geq 1.$$

When f is a homeomorphism, using Proposition 10.1.15 we also get that

$$h(f) \leq \frac{1}{n} \log^+ \sup \|Df^{-n}\| \quad \text{for every } n \geq 1.$$

The following conjecture of Michael Shub [Shu74] is central to the theory of topological entropy:

Conjecture 10.2.9 (Entropy conjecture)**.** If $f : M \to M$ is a diffeomorphism of class C^1 in a Riemannian manifold of dimension d, then

$$h(f) \geq \max_{1\leq k\leq d} \log \rho(f_k), \tag{10.2.4}$$

where each $\rho(f_k)$ denotes the spectral radius of the action $f_k : H_k(M) \to H_k(M)$ induced by f in the real homology of dimension k.

The full statement of the conjecture remains open to date, but several partial answers and related results have been obtained, both positive and negative. Let us summarize what is known in this regard.

It follows from a result of Yano [Yan80] that the inequality (10.2.4) is true for an open and dense subset of the space of homeomorphisms in any manifold of dimension $d \geq 2$. Moreover, it is true for *every* homeomorphism in certain classes of manifolds, such as the spheres or the infranilmanifolds [MP77b, MP77a, MP08]. On the other hand, Shub [Shu74] exhibited a Lipschitz homeomorphism, with zero topological entropy, for which (10.2.4) is false. See Exercise 10.2.7.

A useful way to approach (10.2.4) is by comparing the topological entropy with each one of the spectral radii $\rho(f_k)$. The case $k = d$ is relatively easy. Indeed, for any continuous map f in a manifold of dimension d, the spectral radius $\rho(f_d)$ is equal to the absolute value $|\deg f|$ of the degree of the map. In particular, the inequality $h(f) \geq \log \rho(f_d)$ is trivial for any homeomorphism. For non-invertible continuous maps, the topological entropy may be less than the logarithm of the absolute value of the degree. However, it was shown in [MP77b] that for differentiable maps one always has $h(f) \geq \log|\deg f|$.

Anthony Manning [Man75] proved that the inequality $h(f) \geq \log \rho(f_1)$ is true for every homeomorphism in a manifold of any dimension d. It follows that $h(f) \geq \log \rho(f_{d-1})$, since the duality theorem of Poincaré implies that

$$\rho(f_k) = \rho(f_{d-k}) \quad \text{for every } 0 < k < d.$$

In particular, the theorem of Manning together with the observations in the previous paragraph prove that entropy conjecture is true for every homeomorphisms in any manifold of dimension $d \leq 3$.

Rufus Bowen [Bow78] proved that for any homeomorphism in a manifold the topological entropy $h(f)$ is greater than or equal to the logarithm of the rate of growth of the fundamental group. One can show that this rate of growth is greater than or equal to $\rho(f_1)$. Thus, this result of Bowen implies the theorem of Manning that we have just mentioned.

The main result concerning the entropy conjecture is the theorem of Yosef Yomdin [Yom87], according to which the conjecture is true for every diffeomorphism of class C^∞. The crucial ingredient in the proof is a relation between the topological entropy $h(f)$ and the diffeomorphism's *rate of growth of volume*, which is defined as follows. For each $1 \leq k < d$, let B^k be the unit ball in $\mathbb{R}^k$. Denote by $v(\sigma)$ the k-dimensional volume of the image of any differentiable embedding $\sigma : B^k \to M$. Then, define

$$v_k(f) = \sup_\sigma \limsup_n \frac{1}{n} \log v(f^n \circ \sigma),$$

where the supremum is taken over all the embeddings $\sigma : B^k \to M$ of class C^∞. Define also $v(f) = \max\{v_k(f) : 1 \leq k < d\}$. It is not difficult to check that

$$\log \rho(f_k) \leq v_k(f) \quad \text{for every } 1 \leq k < d. \tag{10.2.5}$$

On the one hand, Sheldon Newhouse [New88] proved that $h(f) \leq v(f)$ for every diffeomorphism of class C^r with $r > 1$. On the other hand, Yomdin [Yom87] proved the opposite inequality:

$$v(f) \leq h(f), \tag{10.2.6}$$

for every diffeomorphism of class C^∞ (this inequality is false, in general, in the C^r case with $r < \infty$). Combining (10.2.5) with (10.2.6), one gets the entropy conjecture (10.2.4) for every diffeomorphism of class C^∞.

Concerning systems of class C^1, it is also known that the inequality (10.2.4) is true for every Axiom A diffeomorphism with no cycles [SW75], for certain partially hyperbolic diffeomorphisms [SX10] and, more generally, for any C^1 diffeomorphism far from homoclinic tangencies [LVY13].

10.2.5 Linear endomorphisms of the torus

In this section we calculate the topological entropy of the linear endomorphisms of the torus:

Proposition 10.2.10. *Let $f_A : \mathbb{T}^d \to \mathbb{T}^d$ be the endomorphism induced on the torus $\mathbb{T}^d$ by some square matrix A of dimension d with integer coefficients and non-zero determinant. Then*

$$h(f_A) = \sum_{j=1}^{d} \log^+ |\lambda_j|, \tag{10.2.7}$$

where $\lambda_1, \ldots, \lambda_d$ are the eigenvalues of A, counted with multiplicity.

We have seen in Proposition 9.4.3 that the entropy of f_A with respect to the Haar measure μ is equal to the expression on the right-hand side of (10.2.7). By the variational principle (Theorem 10.1), whose proof is contained in Section 10.4 below, the topological entropy is greater than or equal to the entropy of the transformation with respect to any invariant probability measure. Thus,

$$h(f_A) \ge h_\mu(f) = \sum_{j=1}^{d} \log^+ |\lambda_j|.$$

In what follows, we focus on proving the opposite inequality:

$$h(f_A) \le \sum_{j=1}^{d} \log^+ |\lambda_j|. \tag{10.2.8}$$

Initially, assume that A is diagonalizable, that is, that there exists a basis $v_1, \ldots, v_d$ of $\mathbb{R}^d$ with $Av_i = \lambda_i v_i$ for each i. Then, clearly, we may take the elements of such a basis to be unit vectors. Moreover, up to renumbering the eigenvalues, we may assume that there exists $u \in \{0, \ldots, d\}$ such that $|\lambda_i| > 1$ for $1 \le i \le u$ and $|\lambda_i| \le 1$ for every $i > u$. Let $e_1, \ldots, e_d$ be the canonical basis of $\mathbb{R}^d$ and $P : \mathbb{R}^d \to \mathbb{R}^d$ be the linear isomorphism defined by $P(e_i) = v_i$ for each i. Then $P^{-1}AP$ is a diagonal matrix. Fix $L > 0$ large enough so that $P((0,L)^d)$ contains some unit cube $\prod_{i=1}^d [b_i, b_i + 1]^d$. See Figure 10.2. Let $\pi : \mathbb{R}^d \to \mathbb{T}^d$ be the canonical projection. Then $\pi P((0,L)^d)$ contains the whole torus $\mathbb{T}^d$.

Given $n \ge 1$ and $\varepsilon > 0$, fix $\delta > 0$ such that $\|P\|\delta\sqrt{d} < \varepsilon$. Moreover, for each $i = 1, \ldots, d$, take

$$\delta_i = \begin{cases} \delta |\lambda_i|^{-n} & \text{if } i \le u \\ \delta & \text{if } i > u. \end{cases}$$

Consider the set

$$E = \pi P\big(\{(k_1\delta_1, \ldots, k_d\delta_d) \in (0,L)^d : k_1, \ldots, k_d \in \mathbb{Z}\}\big).$$

Observe also that, given any $j \ge 0$,

$$f_A^j(E) \subset \pi P\big(\{(k_1\lambda_1^j\delta_1, \ldots, k_d\lambda_d^j\delta_d) : k_1, \ldots, k_d \in \mathbb{Z}\}\big).$$

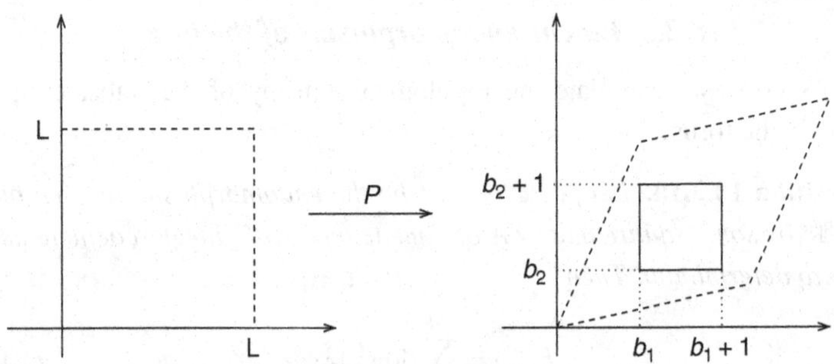

Figure 10.2. Building an (n,ε)-generating set in $\mathbb{T}^d$

Consider $0 \le j < n$. By construction, $|\lambda_i^j \delta_i| \le \delta$ for every $i = 1, \ldots, d$. Therefore, every point of $\mathbb{R}^d$ is at a distance less than or equal to $\delta\sqrt{d}$ from some point of the form $(k_1\lambda_1^j\delta_1, \ldots, k_d\lambda_d^j\delta_d)$. Then (see Figure 10.2), for each $x \in \mathbb{T}^d$ we may find $a \in E$ such that $d(f^j(x), f^j(a)) \le \|P\|\delta\sqrt{d} < \varepsilon$ for every $0 \le j < n$. This shows that E is an (n,ε)-generating set for $\mathbb{T}^d$. On the other hand,

$$\#E \le \prod_{i=1}^{d} \frac{L}{\delta_i} = \left(\frac{L}{\delta}\right)^d \prod_{i=1}^{u} |\lambda_i|^n.$$

These observations show that $g_n(f_A, \varepsilon, \mathbb{T}^d) \le (L/\delta)^d \prod_{i=1}^{u} |\lambda_i|^n$ for every $n \ge 1$ and $\varepsilon > 0$. Hence,

$$h(f) = \lim_{\varepsilon \to 0} \limsup_n \frac{1}{n} g_n(f_A, \varepsilon, \mathbb{T}^d) \le \sum_{i=1}^{u} \log |\lambda_i| = \sum_{i=1}^{d} \log^+ |\lambda_i|.$$

This proves Proposition 10.2.10 in the case when A is diagonalizable.

The general case may be treated in a similar fashion, writing the matrix A in its Jordan canonical form. The reader is invited to carry out the details.

10.2.6 Exercises

10.2.1. Let (M_i, d_i), $i = 1, 2$ be metric spaces and $f_i : M_i \to M_i$, $i = 1, 2$ be continuous transformations. Let $M = M_1 \times M_2$, d be the distance defined in M by

$$d((x_1, x_2), (y_1, y_2)) = \max\{d_1(x_1, y_1), d_2(x_2, y_2)\}$$

and $f : M \to M$ be the transformation defined by $f(x_1, x_2) = (f_1(x_1), f_2(x_2))$. Show that $h(f) \le h(f_1) + h(f_2)$ and the identity holds if at least one of the spaces is compact.

10.2.2. Let $\sigma_A : \Sigma_A \to \Sigma_A$ be a shift of finite type, either one-sided or two-sided. We say that a transition matrix A is *irreducible* if for any $i, j \in X$ there exists $n \ge 1$ such that $A_{i,j}^n > 0$ and that A is *aperiodic* if there exists $n \ge 1$ such that $A_{i,j}^n > 0$ for every $i, j \in X$. Show that:

(a) If A is irreducible then the set of periodic points of σ_A is dense in Σ_A.

(b) σ_A is transitive if and only if A is irreducible.

(c) σ_A is topologically mixing if and only if A is aperiodic.

[Observation: Condition (b) means that the oriented graph G_A is connected: given any $a,b \in X$ there exists some path in G_A starting at a and ending at b.]

10.2.3. Prove Proposition 10.2.6.

10.2.4. Let M be a compact metric space. Show that, given any $\varepsilon > 0$, the restriction of the topological entropy function $f \mapsto h(f)$ to the set of continuous transformations $f : M \to M$ that are ε-expansive is upper semi-continuous (with respect to the topology of uniform convergence).

10.2.5. Show that if A is a transition matrix then, for every $k \geq 1$, no row of A^k is identically zero. The same is true for the columns of A^k, $k \geq 1$, if we assume that A is a transition matrix in the two-sided sense.

10.2.6. (a) Let $f : M \to M$ be a surjective local homeomorphism in a compact metric space and let $d = \inf_y \#f^{-1}(y)$. Prove that $h(f) \geq \log d$.

(b) Let $f : S^1 \to S^1$ be a continuous map in the circle. Show that $h(f)$ is greater than or equal to the logarithm of the absolute value of the degree of f, that is, $h(f) \geq \log|\deg f|$.

[Observation: Misiurewicz and Przytycki [MP77b] proved that $h(f) \geq \log|\deg f|$ for every map $f : M \to M$ of class C^1 in a compact manifold.]

10.2.7. Consider the map $f : \overline{\mathbb{C}} \to \overline{\mathbb{C}}$ defined by $f(z) = z^d/(2|z|^{d-1})$, with $d \geq 2$. Prove that the topological entropy of f is zero, but the degree of f is d. Why is this not in contradiction with Exercise 10.2.6?

10.2.8. Let $f : M \to M$ be a continuous map in a compact metric space M. Given $\varepsilon > 0$, define

$$g_*(f,\varepsilon) = \sup\{g(f, B(x,\infty,\varepsilon)) : x \in M\},$$

where $B(x,\infty,\varepsilon)$ denotes the set of all $y \in M$ such that $d(f^i(x), f^i(y)) \leq \varepsilon$ for every $n \geq 0$. Bowen [Bow72] has shown that, given $b > 0$ and $\delta > 0$, there exists $c > 0$ such that

$$\log g_n(f,\delta,B(x,n,\varepsilon)) < c + (g_*(f,\varepsilon) + b)n \quad \text{for every } x \in M \text{ and } n \geq 1.$$

Using this fact, prove that $h(f) \leq g(f,\varepsilon,M) + g_*(f,\varepsilon)$. One says that f is *h-expansive* if $g_*(f,\varepsilon) = 0$ for some $\varepsilon > 0$. Conclude that in that case $h(f) = g(f,\varepsilon,M)$. [Observation: This generalizes Proposition 10.2.1, since every expansive transformation is also h-expansive.]

10.3 Pressure

In this section we introduce an important extension of the concept of topological entropy, called *(topological) pressure*, and we study its main properties. Throughout, we consider only continuous transformations in compact metric spaces. Related to this, check Exercises 10.3.4 and 10.3.5.

10.3.1 Definition via open covers

Let $f : M \to M$ be a continuous transformation in a compact metric space. We call a *potential* in M any continuous function $\phi : M \to \mathbb{R}$. For each $n \in \mathbb{N}$, define $\phi_n : M \to \mathbb{R}$ by $\phi_n = \sum_{i=0}^{n-1} \phi \circ f^i$. Given an open cover α of M, let

$$P_n(f,\phi,\alpha) = \inf\left\{\sum_{U\in\gamma} \sup_{x\in U} e^{\phi_n(x)} : \gamma \text{ is a finite subcover of } \alpha^n\right\}. \tag{10.3.1}$$

This sequence $\log P_n(f,\phi,\alpha)$ is subadditive (Exercise 10.3.1) and so the limit

$$P(f,\phi,\alpha) = \lim_n \frac{1}{n} \log P_n(f,\phi,\alpha) \tag{10.3.2}$$

exists. Define the *pressure* of the potential ϕ with respect to f to be the limit $P(f,\phi)$ of $P(f,\phi,\alpha)$ when the diameter of α goes to zero. The existence of this limit is guaranteed by the following lemma:

Lemma 10.3.1. *There exists* $\lim_{\operatorname{diam}\alpha\to 0} P(f,\phi,\alpha)$*, that is, there exists some* $P(f,\phi) \in \mathbb{R}$ *such that*

$$\lim_k P(f,\phi,\alpha_k) = P(f,\phi)$$

for every sequence $(\alpha_k)_k$ *of open covers with* $\operatorname{diam}\alpha_k \to 0$.

Proof. Let $(\alpha_k)_k$ and $(\beta_k)_k$ be any sequences of open covers with diameters converging to zero. Given any $\varepsilon > 0$, fix $\delta > 0$ such that $|\phi(x) - \phi(y)| \le \varepsilon$ whenever $d(x,y) \le \delta$. By assumption, $\operatorname{diam}\alpha_k < \delta$ for every k sufficiently large. For fixed k, let $\rho > 0$ be a Lebesgue number for α_k. By assumption, $\operatorname{diam}\beta_l < \rho$ for every l sufficiently large. By the definition of Lebesgue number, it follows that every $B \in \beta_l$ is contained in some $A \in \alpha_k$. Observe also that

$$\sup_{x\in A} \phi_n(x) \le n\varepsilon + \sup_{y\in B} \phi_n(y)$$

for every $n \ge 1$, since $\operatorname{diam}\alpha_k < \delta$. This implies that

$$P_n(f,\phi,\alpha_k) \le e^{n\varepsilon} P_n(f,\phi,\beta_l) \quad \text{for every } n \ge 1$$

and, hence, $P(f,\phi,\alpha_k) \le \varepsilon + P(f,\phi,\beta_l)$. Making $l \to \infty$ and then $k \to \infty$, we get that

$$\limsup_k P(f,\phi,\alpha_k) \le \varepsilon + \liminf_l P(f,\phi,\beta_l).$$

Since $\varepsilon > 0$ is arbitrary, it follows that $\limsup_k P(f,\phi,\alpha_k) \le \liminf_l P(f,\phi,\beta_l)$. Exchanging the roles of the two sequences of covers, we conclude that the limits $\lim_k P(f,\phi,\alpha_k)$ and $\lim_l P(f,\phi,\beta_l)$ exist and are equal.

Before we proceed, let us mention a few simple consequences of the definitions. The first is that the pressure of the zero potential coincides with the topological entropy. Indeed, it is immediate from (10.3.1) that $P_n(f,0,\alpha) = N(\alpha^n)$ for every $n \ge 1$ and, thus, $P(f,0,\alpha) = h(f,\alpha)$ for every open cover α.

Let $(\alpha_k)_k$ be any sequence of open covers with diameters going to zero. Then, by Proposition 10.1.12 and the definition of the pressure,

$$h(f) = \lim_k h(f,\alpha_k) = \lim_k P(f,0,\alpha_k) = P(f,0). \tag{10.3.3}$$

Observe, however, that for general potentials $P(f,\phi)$ need not coincide with the supremum of $P(f,\phi,\alpha)$ over all open covers α (see Exercise 10.3.5).

Given any constant $c \in \mathbb{R}$, we have that $P_n(f,\phi+c,\alpha) = e^{cn}P_n(f,\phi,\alpha)$ for every $n \geq 1$ and, consequently, $P(f,\phi+c,\alpha) = P(f,\phi,\alpha)+c$ for any open cover α. Hence,

$$P(f,\phi+c) = P(f,\phi)+c. \tag{10.3.4}$$

Analogously, if $\phi \leq \psi$ then $P_n(f,\phi,\alpha) \leq P_n(f,\psi,\alpha)$ for every $n \geq 1$, which implies that $P(f,\phi,\alpha) = P(f,\psi,\alpha)$ for every open cover α. That is,

$$\phi \leq \psi \Rightarrow P(f,\phi) \leq P(f,\psi). \tag{10.3.5}$$

In particular, since $\inf\phi \leq \phi \leq \sup\phi$, we have that

$$h(f)+\inf\phi \leq P(f,\phi) \leq h(f)+\sup\phi \tag{10.3.6}$$

for every potential ϕ. An interesting corollary is that if $h(f)$ is finite then $P(f,\phi) < \infty$ for every potential ϕ and, otherwise, $P(f,\phi) = \infty$ for every potential ϕ. An example of this last situation is given in Exercise 10.1.6.

Another simple consequence of the definition is that the pressure is an invariant of topological equivalence:

Proposition 10.3.2. *Let $f: M \to M$ and $g: N \to N$ be continuous transformations in compact metric spaces. If there exists a homeomorphism $h: M \to N$ such that $h \circ f = g \circ h$ then $P(g,\phi) = P(f,\phi\circ h)$ for every potential ϕ in N.*

Proof. The correspondence $\alpha \mapsto h(\alpha)$ is a bijection between the spaces of open covers of M and N, respectively. Moreover, since h and its inverse are (uniformly) continuous, $\operatorname{diam}\alpha_k \to 0$ if and only if $\operatorname{diam} h(\alpha_k) \to 0$. Consider the potential $\psi = \phi \circ h$ in M. Note that $\psi_n = \phi_n \circ h$ and so

$$\sup_{x\in U} \psi_n(x) = \sup_{y\in h(U)} \phi_n(y)$$

for every $U \subset M$ and every $n \geq 1$. Hence, $P_n(f,\psi,\alpha) = P_n(g,\phi,h(\alpha))$ for every n and every open cover α of M. Thus, $P(f,\psi,\alpha) = P(g,\phi,h(\alpha))$ and, taking the limit when the diameter of α goes to zero, $P(f,\psi) = P(g,\phi)$.

One may replace the supremum by the infimum in (10.3.1), that is, replace $P_n(f,\phi,\alpha)$ with

$$Q_n(f,\phi,\alpha) = \inf\left\{\sum_{U\in\gamma} \inf_{x\in U} e^{\phi_n(x)} : \gamma \text{ is a finite subcover of } \alpha^n\right\},$$

although this makes the definition a bit more complicated. In contrast with $\log P_n(f,\phi,\alpha)$, the sequence $\log Q_n(f,\phi,\alpha)$ need not be subadditive. Denote

$$Q^-(f,\phi,\alpha) = \liminf_n \frac{1}{n}\log Q_n(f,\phi,\alpha) \quad \text{and}$$
$$Q^+(f,\phi,\alpha) = \limsup_n \frac{1}{n}\log Q_n(f,\phi,\alpha).$$

Clearly, $Q^-(f,\phi,\alpha) \le Q^+(f,\phi,\alpha)$ for every open cover α of M. Furthermore, $Q_n(f,0,\alpha) = P_n(f,0,\alpha) = N(\alpha^n)$ for every n and so $Q^-(f,0,\alpha) = Q^+(f,0,\alpha) = P(f,0,\alpha) = h(f,\alpha)$.

Corollary 10.3.3. *For any potential $\phi : M \to \mathbb{R}$,*

$$P(f,\phi) = \lim_{\operatorname{diam}\alpha\to 0} Q^+(f,\phi,\alpha) = \lim_{\operatorname{diam}\alpha\to 0} Q^-(f,\phi,\alpha).$$

Proof. Since ϕ is (uniformly) continuous, given any $\varepsilon > 0$ there exists $\delta > 0$ such that

$$\inf_{x\in C}\phi_n(x) \le \sup_{x\in C}\phi_n(x) \le n\varepsilon + \inf_{x\in C}\phi_n(x)$$

whenever $\operatorname{diam} C \le \delta$. So,

$$Q_n(f,\phi,\alpha) \le P_n(f,\phi,\alpha) \le e^{n\varepsilon} Q_n(f,\phi,\alpha)$$

for every open cover α with $\operatorname{diam}\alpha \le \delta$. It follows that

$$\limsup_n \frac{1}{n}\log Q_n(f,\phi,\alpha) \le P(f,\phi,\alpha) \le \varepsilon + \liminf_n \frac{1}{n}\log Q_n(f,\phi,\alpha).$$

As the diameter of α goes to zero, we may take $\varepsilon \to 0$. Thus,

$$\lim_{\operatorname{diam}\alpha\to 0} Q^-(f,\phi,\alpha) = \lim_{\operatorname{diam}\alpha\to 0} Q^+(f,\phi,\alpha) = \lim_{\operatorname{diam}\alpha\to 0} P(f,\phi,\alpha) = P(f,\phi),$$

as claimed.

10.3.2 Generating sets and separated sets

Now we present two alternative definitions of pressure, in terms of generating sets and separated sets. As before, $f : M \to M$ is a continuous transformation in a compact metric space and $\phi : M \to \mathbb{R}$ is a continuous function.

Given $n \ge 1$ and $\varepsilon > 0$, define

$$G_n(f,\phi,\varepsilon) = \inf\left\{\sum_{x\in E} e^{\phi_n(x)} : E \text{ is an } (n,\varepsilon)\text{-generating set for } M\right\} \quad \text{and}$$
$$S_n(f,\phi,\varepsilon) = \sup\left\{\sum_{x\in E} e^{\phi_n(x)} : E \text{ is an } (n,\varepsilon)\text{-separated set in } M\right\}. \tag{10.3.7}$$

Next, define

$$G(f,\phi,\varepsilon) = \limsup_n \frac{1}{n} \log G_n(f,\phi,\varepsilon) \quad \text{and}$$
$$S(f,\phi,\varepsilon) = \limsup_n \frac{1}{n} \log S_n(f,\phi,\varepsilon), \tag{10.3.8}$$

and also

$$G(f,\phi) = \lim_{\varepsilon\to 0} G(f,\phi,\varepsilon) \quad \text{and} \quad S(f,\phi) = \lim_{\varepsilon\to 0} S(f,\phi,\varepsilon) \tag{10.3.9}$$

(these limits exist because the functions are monotonic in ε).

Note that $G_n(f,0,\varepsilon) = g_n(f,\varepsilon)$ and $S_n(f,0,\varepsilon) = s_n(f,\varepsilon)$ for every $n \geq 1$ and every $\varepsilon > 0$. Therefore (Proposition 10.1.6), $G(f,0) = g(f)$ and $S(f,0) = s(f)$ coincide with the topological entropy $h(f)$. In fact,

Proposition 10.3.4. *$P(f,\phi) = G(f,\phi) = S(f,\phi)$ for every potential ϕ in M.*

Proof. Consider $n \geq 1$ and $\varepsilon > 0$. It is clear from the definitions that every maximal (n,ε)-separated set is (n,ε)-generating. Then,

$$\begin{aligned} S_n(f,\phi,\varepsilon) &= \sup\left\{\sum_{x\in E} e^{\phi_n(x)} : E \text{ is } (n,\varepsilon)\text{-separated}\right\} \\ &= \sup\left\{\sum_{x\in E} e^{\phi_n(x)} : E \text{ is } (n,\varepsilon)\text{-separated maximal}\right\} \\ &\geq \inf\left\{\sum_{x\in E} e^{\phi_n(x)} : E \text{ is } (n,\varepsilon)\text{-generating}\right\} = G_n(f,\phi,\varepsilon) \end{aligned} \tag{10.3.10}$$

for every n and every ε. This implies that $G(f,\phi,\varepsilon) \leq S(f,\phi,\varepsilon)$ for every ε and, thus, $G(f,\phi) \leq S(f,\phi)$.

Next, we prove that $S(f,\phi) \leq P(f,\phi)$. Let ε and δ be positive numbers such that $d(x,y) \leq \delta$ implies $|\phi(x) - \phi(y)| \leq \varepsilon$. Let α be any open cover of M with $\operatorname{diam}\alpha < \delta$ and $E \subset M$ be any (n,δ)-separated set. Given any subcover γ of α^n, it is obvious that every point of E is contained in some element of γ. On the other hand, the hypothesis that E is (n,δ)-separated implies that each element of γ contains at most one element of E. Therefore,

$$\sum_{x\in E} e^{\phi_n(x)} \leq \sum_{U\in\gamma} \sup_{y\in U} e^{\phi_n(y)}.$$

Taking the supremum in E and the infimum in γ, we get that

$$S_n(f,\phi,\delta) \leq P_n(f,\phi,\alpha). \tag{10.3.11}$$

It follows that $S(f,\phi,\delta) \leq P(f,\phi,\alpha)$. Making $\delta \to 0$ (hence $\operatorname{diam}\alpha \to 0$), we conclude that $S(f,\phi) \leq P(f,\phi)$, as stated.

Finally, we prove that $P(f,\phi) \leq G(f,\phi)$. Let ε and δ be positive numbers such that $d(x,y) \leq \delta$ implies $|\phi(x) - \phi(y)| \leq \varepsilon$. Let α be any open cover of M with $\operatorname{diam}\alpha < \delta$ and $\rho > 0$ be a Lebesgue number of α. Let $E \subset M$ be any

(n,ρ)-generating set for M. For each $x \in E$ and $i = 0,\dots,n-1$, there exists $A_{x,i} \in \alpha$ such that $B(f^i(x),\rho)$ is contained in $A_{x,i}$. Denote

$$\gamma(x) = \bigcap_{i=0}^{n-1} f^{-i}(A_{x,i}).$$

Observe that $\gamma(x) \in \alpha^n$ and $B(x,n,\rho) \subset \gamma(x)$. Hence, the hypothesis that E is (n,ρ)-generating implies that $\gamma = \{\gamma(x) : x \in E\}$ is a subcover of α. Observe also that

$$\sup_{y \in \gamma(x)} \phi_n(y) \le n\varepsilon + \phi_n(x) \quad \text{for every } x \in E,$$

since $\operatorname{diam} A_{x,i} < \delta$ for every i. It follows that

$$\sum_{U \in \gamma} \sup_{y \in U} e^{\phi_n(y)} \le e^{n\varepsilon} \sum_{x \in E} e^{\phi_n(x)}.$$

This proves that $P_n(f,\phi,\alpha) \le e^{n\varepsilon} G_n(f,\phi,\rho)$ for every $n \ge 1$ and, consequently,

$$P(f,\phi,\alpha) \le \varepsilon + \liminf_n \frac{1}{n} G_n(f,\phi,\rho) \le \varepsilon + G(f,\phi,\rho). \tag{10.3.12}$$

Making $\rho \to 0$ we find that $P(f,\phi,\alpha) \le \varepsilon + G(f,\phi)$. Hence, making ε, δ and $\operatorname{diam}\alpha$ go to zero, $P(f,\phi) \le G(f,\phi)$.

The conclusion of Proposition 10.3.4 may be rewritten as follows:

$$\begin{aligned} P(f,\phi) &= \lim_{s\to 0} \limsup_n \frac{1}{n} \log G_n(f,\phi,s) \\ &= \lim_{s\to 0} \limsup_n \frac{1}{n} \log S_n(f,\phi,s). \end{aligned} \tag{10.3.13}$$

The relations (10.3.12) and (10.3.10) in the proof also give that

$$P(f,\phi) \le \lim_{s\to 0} \liminf_n \frac{1}{n} \log G_n(f,\phi,s) \le \lim_{s\to 0} \liminf_n \frac{1}{n} \log S_n(f,\phi,s).$$

Combining these observations, we get:

$$\begin{aligned} P(f,\phi) &= \lim_{s\to 0} \liminf_n \frac{1}{n} \log G_n(f,\phi,s) \\ &= \lim_{s\to 0} \liminf_n \frac{1}{n} \log S_n(f,\phi,s). \end{aligned} \tag{10.3.14}$$

10.3.3 Properties

Properties of the pressure function in the spirit of Proposition 10.1.10 and Corollary 10.1.13 are stated in Exercise 10.3.3. Let us also extend Propositions 10.1.14 and 10.1.15 to the present context:

Proposition 10.3.5. *Let $f : M \to M$ be a continuous transformation in a compact metric space and ϕ be a potential in M. Then:*

(1) $P(f^k,\phi_k) = kP(f,\phi)$ for every $k \ge 1$.
(2) If f is a homeomorphism then $P(f^{-1},\phi) = P(f,\phi)$.

Proof. Given a potential $\phi : M \to \mathbb{R}$ and an open cover α, denote $\psi = \sum_{i=0}^{k-1} \phi \circ f^i$ and $\beta = \bigvee_{i=0}^{k-1} f^{-i}(\alpha)$. Let $g = f^k$. It is clear that

$$\sum_{j=0}^{n-1} \psi \circ g^j = \sum_{l=0}^{kn-1} \phi \circ f^l \quad \text{and} \quad \bigvee_{j=0}^{n-1} g^{-j}(\beta) = \bigvee_{l=0}^{nk-1} f^{-l}(\alpha).$$

Then,

$$P_n(g,\psi,\beta) = \inf\left\{ \sum_{U\in\gamma} \sup_{x\in U} e^{\sum_{j=0}^{n-1} \psi(g^j(x))} : \gamma \subset \bigvee_{j=0}^{n-1} g^{-j}(\beta) \right\}$$

$$= \inf\left\{ \sum_{U\in\gamma} \sup_{x\in U} e^{\sum_{l=0}^{kn-1} \phi(f^l(x))} : \gamma \subset \bigvee_{l=0}^{nk-1} f^{-l}(\alpha) \right\} = P_{kn}(f,\phi,\alpha).$$

Consequently, $P(f^k,\psi,\beta) = kP(f,\phi,\alpha)$ for any α. Making $\operatorname{diam}\alpha \to 0$ (note that $\operatorname{diam}\beta \to 0$), we deduce that $P(f^k,\psi) = kP(f,\phi)$. This proves part (1).

Suppose that f is a homeomorphism. Given an open cover α and an integer number $n \ge 1$, denote

$$\phi_n^- = \sum_{j=0}^{n-1} \phi \circ f^{-j} \quad \text{and} \quad \alpha_-^n = \alpha \vee f(\alpha) \vee \cdots \vee f^{n-1}(\alpha).$$

It is clear that $\phi_n^- = \phi_n \circ f^{n-1}$ and $\alpha_-^n = f^{n-1}(\alpha^n)$. Moreover, γ is a subcover of α^n if and only if $\delta = f^{n-1}(\gamma)$ is a subcover of α_-^n. Combining these facts, we find that

$$P_n(f^{-1},\phi,\alpha) = \inf\left\{ \sum_{U\in\delta} \sup_{x\in U} e^{\phi_n^-(x)} : \gamma \subset \alpha_-^n \right\}$$

$$= \inf\left\{ \sum_{V\in\gamma} \sup_{y\in V} e^{\phi_n(y)} : \delta \subset \alpha^n \right\} = P_n(f,\phi,\alpha)$$

for every $n \ge 1$. Hence, $P(f^{-1},\phi,\alpha) = P(f,\phi,\alpha)$ for every open cover α. Making $\operatorname{diam}\alpha \to 0$, we reach the conclusion in part (2).

Next, we fix the transformation $f : M \to M$ and we consider $P(f,\cdot)$ as a function in the space $C^0(M)$ of all continuous functions, with the norm defined by

$$\|\varphi\| = \sup\{|\varphi(x)| : x \in M\}.$$

We have seen in (10.3.6) that if the topological entropy $h(f)$ is infinite then the pressure function is constant and equal to ∞. In what follows we assume that $h(f)$ is finite. Then, $P(f,\phi)$ is finite for every potential ϕ.

Proposition 10.3.6. *The pressure function is Lipschitz, with Lipschitz constant equal to 1: $|P(f,\phi) - P(f,\psi)| \le \|\phi - \psi\|$ for any potentials ϕ and ψ.*

Proof. Clearly, $\phi \le \psi + \|\phi - \psi\|$. Hence, by (10.3.4) and (10.3.5), we have that $P(f,\phi) \le P(f,\psi) + \|\phi - \psi\|$. Exchanging the roles of ϕ and ψ, one gets the other inequality.

Proposition 10.3.7. *The pressure function is convex:*

$$P(f,(1-t)\phi + t\psi) \le (1-t)P(f,\phi) + tP(f,\psi)$$

for any potentials ϕ and ψ in M and any $0 \le t \le 1$.

Proof. Write $\xi = (1-t)\phi + t\psi$. Then $\xi_n = (1-t)\phi_n + t\psi_n$ for every $n \ge 1$ and, thus, $\sup(\xi_n \mid U) \le (1-t)\sup(\phi_n \mid U) + t\sup(\psi_n \mid U)$ for every $U \subset M$. Then, by the Hölder inequality (Theorem A.5.5),

$$\sum_{U\in\gamma} \sup_{x\in U} e^{\xi_n(x)} \le \Big(\sum_{U\in\gamma} \sup_{x\in U} e^{\phi_n(x)}\Big)^{1-t} \Big(\sum_{U\in\gamma} \sup_{x\in U} e^{\psi_n(x)}\Big)^{t}$$

for any finite family γ of subsets of M. This implies that, given any open cover α,

$$P_n(f,\xi,\alpha) \le P_n(f,\phi,\alpha)^{1-t} P_n(f,\psi,\alpha)^t$$

for every $n \ge 1$ and, hence, $P(f,\xi,\alpha) \le (1-t)P(f,\phi,\alpha) + tP(f,\psi,\alpha)$. Passing to the limit when $\operatorname{diam}\alpha \to 0$, we get the conclusion of the proposition.

We say that two potentials $\phi, \psi : M \to \mathbb{R}$ are *cohomologous* if there exists a continuous function $u : M \to \mathbb{R}$ such that $\phi = \psi + u \circ f - u$. Note that this is an equivalence relation in the space of potentials (Exercise 10.3.6).

Proposition 10.3.8. *Let $f : M \to M$ be a continuous transformation in a compact topological space. If $\phi, \psi : M \to \mathbb{R}$ are cohomologous potentials then $P(f,\phi) = P(f,\psi)$.*

Proof. If $\psi = \phi + u \circ f - u$ then $\psi_n(x) = \phi_n(x) + u(f^n(x)) - u(x)$ for every $n \in \mathbb{N}$. Let $K = \sup|u|$. Then $|\sup_{x\in C} \psi_n(x) - \sup_{x\in C} \phi_n(x)| \le 2K$ for every set $C \subset M$. Hence, for any open cover γ,

$$e^{-2K} \sum_{U\in\gamma} \sup_{x\in U} e^{\phi_n(x)} \le \sum_{U\in\gamma} \sup_{x\in U} e^{\psi_n(x)} \le e^{2K} \sum_{U\in\gamma} \sup_{x\in U} e^{\phi_n(x)}.$$

This implies that, given any open cover α of M,

$$e^{-2K} P_n(f,\phi,\alpha) \le P_n(f,\psi,\alpha) \le e^{2K} P_n(f,\phi,\alpha)$$

for every n. Therefore, $P(f,\phi,\alpha) = P(f,\psi,\alpha)$ for every α and, consequently, $P(f,\phi) = P(f,\psi)$.

10.3.4 Comments in statistical mechanics

Let us take a pause to explain the relation between the mathematical concept of pressure and the issues in physics that originated it. This also serves as a preview to Chapter 12, where this theory will be developed in the context of expanding maps in metric spaces. The discussion that follows is a combination of mathematical results and physical considerations, not necessarily rigorous, and is quite brief: we refer the reader to the classical works of David Ruelle [Rue04] and Oscar Lanford [Lan73] for actual presentations of the subject.

The goal of statistical mechanics is to describe the properties of physical systems consisting of a large number of units that interact with each other. For example, these units may be particles, such as molecules of a gas, or sites in a crystal grid, which may or may not be occupied by particles. The *constant of Avogadro* 6.022×10^{23} illustrates what one means by "large" in specific situations in this context.

The main challenge in this area of mathematical physics is to understand the phenomena of *phase transitions*, that is, sudden changes from one physical state to another: for example, what happens when liquid water turns into ice? Why does this occur suddenly, at a given freezing temperature? Mathematical methods developed for tackling this kind of question turn out to be very useful in other areas of science, such as quantum field theory and, closer to the scope of this book, the ergodic theory of hyperbolic dynamical systems (Bowen [Bow75a]).

In order to formulate these problems in mathematical terms, it is convenient to assume that the set L of units in the system is actually infinite, because finite systems do not have genuine phase transitions. The best-studied examples are the *lattice systems*, for which $L = \mathbb{Z}^d$ with $d \geq 1$. It is assumed that each unit has a finite set F of possible values (or "states"). For example, $F = \{-1, +1\}$ in the case of *spin systems*, with ± 1 representing the two possible orientations of the particle's "spin", and $F = \{0, 1\}$ in the case of *lattice gases*: 1 means that the site $k \in L$ is occupied by a gas molecule, whereas 0 means that the site is empty.

Then, the system's *configuration space* is a subset Ω of the product space F^L. We assume Ω to be closed in F^L and invariant under the shift map

$$\sigma^n : F^L \to F^L, \quad (\xi_k)_{k \in L} \mapsto (\xi_{k+n})_{k \in L}$$

for every $n \in L$. A *state* of the system is a probability measure μ on Ω: intuitively, one presumes that at the microscopic level the system oscillates randomly between different configurations $\xi \in \Omega$ (for example, different positions or velocities of the molecules), all corresponding to the same macroscopic parameters (same temperature, etc.); then, the measure μ describes the probability distribution of these microscopic configurations.

States corresponding to macroscopic configurations that can be physically observed, that is, that actually occur in Nature, are called *equilibrium states*. This notion has a central role in the theory, in particular, because *phase transitions are associated with the coexistence of more than one equilibrium state*. Under our hypotheses on Ω, one can show that every equilibrium state μ is invariant under the shift maps σ^n, $n \in L$. Thus, the study of lattice systems is naturally inserted in the scope of ergodic theory.

According to the *variational principle* of statistical mechanics, which goes back to the *principle of least action* of Maupertuis, the equilibrium states are characterized by the fact that they minimize a certain fundamental quantity, called *Gibbs free energy*, whose definition involves the energy E, the temperature T and the entropy S of the system's state. The *pressure* of that state is, simply, the product of the Gibbs free energy and a negative factor $-\beta$ whose nature will be explained shortly.[1] Therefore, the equilibrium states are also characterized by the fact that they maximize the pressure among all probability measures invariant under the shift maps σ^n, $n \in L$.

From these facts, one can obtain a rather explicit description of the equilibrium states for lattice systems: under suitable hypotheses, *the equilibrium states are precisely the Gibbs states invariant under the shift maps*. In the remainder of this section we are going to motivate and define this concept of *Gibbs state*, which will also allow us to illustrate the ideas outlined in the previous paragraphs. By the end of the section we briefly comment on the case of one-dimensional lattice systems, that is, the case $d = 1$, whose theory is much simpler and which is more closely related to the topics treated in this book.

Let us start by considering the particularly simple case of finite systems, that is, such that the configuration space Ω is finite. The *entropy* of a state μ in Ω is the number

$$S(\mu) = \sum_{\xi \in \Omega} -\mu(\{\xi\}) \log \mu(\{\xi\}).$$

To each configuration $\xi \in \Omega$ corresponds a value $E(\xi)$ for the energy of the system. Denote by $E(\mu)$ the energy of the state μ, that is, the mean

$$E(\mu) = \sum_{\xi \in \Omega} \mu(\{\xi\}) E(\{\xi\}).$$

Take the system's absolute temperature T to be constant in time. Then, the *Gibbs free energy* is defined by

$$G(\mu) = E(\mu) - \kappa T S(\mu),$$

[1] From the mathematical point of view, the two quantities are equivalent. Preference for one denomination or the other has mostly to do with the physical interpretation of the set F: for spin systems one usually refers to the Gibbs free energy, whereas for lattice gases, where the elements of F describe the rate of occupation of each site, it is more natural to refer to the pressure.

where $\kappa = 1.380 \times 10^{-23}\,\mathrm{m}^2\,\mathrm{kg}\,\mathrm{s}^{-2}\,\mathrm{K}^{-1}$ is the *Boltzmann constant*. In other words, denoting $\beta = 1/(\kappa T)$,

$$-\beta G(\mu) = \sum_{\xi\in\Omega} \mu(\{\xi\})\big[-\beta E(\xi) - \log\mu(\{\xi\})\big]. \tag{10.3.15}$$

This expression is denoted by $P(\mu)$ and is called *pressure* of the state μ.

It is easy to check that the pressure (10.3.15) is maximum (hence, the Gibbs free energy $G(\mu)$ is minimum) if and only if

$$\mu(\{\xi\}) = \frac{e^{-\beta E(\xi)}}{\sum_{\eta\in\Omega} e^{-\beta E(\eta)}} \quad \text{for every } \xi\in\Omega \tag{10.3.16}$$

(see Lemma 10.4.4 below). Therefore, the *Gibbs distribution* μ given by (10.3.16) is the unique equilibrium state of the system. In particular, in this simple context there are no phase transitions.

Now we sketch how this analysis can be extended to infinite lattice systems, assuming that the interaction between sites that are far apart is sufficiently weak. It is part of the hypotheses that the energy associated with each configuration $\xi\in\Omega$ comes from the pairwise interactions between the different sites in the lattice (including self-interactions) and that this interaction is invariant under the shift maps σ^n, $n\in L$. Then, the energy $E_{k,l}$ resulting from the action of any site $k\in L$ on any other site $l\in L$ depends only on their relative position and on the values of ξ_k and ξ_l. In other words, there exists a function $\Psi : L\times F\times F \to \bar{\mathbb{R}}$ such that

$$E_{k,l} = \Psi(k-l,\xi_k,\xi_l) \quad \text{for any } k,l\in L.$$

It is also assumed that the strength of the interaction decays exponentially with the distance between the sites, in the following sense: there exist constants $K>0$ and $\theta>0$ such that

$$|\Psi(m,a,b)| \le Ke^{-\theta|m|} \quad \text{for any } m\in L \text{ and } a,b\in F, \tag{10.3.17}$$

where $|m| = \max\{|m_1|,\dots,|m_d|\}$. In particular (Exercise 10.3.9), the energy

$$\varphi(\xi) = \sum_{k\in L} \Psi(k,\xi_k,\xi_0)$$

resulting from the action of all the sites on the site 0 at the origin is uniformly bounded.

Initially, given any finite set $\Lambda\subset L$, let us consider the system one obtains by observing only the sites $k\in\Lambda$ and "switching off" their interactions with the sites in the complement of Λ. This is a finite system, as the configuration space is contained in F^Λ, with energy function given by

$$E_\Lambda(x) = \sum_{l\in\Lambda}\sum_{k\in\Lambda} \Psi(k-l,x_k,x_l) \quad \text{for every } x\in F^\Lambda.$$

Hence, according to (10.3.16), its Gibbs distribution μ_Λ is given by

$$\mu_\Lambda(\{x\}) = \frac{e^{-\beta E_\Lambda(x)}}{\sum_{y\in F^\Lambda} e^{-\beta E_\Lambda(y)}} \quad \text{for each } x \in F^\Lambda. \tag{10.3.18}$$

The notion of Gibbs state is obtained from this one by "switching back on" the interaction with the sites outside Λ, in the way we are going to explain. Denote by $r_\Lambda(x)$ the expression on the right-hand side of (10.3.18). Observe that

$$r_\Lambda(x) = \left[\sum_{y\in F^\Lambda} e^{\beta E_\Lambda(x) - \beta E_\Lambda(y)}\right]^{-1}$$

and recall that

$$E_\Lambda(x) - E_\Lambda(y) = \sum_{l\in\Lambda}\sum_{k\in\Lambda} \Psi(k-l, x_k, x_l) - \Psi(k-l, y_k, y_l).$$

For $\xi, \eta \in F^L$, define

$$\begin{aligned}\mathcal{E}(\xi,\eta) &= \sum_{l\in L}\sum_{k\in L} \Psi(k-l,\xi_k,\xi_l) - \Psi(k-l,\eta_k,\eta_l)\\ &= \sum_{l\in L}\sum_{j\in L} \Psi(j,\xi_{j+l},\xi_l) - \Psi(j,\eta_{j+l},\eta_l) = \sum_{l\in L} \varphi(\sigma^l(\xi)) - \varphi(\sigma^l(\eta)).\end{aligned}$$

It follows from the condition (10.3.17) that this sum converges whenever the two configurations are such that $\xi_k = \eta_k$ for every k in the complement of some finite set (Exercise 10.3.9). Then,

$$\rho_\Lambda(\xi) = \left[\beta \sum_{\eta|\Lambda^c = \xi|\Lambda^c} e^{\beta\mathcal{E}(\xi,\eta)}\right]^{-1}$$

is well defined for every $\xi \in F^\Lambda$.

A probability measure μ supported in $\Omega \subset F^L$ is called a *Gibbs state* if, for every finite set $\Lambda \subset L$, the disintegration $\{\mu_{\Lambda,\theta} : \theta \in F^{\Lambda^c}\}$ of μ relative to the partition $\{F^\Lambda \times \{\theta\} : \theta \in F^{\Lambda^c}\}$ of the space $F^L = F^\Lambda \times F^{\Lambda^c}$ is given by

$$\mu_{\Lambda,\theta}(\{x\}\times\{\theta\}) = \begin{cases} \rho_\Lambda(x,\theta) & \text{if } (x,\theta)\in\Omega \\ 0 & \text{otherwise.}\end{cases}$$

To conclude this section, we state one of the main results of this formalism that we have been describing: *one-dimensional lattice systems exhibit no phase transitions*. More precisely:

Theorem 10.3.9 (Ruelle). *If $d = 1$ and the interactions decay exponentially with the distance, then there exists a unique Gibbs state and it is also the unique equilibrium state.*

The arguments in the proof of this theorem (Ruelle [Rue04]) are at the basis of the thermodynamic formalism of expanding maps, which we are going to present in Chapter 12. Let us point out that the theorem is false in dimension $d \geq 2$. Exercise 10.3.10 highlights one of the specificities of the one-dimensional case that are behind this result.

10.3.5 Exercises

10.3.1. Check that the sequence $\log P_n(f,\phi,\alpha)$ is subadditive.

10.3.2. Show that if f is a homeomorphism then $P(f,\phi,f(\alpha)) = P(f,\phi,\alpha)$, $Q^+(f,\phi,f(\alpha)) = Q^+(f,\phi,\alpha)$ and $Q^-(f,\phi,f(\alpha)) = Q^-(f,\phi,\alpha)$ for every open cover α.

10.3.3. Show that, for any potential $\phi : M \to \mathbb{R}$:

(a) If α, β are open covers with $\alpha \prec \beta$ then $Q^+(f,\phi,\alpha) \leq Q^+(f,\phi,\beta)$ and $Q^-(f,\phi,\alpha) \leq Q^-(f,\phi,\beta)$.

(b) $Q^+(f,\phi,\alpha) = Q^+(f,\phi,\alpha^k)$ and $Q^-(f,\phi,\alpha) = Q^-(f,\phi,\alpha^k)$ for every $k \geq 1$ and every open cover α.

(c) $Q^+(f,\phi,\alpha) = P(f,\phi) = Q^-(f,\phi,\alpha)$ for any open cover α such that $\operatorname{diam}\alpha^k \to 0$.

(d) $P(f,\phi,\alpha) = P(f,\phi,\alpha^k)$ for every $k \geq 1$ and any open cover α whose elements are pairwise disjoint.

(e) $P(f,\phi,\alpha) = P(f,\phi)$ for any open cover α such that $\operatorname{diam}\alpha^k \to 0$ and whose elements are pairwise disjoint.

(f) If f is a homeomorphism, one may replace α^k by $\alpha^{\pm k}$ in statements (b), (c), (d) and (e).

10.3.4. (Walters). Prove that

$$P(f,\phi) = \sup\{Q^-(f,\phi,\alpha) : \alpha \text{ is an open cover of } M\}$$
$$= \sup\{Q^+(f,\phi,\alpha) : \alpha \text{ is an open cover of } M\}.$$

[Observation: In particular, the pressure depends only on the topology of M, not the distance. This also provides a way to extend the definition to continuous transformations in compact topological spaces.]

10.3.5. Exhibit a homeomorphism $f : M \to M$, a potential $\phi : M \to \mathbb{R}$ and open covers α and β of a compact metric space M such that $\alpha \prec \beta$ and $P(f,\phi,\alpha) > P(f,\phi,\beta) = P(f,\phi)$. [Observation: Thus, the conclusions of Exercise 10.3.3(a) and Exercise 10.3.4 are no longer valid if one replaces $Q^{\pm}(f,\phi,\alpha)$ by $P(f,\phi,\alpha)$.]

10.3.6. Check that the cohomology relation

$$\phi \sim \psi \Leftrightarrow \psi = \phi + u \circ f - u \text{ for some continuous function } u : M \to \mathbb{R}$$

is an equivalence relation.

10.3.7. Let $f_i : M_i \to M_i$, $i = 1,2$ be continuous transformations in compact metric spaces and, for each i, let ϕ_i be a potential in M_i. Define

$$f_1 \times f_2 : M_1 \times M_2 \to M_1 \times M_2, \quad f_1 \times f_2(x_1,x_2) = (f_1(x_1), f_2(x_2))$$
$$\phi_1 \times \phi_2 : M_1 \times M_2 \to \mathbb{R}, \quad \phi_1 \times \phi_2(x_1,x_2) = \phi_1(x_1) + \phi_2(x_2).$$

Show that $P(f_1 \times f_2, \phi_1 \times \phi_2) = P(f_1,\phi_1) + P(f_2,\phi_2)$.

10.3.8. Consider the transformation $f : S^1 \to S^1$ defined by $f(x) = 2x \mod \mathbb{Z}$. Prove that if $\phi : S^1 \to \mathbb{R}$ is a Hölder function then

$$P(f,\phi) = \lim_n \frac{1}{n} \log \sum_{p \in \operatorname{Fix}(f^n)} e^{\phi_n(p)}.$$

[Observation: We will get a more general result in Exercise 11.3.4.]

10.3.9. Assuming the conditions in Section 10.3.4, prove that:

(a) There exists $C > 0$ such that $|\varphi(\xi)| \le C$ and $|\varphi(\xi) - \varphi(\eta)| \le Ce^{-\theta N/2}$ for any $\xi, \eta \in F^L$ such that $\xi_k = \eta_k$ whenever $|k| < N$.

(b) For any finite set $\Lambda \subset L$, there exists $C_\Lambda > 0$ such that $|\mathcal{E}(\xi,\eta)| \le C_\Gamma$ for any $\xi, \eta \in F^L$ such that $\xi_k = \eta_k$ for every $k \in \Lambda^c$.

10.3.10. Assuming the conditions in Section 10.3.4, prove that if $d = 1$ then there exists $C > 0$ such that $|\mathcal{E}(\xi,\eta) - E_\Lambda(\xi) + E_\Lambda(\eta)| \le C$ for every finite interval $\Lambda \subset \mathbb{Z}$ and any $\xi, \eta \in F^{\mathbb{Z}}$ with $\xi \mid \Lambda^c = \eta \mid \Lambda^c$. Consequently,

$$e^{-\beta C} \le \frac{r_\Lambda(\xi \mid \Lambda)}{\rho_\Lambda(\xi)} \le e^{\beta C}$$

for every $\xi \in F^{\mathbb{Z}}$ and every finite interval Λ. [Observation: Therefore, the probability distribution of the configurations restricted to each finite interval is not affected in a significant way by the interactions with the sites outside that interval.]

10.4 Variational principle

The variational principle for the pressure, which we state below, was proved originally by Ruelle [Rue73], under more restrictive assumptions, and was then extended by Walters [Wal75] to the context we consider here:

Theorem 10.4.1 (Variational principle). *Let $f : M \to M$ be a continuous transformation in a compact metric space and $\mathcal{M}_1(f)$ denote the set of probability measures invariant under f. Then, for every continuous function $\phi : M \to \mathbb{R}$,*

$$P(\phi,f) = \sup\{h_\nu(f) + \int \phi\, d\nu : \nu \in \mathcal{M}_1(f)\}.$$

Theorem 10.1 corresponds to the special case $\phi \equiv 0$. In particular, it follows that the topological entropy of f is zero if and only if $h_\nu(f) = 0$ for every invariant probability measure ν. That is the case, for example, for every circle homeomorphism (Example 10.1.1) and every translation in a compact metrizable group (Example 10.1.7). The compactness hypothesis is crucial: as observed in Exercise 10.4.4, there exist transformations (in non-compact spaces) without invariant measures and whose topological entropy is positive.

In Sections 10.4.1 and 10.4.2 we present a proof of Theorem 10.4.1 that is due to Misiurewicz [Mis76]. Before that, let us mention a couple of consequences.

Corollary 10.4.2. *Let $f : M \to M$ be a continuous transformation in a compact metric space and $\mathcal{M}_e(f)$ denote the set of probability measures invariant and ergodic. Then*

$$P(\phi,f) = \sup\{h_\nu(f) + \int \phi\, d\nu : \nu \in \mathcal{M}_e(f)\}.$$

Proof. Given any $\nu \in \mathcal{M}_1(f)$, let $\{\nu_P : P \in \mathcal{P}\}$ be its ergodic decomposition. By Theorems 5.1.3 and 9.6.2,

$$h_\nu(f) + \int \phi \, d\nu = \int \left(h_{\nu_P}(f) + \int \phi \, d\nu_P \right) d\hat{\mu}(P).$$

This implies that

$$\sup\{h_\nu(f) + \int \phi \, d\nu : \nu \in \mathcal{M}_1(f)\} \le \sup\{h_\nu(f) + \int \phi \, d\nu : \nu \in \mathcal{M}_e(f)\}.$$

The converse inequality is trivial, since $\mathcal{M}_e(f) \subset \mathcal{M}_1(f)$. Now it suffices to apply Theorem 10.4.1.

Another interesting consequence is that for transformations with finite topological entropy the pressure function determines the set of all invariant probability measures:

Corollary 10.4.3 (Walters). *Let $f : M \to M$ be a continuous transformation in a compact metric space with topological entropy $h(f) < \infty$. Let η be any finite signed measure on M. Then, η is a probability measure invariant under f if and only if $\int \phi \, d\eta \le P(f, \phi)$ for every continuous function $\phi : M \to \mathbb{R}$.*

Proof. The "only if" claim is an immediate consequence of Theorem 10.4.1: if η is an invariant probability measure then

$$P(f, \phi) \ge h_\eta(f) + \int \phi \, d\eta \ge \int \phi \, d\eta$$

for every continuous function ϕ. In what follows we prove the converse.

Let η be a finite signed measure such that $\int \phi \, d\eta \le P(f, \phi)$ for every ϕ. Consider any $\phi \ge 0$. For any $c > 0$ and $\varepsilon > 0$,

$$c \int (\phi + \varepsilon) \, d\eta = -\int -c(\phi + \varepsilon) \, d\eta \ge -P(f, -c(\phi + \varepsilon)).$$

By the relation (10.3.6), we also have that

$$P(f, -c(\phi + \varepsilon)) \le h(f) + \sup\big(-c(\phi + \varepsilon)\big) = h(f) - c \inf(\phi + \varepsilon).$$

Therefore, $c \int (\phi + \varepsilon) \, d\eta \ge -h(f) + c \inf(\phi + \varepsilon)$. When $c > 0$ is sufficiently large, the right-hand side of this inequality is positive. Hence, $\int (\phi + \varepsilon) \, d\eta > 0$. Since $\varepsilon > 0$ is arbitrary, this implies that $\int \phi \, d\eta \ge 0$ for every $\phi \ge 0$. So, η is a positive measure.

The next step is to show that η is a probability measure. By assumption,

$$\int c \, d\eta \le P(f, c) = h(f) + c$$

for every $c \in \mathbb{R}$. For $c > 0$, this implies that $\eta(M) \le 1 + h(f)/c$. Passing to the limit when $c \to +\infty$, we get that $\eta(M) \le 1$. Analogously, considering $c < 0$ and taking the limit when $c \to -\infty$, we get $\eta(M) \ge 1$. Therefore, η is a probability measure, as stated.

We are left to prove that η is invariant under f. By assumption, given any $c \in \mathbb{R}$ and any potential ϕ,

$$c\int(\phi\circ f-\phi)\,d\eta \leq P(f,c(\phi\circ f-\phi)).$$

By Proposition 10.3.8, the expression on the right-hand side is equal to $P(f,0)=h(f)$. For $c>0$, this implies that

$$\int(\phi\circ f-\phi)\,d\eta \leq \frac{h(f)}{c}$$

and, taking the limit when $c\to+\infty$, it follows that $\int(\phi\circ f-\phi)\,d\eta\leq 0$. The same argument, applied to the function $-\phi$, shows that $\int(\phi\circ f-\phi)\,d\eta\geq 0$. Hence, $\int\phi\circ f\,d\eta=\int\phi\,d\eta$ for every ϕ. By Proposition A.3.3, this implies that $f_*\eta=\eta$.

10.4.1 Proof of the upper bound

In this section we prove that, given any invariant probability measure ν,

$$h_\nu(f)+\int\phi\,d\nu\leq P(f,\phi). \tag{10.4.1}$$

To do this, let $\mathcal{P}=\{P_1,\dots,P_s\}$ be any finite partition. We are going to show that if α is an open cover of M with sufficiently small diameter, depending only on $\mathcal{P}$, then

$$h_\nu(f,\mathcal{P})+\int\phi\,d\nu\leq\log 4+P(f,\phi,\alpha). \tag{10.4.2}$$

Making $\operatorname{diam}\alpha\to 0$, it follows that $h_\nu(f,\mathcal{P})+\int\phi\,d\nu\leq\log 4+P(f,\phi)$ for every finite partition $\mathcal{P}$. Hence, $h_\nu(f)+\int\phi\,d\nu\leq\log 4+P(f,\phi)$. Now replace the transformation f by f^k and the potential ϕ by ϕ_k. Note that $\int\phi_k\,d\nu=k\int\phi\,d\nu$, since ν is invariant under f. Using Propositions 9.1.14 and 10.3.5, we get that

$$kh_\nu(f,\mathcal{P})+k\int\phi\,d\nu\leq\log 4+kP(f,\phi)$$

for every $k\geq 1$. Dividing by k and taking the limit when $k\to\infty$, we get the inequality (10.4.1).

For proving (10.4.2) we need the following elementary fact:

Lemma 10.4.4. *Let $a_1,\dots,a_k$ be real numbers and $p_1,\dots,p_k$ be non-negative numbers such that $p_1+\cdots+p_k=1$. Let $A=\sum_{i=1}^k e^{a_i}$. Then*

$$\sum_{i=1}^k p_i(a_i-\log p_i)\leq\log A.$$

Moreover, the identity holds if and only if $p_j=e^{a_j}/A$ for every j.

Proof. Write $t_i = e^{a_i}/A$ and $x_i = p_i/e^{a_i}$. Note that $\sum_{i=1}^k t_i = 1$. By the concavity property (9.1.8) of the function $\phi(x) = -x\log x$,

$$\sum_{i=1}^{k} t_i\phi(x_i) \le \phi(\sum_{i=1}^{k} t_i x_i).$$

Note that $t_i\phi(x_i) = (p_i/A)(a_i - \log p_i)$ and $\sum_{i=1}^k t_i x_i = 1/A$. So, the previous inequality may be rewritten as follows:

$$\sum_{i=1}^{k} \frac{p_i}{A}(a_i - \log p_i) \le \frac{1}{A}\log A.$$

Multiplying by A we get the inequality in the statement of the lemma. Moreover, the identity holds if and only if the x_i are all equal, that is, if and only if there exists c such that $p_i = ce^{a_i}$ for every i. Summing over $i = 1,\dots,k$ we see that in that case $c = 1/A$, as stated.

Since the measure ν is regular (Proposition A.3.2), given $\varepsilon > 0$ we may find compact sets $Q_i \subset P_i$ such that $\nu(P_i \setminus Q_i) < \varepsilon$ for every $i = 1,\dots,s$. Let Q_0 be the complement of $\bigcup_{i=1}^s Q_i$ and let $P_0 = \emptyset$. Then $\mathcal{Q} = \{Q_0, Q_1, \dots, Q_s\}$ is a finite partition of M such that $\nu(P_i \Delta Q_i) < s\varepsilon$ for every $i = 0,1,\dots,s$. Hence, by Lemma 9.1.6,

$$H_\nu(\mathcal{P}/\mathcal{Q}) \le \log 2$$

as long as $\varepsilon > 0$ is sufficiently small (depending only on s). Let ε and $\mathcal{Q}$ be fixed from now on and assume that the open cover α satisfies

$$\operatorname{diam}\alpha < \min\{d(Q_i, Q_j) : 1 \le i < j \le s\}. \tag{10.4.3}$$

By Lemma 9.1.11, we have that $h_\nu(f,\mathcal{P}) \le h_\nu(f,\mathcal{Q}) + H_\nu(\mathcal{P}/\mathcal{Q}) \le h_\nu(f,\mathcal{Q}) + \log 2$. Hence, to prove (10.4.2) it suffices to show that

$$h_\nu(f,\mathcal{Q}) + \int \phi\, d\nu \le \log 2 + P(f,\phi,\alpha). \tag{10.4.4}$$

To that end, observe that

$$H_\nu(\mathcal{Q}^n) + \int \phi_n\, d\nu \le \sum_{Q\in\mathcal{Q}^n} \nu(Q)\big(-\log\nu(Q) + \sup_{x\in Q}\phi_n(x)\big)$$

for every $n \ge 1$. Then, by Lemma 10.4.4,

$$H_\nu(\mathcal{Q}^n) + \int \phi_n\, d\nu \le \log\Big(\sum_{Q\in\mathcal{Q}^n} \sup_{x\in Q} e^{\phi_n(x)}\Big). \tag{10.4.5}$$

Let γ be any finite subcover of α^n. For each $Q \in \mathcal{Q}^n$, consider any point x_Q in the closure of Q such that $\phi_n(x_Q) = \sup_{x\in Q}\phi_n(x)$. Pick $U_Q \in \gamma$ such that $x_Q \in U_Q$. Then $\sup_{x\in Q}\phi_n(x) \le \sup_{y\in U_Q}\phi_n(y)$ for every $Q \in \mathcal{Q}^n$. The condition (10.4.3) implies that each element of α intersects the closure of not more than two elements of $\mathcal{Q}$. Therefore, each element of α^n intersects the closure of not more than 2^n elements of $\mathcal{Q}^n$. In particular, for each $U \in \gamma$ there exist not more than 2^n sets $Q \in \mathcal{Q}^n$ such that $U_Q = U$. Therefore,

$$\sum_{Q \in \mathcal{Q}^n} \sup_{x \in Q} e^{\phi_n(x)} \leq 2^n \sum_{U \in \gamma} \sup_{y \in U} e^{\phi_n(y)}, \tag{10.4.6}$$

for any finite subcover γ of α^n. Combining (10.4.5) and (10.4.6),

$$H_\nu(\mathcal{Q}^n) + \int \phi_n \, d\nu \leq n \log 2 + \log P_n(f, \phi, \alpha).$$

Dividing by n and taking the limit when $n \to \infty$, we get (10.4.4). This completes the proof of the upper bound (10.4.1).

10.4.2 Approximating the pressure

To finish the proof of Theorem 10.4.1, we now show that for every $\varepsilon > 0$ there exists a probability measure μ invariant under f and such that

$$h_\mu(f) + \int \phi \, d\mu \geq S(f, \phi, \varepsilon). \tag{10.4.7}$$

Clearly, this implies that the supremum of the values of $h_\nu(f) + \int \phi \, d\nu$ when ν varies in $\mathcal{M}_1(f)$ is greater than or equal to $S(f, \phi) = P(f, \phi)$.

For each $n \geq 1$, let E be an (n, ε)-separated set such that

$$\sum_{y \in E} e^{\phi_n(y)} \geq \frac{1}{2} S_n(f, \phi, \varepsilon). \tag{10.4.8}$$

Denote by A the expression on the left-hand side of this inequality. Consider the probability measures ν_n and μ_n defined on M by

$$\nu_n = \frac{1}{A} \sum_{x \in E} e^{\phi_n(x)} \delta_x \quad \text{and} \quad \mu_n = \sum_{j=0}^{n-1} f_*^j \nu_n.$$

By the definition (10.3.8), recalling also that the space of probability measures is compact (Theorem 2.1.5), we may choose a subsequence $(n_j)_j \to \infty$ such that

1. $\dfrac{1}{n_j} \log S_{n_j}(f, \phi, \varepsilon)$ converges to $S(f, \phi, \varepsilon)$, and
2. μ_{n_j} converges, in the weak* topology, to some probability measure μ.

We are going to check that such a measure μ is invariant under f and satisfies (10.4.7). For the reader's convenience, we split the argument into four steps.

Step 1: First, we prove that μ is invariant. Let $\varphi : M \to \mathbb{R}$ be any continuous function. For each $n \geq 1$,

$$\int \varphi \, d(f_* \mu_n) = \frac{1}{n} \sum_{j=1}^{n} \int \varphi \circ f^j \, d\nu_n = \int \varphi \, d\mu_n + \frac{1}{n} \Big(\int \varphi \circ f^n \, d\nu_n - \int \varphi \, d\nu_n \Big)$$

and, consequently,

$$\left| \int \varphi \, d(f_* \mu_n) - \int \varphi \, d\mu_n \right| \leq \frac{2}{n} \sup |\varphi|.$$

Restricting to $n = n_j$ and taking the limit when $j \to \infty$, we see that $\int \varphi \, df_* \mu = \int \varphi \, d\mu$ for every continuous function $\varphi : M \to \mathbb{R}$. This proves (recall Proposition A.3.3) that $f_* \mu = \mu$, as claimed.

Step 2: Next, we estimate the entropy with respect to ν_n. Let $\mathcal{P}$ be any finite partition of M such that $\operatorname{diam} \mathcal{P} < \varepsilon$ and $\mu(\partial \mathcal{P}) = 0$, where $\partial \mathcal{P}$ denotes the union of the boundaries ∂P of all sets $P \in \mathcal{P}$. The first condition implies that each element of $\mathcal{P}^n$ contains at most one element of E. On the other hand, it is clear that every element of E is contained in some element of $\mathcal{P}^n$. Hence,

$$\begin{aligned} H_{\nu_n}(\mathcal{P}^n) &= \sum_{x \in E} -\nu_n(\{x\}) \log \nu_n(\{x\}) = \sum_{x \in E} -\frac{1}{A} e^{\phi_n(x)} \log \left(\frac{1}{A} e^{\phi_n(x)} \right) \\ &= \log A - \frac{1}{A} \sum_{x \in E} e^{\phi_n(x)} \phi_n(x) = \log A - \int \phi_n \, d\nu_n \end{aligned} \tag{10.4.9}$$

(the last equality follows directly from the definition of ν_n).

Step 3: Now we calculate the entropy with respect to μ_n. Consider $1 \leq k < n$. For each $r \in \{0, \ldots, k-1\}$, let $q_r \geq 0$ be the largest integer number such that $r + kq_r \leq n$. In other words, $q_r = [(n-r)/k]$. Then,

$$\mathcal{P}^n = \mathcal{P}^r \vee \left[\bigvee_{j=0}^{q_r - 1} f^{-(kj+r)}(\mathcal{P}^k) \right] \vee f^{-(kq_r+r)}(\mathcal{P}^{n-(kq_r+r)})$$

(the first term is void if $r = 0$ and the third one is void if $n = kq_r + r$). Therefore,

$$H_{\nu_n}(\mathcal{P}^n) \leq \sum_{j=0}^{q_r-1} H_{\nu_n}(f^{-(kj+r)}(\mathcal{P}^k)) + H_{\nu_n}(\mathcal{P}^r) + H_{\nu_n}(f^{-(kq_r+r)}(\mathcal{P}^{n-(kq_r+r)})).$$

Clearly, $\#\mathcal{P}^r \leq (\#\mathcal{P})^k$. Using Lemma 9.1.3, we find that $H_{\nu_n}(\mathcal{P}^r) \leq k \log \#\mathcal{P}$. For the same reason, the last term in the previous inequality is also bounded by $k \log \#\mathcal{P}$. Then, using the property (9.1.12),

$$H_{\nu_n}(\mathcal{P}_n) \leq \sum_{j=0}^{q_r-1} H_{f_*^{(kj+r)} \nu_n}(\mathcal{P}^k) + 2k \log \#\mathcal{P} \tag{10.4.10}$$

for every $r \in \{0, \ldots, k-1\}$. Now, it is clear that every number $i \in \{0, \ldots, n-1\}$ may be written in a unique way as $i = kj + r$ with $0 \leq j \leq q_r - 1$. Then, summing (10.4.10) over all the values of r,

$$kH_{\nu_n}(\mathcal{P}_n) \leq \sum_{i=0}^{n-1} H_{f_*^i \nu_n}(\mathcal{P}^k) + 2k^2 \log \#\mathcal{P}. \tag{10.4.11}$$

The concavity property (9.1.8) of the function $\phi(x) = -x\log x$ implies that

$$\frac{1}{n}\sum_{i=0}^{n-1} H_{f_*^i\nu_n}(\mathcal{P}^k) \le H_{\mu_n}(\mathcal{P}^k).$$

Combining this inequality with (10.4.11), we see that

$$\frac{1}{n}H_{\nu_n}(\mathcal{P}_n) \le \frac{1}{k}H_{\mu_n}(\mathcal{P}^k) + \frac{2k}{n}\log\#\mathcal{P}.$$

On the other hand, by the definition of μ_n,

$$\frac{1}{n}\int \phi_n\,d\nu_n = \frac{1}{n}\sum_{j=0}^{n-1}\int \phi\circ f^j\,d\nu_n = \int \phi\,d\mu_n.$$

Thus, the previous inequality yields

$$\frac{1}{n}H_{\nu_n}(\mathcal{P}_n) + \frac{1}{n}\int \phi_n\,d\nu_n \le \frac{1}{k}H_{\mu_n}(\mathcal{P}^k) + \int \phi\,d\mu_n + \frac{2k}{n}\log\#\mathcal{P}. \quad (10.4.12)$$

Step 4: Finally, we translate the previous estimates to the limit measure μ. From (10.4.9) and (10.4.12), we get

$$\frac{1}{k}H_{\mu_n}(\mathcal{P}^k) + \int \phi\,d\mu_n \ge \frac{1}{n}\log A - \frac{2k}{n}\log\#\mathcal{P}.$$

By the choice of E in (10.4.8), it follows that

$$\frac{1}{k}H_{\mu_n}(\mathcal{P}^k) + \int \phi\,d\mu_n \ge \frac{1}{n}\log S_n(f,\phi,\varepsilon) - \frac{1}{n}\log 2 - \frac{2k}{n}\log\#\mathcal{P}. \quad (10.4.13)$$

The choice of the partition $\mathcal{P}$ with $\mu(\partial\mathcal{P}) = 0$ implies that $\mu(\partial\mathcal{P}^k) = 0$ for every $k \ge 1$, since

$$\partial\mathcal{P}^k \subset \partial\mathcal{P} \cup f^{-1}(\partial\mathcal{P}) \cup \cdots \cup f^{-k+1}(\partial\mathcal{P}).$$

In other words, every element of $\mathcal{P}^k$ is a continuity set for the measure μ. According to Exercise 2.1.1, it follows that $\mu(P) = \lim_j \mu_{n_j}(P)$ for every $P \in \mathcal{P}^k$ and, hence,

$$H_\mu(\mathcal{P}^k) = \lim_j H_{\mu_{n_j}}(\mathcal{P}^k) \quad \text{for every } k \ge 1.$$

Since the function ϕ is continuous, we also have that $\int \phi\,d\mu = \lim_j \int \phi\,d\mu_{n_j}$. Therefore, restricting (10.4.13) to the subsequence $(n_j)_j$ and taking the limit when $j \to \infty$,

$$\frac{1}{k}H_\mu(\mathcal{P}^k) + \int \phi\,d\mu \ge S(f,\phi,\varepsilon).$$

Taking the limit when $k \to \infty$, we find that

$$h_\mu(f,\mathcal{P}) + \int \phi\,d\mu \ge S(f,\phi,\varepsilon).$$

Then, making $\varepsilon \to 0$ (and, consequently, $\operatorname{diam}\mathcal{P} \to 0$), we get (10.4.7).

This completes the proof of the variational principle (Theorem 10.4.1).

10.4.3 Exercises

10.4.1. Let $f : M \to M$ be a continuous transformation in a compact metric space M. Check that $P(f,\varphi) \le h(f) + \sup\varphi$ for every continuous function $\varphi : M \to \mathbb{R}$.

10.4.2. Show that if $f : M \to M$ is a continuous transformation in a compact metric space and $X \subset M$ is a forward invariant set, meaning that $f(X) \subset X$, then $P(f \mid X, \varphi \mid X) \le P(f,\varphi)$.

10.4.3. Give an alternative proof of Proposition 10.3.8, using the variational principle.

10.4.4. Exhibit a continuous transformation $f : M \to M$ in a non-compact metric space M such that f has no invariant probability measure and yet the topological entropy $h(f)$ is positive. [Observation: Thus, the variational principle need not hold when the ambient space is not compact.]

10.4.5. Given numbers $\alpha, \beta > 0$ such that $\alpha + \beta < 1$, define

$$g : [0,\alpha] \cup [1-\beta, 1] \to [0,1] \quad g(x) = \begin{cases} x/\alpha & \text{if } x \in [0,\alpha] \\ (x-1)/\beta + 1 & \text{if } x \in [1-\beta, 1]. \end{cases}$$

Let $K \subset [0,1]$ be the Cantor set formed by the points x such that $g^n(x)$ is defined for every $n \ge 0$ and $f : K \to K$ be the restriction of g. Calculate the function $\psi : \mathbb{R} \to \mathbb{R}$ defined by $\psi(t) = P(f, -t \log g')$. Check that ψ is convex and decreasing and admits a (unique) zero in $(0,1)$. Show that $h_\mu(f) < \int \log g' \, d\mu$ for every probability measure μ invariant under f.

10.4.6. Let $f : M \to M$ be a continuous transformation in a compact metric space, such that the set of ergodic invariant probability measures is finite. Show that for every potential $\varphi : M \to \mathbb{R}$ there exists some invariant probability measure that realizes the supremum in (10.0.1).

10.5 Equilibrium states

Let $f : M \to M$ be a continuous transformation and $\phi : M \to M$ be a potential in a compact metric space. In this section we study the fundamental properties of the set $\mathcal{E}(f,\phi)$ formed by the *equilibrium states*, that is, the invariant probability measures μ such that

$$h_\mu(f) + \int \phi \, d\mu = P(\phi, f) = \sup\{h_\nu(f) + \int \phi \, d\nu : \nu \in \mathcal{M}_1(f)\}.$$

In the special case $\phi \equiv 0$ the elements of $\mathcal{E}(f,\phi)$ are also called *measures of maximal entropy*. Let us start with a few simple examples.

Example 10.5.1. If $f : M \to M$ has zero topological entropy then every invariant probability measure μ is a measure of maximal entropy: $h_\mu(f) = 0 = h(f)$. For any potential $\phi : M \to \mathbb{R}$,

$$P(f,\phi) = \sup\{\int \phi \, d\nu : \nu \in \mathcal{M}_1(f)\}.$$

Hence, ν is an equilibrium state if and only if ν maximizes the integral of ϕ. Since the function $\nu \mapsto \int \phi \, d\nu$ is continuous and $\mathcal{M}_1(f)$ is compact, with respect to the weak* topology, maxima do exist for every potential ϕ.

Example 10.5.2. Let $f_A : M \to M$ be the linear endomorphism induced in $\mathbb{T}^d$ by some matrix A with integer coefficients and non-zero determinant. Let μ be the Haar measure on $\mathbb{T}^d$. By Propositions 9.4.3 and 10.2.10,

$$h_\mu(f_A) = \sum_{i=1}^{d} \log^+ |\lambda_i| = h(f),$$

where $\lambda_1, \dots, \lambda_d$ are the eigenvalues of A. In particular, the Haar measure is a measure of maximal entropy for f.

Example 10.5.3. Let $\sigma : \Sigma \to \Sigma$ be the shift map in $\Sigma = \{1, \dots, d\}^{\mathbb{N}}$ and μ be the Bernoulli measure associated with a probability vector $p = (p_1, \dots, p_d)$. As observed in Example 9.1.10,

$$h_\mu(\sigma, \mathcal{P}) = \lim_n \frac{1}{n} H_\mu(\mathcal{P}^n) = \sum_{i=1}^{d} -p_i \log p_i.$$

We leave it to the reader (Exercise 10.5.1) to check that this function attains its maximum precisely when the coefficients p_i are all equal to $1/d$. Moreover, in that case $h_\mu(\sigma) = \log d$. Recall also (Example 10.1.2) that $h(\sigma) = \log d$. Therefore, the Bernoulli measure associated with the vector $p = (1/d, \dots, 1/d)$ is the only measure of maximal entropy among the Bernoulli measures. In fact, it follows from the theory that we develop in Chapter 12 that μ is the unique measure of maximal entropy among *all* invariant measures.

Let us start with the following extension of the variational principle:

Proposition 10.5.4. *For every potential* $\phi : M \to \mathbb{R}$,

$$P(f, \phi) = \sup\{h_\mu(f) + \int \phi \, d\mu : \mu \text{ invariant and ergodic for } f\}.$$

Proof. Consider the function $\Psi : \mathcal{M}_1(f) \to \mathbb{R}$ given by $\Psi(\mu) = h_\mu(f) + \int \phi \, d\mu$. For each invariant probability measure μ, let $\{\mu_P : P \in \mathcal{P}\}$ be the corresponding ergodic decomposition. It follows from Theorem 9.6.2 that

$$\Psi(\mu) = \int \Psi(\mu_P) \, d\hat{\mu}(P). \tag{10.5.1}$$

This implies that the supremum of Ψ over all the invariant probability measures is less than or equal to the supremum of Ψ over the ergodic invariant probability measures. Since the opposite inequality is obvious, it follows that the two suprema coincide. By the variational principle (Theorem 10.4.1), the supremum of Ψ over all the invariant probability measures is equal to $P(f, \phi)$.

Proposition 10.5.5. *Assume that $h(f) < \infty$. Then the set of equilibrium states for any potential $\phi : M \to \mathbb{R}$ is a convex subset of $\mathcal{M}_1(f)$: more precisely, given $t \in (0,1)$ and $\mu_1, \mu_2 \in \mathcal{M}_1(f)$,*

$$(1-t)\mu_1 + t\mu_2 \in \mathcal{E}(f,\phi) \quad \Leftrightarrow \quad \{\mu_1, \mu_2\} \subset \mathcal{E}(f,\phi).$$

Moreover, an invariant probability measure μ is in $\mathcal{E}(f,\phi)$ if and only if almost every ergodic component of μ is in $\mathcal{E}(f,\phi)$.

Proof. As we have seen in (10.3.6), the hypothesis that the topological entropy is finite ensures that $P(f,\phi) < \infty$ for every potential ϕ. Let us consider the functional $\Psi(\mu) = h_\mu(f) + \int \phi \, d\mu$ introduced in the proof of the previous result. By Proposition 9.6.1, this functional is convex:

$$\Psi((1-t)\mu_1 + t\mu_2) = (1-t)\Psi(\mu_1) + t\Psi(\mu_2)$$

for every $t \in (0,1)$ and any $\mu_1, \mu_2 \in \mathcal{M}_1(f)$. Then, $\Psi((1-t)\mu_1 + t\mu_2)$ is equal to the supremum of Ψ if and only if both $\Psi(\mu_1)$ and $\Psi(\mu_2)$ are. This proves the first part of the proposition. The proof of the second part is analogous: the relation (10.5.1) implies that $\Psi(\mu) = \sup \Psi$ if and only if $\Psi(\mu_P) = \sup \Psi$ for $\hat{\mu}$-almost every P.

Corollary 10.5.6. *If $\mathcal{E}(f,\phi)$ is non-empty then it contains ergodic invariant probability measures. Moreover, the extremal elements of the convex set $\mathcal{E}(f,\phi)$ are precisely the ergodic measures contained in it.*

Proof. To get the first claim it suffices to consider the ergodic components of any element of $\mathcal{E}(f,\phi)$. Let us move on to proving the second claim. If $\mu \in \mathcal{E}(f,\phi)$ is ergodic then (Proposition 4.3.2) μ is an extremal element of $\mathcal{M}_1(f)$ and so it must be an extremal element of $\mathcal{E}(f,\phi)$. Conversely, if $\mu \in \mathcal{E}(f,\phi)$ is not ergodic then we may write

$$\mu = (1-t)\mu_1 + t\mu_2, \quad \text{with } 0 < t < 1 \text{ and } \mu_1, \mu_2 \in \mathcal{M}_1(f).$$

By Proposition 10.5.5 we have that $\mu_1, \mu_2 \in \mathcal{E}(f,\phi)$, which implies that μ is not an extremal element of the set $\mathcal{E}(f,\phi)$.

In general, the set of equilibrium states *may* be empty. The first example of this kind was given by Gurevič. The following construction is taken from Walters [Wal82]:

Example 10.5.7. Let $f_n : M_n \to M_n$ be a sequence of homeomorphisms in compact metric spaces such that the sequence $(h(f_n))_n$ is increasing and bounded. We are going to build a metric space M and a homeomorphism $f : M \to M$ with the following features:

- M is the union of all M_n with an additional point, denoted as ∞, endowed with a distance function relative to which $(M_n)_n$ converges to ∞.

- f fixes the point ∞ and its restriction to each M_n coincides with f_n.

Then we are going to check that $f : M \to M$ has no measure of maximal entropy. Let us explain how this is done.

Denote by d_n the distance in each metric space M_n. It is no restriction to assume that $d_n \leq 1$ for every n. Define $M = \bigcup_n M_n \cup \{\infty\}$ and consider the distance d defined in M by:

$$d(x,y) = \begin{cases} n^{-2} d_n(x,y) & \text{if } x \in X_n \text{ and } y \in X_n \text{ with } n \geq 1 \\ \sum_{i=n}^{m} i^{-2} & \text{if } x \in X_n \text{ and } y \in X_m \text{ with } n < m \\ \sum_{i=n}^{\infty} i^{-2} & \text{if } x \in X_n \text{ and } y = \infty. \end{cases}$$

We leave it to the reader to check that d is indeed a distance in M and that (M,d) is a compact space. Let $\beta = \sup_n h(f_n)$. Since the sets $\{\infty\}$ and M_n, $n \geq 1$ are invariant and cover the whole of M, any ergodic probability measure μ of f satisfies either $\mu(\{\infty\}) = 1$ or $\mu(M_n) = 1$ for some $n \geq 1$. In the first case, $h_\mu(f) = 0$. In the second, μ may be viewed as a probability measure invariant under f_n and, consequently, $h_\mu(f) \leq h(f_n)$. In particular, $h_\mu(f) < \beta$ for every ergodic probability measure μ of f. The previous observation also shows that

$$\begin{aligned} &\sup\{h_\mu(f) : \mu \text{ invariant and ergodic for } f\} \\ &\qquad = \sup_n \sup\{h_\mu(f) : \mu \text{ invariant and ergodic for } f_n\}. \end{aligned}$$

According to Proposition 10.5.4, this means that $h(f) = \sup_n h(f_n) = \beta$. Thus, no ergodic invariant measure of f realizes the topological entropy. By Proposition 10.5.5, it follows that f has no measure of maximal entropy.

Nevertheless, there is a broad class of transformations for which equilibrium states do exist for every potential:

Lemma 10.5.8. *If the entropy function $\nu \mapsto h_\nu(f)$ is upper semi-continuous then $\mathcal{E}(f,\phi)$ is compact, relative to the weak* topology, and non-empty, for any potential $\phi : M \to \mathbb{R}$.*

Proof. Let $(\mu_n)_n$ be a sequence in $\mathcal{M}_1(f)$ such that

$$h_{\mu_n}(f) + \int \phi \, d\mu_n \text{ converges to } P(f,\phi).$$

Since $\mathcal{M}_1(f)$ is compact (Theorem 2.1.5), there exists some accumulation point μ. The assumption implies that $\nu \mapsto h_\nu(f) + \int \phi \, d\nu$ is upper semi-continuous. Consequently,

$$h_\mu(f) + \int \phi \, d\mu \geq \liminf_n h_{\mu_n}(f) + \int \phi \, d\mu_n = P(f,\phi)$$

and so μ is an equilibrium state, as stated. Analogously, taking any sequence $(\nu_n)_n$ in $\mathcal{E}(f,\phi)$ we see that every accumulation point ν is an equilibrium state. This shows that $\mathcal{E}(f,\phi)$ is closed and, thus, compact.

Corollary 10.5.9. *Assume that $f : M \to M$ is an expansive continuous transformation in a compact metric space M. Then every potential $\phi : M \to \mathbb{R}$ admits some equilibrium state.*

Proof. Just combine Corollary 9.2.17 with Lemma 10.5.8.

The conclusions of Corollaries 9.2.17 and 10.5.9 remain valid when f is just h-expansive, in the sense of Exercise 10.2.8. See Bowen [Bow72]. Misiurewicz [Mis73] noted that the same is still true when f is just *asymptotically h-expansive*, meaning that $g_*(f,\varepsilon) \to 0$ when $\varepsilon \to 0$. Buzzi [Buz97] proved that every C^∞ diffeomorphism is asymptotically h-expansive. The corresponding statement for h-expansivity is false: Burguet, Liao and Yang [BLY] found open sets, in the C^2 topology, formed by diffeomorphisms that are not h-expansive; C^∞ diffeomorphisms are dense in such sets.

Combining these results of Misiurewicz and Buzzi, one gets the following theorem of Newhouse [New90]: for every C^∞ diffeomorphism f, the entropy function $\nu \mapsto h_\nu(f)$ is upper semi-continuous and so equilibrium states always exist. Yomdin [Yom87] also proved that the topological entropy function $f \mapsto h(f)$ is upper semi-continuous in the realm of C^∞ diffeomorphisms. Both conclusions, Newhouse's and Yomdin's, are usually false for C^r diffeomorphisms with $r < \infty$, according to Misiurewicz [Mis73]. But Liao, Viana and Yang [LVY13] proved that they both extend to C^1 diffeomorphisms away from homoclinic tangencies and any such diffeomorphism is h-expansive. In particular, equilibrium states always exist in that generality.

Uniqueness of equilibrium states is a much more delicate problem. It is very easy to exhibit transformations with infinitely many ergodic equilibrium states. For example, let $f : S^1 \to S^1$ be a circle homeomorphism with infinitely many fixed points. The Dirac measures on those points are ergodic invariant probability measures. Since the topological entropy $h(f)$ is equal to zero (Example 10.5.1), each of those measures is an equilibrium state for any potential that attains its maximum at the corresponding point.

This type of example is trivial, of course, because the transformation is not transitive. A more interesting question is whether an indivisibility property, such as transitivity or topological mixing, ensures uniqueness of the equilibrium state. It turns out that this is also not true. The first counter-example (called Dyck shift) was exhibited by Krieger [Kri75]. Next, we present a particularly transparent and flexible construction, due to Haydn [Hay]. Other interesting examples were studied by Hofbauer [Hof77].

Example 10.5.10. Let $X = \{*, 1, 2, 3, 4\}$ and consider the subsets $E = \{2, 4\}$ (the *even* symbols) and $O = \{1, 3\}$ (the *odd* symbols). We are going to exhibit a compact $\mathcal{H} \subset X^{\mathbb{Z}}$ invariant under the shift map in $X^{\mathbb{Z}}$ such that the restriction $\sigma : \mathcal{H} \to \mathcal{H}$ is topologically mixing and yet admits two mutually singular invariant measures, μ_v and μ_a, such that

$$h_{\mu_v}(\sigma) = h_{\mu_a}(\sigma) = \log 2 = h(\sigma).$$

Let us describe this example. By definition, $\mathcal{H} = E^{\mathbb{Z}} \cup O^{\mathbb{Z}} \cup \mathcal{H}_*$, where $\mathcal{H}_*$ consists of the sequences $x \in X^{\mathbb{Z}}$ that satisfy the following rule: Whenever one block with m symbols of one type, even or odd, is followed by another block with n symbols of the other type, odd or even, the two of them are separated by a block of no less than $m+n$ symbols $*$. In other words, the following configurations are admissible in sequences $x \in \mathcal{H}_*$:

$$x = (\dots, *, e_1, \dots, e_m, \underbrace{*, \dots, *}_{k}, o_1, \dots, o_n, * \dots) \quad \text{or}$$

$$x = (\dots, *, o_1, \dots, o_m, \underbrace{*, \dots, *}_{k}, e_1, \dots, e_n, * \dots),$$

with $e_i \in E$, $o_j \in O$ and $k \geq m+n$. Observe that a sequence $x \in \mathcal{H}_*$ may start and/or end with an infinite block of $*$ but it can neither start nor end with an infinite block of either even or odd type. It is clear that $\mathcal{H}$ is invariant under the shift map. Haydn [Hay] proved that (see Exercise 10.5.6):

(i) the shift map $\sigma : \mathcal{H} \to \mathcal{H}$ is topologically mixing;
(ii) $h(\sigma) = \log 2$.

We know that $E^{\mathbb{Z}}$ and $O^{\mathbb{Z}}$ support Bernoulli measures μ_v and μ_a with entropy equal to $\log 2$. Then, μ_v and μ_a are measures of maximal entropy for $\sigma : \mathcal{H} \to \mathcal{H}$ and they are mutually singular.

Clearly, this construction may be modified to yield transformations with any given number of ergodic measures of maximal entropy. Haydn [Hay] has also shown how to adapt it to construct examples with multiple equilibrium states for other potentials as well.

In Chapters 11 and 12 we study a class of transformations, called *expanding*, for which every Hölder potential admits exactly one equilibrium state. In particular, these transformations are *intrinsically ergodic*, that is, they have a unique measure of maximal entropy.

10.5.1 Exercises

10.5.1. Show that, among the Bernoulli measures of the shift map $\sigma : \Sigma \to \Sigma$ in the space $\Sigma = \{1, \dots, d\}^{\mathbb{Z}}$, the one with the largest entropy is given by the probability vector $(1/d, \dots, 1/d)$.

10.5.2. Let $\sigma : \Sigma \to \Sigma$ be the shift map in $\Sigma = \{1, \dots, d\}^{\mathbb{Z}}$ and $\phi : \Sigma \to \mathbb{R}$ be a locally constant potential, that is, such that ϕ is constant on each cylinder $[0; i]$. Calculate $P(f, \phi)$ and show that there exists some equilibrium state that is a Bernoulli measure.

10.5.3. Let $\sigma : \Sigma \to \Sigma$ be the shift map in $\Sigma = \{1,\dots,d\}^{\mathbb{N}}$. An invariant probability measure μ is called a *Gibbs state* for a potential $\varphi : \Sigma \to \mathbb{R}$ if there exist $P \in \mathbb{R}$ and $K > 0$ such that

$$K^{-1} \le \frac{\mu(C)}{\exp(\varphi_n(x) - nP)} \le K \tag{10.5.2}$$

for every cylinder $C = [0; i_0, \dots, i_{n-1}]$ and any $x \in C$. Prove that if μ is a Gibbs state then $h_\mu(\sigma) + \int \varphi \, d\mu$ coincides with the constant P in (10.5.2). Therefore, μ is an equilibrium state if and only if $P = P(\sigma, \varphi)$. Prove that for each choice of the constant P there exists at most one ergodic Gibbs state.

10.5.4. Let $f : M \to M$ be a continuous transformation in a compact metric space and $\phi : M \to \mathbb{R}$ be a continuous function. If μ is an equilibrium state for ϕ, then the functional $F_\mu : C^0(M) \to \mathbb{R}$ defined by $F_\mu(\psi) = \int \psi \, d\mu$ is such that $F_\mu(\psi) \le P(f, \phi + \psi) - P(f, \phi)$ for every $\psi \in C^0(M)$. Conclude that if the pressure function $P(f, \cdot) : C^0(M) \to \mathbb{R}$ is differentiable in every direction at a point ϕ then ϕ admits at most one equilibrium state.

10.5.5. Let $f : M \to M$ be a continuous transformation in a compact metric space. Show that the subset of functions $\phi : M \to \mathbb{R}$ for which there exists a unique equilibrium state is residual in $C^0(M)$.

10.5.6. Check the claims (i) and (ii) in Example 10.5.10.

11
Expanding maps

The distinctive feature of the transformations $f : M \to M$ that we study in the last two chapters of this book is that they expand the distance between nearby points: there exists a constant $\sigma > 1$ such that

$$d(f(x), f(y)) \geq \sigma\, d(x,y)$$

whenever the distance between x and y is small (a precise definition will be given shortly). There is more than one reason why this class of transformations has an important role in ergodic theory.

On the one hand, as we are going to see, expanding maps exhibit very rich dynamical behavior, from the metric and topological point of view as well as from the ergodic point of view. Thus, they provide a natural and interesting context for utilizing many of the ideas and methods that have been introduced so far.

On the other hand, expanding maps lead to paradigms that are useful for understanding many other systems, technically more complex. A good illustration of this is the ergodic theory of uniformly hyperbolic systems, for which an excellent presentation can be found in Bowen [Bow75a].

An important special case of expanding maps are the differentiable transformations on manifolds such that

$$\|Df(x)v\| \geq \sigma\, \|v\|$$

for every $x \in M$ and every vector v tangent to M at the point x. We focus on this case in Section 11.1. The main result (Theorem 11.1.2) is that, under the hypothesis that the Jacobian $\det Df$ is Hölder, the transformation f admits a unique invariant probability measure absolutely continuous with respect to the Lebesgue measure. Moreover, that probability measure is ergodic and positive on the open subsets of M.

In Section 11.2 we extend the notion of an expanding map to metric spaces and we give a global description of the topological dynamics of such maps, starting from the study of their periodic points. The main objective is to show that the global dynamics may always be reduced to the topologically exact case (Theorem 11.2.15). In Section 11.3 we complement this analysis by showing that for these transformations the topological entropy coincides with the growth rate of the number of periodic points.

The study of expanding maps will proceed in Chapter 12, where we will develop the so-called *thermodynamic formalism* for such systems.

11.1 Expanding maps on manifolds

Let M be a compact manifold and $f : M \to M$ be a map of class C^1. We say that f is *expanding* if there exists $\sigma > 1$ and some Riemannian metric on M such that

$$\|Df(x)v\| \geq \sigma \|v\| \quad \text{for every } x \in M \text{ and every } v \in T_xM. \tag{11.1.1}$$

In particular, f is a local diffeomorphism: the condition (11.1.1) implies that $Df(x)$ is an isomorphism for every $x \in M$. In what follows, we call *Lebesgue measure* on M the volume measure m induced by such a Riemannian metric. The precise choice of the metric is not very important, since the volume measures induced by different Riemannian metrics are all equivalent.

Example 11.1.1. Let $f_A : \mathbb{T}^d \to \mathbb{T}^d$ be the linear endomorphism of the torus induced by some matrix A with integer coefficients and determinant different from zero. Assume that all the eigenvalues $\lambda_1, \dots, \lambda_d$ of A are larger than 1 in absolute value. Then, given any $1 < \sigma < \inf_i |\lambda_i|$, there exists an inner product in $\mathbb{R}^d$ relative to which $\|Av\| \geq \sigma \|v\|$ for every v. Indeed, suppose that the eigenvalues are real. Consider any basis of $\mathbb{R}^d$ that sets A in canonical Jordan form: $A = D + \varepsilon N$ where N is nilpotent and D is diagonal with respect to that basis. The inner product relative to which such basis is orthonormal has the required property, as long as $\varepsilon > 0$ is small enough. The reader should have no difficulty extending this argument to the case when there are complex eigenvalues. This shows that the transformation f_A is expanding.

It is clear from the definition that any map sufficiently close to an expanding one, relative to the C^1 topology, is still expanding. Thus, the observation in Example 11.1.1 provides a whole open set of examples of expanding maps. A classical result of Michael Shub [Shu69] asserts a (much deeper) kind of converse: every expanding map on the torus $\mathbb{T}^d$ is topologically conjugate to an expanding linear endomorphism f_A.

Given a probability measure μ invariant under a transformation $f : M \to M$, we call the *basin* of μ the set $B(\mu)$ of all points $x \in M$ such that

$$\lim_{n\to\infty} \frac{1}{n} \sum_{j=0}^{n-1} \varphi(f^j(x)) = \int \varphi \, d\mu$$

for every continuous function $\varphi : M \to \mathbb{R}$. Note that the basin is always an invariant set. If μ is ergodic then $B(\mu)$ is a full measure set (Exercise 4.1.5).

Theorem 11.1.2. *Let $f : M \to M$ be an expanding map on a compact (connected) manifold M and assume that the Jacobian $x \mapsto \det Df(x)$ is Hölder. Then f admits a unique invariant probability measure μ absolutely continuous with respect to Lebesgue measure m. Moreover, μ is ergodic, its support coincides with M and its basin has full Lebesgue measure in M.*

First, let us outline the strategy of the proof of Theorem 11.1.2. The details will be given in the forthcoming sections. The conclusion is generally false if one omits the hypothesis of Hölder continuity: see Quas [Qua99].

It is easy to check (Exercise 11.1.1) that the pre-image under f of any set with zero Lebesgue measure m also has zero Lebesgue measure. This means that the image $f_*\nu$ under f of any measure ν absolutely continuous with respect to m is also absolutely continuous with respect to m. In particular, the n-th image f^n_*m is always absolutely continuous with respect to m.

In Proposition 11.1.7 we prove that the *density* (that is, the Radon–Nikodym derivative) of each f^n_*m with respect to m is bounded by some constant independent of $n \geq 1$. We deduce from this fact that every accumulation point of the sequence

$$\frac{1}{n}\sum_{j=0}^{n-1} f^j_* m,$$

with respect to the weak* topology, is an invariant probability measure absolutely continuous with respect to Lebesgue measure, with density bounded by that same constant.

An additional argument, using the fact that M is connected, proves that the accumulation point is unique and has all the properties in the statement of the theorem.

11.1.1 Distortion lemma

Starting the proof of Theorem 11.1.2, let us prove the following elementary fact:

Lemma 11.1.3. *Let $f : M \to M$ be a local diffeomorphism of class C^r, $r \geq 1$ on a compact Riemannian manifold M and $\sigma > 0$ be such that $\|Df(x)v\| \geq \sigma \|v\|$ for every $x \in M$ and every $v \in T_xM$. Then there exists $\rho > 0$ such that, for any pre-image x of any point $y \in M$, there exists a map $h : B(y,\rho) \to M$ of class C^r such that $f \circ h = \mathrm{id}$, $h(y) = x$ and*

$$d(h(y_1),h(y_2)) \leq \sigma^{-1} d(y_1,y_2) \quad \textit{for every } y_1,y_2 \in B(y,\rho). \qquad (11.1.2)$$

Proof. By the inverse function theorem, for every $\xi \in M$ there exist open neighborhoods $U(\xi) \subset M$ of ξ and $V(\xi) \subset N$ of $f(\xi)$ such that f maps $U(\xi)$

diffeomorphically onto $V(\xi)$. Since M is compact, it follows that there exists $\delta > 0$ such that $d(\xi,\xi') \geq \delta$ whenever $f(\xi) = f(\xi')$. In particular, we may choose these neighborhoods in such a way that $U(\xi) \cap U(\xi') = \emptyset$ whenever $f(\xi) = f(\xi')$. For each $\eta \in M$, let

$$W(\eta) = \bigcap_{\xi \in f^{-1}(\eta)} V(\xi).$$

Since $f^{-1}(\eta)$ is finite (Exercise A.4.6), every $W(\eta)$ is an open set. Fix $\rho > 0$ such that 2ρ is a Lebesgue number for the open cover $\{W(\eta) : \eta \in M\}$ of M. In particular, for every $y \in M$ there exists $\eta \in M$ such that $B(y,\rho)$ is contained in $W(\eta)$, that is, it is contained in $V(\xi)$ for all $\xi \in f^{-1}(\eta)$. Since the $U(\xi)$ are pairwise disjoints and $\#f^{-1}(y) = \text{degree}(f) = \#f^{-1}(\eta)$, given any $x \in f^{-1}(y)$ there exists exactly one $\xi \in f^{-1}(\eta)$ such that $x \in U(\xi)$. Let h be the restriction to $B(y,\rho)$ of the inverse of $f : U(\xi) \to V(\xi)$. By construction, $f \circ h = \text{id}$ and $h(y) = x$. Moreover, $\|Dh(z)\| = \|Df(h(z))^{-1}\| \leq \sigma^{-1}$ for every z in the domain of h. By the mean value theorem, this implies that h has the property (11.1.2).

Transformations h as in this statement are called *inverse branches* of the local diffeomorphism f. Now assume that f is an expanding map. The condition (11.1.1) means that in this case we may take $\sigma > 1$ in the hypothesis of the lemma. Then the conclusion (11.1.2) implies that the inverse branches are contractions, with uniform contraction rate.

In particular, we may define inverse branches h^n of any iterate f^n, $n \geq 1$, as follows. Given $y \in M$ and $x \in f^{-n}(y)$, let $h_1, \dots, h_n$ be inverse branches of f with

$$h_j(f^{n-j+1}(x)) = f^{n-j}(x)$$

for every $1 \leq j \leq n$. Since every h_j is a contraction, its image is contained in a ball around $f^{n-j}(x)$ with radius smaller than ρ. Then $h^n = h_n \circ \cdots \circ h_1$ is well defined on the closure of the ball of radius ρ around y. It is clear that $f^n \circ h^n = \text{id}$ and $h^n(y) = x$. Moreover, each h^n is a contraction:

$$d(h^n(y_1), h^n(y_2)) \leq \sigma^{-n} d(y_1, y_2) \quad \text{for every } y_1, y_2 \in B(y,\rho).$$

Lemma 11.1.4. *If $f : M \to M$ is a C^1 expanding map on a compact manifold then f is expansive.*

Proof. By Lemma 11.1.3, there exists $\rho > 0$ such that, for any pre-image x of a point $y \in M$, there exists a map $h : B(y,\rho) \to M$ of class C^1 such that $f \circ h = \text{id}$, $h(y) = x$ and

$$d(h(y_1), h(y_2)) \leq \sigma^{-1} d(y_1, y_2) \quad \text{for every } y_1, y_2 \in B(y,\rho).$$

Hence, if $d(f^n(x), f^n(y)) \leq \rho$ for every $n \geq 0$ then

$$d(x,y) \leq \sigma^{-n} d(f^n(x), f^n(y)) \leq \sigma^{-n}\rho,$$

which immediately implies that $x = y$.

The next result provides a good control of the distortion of the iterates of f and their inverse branches, which is crucial for the proof of Theorem 11.1.2. This is the only step of the proof where we use the hypothesis that the Jacobian $x \mapsto \det Df(x)$ is Hölder. Note that, since f is a local diffeomorphism and M is compact, the Jacobian is bounded from zero and infinity. Hence, the logarithm $\log|\det Df|$ is also Hölder: there exist $C_0 > 0$ and $\nu > 0$ such that

$$\big|\log|\det Df(x)| - \log|\det Df(y)|\big| \le C_0 d(x,y)^\nu \quad \text{for any } x,y \in M.$$

Proposition 11.1.5 (Distortion lemma)**.** *There exists $C_1 > 0$ such that, given any $n \ge 1$, any $y \in M$ and any inverse branch $h^n : B(y,\rho) \to M$ of f^n,*

$$\log \frac{|\det Dh^n(y_1)|}{|\det Dh^n(y_2)|} \le C_1 d(y_1,y_2)^\nu \le C_1(2\rho)^\nu$$

for every $y_1, y_2 \in B(y,\rho)$.

Proof. Write h^n as a composition $h^n = h_n \circ \cdots \circ h_1$ of inverse branches of f. Analogously, $h^i = h_i \circ \cdots \circ h_1$ for $1 \le i < n$ and $h^0 = \mathrm{id}$. Then,

$$\log \frac{|\det Dh^n(y_1)|}{|\det Dh^n(y_2)|} = \sum_{i=1}^{n} \log|\det Dh_i(h^{i-1}(y_1))| - \log|\det Dh_i(h^{i-1}(y_2))|.$$

Note that $\log|\det Dh_i| = -\log|\det Df| \circ h_i$ and recall that every h_j is a σ^{-1}-contraction. Hence,

$$\log \frac{|\det Dh^n(y_1)|}{|\det Dh^n(y_2)|} \le \sum_{i=1}^{n} C_0 d(h^i(y_1), h^i(y_2))^\nu \le \sum_{i=1}^{n} C_0 \sigma^{-i\nu} d(y_1,y_2)^\nu.$$

Therefore, to prove the lemma it suffices to take $C_1 = C_0 \sum_{i=1}^{\infty} \sigma^{-i\nu}$.

The geometric meaning of this proposition is made even more transparent by the following corollary:

Corollary 11.1.6. *There exists $C_2 > 0$ such that, for every $y \in M$ and any measurable sets $B_1, B_2 \subset B(y,\rho)$,*

$$\frac{1}{C_2}\frac{m(B_1)}{m(B_2)} \le \frac{m(h^n(B_1))}{m(h^n(B_2))} \le C_2 \frac{m(B_1)}{m(B_2)}.$$

Proof. Take $C_2 = \exp(2C_1(2\rho)^\nu)$. It follows from the Proposition 11.1.5 that

$$m(h^n(B_1)) = \int_{B_1} |\det Dh^n|\, dm \le \exp(C_1(2\rho)^\nu)|\det Dh^n(y)| m(B_1) \quad \text{and}$$

$$m(h^n(B_2)) = \int_{B_2} |\det Dh^n|\, dm \ge \exp(-C_1(2\rho)^\nu)|\det Dh^n(y)| m(B_2).$$

Dividing the two inequalities, we get that

$$\frac{m(h^n(B_1))}{m(h^n(B_2))} \le C_2 \frac{m(B_1)}{m(B_2)}.$$

Inverting the roles of B_1 and B_2 we get the other inequality.

The next result, which is also a consequence of the distortion lemma, asserts that the iterates $f_*^n m$ of the Lebesgue measure have uniformly bounded densities:

Proposition 11.1.7. *There exists $C_3 > 0$ such that $(f_*^n m)(B) \le C_3 m(B)$ for every measurable set $B \subset M$ and every $n \ge 1$.*

Proof. It is no restriction to suppose that B is contained in a ball $B_0 = B(z,\rho)$ of radius ρ around some point in the pre-image of $z \in M$. Using Corollary 11.1.6, we see that

$$\frac{m(h^n(B))}{m(h^n(B_0))} \le C_2 \frac{m(B)}{m(B_0)}$$

for every inverse branch h^n of f^n at the point z. Moreover, $(f_*^n m)(B) = m(f^{-n}(B))$ is the sum of $m(h^n(B))$ over all the inverse branches, and analogously for B_0. In this way, we find that

$$\frac{(f_*^n m)(B)}{(f_*^n m)(B_0)} \le C_2 \frac{m(B)}{m(B_0)}.$$

It is clear that $(f_*^n m)(B_0) \le (f_*^n m)(M) = 1$. Moreover, the Lebesgue measure of the balls with a fixed radius ρ is bounded from zero for some constant $\alpha_0 > 0$ that depends only on ρ. So, to get the conclusion of the proposition it suffices to take $C_3 = C_2 \alpha_0$.

We also need the auxiliary result that follows. Recall that, given a function φ and a measure ν, we denote by $\varphi\nu$ the measure defined by $(\varphi\nu)(B) = \int_B \varphi\, d\nu$.

Lemma 11.1.8. *Let ν be a probability measure on a compact metric space X and $\varphi : X \to [0,+\infty)$ be an integrable function with respect to ν. Let μ_i, $i \ge 1$, be a sequence of probability measures on X converging, in the weak* topology, to a probability measure μ. If $\mu_i \le \varphi\nu$ for every $i \ge 1$ then $\mu \le \varphi\nu$.*

Proof. Let B be any measurable set. For each $\varepsilon > 0$, let K_ε be a compact subset of B such that $\mu(B \setminus K_\varepsilon)$ and $(\varphi\nu)(B \setminus K_\varepsilon)$ are both less than ε (such a compact set does exist, by Proposition A.3.2). Fix $r > 0$ small enough that the measure of $A_\varepsilon \setminus K_\varepsilon$ is also less than ε, for both μ and $\varphi\nu$, where $A_\varepsilon = \{z : d(z, K_\varepsilon) < r\}$. The set of values of r for which the boundary of A_ε has positive μ-measure is at most countable (Exercise A.3.2). Hence, up to changing r slightly if necessary, we may suppose that the boundary of A_ε has measure zero. Then $\mu = \lim_i \mu_i$ implies that $\mu(A_\varepsilon) = \lim_i \mu_i(A_\varepsilon) \le (\varphi\nu)(A_\varepsilon)$. Making $\varepsilon \to 0$, we conclude that $\mu(B) \le (\varphi\nu)(B)$.

Applying this lemma to our present situation, we obtain

Corollary 11.1.9. *Every accumulation point μ of the sequence $n^{-1}\sum_{j=0}^{n-1} f_*^j m$ is an invariant probability measure for f absolutely continuous with respect to the Lebesgue measure.*

Proof. Take φ constant equal to C_3 and let $\nu = m$. Choose a subsequence $(n_i)_i$ such that $\mu_i = n_i^{-1} \sum_{j=0}^{n_i-1} f_*^j m$ converges to some probability measure μ. Proposition 11.1.7 ensures that $\mu_i \le \varphi\nu$. Then, by Lemma 11.1.8, we also have $\mu \le \varphi\nu = C_3 m$. This implies that $\mu \ll m$ with density bounded by C_3.

11.1.2 Existence of ergodic measures

Next, we show that the measure μ we have just constructed is the unique invariant probability measure absolutely continuous with respect to the Lebesgue measure and, moreover, it is ergodic for f.

Start by fixing a finite partition $\mathcal{P}_0 = \{U_1, \dots, U_s\}$ of M into regions with non-empty interior and diameter less than ρ. Then, for each $n \ge 1$, define $\mathcal{P}_n$ to be the partition of M into the images of the U_i, $1 \le i \le s$, under the inverse branches of f^n. The diameter of each $\mathcal{P}_n$, that is, the supremum of the diameters of its elements, is less than $\rho\sigma^{-n}$.

Lemma 11.1.10. *Let $\mathcal{P}_n$, $n \ge 1$, be a sequence of partitions in a compact metric space M, with diameters converging to zero. Let ν be a probability measure on M and B be any measurable set with $\nu(B) > 0$. Then there exist $V_n \in \mathcal{P}_n$, $n \ge 1$, such that*

$$\nu(V_n) > 0 \quad \text{and} \quad \frac{\nu(B \cap V_n)}{\nu(V_n)} \to 1 \quad \text{when } n \to \infty.$$

Proof. Given any $0 < \varepsilon < \nu(B)$, let $K_\varepsilon \subset B$ be a compact set with $\nu(B \setminus K_\varepsilon) < \varepsilon$. Let $K_{\varepsilon,n}$ be the union of all the elements of $\mathcal{P}_n$ that intersect K_ε. Since the diameters of the partitions converge to zero, $\nu(K_{\varepsilon,n} \setminus K_\varepsilon) < \varepsilon$ for every n sufficiently large. By contradiction, suppose that

$$\nu\big(K_\varepsilon \cap V_n\big) \le \frac{\nu(B) - \varepsilon}{\nu(B) + \varepsilon}\,\nu(V_n)$$

for every $V_n \in \mathcal{P}_n$ that intersects K_ε. It would follow that

$$\nu(K_\varepsilon) \le \sum_{V_n} \nu\big(K_\varepsilon \cap V_n\big) \le \sum_{V_n} \frac{\nu(B) - \varepsilon}{\nu(B) + \varepsilon}\,\nu(V_n) = \frac{\nu(B) - \varepsilon}{\nu(B) + \varepsilon}\,\nu(K_{\varepsilon,n})$$
$$\le \frac{\nu(B) - \varepsilon}{\nu(B) + \varepsilon}\,(\nu(K_\varepsilon) + \varepsilon) \le \nu(B) - \varepsilon < \nu(K_\varepsilon).$$

This contradiction shows that there must exist some $V_n \in \mathcal{P}_n$ such that

$$\nu(V_n) \ge \nu\big(B \cap V_n\big) \ge \nu\big(K_\varepsilon \cap V_n\big) > \frac{\nu(B) - \varepsilon}{\nu(B) + \varepsilon}\,\nu(V_n)$$

and, consequently, $\nu(V_n) > 0$. Making $\varepsilon \to 0$ we get the claim.

In the statements that follow, we say that a measurable set $A \subset M$ is invariant under $f : M \to M$ if $f^{-1}(A) = A$ up to zero Lebesgue measure. According to Exercise 11.1.1, then we also have that $f(A) = A$ up to zero Lebesgue measure.

Lemma 11.1.11. *If $A \subset M$ satisfies $f(A) \subset A$ and has positive Lebesgue measure then A has full Lebesgue measure inside some $U_i \in \mathcal{P}_0$, that is, there exists $1 \le i \le s$ such that $m(U_i \setminus A) = 0$.*

Proof. By Lemma 11.1.10, we may choose $V_n \in \mathcal{P}_n$ so that $m(V_n \setminus A)/m(V_n)$ converges to zero when $n \to \infty$. Let $U_{i(n)} = f^n(V_n)$. By Proposition 11.1.5 applied to the inverse branch of f^n that maps $U_{i(n)}$ to V_n, we get that

$$\frac{m(U_{i(n)} \setminus A)}{m(U_{i(n)})} \le \frac{m(f^n(V_n \setminus A))}{m(f^n(V_n))} \le \exp\left(C_1 (2\rho)^\nu\right) \frac{m(V_n \setminus A)}{m(V_n)}$$

also converges to zero. Since $\mathcal{P}_0$ is finite, there must be $1 \le i \le s$ such that $i(n) = i$ for infinitely many values of n. Then $m(U_i \setminus A) = 0$.

Corollary 11.1.12. *The transformation $f : M \to M$ admits some ergodic invariant probability measure absolutely continuous with respect to the Lebesgue measure.*

Proof. It follows from the previous lemma there exist at most $s = \#\mathcal{P}_0$ pairwise disjoint invariant sets with positive Lebesgue measure. Therefore, M may be partitioned into a finite number of minimal invariant sets $A_1, \dots, A_r$, $r \le s$ with positive Lebesgue measure, where by *minimal* we mean that there are no invariant sets $B_i \subset A_i$ with $0 < m(B_i) < m(A_i)$. Given any absolutely continuous invariant probability measure μ, there exists some i such that $\mu(A_i) > 0$. The normalized restriction

$$\mu_i(B) = \frac{\mu(B \cap A_i)}{\mu(A_i)}$$

of μ to any such A_i is invariant and absolutely continuous. Moreover, the assumption that A_i is minimal implies that μ_i is ergodic.

11.1.3 Uniqueness and conclusion of the proof

The previous argument also shows that there exist only a finite number of absolutely continuous ergodic probability measures. The last step is to show that, in fact, such a probability measure is unique. For that we use the fact that f is *topologically exact:*

Lemma 11.1.13. *Given any non-empty open set $U \subset M$, there exists $N \ge 1$ such that $f^N(U) = M$.*

Proof. Let $x \in U$ and $r > 0$ be such that the ball of radius r around x is contained in U. Given any $n \ge 1$, suppose that $f^n(U)$ does not cover the whole manifold. Then there exists some curve γ connecting $f^n(x)$ to a point $y \in M \setminus f^n(U)$, and that curve may be taken with length smaller than $\operatorname{diam} M + 1$.

Lifting[1] γ by the local diffeomorphism f^n, we obtain a curve γ_n connecting x to some point $y_n \in M \setminus U$. Then $r \le \circ(\gamma_n) \le \sigma^{-n}(\operatorname{diam} M + 1)$. This provides an upper bound on the possible value of n. Hence, $f^n(U) = M$ for every n sufficiently large, as claimed.

Corollary 11.1.14. *If $A \subset M$ has positive Lebesgue measure and satisfies $f(A) \subset A$ then A has full Lebesgue measure in the whole manifold M.*

Proof. Let U be the interior of a set U_i as in Lemma 11.1.11 and $N \ge 1$ be such that $f^N(U) = M$. Then $m(U \setminus A) = 0$ and, using the fact that f is a local diffeomorphism, it follows that $M \setminus A = f^N(U) \setminus f^N(A) \subset f^N(U \setminus A)$ also has Lebesgue measure zero.

The next statement completes the proof of Theorem 11.1.2:

Corollary 11.1.15. *Let μ be any absolutely continuous invariant probability measure. Then μ is ergodic and its basin $B(\mu)$ has full Lebesgue measure in M. Consequently, μ is unique. Moreover, its support is the whole manifold M.*

Proof. If A is an invariant set then, by Corollary 11.1.14, either A or its complement A^c has Lebesgue measure zero. Since μ is absolutely continuous, it follows that either $\mu(A) = 0$ or $\mu(A^c) = 0$. This proves that μ is ergodic. Then $\mu(B(\mu)) = 1$ and, in particular, $m(B(\mu)) > 0$. Since $B(\mu)$ is an invariant set, it follows that it has full Lebesgue measure. Analogously, since the support of μ is a compact set with positive Lebesgue measure and $f(\operatorname{supp}\mu) \subset \operatorname{supp}\mu$, it must coincide with M.

Finally, let μ and ν be any two absolutely continuous invariant probability measures. It follows from what we have just said that the two measures are ergodic and their basins intersect each other. Given any point x in $B(\mu) \cap B(\nu)$, the sequence

$$\frac{1}{n}\sum_{j=0}^{n-1} \delta_{f^j(x)}$$

converges to both μ and ν in the weak* topology. Thus, $\mu = \nu$.

In general, we say that an invariant probability measure μ of a local diffeomorphism $f : M \to M$ is a *physical measure* if its basin has positive Lebesgue measure. It follows from Corollary 11.1.15 that in the present context there exists a unique physical measure, which is the absolutely continuous invariant measure μ, and its basin has full Lebesgue measure. This last fact may be expressed as follows:

$$\frac{1}{n}\sum_{j=0}^{n-1} \delta_{f^j(x)} \to \mu \quad \text{for Lebesgue almost every } x.$$

[1] Note that any local diffeomorphism from a compact manifold to itself is a covering map.

In Chapter 12 we will find this absolutely continuous invariant probability measure μ through a different approach (Proposition 12.1.20) that also shows that the density $h = d\mu/dm$ is Hölder and bounded away from zero. In particular, μ is equivalent to the Lebesgue measure m, not just absolutely continuous. Moreover (Section 12.1.7), the system (f,μ) is exact, not just ergodic. In addition (Lemma 12.1.12), its Jacobian is given by $J_\mu f = |\det Df|(h \circ f)/h$. Hence, by the Rokhlin formula (Theorem 9.7.3),

$$h_\mu(f) = \int \log J_\mu f \, d\mu = \int \log|\det Df| \, d\mu + \int \log(h \circ f) \, d\mu - \int \log h \, d\mu.$$

Since μ is invariant, this means that

$$h_\mu(f) = \int \log|\det Df| \, d\mu.$$

Actually, the facts stated in the previous paragraph can already be proven with the methods available at this point. We invite the reader to do just that (Exercises 11.1.3 through 11.1.6), in the context of expanding maps of the interval, which are technically a bit simpler than expanding maps on a general manifold.

Example 11.1.16. We say that a transformation $f : [0,1] \to [0,1]$ is an *expanding map of the interval* if there exists a countable (possibly finite) family $\mathcal{P}$ of pairwise disjoint open subintervals whose union has full Lebesgue measure in $[0,1]$ and which satisfy:

(i) The restriction of f to each $P \in \mathcal{P}$ is a diffeomorphism onto $(0,1)$; denote by $f_P^{-1} : (0,1) \to P$ its inverse.

(ii) There exist $C > 0$ and $\theta > 0$ such that, for every x, y and every $P \in \mathcal{P}$,

$$\big|\log|D(f_P^{-1})(x)| - \log|D(f_P^{-1})(y)|\big| \le C|x-y|^\theta.$$

(iii) There exist $c > 0$ and $\sigma > 1$ such that, for every n and every x,

$$|Df^n(x)| \ge c\sigma^n \quad \text{(whenever the derivative is defined.)}$$

This class of transformations includes the decimal expansion and the Gauss map as special cases. Its properties are analyzed in Exercises 11.1.3 through 11.1.5.

Exercise 11.1.6 deals with a slightly more general class of transformations, where we replace condition (i) by

(i′) There exists $\delta > 0$ such that the restriction of f to each $P \in \mathcal{P}$ is a diffeomorphism onto some interval $f(P)$ of length larger than δ that contains every element of $\mathcal{P}$ that it intersects.

11.1.4 Exercises

11.1.1. Let $f : M \to M$ be a local diffeomorphism in a compact manifold and m be the Lebesgue measure on M. Check the following facts:

(a) If $m(B) = 0$ then $m(f^{-1}(B)) = 0$.
(b) If B is measurable then $f(B)$ is measurable.
(c) If $m(B) = 0$ then $m(f(B)) = 0$.
(d) If $A = B$ up to zero Lebesgue measure zero then $f(A) = f(B)$ and $f^{-1}(A) = f^{-1}(B)$ up to zero Lebesgue measure.
(e) If A is an invariant set then $f(A) = A$ up to zero Lebesgue measure.

11.1.2. Let $f : M \to M$ be a transformation of class C^1 such that there exist $\sigma > 1$ and $k \geq 1$ satisfying $\|Df^k(x)v\| \geq \sigma \|v\|$ for every $x \in M$ and every $v \in T_xM$. Show that there exists $\theta > 1$ and a Riemannian norm $\langle \cdot \rangle$ equivalent to $\| \cdot \|$ such that $\langle Df(x)v \rangle \geq \theta \langle v \rangle$ for every $x \in M$ and every $v \in T_xM$.

11.1.3. Show that if $f : [0,1] \to [0,1]$ is an expanding map of the interval and m is the Lebesgue measure on $[0,1]$ then there exists a function $\rho : (0,1) \to (0,\infty)$ such that $\log \rho$ is bounded and Hölder and $\mu = \rho m$ is a probability measure invariant under f.

11.1.4. Show that the measure μ in Exercise 11.1.3 is exact and is the unique invariant probability measure of f absolutely continuous with respect to the Lebesgue measure m.

11.1.5. Show that the measure μ in Exercise 11.1.3 satisfies the Rokhlin formula: assuming that $\log |f'| \in L^1(\mu)$, we have that $h_\mu(f) = \int \log |f'| \, d\mu$.

11.1.6. Prove the following generalization of Exercises 11.1.3 and 11.1.4: if f satisfies the conditions (i′), (ii) and (iii) in Example 11.1.16 then there exists a finite (non-empty) family of absolutely continuous invariant probability measures ergodic for f and such that every absolutely continuous invariant probability measure is a convex combination of those ergodic probability measures.

11.2 Dynamics of expanding maps

In this section we extend the notion of an expanding map to compact metric spaces and we mention a few interesting examples. In this general setup, an expanding map need not be transitive, let alone topologically exact (compare Lemma 11.1.13). However, Theorems 11.2.14 and 11.2.15 assert that the dynamics may always be reduced to the topologically exact case. This is relevant because for the main results in this section we need the transformation to be topologically exact or, equivalently (Exercise 11.2.2), topologically mixing.

A continuous transformation $f : M \to M$ in a compact metric space M is an *expanding map* if there exist constants $\sigma > 1$ and $\rho > 0$ such that for every $p \in M$ the image of the ball $B(p,\rho)$ contains a neighborhood of the closure of $B(f(p),\rho)$ and

$$d(f(x),f(y)) \geq \sigma d(x,y) \quad \text{for every } x,y \in B(p,\rho). \tag{11.2.1}$$

Every expanding map on a manifold, in the sense of Section 11.1, is also expanding in the present sense:

Example 11.2.1. Let M be a compact Riemannian manifold and $f : M \to M$ be a map of class C^1 such that $\|Df(x)v\| \geq \sigma \|v\|$ for every $x \in M$ and every $v \in T_xM$, where σ is a constant larger than 1. Denote $K = \sup \|Df\|$ (observe that $K > 1$). Fix $\rho > 0$ small enough that the restriction of f to every ball $B(p, 2K\rho)$ is a diffeomorphism onto its image. Consider any $y \in B(f(p), \sigma\rho)$ and let $\gamma : [0,1] \to B(f(p), \sigma\rho)$ be a minimizing geodesic (that is, such that it realizes the distance between points) with $\gamma(0) = f(p)$ and $\gamma(1) = y$. By the choice of ρ, there exists a differentiable curve $\beta : [0, \delta] \to B(p, \rho)$ such that $\beta(0) = p$ and $f(\beta(t)) = \gamma(t)$ for every t. Observe that (using $\ell(\cdot)$ to denote the length of a curve),

$$d(p, \beta(t)) \leq \ell(\beta \mid [0,t]) \leq \sigma^{-1}\ell(\gamma \mid [0,t]) = \sigma^{-1} t d(f(p), y) < t\rho$$

for every t. This shows that we may take $\delta = 1$. Then, $\beta(1) \in B(p, \rho)$ and $f(\beta(1)) = \gamma(1) = y$. In this way, we have shown that $f(B(p, \rho))$ contains $B(f(p), \sigma\rho)$, which is a neighborhood of the closure of $B(f(p), \rho)$. Now consider any $x, y \in B(p, \rho)$. Note that $d(f(x), f(y)) < 2K\rho$. Let $\gamma : [0,1] \to B(f(x), 2K\rho)$ be a minimizing geodesic connecting $f(x)$ to $f(y)$. Arguing as in the previous paragraph, we find a differentiable curve $\beta : [0,1] \to B(x, 2K\rho)$ connecting x to y and such that $f(\beta(t)) = \gamma(t)$ for every t. Then,

$$d(f(x), f(y)) = \ell(\gamma) \geq \sigma \ell(\beta) \geq \sigma d(x, y).$$

This completes the proof that f is an expanding map.

The following fact is useful for constructing further examples:

Lemma 11.2.2. *Assume that $f : M \to M$ is an expanding map and $\Lambda \subset M$ is a compact set such that $f^{-1}(\Lambda) = \Lambda$. Then the restriction $f : \Lambda \to \Lambda$ is also an expanding map.*

Proof. It is clear that the condition (11.2.1) remains valid for the restriction. We are left to check that $f(\Lambda \cap B(p, \rho))$ contains a neighborhood of $\Lambda \cap \overline{B(f(p), \rho)}$ inside Λ. By assumption, $f(B(p, \rho))$ contains some neighborhood V of the closure of $B(f(p), \rho)$. Then $\Lambda \cap V$ is a neighborhood of $\Lambda \cap \overline{B(f(p), \rho)}$. Moreover, given any $y \in \Lambda \cap V$ there exists $x \in B(p, \rho)$ such that $f(x) = y$. Since $f^{-1}(\Lambda) = \Lambda$, this point is necessarily in Λ. This proves that $\Lambda \cap V$ is contained in the image $f(\Lambda \cap B(p, \rho))$. Hence, the restriction of f to the set Λ is an expanding map, as stated.

It is not possible to replace the hypothesis of Lemma 11.2.2 by $f(\Lambda) = \Lambda$. See Exercise 11.2.4.

Example 11.2.3. Let $J \subset [0,1]$ be a finite union of (two or more) pairwise disjoint compact intervals. Consider a map $f : J \to [0,1]$ such that the

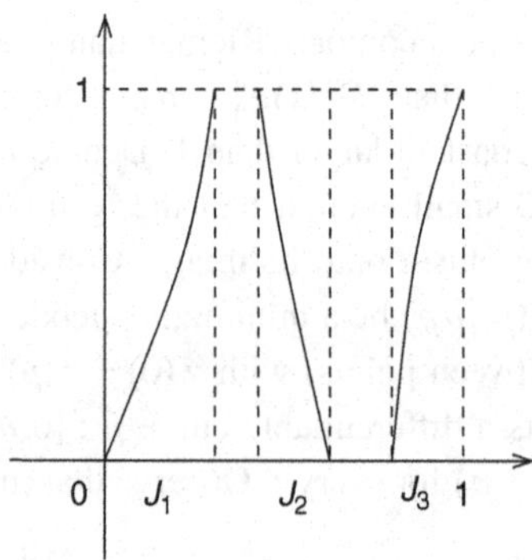

Figure 11.1. Expanding map on a Cantor set

restriction of f to each connected component of J is a diffeomorphism onto $[0,1]$. See Figure 11.1. Assume that there exists $\sigma > 1$ such that

$$|f'(x)| \geq \sigma \quad \text{for every } x \in J. \tag{11.2.2}$$

Denote $\Lambda = \bigcap_{n=0}^{\infty} f^{-n}(J)$. In other words, Λ is the set of all points x whose iterates $f^n(x)$ are defined for every $n \geq 0$. It follows from the definition that Λ is compact (one can show that K is a Cantor set) and $f^{-1}(\Lambda) = \Lambda$. The restriction $f : \Lambda \to \Lambda$ is an expanding map. Indeed, fix any $\rho > 0$ smaller than the distance between any two connected components of J. Then every ball of radius ρ inside Λ is contained in a unique connected component of J and so, by (11.2.2), it is dilated by a factor greater than or equal to σ.

Example 11.2.4. Let $\sigma : \Sigma_A \to \Sigma_A$ be the one-sided shift of finite type associated with a transition matrix A (these notions were introduced in Section 10.2.2). Consider in Σ_A the distance defined by

$$d\big((x_n)_n, (y_n)_n\big) = 2^{-N}, \quad N = \inf\{n \in \mathbb{N} : x_n \neq y_n\}. \tag{11.2.3}$$

Then σ_A is an expanding map. Indeed, fix $\rho \in (1/2, 1)$ and $\sigma = 2$. The open ball of radius ρ around any point $(p_n)_n \in \Sigma_A$ is just the cylinder $[0; p_0]_A$ that contains the point. The definition (11.2.3) yields

$$d\big((x_{n+1})_n, (y_{n+1})_n\big) = 2d\big((x_n)_n, (y_n)_n\big)$$

for any $(x_n)_n$ and $(y_n)_n$ in the cylinder $[0; p_0]_A$. Moreover, $\sigma_A([0; p_0]_A)$ is the union of all cylinders $[0; q]$ such that $A_{p_0,q} = 1$. In particular, it contains the cylinder $[0; p_1]_A$. Since the cylinders are both open and closed in Σ_A, this shows that the image of the ball of radius ρ around $(p_n)_n$ contains a neighborhood of the closure of the ball of radius ρ around $(p_{n+1})_n$. This completes the proof that every shift of finite type is an expanding map.

Example 11.2.5. Let $f : S^1 \to S^1$ be a local diffeomorphism of class C^2 with degree larger than 1. Assume that all the periodic points of f are hyperbolic, that is, $|(f^n)'(x)| \neq 1$ for every $x \in \mathrm{Fix}(f^n)$ and every $n \geq 1$. Let Λ be the complement of the union of the basins of attraction of all the attracting periodic points of f. Then the restriction $f : \Lambda \to \Lambda$ is an expanding map: this is a consequence of a deep theorem of Ricardo Mañé [Mañ85].

For expanding maps $f : M \to M$ in a metric space M, the number of pre-images of a point $y \in M$ may vary with y (unless M is connected; see Exercises 11.2.1 and A.4.6). For example, for a shift of finite type $\sigma : \Sigma_A \to \Sigma_A$ (Example 11.2.4) the number of pre-images of a point $y = (y_n)_n \in \Sigma_A$ is equal to the number of symbols i such that $A_{i,y_0} = 1$; in general, this number depends on y_0.

On the other hand, it is easy to see that the number of pre-images is always finite, and even bounded: it suffices to consider a finite cover of M by balls of radius ρ and to notice that every point has at most one pre-image in each of those balls. By a slight abuse of language, we call the *degree* of an expanding map $f : M \to M$ the maximum number of pre-images of any point, that is,

$$\mathrm{degree}(f) = \max\{\#f^{-1}(y) : y \in M\}. \tag{11.2.4}$$

11.2.1 Contracting inverse branches

Let $f : M \to M$ be an expanding map. By definition, the restriction of f to each ball $B(p,\rho)$ of radius ρ is injective and its image contains the closure of $B(f(p),\rho)$. Thus, the restriction to $B(p,\rho) \cap f^{-1}(B(f(p),\rho))$ is a homeomorphism onto $B(f(p),\rho)$. We denote by

$$h_p : B(f(p),\rho) \to B(p,\rho)$$

its inverse and call it the *inverse branch of f* at p. It is clear that $h_p(f(p)) = p$ and $f \circ h_p = \mathrm{id}$. The condition (11.2.1) implies that h_p is a σ^{-1}-contraction:

$$d(h_p(z),h_p(w)) \leq \sigma^{-1} d(z,w) \quad \text{for every } z,w \in B(f(p),\rho). \tag{11.2.5}$$

Lemma 11.2.6. *If $f : M \to M$ is expanding then, for every $y \in M$,*

$$f^{-1}(B(y,\rho)) = \bigcup_{x \in f^{-1}(y)} h_x(B(y,\rho)).$$

Proof. The relation $f \circ h_x = \mathrm{id}$ implies that $h_x(B(y,\rho))$ is contained in the pre-image of $B(y,\rho)$ for every $x \in f^{-1}(y)$. To prove the other inclusion, let z be any point such that $f(z) \in B(y,\rho)$. By the definition of an expanding map, $f(B(z,\rho))$ contains $B(f(z),\rho)$ and, hence, contains y. Let $h_z : B(f(z),\rho) \to M$ be the inverse branch of f at z and let $x = h_z(y)$. Both z and $h_x(f(z))$ are in $B(x,\rho)$. Since f is injective on every ball of radius ρ and $f(h_x(f(z))) = f(z)$, it follows that $z = h_x(f(z))$. This completes the proof.

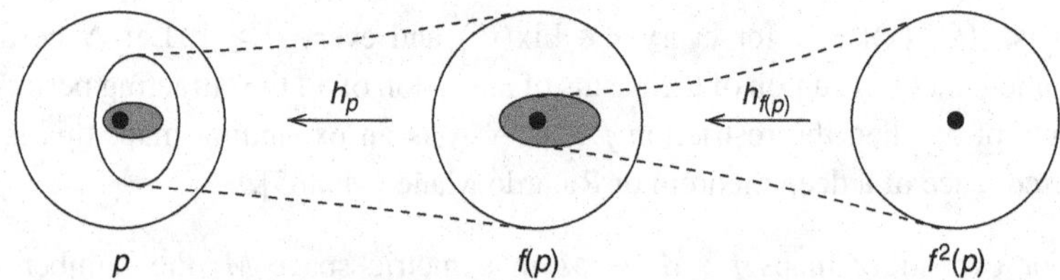

Figure 11.2. Inverse branches of f^n

More generally, for any $n \geq 1$, we call the *inverse branch of* f^n at p the composition

$$h_p^n = h_p \circ \cdots \circ h_{f^{n-1}(p)} : B(f^n(p), \rho) \to B(p, \rho)$$

of the inverse branches of f at the iterates of p. See Figure 11.2. Observe that $h_p^n(f^n(p)) = p$ and $f^n \circ h_p^n = \mathrm{id}$. Moreover, $f^j \circ h_p^n = h_{f^j(p)}^{n-j}$ for each $0 \leq j < n$. Hence,

$$d(f^j \circ h_p^n(z), f^j \circ h_p^n(w)) \leq \sigma^{j-n} d(z, w) \tag{11.2.6}$$

for every $z, w \in B(f^n(p), \rho)$ and every $0 \leq j \leq n$.

Lemma 11.2.7. *If $f : M \to M$ is an expanding map then $f^n(B(p, n+1, \varepsilon)) = B(f^n(p), \varepsilon)$ for every $p \in M$, $n \geq 0$ and $\varepsilon \in (0, \rho]$.*

Proof. The inclusion $f^n(B(p, n+1, \varepsilon)) \subset B(f^n(p), \varepsilon)$ is an immediate consequence of the definition of a dynamical ball. To prove the converse, consider the inverse branch $h_p^n : B(f^n(p), \rho) \to B(p, \rho)$. Given any $y \in B(f^n(p), \varepsilon)$, let $x = h_p^n(y)$. Then $f^n(x) = y$ and, by (11.2.6),

$$d(f^j(x), f^j(p)) \leq \sigma^{j-n} d(f^n(x), f^n(p)) \leq d(y, f^n(p)) < \varepsilon$$

for every $0 \leq j \leq n$. This shows that $x \in B(p, n+1, \varepsilon)$.

Corollary 11.2.8. *Every expanding map is expansive.*

Proof. Assume that $d(f^n(z), f^n(w)) < \rho$ for every $n \geq 0$. This implies that $z = h_w^n(f^n(z))$ for every $n \geq 0$. Then, the property (11.2.6) gives that

$$d(z, w) \leq \sigma^{-n} d(f^n(z), f^n(w)) < \rho\sigma^{-n}.$$

Making $n \to \infty$, we get that $z = w$. So, ρ is a constant of expansivity for f.

11.2.2 Shadowing and periodic points

Given $\delta > 0$, we call a *δ-pseudo-orbit* of a transformation $f : M \to M$ any sequence $(x_n)_{n \geq 0}$ such that

$$d(f(x_n), x_{n+1}) < \delta \quad \text{for every } n \geq 0.$$

We say that the δ-pseudo-orbit is *periodic* if there is $\kappa \geq 1$ such that $x_n = x_{n+\kappa}$ for every $n \geq 0$. It is clear that every orbit is a δ-pseudo-orbit, for every

$\delta > 0$. For expanding maps we have a kind of converse: every pseudo-orbit is uniformly close to (we say that it is *shadowed* by) some orbit of the transformation:

Proposition 11.2.9 (Shadowing lemma). *Assume that $f : M \to M$ is an expanding map. Then, given any $\varepsilon > 0$ there exists $\delta > 0$ such that for every δ-pseudo-orbit $(x_n)_n$ there exists $x \in M$ such that $d(f^n(x), x_n) < \varepsilon$ for every $n \geq 0$.*

If ε is small enough, so that 2ε is a constant of expansivity for f, then the point x is unique. If, in addition, the pseudo-orbit is periodic then x is a periodic point.

Proof. It is no restriction to suppose that ε is less than ρ. Fix $\delta > 0$ so that $\sigma^{-1}\varepsilon + \delta < \varepsilon$. For each $n \geq 0$, let $h_n : B(f(x_n), \rho) \to B(x_n, \rho)$ be the inverse branch of f at x_n. The property (11.2.5) ensures that

$$h_n\big(B(f(x_n), \varepsilon)\big) \subset B(x_n, \sigma^{-1}\varepsilon) \quad \text{for every } n \geq 1. \tag{11.2.7}$$

Since $d(x_n, f(x_{n-1})) < \delta$, it follows that

$$h_n\big(B(f(x_n), \varepsilon)\big) \subset B(f(x_{n-1}), \varepsilon) \quad \text{for every } n \geq 1. \tag{11.2.8}$$

Then, we may consider the composition $h^{n+1} = h_0 \circ \cdots \circ h_n$, and (11.2.8) implies that the sequence of compact sets $K_{n+1} = h^{n+1}\big(B(f(x_n), \varepsilon)\big)$ is nested. Take x in the intersection. For every $n \geq 0$, we have that $x \in K_{n+1}$ and so $f^n(x)$ belongs to

$$f^n \circ h^{n+1}\big(B(f(x_n), \varepsilon)\big) = h_n\big(B(f(x_n), \varepsilon)\big).$$

By (11.2.7), this implies that $d(f^n(x), x_n) \leq \sigma^{-1}\varepsilon < \varepsilon$ for every $n \geq 0$.

The other claims in the proposition are simple consequences, as we now explain. If x' is another point as in the conclusion of the proposition then

$$d(f^n(x), f^n(x')) \leq d(f^n(x), x_n) + d(f^n(x'), x_n) < 2\varepsilon \quad \text{for every } n \geq 0.$$

Since 2ε is an expansivity constant, it follows that $x = x'$. Moreover, if the pseudo-orbit is periodic, with period $\kappa \geq 1$, then

$$d(f^n(f^\kappa(x)), x_n) = d(f^{n+\kappa}(x), x_{n+\kappa}) < \varepsilon \quad \text{for every } n \geq 0.$$

By uniqueness, it follows $f^\kappa(x) = x$.

It is worthwhile pointing out that δ depends linearly on ε: the proof of Proposition 11.2.9 shows that we may take $\delta = c\varepsilon$, where $c > 0$ depends only on σ.

We call *pre-orbit* of a point $x \in M$ any sequence $(x_{-n})_{n \geq 0}$ such that $x_0 = x$ and $f(x_{-n}) = x_{-n+1}$ for every $n \geq 1$. If x is a periodic point, with period $l \geq 1$, then it admits a distinguished *periodic pre-orbit* $(\bar{x}_{-n})_n$, such that $\bar{x}_{-kl} = x$ for every integer $k \geq 1$.

Lemma 11.2.10. *If $d(x,y) < \rho$ then, given any pre-orbit $(x_{-n})_n$ of x, there exists a pre-orbit $(y_{-n})_n$ of y asymptotic to $(x_{-n})_n$, in the sense that $d(x_{-n}, y_{-n})$ converges to 0 when $n \to \infty$.*

Proof. For each $n \geq 1$, let $h_n : B(x,\rho) \to M$ be the inverse branch of f^n with $h_n(x) = x_{-n}$. Define $y_{-n} = h_n(y)$. It is clear that $d(x_{-n}, y_{-n}) \leq \sigma^{-n} d(x,y)$. This implies the claim.

Theorem 11.2.11. *Let $f : M \to M$ be an expanding map in a compact metric space and $\Lambda \subset M$ be the closure of the set of all periodic points of f. Then $f(\Lambda) = \Lambda$ and the restriction $f : \Lambda \to \Lambda$ is an expanding map.*

Proof. On the one hand, it is clear that $f(\Lambda)$ is contained in Λ: if a point x is accumulated by periodic points p_n then $f(x)$ is accumulated by the images $f(p_n)$, which are also periodic points. On the other hand, since $f(\Lambda)$ is a compact set that contains all the periodic points, it must contain Λ. This shows that $f(\Lambda) = \Lambda$.

Next, we prove that the restriction $f : \Lambda \to \Lambda$ is an expanding map. It is clear that the property (11.2.1) remains valid for the restriction. So, we only have to show that there exists $r \leq \rho$ (we are going to take $r = \sigma^{-1}\rho$) such that, for every $x \in \Lambda$, the image $f(\Lambda \cap B(x,r))$ contains a neighborhood of $\Lambda \cap \overline{B(x,r)}$ inside Λ. The main ingredient is the following lemma:

Lemma 11.2.12. *Let p be a periodic point and $h_p : B(f(p),\rho) \to B(p,\rho)$ be the inverse branch of f at p. If $y \in B(f(p),\rho)$ is a periodic point then $h_p(y) \in \Lambda$.*

Proof. Write $x = h_p(y)$ and $q = f(p)$. Consider any $\varepsilon > 0$ such that 2ε is a constant of expansivity for f. Take $\delta > 0$ given by the shadowing lemma (Proposition 11.2.9). By Lemma 11.2.10, there exists a pre-orbit $(x_{-n})_n$ of x asymptotic to the periodic pre-orbit $(\bar{p}_{-n})_n$ of p. In particular,

$$d(x_{-k+1}, q) = d(x_{-k+1}, \bar{p}_{-k+1}) < \delta \tag{11.2.9}$$

for every large multiple k of the period of p. Analogously, there exists a pre-orbit $(q_{-n})_n$ of q asymptotic to the periodic pre-orbit $(\bar{y}_{-n})_n$ of y. Fix any multiple l of the period of y such that

$$d(q_{-l}, f(x)) = d(q_{-l}, y) < \delta. \tag{11.2.10}$$

Now consider the periodic sequence $(z_n)_n$, with period $k+l$, given by

$$z_0 = x, z_1 = q_{-l}, \ldots, z_l = q_{-1}, z_{l+1} = x_{-k+1}, \ldots, z_{l+k-1} = x_{-1}, z_{k+l} = x.$$

See Figure 11.3. We claim that $(z_n)_n$ is a δ-pseudo-orbit. Indeed, if n is a multiple of $k+l$ then, by (11.2.10),

$$d(f(z_n), z_{n+1}) = d(f(x), q_{-1}) = d(y, q_{-1}) < \delta.$$

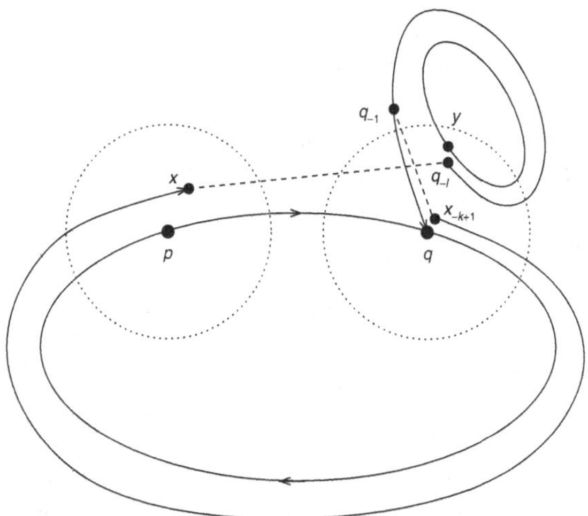

Figure 11.3. Constructing periodic orbits

If $n-l$ is a multiple of $k+l$ then, by (11.2.9),

$$d(f(z_n), z_{n+1}) = d(f(q_{-1}), x_{-k+1}) = d(q, x_{-k+1}) < \delta.$$

In all the other cases, $f(z_n) = z_{n+1}$. This proves our claim. Now we may use Proposition 11.2.9 to conclude that there exists a periodic point z such that $d(f^n(z), z_n) < \varepsilon$ for every $n \geq 0$. In particular, $d(z, x) < \varepsilon$. Since $\varepsilon > 0$ is arbitrary, this shows that x is in the closure of the set of periodic points, as stated.

Corollary 11.2.13. *Let $z \in \Lambda$ and $h_z : B(f(z), \rho) \to B(z, \rho)$ be the inverse branch of f at z. If $w \in \Lambda \cap B(f(z), \rho)$ then $h_z(w) \in \Lambda$.*

Proof. Since $z \in \Lambda$, we may find some periodic point p close enough to z that $w \in B(f(p), \rho)$ and $h_p(w) = h_z(w)$. Since $w \in \Lambda$, we may find periodic points $y_n \in B(f(p), \rho)$ converging to w. By Lemma 11.2.12, we have that $h_p(y_n) \in \Lambda$ for every n. Passing to the limit, we conclude that $h_p(w) \in \Lambda$.

We are ready to conclude the proof of Theorem 11.2.11. Take $r = \sigma^{-1}\rho$. The property (11.2.6) implies that $h_z(B(f(z), \rho))$ is contained in $B(z, r)$, for every $z \in \Lambda$. Then, Corollary 11.2.13 implies that $f(\Lambda \cap B(z, r))$ contains $\Lambda \cap B(f(z), \rho)$, which is a neighborhood of $\Lambda \cap \overline{B(f(z), r)}$ in Λ. Thus the argument is complete.

Theorem 11.2.14. *Let $f : M \to M$ be an expanding map in a compact metric space and $\Lambda \subset M$ be the closure of the set of periodic points of f. Then*

$$M = \bigcup_{k=0}^{\infty} f^{-k}(\Lambda).$$

Proof. Given any $x \in M$, let $\omega(x)$ denote its ω-limit set, that is, the set of accumulation points of the iterates $f^n(x)$ when $n \to \infty$. First, we show that $\omega(x) \subset \Lambda$. Then, we deduce that $f^k(x) \in \Lambda$ for some $k \geq 0$.

Let $\varepsilon > 0$ be such that 2ε is a constant of expansivity for f. Take $\delta > 0$ given by the shadowing lemma (Proposition 11.2.9) and let $\alpha \in (0,\delta)$ be such that $d(f(z),f(w)) < \delta$ whenever $d(z,w) < \alpha$. Let y be any point in $\omega(x)$. The definition of the ω-limit set implies that there exist $r \geq 0$ and $s \geq 1$ such that

$$d(f^r(x),y) < \alpha \quad \text{and} \quad d(f^{r+s}(x),y) < \alpha.$$

Consider the periodic sequence (z_n), with period s, given by

$$z_0 = y, z_1 = f^{r+1}(x), \ldots, z_{s-1} = f^{r+s-1}(x), z_s = y.$$

Observe that $d(f(z_0),z_1) = d(f(y),f^{r+1}(x)) < \delta$ (because $d(y,f^r(x)) < \alpha$), $d(f(z_{s-1}),z_s) = d(f^{r+s}(x),y) < \alpha < \delta$ and $f(z_n) = z_{n+1}$ in all the other cases. In particular, $(z_n)_n$ is a δ-pseudo-orbit. Then, by Proposition 11.2.9, there exists some periodic point z such that $d(y,z) < \varepsilon$. Making $\varepsilon \to 0$, we conclude that y is accumulated by periodic points, that is, $y \in \Lambda$.

Let $\varepsilon > 0$ and $\delta > 0$ be as before. It is no restriction to suppose that $\delta < \varepsilon$. Take $\beta \in (0,\delta/2)$ such that $d(f(z),f(w)) < \delta/2$ whenever $d(z,w) < \beta$. Since $\omega(x)$ is contained in Λ, there exist $k \geq 1$ and points $w_n \in \Lambda$ such that $d(f^{n+k}(x),w_n) < \beta$ for every $n \geq 0$. Observe that

$$d(f(w_n),w_{n+1}) \leq d(f(w_n),f^{n+k+1}(x)) + d(f^{n+k+1}(x),w_{n+1}) < \delta/2 + \beta < \delta$$

for every $n \geq 0$. Therefore, $(w_n)_n$ is a δ-pseudo-orbit in Λ. Since the restriction of f to Λ is an expanding map (Theorem 11.2.11), it follows from Proposition 11.2.9 *applied to the restriction* that there exists $w \in \Lambda$ such that $d(f^n(w),w_n) < \varepsilon$ for every $n \geq 0$. Then,

$$d(f^n(f^k(x)),f^n(w)) \leq d(f^{n+k}(x),w_n) + d(w_n,f^n(w)) < \beta + \varepsilon < 2\varepsilon$$

for every $n \geq 0$. Then, by expansivity, $f^k(x) = w$.

11.2.3 Dynamical decomposition

Theorem 11.2.14 shows that for expanding maps the interesting dynamics are localized in the closure Λ of the set of periodic points. In particular, $\operatorname{supp}\mu \subset \Lambda$ for every invariant probability measure f. Moreover (Theorem 11.2.11), the restriction of f to Λ is still an expanding map. Thus, up to replacing M by Λ, it is no restriction to suppose that the set of periodic points is dense in M.

Theorem 11.2.15 (Dynamical decomposition). *Let $f : M \to M$ be an expanding map whose set of periodic points is dense in M. Then there exists a partition of M into non-empty compact sets $M_{i,j}$, with $1 \leq i \leq k$ and $1 \leq j \leq m(i)$, such that:*

(i) $M_i = \bigcup_{j=1}^{m(i)} M_{i,j}$ *is invariant under* f, *for every* i;
(ii) $f(M_{i,j}) = M_{i,j+1}$ *if* $j < m(i)$ *and* $f(M_{i,m(i)}) = M_{i,1}$, *for every* i,j;
(iii) *each restriction* $f : M_i \to M_i$ *is a transitive expanding map;*
(iv) *each* $f^{m(i)} : M_{i,j} \to M_{i,j}$ *is a topologically exact expanding map.*

Moreover, the number k, *the numbers* $m(i)$ *and the sets* $M_{i,j}$ *are unique up to renumbering.*

Proof. Consider the relation $\sim$ defined in the set of periodic points of f as follows. Given two periodic points p and q, let $(\bar{p}_{-n})_n$ and $(\bar{q}_{-n})_n$, respectively, be their periodic pre-orbits. By definition, $p \sim q$ if and only if there exist pre-orbits $(p_{-n})_n$ of p and $(q_{-n})_n$ of q such that

$$d(p_{-n}, \bar{q}_{-n}) \to 0 \quad \text{and} \quad d(\bar{p}_{-n}, q_{-n}) \to 0. \tag{11.2.11}$$

We claim that $\sim$ is an equivalence relation. It is clear from the definition that the relation $\sim$ is reflexive and symmetric. To prove that it is also transitive, suppose that $p \sim q$ and $q \sim r$. Then there exist pre-orbits $(q_{-n})_n$ of q and $(r_{-n})_n$ of r asymptotic to the periodic pre-orbits $(\bar{p}_{-n})_n$ of p and $(\bar{q}_{-n})_n$ of q, respectively. Let $k \geq 1$ be a multiple of the periods of p and q such that $d(r_{-k}, \bar{q}_{-k}) < \rho$. Note that $\bar{q}_{-k} = q$, since k is a multiple of the period of q. Then, by Lemma 11.2.10, there exists a pre-orbit $(r'_{-n})_n$ of the point $r' = r_{-k}$ such that $d(r'_{-n}, q_{-n}) \to 0$. Then $d(r'_{-n}, \bar{p}_{-n}) \to 0$. Consider the pre-orbit $(r''_{-n})_n$ of r defined by

$$r''_{-n} = \begin{cases} r_{-n} & \text{if } n \leq k \\ r'_{-n+k} & \text{if } n > k. \end{cases}$$

Since k is a multiple of the period of p, we have $d(r''_{-n}, \bar{p}_{-n}) = d(r'_{-n+k}, \bar{p}_{-n+k})$ for every $n > k$. Therefore, $(r''_{-n})_n$ is asymptotic to $(\bar{p}_{-n})_n$. Analogously, one can find a pre-orbit (p''_{-n}) of p asymptotic to the periodic pre-orbit $(\bar{r}_{-n})_n$ of r. Therefore, $p \sim r$ and this proves that the relation $\sim$ is indeed transitive.

Next, we claim that $p \sim q$ if and only if $f(p) \sim f(q)$. Start by supposing that $p \sim q$, and let $(p_{-n})_n$ and $(q_{-n})_n$ be pre-orbits of p and q as in (11.2.11). The periodic pre-orbits of $p' = f(p)$ and $q' = f(q)$ are given by, respectively,

$$\bar{p}'_{-n} = \begin{cases} f(p) & \text{if } n = 0 \\ \bar{p}_{-n+1} & \text{if } n \geq 1 \end{cases} \quad \text{and} \quad \bar{q}'_{-n} = \begin{cases} f(q) & \text{if } n = 0 \\ \bar{q}_{-n+1} & \text{if } n \geq 1. \end{cases}$$

Consider the pre-orbits of p and q given by, respectively,

$$p'_{-n} = \begin{cases} f(p) & \text{if } n = 0 \\ p_{-n+1} & \text{if } n \geq 1 \end{cases} \quad \text{and} \quad q'_{-n} = \begin{cases} f(q) & \text{if } n = 0 \\ q_{-n+1} & \text{if } n \geq 1. \end{cases}$$

It is clear that $(p'_{-n})_n$ is asymptotic to $(\bar{q}'_{-n})_n$ and $(q'_{-n})_n$ is asymptotic to $(\bar{p}'_{-n})_n$. Hence, $f(p) \sim f(q)$. Now suppose that $f(p) \sim f(q)$. The previous argument shows that $f^k(p) \sim f^k(q)$ for every $k \geq 1$. When k is a common multiple of the periods of p and q this means that $p \sim q$. This completes the proof of our claim. Note that the statement means that the image and the pre-image of any equivalence class are both equivalence classes.

Observe also that if $d(p,q) < \rho$ then $p \sim q$. Indeed, by Lemma 11.2.10 we may find a pre-orbit of q asymptotic to the periodic pre-orbit of p and, analogously, a pre-orbit of p asymptotic to the periodic pre-orbit of q. It follows that the equivalence classes are open sets and, since M is compact, they are finite in number. Moreover, if A and B are two different equivalence classes, then their closures $\bar{A}$ and $\bar{B}$ are disjoint: the distance between them is at least ρ. Since $p \sim q$ if and only if $f(p) \sim f(q)$, it follows that the transformation f permutes the closures of the equivalence classes.

Thus, we may enumerate the closures of the equivalence classes as $M_{i,j}$, with $1 \le i \le k$ and $1 \le j \le m(i)$, in such a way that

$$f(M_{i,j}) = M_{i,j+1} \text{ for } j < m(i) \quad \text{and} \quad f(M_{i,m(i)}) = M_{i,1}. \tag{11.2.12}$$

The properties (i) and (ii) in the statement of the theorem are immediate consequences.

Let us prove property (iii). Since the M_i are pairwise disjoint, it follows from (11.2.12) that $f^{-1}(M_i) = M_i$ for every i. Hence, Lemma 11.2.2 implies that $f : M_i \to M_i$ is an expanding map. By Lemma 4.3.4, to show that this map is transitive it suffices to show that given any open subsets U and V of M_i there exists $n \ge 1$ such that $f^n(U)$ intersects V. It is no restriction to assume that $U \subset M_{i,j}$ for some j. Moreover, up to replacing V by some pre-image $f^{-k}(V)$, we may suppose that V is contained in the same $M_{i,j}$. Choose periodic points $p \in U$ and $q \in V$. By the definition of equivalence classes, there exists some pre-orbit $(q_{-n})_n$ of q asymptotic to the periodic pre-orbit $(\bar{p}_{-n})_n$ of p. In particular, we may find n arbitrarily large such that $q_{-n} \in U$. Then $q \in f^n(U) \cap V$. Therefore, $f : M_i \to M_i$ is transitive.

Next, we prove property (iv). Since the $M_{i,j}$ are pairwise disjoint, it follows from (11.2.12) that $f^{-m(i)}(M_{i,j}) = M_{i,j}$ for every i. Hence (Lemma 11.2.2), $g = f^{m(i)} : M_{i,j} \to M_{i,j}$ is an expanding map. We also want to prove that g is topologically exact. Let U be a non-empty open subset of $M_{i,j}$ and p be a periodic point of f in U. By (11.2.12), the period κ is a multiple of $m(i)$, say $\kappa = sm(i)$. Let q be any periodic point of f in $M_{i,j}$. By the definition of the equivalence relation $\sim$, there exists some pre-orbit $(q_{-n})_n$ of q asymptotic to the periodic pre-orbit $(\bar{p}_{-n})_n$ of p. In particular, $d(q_{-\kappa n}, p) \to 0$ when $n \to \infty$. Then $h_q^{\kappa n}(B(q,\rho))$ is contained in U for every n sufficiently large. This implies that $g^{sn}(U) = f^{\kappa n}(U)$ contains $B(q,\rho)$ for every n sufficiently large. Since $M_{i,j}$ is compact, we may find a finite cover by balls of radius ρ around periodic points. Applying the previous argument to each of those periodic points, we deduce that $g^{sn}(U)$ contains $M_{i,j}$ for every n sufficiently large. Therefore, g is topologically exact.

We are left to prove that k, the $m(i)$ and the $M_{i,j}$ are unique. Let $N_{r,s}$, with $1 \le r \le l$ and $1 \le s \le n(r)$, be another partition as in the statement. Initially, let us consider the partitions $\mathcal{M} = \{M_i : 1 \le i \le k\}$ and $\mathcal{N} = \{N_r : 1 \le r \le l\}$, where $N_r = \bigcup_{s=1}^{n(r)} N_{r,s}$. Given any i and r, the sets M_i and N_r are open, closed, invariant and transitive. We claim that either $M_i \cap N_r = \emptyset$ or $M_i = N_r$. Indeed, since the intersection is open, if it is non-empty then it intersects any orbit that is dense in M_i (or in N_r). Since the intersection is also closed and invariant, it follows that it contains M_i (and N_r). In other words, $M_i = N_r$. This proves our claim. It follows that the partitions $\mathcal{M}$ and $\mathcal{N}$ coincide, that is, $k = l$ and $M_i = N_i$ up to renumbering. Now, fix i. The transformation f permutes the $M_{i,j}$ and the $N_{i,s}$ cyclically, with periods $m(i)$ and $n(i)$. Since $f^{m(i)n(i)}$ is transitive on each $M_{i,j}$ and each $N_{i,s}$, the same argument as in the first part of this paragraph shows that, given any j and s, either $M_{i,j} \cap N_{i,s} = \emptyset$ or $M_{i,j} = N_{i,s}$. It follows that $m(i) = n(i)$ and the families $M_{i,j}$ and $N_{i,s}$ coincide, up to cyclic renumbering.

The following consequence of the theorem contains Lemma 11.1.13:

Corollary 11.2.16. *If M is connected and $f : M \to M$ is an expanding map then the set of periodic points is dense in M and f is topologically exact.*

Proof. We claim that Λ is an open subset of $f^{-1}(\Lambda)$. Indeed, consider $\delta \in (0, \rho)$ such that $d(x,y) < \delta$ implies $d(f(x),f(y)) < \rho$. Assume that $x \in f^{-1}(\Lambda)$ is such that $d(x, \Lambda) < \delta$. Then there exists $z \in \Lambda$ such that $d(x,z) < \delta < \rho$ and so $d(f(x),f(z)) < \rho$. Applying Corollary 11.2.13 with $w = f(x)$, we get that $x = h_z(w) \in \Lambda$. Therefore, Λ contains its δ-neighborhood inside $f^{-1}(\Lambda)$. This implies our claim.

Then, the set $S = f^{-1}(\Lambda) \setminus \Lambda$ is closed in $f^{-1}(\Lambda)$ and, consequently, it is closed in M, so $f^{-n}(S)$ is closed in M for every $n \ge 0$. By Theorem 11.2.14, the space M is a countable pairwise disjoint union of closed sets Λ and $f^{-n}(S)$, $n \ge 0$. By the Baire theorem, some of these closed sets have a non-empty interior. Since f is an open map, it follows that Λ has a non-empty interior.

Now consider the restriction $f : \Lambda \to \Lambda$. By Theorem 11.2.11, this is an expanding map. Let $\{\Lambda_{i,j} : 1 \le i \le k, 1 \le j \le m(i)\}$ be the partition of the domain Λ given by Theorem 11.2.15. Then some $\Lambda_{i,j}$ contains an open subset V of M. Since $f^{m(i)}$ is topologically exact, $f^{nm(i)}(V) = \Lambda_{i,j}$ for some $n \ge 1$. Using once more the fact that f is an open map, it follows that the compact set $M_{i,j}$ is an open subset of M. By connectivity, it follows that $M = \Lambda_{i,j}$. This implies that $\Lambda = M$ and $f : M \to M$ is topologically exact.

11.2.4 Exercises

11.2.1. Show that if $f : M \to M$ is a local homeomorphism in a compact connected metric space then the number of pre-images $\#f^{-1}(y)$ is the same for every $y \in M$.

11.2.2. Show that if an expanding map is topologically mixing then it is topologically exact.

11.2.3. Let $f : M \to M$ be a topologically exact transformation in a compact metric space. Show that for every $r > 0$ there exists $N \geq 1$ such that $f^N(B(x,r)) = M$ for every $x \in M$.

11.2.4. Consider the expanding map $f : S^1 \to S^1$ given by $f(x) = 2x \mod \mathbb{Z}$. Give an example of a compact set $\Lambda \subset S^1$ such that $f(\Lambda) = \Lambda$ but the restriction $f : \Lambda \to \Lambda$ is not an expanding map.

11.2.5. Let $f : M \to M$ be an expanding map and Λ be the closure of the set of periodic points of f. Show that $h(f) = h(f \mid \Lambda)$.

11.2.6. Let $f : M \to M$ be an expanding map such that the set of periodic points is dense in M and let M_i, $M_{i,j}$ be the compact subsets given by Theorem 11.2.15. Show that $h(f) = \max_i h(f \mid M_i)$ and

$$h(f \mid M_i) = \frac{1}{m(i)} h(f^{m(i)} \mid M_{i,j}) \quad \text{for any } i, j.$$

11.2.7. Let $\sigma_A : \Sigma_A \to \Sigma_A$ be a shift of finite type. Interpret the decomposition given by Theorem 11.2.15 in terms of the matrix A.

11.3 Entropy and periodic points

In this section we analyze the distribution of periodic points of an expanding map $f : M \to M$ from a quantitative point of view. We show (Section 11.3.1) that the rate of growth of the number of periodic points is equal to the topological entropy; compare this statement with the discussion in Section 10.2.1. Another interesting conclusion (Section 11.3.2) is that every invariant probability measure may be approximated, in the weak* topology, by invariant probability measures supported on periodic orbits. These results are based on the following property:

Proposition 11.3.1. *Let $f : M \to M$ be a topologically exact expanding map. Then, given any $\varepsilon > 0$ there exists $\kappa \geq 1$ such that, given any $x_1, \dots, x_s \in M$, any $n_1, \dots, n_s \geq 1$ and any $k_1, \dots, k_s \geq \kappa$, there exists a point $p \in M$ such that, denoting $m_j = \sum_{i=1}^{j} n_i + k_i$ for $j = 1, \dots, s$ and $m_0 = 0$,*

(i) $d(f^{m_{j-1}+i}(p), f^i(x_j)) < \varepsilon$ for $0 \leq i < n_j$ and $1 \leq j \leq s$, and
(ii) $f^{m_s}(p) = p$.

Proof. Given $\varepsilon > 0$, take $\delta > 0$ as in the shadowing lemma (Proposition 11.2.9). Without loss of generality, we may suppose that $\delta < \varepsilon$ and 2ε is a constant of expansivity for f (recall Corollary 11.2.8). Since f is topologically exact, given any $z \in M$ there exists $\kappa \geq 1$ such that $f^k(B(z,\delta)) = M$ for every $k \geq \kappa$. Moreover (see Exercise 11.2.3), since M is compact, we may choose κ depending only on δ. Let $x_j, n_j, k_j \geq \kappa$, $j = 1, \dots, s$ be as in the statement. In particular, for each $j = 1, \dots, s-1$ there exists $y_j \in B(f^{n_j}(x_j), \delta)$ such that $f^{k_j}(y_j) = x_{j+1}$. Analogously, there exists $y_s \in B(f^{n_s}(x_s), \delta)$ such that $f^{k_s}(y_s) = x_1$. Consider the periodic δ-pseudo-orbit $(z_n)_{n\geq 0}$ defined by

$$z_n = \begin{cases} f^{n-m_{j-1}}(x_j) & \text{for } 0 \leq n - m_{j-1} < n_j \text{ and } j = 1, \dots s \\ f^{n-m_{j-1}-n_j}(y_j) & \text{for } 0 \leq n - m_{j-1} - n_j < k_j \text{ and } j = 1, \dots, s \\ z_{n-m_s} & \text{for } n \geq m_s. \end{cases}$$

By the second part of the shadowing lemma, there exists some periodic point $p \in M$, with period m_s, whose trajectory ε-shadows this periodic pseudo-orbit $(z_n)_n$. In particular, the conditions (i) and (ii) in the statement hold.

The property in the conclusion of Proposition 11.3.1 was introduced by Rufus Bowen [Bow71] and is called *specification by periodic points*. When condition (i) holds, but the point p is not necessarily periodic, we say that f has the property of *specification*.

11.3.1 Rate growth of periodic points

Let $f : M \to M$ be an expanding map. Then f is expansive (by Lemma 11.1.4) and so it follows from Proposition 10.2.2 that the rate of growth of the number of periodic points is bounded above by the topological entropy:

$$\limsup_n \frac{1}{n} \log \#\mathrm{Fix}(f^n) \leq h(f). \tag{11.3.1}$$

In this section we prove that, in fact, the identity holds in (11.3.1). We start with the topologically exact case, where one may even replace the limit superior by a limit:

Proposition 11.3.2. *For any topologically exact expanding map $f : M \to M$,*

$$\lim_n \frac{1}{n} \log \#\mathrm{Fix}(f^n) = h(f).$$

Proof. Given $\varepsilon > 0$, fix $\kappa \geq 1$ satisfying the conclusion of Proposition 11.3.1 with $\varepsilon/2$ instead of ε. For each $n \geq 1$, let E be a maximal (n,ε)-separated set. According to Proposition 11.3.1, for each $x \in E$ there exists $p(x) \in B(x,n,\varepsilon/2)$ with $f^{n+\kappa}(p(x)) = p(x)$. We claim that the map $x \mapsto p(x)$ is injective. Indeed, consider any $y \in E \setminus \{x\}$. Since the set E was chosen to be (n,ε)-separated, $B(x,n,\varepsilon/2) \cap B(y,n,\varepsilon/2) = \emptyset$. This implies that $p(x) \neq p(y)$, which proves our claim. In particular, it follows that

$$\#\mathrm{Fix}(f^{n+\kappa}) \geq \#E = s_n(f,\varepsilon,M) \quad \text{for every } n \geq 1$$

(recall the definition (10.1.9) in Section 10.2.1). Then,

$$\liminf_n \frac{1}{n} \log \#\mathrm{Fix}(f^{n+\kappa}) \geq \liminf_n \frac{1}{n} \log s_n(f,\varepsilon,M).$$

Making $\varepsilon \to 0$ and using Corollary 10.1.8, we find that

$$\liminf_n \frac{1}{n} \log \#\operatorname{Fix}(f^{n+\kappa}) \geq \lim_{\varepsilon \to 0} \liminf_n \frac{1}{n} \log s_n(f, \varepsilon, M) = h(f). \tag{11.3.2}$$

Together with (11.3.1), this implies the claim in the proposition.

Proposition 11.3.2 is not true, in general, if f is not topologically exact. For example, given an arbitrary expanding map $g : M \to M$, consider $f : M \times \{0,1\} \to M \times \{0,1\}$ defined by $f(x,i) = (g(x), 1-i)$. Then f is an expanding map but all its periodic points have even period. In particular, in this case,

$$\liminf_n \frac{1}{n} \log \#\operatorname{Fix}(f^n) = 0.$$

However, the next proposition shows that this type of example is the worst that can happen. The proof also makes it clear when and how the limit may fail to exist.

Proposition 11.3.3. *For any expanding map $f : M \to M$,*

$$\limsup_n \frac{1}{n} \log \#\operatorname{Fix}(f^n) = h(f).$$

Proof. By Theorem 11.2.11, the restriction f to the set of periodic points is an expanding map. According to Exercise 11.2.5 this restriction has the same entropy as f. Obviously, the two transformations have the same periodic points. Therefore, up to replacing f by this restriction, we may suppose that the set of periodic points is dense in M. Then, by the theorem of dynamical decomposition (Theorem 11.2.11), one may write M as a finite union of compact sets $M_{i,j}$, with $1 \leq i \leq k$ and $1 \leq j \leq m(i)$, such that each $f^{m(i)} : M_{i,j} \to M_{i,j}$ is a topologically exact expanding map. According to Exercise 11.2.6, there exists $1 \leq i \leq k$ such that

$$h(f) = \frac{1}{m(i)} h(f^{m(i)} \mid M_{i,1}). \tag{11.3.3}$$

It is clear that

$$\begin{aligned} \limsup_n \frac{1}{n} \log \#\operatorname{Fix}\left(f^n\right) &\geq \limsup_n \frac{1}{nm(i)} \log \#\operatorname{Fix}\left(f^{nm(i)}\right) \\ &\geq \frac{1}{m(i)} \limsup_n \frac{1}{n} \log \#\operatorname{Fix}\left((f^{m(i)} \mid M_{i,1})^n\right). \end{aligned} \tag{11.3.4}$$

Moreover, Proposition 11.3.2 applied to $f^{m(i)} : M_{i,1} \to M_{i,1}$ yields

$$\lim_n \frac{1}{n} \log \#\operatorname{Fix}\left((f^{m(i)} \mid M_{i,1})^n\right) = h(f^{m(i)} \mid M_{i,1}). \tag{11.3.5}$$

Combining (11.3.3)–(11.3.5), we find that

$$\limsup_n \frac{1}{n} \log \#\operatorname{Fix}\left(f^n\right) \geq h(f), \tag{11.3.6}$$

as we wanted to prove.

11.3.2 Approximation by atomic measures

Given a periodic point p, with period $n \geq 1$, consider the probability measure μ_p defined by

$$\mu_p = \frac{1}{n}\big(\delta_p + \delta_{f(p)} + \cdots + \delta_{f^{n-1}(p)}\big).$$

Clearly, μ_p is invariant and ergodic for f. We are going to show that if f is expanding then the set of measures of this form is dense in the space $\mathcal{M}_1(f)$ of all invariant probability measures:

Theorem 11.3.4. *Let $f : M \to M$ be a topologically exact expanding map. Then every probability measure μ invariant under f can be approximated, in the weak* topology, by invariant probability measures supported on periodic orbits.*

Proof. Let $\varepsilon > 0$ and $\Phi = \{\phi_1, \dots, \phi_N\}$ be a finite family of continuous functions in M. We want to show that the neighborhood $V(\mu, \Phi, \varepsilon)$ defined in (2.1.1) contains some measure μ_p supported on a periodic orbit. By the theorem of Birkhoff, for μ-almost every point $x \in M$,

$$\tilde{\phi}_i(x) = \lim_n \frac{1}{n}\sum_{t=0}^{n-1} \phi_i(f^t(x)) \quad \text{exists for every } i. \tag{11.3.7}$$

Fix $C > \sup|\phi_i| \geq \sup|\tilde{\phi}_i|$ and take $\delta > 0$ such that

$$d(x,y) < \delta \quad \Rightarrow \quad |\phi_i(x) - \phi_i(y)| < \frac{\varepsilon}{5} \quad \text{for every } i. \tag{11.3.8}$$

Fix $\kappa = \kappa(\delta) \geq 1$ given by the property of specification (Proposition 11.3.1). Choose points $x_j \in M$, $1 \leq j \leq s$ satisfying (11.3.7) and positive numbers α_j, $1 \leq j \leq s$ such that $\sum_j \alpha_j = 1$ and

$$\left| \int \tilde{\phi}_i \, d\mu - \sum_{j=1}^{s} \alpha_j \tilde{\phi}_i(x_j) \right| < \frac{\varepsilon}{5} \quad \text{for every } i \tag{11.3.9}$$

(use Exercise A.2.6). Take $k_j \equiv \kappa$ and choose integer numbers n_j much bigger than κ, in such a way that

$$\left| \frac{n_j}{m_s} - \alpha_j \right| < \frac{\varepsilon}{5Cs} \tag{11.3.10}$$

(recall that $m_s = \sum_j (n_j + k_j) = s\kappa + \sum_j n_j$) and, using (11.3.8),

$$\left| \sum_{t=0}^{n_j - 1} \phi_i(f^t(x_j)) - n_j \tilde{\phi}_i(x_j) \right| < \frac{\varepsilon}{5} n_j \quad \text{for } 1 \leq i \leq N. \tag{11.3.11}$$

Combining (11.3.9) and (11.3.10) with the fact that $\int \tilde{\phi}_i \, d\mu = \int \phi_i \, d\mu$, we get

$$\left| \int \phi_i \, d\mu - \sum_{j=1}^{s} \frac{n_j}{m_s} \tilde{\phi}_i(x_j) \right| < \frac{\varepsilon}{5} + \frac{\varepsilon}{5Cs} s \sup|\tilde{\phi}_i| < \frac{2\varepsilon}{5}. \tag{11.3.12}$$

By Proposition 11.3.1, there exists some periodic point $p \in M$, with period m_s, such that $d(f^{m_{j-1}+t}(p), f^t(x_j)) < \delta$ for $0 \le t < n_j$ and $1 \le j \le s$. Then, the property (11.3.8) implies that

$$\left| \sum_{t=0}^{n_j-1} \phi_i(f^{m_{j-1}+t}(p)) - \sum_{t=0}^{n_j-1} \phi_i(f^t(x_j)) \right| < \frac{\varepsilon}{5} n_j \quad \text{for } 1 \le j \le s.$$

Combining this relation with (11.3.11), we obtain

$$\left| \sum_{t=0}^{n_j-1} \phi_i(f^{m_{j-1}+t}(p)) - n_j \tilde{\phi}_i(x_j) \right| < \frac{2\varepsilon}{5} n_j \quad \text{for } 1 \le j \le s. \tag{11.3.13}$$

Since $\sum_j \alpha_j = 1$, the condition (11.3.10) implies that

$$s\kappa = m_s - \sum_{j=1}^{s} n_j < \frac{\varepsilon}{5C} m_s.$$

Then (11.3.13) implies that

$$\left| \sum_{t=0}^{m_s-1} \phi_i(f^t(p)) - \sum_{j=1}^{s} n_j \tilde{\phi}_i(x_j) \right| < \frac{2\varepsilon}{5} \sum_{j=1}^{s} n_j + s\kappa \sup |\tilde{\phi}_i| < \frac{3\varepsilon}{5} m_s. \tag{11.3.14}$$

Let μ_p be the invariant probability measure supported on the orbit of p. The first term in (11.3.14) coincides with $m_s \int \phi_i \, d\mu_p$. Therefore, adding the inequalities (11.3.12) and (11.3.14), we conclude that

$$\left| \int \phi_i \, d\mu_p - \int \phi_i \, d\mu \right| < \frac{2\varepsilon}{5} + \frac{3\varepsilon}{5} = \varepsilon \quad \text{for every } 1 \le i \le N.$$

This means that $\mu_p \in V(\mu, \Phi, \varepsilon)$, as we wanted to prove.

11.3.3 Exercises

11.3.1. Let $f : M \to M$ be a continuous transformation in a compact metric space M. Check that if some iterate f^l, $l \ge 1$ has the property of specification, or specification by periodic points, then so does f.

11.3.2. Let $f : M \to M$ be a continuous transformation in a metric space with the property of specification. Show that f is topologically mixing.

11.3.3. Let $f : M \to M$ be a topologically mixing expanding map and $\varphi : M \to \mathbb{R}$ be a continuous function. Assume that there exist probability measures μ_1, μ_2 invariant under f and such that $\int \varphi \, d\mu_1 \neq \int \varphi \, d\mu_2$. Show that there exists $x \in M$ such that the time average of φ on the orbit of x does not converge. [Observation: One can show (see [BS00]) that the set M_φ of points where the time average of φ does not converge has full entropy and full Hausdorff dimension.]

11.3.4. Prove the following generalization of Proposition 11.3.2: if $f : M \to M$ is a topologically exact expanding map then

$$P(f, \phi) = \lim_k \frac{1}{k} \log \sum_{p \in \mathrm{Fix}(f^k)} e^{\phi_k(p)} \quad \text{for every Hölder function } \phi : M \to \mathbb{R}.$$

11.3.5. Let $f : M \to M$ be an expanding map of class C^1 on a compact manifold M. Show that f admits:

(a) A neighborhood $\mathcal{U}_0$ in the C^0 topology (that is, the topology of uniform convergence) such that f is a topological factor of every $g \in \mathcal{U}_0$. In particular, $h(g) \geq h(f)$ for every $g \in \mathcal{U}_0$.

(b) A neighborhood $\mathcal{U}_1$ in the C^1 topology such that every $g \in \mathcal{U}_1$ is topologically conjugate to f. In particular, $g \mapsto h(g)$ is constant on $\mathcal{U}_1$.

12

Thermodynamic formalism

In this chapter we develop the ergodic theory of expanding maps on compact metric spaces. This theory evolved from the kind of ideas in statistical mechanics that we discussed in Section 10.3.4 and, for that reason, is often called thermodynamic formalism. We point out, however, that this last expression is much broader, encompassing not only the original setting of mathematical physics but also applications to other mathematical systems, such as the so-called uniformly hyperbolic diffeomorphisms and flows (in this latter regard, see the excellent monograph of Rufus Bowen [Bow75a]).

The main result in this chapter is the following theorem of David Ruelle, which we prove in Section 12.1 (the notion of Gibbs state is also introduced in Section 12.1):

Theorem 12.1 (Ruelle). *Let $f : M \to M$ be a topologically exact expanding map on a compact metric space and $\varphi : M \to \mathbb{R}$ be a Hölder function. Then there exists a unique equilibrium state μ for φ. Moreover, the measure μ is exact, it is supported on the whole of M and is a Gibbs state.*

Recall that an expanding map is topologically exact if (and only if) it is topologically mixing (Exercise 11.2.2). Moreover, a topologically exact map is necessarily surjective.

In the particular case when M is a Riemannian manifold and f is differentiable, the equilibrium state of the potential $\varphi = -\log|\det Df|$ coincides with the absolutely continuous invariant measure given by Theorem 11.1.2. In particular, it is the unique physical measure of f. These facts are proved in Section 12.1.8.

The theorem of Livšic that we present in Section 12.2 complements the theorem of Ruelle in a very elegant way. It asserts that two potentials φ and ψ have the same equilibrium state if and only if the difference between them is cohomologous to a constant. In other words, this happens if and only if $\varphi - \psi = c + u \circ f - u$ for some $c \in \mathbb{R}$ and some continuous function u. Moreover, and remarkably, it suffices to check this condition on the periodic orbits of f.

In Section 12.3 we show that the system (f,μ) exhibits exponential decay of correlations in the space of Hölder functions, for every equilibrium state μ of any Hölder potential.

We close this chapter (Section 12.4) with an application of these ideas to a class of geometric and dynamical objects called *conformal repellers*. We prove the *Bowen–Manning formula* according to which the Hausdorff dimension of the repeller is given by the unique zero of the function $t \mapsto P(f, t\varphi_u)$.

12.1 Theorem of Ruelle

Let $f: M \to M$ be a topologically exact expanding map and φ be a Hölder potential. In what follows, $\rho > 0$ and $\sigma > 1$ are the same constants as in the definition (11.2.1). Recall that we denote by φ_n the orbital sums of φ:

$$\varphi_n(x) = \sum_{j=0}^{n-1} \varphi(f^j(x)) \quad \text{for } x \in M.$$

Before getting into the details of the proof of Theorem 12.1 let us outline the main points. The arguments in the proof turn around the *transfer operator* (or *Ruelle–Perron–Frobenius operator*), the linear operator $\mathcal{L}: C^0(M) \to C^0(M)$ defined in the Banach space $C^0(M)$ of continuous complex functions by

$$\mathcal{L}g(y) = \sum_{x \in f^{-1}(y)} e^{\varphi(x)} g(x). \tag{12.1.1}$$

Observe that $\mathcal{L}$ is well defined: $\mathcal{L}g \in C^0(M)$ whenever $g \in C^0(M)$. Indeed, as we saw in Lemma 11.2.6, for each $y \in M$ there exist inverse branches

$$h_i : B(y,\rho) \to M, \quad i = 1, \dots, k$$

of the transformation f such that $\bigcup_{i=1}^k h_i(B(y,\rho))$ coincides with the pre-image of the ball $B(y,\rho)$. Then,

$$\mathcal{L}g = \sum_{i=1}^{k} \left(e^{\varphi} g\right) \circ h_i \tag{12.1.2}$$

restricted to $B(y,\rho)$ and, clearly, this expression defines a continuous function.

It is clear from the definition that $\mathcal{L}$ is a positive operator: if $g(x) \ge 0$ for every $x \in M$ then $\mathcal{L}g(y) \ge 0$ for every $y \in M$. It is also easy to check that $\mathcal{L}$ is a continuous operator: indeed,

$$\|\mathcal{L}g\| = \sup|\mathcal{L}g| \le \operatorname{degree}(f) e^{\sup\varphi} \sup|g| = \operatorname{degree}(f) e^{\sup\varphi} \|g\| \tag{12.1.3}$$

for every $g \in C^0(M)$, and that means that $\|\mathcal{L}\| \le \operatorname{degree}(f)\, e^{\sup\varphi}$; recall that $\operatorname{degree}(f)$ was defined in (11.2.4).

According to the theorem of Riesz–Markov (Theorem A.3.12), the dual space of the Banach space $C^0(M)$ may be identified with the space $\mathcal{M}(M)$

of all complex Borel measures. Then, the *dual* of the transfer operator is the linear operator $\mathcal{L}^* : \mathcal{M}(M) \to \mathcal{M}(M)$ defined by

$$\int g\, d(\mathcal{L}^*\eta) = \int (\mathcal{L}g)\, d\eta \quad \text{for every } g \in C^0(M) \text{ and } \eta \in \mathcal{M}(M). \qquad (12.1.4)$$

This operator is positive, in the sense that if η is a positive measure then $\mathcal{L}^*\eta$ is also a positive measure.

The first step in the proof (Section 12.1.1) is to show that $\mathcal{L}^*$ admits a positive eigenmeasure ν associated with a positive eigenvalue λ. We will see that such a measure admits a positive Jacobian which is Hölder and whose support is the whole space M. Moreover (Section 12.1.2), the eigenmeasure ν is a *Gibbs state*: there exists a constant $P \in \mathbb{R}$ and for each $\varepsilon > 0$ there exists $K \geq 1$ such that

$$K^{-1} \leq \frac{\nu(B(x,n,\varepsilon))}{\exp\left(\varphi_n(x) - nP\right)} \leq K \quad \text{for every } x \in M \text{ and every } n \geq 1, \qquad (12.1.5)$$

where $B(x,n,\varepsilon)$ is the dynamical ball defined in (9.3.2). Actually, $P = \log\lambda$.

Behind the proof of the Gibbs property are certain results about distortion control that are also crucial to show (Section 12.1.3) that the transfer operator itself $\mathcal{L}$ admits an eigenfunction associated with the eigenvalue λ. This function is strictly positive and Hölder. The measure $\mu = h\nu$ is the equilibrium state we are looking for, although that will take a little while to prove.

It follows easily from the properties of h (Section 12.1.4) that μ is invariant and a Gibbs state, and its support is the whole of M. Moreover, $h_\mu(f) + \int \varphi\, d\mu = P$. To conclude that μ is indeed an equilibrium state, we need to check that P is equal to the pressure $P(f,\varphi)$. This is done (Section 12.1.5) with the help of the Rokhlin formula (Theorem 9.7.3), which also allows us to conclude that if η is an equilibrium state then η/h is an eigenmeasure of $\mathcal{L}^*$ associated with the eigenvalue $\lambda = \log P(f,\varphi)$. This last result is the key ingredient for proving that the equilibrium state is unique (Section 12.1.6).

The distortion control is, again, crucial for checking (Section 12.1.7) that the system (f,μ) is exact. Finally, in Section 12.1.8 we comment on the special case $\varphi = -\log|\det f|$, when f is an expanding map on a Riemannian manifold. In this case, the reference measure ν is the Lebesgue measure on the manifold itself. Thus, the equilibrium state μ is an invariant measure equivalent to the Lebesgue measure, and so it coincides with the invariant measure constructed in Section 11.1.

Before we start to detail these steps, it is convenient to make a couple of quick comments. First, note that the existence of an equilibrium state follows immediately from Corollary 10.5.9, since Lemma 11.1.4 asserts that every expanding map is expansive. However, this fact is not used in the proof: instead, in Section 12.1.4 we present a much more explicit construction of the equilibrium state.

The other comment concerns the Rokhlin formula. Let $\mathcal{P}$ be any finite partition of M with $\operatorname{diam}\mathcal{P} < \rho$. For each $n \geq 1$, every element of the partition $\mathcal{P}^n = \bigvee_{j=0}^{n-1} f^{-j}(\mathcal{P})$ is contained in the image $h^{n-1}(P)$ of some $P \in \mathcal{P}$ by an inverse branch h^{n-1} of the iterate f^{n-1}. In particular, $\operatorname{diam}\mathcal{P}^n < \sigma^{-n+1}\rho$ for every n. Then, $\mathcal{P}$ satisfies the hypotheses of Theorem 9.7.3 *at every point*. Hence, the Rokhlin formula holds for every invariant probability measure.

12.1.1 Reference measure

Recall that $C^0_+(M)$ denotes the cone of positive continuous functions. As observed previously, this cone is preserved by the transfer operator $\mathcal{L}$. The dual cone (recall Example 2.3.3) is defined by

$$C^0_+(M)^* = \{\eta \in C^0(M)^* : \eta(\psi) \geq 0 \text{ for every } \psi \in C^0_+(M)\}$$

and may be seen as the cone of finite positive Borel measures. It follows directly from (12.1.4) that $C^0_+(M)^*$ is preserved by the dual operator $\mathcal{L}^*$.

Lemma 12.1.1. *Consider the spectral radius $\lambda = \rho(\mathcal{L}^*) = \rho(\mathcal{L})$. Then there exists some probability measure ν on M such that $\mathcal{L}^*\nu = \lambda\nu$.*

Proof. As we saw in Exercise 2.3.3, the cone $C^0_+(M)$ is normal. Hence, we may apply Theorem 2.3.4 with $E = C^0(M)$ and $C = C^0_+(M)$ and $T = \mathcal{L}$. The conclusion of the theorem means that $\mathcal{L}^*$ admits some eigenvector $\nu \in C^0_+(M)^*$ associated with the eigenvalue λ. As we have just explained, ν may be identified with a finite positive measure. Normalizing ν, we may take it to be a probability measure.

In Exercise 12.1.2 we propose an alternative proof of Lemma 12.1.1, based on the Tychonoff–Schauder theorem (Theorem 2.2.3).

Example 12.1.2. Let $f : M \to M$ be a local diffeomorphism on a compact Riemannian manifold M. Consider the transfer operator $\mathcal{L}$ associated with the potential $\varphi = -\log|\det Df|$. The Lebesgue measure m (that is, the volume measure induced by the Riemannian metric) of M is an eigenmeasure of the transfer operator associated with the eigenvalue $\lambda = 1$:

$$\mathcal{L}^* m = m. \tag{12.1.6}$$

To check this fact, it is enough to show that $\mathcal{L}^*m(E) = m(E)$ for every measurable set E contained in the image of some inverse branch $h_j : B(y,\rho) \to M$ (because, M being compact, every measurable set may be written as a finite disjoint union of subsets E of this kind). Now, using the expression (12.1.2),

$$\mathcal{L}^*m(E) = \int \mathcal{X}_E \, d(\mathcal{L}^*m) = \int (\mathcal{L}\mathcal{X}_E)\, dm = \int \sum_{i=1}^{k} \frac{\mathcal{X}_E}{|\det Df|} \circ h_i \, dm.$$

Hence, by the choice of E and the formula of change of variables,

$$\mathcal{L}^* m(E) = \int \sum_{i=1}^{k} \frac{\mathcal{X}_E}{|\det Df|} \circ h_i \, dm = \int \mathcal{X}_E \, dm = m(E).$$

This proves that m is, indeed, a fixed point of $\mathcal{L}^*$.

Exercise 12.1.3 gives a similar conclusion for Markov measures.

From now on, we always take ν to be a *reference measure*, that is, a probability measure such that $\mathcal{L}^*\nu = \lambda\nu$ for *some* $\lambda > 0$. By the end of the proof of Theorem 12.1 we will find that λ is uniquely determined (in view of Lemma 12.1.1, that means that λ is necessarily equal to the spectral radius of $\mathcal{L}$ and $\mathcal{L}^*$) and the measure ν itself is also unique.

Initially, we show that f admits a Jacobian with respect to ν, which may be written explicitly in terms of the eigenvalue λ and the potential φ:

Lemma 12.1.3. *The transformation $f : M \to M$ admits a Jacobian with respect to ν, given by $J_\nu f = \lambda e^{-\varphi}$.*

Proof. Let A be any domain of invertibility of f. Let $(g_n)_n$ be a sequence of continuous functions converging to the characteristic function of A at ν-almost every point and such that $\sup |g_n| \le 1$ for every n (see Exercise A.3.5). Observe that

$$\mathcal{L}(e^{-\varphi} g_n)(y) = \sum_{x \in f^{-1}(y)} g_n(x).$$

The expression on the right-hand side is bounded by the degree of f, as defined in (11.2.4), and it converges to $\mathcal{X}_{f(A)}(y)$ at ν-almost every point. Hence, using the dominated convergence theorem, the sequence

$$\int \lambda e^{-\varphi} g_n \, d\nu = \int e^{-\varphi} g_n \, d(\mathcal{L}^*\nu) = \int \mathcal{L}(e^{-\varphi} g_n) \, d\nu$$

converges to $\nu(f(A))$. Since the expression on the left-hand side converges to $\int_A \lambda e^{-\varphi} d\nu$, we conclude that

$$\nu(f(A)) = \int_A \lambda e^{-\varphi} d\nu,$$

which proves the claim.

The next lemma applies, in particular, to the reference measure ν:

Lemma 12.1.4. *Let $f : M \to M$ be a topologically exact expanding map and η be any Borel probability measure such that there exists Jacobian of f with respect to η. Then η it is supported on the whole of M.*

Proof. Suppose, by contradiction, that there exists some open set $U \subset M$ such that $\eta(U) = 0$. Note that f is an open map, since it is a local homeomorphism.

Thus, the image $f(U)$ is also an open set. Moreover, we may write U as a finite disjoint union of domains of invertibility A. For each one of them,

$$\eta(f(A)) = \int_A J_\eta f \, d\eta = 0.$$

Therefore, $\eta(f(U)) = 0$. By induction, it follows that $\eta(f^n(U)) = 0$ for every $n \geq 0$. Since we take f to be topologically exact, there exists $n \geq 1$ such that $f^n(U) = M$. This contradicts the fact that $\eta(M) = 1$.

12.1.2 Distortion and the Gibbs property

In this section we prove certain distortion bounds that have a central role in the proof of Theorem 12.1. The hypothesis that φ is Hölder is critical at this stage: most of what follows is false, in general, if the potential is only continuous. As a first application of this distortion control, we prove that every reference measure ν is a Gibbs state.

Fix constants $K_0 > 0$ and $\alpha > 0$ such that $|\varphi(z) - \varphi(w)| \leq K_0 d(z,w)^\alpha$ for any $z, w \in M$.

Lemma 12.1.5. *There exists $K_1 > 0$ such that for every $n \geq 1$, every $x \in M$ and every $y \in B(x, n+1, \rho)$,*

$$|\varphi_n(x) - \varphi_n(y)| \leq K_1 d(f^n(x), f^n(y))^\alpha.$$

Proof. By hypothesis, $d(f^i(x), f^i(y)) < \rho$ for every $0 \leq i \leq n$. Then, for each $j = 1, \dots, n$, the inverse branch $h_j : B(f^n(x), \rho) \to M$ of f^j at the point $f^{n-j}(x)$, which maps $f^n(x)$ to $f^{n-j}(x)$, also maps $f^n(y)$ to $f^{n-j}(y)$. Hence, recalling (11.2.6), $d(f^{n-j}(x), f^{n-j}(y)) \leq \sigma^{-j} d(f^n(x), f^n(y))$ for every $j = 1, \dots, n$. Then,

$$\begin{aligned} |\varphi_n(x) - \varphi_n(y)| &\leq \sum_{j=1}^{n} |\varphi(f^{n-j}(x)) - \varphi(f^{n-j}(y))| \\ &\leq \sum_{j=1}^{n} K_0 \sigma^{-j\alpha} d(f^n(x), f^n(y))^\alpha. \end{aligned}$$

Therefore, we may take any $K_1 \geq K_0 \sum_{j=0}^{\infty} \sigma^{-j\alpha}$.

As a consequence of Lemma 12.1.5, we obtain the following variation of Proposition 11.1.5 where the usual Jacobian with respect to the Lebesgue measure is replaced by the Jacobian with respect to any reference measure ν:

Corollary 12.1.6. *There exists $K_2 > 0$ such that for every $n \geq 1$, every $x \in M$ and every $y \in B(x, n+1, \rho)$,*

$$K_2^{-1} \leq \frac{J_\nu f^n(x)}{J_\nu f^n(y)} \leq K_2.$$

Proof. From the expression of the Jacobian in Lemma 12.1.3 it follows that (recall Exercise 9.7.5)

$$J_\nu f^n(z) = \lambda^n e^{-\varphi_n(z)} \quad \text{for every } z \in M \text{ and every } n \geq 1. \tag{12.1.7}$$

Then, Lemma 12.1.5 yields

$$\left|\log \frac{J_\nu f^n(x)}{J_\nu f^n(y)}\right| = \left|\varphi_n(x) - \varphi_n(y)\right| \leq K_1 d(f^n(x), f^n(y))^\alpha \leq K_1 \rho^\alpha.$$

So, it suffices to take $K_2 = \exp(K_1\rho^\alpha)$.

Now we may show that every reference measure ν is a Gibbs state:

Lemma 12.1.7. *For every small $\varepsilon > 0$, there exists $K_3 = K_3(\varepsilon) > 0$ such that, denoting $P = \log\lambda$,*

$$K_3^{-1} \leq \frac{\nu(B(x,n,\varepsilon))}{\exp(\varphi_n(x) - nP)} \leq K_3 \quad \textit{for every } x \in M \textit{ and every } n \geq 1.$$

Proof. Consider $\varepsilon < \rho$. Then, $f \mid B(y,\varepsilon)$ is injective for every $y \in M$ and, consequently, $f^n \mid B(x,n,\varepsilon)$ is injective for every $x \in M$ and every n. Then,

$$\nu(f^n(B(x,n,\varepsilon))) = \int_{B(x,n,\varepsilon)} J_\nu f^n(y)\, d\nu(y).$$

Up to reducing ε, we may assume that $d(f(x), f(y)) < \rho$ whenever $d(x,y) < \varepsilon$. This implies that $B(x,n,\varepsilon) \subset B(x,n+1,\rho)$ for every $x \in M$ and $n \geq 1$. Then, by Corollary 12.1.6, the value of $J_\nu f^n$ at any point $y \in B(x,n,\varepsilon)$ differs from $J_\nu f^n(x)$ by a factor bounded by the constant K_2. It follows that

$$K_2^{-1}\nu(f^n(B(x,n,\varepsilon))) \leq J_\nu f^n(x)\nu(B(x,n,\varepsilon)) \leq K_2\nu(f^n(B(x,n,\varepsilon))). \tag{12.1.8}$$

Now, $J_\nu f^n(x) = \lambda^n e^{-\varphi_n(x)} = \exp(nP - \varphi_n(x))$, as we saw in (12.1.7). By Lemma 11.2.7 we also have that $f^n(B(x,n,\varepsilon)) = f(B(f^{n-1}(x),\varepsilon))$, and so

$$\nu(f^n(B(x,n,\varepsilon))) = \int_{B(f^{n-1}(x),\varepsilon)} J_\nu f\, d\nu \tag{12.1.9}$$

for every $x \in M$ and every n. It is clear that the left-hand side of (12.1.9) is bounded above by 1. Moreover, $J_\nu f = \lambda e^{-\varphi}$ is bounded from zero and (by Exercise 12.1.1 and Lemma 12.1.4) the set $\{\nu(B(y,\varepsilon)) : y \in M\}$ is also bounded from zero. Therefore, the right-hand side of (12.1.9) is bounded below by some number $a > 0$. Using these observations in (12.1.8), we obtain

$$K_2^{-1} a \leq \frac{\nu(B(x,n,\varepsilon))}{\exp(\varphi_n(x) - nP)} \leq K_2.$$

Now it suffices to take $K_3 = \max\{K_2/a, K_2\}$.

12.1.3 Invariant density

Next, we show that the transfer operator $\mathcal{L}$ admits some positive eigenfunction h associated with the eigenvalue λ. We are going to find h as a Cesàro accumulation point of the sequence of functions $\lambda^{-n}\mathcal{L}^n 1$. To show that there does exist some accumulation point, we start by proving that this sequence is uniformly bounded and equicontinuous.

Lemma 12.1.8. *There exists $K_4 > 0$ such that*

$$-K_4 d(y_1, y_2)^\alpha \le \log \frac{\mathcal{L}^n 1(y_1)}{\mathcal{L}^n 1(y_2)} \le K_4 d(y_1, y_2)^\alpha$$

for every $n \ge 1$ and any $y_1, y_2 \in M$ with $d(y_1, y_2) < \rho$.

Proof. It follows from (12.1.2) that, given any continuous function g,

$$\mathcal{L}^n g = \sum_i \left(e^{\varphi_n} g\right) \circ h_i^n \quad \text{restricted to each ball } B(y, \rho),$$

where the sum is over all inverse branches $h_i^n : B(y, \rho) \to M$ of the iterate f^n. In particular,

$$\frac{\mathcal{L}^n 1(y_1)}{\mathcal{L}^n 1(y_2)} = \frac{\sum_i e^{\varphi_n(h_i^n(y_1))}}{\sum_i e^{\varphi_n(h_i^n(y_2))}}.$$

By Lemma 12.1.5, for each of these inverse branches h_i^n one has

$$|\varphi_n(h_i^n(y_1)) - \varphi_n(h_i^n(y_2))| \le K_1 d(y_1, y_2)^\alpha.$$

Consequently,

$$e^{-K_1 d(y_1, y_2)^\alpha} \le \frac{\mathcal{L}^n 1(y_1)}{\mathcal{L}^n 1(y_2)} \le e^{K_1 d(x_1, x_2)^\alpha}.$$

Therefore, one may take any $K_4 \ge K_1$.

It follows that the sequence $\lambda^{-n}\mathcal{L}^n 1$ is bounded from zero and infinity:

Corollary 12.1.9. *There exists $K_5 > 0$ such that $K_5^{-1} \le \lambda^{-n}\mathcal{L}^n 1(x) \le K_5$ for every $n \ge 1$ and any $x \in M$.*

Proof. Start by observing that, for every $n \ge 1$,

$$\int \mathcal{L}^n 1 \, d\nu = \int 1 \, d(\mathcal{L}^{*n}\nu) = \int \lambda^n \, d\nu = \lambda^n.$$

In particular, for every $n \ge 1$,

$$\min_{y \in M} \lambda^{-n}\mathcal{L}^n 1(y) \le 1 \le \max_{y \in M} \lambda^{-n}\mathcal{L}^n 1(y). \tag{12.1.10}$$

Since f is topologically exact, there exists $N \ge 1$ such that $f^N(B(x, \rho)) = M$ for every $x \in M$ (check Exercise 11.2.3). Now, given any $x, y \in M$, we may find $x' \in B(x, \rho)$ such that $f^N(x') = y$. Then, on the one hand,

$$\mathcal{L}^{n+N} 1(y) = \sum_{z \in f^{-N}(y)} e^{\varphi_N(z)} \mathcal{L}^n 1(z) \ge e^{\varphi_N(x')} \mathcal{L}^n 1(x') \ge e^{-cN} \mathcal{L}^n 1(x').$$

On the other hand, Lemma 12.1.8 gives that $\mathcal{L}^n 1(x') \geq \mathcal{L}^n 1(x)\exp(-K_4\rho^\alpha)$. Take $c = \sup|\varphi|$ and $K \geq \exp(K_4\rho^\alpha)e^{cN}\lambda^N$. Combining the previous inequalities, we get that

$$\mathcal{L}^{n+N}1(y) \geq \exp(-K_4\rho^\alpha)e^{-cN}\mathcal{L}^n 1(x) \geq K^{-1}\lambda^N \mathcal{L}^n 1(x)$$

for every $x, y \in M$. Therefore, for every $n \geq 1$,

$$\min \lambda^{-(n+N)}\mathcal{L}^{n+N}1 \geq K^{-1}\max \lambda^{-n}\mathcal{L}^n 1. \tag{12.1.11}$$

Combining (12.1.10) and (12.1.11), we get:

$$\max \lambda^{-n}\mathcal{L}^n 1 \leq K \min \lambda^{-(n+N)}\mathcal{L}^{n+N}1 \leq K \quad \text{for every } n \geq 1,$$

$$\min \lambda^{-n}\mathcal{L}^n 1 \geq K^{-1}\max \lambda^{-n+N}\mathcal{L}^{n-N}1 \geq K^{-1} \quad \text{for every } n > N.$$

To conclude the proof, we only have to extend this last estimate to the values $n = 1, \ldots, N$. For that, observe that each $\mathcal{L}^n 1$ is a positive continuous function. Since M is compact, it follows that the minimum of $\mathcal{L}^n 1$ is positive for every n. Then, we may take $K_5 \geq K$ such that $\min \lambda^{-n}\mathcal{L}^n 1 \geq K_5^{-1}$ for every $n = 1, \ldots, N$.

It follows immediately from Corollary 12.1.9 that the positive eigenvalue λ is uniquely determined. By Lemma 12.1.1, this implies that $\lambda = \rho(\mathcal{L}) = \rho(\mathcal{L}^*)$. We are also going to see, in a while, that $\lambda = e^{P(f,\varphi)}$.

Lemma 12.1.10. *There exists $K_6 > 0$ such that*

$$|\lambda^{-n}\mathcal{L}^n 1(x) - \lambda^{-n}\mathcal{L}^n 1(y)| \leq K_6 d(x,y)^\alpha \quad \textit{for any } n \geq 1 \textit{ and } x, y \in M.$$

In particular, the sequence $\lambda^{-n}\mathcal{L}^n 1$ is equicontinuous.

Proof. Initially, suppose that $d(x,y) < \rho$. By Lemma 12.1.8,

$$\mathcal{L}^n 1(x) \leq \mathcal{L}^n 1(y)\exp(K_4 d(x,y)^\alpha)$$

and, hence,

$$\lambda^{-n}\mathcal{L}^n 1(x) - \lambda^{-n}\mathcal{L}^n 1(y) \leq \left[\exp(K_4 d(x,y)^\alpha) - 1\right]\lambda^{-n}\mathcal{L}^n 1(y).$$

Take $K > 0$ such that $|\exp(K_4 t) - 1| \leq K|t|$ whenever $|t| \leq \rho^\alpha$. Then, using Corollary 12.1.9,

$$\lambda^{-n}\mathcal{L}^n 1(x) - \lambda^{-n}\mathcal{L}^n 1(y) \leq KK_5 d(x,y)^\alpha.$$

Reversing the roles of x and y, we conclude that

$$|\lambda^{-n}\mathcal{L}^n 1(x) - \lambda^{-n}\mathcal{L}^n 1(y)| \leq KK_5 d(x,y)^\alpha \quad \text{whenever } d(x,y) < \rho.$$

When $d(x,y) \geq \rho$, Corollary 12.1.9 gives that

$$|\lambda^{-n}\mathcal{L}^n 1(x) - \lambda^{-n}\mathcal{L}^n 1(y)| \leq 2K_5 \leq 2K_5\rho^{-\alpha}d(x,y)^\alpha.$$

Hence, it suffices to take $K_6 \geq \max\{KK_5, 2K_5\rho^{-\alpha}\}$ to get the first part of the statement. The second part is an immediate consequence.

We are ready to show that the transfer operator $\mathcal{L}$ admits some eigenfunction associated with the eigenvalue λ. Corollary 12.1.9 and Lemma 12.1.10 imply that the time average

$$h_n = \frac{1}{n}\sum_{i=0}^{n-1} \lambda^{-i}\mathcal{L}^i 1$$

defines an equicontinuous bounded sequence. Then, by the theorem of Ascoli–Arzelá, there exists some subsequence $(h_{n_i})_i$ converging uniformly to a continuous function h.

Lemma 12.1.11. *The function h satisfies $\mathcal{L}h = \lambda h$. Moreover, $\int h\,d\nu = 1$ and*

$$K_5^{-1} \le h(x) \le K_5 \quad \textit{and} \quad |h(x) - h(y)| \le K_6 d(x,y)^\alpha \quad \textit{for every } x,y \in M.$$

Proof. Consider any subsequence $(h_{n_i})_i$ converging to h. As the transfer operator $\mathcal{L}$ is continuous,

$$\mathcal{L}h = \lim_i \mathcal{L}h_{n_i} = \lim_i \frac{1}{n_i}\sum_{k=0}^{n_i-1} \lambda^{-k}\mathcal{L}^{k+1}1 = \lim_i \frac{\lambda}{n_i}\sum_{k=1}^{n_i} \lambda^{-k}\mathcal{L}^k 1$$

$$= \lim_i \frac{\lambda}{n_i}\sum_{k=0}^{n_i-1} \lambda^{-k}\mathcal{L}^k 1 + \frac{\lambda}{n_i}\big(\lambda^{-n_i}\mathcal{L}^{n_i}1 - 1\big).$$

The first term on the right-hand side converges to λh whereas the second one converges to zero, because the sequence $\lambda^{-n}\mathcal{L}^n 1$ is uniformly bounded. It follows that $\mathcal{L}h = \lambda h$, as we stated.

Note that $\int \lambda^{-n}\mathcal{L}^n 1\,d\nu = \int \lambda^{-n} d(\mathcal{L}^{*n}\nu) = \int 1\,d\nu = 1$ for every $n \in \mathbb{N}$, by the definition of ν. It follows that $\int h_n\,d\nu = 1$ for every n and, using the dominated convergence theorem, $\int h\,d\nu = 1$. All the other claims in the statement follow, in an entirely analogous way, from Corollary 12.1.9 and Lemma 12.1.10.

12.1.4 Construction of the equilibrium state

Consider the measure defined by $\mu = h\nu$, that is,

$$\mu(A) = \int_A h\,d\nu \quad \text{for each measurable set } A \subset M.$$

We are going to see that μ is an equilibrium state for the potential φ and satisfies all the other conditions in Theorem 12.1.

From Lemma 12.1.11 we get that $\mu(M) = \int h\,d\nu = 1$ and so μ is a probability measure. Moreover,

$$K_5^{-1}\nu(A) \le \mu(A) \le K_5\nu(A) \tag{12.1.12}$$

for every measurable set $A \subset M$. In particular, μ is equivalent to the reference measure ν. This fact, together with Lemma 12.1.4, gives that $\operatorname{supp}\mu = M$. It also follows from the relation (12.1.12), together with Lemma 12.1.7, that μ is a Gibbs state: taking $L = K_5 K$, we find that

$$L^{-1} \le \frac{\mu(B(x,n,\varepsilon))}{\exp(\varphi_n(x) - nP)} \le L, \tag{12.1.13}$$

for every $x \in M$ and every $n \ge 1$. Recall that $P = \log\lambda$.

Lemma 12.1.12. *The probability measure μ is invariant under f. Moreover, f admits a Jacobian with respect to μ, given by $J_\mu f = \lambda e^{-\varphi}(h\circ f)/h$.*

Proof. Start by noting that $\mathcal{L}\big((g_1\circ f)g_2\big) = g_1\mathcal{L}g_2$, for any continuous functions $g_1, g_2 : M \to \mathbb{R}$. Indeed, for every $y \in M$,

$$\begin{aligned}\mathcal{L}\big((g_1\circ f)g_2\big)(y) &= \sum_{x\in f^{-1}(y)} e^{\varphi(x)} g_1(f(x))g_2(x)\\ &= g_1(y)\sum_{x\in f^{-1}(y)} e^{\varphi(x)} g_2(x) = g_1(y)\mathcal{L}g_2(y).\end{aligned} \tag{12.1.14}$$

Thus, for every continuous function $g : M \to \mathbb{R}$,

$$\begin{aligned}\int (g\circ f)\,d\mu &= \lambda^{-1}\int (g\circ f)h\,d(\mathcal{L}^*\nu) = \lambda^{-1}\int \mathcal{L}\big((g\circ f)h\big)\,d\nu\\ &= \lambda^{-1}\int g\mathcal{L}h\,d\nu = \int gh\,d\nu = \int g\,d\mu.\end{aligned}$$

In view of Proposition A.3.3, this proves that the probability measure μ is invariant under f.

To prove the second claim, consider any domain of invertibility A of f. Then, using Lemma 9.7.4(i),

$$\mu(f(A)) = \int_{f(A)} 1\,d\mu = \int_{f(A)} h\,d\nu = \int_A J_\nu f(h\circ f)\,d\nu = \int_A J_\nu f\frac{h\circ f}{h}\,d\mu.$$

By Lemma 12.1.3, this means that

$$J_\mu f = J_\nu f\frac{h\circ f}{h} = \lambda e^{-\varphi}\frac{h\circ f}{h},$$

as stated.

Corollary 12.1.13. *The invariant probability measure $\mu = h\nu$ satisfies*

$$h_\mu(f) + \int \varphi\,d\mu = P.$$

Proof. Combining the Rokhlin formula (Theorem 9.7.3) with the second part of Lemma 12.1.12,

$$h_\mu(f) = \int \log J_\mu f\,d\mu = \log\lambda - \int\varphi\,d\mu + \int(\log h\circ f - \log h)\,d\mu.$$

Since μ is invariant and $\log h$ is bounded (Corollary 12.1.9), the last term is equal to zero. This shows that $h_\mu(f) = P - \int\varphi\,d\mu$, as stated.

To complete the proof that $\mu = h\nu$ is an equilibrium state, all that we need to do is to check that $P = \log\lambda$ is equal to the pressure $P(f,\varphi)$. This is done in Corollary 12.1.15 below.

12.1.5 Pressure and eigenvalues

Let η be any probability measure invariant under f and such that

$$h_\eta(f) + \int \varphi\, d\eta \geq P \tag{12.1.15}$$

(for example: the probability measure μ constructed in the previous section). Let $g_\eta = 1/J_\eta f$ (the Jacobian $J_\eta f$ does exist, by Exercise 9.7.8) and consider also the function $g = \lambda^{-1}e^{\varphi}h/(h\circ f)$. Observe that

$$\sum_{x\in f^{-1}(y)} g(x) = \frac{1}{\lambda h(y)} \sum_{x\in f^{-1}(y)} e^{\varphi(x)}h(x) = \frac{\mathcal{L}h(y)}{\lambda h(y)} = 1 \tag{12.1.16}$$

for every $y \in M$. Moreover, since η is invariant under f, Exercise 9.7.4 gives that

$$\sum_{x\in f^{-1}(y)} g_\eta(x) = 1 \quad \text{for } \eta\text{-almost every } y \in M. \tag{12.1.17}$$

Using (12.1.15) and the Rokhlin formula (Theorem 9.7.3),

$$0 \leq h_\eta(f) + \int \varphi\, d\eta - P = \int (-\log g_\eta + \varphi - \log\lambda)\, d\eta. \tag{12.1.18}$$

By the definition of g and the hypothesis that η is invariant, the integral on the right-hand side of (12.1.18) is equal to

$$\int (-\log g_\eta + \log g + \log h\circ f - \log h)\, d\eta = \int \log\frac{g}{g_\eta}\, d\eta. \tag{12.1.19}$$

Recalling the definition of g_η, Exercise 9.7.3 gives that

$$\int \log\frac{g}{g_\eta}\, d\eta = \int \Big(\sum_{x\in f^{-1}(y)} g_\eta(x) \log\frac{g}{g_\eta}(x) \Big)\, d\eta(y). \tag{12.1.20}$$

At this point we need the following elementary fact:

Lemma 12.1.14. *Let p_i, b_i, $i = 1,\dots,k$ be positive real numbers such that $\sum_{i=1}^k p_i = 1$. Then*

$$\sum_{i=1}^k p_i \log b_i \leq \log(\sum_{i=1}^k p_i b_i),$$

and the identity holds if and only if the numbers b_j are all equal to $\sum_{i=1}^k p_i b_i$.

Proof. Take $a_i = \log(p_i b_i)$ in Lemma 10.4.4. Then the inequality in the conclusion of Lemma 10.4.4 corresponds exactly to the inequality in the present lemma. Moreover, the identity holds if and only if

$$p_j = \frac{e^{a_j}}{\sum_i e^{a_i}} \Leftrightarrow p_j = \frac{p_j b_j}{\sum_i p_i b_i} \Leftrightarrow b_j = \sum_i p_i b_i$$

for every $j = 1, \ldots, n$.

For each $y \in M$, take $p_i = g_\eta(x_i)$ and $b_i = g(x_i)/g_\eta(x_i)$, where the x_i are the pre-images of y. The identity (12.1.17) means that $\sum_i p_i = 1$ for η-almost every y. Then, we may apply Lemma 12.1.14:

$$\begin{aligned} \sum_{x \in f^{-1}(y)} g_\eta(x) \log \frac{g}{g_\eta}(x) &\leq \log \sum_{x \in f^{-1}(y)} g_\eta(x) \frac{g}{g_\eta}(x) \\ &= \log \sum_{x \in f^{-1}(y)} g(x) = 0 \end{aligned} \tag{12.1.21}$$

for η-almost every y; in the last step we used (12.1.16). Combining the relations (12.1.18) through (12.1.21), we find:

$$h_\eta(f) + \int \varphi \, d\eta - P = \int \log \frac{g}{g_\eta} \, d\eta = 0. \tag{12.1.22}$$

Corollary 12.1.15. $P(f, \varphi) = P = \log \rho(\mathcal{L})$.

Proof. By (12.1.22), we have that $h_\eta(f) + \int \varphi \, d\eta = P$ for every invariant probability measure η such that $h_\eta(f) + \int \varphi \, d\eta \geq P$. By the variational principle (Theorem 10.4.1), it follows that $P(f, \varphi) = P$. The second identity has been observed before, right after Corollary 12.1.9.

At this point we have completed the proof that the measure $\mu = h\nu$ constructed in the previous section is an equilibrium state for φ. The statement that follows arises from the same kind of ideas and is the basis for proving that this equilibrium state is unique:

Corollary 12.1.16. *If η is an equilibrium state for φ then* $\operatorname{supp} \eta = M$ *and*

$$J_\eta f = \lambda e^{-\varphi} (h \circ f)/h \quad \text{and} \quad \mathcal{L}^*(\eta/h) = \lambda(\eta/h).$$

Proof. The first claim is an immediate consequence of the second one and Lemma 12.1.4.

Note that the identity in (12.1.22) also implies that the identity in (12.1.21) holds for η-almost every $y \in M$. According to Lemma 12.1.14, that happens if and only if the numbers $b_i = \log(g(x_i)/g_\eta(x_i))$ are all equal. In other words, for η-almost every $y \in M$ there exists a number $c(y)$ such that

$$\frac{g(x)}{g_\eta(x)} = c(y) \quad \text{for every } x \in f^{-1}(y).$$

Moreover, recalling the identities (12.1.18) and (12.1.19),

$$c(y) = \sum_{x\in f^{-1}(y)} c(y)g_\eta(x) = \sum_{x\in f^{-1}(y)} g(x) = 1$$

for η-almost every y. It follows that $g_\eta = g$ at η-almost every point, and so the function $1/g = \lambda e^{-\varphi}(h\circ f)/h$ is a Jacobian of f with respect to η. This proves the second claim.

To prove the third claim, let $\xi : M \to \mathbb{R}$ be any continuous function. On the one hand, by the definition of the transfer operator,

$$\int \xi \, d\mathcal{L}^*\left(\frac{\eta}{h}\right) = \int \frac{1}{h}(\mathcal{L}\xi)\, d\eta = \int \frac{1}{h(y)}\left(\sum_{x\in f^{-1}(y)} e^{\varphi(x)}\xi(x)\right) d\eta(y). \quad (12.1.23)$$

By the definition of the function g,

$$\frac{e^{\varphi(x)}}{h(y)} = \frac{\lambda g(x)}{h(x)}.$$

Replacing this identity in (12.1.23), we obtain:

$$\int \xi \, d\mathcal{L}^*\left(\frac{\eta}{h}\right) = \int \left(\sum_{x\in f^{-1}(y)} \frac{\lambda g\xi}{h}(x)\right) d\eta(y). \quad (12.1.24)$$

Then, recalling that $g = g_\eta = 1/J_\eta f$, we may use Exercise 9.7.3 to conclude that

$$\int \xi \, d\mathcal{L}^*\left(\frac{\eta}{h}\right) = \int \left(\sum_{x\in f^{-1}(y)} \frac{\lambda g\xi}{h}(x)\right) d\eta(y) = \int \frac{\lambda\xi}{h}\, d\eta.$$

Since the continuous function ξ is arbitrary, this shows that $\mathcal{L}^*(\eta/h) = \lambda(\eta/h)$, as stated.

12.1.6 Uniqueness of the equilibrium state

Let us start by proving the following distortion bound:

Corollary 12.1.17. *There exists $K_7 > 0$ such that for every equilibrium state η, every $n \geq 1$, every $x \in M$ and every $y \in B(x, n+1, \rho)$,*

$$K_7^{-1} \leq \frac{J_\eta f^n(x)}{J_\eta f^n(y)} \leq K_7.$$

Proof. By Corollary 12.1.16,

$$J_\eta f^n = \lambda e^{-\varphi_n}\frac{h\circ f^n}{h} = J_\nu f^n \frac{h\circ f^n}{h}$$

for each $n \geq 1$. Then, using Corollary 12.1.6 and Lemma 12.1.11,

$$K_2^{-1}K_5^{-4} \leq \frac{J_\eta f^n(x)}{J_\eta f^n(y)} = \frac{J_\nu f^n(x)}{J_\nu f^n(y)}\frac{h(f^n(x))h(y)}{f(f^n(y))h(x)} \leq K_2 K_5^4.$$

Therefore, it suffices to take $K_7 = K_2 K_5^4$.

Lemma 12.1.18. *All the equilibrium states of φ are equivalent measures.*

Proof. Let η_1 and η_2 be equilibrium states. Consider any finite partition $\mathcal{P}$ of M such that every $P \in \mathcal{P}$ has non-empty interior and diameter less than ρ. Since $\operatorname{supp} \eta_1 = \operatorname{supp} \eta_2 = M$ (by Corollary 12.1.16), the set $\{\eta_i(P) : i = 1, 2 \text{ and } P \in \mathcal{P}\}$ is bounded from zero. Consequently, there exists $C_1 > 0$ such that

$$\frac{1}{C_1} \le \frac{\eta_1(P)}{\eta_2(P)} \le C_1 \quad \text{for every } P \in \mathcal{P}. \tag{12.1.25}$$

We are going to show that this relation extends to every measurable subset of M, up to replacing C_1 by a convenient constant $C_2 > C_1$.

For each $n \ge 1$, let $\mathcal{Q}_n$ be the partition of M formed by the images $h^n(P)$ of the elements of P under the inverse branches h^n of the iterate f^n. By the definition of Jacobian, $\eta_i(P) = \int_{h^n(P)} J_{\eta_i} f^n \, d\eta_i$. Hence, using Corollary 12.1.17,

$$K_7^{-1} J_{\eta_i} f^n(x) \le \frac{\eta_i(P)}{\eta_i(h^n(P))} \le K_7 J_{\eta_i} f^n(x)$$

for any $x \in h^n(P)$. Recalling that $J_{\eta_1} f = J_{\eta_2} f$ (Corollary 12.1.16), it follows that

$$K_7^{-2} \le \frac{\eta_2(P)\eta_1(h^n(P))}{\eta_1(P)\eta_2(h^n(P))} \le K_7^2. \tag{12.1.26}$$

Combining (12.1.25) and (12.1.26), and taking $C_2 = C_1 K_7^2$, we get that

$$\frac{1}{C_2} \le \frac{\eta_1(h^n(P))}{\eta_2(h^n(P))} \le C_2 \tag{12.1.27}$$

for every $P \in \mathcal{P}$, every inverse branch h^n of f^n and every $n \ge 1$. In other words, the property in (12.1.25) holds for every element of $\mathcal{Q}_n$, with C_2 in the place of C_1.

Now observe that $\operatorname{diam} \mathcal{Q}_n \le 2\sigma^{-n}\rho$ for every n. Given any measurable set B and any $\delta > 0$, we may use Proposition A.3.2 to find a compact set $F \subset B$ and an open set $A \supset B$ such that $\eta_i(A \setminus F) < \delta$ for $i = 1, 2$. Let Q_n be the union of all the elements of the partition $\mathcal{Q}_n$ that intersect F. It is clear that $Q_n \supset F$ and, assuming that n is large enough, $Q_n \subset A$. Then,

$$\eta_1(B) \le \eta_1(A) < \eta_1(Q_n) + \delta \quad \text{and} \quad \eta_2(B) \ge \eta_2(F) > \eta_2(Q_n) - \delta.$$

The relation (12.1.27) gives that $\eta_1(Q_n) \le C_2 \eta_2(Q_n)$, since Q_n is a (disjoint) union of elements of $\mathcal{Q}_n$. Combining these three inequalities, we obtain

$$\eta_1(B) < C_2\big(\eta_2(B) + \delta\big) + \delta.$$

Since δ is arbitrary, we conclude that $\eta_1(B) \le C_2 \eta_2(B)$ for every measurable set $B \subset M$. Reversing the roles of the two measures, we also get $\eta_2(B) \le C_2 \eta_2(B)$ for every measurable set $B \subset M$.

These inequalities prove that any two equilibrium states are equivalent measures, with Radon–Nikodym derivatives bounded from zero and infinity.

Combining Lemmas 4.3.3 and 12.1.18 we get that all the *ergodic* equilibrium states are equal. Now, by Proposition 10.5.5, the connected components of any equilibrium state are also equilibrium states (ergodic, of course). It follows that there exists a unique equilibrium state, as stated.

There is an alternative proof of the fact that the equilibrium state is unique that does not use Proposition 10.5.5 and, thus, does not require the theorem of Jacobs. Indeed, the results in the next section imply that the equilibrium state $\mu = h\nu$ in Section 12.1.4 is ergodic. By Lemma 12.1.18, that implies that all the equilibrium states are ergodic. Using Lemma 4.3.3, it follows that all the equilibrium states must coincide.

As a consequence, the reference measure ν is also unique: if there were two distinct reference measures, ν_1 and ν_2, then $\mu_1 = h\nu_1$ and $\mu_2 = h\nu_2$ would be distinct equilibrium states. Analogously, the positive eigenfunction h is unique up to multiplication by a positive constant.

12.1.7 Exactness

Finally, let us prove that the system (f, μ) is exact. Recall that this means that if $B \subset M$ is such that there exist measurable sets B_n satisfying $B = f^{-n}(B_n)$ for every $n \geq 1$, then B has measure 0 or measure 1.

Let B be such a subset of M and assume that $\mu(B) > 0$. Let $\mathcal{P}$ be a finite partition of M by subsets with non-empty interior and diameter less than ρ. For each n, let $\mathcal{Q}_n$ be the partition of M whose elements are the images $h^n(P)$ of the sets $P \in \mathcal{P}$ under the inverse branches h^n of the iterate f^n.

Lemma 12.1.19. *For every $\varepsilon > 0$ and every $n \geq 1$ sufficiently large there exists some $h^n(P) \in \mathcal{Q}_n$ such that*

$$\mu\big(B \cap h^n(P)\big) > (1-\varepsilon)\mu(h^n(P)). \tag{12.1.28}$$

Proof. Fix $\varepsilon > 0$. Since the measure μ is regular (Proposition A.3.2), given any $\delta > 0$ there exist some compact set $F \subset B$ and some open set $A \supset B$ satisfying $\mu(A \setminus F) < \delta$. Since we assume that $\mu(B) > 0$, this inequality implies that $\mu(F) > (1-\varepsilon)\mu(A)$, as long as $\delta > 0$ is sufficiently small. Fix δ from now on. Note that $\operatorname{diam} \mathcal{Q}_n < \sigma^{-n}\rho$. Then, for every n sufficiently large, any element $h^n(P)$ of $\mathcal{Q}_n$ that intersects F is contained in A. By contradiction, suppose that (12.1.28) is false for every $h^n(P)$. Then, adding over all the $h^n(P)$ that intersect F,

$$\begin{aligned}\mu(F) &\leq \sum_{P,h^n} \mu\big(F \cap h^n(P)\big) \leq \sum_{P,h^n} \mu\big(B \cap h^n(P)\big)\\ &\leq (1-\varepsilon)\sum_{P,h^n} \mu(h^n(P)) \leq (1-\varepsilon)\mu(A).\end{aligned}$$

This contradiction proves that (12.1.28) is valid for some $h^n(P) \in \mathcal{Q}_n$.

Consider any $h^n(P) \in \mathcal{Q}_n$ such that (12.1.28). Since $B = f^{-n}(B_n)$ and $f^n \circ h^n = \mathrm{id}$ in its domain, we have that $f^n\big(h^n(P) \setminus B\big) = P \setminus B_n$. Then, applying Corollary 12.1.17 to the measure $\eta = \mu$,

$$\begin{aligned} \mu(P \setminus B_n) &= \int_{h^n(P)\setminus B} J_\mu f^n \, d\mu \le K_7 \mu(h^n(P) \setminus B) J_\mu f^n(x) \\ \text{and} \quad \mu(P) &= \int_{h^n(P)} J_\mu f^n \, d\mu \ge K_7^{-1} \mu(h^n(P)) J_\mu f^n(x) \end{aligned} \tag{12.1.29}$$

for any $x \in h^n(P)$. Combining (12.1.28) and (12.1.29),

$$\frac{\mu(P \setminus B_n)}{\mu(P)} \le K_7^2 \frac{\mu(h^n(P) \setminus B)}{\mu(h^n(P))} \le K_7^2 \varepsilon.$$

In this way we have shown that, given any $\varepsilon > 0$ and any $n \ge 1$ sufficiently large, there exists some $P \in \mathcal{P}$ such that $\mu(P \setminus B_n) \le K_7^2 \varepsilon \mu(P)$.

Since the partition $\mathcal{P}$ is finite, it follows that there exist some $P \in \mathcal{P}$ and some sequence $(n_j)_j \to \infty$ such that

$$\mu(P \setminus B_{n_j}) \to 0 \quad \text{when} \quad j \to \infty. \tag{12.1.30}$$

Let P be fixed from now on. Since, by assumption, P has non-empty interior and f is topologically exact, there exists $N \ge 1$ such that $f^N(P) = M$. Let $P = P_1 \cup \cdots \cup P_s$ be a finite partition of P into domains of invertibility of f^N. Corollaries 12.1.9 and 12.1.16 give that $J_\mu f^N = \lambda^N e^{-\varphi_N} (h \circ f^N)/f$ is bounded from zero and infinity. Note also that $f^N(P_i \setminus B_{n_j}) = f^N(P_i) \setminus B_{n_j+N}$, because $f^{-n}(B_n) = B$ for every n. Combining these two observations with (12.1.30), we find that, given any $i = 1, \dots, s$, the sequence

$$\mu(f^N(P_i) \setminus B_{n_j+N}) = \mu(f^N(P_i \setminus B_{n_j})) = \int_{P_i \setminus B_{n_j}} J_\mu f^N \, d\mu$$

converges to zero when $j \to \infty$. Now, $\{f^N(P_i) : i = 1, \dots, s\}$ is a finite cover of M by measurable sets. Therefore, this last conclusion implies that $\mu(M \setminus B_{n_j+N})$ converges to zero, that is, $\mu(B) = \mu(B_{n_j+N})$ converges to 1 when $j \to \infty$. That means that $\mu(B) = 1$, of course.

The proof of Theorem 12.1 is complete.

12.1.8 Absolutely continuous measures

In this last section on the theorem of Ruelle we briefly discuss the special case when $f : M \to M$ is a local diffeomorphism on a compact Riemannian manifold and $\varphi = -\log|\det Df|$. It is assumed that f is such that this potential φ is Hölder. The first goal is to compare the conclusions of the theorem of Ruelle in this case with the results in Section 11.1:

Proposition 12.1.20. *The invariant absolutely continuous probability measure coincides with the equilibrium state μ of the potential $\varphi = -\log|\det Df|$.*

Consequently, it is equivalent to the Lebesgue measure m, with density $d\mu/dm$ Hölder and bounded from zero and infinity, and it is exact.

Proof. We saw in Example 12.1.2 that the Lebesgue measure m is an eigenvector of the dual $\mathcal{L}^*$ of the transfer operator associated with the potential $\varphi = -\log|\det Df|$: more precisely,

$$\mathcal{L}^* m = m.$$

Applying the previous theory (from Lemma 12.1.3 on) with $\lambda = 1$ and $\nu = m$, we find a Hölder function $h : M \to \mathbb{R}$, bounded from zero and infinity, such that $\mathcal{L}h = h$ and the measure $\mu = hm$ is the equilibrium state of the potential φ. Recalling Corollary 11.1.15, it follows that μ is also the unique probability measure invariant under f and absolutely continuous with respect to m. The fact that h is positive implies that μ and m are equivalent measures. Exactness was proven in Section 12.1.7.

It is worthwhile pointing out that, while the absolutely continuous invariant measure is unique (Theorem 11.1.2), the potential $\varphi = -\log|\det Df|$ depends on the choice of the Riemannian metric on M, because the determinant does. So, Proposition 12.1.20 also implies that all the potentials of this form, corresponding to different choices of the Riemannian metric, have the same equilibrium state. This type of situation is the subject of Section 12.2 and, in particular, Exercise 12.2.3.

It also follows from the proof of Theorem 12.1 that

$$h_\mu(f) - \int \log|\det Df|\,d\mu = P(f, -\log|\det Df|) = \log\lambda = 0. \tag{12.1.31}$$

Let $\tilde{\varphi}$ be the time average of the function φ, given by the Birkhoff ergodic theorem. Then,

$$\int \log|\det Df|\,d\mu = \int -\varphi\,d\mu = \int -\tilde{\varphi}\,d\mu. \tag{12.1.32}$$

Moreover,

$$-\tilde{\varphi}(x) = \lim_n \frac{1}{n}\sum_{j=0}^{n-1}\log|\det Df(f^j(x))| = \lim_n \frac{1}{n}\log|\det Df^n(x)| \tag{12.1.33}$$

at μ-almost every point. In the context of our comments about the Oseledets theorem (see the relation (c1) in Section 3.3.5) we mentioned that

$$\lim_n \frac{1}{n}\log|\det Df^n(x)| = \sum_{i=1}^{k(x)} d_i(x)\lambda_i(x), \tag{12.1.34}$$

where $\lambda_1(x), \ldots, \lambda_{k(x)}(x)$ are the Lyapunov exponents of the transformation f at the point x and $d_1(x), \ldots, d_{k(x)}(x)$ are the corresponding multiplicities.

Combining the relations (12.1.31)–(12.1.34), we find that

$$h_\mu(f) = \int \left(\sum_{i=1}^{k(x)} d_i(x)\lambda_i(x) \right) d\mu(x). \qquad (12.1.35)$$

Since these functions are invariant (see the relation (a1) in Section 3.3.5) and the measure μ is ergodic, the functions $k(x)$, $\lambda_i(x)$ and $d_i(x)$ are constant at μ-almost every point. Let us denote by k, λ_i and d_i these constants. Then (12.1.35) translates into the following theorem:

Theorem 12.1.21. *Let $f : M \to M$ be an expanding map on a compact Riemannian manifold, such that the derivative Df is Hölder. Let μ be the unique invariant probability measure absolutely continuous with respect to the Lebesgue measure on M. Then*

$$h_\mu(f) = \sum_{i=1}^{k} d_i\lambda_i, \qquad (12.1.36)$$

where λ_i, $i = 1,\dots,k$ are the Lyapunov exponents of f at μ-almost every point and d_i, $i = 1,\dots,k$ are the corresponding multiplicities.

As we pointed out before, in Section 9.4.4, this is a special instance of the Pesin entropy formula (Theorem 9.4.5).

12.1.9 Exercises

12.1.1. Show that if η is a Borel measure on a compact metric space then for every $\varepsilon > 0$ there exists $b > 0$ such that $\eta(B(y,\varepsilon)) > b$ for every $y \in \operatorname{supp}\eta$.

12.1.2. Let $f : M \to M$ be an expanding map. Consider the non-linear operator $G : \mathcal{M}_1(M) \to \mathcal{M}_1(M)$ defined in the space $\mathcal{M}_1(M)$ of all Borel probability measures by

$$G(\eta) = \frac{\mathcal{L}^*(\eta)}{\int \mathcal{L}1\, d\eta}.$$

Use the Tychonoff–Schauder theorem (Theorem 2.2.3) to show that G admits some fixed point and deduce Lemma 12.1.1.

12.1.3. Let $\sigma : \Sigma_A \to \Sigma_A$ be the one-sided shift of finite type associated with a given transition matrix A (recall Section 10.2.2). Let P be a stochastic matrix such that $P_{i,j} = 0$ whenever $A_{i,j} = 0$ and p be a probability vector with positive coefficients such that $P^*p = p$. Consider the transfer operator $\mathcal{L}$ associated with the locally constant potential

$$\varphi(i_0, i_1, \dots, i_n, \dots) = -\log \frac{p_{i_1}}{p_{i_0} P_{i_0,i_1}}.$$

Show that the Markov measure μ associated with the matrix P and the vector p satisfies $\mathcal{L}^*\mu = \mu$.

12.1.4. Let λ be any positive number and ν be a Borel probability measure such that $\mathcal{L}^*\nu = \lambda\nu$. Show that, given any $u \in L^1(\nu)$ and any continuous function $v : M \to \mathbb{R}$,

$$\int (u \circ f) v \, d\nu = \int u(\lambda^{-1} \mathcal{L} v) \, d\nu.$$

12.2 Theorem of Livšic

Now we discuss the following issue: when is it the case that two different Hölder potentials ϕ and ψ have the same equilibrium state? Observe that, since these are ergodic measures, the two equilibrium states μ_ϕ and μ_ψ either coincide or are mutually singular (by Lemma 4.3.3).

Recall that two potentials are said to be cohomologous (with respect to f) if the difference between them may be written as $u \circ f - u$ for some continuous function $u : M \to \mathbb{R}$.

Theorem 12.2.1 (Livšic). *A potential $\varphi : M \to \mathbb{R}$ is cohomologous to zero if and only if $\varphi_n(x) = 0$ for every $x \in \mathrm{Fix}(f^n)$ and every $n \geq 1$.*

Proof. It is clear that if $\varphi = u \circ f - u$ for some u then

$$\varphi_n(x) = \sum_{j=1}^{n} u(f^j(x)) - \sum_{j=0}^{n-1} u(f^j(x)) = 0$$

for every $x \in M$ such that $f^n(x) = x$. The converse is a lot more interesting.

Suppose that $\varphi_n(x) = 0$ for every $x \in \mathrm{Fix}(f^n)$ and every $n \geq 1$. Consider any point $z \in M$ whose orbit is dense in M; such a point exists because f is topologically exact and, consequently, transitive. Define the function u on the orbit of z through the following relation:

$$u(f^n(z)) = u(z) + \varphi_n(z), \tag{12.2.1}$$

where $u(z)$ is arbitrary. Observe that

$$u(f^{n+1}(z)) - u(f^n(z)) = \varphi_{n+1}(z) - \varphi_n(z) = \varphi(f^n(z)) \tag{12.2.2}$$

for every $n \geq 0$. In other words, the cohomology relation

$$\phi - \psi = u \circ f - u \tag{12.2.3}$$

holds on the orbit of z. To extend this relation to the whole of M, we use the following fact:

Lemma 12.2.2. *The function u is uniformly continuous on the orbit of z.*

Proof. Given $\varepsilon \in (0, \rho)$, take $\delta > 0$ given by the shadowing lemma (Proposition 11.2.9). Suppose that $k \geq 0$ and $l \geq 1$ are such that $d(f^k(z), f^{k+l}(z)) < \delta$. Then the periodic sequence $(x_n)_n$ of period l given by

$$x_0 = f^k(z), x_1 = f^{k+1}(z), \ldots, x_{l-1} = f^{k+l-1}(z), x_l = f^k(z)$$

is a δ-pseudo-orbit. Hence, by Proposition 11.2.9, there exists $x \in \text{Fix}(f^l)$ such that $d(f^j(x), f^{k+j}(z)) < \varepsilon$ for every $j \geq 0$. Since we took $\varepsilon < \rho$, this also implies that $x = h_l(f^l(x))$, where $h_l : B(f^{k+l}(z), \rho) \to M$ denotes the inverse branch of f^l that maps $f^{k+l}(z)$ to $f^k(z)$. By (11.2.6), it follows that

$$d(f^j(x), f^{k+j}(z)) \leq \sigma^{j-l} d(f^l(x), f^{k+l}(z)) \quad \text{for every } 0 \leq j \leq l. \tag{12.2.4}$$

By the definition (12.2.1),

$$u(f^{k+l}(z)) - u(f^k(z)) = \varphi_{k+l}(z) - \varphi_k(z) = \varphi_l(f^k(z)). \tag{12.2.5}$$

Fix constants $C > 0$ and $\nu > 0$ such that $|\varphi(x) - \varphi(y)| \leq Cd(x,y)^\nu$ for any $x, y \in M$. Then,

$$\left|\varphi_l(f^k(z)) - \varphi_l(x)\right| \leq \sum_{j=0}^{j-1} \left|\varphi(f^{k+j}(z)) - \varphi(f^j(x))\right| \leq \sum_{j=0} Cd(f^j(x), f^{k+j}(z))^\nu.$$

Using (12.2.4), it follows that

$$|\varphi_l(f^k(z)) - \varphi_l(x)| \leq \sum_{j=0} C\sigma^{\nu(j-l)} d(x, f^{k+l}(z))^\nu \leq C_1 \varepsilon^\nu, \tag{12.2.6}$$

where $C_1 = C\sum_{i=0}^\infty \sigma^{-i\nu}$. Recall that, by assumption, $\psi_l(x) = 0$. Hence, combining (12.2.5) and (12.2.6), we find that $|u(f^{k+l}(z)) - u(f^k(z))| \leq C_1\varepsilon^\nu$. This completes the proof of the lemma.

It follows from Lemma 12.2.2 that u admits a (unique) continuous extension to the closure of the orbit of z, that is, the ambient space M. Then, by continuity of φ and u, the cohomology relation (12.2.3) extends to the whole M. This proves Theorem 12.2.1.

Theorem 12.2.3. *Let $f : M \to M$ be a topologically exact expanding map on a compact metric space and ϕ and ψ be two Hölder potentials in M. The following conditions are equivalent:*

(i) $\mu_\phi = \mu_\psi$;
(ii) there exist $c \in \mathbb{R}$ and an arbitrary function $u : M \to \mathbb{R}$ such that $\phi - \psi = c + u \circ f - u$;
(iii) $\phi - \psi$ is cohomologous to some constant $c \in \mathbb{R}$;
(iv) there exist $c \in \mathbb{R}$ and a Hölder function $u : M \to \mathbb{R}$ such that $\phi - \psi = c + u \circ f - u$;
(v) there exists $c \in \mathbb{R}$ such that $\phi_n(x) - \psi_n(x) = cn$ for every $x \in \text{Fix}(f^n)$ and $n \geq 1$.

Moreover, the constants $c \in \mathbb{R}$ in (ii), (iii), (iv) and (v) coincide; indeed, they are all equal to $P(f, \phi) - P(f, \psi)$.

Proof. It is clear that (iv) implies (iii) and (iii) implies (ii).

If $\phi - \psi = c + u \circ f - u$ for some function u then, given any $x \in \operatorname{Fix}(f^n)$,

$$\phi_n(x) - \psi_n(x) = \sum_{j=0}^{n-1} (\phi - \psi)(f^j(x)) = \sum_{j=0}^{n-1} \big(c + u(f^{j+1}(x)) - u(f^j(x))\big).$$

Since $f^n(x) = x$, the sum of the last two terms over every $j = 0, \dots, n-1$ vanishes. Therefore, $\phi_n(x) - \psi_n(x) = cn$. This proves that (ii) implies (v).

Suppose that $\phi_n(x) - \psi_n(x) = cn$ for every $x \in \operatorname{Fix}(f^n)$ and every $n \geq 0$. That means that the function $\varphi = \phi - \psi - c$ satisfies $\varphi_n(x) = 0$ for every $x \in \operatorname{Fix}(f^n)$ and every $n \geq 0$. Note also that φ is Hölder. Hence, by Theorem 12.2.1, there exists a continuous function $u : M \to \mathbb{R}$ such that $\varphi = u \circ f - u$. In other words, $\phi - \psi$ is cohomologous to c. This shows that (v) implies (iii).

It follows from (10.3.4) and Proposition 10.3.8 that if ϕ is cohomologous to $\psi + c$ then

$$P(f, \phi) = P(f, \psi + c) = P(f, \psi) + c.$$

On the other hand, given any invariant probability measure ν,

$$h_\nu(f) + \int \phi\, d\nu = h_\nu(f) + \int (\psi + c)\, d\nu = h_\nu(f) + \int \psi\, d\nu + c.$$

Therefore, ν is an equilibrium state for ϕ if and only if ν is an equilibrium state for ψ. This shows that (iii) implies (i).

If μ_ϕ and μ_ψ coincide then they have the same Jacobian, of course. By Lemma 12.1.12, this means that

$$\lambda_\phi e^{-\phi} \frac{h_\phi \circ f}{h_\phi} = \lambda_\psi e^{-\psi} \frac{h_\psi \circ f}{h_\psi}. \tag{12.2.7}$$

Let $c = \log \lambda_\phi - \log \lambda_\psi$ and $u = \log h_\phi - \log h_\psi$. Both are well defined, since λ_ϕ, λ_ψ, h_ϕ and h_ψ are all positive. Moreover, since the functions h_ϕ and h_ψ are Hölder and bounded from zero and infinity (Corollary 12.1.9), the function u is Hölder. Finally, (12.2.7) may be rewritten as follows:

$$\phi - \psi = c + \log u \circ f - u.$$

This shows that (i) implies (iv). The proof of the theorem is complete.

Here is an interesting consequence in the differentiable setting:

Corollary 12.2.4. *Let $f : M \to M$ be a differentiable expanding map on a compact Riemannian manifold such that the Jacobian $\det Df$ is Hölder. The absolutely continuous invariant probability measure μ coincides with the measure of maximum entropy if and only if there exists $c \in \mathbb{R}$ such that*

$$|\det Df^n(x)| = e^{cn} \quad \textit{for every } x \in \operatorname{Fix}(f^n) \textit{ and every } n \geq 1.$$

Proof. As we saw in Proposition 12.1.20, μ is the equilibrium state of the potential $\varphi = -\log|\det Df|$. It is clear that the measure of maximum entropy

μ_0 is the equilibrium state of the zero function. Observe that

$$\varphi_n(x) = \sum_{j=0}^{n-1} \log|\det Df(f^j(x))| = \log|\det Df^n(x)|.$$

Therefore, Theorem 12.2.3 gives that $\mu = \mu_0$ if and only if there exists some number $c \in \mathbb{R}$ such that $\log|\det Df^n(x)| = 0 + cn$ for every $x \in \operatorname{Fix}(f^n)$ and every $n \geq 1$.

12.2.1 Exercises

12.2.1. Consider the two-sided shift $\sigma : \Sigma \to \Sigma$ in $\Sigma = \{1,\dots,d\}^{\mathbb{Z}}$. Show that for every Hölder function $\varphi : \Sigma \to \mathbb{R}$, there exists a Hölder function $\varphi^+ : \Sigma \to \mathbb{R}$, cohomologous to φ and such that $\varphi^+(x) = \varphi^+(y)$ whenever $x = (x_i)_{i\in\mathbb{Z}}$ and $y = (y_i)_{i\in\mathbb{Z}}$ are such that $x_i = y_i$ for $i \geq 0$.

12.2.2. Prove that if the functions $\varphi, \psi : M \to \mathbb{R}$ are such that there exist constants C, L satisfying $|\varphi_n(x) - \psi_n(x) - nC| \leq L$ for every $x \in M$, then $P(f,\varphi) = P(f,\psi) + C$ and φ is cohomologous to $\psi + C$.

12.2.3. Let $f : M \to M$ be a differentiable expanding map on a compact manifold, with Hölder derivative. Check that any two potentials of the form $\varphi = -\log|\det Df|$, for two different choices of a Riemannian metric on M, are cohomologous. [Observation: In particular, all such potentials have the same equilibrium state, namely, the absolutely continuous invariant probability measure. This was observed before, in Section 12.1.8.]

12.2.4. Given $k \geq 2$, let $f : S^1 \to S^1$ be the (expanding) map given by $f(x) = kx \mod \mathbb{Z}$. Let $g : S^1 \to S^1$ be a differentiable expanding map of degree k. Show that f and g are topologically conjugate.

12.2.5. Given $k \geq 2$, let $f : S^1 \to S^1$ be the map given by $f(x) = kx \mod \mathbb{Z}$. Let $g : S^1 \to S^1$ be a differentiable expanding map of degree k, with Hölder derivative. Show that the following conditions are equivalent:
  (a) f and g are conjugated by some diffeomorphism;
  (b) f and g are conjugated by some absolutely continuous homeomorphism whose inverse is also absolutely continuous;
  (c) $(g^n)'(p) = k^n$ for every $p \in \operatorname{Fix}(f^n)$.

12.3 Decay of correlations

Let $f : M \to M$ be a topologically exact expanding map and $\varphi : M \to \mathbb{R}$ be a Hölder potential. As before, we denote by ν the reference measure (Section 12.1.1) and by μ the equilibrium state (Section 12.1.4) of the potential φ. Recall that $\mu = h\nu$, where the function h is bounded from zero and infinity (Corollary 12.1.9). In particular, $L^1(\mu) = L^1(\nu)$.

Given $b > 0$ and $\beta > 0$, we say that a function $g : M \to \mathbb{R}$ is (b,β)-Hölder if

$$|g(x) - g(y)| \leq b\, d(x,y)^{\beta} \quad \text{for any } x, y \in M. \tag{12.3.1}$$

We say that g is β-Hölder if it is (b,β)-Hölder for some $b > 0$. Then we denote by $H_\beta(g)$ the smallest of such constants b. Moreover, fixing $\rho > 0$ as in (11.2.1), we denote by $H_{\beta,\rho}(g)$ the smallest constant b such that the inequality in (12.3.1) holds for any $x,y \in M$ with $d(x,y) < \rho$.

The correlations sequence of two functions g_1 and g_2, with respect to the invariant measure μ, was defined in (7.1.1):

$$C_n(g_1,g_2) = \left| \int (g_1 \circ f^n) g_2 \, d\mu - \int g_1 \, d\mu \int g_2 \, d\mu \right|.$$

We also consider a similar notion for the reference measure ν:

$$B_n(g_1,g_2) = \left| \int (g_1 \circ f^n) g_2 \, d\nu - \int g_1 \, d\mu \int g_2 \, d\nu \right|.$$

In this section we prove that these sequences decay exponentially.

Theorem 12.3.1 (Exponential convergence to equilibrium). *Given $\beta \in (0,\alpha]$, there exists $\Lambda < 1$ and for every β-Hölder function $g_2 : M \to \mathbb{C}$ there exists $K_1(g_2) > 0$ such that*

$$B_n(g_1,g_2) \le K_1(g_2)\Lambda^n \int |g_1| \, d\nu \quad \textit{for every } g_1 \in L_1(\nu) \textit{ and every } n \ge 1.$$

The proof is presented in Sections 12.3.1 through 12.3.3. It provides an explicit expression for the factor $K_1(g_2)$. Observe also that

$$B_n(g_1,g_2) = \left| \int g_1 \, d(f_*^n(g_2\nu)) - \int g_1 \, d\left(\mu \int g_2 \, d\nu\right) \right|.$$

Then, the conclusion of Theorem 12.3.1 may be interpreted as follows: *the iterates of any measure of the form $g_2\nu$ converge to the invariant measure $\mu \int g_2 \, d\nu$ exponentially fast.*

Theorem 12.3.2 (Exponential decay of correlations). *For every $\beta \in (0,\alpha]$ there exists $\Lambda < 1$ and for every β-Hölder function $g_2 : M \to \mathbb{C}$ there exists $K_2(g_2) > 0$ such that*

$$C_n(g_1,g_2) \le K_2(g_2)\Lambda^n \int |g_1| \, d\mu \quad \textit{for every } g_1 \in L_1(\mu) \textit{ and every } n \ge 1.$$

In particular, given any pair g_1 and g_2 of β-Hölder functions, there exists $K(g_1,g_2) > 0$ such that $C_n(g_1,g_2) \le K(g_1,g_2)\Lambda^n$ for every $n \ge 1$.

Proof. Recall that $\mu = h\nu$ and, according to Corollary 12.1.9, the function h is α-Hölder and satisfies $K_5^{-1} \le h \le K_5$ for some $K_5 > 0$. Hence (see Exercise 12.3.5), g_2 is β-Hölder if and only if $g_2 h$ is β-Hölder. Moreover,

$$\begin{aligned} C_n(g_1,g_2) &= \int (g_1 \circ f^n) g_2 \, d\mu - \int g_1 \, d\mu \int g_2 \, d\mu \\ &= \int (g_1 \circ f^n)(g_2 h) \, d\nu - \int g_1 \, d\mu \int (g_2 h) \, d\nu = B_n(g_1,g_2 h). \end{aligned}$$

Therefore, it follows from Theorem 12.3.1 that

$$C_n(g_1,g_2) \le K_1(g_2 h)\Lambda^n \int |g_1|\, d\nu \le K_1(g_2 h)/K_5 \Lambda^n \int |g_1|\, d\mu.$$

This proves the first part of the theorem, with $K_2(g_2) = K_1(g_2 h)/K_5$. The second part is an immediate consequence: if g_1 is β-Hölder then $g_1 \in L^1(\mu)$ and it suffices to take $K(g_1,g_2) = K_2(g_2)\int |g_1|\, d\mu$.

Before we move to prove Theorem 12.3.1, let us make a few quick comments. The issue of decay of correlations was already discussed in Section 7.4, from the viewpoint of the spectral gap property. Here we introduce a different approach. The proof of the theorem that we are going to present is based on the notion of *projective distance* associated with a cone, which was introduced by Garret Birkhoff [Bir67]. This tool allows us to obtain exponential convergence to equilibrium (which yields exponential decay of correlations, as we have just shown) without having to analyze the spectrum of the transfer operator. Incidentally, this can also be used to deduce that the spectral gap property does hold in the present context. We will come back to this theme near the end of Section 12.3.

12.3.1 Projective distances

Let E be a Banach space. We call a *cone* any convex subset C of E such that

$$tC \subset C \text{ for every } t > 0 \quad \text{and} \quad \bar{C} \cap (-\bar{C}) = \{0\}, \tag{12.3.2}$$

where $\bar{C}$ denotes the closure of C (previously we considered only closed cones but at this point it is convenient to loosen that requirement). Given $v_1, v_2 \in C$, define

$$\alpha(v_1,v_2) = \sup\{t > 0 : v_2 - t v_1 \in C\} \text{ and } \beta(v_1,v_2) = \inf\{s > 0 : s v_1 - v_2 \in C\}.$$

Figure 12.1 helps illustrate the geometric meaning of these numbers. By convention, $\alpha(v_1,v_2) = 0$ if $v_2 - t v_1 \notin C$ for every $t > 0$ and $\beta(v_1,v_2) = +\infty$ if $s v_1 - v_2 \notin C$ for every $s > 0$.

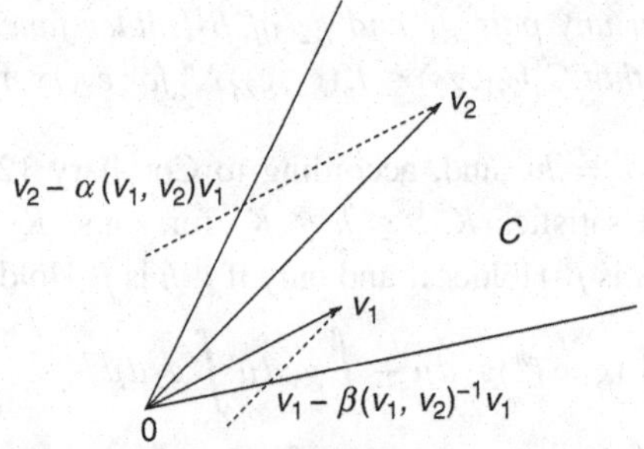

Figure 12.1. Defining the projective distance in a cone C

Note that $\alpha(v_1,v_2)$ is always finite. Indeed, $\alpha(v_1,v_2) = +\infty$ would mean that there exists a sequence $(t_n)_n \to +\infty$ with $v_2 - t_n v_1 \in C$ for every n. Then, $s_n = 1/t_n$ would be a sequence of positive numbers converging to zero and such that $s_n v_2 - v_1 \in C$ for every n. This would imply that $-v_1 \in \bar{C}$, which would contradict the second condition in (12.3.2). An analogous argument shows that $\beta(v_1,v_2)$ is always positive: $\beta(v_1,v_2) = 0$ would imply $-v_2 \in \bar{C}$.

Given any cone $C \subset E$ and any $v_1, v_2 \in C \setminus \{0\}$, we define

$$\theta(v_1,v_2) = \log \frac{\beta(v_1,v_2)}{\alpha(v_1,v_2)}, \tag{12.3.3}$$

with $\theta(v_1,v_2) = +\infty$ whenever $\alpha(v_1,v_2) = 0$ or $\beta(v_1,v_2) = +\infty$. The remarks in the previous paragraph ensure that $\theta(v_1,v_2)$ is always well defined. We call θ the *projective distance* associated with the cone C. This terminology is justified by the proposition that follows, which shows that θ defines a distance in the projective quotient of $C \setminus \{0\}$, that is, in the set of equivalence classes of the relation $\sim$ defined by $v_1 \sim v_2 \Leftrightarrow v_1 = tv_2$ for some $t > 0$.

Proposition 12.3.3. *If C is a cone then*

(i) $\theta(v_1,v_2) = \theta(v_2,v_1)$ *for any* $v_1, v_2 \in C$;
(ii) $\theta(v_1,v_2) + \theta(v_2,v_3) \geq \theta(v_1,v_3)$ *for any* $v_1, v_2, v_3 \in C$;
(iii) $\theta(v_1,v_2) \geq 0$ *for any* $v_1, v_2 \in C$;
(iv) $\theta(v_1,v_2) = 0$ *if and only if there exists* $t > 0$ *such that* $v_1 = tv_2$;
(v) $\theta(t_1 v_1, t_2 v_2) = \theta(v_1,v_2)$ *for any* $v_1, v_2 \in C$ *and* $t_1, t_2 > 0$.

Proof. If $\alpha(v_2,v_1) > 0$ then

$$\begin{aligned}\alpha(v_2,v_1) &= \sup\{t > 0 : v_1 - tv_2 \in C\} = \sup\left\{t > 0 : \frac{1}{t} v_1 - v_2 \in C\right\} \\ &= \left(\inf\{s > 0 : sv_1 - v_2 \in C\}\right)^{-1} = \beta(v_1,v_2)^{-1}.\end{aligned}$$

Moreover,

$$\begin{aligned}\alpha(v_2,v_1) = 0 &\Leftrightarrow v_1 - tv_2 \notin C \text{ for every } t > 0 \\ &\Leftrightarrow sv_1 - v_2 \notin C \text{ for every } s > 0 \Leftrightarrow \beta(v_1,v_2) = +\infty.\end{aligned}$$

Therefore, $\alpha(v_2,v_1) = \beta(v_1,v_2)^{-1}$ in all cases. Exchanging the roles of v_1 and v_2, we also get that $\beta(v_2,v_1) = \alpha(v_1,v_2)^{-1}$ for any $v_1, v_2 \in C$. Part (i) of the proposition is an immediate consequence of these observations.

Next, we claim that $\alpha(v_1,v_2)\alpha(v_2,v_3) \leq \alpha(v_1,v_3)$ for any $v_1, v_2, v_3 \in C$. This is obvious if $\alpha(v_1,v_2) = 0$ or $\alpha(v_2,v_3) = 0$; therefore, we may suppose that $\alpha(v_1,v_2) > 0$ and $\alpha(v_2,v_3) > 0$. Then, by definition, there exist increasing sequences of positive numbers $(r_n)_n \to \alpha(v_1,v_2)$ and $(s_n)_n \to \alpha(v_2,v_3)$ such that

$$v_2 - r_n v_1 \in C \quad \text{and} \quad v_3 - s_n v_2 \in C \quad \text{for every } n \geq 1.$$

Since C is convex, it follows that $v_3 - s_n r_n v_1 \in C$ and so $s_n r_n \le \alpha(v_1, v_3)$, for every $n \ge 1$. Passing to the limit as $n \to +\infty$, we get the claim. An analogous argument shows that $\beta(v_1, v_2)\beta(v_2, v_3) \ge \beta(v_1, v_3)$ for any $v_1, v_2, v_3 \in C$. Part (ii) of the proposition follows immediately from these inequalities.

Part (iii) means, simply, that $\alpha(v_1, v_2) \le \beta(v_1, v_2)$ for any $v_1, v_2 \in C$. To prove this fact, consider $t > 0$ and $s > 0$ such that $v_2 - tv_1 \in C$ and $sv_1 - v_2 \in C$. Then, by convexity, $(s-t)v_1 \in C$. If $s-t$ were negative, then we would have $-v_1 \in C$, which would contradict the last part of (12.3.2). Therefore, $s \ge t$ for any t and s as above. This implies that $\alpha(v_1, v_2) \le \beta(v_1, v_2)$.

Let $v_1, v_2 \in C$ be such that $\theta(v_1, v_2) = 0$. Then, $\alpha(v_1, v_2) = \beta(v_1, v_2) = \gamma$ for some $\gamma \in (0, +\infty)$. Hence, there exist an increasing sequence $(t_n)_n \to \gamma$ and a decreasing sequence $(s_n)_n \to \gamma$ with

$$v_2 - t_n v_1 \in C \quad \text{and} \quad s_n v_1 - v_2 \in C \quad \text{for every } n \ge 1.$$

Writing $v_2 - t_n v_1 = (v_2 - \gamma v_1) + (\gamma - t_n)v_1$, we conclude that $v_2 - \gamma v_1$ is in the closure $\bar{C}$ of C. Analogously, $\gamma v_1 - v_2 \in \bar{C}$. By the second part of (12.3.2), it follows that $v_2 - \gamma v_1 = 0$. This proves part (iv) of the proposition.

Finally, consider any $t_1, t_2 > 0$ and $v_1, v_2 \in C$. By definition,

$$\alpha(t_1 v_1, t_2 v_2) = \frac{t_2}{t_1}\alpha(v_1, v_2) \quad \text{and} \quad \beta(t_1 v_1, t_2 v_2) = \frac{t_2}{t_1}\beta(v_1, v_2).$$

Hence, $\theta(t_1 v_1, t_2 v_2) = \theta(v_1, v_2)$, as stated in part (v) of the proposition.

Example 12.3.4. Consider the cone $C = \{(x,y) \in E : y > |x|\}$ in $E = \mathbb{R}^2$. The projective quotient of C may be identified with the interval $(-1, 1)$, through $(x, 1) \mapsto x$. Given $-1 < x_1 \le x_2 < 1$, we have:

$$\begin{aligned} \alpha((x_1, 1), (x_2, 1)) &= \sup\{t > 0 : (x_2, 1) - t(x_1, 1) \in C\} \\ &= \sup\{t > 0 : 1 - t \ge |x_2 - tx_1|\} = \frac{1 - x_2}{1 - x_1}, \end{aligned}$$

$$\text{and} \quad \beta((x_1, 1), (x_2, 1)) = \frac{x_2 + 1}{x_1 + 1}.$$

Therefore,

$$\theta((x_1, 1), (x_2, 1)) = \log R(-1, x_1, x_2, 1), \tag{12.3.4}$$

where

$$R(a, b, c, d) = \frac{(c-a)(d-b)}{(b-a)(d-c)}$$

denotes the *cross-ratio* of four positive numbers $a < b \le c < d$.

In Exercise 12.3.2 we invite the reader to check a similar fact when the interval is replaced by the unit disk $\mathbb{D} = \{z \in \mathbb{C} : |z| < 1\}$.

Example 12.3.5. Let $E = C^0(M)$ be the space of continuous functions on a compact metric space M. Consider the cone $C_+ = \{g \in E : g(x) > 0 \text{ for } x \in M\}$. For any $g_1, g_2 \in C_+$,

$$\begin{aligned}\alpha(g_1,g_2) &= \sup\left\{t > 0 : (g_2 - tg_1)(x) > 0 \text{ for every } x \in M\right\}\\ &= \inf\left\{\frac{g_2}{g_1}(x) : x \in M\right\}\end{aligned}$$

$$\text{and} \quad \beta(g_1,g_2) = \sup\left\{\frac{g_2}{g_1}(x) : x \in M\right\}.$$

Therefore,

$$\theta(g_1,g_2) = \log\frac{\sup(g_2/g_1)}{\inf(g_2/g_1)} = \log\sup\left\{\frac{g_2(x)g_1(y)}{g_1(x)g_2(y)} : x,y \in M\right\}. \tag{12.3.5}$$

This projective distance is complete (Exercise 12.3.3) but that is not always the case (Exercise 12.3.4).

Next, we observe that the projective distance depends monotonically on the cone. Indeed, let C_1 and C_2 be two cones with $C_1 \subset C_2$ and let $\alpha_i(\cdot,\cdot)$, $\beta_i(\cdot,\cdot)$, $\theta_i(\cdot,\cdot)$, $i = 1,2$ be the corresponding functions, as defined previously. It is clear from the definitions that, given any $v_1, v_2 \in C_2$,

$$\alpha_1(v_1,v_2) \le \alpha_2(v_1,v_2) \quad \text{and} \quad \beta_1(v_1,v_2) \ge \beta_2(v_1,v_2)$$

and, consequently, $\theta_1(v_1,v_2) \ge \theta_2(v_1,v_2)$.

More generally, let C_1 and C_2 be cones in Banach spaces E_1 and E_2, respectively, and let $L : E_1 \to E_2$ be a linear operator such that $L(C_1) \subset C_2$. Then,

$$\begin{aligned}\alpha_1(v_1,v_2) &= \sup\{t > 0 : v_2 - tv_1 \in C_1\}\\ &\le \sup\{t > 0 : L(v_2 - tv_1) \in C_2\}\\ &= \sup\{t > 0 : L(v_2) - tL(v_1) \in C_2\} = \alpha_2(L(v_1),L(v_2))\end{aligned}$$

and, analogously, $\beta_1(v_1,v_2) \ge \beta_2(L(v_1),L(v_2))$. Consequently,

$$\theta_2(L(v_1),L(v_2)) \le \theta_1(v_1,v_2) \quad \text{for any } v_1,v_2 \in C_1. \tag{12.3.6}$$

Of course, the inequality (12.3.6) is not strict, in general. However, according to the next proposition, one does have a strict inequality whenever $L(C_1)$ has finite θ_2-diameter in C_2; actually, in that case L is a contraction with respect to the projective distances θ_1 and θ_2. Recall that the hyperbolic tangent is defined by

$$\tanh x = \frac{1 - e^{-2x}}{1 + e^{-2x}} \quad \text{for every } x \in \mathbb{R}.$$

Keep in mind that the function tanh takes values in the interval $(0,1)$.

Proposition 12.3.6. *Let C_1 and C_2 be cones in Banach spaces E_1 and E_2, respectively, and let $L : E_1 \to E_2$ be a linear operator such that $L(C_1) \subset C_2$. Suppose that $D = \sup\{\theta_2(L(v_1), L(v_2)) : v_1, v_2 \in C_1\}$ is finite. Then*

$$\theta_2(L(v_1), L(v_2)) \le \tanh\left(\frac{D}{4}\right)\theta_1(v_1, v_2) \quad \textit{for any } v_1, v_2 \in C.$$

Proof. Let $v_1, v_2 \in C_1$. It is no restriction to suppose that $\alpha_1(v_1, v_2) > 0$ and $\beta_1(v_1, v_2) < +\infty$, otherwise $\theta_1(v_1, v_2) = +\infty$ and there is nothing to prove. Then there exist an increasing sequence $(t_n)_n \to \alpha_1(v_1, v_2)$ and a decreasing sequence $(s_n)_n \to \beta_1(v_1, v_2)$ such that

$$v_2 - t_n v_1 \in C_1 \quad \text{and} \quad s_n v_1 - v_2 \in C_1.$$

In particular, $\theta_2(L(v_2 - t_n v_1), L(s_n v_1 - v_2)) \le D$ for every $n \ge 1$. Fix any $D_0 > D$. Then we may choose positive numbers T_n and S_n such that

$$\begin{aligned} &L(s_n v_1 - v_2) - T_n L(v_2 - t_n v_1) \in C_2 \quad \text{and} \\ &S_n L(v_2 - t_n v_1) - L(s_n v_1 - v_2) \in C_2, \end{aligned} \tag{12.3.7}$$

and $\log(S_n/T_n) \le D_0$ for every $n \ge 1$. The first part of (12.3.7) gives that

$$(s_n + t_n T_n) L(v_1) - (1 + T_n) L(v_2) \in C_2$$

and, by definition of $\beta_2(\cdot, \cdot)$, this implies that

$$\beta_2(L(v_1), L(v_2)) \le \frac{s_n + t_n T_n}{1 + T_n}.$$

Analogously, the second part of (12.3.7) implies that

$$\alpha_2(L(v_1), L(v_2)) \ge \frac{s_n + t_n S_n}{1 + S_n}.$$

Therefore, $\theta_2(L(v_1), L(v_2))$ cannot exceed

$$\log\left(\frac{s_n + t_n T_n}{1 + T_n} \cdot \frac{1 + S_n}{s_n + t_n S_n}\right) = \log\left(\frac{s_n/t_n + T_n}{1 + T_n} \cdot \frac{1 + S_n}{s_n/t_n + S_n}\right).$$

The last term may be rewritten as

$$\begin{aligned} \log\left(\frac{s_n}{t_n} + T_n\right) - \log(1 + T_n) - \log\left(\frac{s_n}{t_n} + S_n\right) + \log(1 + S_n) = \\ = \int_0^{\log(s_n/t_n)} \left(\frac{e^x\,dx}{e^x + T_n} - \frac{e^x\,dx}{e^x + S_n}\right), \end{aligned}$$

and this expression is less than or equal to

$$\sup_{x>0} \frac{e^x(S_n - T_n)}{(e^x + T_n)(e^x + S_n)} \log\left(\frac{s_n}{t_n}\right).$$

Now we use the following elementary facts:

$$\sup_{y>0} \frac{y(S_n - T_n)}{(y + T_n)(y + S_n)} = \frac{1 - \sqrt{T_n/S_n}}{1 + \sqrt{T_n/S_n}} \le \frac{1 - e^{-D_0/2}}{1 + e^{-D_0/2}} = \tanh\frac{D_0}{4}.$$

Indeed, the supremum is attained for $y = \sqrt{S_n T_n}$ and the inequality is a consequence of the fact that $\log(S_n/T_n) \leq D_0$. This proves that

$$\theta_2(L(v_1), L(v_2)) \leq \tanh\left(\frac{D_0}{4}\right) \log\left(\frac{s_n}{t_n}\right).$$

Note also that $\theta(v_1, v_2) = \lim_n \log(s_n/t_n)$, due to our choice of s_n and t_n. Hence, taking the limit when $n \to \infty$ and then making $D_0 \to D$, we obtain the conclusion of the proposition.

Example 12.3.7. Let C_+ be the cone of positive continuous functions in M. For each $L > 1$, let $C(L) = \{g \in C_+ : \sup|g| \leq L \inf|g|\}$. Then, $C(L)$ has finite diameter in C_+, for every $L > 1$. Indeed, we have seen in Example 12.3.5 that the projective distance θ associated with C_+ is given by

$$\theta(g_1, g_2) = \log\sup\left\{\frac{g_2(x)g_1(y)}{g_1(x)g_2(y)} : x, y \in M\right\}.$$

In particular, $\theta(g_1, g_2) \leq 2\log L$ for any $g_1, g_2 \in C(L)$.

12.3.2 Cones of Hölder functions

Let $f : M \to M$ be a topologically exact expanding map and $\rho > 0$ and $\sigma > 1$ be the constants in the definition (11.2.1). Let $\mathcal{L} : C^0(M) \to C^0(M)$ be the transfer operator associated with a Hölder potential $\varphi : M \to M$. Fix constants $K_0 > 0$ and $\alpha > 0$ such that

$$|\varphi(x) - \varphi(y)| \leq K_0 d(x, y)^\alpha \quad \text{for any } x, y \in M.$$

Given $b > 0$ and $\beta > 0$, we denote by $C(b, \beta)$ the set of positive functions $g \in C^0(M)$ whose logarithm is (b, β)-Hölder on balls of radius ρ, that is, such that

$$|\log g(x) - \log g(y)| \leq b d(x, y)^\beta \quad \text{whenever } d(x, y) < \rho. \tag{12.3.8}$$

Lemma 12.3.8. *For any $b > 0$ and $\beta > 0$, the set $C(b, \beta)$ is a cone in the space $E = C^0(M)$ and the corresponding projective distance is given by*

$$\theta(g_1, g_2) = \log\frac{\beta(g_1, g_2)}{\alpha(g_1, g_2)},$$

where $\alpha(g_1, g_2)$ is the infimum and $\beta(g_1, g_2)$ is the supremum of the set

$$\left\{\frac{g_2}{g_1}(x), \frac{\exp(bd(x,y)^\beta)g_2(x) - g_2(y)}{\exp(bd(x,y)^\beta)g_1(x) - g_1(y)} : x \neq y \text{ and } d(x, y) < \rho\right\}.$$

Proof. It is clear that $g \in C$ implies $tg \in C$ for every $t > 0$. Moreover, the closure of C is contained in the set of non-negative functions and so $-\bar{C} \cap \bar{C}$ contains only the zero function. Now, to conclude that C is a cone, we only have to check that it is convex. Consider any $g_1, g_2 \in C(b,\beta)$. The definition (12.3.8) means that

$$\exp(-bd(x,y)^{\beta}) \leq \frac{g_i(x)}{g_i(y)} \leq \exp(bd(x,y)^{\beta})$$

for $i = 1,2$ and any $x, y \in M$ with $d(x,y) < \rho$. Then, given $t_1, t_2 > 0$,

$$\exp(-bd(x,y)^{\beta}) \leq \frac{t_1 g_1(x) + t_2 g_2(x)}{t_1 g_1(y) + t_2 g_2(y)} \leq \exp(bd(x,y)^{\beta})$$

for any $x, y \in M$ with $d(x,y) < \rho$. Hence, $t_1 g_1 + t_2 g_2$ is in $C(b,\beta)$.

We proceed to calculate the projective distance. By definition, $\alpha(g_1, g_2)$ is the supremum of all the numbers $t > 0$ satisfying the following three conditions:

$$(g_2 - tg_1)(x) > 0 \Leftrightarrow t < \frac{g_2}{g_1}(x)$$

$$\frac{(g_2 - tg_1)(x)}{(g_2 - tg_1)(y)} \leq \exp(bd(x,y)^{\beta}) \Leftrightarrow t \leq \frac{\exp(bd(x,y)^{\beta})g_2(y) - g_2(x)}{\exp(bd(x,y)^{\beta})g_1(y) - g_1(x)}$$

$$\frac{(g_2 - tg_1)(x)}{(g_2 - tg_1)(y)} \geq \exp(-bd(x,y)^{\beta}) \Leftrightarrow t \leq \frac{\exp(bd(x,y)^{\beta})g_2(x) - g_2(y)}{\exp(bd(x,y)^{\beta})g_1(x) - g_1(y)}$$

for any $x, y \in M$ with $x \neq y$ and $d(x,y) < \rho$. Hence, $\alpha(g_1, g_2)$ is equal to

$$\inf\left\{\frac{g_2(x)}{g_1(x)}, \frac{\exp(bd(x,y)^{\beta})g_2(x) - g_2(y)}{\exp(bd(x,y)^{\beta})g_1(x) - g_1(y)} : x \neq y \text{ and } d(x,y) < \rho\right\}.$$

Analogously, $\beta(g_1, g_2)$ is the supremum of this same set.

The crucial fact that makes the proof of Theorem 12.3.1 work is that the transfer operator tends to improve the regularity of functions or, more precisely, their Hölder constants. The next proposition is a concrete manifestation of this fact:

Lemma 12.3.9. *For each $\beta \in (0, \alpha]$ there exists a constant $\lambda_0 \in (0,1)$ such that $\mathcal{L}(C(b,\beta)) \subset C(\lambda_0 b, \beta)$ for every b sufficiently large (depending on β).*

Proof. It follows directly from the expression of the transfer operator in (12.1.1) that $\mathcal{L}g$ is positive whenever g is positive. Therefore, we only have to check the second condition in the definition of $C(\lambda_0 b, \beta)$. Consider $y_1, y_2 \in M$ with $d(y_1, y_2) < \rho$. The expression (12.1.2) gives that

$$\mathcal{L}g(y_i) = \sum_{j=1}^{k} e^{\varphi(x_{i,j})} g(x_{i,j})$$

for $i = 1,2$, where the points $x_{i,j} \in f^{-1}(y_i)$ satisfy $d(x_{1i}, x_{2i}) \leq \sigma^{-1} d(y_1, y_2)$ for every $1 \leq j \leq k$. By hypothesis, φ is (K_0, α)-Hölder. Since we suppose that

$\beta \le \alpha$, it follows that φ is (K,β)-Hölder, with $K = K_0(\operatorname{diam} M)^{\alpha-\beta}$. Therefore,

$$(\mathcal{L}g)(y_1) = \sum_{i=1}^{k} e^{\varphi(x_{1,i})} g(x_{1,i}) = \sum_{i=1}^{k} e^{\varphi(x_{2,i})} g(x_{2,i}) \frac{g(x_{1,i}) e^{\varphi(x_{1,i})}}{g(x_{2,i}) e^{\varphi(x_{2,i})}}$$

$$\le \sum_{i=1}^{k} e^{\varphi(x_{2,i})} g(x_{2,i}) \exp\left(bd(x_{1,i},x_{2,i})^{\beta} + Kd(x_{1,i},x_{2,i})^{\beta}\right)$$

$$\le (\mathcal{L}g)(y_2) \exp\left((b+K)\sigma^{-\beta} d(y_1,y_2)^{\beta}\right)$$

for every $g \in C(b,\beta)$. Fix $\lambda_0 \in (\sigma^{-\beta},1)$. For every b sufficiently large, $(b+K)\sigma^{-\beta} \le b\lambda_0$. Then the previous relation gives that

$$(\mathcal{L}g)(y_1) \le (\mathcal{L}g)(y_2) \exp(\lambda_0 bd(y_1,y_2)^{\beta}),$$

for any $y_1, y_2 \in M$ with $d(y_1,y_2) < \rho$. Exchanging the roles of y_1 and y_2, we obtain the other inequality.

Next, we use the family of cones $C(L)$ introduced in Example 12.3.7:

Lemma 12.3.10. *There exists $N \ge 1$ and for every $\beta > 0$ and every $b > 0$ there exists $L > 1$ satisfying $\mathcal{L}^N(C(b,\beta)) \subset C(L)$.*

Proof. By hypothesis, f is topologically exact. Hence, there exists $N \ge 1$ such that $f^N(B(z,\rho)) = M$ for every $z \in M$. Fix N once and for all. Given $g \in C(b,\beta)$, consider any point $z \in M$ such that $g(z) = \sup g$. Consider $y_1, y_2 \in M$. On the one hand,

$$\mathcal{L}^N g(y_1) = \sum_{x \in f^{-N}(y_1)} e^{\varphi_N(x)} g(x) \le \operatorname{degree}(f^N) e^{N \sup|\varphi|} g(z).$$

On the other hand, by the choice of N, there exists $x \in B(z,\rho)$ such that $f^N(x) = y_2$. Then,

$$\mathcal{L}^N g(y_2) \ge e^{\varphi_N(x)} g(x) \ge e^{-N\sup|\varphi|} e^{-bd(x,z)^{\beta}} g(z) \ge e^{-N\sup|\varphi| - b\rho^{\beta}} g(z).$$

Since y_1 and y_2 are arbitrary, this proves that

$$\frac{\sup \mathcal{L}^N g}{\inf \mathcal{L}^N g} \le \operatorname{degree}(f^N) e^{2N\sup|\varphi| + b\rho^{\beta}}.$$

Now it suffices to take L equal to the expression on the right-hand side of this inequality.

Combining Lemmas 12.3.9 and 12.3.10, we get that there exists $N \ge 1$ and, given $\beta \in (0,\alpha]$ there exists $\lambda_0 \in (0,1)$ such that, for every $b > 0$ sufficiently large (depending on N and β) there exists $L > 1$, satisfying

$$\mathcal{L}^N(C(b,\beta)) \subset C(\lambda_0^N b, \beta) \cap C(L). \tag{12.3.9}$$

In what follows, we write $C(c,\beta,R) = C(c,\beta) \cap C(R)$ for any $c > 0$, $\beta > 0$ and $R > 1$.

Lemma 12.3.11. *For every $c \in (0,b)$ and $R > 1$, the set $C(c,\beta,R) \subset C(b,\beta)$ has finite diameter with respect to the projective distance of the cone $C(b,\beta)$.*

Proof. We use the expression of θ given by Lemma 12.3.8. On the one hand, the hypothesis that $g_1, g_2 \in C(c,\beta)$ ensures that

$$\begin{aligned}\frac{\exp\left(bd(x,y)^\beta\right)g_2(x) - g_2(y)}{\exp\left(bd(x,y)^\beta\right)g_1(x) - g_1(y)} &= \frac{g_2(x)}{g_1(x)}\frac{1-\exp\left(-bd(x,y)^\beta\right)\left(g_2(y)/g_2(x)\right)}{1-\exp\left(-bd(x,y)^\beta\right)\left(g_1(y)/g_1(x)\right)}\\ &\geq \frac{g_2(x)}{g_1(x)}\frac{1-\exp\left(-(b-c)d(x,y)^\beta\right)}{1-\exp\left(-(b+c)d(x,y)^\beta\right)}\\ &\geq \frac{g_2(x)}{g_1(x)}\frac{1-\exp\left(-(b-c)\rho^\beta\right)}{1-\exp\left(-(b+c)\rho^\beta\right)}\end{aligned}$$

for any $x,y \in M$ with $d(x,y) < \rho$. Denote by r the value of the last fraction on the right-hand side. Then, observing that $r \in (0,1)$,

$$\alpha(g_1,g_2) \geq \inf\left\{\frac{g_2(x)}{g_1(x)}, r\frac{g_2(x)}{g_1(x)} : x \in M\right\} = r\inf\left\{\frac{g_2(x)}{g_1(x)} : x \in M\right\} \geq r\frac{\inf g_2}{\sup g_1}.$$

Analogously,

$$\beta(g_1,g_2) \leq \sup\left\{\frac{g_2(x)}{g_1(x)}, \frac{1}{r}\frac{g_2(x)}{g_1(x)} : x \in M\right\} = \frac{1}{r}\sup\left\{\frac{g_2(x)}{g_1(x)} : x \in M\right\} \leq \frac{1}{r}\frac{\sup g_2}{\inf g_1}.$$

On the other hand, the hypothesis that $g_1, g_2 \in C(R)$ gives that

$$\frac{\sup g_2}{\inf g_1} \leq R^2\frac{\inf g_2}{\sup g_1}.$$

Combining these three inequalities, we conclude that $\theta(g_1,g_2) \leq \log(R^2/r^2)$ for any $g_1, g_2 \in C(c,\beta,R)$.

Corollary 12.3.12. *There exists $N \geq 1$ such that for every $\beta \in (0,\alpha]$ and every $b > 0$ sufficiently large there exists $\Lambda_0 < 1$ satisfying*

$$\theta(\mathcal{L}^N g_1, \mathcal{L}^N g_2) \leq \Lambda_0 \theta(g_1,g_2) \quad \textit{for any } g_1,g_2 \in C(b,\beta).$$

Proof. Take $N \geq 1$, $\lambda_0 \in (0,1)$ and $L > 1$ as in (12.3.9) and consider

$$c = \lambda_0^N b \quad \text{and} \quad R = L. \tag{12.3.10}$$

Then $\mathcal{L}^N(C(b,\beta)) \subset C(c,\beta,R)$ and it follows from Lemma 12.3.11 that the diameter D of the image $\mathcal{L}^N(C(b,\beta))$ with respect to the projective distance θ is finite. Take $\Lambda_0 = \tanh(D/4)$. Now the conclusion of the corollary is an immediate application of Proposition 12.3.6.

12.3.3 Exponential convergence

Fix $N \geq 1$, $\beta \in (0,\alpha]$, $b > 0$ and $L > 1$ as in Corollary 12.3.12, and then consider $c > 0$ and $R > 1$ given by (12.3.10). As before, we denote by h the positive eigenfunction (Lemma 12.1.11) and by λ the spectral radius (Corollary 12.1.15) of the transfer operator $\mathcal{L}$. Recall that h is α-Hölder and bounded from zero and infinity. Therefore, up to increasing the constants b and L, if necessary, we may assume that $h \in C(c,\beta,R)$.

The next lemma follows directly from the previous considerations and is a significant step toward the estimate in Theorem 12.3.1. We continue denoting by $\|\cdot\|$ the norm defined in $C^0(M)$ by $\|\phi\| = \sup\{|\phi(x)| : x \in M\}$.

Lemma 12.3.13. *There exists $C > 0$ and $\Lambda \in (0,1)$ such that*

$$\|\lambda^{-n}\mathcal{L}^n g - h\int g\,d\nu\| \leq C\Lambda^n \int g\,d\nu \quad \textit{for } g \in C(c,\beta,R) \textit{ and } n \geq 1.$$

Proof. Let $g \in C(c,\beta,R)$. In particular, $g > 0$ and so $\int g\,d\nu > 0$. The conclusion of the lemma is not affected when we multiply g by any positive number. Hence, it is no restriction to suppose that $\int g\,d\nu = 1$. Then,

$$\int \lambda^{-n}\mathcal{L}^n g\,d\nu = \int \lambda^{-n} g\,d(\mathcal{L}^{*n}\nu) = \int g\,d\nu = 1 = \int h\,d\nu$$

and, hence, $\inf(\lambda^{-n}\mathcal{L}^n g/h) \leq 1 \leq \sup(\lambda^{-n}\mathcal{L}^n g/h)$ for every $n \geq 1$. Now, it follows from the expressions in Lemma 12.3.8 that

$$\alpha(\lambda^{-jN}\mathcal{L}^{jN}g,h) \leq \inf\frac{\lambda^{-jN}\mathcal{L}^{jN}g}{h} \leq 1,$$
$$\beta(\lambda^{-jN}\mathcal{L}^{jN}g,h) \geq \sup\frac{\lambda^{-jN}\mathcal{L}^{jN}g}{h} \geq 1.$$

Consequently,

$$\theta(\lambda^{-jN}\mathcal{L}^{jN}g,h) \geq \log\beta(\lambda^{-jN}\mathcal{L}^{jN}g,h) \geq \log\sup\frac{\lambda^{-jN}\mathcal{L}^{jN}g}{h},$$
$$\theta(\lambda^{-jN}\mathcal{L}^{jN}g,h) \geq -\log\alpha(\lambda^{-jN}\mathcal{L}^{jN}g,h) \geq -\log\inf\frac{\lambda^{-jN}\mathcal{L}^{jN}g}{h}$$

for every $j \geq 0$. Now let D be the diameter of $C(c,\beta,R)$ with respect to the projective distance θ (Lemma 12.3.11). By Proposition 12.3.3 and Corollary 12.3.12,

$$\theta(\lambda^{-jN}\mathcal{L}^{jN}g,h) = \theta(\mathcal{L}^{jN}g,\mathcal{L}^{jN}h) \leq \Lambda_0^j\theta(g,h) \leq \Lambda_0^j D$$

for every $j \geq 0$. Combining this with the previous two inequalities,

$$\exp(-\Lambda_0^j D) \leq \inf\frac{\lambda^{-jN}\mathcal{L}^{jN}g}{h} \leq \sup\frac{\lambda^{-jN}\mathcal{L}^{jN}g}{h} \leq \exp(\Lambda_0^j D)$$

for every $j \geq 0$. Fix $C_1 > 0$ such that $|e^x - 1| \leq C_1|x|$ whenever $|x| \leq D$. Then the previous relation implies that

$$\left|\lambda^{-jN}\mathcal{L}^{jN}g(x) - h(x)\right| \leq h(x)C_1\Lambda_0^j D \quad \text{for every } x \in M \text{ and } j \geq 0. \quad (12.3.11)$$

Take $C_2 = C_1 D \sup h$ and $\Lambda = \Lambda_0^{1/N}$. The inequality (12.3.11) means that

$$\|\lambda^{-jN}\mathcal{L}^{jN}g - h\| \leq C_2\Lambda^{jN} \quad \text{for every } j \geq 1.$$

Given any $n \geq 1$, write $n = jN + r$ with $j \geq 0$ and $0 \leq r < N$. Since the transfer operator $\mathcal{L} : C^0(M) \to C^0(M)$ is continuous and $\mathcal{L}h = \lambda h$,

$$\|\lambda^{-n}\mathcal{L}^n g - h\| = \|\lambda^{-r}\mathcal{L}^r(\lambda^{-jN}\mathcal{L}^{jN}g - h)\| \leq (\|\mathcal{L}\|/\lambda)^r\|\lambda^{-jN}\mathcal{L}^{jN}g - h\|.$$

Combining the last two inequalities,

$$\|\lambda^{-n}\mathcal{L}^n g - h\| \leq (\|\mathcal{L}\|/\lambda)^r C_2\Lambda^{n-r}.$$

This proves the conclusion of the lemma, as long as we take $C \geq C_2(\|\mathcal{L}\|/(\lambda\Lambda))^r$ for every $0 \leq r < N$.

Now we are ready to prove Theorem 12.3.1:

Proof. Start by considering $g_2 \in C(c,\beta,R)$. Using the identity in Exercise 12.1.4 and recalling that $\mu = h\nu$,

$$\begin{aligned} B_n(g_1,g_2) &= \left|\int g_1\left(\lambda^{-n}\mathcal{L}^n g_2 - h\int g_2\,d\nu\right)d\nu\right| \\ &\leq \left\|\lambda^{-n}\mathcal{L}^n g_2 - h\int g_2\,d\nu\right\|\int |g_1|\,d\nu. \end{aligned}$$

Therefore, using Lemma 12.3.13,

$$B_n(g_1,g_2) \leq C\Lambda^n\int |g_1|\,d\nu\int g_2\,d\nu. \quad (12.3.12)$$

Now let $g_2 : M \to \mathbb{R}$ be any β-Hölder function and $H = H_\beta(g_2)$. Write $g_2 = g_2^+ - g_2^-$ with

$$g_2^+ = \frac{1}{2}(|g_2| + g_2) + B \quad \text{and} \quad g_2^- = \frac{1}{2}(|g_2| - g_2) + B,$$

with B defined by $B = \max\{H/c, \sup|g_2|/(R-1)\}$. It is clear that the functions $g_2^\pm$ are positive: $g_2^\pm \geq B > 0$. Moreover, they are (H,β)-Hölder:

$$|g_2^\pm(x) - g_2^\pm(y)| \leq |g_2(x) - g_2(y)| \leq Hd(x,y)^\beta,$$

for $x, y \in M$. Hence, using the mean value theorem and the fact that $B \geq H/c$,

$$\left|\log g_2^\pm(x) - \log g_2^\pm(y)\right| \leq \frac{|g_2^\pm(x) - g_2^\pm(y)|}{B} \leq \frac{Hd(x,y)^\beta}{B} \leq cd(x,y)^\beta.$$

Moreover, since $B \geq \sup|g_2|/(R-1)$,

$$\sup g_2^\pm \leq \sup|g_2| + B \leq RB \leq R\inf g_2^\pm.$$

Together with the previous relation, this means that $g_2^{\pm} \in C(c,\beta,R)$. Then, we may apply (12.3.12) to both functions:

$$B_n(g_1,g_2^{\pm}) \le C\Lambda^n \int |g_1|\,d\nu \int g_2^{\pm}\,d\nu$$

and, consequently,

$$\begin{aligned} B_n(g_1,g_2) &\le B_n(g_1,g_2^{+}) + B_n(g_1,g_2^{-}) \\ &\le C\Lambda^n \int |g_1|\,d\nu \int (g_2^{+} + g_2^{-})\,d\nu. \end{aligned} \tag{12.3.13}$$

Moreover, by the definition of $g_2^{\pm}$,

$$\begin{aligned} \int (g_2^{+} + g_2^{-})\,d\nu &= \int |g_2|\,d\nu + 2B \le \int |g_2|\,d\nu + \frac{2H}{c} + \frac{2\sup|g_2|}{R-1} \\ &\le \frac{2}{c}H_\beta(g_2) + \frac{R+1}{R-1}\sup|g_2|. \end{aligned} \tag{12.3.14}$$

Take $C_1 = C\max\{2/c,(R+1)/(R-1)\}$ and define

$$K_1(g_2) = 2C_1\big(\sup|g_2| + H_\beta(g_2)\big).$$

The relations (12.3.13) and (12.3.14) give that

$$B_n(g_1,g_2) \le C_1\Lambda^n \int |g_1|\,d\nu\big(H_\beta(g_2) + \sup|g_2|\big) \le \frac{1}{2}K_1(g_2)\Lambda^n \int |g_1|\,d\nu.$$

This closes the proof of the theorem in the case when g_2 is a real function.

The general (complex) case follows easily. Note that $K_1(\Re g_2) \le K_1(g_2)$, because $\sup|\Re g_2| \le \sup|g_2|$ and $H_\beta(\Re g_2) \le H_\beta(g_2)$. Analogously, $K_1(\Im g_2) \le K_1(g_2)$. Therefore, the previous argument yields

$$\begin{aligned} B_n(g_1,g_2) &\le B_n(g_1,\Re g_2) + B_n(g_1,\Im g_2) \le \frac{1}{2}\big(K_1(\Re g_2) + K_1(\Im g_2)\big)\Lambda^n \int |g_1|\,d\nu \\ &\le K_1(g_2)\Lambda^n \int |g_1|\,d\nu. \end{aligned}$$

This completes the proof of Theorem 12.3.1.

We close this section with a few comments about the spectral gap property. Let $C^\beta(M)$ be the vector space of β-Hölder functions $g: M \to \mathbb{C}$. We leave it to the reader (Exercise 12.3.6) to check the following facts:

(i) The function $\|g\|_{\beta,\rho} = \sup|g| + H_{\beta,\rho}(g)$ is a complete norm in $C^\beta(M)$.
(ii) $C^\beta(M)$ is invariant under the transfer operator: $\mathcal{L}(C^\beta(M)) \subset C^\beta(M)$.
(iii) The restriction $\mathcal{L}: C^\beta(M) \to C^\beta(M)$ is continuous with respect to the norm $\|\cdot\|_{\beta,\rho}$.

Note that $h \in C^\beta(M)$, since $\beta \le \alpha$. Define $V = \{g \in C^\beta(M) : \int g\,d\nu = 0\}$. Then $C^\beta(M) = V \oplus \mathbb{C}h$, because every function $g \in C^\beta(M)$ may be decomposed, in a unique way, as the sum of a function in V with a multiple of h:

$$g = \left(g - h \int g \, d\nu\right) + h \int g \, d\nu.$$

Moreover, the direct sum $C^\beta(M) = V \oplus \mathbb{C}h$ is invariant under the transfer operator. Indeed, if $g \in V$ then

$$g \in V \Rightarrow \int \mathcal{L} g \, d\nu = \int g \, d\mathcal{L}^* \nu = \lambda \int g \, d\nu = 0 \Rightarrow \mathcal{L} g \in V.$$

It follows that the spectrum of $\mathcal{L} : C^\beta(M) \to C^\beta(M)$ is the union of $\{\lambda\}$ with the restriction of $\mathcal{L}$ to the hyperplane V. In Exercise 12.3.8 we invite the reader to show that the spectral radius of $\mathcal{L} \mid V$ is strictly less than λ. Consequently, $\mathcal{L} : C^\beta(M) \to C^\beta(M)$ has the spectral gap property.

The book of Viviane Baladi [Bal00] contains an in-depth presentation of the spectral theory of transfer operators and its connections to the issue of decay of correlations, for differentiable (or piecewise differentiable) expanding maps and also for uniformly hyperbolic diffeomorphisms.

12.3.4 Exercises

12.3.1. Show that the cross-ratio $R(a,x,y,b)$ is invariant under every Möbius automorphism of the real line, that is, $R(\phi(a),\phi(b),\phi(c),\phi(d)) = R(a,b,c,d)$ for any $a < b \le c < d$ and every transformation of the form $\phi(x) = (\alpha x + \beta)/(\gamma x + \delta)$ with $\alpha\delta - \beta\gamma \neq 0$.

12.3.2. Consider the cone $C = \{(z,s) \in \mathbb{C} \times \mathbb{R} : s > |z|\}$. Its projective quotient may be identified with the unit disk $\mathbb{D} = \{z \in \mathbb{C} : |z| < 1\}$ through $(z,1) \mapsto z$. Let d be the distance induced in $\mathbb{D}$, through this identification, by the projective distance of C. Show that d coincides with the *Cayley–Klein distance* Δ, which is defined by

$$\Delta(p,q) = \log \frac{|aq|\,|pb|}{|ap|\,|bq|}, \quad \text{for } p,q \in \mathbb{D},$$

where a and b are the points where the straight line through p and q intersects the boundary of the disk, denoted in such a way that p is between a and q and q is between p and b. [Observation: The Cayley–Klein distance is related to the Poincaré distance in the disk through the map $z \mapsto (2z)/1 + |z|^2$.]

12.3.3. Show that the projective distance associated with the cone C_+ in Example 12.3.7 is complete, in the following sense: with respect to the projective distance, every Cauchy sequence $(g_n)_n$ converges to some element of C_+. Moreover, if we normalize the functions (for example, fixing any probability measure η on M and requiring that $\int g_n \, d\eta = 1 = \int g \, d\eta$ for every n), then $(g_n)_n$ converges uniformly to g.

12.3.4. Let M be a compact manifold and C_1 be the cone of positive differentiable functions in M. Show that the corresponding projective distance θ_1 is *not* complete.

12.3.5. Check that if $g_1, g_2 : M \to \mathbb{R}$ are β-Hölder functions, $\theta : M \to M$ is an L-Lipschitz transformation and η is a probability measure on M then:

(a) $H_\beta(g_1g_2) \le \sup|g_1|H_\beta(g_2) + \sup|g_2|H_\beta(g_1)$;
(b) $\int |g_1|\,d\eta \le \sup|g_1| \le \int |g_1|\,d\eta + H_\beta(g_1)(\operatorname{diam} M)^\beta$;
(c) $H_\beta(g \circ \theta) \le L^\beta H_\beta(g)$.

Moreover, the claim in (a) remains true if we replace H_β by $H_{\beta,\rho}$. The same holds for the claim in (c), as long as $L \le 1$.

12.3.6. Let $C^\beta(M)$ be the vector space of β-Hölder functions on a compact metric space M. Prove the properties (i), (ii), (iii) stated at the end of Section 12.3.

12.3.7. Endow $C^\beta(M)$ with the norm $\|\cdot\|_{\beta,\rho}$. Let $\mathcal{L} : C^\beta(M) \to C^\beta(M)$ be the transfer operator associated with an α-Hölder potential $\varphi : M \to \mathbb{R}$, with $\alpha \ge \beta$. Let λ be the spectral radius, ν be the reference measure, h be the eigenfunction and $\mu = h\nu$ be the equilibrium state of the potential φ. Consider the transfer operator $\mathcal{P} : C^\beta(M) \to C^\beta(M)$ associated with the potential $\psi = \varphi + \log h - \log h \circ f - \log \lambda$.

(a) Check that $\mathcal{L}$ is linearly conjugate to $\lambda\mathcal{P}$, and so $\operatorname{spec}(\mathcal{L}) = \lambda \operatorname{spec}(\mathcal{P})$. Moreover, $\mathcal{P}1 = 1$ and $\mathcal{P}^*\mu = \mu$.
(b) Show that $\int |\mathcal{P}^n g|\,d\mu \le \int |g|\,d\mu$ and $\sup|\mathcal{P}^n g| \le \sup|g|$ and there exist constants $C > 0$ and $\tau < 1$ such that $H_{\beta,\rho}(\mathcal{P}^n g) \le \tau^n H_{\beta,\rho}(g) + C \sup|g|$ for every $g \in C^\beta(M)$ and every $n \ge 1$.

12.3.8. The goal of this exercise is to prove that the spectral radius of the restriction of $\mathcal{L}$ to the hyperplane $V = \{g \in C^\beta(M) : \int g\,d\nu = 0\}$ is strictly less than λ. By part (a) of Exercise 12.3.7, it is enough to consider the case $\mathcal{L} = \mathcal{P}$ (with $\lambda = 1$ and $\nu = \mu$ and $h = 1$). Fix b, β, R as in Corollary 12.3.12.

(a) Show that there exist $K > 1$ and $r > 0$ such that, for every $v \in V$ with $\|v\|_{\beta,\rho} \le r$, the function $g = 1 + v$ is in the cone $C(b,\beta,R)$ and satisfies

$$K^{-1}\|v\|_{\beta,\rho} \le \theta(1,g) \le K\|v\|_{\beta,\rho}.$$

(b) Use Corollary 12.3.12 and the previous item to find $C > 0$ and $\tau < 1$ such that $\|\mathcal{P}^n v\|_{\beta,\rho} \le C\tau^n\|v\|_{\beta,\rho}$ for every $v \in V$. Deduce that the spectral radius of $\mathcal{P} \mid V$ is less or equal than $\tau < 1$.

12.4 Dimension of conformal repellers

In this section we present an application of the theory developed previously to the calculation of the Hausdorff dimension of certain invariant sets of expanding maps, that we call conformal repellers. The main result (Theorem 12.4.3) contains a formula for the value of the Hausdorff dimension of the repeller in terms of the pressure of certain potentials.

Detailed presentations of the theory of fractal dimensions and its many applications can be found in the books of Falconer [Fal90], Palis and Takens [PT93, Chapter 4], Pesin [Pes97] and Bonatti, Díaz and Viana [BDV05, Chapter 3].

12.4.1 Hausdorff dimension

Let M be a metric space. In this section, we call a *cover* of M any countable (possibly finite) family of subsets of M, not necessarily open, whose union is the whole of M. The *diameter* of a cover $\mathcal{U}$ is the supremum of the diameters of its elements. For each $d > 0$ and $\delta > 0$, define

$$m_d(M,\delta) = \inf\left\{\sum_{U\in\mathcal{U}} \operatorname{diam}(U)^d : \mathcal{U} \text{ cover with } \operatorname{diam}\mathcal{U} < \delta\right\}. \tag{12.4.1}$$

That is, we consider all possible covers of M by subsets with diameter less than δ and we try to minimize the sum of the diameters raised to the power d. This number varies with δ in a monotonic fashion: when δ decreases, the class of admissible covers decreases and, thus, the infimum can only increase. We call *Hausdorff measure of M in dimension d* the limit

$$m_d(M) = \lim_{\delta\to 0} m_d(M,\delta). \tag{12.4.2}$$

Note that $m_d(M) \in [0,\infty]$. Moreover, it follows directly from the definition that

$$m_{d_1}(M,\delta) \le \delta^{d_1-d_2} m_{d_2}(M,\delta) \quad \text{for every } \delta > 0 \text{ and any } d_1 > d_2 > 0.$$

Making $\delta \to 0$, it follows that $m_{d_1}(M) = 0$ or $m_{d_2}(M) = \infty$ or both. Therefore, there exists a unique $d(M) \in [0,\infty]$ such that $m_d(M) = \infty$ for every $d < d(M)$ and $m_d(M) = 0$ for every $d > d(M)$. We call $d(M)$ the *Hausdorff dimension* of the metric space M.

Example 12.4.1. Consider the usual Cantor set K in the real line. That is,

$$K = \bigcap_{n=0}^{\infty} K^n$$

where $K^0 = [0,1]$ and every K^n, $n \ge 1$ is obtained by removing from each connected component of K^{n-1} the central open subinterval with relative length $1/3$. Let $d_0 = \log 2/\log 3$. We are going to show that $m_{d_0}(M) = 1$, which implies that $d(M) = d_0$.

To prove the upper bound, consider, for each $n \ge 0$, the cover $\mathcal{V}^n$ of K whose elements are the intersections of K with each of the connected components of K^n. It is clear that the sequence $(\mathcal{V}^n)_n$ is increasing: $\mathcal{V}^{n-1} \prec \mathcal{V}^n$ for every $n \ge 1$. Note also that $\mathcal{V}^n$ has exactly 2^n elements, all with diameter equal to 3^{-n}. Therefore,

$$\sum_{V\in\mathcal{V}^n} (\operatorname{diam} V)^{d_0} = 2^n 3^{-nd_0} = 1 \quad \text{for every } n. \tag{12.4.3}$$

Since $\operatorname{diam}\mathcal{V}^n \to 0$ when $n \to \infty$, it follows that $m_{d_0}(M) \le 1$ and so $d(M) \le d_0$.

The lower bound is a bit more difficult, because one needs to deal with arbitrary covers. We are going to show that, given any cover $\mathcal{U}$ of M,

$$\sum_{U\in\mathcal{U}} (\operatorname{diam} U)^{d_0} \geq 1. \tag{12.4.4}$$

Clearly, this implies that $m_{d_0}(M) \geq 1$ and so $d(M) \geq d_0$.

Let us call an *open segment* the intersection of K with any interval of the real line whose endpoints are not in K. It is clear that every subset of K is contained in some open segment whose diameter is only slightly larger. Hence, given any cover $\mathcal{U}$, we can always find covers $\mathcal{U}'$ whose elements are open segments such that $\sum_{U'\in\mathcal{U}'}(\operatorname{diam} U')^{d_0}$ is as close to $\sum_{U\in\mathcal{U}}(\operatorname{diam} U)^{d_0}$ as we want. So, it is no restriction to assume from the start that the elements of $\mathcal{U}$ are open segments. Then, since K is compact, we may also assume that $\mathcal{U}$ is finite. For any open segment U there exists $n \geq 0$ such that every element of $\mathcal{V}^m$, $m \geq n$ that intersects U is contained in U. Since $\mathcal{U}$ is finite, we may choose the same n for all its elements. We claim that

$$\sum_{U\in\mathcal{U}} (\operatorname{diam} U)^{d_0} \geq \sum_{V\in\mathcal{V}^n} (\operatorname{diam} V)^{d_0}. \tag{12.4.5}$$

Clearly, (12.4.3) and (12.4.5) imply (12.4.4). We are left to prove (12.4.5).

The strategy is to modify the cover $\mathcal{U}$ successively, in such a way that the expression on the left-hand side of (12.4.5) never increases and one reaches the cover $\mathcal{V}^n$ after finitely many modifications. For each $U \in \mathcal{U}$, let $k \geq 0$ be minimum such that U intersects a unique element V of $\mathcal{V}^k$. The choice of n implies that $k \leq n$: for $k > n$, if U intersects an element of $\mathcal{V}^k$ then U contains all the 2^{k-n} elements of $\mathcal{V}^k$ inside the same element of $\mathcal{V}^n$. Suppose that $k < n$. By the choice of k, the set U intersects exactly two elements of $\mathcal{V}^{k+1}$. Let them be denoted V_1 and V_2 and let U_1 and U_2 be their intersections with U. Then

$$\operatorname{diam} U_i \leq \operatorname{diam} V_i = 3^{-k-1} \quad \text{and} \quad \operatorname{diam} U = \operatorname{diam} U_1 + 3^{-k-1} + \operatorname{diam} U_2.$$

Hence (Exercise 12.4.1),

$$(\operatorname{diam} U)^{d_0} \geq (\operatorname{diam} U_1)^{d_0} + (\operatorname{diam} U_2)^{d_0}.$$

This means that the value on the left-hand side of (12.4.5) does not increase when we replace U by U_1 and U_2 in the cover $\mathcal{U}$. On the one hand, the new cover satisfies the same conditions as the original: U_1 and U_2 are open segments (because V_1, V_2 and U are open segments) and they contain every element of $\mathcal{V}^n$ that they intersect. On the other hand, by construction, each one of them intersects a unique element of $\mathcal{V}^{k+1}$. Therefore, after finitely many repetitions of this procedure we reduce the initial situation to the case where $k = n$ for every $U \in \mathcal{U}$. Now, the choice of n implies that in that case each $U \in \mathcal{U}$ contains the unique $V \in \mathcal{V}^n$ that it intersects. Observe that this means that $U = V$. In particular, any elements of $\mathcal{U}$ that correspond to the same

$V \in \mathcal{V}^m$ must coincide. Eliminating such repetitions we obtain the cover $\mathcal{V}^n$. This completes the calculation.

In general, the Hausdorff measure of a metric space M in its dimension $d(M)$ may take any value. In many interesting situations, including Example 12.4.1 and the much more general construction that we treat next, this measure is positive and finite. But there are many other cases where it is either zero or infinity.

12.4.2 Conformal repellers

Let $D, D_1, \ldots, D_N$ be compact convex subsets of an Euclidean space $\mathbb{R}^\ell$ such that $D_i \subset D$ for every i and $D_i \cap D_j = \emptyset$ whenever $i \neq j$. Assume that

$$\mathrm{vol}(D \setminus D_*) > 0, \tag{12.4.6}$$

where $D_* = D_1 \cup \cdots \cup D_N$ and vol denote the volume measure on $\mathbb{R}^\ell$. Assume also that there exists a map $f : D_* \to D$ such that the restriction to each D_i is a homeomorphism onto D. See Figure 12.2. Note that the sequence of pre-images $f^{-n}(D)$ is decreasing. Their intersection

$$\Lambda = \bigcap_{n=0}^{\infty} f^{-n}(D) \tag{12.4.7}$$

is called a *repeller* of f. In other words, Λ is the set of points x whose iterates $f^n(x)$ are defined for every $n \geq 1$. It is clear that Λ is compact and $f^{-1}(\Lambda) = \Lambda$.

Example 12.4.2. The Cantor set K in Example 12.4.1 is the repeller of the transformation $f : [0,1/3] \cup [2/3,1] \to [0,1]$ given by $f(x) = 3x$ if $x \in [0,1/3]$ and $f(x) = 3x - 2$ if $x \in [2/3,1]$. A more general class of examples in dimension 1 was introduced in Example 11.2.3.

In what follows we take the map $f : D_* \to D$ to be of class C^1; for points on the boundary of the domain this just means that f admits a C^1 extension to some neighborhood. We also make the following additional hypotheses.

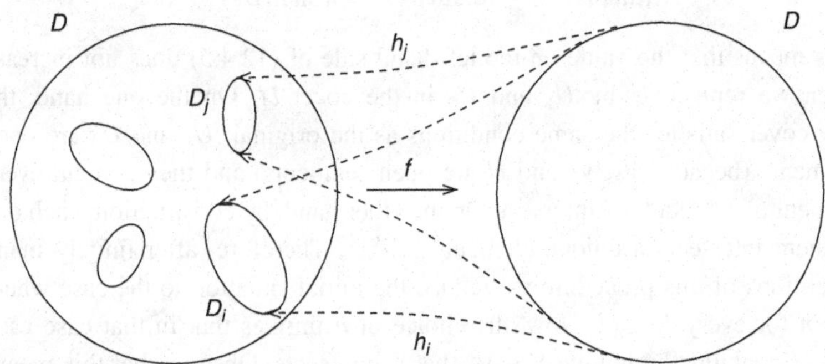

Figure 12.2. A repeller

The first hypothesis is that the map f is expanding: there exists $\sigma > 1$ such that

$$\|Df(x)v\| \geq \sigma \quad \text{for every } x \in D_* \text{ and every } v \in \mathbb{R}^\ell. \tag{12.4.8}$$

Then it is not difficult to check that the restriction $f : \Lambda \to \Lambda$ to the repeller is an expanding map in the sense of Section 11.2.

The second hypothesis is that the logarithm of the Jacobian of f is Hölder: there exist $C > 0$ and $\theta > 0$ such that

$$\log \frac{|\det Df(x)|}{|\det Df(y)|} \leq C\|x-y\|^\theta \quad \text{for every } x, y \in D_*. \tag{12.4.9}$$

Up to choosing C sufficiently large, the inequality is automatically satisfied when x and y belong to distinct subdomains D_i and D_j, because $d(D_i, D_j) > 0$.

The third and last hypothesis is that the map f is *conformal*:

$$\|Df(x)\| \, \|Df(x)^{-1}\| = 1 \quad \text{for every } x \in D_*. \tag{12.4.10}$$

It is important to note that this condition is automatic when $\ell = 1$. For $\ell = 2$, it holds if and only if the map f is analytic.

All these conditions are satisfied in the case of the Cantor set (Examples 12.4.1 and 12.4.2). They are also satisfied in Example 11.2.3, as long as we take the derivative of the corresponding map f to be Hölder.

Theorem 12.4.3 (Bowen–Manning formula). *Suppose that $f : D_* \to D$ satisfies the conditions* (12.4.8), (12.4.9) *and* (12.4.10). *Then the Hausdorff dimension of the repeller is given by*

$$d(\Lambda) = d_0 \ell,$$

where $d_0 \in (0,1)$ is the unique number such that $P(f, -d_0 \log|\det Df|) = 0$.

The reader should be warned that we allow ourselves a slight abuse of language, in order not to overload the notations: throughout this section, $P(f, \psi)$ always denotes the pressure of a potential $\psi : \Lambda \to \mathbb{R}$ *with respect to the restriction* $f : \Lambda \to \Lambda$ to the repeller, even if at other points of the arguments we consider the map f defined on the whole domain D_*.

Before we start proving the theorem, let us mention the following interesting special case:

Example 12.4.4. Let $f : J \to [0,1]$ be a map as in Example 11.2.3 and assume that the restriction of f to each connected component J_i of J is affine: the absolute value of the derivative is constant, equal to the inverse of the length $|J_i|$. Then the Hausdorff dimension of the repeller K of the map f is the unique number τ such that

$$\sum_i |J_i|^\tau = 1. \tag{12.4.11}$$

To obtain this conclusion from Theorem 12.4.3 it suffices to note that

$$P(f, -t\log|f'|) = \log\sum_i |J_i|^t \quad \text{for every } t. \tag{12.4.12}$$

We let the reader check this last claim (Exercise 12.4.6).

12.4.3 Distortion and conformality

Let us call an inverse branch of f the inverse $h_i : D \to D_i$ of the restriction of f to each domain D_i. More generally, we call an inverse branch of f^n any composition

$$h^n = h_{i_0} \circ \cdots \circ h_{i_{n-1}} \tag{12.4.13}$$

with $i_0, \ldots, i_{n-1} \in \{1, \ldots, N\}$. For each $n \geq 1$, denote by $\mathcal{I}^n$ the family of all inverse branches h^n of f^n. By construction, the images $h^n(D)$, $h^n \in \mathcal{I}^n$ are pairwise disjoint and their union contains Λ.

The principal goal in this section is to prove the following geometric estimate, which is at the heart of the proof of Theorem 12.4.3:

Proposition 12.4.5. *There exists $C_0 > 1$ such that for every $n \geq 1$, every $h^n \in \mathcal{I}^n$, every $E \subset h^n(D)$ and every $x \in h^n(D)$:*

$$\frac{1}{C_0}[\operatorname{diam} f^n(E)]^\ell \leq [\operatorname{diam} E]^\ell |\det Df^n(x)| \leq C_0[\operatorname{diam} f^n(E)]^\ell. \tag{12.4.14}$$

Starting the proof of this proposition, observe that our hypotheses imply that every inverse branch h_i of f is a diffeomorphism with $\|Dh_i\| \leq \sigma^{-1}$. Then, since D is convex, we may use the mean value theorem to conclude that

$$\|h_i(z) - h_i(w)\| \leq \sigma^{-1}\|z - w\| \quad \text{for every } z, w \in D. \tag{12.4.15}$$

For each inverse branch h^n as in (12.4.13), let us consider the sequence of inverse branches

$$h^{n-k} = h_{i_k} \circ \cdots \circ h_{i_{n-1}}, \quad k = 0, \ldots, n-1. \tag{12.4.16}$$

Note that $h^{n-k}(D) \subset D_{i_k}$ for each k. It follows from (12.4.15) that each h^{n-k} is a σ^{k-n}-contraction. In particular,

$$\operatorname{diam} h^{n-k}(D) \leq \sigma^{k-n} \operatorname{diam} D \quad \text{for every } k = 0, \ldots, n-1. \tag{12.4.17}$$

Recall that the *convex hull* of a set $X \subset \mathbb{R}^\ell$ is the union of all the line segments whose endpoints are in X. It is clear that the convex hull has the same diameter as the set itself. Since D_i is convex for every i, the convex hull of each $h^{n-k}(D)$ is contained in D_{i_k}. In particular, the derivative Df is defined at every point in the convex hull of every $h^{n-k}(D)$.

Lemma 12.4.6. *There exists $C_1 > 1$ such that, for every $n \geq 1$ and every inverse branch $h^n \in \mathcal{I}^n$,*

$$\prod_{k=0}^{n-1} \frac{|\det Df(z_k)|}{|\det Df(w_k)|} \leq C_1$$

for any z_k, w_k in the convex hull of $h^{n-k}(D)$ for $k = 0, \dots, n-1$.

Proof. The condition (12.4.9) gives that

$$\log \frac{|\det Df(z_k)|}{|\det Df(w_k)|} \leq C\|z_k - w_k\|^\theta \leq C[\operatorname{diam} h^{n-k}(D)]^\theta$$

for each $k = 0, \dots, n-1$. Then, using (12.4.17),

$$\log \prod_{k=0}^{n-1} \frac{|\det Df(z_k)|}{|\det Df(w_k)|} \leq \sum_{k=0}^{n-1} C[\operatorname{diam} h^{n-k}(D)]^\theta \leq C[\operatorname{diam} D]^\theta \sum_{k=0}^{n-1} \sigma^{(k-n)\theta}.$$

Therefore, it suffices to take $C_1 = \exp\left(C(\operatorname{diam} D)^\theta \sum_{j=1}^\infty \sigma^{-j\theta}\right)$.

The time has come for us to exploit the conformality hypothesis (12.4.10). Given any linear isomorphism $L : \mathbb{R}^\ell \to \mathbb{R}^\ell$, it is clear that $|\det L| \leq \|L\|^\ell$, and analogously for the inverse. Therefore,

$$1 = |\det L|\,|\det L^{-1}| \leq \left(\|L\|\,\|L^{-1}\|\right)^\ell.$$

Hence, $\|L\|\,\|L^{-1}\| = 1$ implies that $|\det L| = \|L\|^\ell$, and analogously for the inverse. Therefore, (12.4.10) implies that

$$|\det Df(y)| = \|Df(y)\|^\ell \quad \text{for every } y \in D_*. \tag{12.4.18}$$

Now we are ready to prove Proposition 12.4.5:

Proof of Proposition 12.4.5. Let n, h^n, E and x be as in the statement. Let w be a point of maximum for the norm of Dh^n in the domain D. By the mean value theorem,

$$\|x_1 - x_2\| \leq \|Dh^n(w)\|\,\|f^n(x_1) - f^n(x_2)\| \tag{12.4.19}$$

for any x_1, x_2 in E. Observe that $Dh^n(w)$ is the inverse of $Df^n(z)$, with $z = h^n(w)$. Hence, by conformality, $\|Dh^n(w)\| = \|Df^n(z)\|^{-1}$. Moreover, using Lemma 12.4.6 and (12.4.18),

$$|\det Df^n(x)| \leq C_1|\det Df^n(z)| = C_1\|Df^n(z)\|^\ell. \tag{12.4.20}$$

Combining (12.4.19) and (12.4.20), we obtain

$$\|x_1 - x_2\|^\ell \leq C_1|\det Df^n(x)|^{-1}\|f^n(x_1) - f^n(x_2)\|^\ell.$$

Varying $x_1, x_2 \in E$, it follows that

$$[\operatorname{diam} E]^\ell \leq C_1|\det Df^n(x)|^{-1}[\operatorname{diam} f^n(E)]^\ell.$$

This proves the second inequality in (12.4.14), as long as we take $C_0 \geq C_1$.

The proof of the other inequality is similar. For each $k = 0,\dots,n-1$, let z_k be a point of maximum for the norm of Df restricted to the convex hull of $h^{n-k}(D)$. Then,

$$\begin{aligned}\|f^{k+1}(x_1) - f^{k+1}(x_2)\| &\le \|Df(z_k)\| \, \|f^k(x_1) - f^k(x_2)\| \\ &= |\det Df(z_k)|^{1/\ell} \, \|f^k(x_1) - f^k(x_2)\|\end{aligned}$$

for every k and any $x_1, x_2 \in E$. Hence,

$$\|f^n(x_1) - f^n(x_2)\|^\ell \le \prod_{k=0}^{n-1} |\det Df(z_k)| \, \|x_1 - x_2\|^\ell. \tag{12.4.21}$$

By Lemma 12.4.6,

$$\prod_{k=0}^{n-1} |\det Df(z_k)| \le C_1 |\det Df^n(x)|. \tag{12.4.22}$$

Combining (12.4.21) and (12.4.22), we obtain

$$\|y_1 - y_2\|^\ell \le C_1 |\det Df^n(x)| \, \|x_1 - x_2\|^\ell.$$

Varying y_1, y_2, we conclude that

$$[\operatorname{diam} f^n(E)]^\ell \le C_1 |\det Df^n(x)| [\operatorname{diam} E]^\ell.$$

This proves the first inequality in (12.4.14), for any $C_0 \ge C_1$.

12.4.4 Existence and uniqueness of d_0

Now we prove the existence and uniqueness of the number d_0 in the statement of Theorem 12.4.3. Denote $\phi = -\log|\det Df|$ and consider the function

$$\Psi : \mathbb{R} \to \mathbb{R}, \quad \Psi(t) = P(f, t\phi).$$

We want to show that there exists a unique d_0 such that $\Psi(d_0) = 0$.

Uniqueness is easy to prove. Indeed, the hypotheses (12.4.8) and (12.4.10) imply that

$$\phi = \log|\det Df^{-1} \circ f| = \ell \log \|Df^{-1} \circ f\| \le -\ell \log \sigma.$$

Then, given any $s < t$, we have $t\phi \le s\phi - (t-s)\ell \log\sigma$. Using (10.3.4) and (10.3.5), it follows that

$$P(f, t\phi) \le P(f, s\phi) - (t-s)\ell \log\sigma < P(f, s\phi).$$

This proves that Ψ is strictly decreasing, and so there exists at most one $d_0 \in \mathbb{R}$ such that $\Psi(d_0) = 0$.

On the other hand, it follows from Proposition 10.3.6 that Ψ is continuous. Hence, to prove the existence of d_0 it is enough to show that $\Psi(0) > 0 > \Psi(1)$. This may be done as follows.

Let $\mathcal{L}$ be the open cover of Λ whose elements are the images $h(\Lambda)$ of Λ under all the inverse branches of f. For each $n \geq 1$, the iterated sum $\mathcal{L}^n$ is formed by the images $h^n(\Lambda)$ of Λ under the inverse branches of f^n. It follows from (12.4.17) that $\operatorname{diam} \mathcal{L}^n \leq \sigma^{-n} \operatorname{diam} D$ for every n, and so $\operatorname{diam} \mathcal{L}^n \to 0$. Then, since the elements of $\mathcal{L}$ are pairwise disjoint, we may use Exercise 10.3.3 to conclude that

$$P(f,\psi) = P(f,\psi,\mathcal{L}) \quad \text{for every potential } \psi. \tag{12.4.23}$$

In particular, $\Psi(0) = P(f,0,\mathcal{L}) = h(f,\mathcal{L})$. Note that each family $\mathcal{L}^n$ is a minimal cover of the repeller, that is, no proper subfamily covers Λ. Therefore, $H(\mathcal{L}^n) = \log \#\mathcal{L}^n = n \log N$ for every n and, consequently, $h(f,\mathcal{L}) = \log N$. This proves that $\Psi(0)$ is positive.

Proposition 12.4.7. $\Psi(1) = \lim_n \frac{1}{n} \log \operatorname{vol}\left(f^{-n}(D)\right) < 0$.

Proof. By (12.4.23), we have that $\Psi(1) = P(f,\phi,\mathcal{L})$. In other words,

$$\Psi(1) = \lim_n \frac{1}{n} \log P_n(f,\phi,\mathcal{L}) = \lim_n \frac{1}{n} \log \sum_{h^n \in \mathcal{I}^n} e^{\phi_n(h^n(\Lambda))}.$$

Since $\phi = -\log |\det Df|$, this means that

$$\Psi(1) = \lim_n \frac{1}{n} \log \sum_{h^n \in \mathcal{I}^n} \sup_{h^n(D)} \frac{1}{|\det Df^n|}. \tag{12.4.24}$$

On the other hand, by the formula of change of variables,

$$\operatorname{vol}\left(f^{-n}(D)\right) = \sum_{h^n \in \mathcal{I}^n} \operatorname{vol}(h^n(D)) = \sum_{h^n \in \mathcal{I}^n} \int_D \frac{1}{|\det Df^n|} \circ h^n \, dx.$$

It follows from Lemma 12.4.6 that

$$\inf_{h^n(D)} |\det Df^n| \leq |\det Df^n|(h^n(z)) \leq C_1 \inf_{h^n(D)} |\det Df^n|$$

for every $z \in h^n(D)$ and every $h^n \in \mathcal{I}^n$. Consequently,

$$\operatorname{vol}\left(f^{-n}(D)\right) \leq \operatorname{vol}(D) \sum_{h^n \in \mathcal{I}^n} \sup_{h^n(D)} \frac{1}{|\det Df^n|} \leq C_1 \operatorname{vol}\left(f^{-n}(D)\right).$$

Combining these inequalities with (12.4.24), we conclude that

$$\limsup_n \frac{1}{n} \log \operatorname{vol}\left(f^{-n}(D)\right) \leq \Psi(1) \leq \liminf_n \frac{1}{n} \log \operatorname{vol}\left(f^{-n}(D)\right).$$

This proves the identity in the statement of the proposition.

It remains to prove that the volume of the pre-images $f^{-n}(D)$ decays exponentially fast. For that, observe that $f^{-(n+1)}(D) = f^{-n}(D_*)$ is the disjoint union of the images $h^n(D_*)$, with $h^n \in \mathcal{I}^n$. Therefore,

$$\frac{\operatorname{vol}\left(f^{-(n+1)}(D)\right)}{\operatorname{vol}\left(f^{-n}(D)\right)} = \frac{\sum_{h^n \in \mathcal{I}^n} \operatorname{vol}\left(h^n(D_*)\right)}{\sum_{h^n \in \mathcal{I}^n} \operatorname{vol}\left(h^n(D)\right)} \leq \max_{h^n \in \mathcal{I}^n} \frac{\operatorname{vol}\left(h^n(D*)\right)}{\operatorname{vol}\left(h^n(D)\right)}. \tag{12.4.25}$$

By the formula of change of variables,

$$\operatorname{vol}\left(h^n(D)\right) = \int_D \frac{1}{|\det Df^n|} \circ h^n \, dx \quad \text{and}$$

$$\operatorname{vol}\left(h^n(D \setminus D_*)\right) = \int_{D \setminus D_*} \frac{1}{|\det Df^n|} \circ h^n \, dx.$$

Hence, using Lemma 12.4.6,

$$\frac{\operatorname{vol}\left(h^n(D \setminus D_*)\right)}{\operatorname{vol}\left(h^n(D)\right)} \geq \frac{1}{C_1} \frac{\operatorname{vol}\left(D \setminus D_*\right)}{\operatorname{vol}\left(D\right)} \tag{12.4.26}$$

for every $h^n \in \mathcal{I}^n$. By the hypothesis (12.4.6), the expression on the right-hand side of (12.4.26) is positive. Fix $\beta > 0$ close enough to zero that $1 - e^{-\beta}$ is smaller than that expression. Then

$$\frac{\operatorname{vol}\left(h^n(D) \setminus h^n(D_*)\right)}{\operatorname{vol}\left(h^n(D)\right)} \geq 1 - e^{-\beta}$$

for every $h^n \in \mathcal{I}^n$. Combining this inequality with (12.4.25) and the fact that $\operatorname{vol}(h^n(D) \setminus h^n(D_*)) = \operatorname{vol} h^n(D) - \operatorname{vol} h^n(D_*)$, we obtain that

$$\frac{\operatorname{vol}\left(f^{-(n+1)}(D)\right)}{\operatorname{vol}\left(f^{-n}(D)\right)} \leq e^{-\beta} \quad \text{for every } n \geq 0$$

(the case $n = 0$ follows directly from the hypothesis (12.4.6)). Hence,

$$\lim_n \frac{1}{n} \log \operatorname{vol}\left(f^{-n}(D)\right) \leq -\beta < 0.$$

This concludes the proof of the proposition.

Figure 12.3 summarizes the conclusions in this section. Recall that the function defined by $\Psi(t) = P(f, -t \log|\det Df|)$ is convex, by Proposition 10.3.5.

12.4.5 Upper bound

Here we show that $d(\Lambda) \leq b\ell$ for every $b > 0$ such that $P(f, b\phi) < 0$. In view of the observations in the previous section, this proves that $d(\Lambda) \leq d_0\ell$.

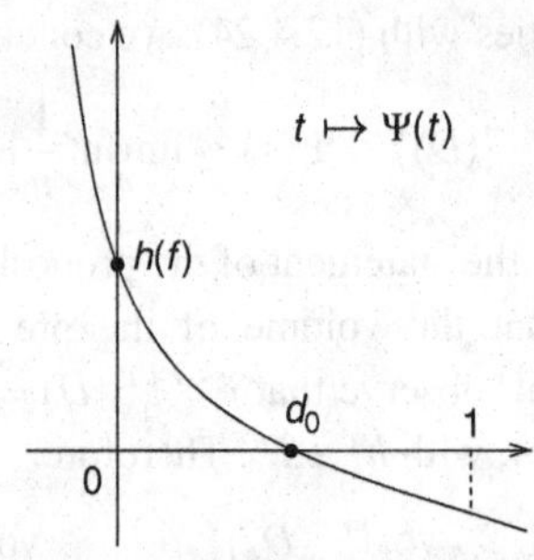

Figure 12.3. Pressure and Hausdorff dimension

Let $\mathcal{L}$ be the open cover of Λ introduced in the previous section and let $b > 0$ be such that $P(f, b\phi) < 0$. The property (12.4.23) implies that

$$P(f, b\phi, \mathcal{L}) = P(f, b\phi) < -\kappa$$

for some $\kappa > 0$. By the definition (10.3.2), it follows that

$$P_n(f, b\phi, \mathcal{L}) \leq e^{-\kappa n} \quad \text{for every } n \text{ sufficiently large.} \tag{12.4.27}$$

It is clear that $\mathcal{L}^n$ is a minimal cover of Λ: no proper subfamily covers Λ. Hence, recalling the definition (10.3.1), the inequality (12.4.27) implies that

$$\sum_{L \in \mathcal{L}^n} e^{b\phi_n(L)} \leq e^{-\kappa n} \quad \text{for every } n \text{ sufficiently large.} \tag{12.4.28}$$

It is clear that every $L \in \mathcal{L}^n$ is compact. Hence, by continuity of the Jacobian,

$$e^{\phi_n(L)} = \sup_L |\det Df^n|^{-1} = |\det Df^n(x)|^{-1}$$

for some $x \in L$. It is also clear that $f^n(L) = \Lambda$ for every $L \in \mathcal{L}^n$. Then, taking $E = L$ in Proposition 12.4.5,

$$[\operatorname{diam} L]^\ell e^{-\phi_n(L)} \leq C_0 [\operatorname{diam} \Lambda]^\ell.$$

Combining this inequality with (12.4.28), we obtain that

$$\sum_{L \in \mathcal{L}^n} [\operatorname{diam} L]^{b\ell} \leq C_0^b [\operatorname{diam} \Lambda]^{b\ell} \sum_{L \in \mathcal{L}^n} e^{b\phi_n(L)} \leq C_0^b [\operatorname{diam} \Lambda]^{b\ell} e^{-\kappa n}$$

for every n sufficiently large. Since the expression on the right-hand side converges to zero, and the diameter of the covers $\mathcal{L}^n$ also converges to zero, it follows that $m_{b\ell}(M) = 0$. Therefore, $d(M) \leq b\ell$.

12.4.6 Lower bound

Now we show that $d(\Lambda) \geq a\ell$ for every a such that $P(f, a\phi) > 0$. This implies that $d(\Lambda) \geq d_0\ell$, which completes the proof of Theorem 12.4.3.

As observed in the previous section, the cover $\mathcal{L}$ realizes the pressure and all its iterated sums $\mathcal{L}^n$ are minimal covers of Λ. Hence, the choice of a implies that there exists $\kappa > 0$ such that

$$P_n(f, a\phi, \mathcal{L}) = \sum_{L \in \mathcal{L}^n} e^{a\phi_n(L)} \geq e^{\kappa n} \quad \text{for every } n \text{ sufficiently large.} \tag{12.4.29}$$

Fix such an n. Let $\varepsilon > 0$ be a lower bound for the distance between any two elements of $\mathcal{L}^n$: a lower bound does exist because the elements of $\mathcal{L}^n$ are compact and pairwise disjoint. Fix $\rho \in (0, \varepsilon^{a\ell})$. The reason for this choice will be clear soon. We claim that

$$\sum_{U \in \mathcal{U}} [\operatorname{diam} U]^{al} \geq 2^{-al} \rho \tag{12.4.30}$$

for every cover $\mathcal{U}$ of Λ. By definition, this implies that $m_{a\ell}(\Lambda) \geq 2^{-al}\rho > 0$ and, consequently, $d(\Lambda) \geq a\ell$. Therefore, to end the proof of Theorem 12.4.3 it suffices to prove this claim.

Let us suppose that there exists some open cover of Λ which does not satisfy (12.4.30). Then, using Exercise 12.4.3, there exists some open cover $\mathcal{U}$ of Λ with

$$\sum_{U \in \mathcal{U}} [\operatorname{diam} U]^{al} < \rho < \varepsilon^{al}. \tag{12.4.31}$$

By compactness, we may suppose that this open cover $\mathcal{U}$ is finite. The relation (12.4.31) implies that every $U \in \mathcal{U}$ has diameter less than ε. Hence, each $U \in \mathcal{U}$ intersects at most one $L \in \mathcal{L}^n$. Since $\mathcal{L}^n$ covers Λ and U is a non-empty subset of Λ, we also have that U intersects some $L \in \mathcal{L}^n$. This means that $\mathcal{U}$ is the disjoint union of the families

$$\mathcal{U}_L = \{U \in \mathcal{U} : U \cap L \neq \emptyset\}, \qquad L \in \mathcal{L}^n.$$

If $U \in \mathcal{U}_L$ then $U \subset L$. Let us consider the families $f^n(\mathcal{U}_L) = \{f^n(U) : U \in \mathcal{U}_L\}$. Observe that each one of them is a cover of Λ. Moreover, using Proposition 12.4.5,

$$\sum_{V \in f^n(\mathcal{U}_L)} [\operatorname{diam} V]^{a\ell} = \sum_{U \in \mathcal{U}_L} [\operatorname{diam} f^n(U)]^{a\ell} \leq C_0 e^{-a\phi_n(L)} \sum_{U \in \mathcal{U}_L} [\operatorname{diam} U]^{a\ell}. \tag{12.4.32}$$

Therefore,

$$\sum_{U \in \mathcal{U}} [\operatorname{diam} U]^{a\ell} = \sum_{L \in \mathcal{L}^n} \sum_{U \in \mathcal{U}_L} [\operatorname{diam} U]^{a\ell} \geq \sum_{L \in \mathcal{L}^n} C_0^{-1} e^{a\phi_n(L)} \sum_{V \in f^n(\mathcal{U}_L)} [\operatorname{diam} V]^{a\ell}.$$

Let us suppose that

$$\sum_{V \in f^n(\mathcal{U}_L)} [\operatorname{diam} V]^{a\ell} \geq \sum_{U \in \mathcal{U}} [\operatorname{diam} U]^{a\ell} \quad \text{for every } L \in \mathcal{L}^n.$$

Then, the previous inequality implies

$$\sum_{U \in \mathcal{U}} [\operatorname{diam} U]^{a\ell} \geq \sum_{L \in \mathcal{L}^n} C_0^{-1} e^{a\phi_n(L)} \sum_{U \in \mathcal{U}} [\operatorname{diam} U]^{a\ell} \geq C_0^{-1} e^{\kappa n} \sum_{U \in \mathcal{U}} [\operatorname{diam} U]^{a\ell}.$$

This is a contradiction, because $e^{\kappa n} > C_0$. Hence, there exists $L \in \mathcal{L}^n$ such that

$$\sum_{V \in f^n(\mathcal{U}_L)} [\operatorname{diam} V]^{a\ell} \leq \sum_{U \in \mathcal{U}} [\operatorname{diam} U]^{a\ell} < \rho.$$

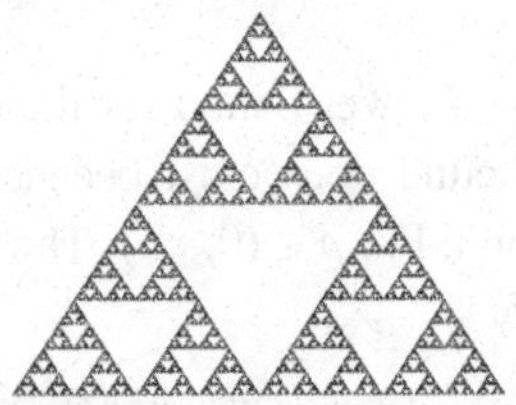

Figure 12.4. Sierpinski triangle

Thus, we may repeat the previous procedure with $f^n(\mathcal{U}_L)$ in the place of $\mathcal{U}$. Observe, however, that $\#f^n(\mathcal{U}_L) = \#\mathcal{U}_L$ is strictly less than $\#\mathcal{U}$. Therefore, this process must stop after a finite number of steps. This contradiction proves the claim 12.4.30.

The proof of Theorem 12.4.3 is complete. However, it is possible to prove an even stronger result: in the conditions of the theorem, the Hausdorff measure of Λ in dimension $d(M)$ is positive and finite. We leave this statement as a special challenge (Exercise 12.4.7) for the reader who remained with us till the end of this book!

12.4.7 Exercises

12.4.1. Let $d = \log 2/\log 3$. Show that $(x_1 + 1 + x_2)^d \geq x_1^d + x_2^d$ for every $x_1, x_2 \in [0,1]$. Moreover, the identity holds if and only if $x_1 = x_2 = 1$.

12.4.2. Let $f : M \to N$ be a Lipschitz map, with Lipschitz constant L. Show that

$$m_d(f(A)) \leq L^d m_d(A)$$

for any $d \in (0,\infty)$ and any $A \subset M$. Use this fact to show that if $A \subset \mathbb{R}^n$ and $t > 0$, then $m_d(tA) = t^d m_d(A)$, where $tA = \{tx : x \in A\}$.

12.4.3. Represent by $m_d^o(M)$ and $m_d^c(M)$ the numbers defined in the same ways as the Hausdorff measure $m_d(M)$ but considering only covers by open sets and covers by closed sets, respectively. Show that $m_d^o(M) = m_d^c(M) = m_d(M)$.

12.4.4. (Mass distribution principle) Let μ be a finite measure on a compact metric space M and assume that there exist numbers d, K, $\rho > 0$ such that $\mu(B) \leq K(\operatorname{diam} U)^d$ for every set $B \subset M$ with diameter less than ρ. Show that if $A \subset M$ is such that $\mu(A) > 0$ then $m_d(A) > 0$ and so $d(A) \geq d$.

12.4.5. Use the mass distribution principle to show that the Hausdorff dimension of the *Sierpinski triangle* (Figure 12.4) is equal to $d_0 = \log 3/\log 2$ and the Hausdorff measure in dimension d_0 is positive and finite.

12.4.6. Check the pressure formula (12.4.12).

12.4.7. Adapting arguments from Exercise 12.4.5, show that in the conditions of Theorem 12.4.3 one has $0 < m_{d(\Lambda)}(\Lambda) < \infty$.

Appendix A
Topics in measure theory, topology and analysis

In this series of appendices we recall several basic concepts and facts in measure theory, topology and functional analysis that are useful throughout the book. Our purpose is to provide the reader with a quick, accessible source of references to measure and integration, general and differential topology and spectral theory, to try and make this book as self-contained as possible. We have not attempted to make the material in these appendices completely sequential: it may happen that a notion mentioned in one section is defined or discussed in more depth in a later one (check the index).

As a general rule, we omit the proofs. For Appendices A.1, A.2 and A.5, the reader may find detailed information in the books of Castro [Cas04], Fernandez [Fer02], Halmos [Hal50], Royden [Roy63] and Rudin [Rud87]. The presentation in Appendix A.3 is a bit more complete, including the proofs of most results, but the reader may find additional relevant material in the books of Billingsley [Bil68, Bil71]. We recommend the books of Hirsch [Hir94] and do Carmo [dC79] to all those interested in going further into the topics in Appendix A.4. For more information on the subjects of Appendices A.6 and A.7, including proofs of the results quoted here, check the book of Halmos [Hal51] and the treatise of Dunford and Schwarz [DS57, DS63], especially Section IV.4 of the first volume and the initial sections of the second volume.

A.1 Measure spaces

Measure spaces are the natural environment for the definition of the Lebesgue integral, which is the main topic to be presented in Appendix A.2. We begin by introducing the notions of algebra and σ-algebra of subsets of a set, which lead to the concept of measurable space. Next, we present the notion of measure on a σ-algebra and we analyze some of its properties. In particular, we mention a few results on the construction of measures, including Lebesgue measures in

Euclidean spaces. The last part is dedicated to measurable maps, which are the maps that preserve the structure of measurable spaces.

A.1.1 Measurable spaces

Given a set X, we often denote by A^c the complement $X \setminus A$ of each subset A.

Definition A.1.1. An *algebra* of subsets of a set X is a family $\mathcal{B}$ of subsets of X that contains the empty set and is closed under the elementary operations of set theory:

(i) $\emptyset \in \mathcal{B}$;
(ii) $A \in \mathcal{B}$ implies $A^c \in \mathcal{B}$;
(iii) $A \in \mathcal{B}$ and $B \in \mathcal{B}$ implies $A \cup B \in \mathcal{B}$;
(iv) $A \in \mathcal{B}$ and $B \in \mathcal{B}$ implies $A \cap B \in \mathcal{B}$;
(v) $A \in \mathcal{B}$ and $B \in \mathcal{B}$ implies $A \setminus B \in \mathcal{B}$.

The two last properties are immediate consequences of the previous ones, since $A \cap B = (A^c \cup B^c)^c$ and $A \setminus B = A \cap B^c$. Moreover, by associativity, properties (iii) and (iv) imply that the union and the intersection of any finite family of elements of $\mathcal{B}$ are also in $\mathcal{B}$.

Definition A.1.2. A σ*-algebra* of subsets of a set X is an algebra $\mathcal{B}$ of subsets of X that is also closed under countable unions:

$$A_j \in \mathcal{B} \text{ for } j = 1, \dots, n, \dots \quad \text{implies} \quad \bigcup_{j=1}^{\infty} A_j \in \mathcal{B}.$$

Then $\mathcal{B}$ is also closed under countable intersections:

$$A_j \in \mathcal{B} \text{ for } j = 1, \dots, n, \dots \quad \text{implies} \quad \bigcap_{j=1}^{\infty} A_j = \Big(\bigcup_{j=1}^{\infty} A_j^c\Big)^c \in \mathcal{B}.$$

Definition A.1.3. A *measurable space* is a pair $(X, \mathcal{B})$ where X is a set and $\mathcal{B}$ is a σ-algebra of subsets of X. The elements of $\mathcal{B}$ are called *measurable sets*.

Next, we describe a few examples of constructions of σ-algebras.

Example A.1.4. For any set X, the following families of subsets are σ-algebras:

$$\{\emptyset, X\} \quad \text{and} \quad 2^X = \{\text{ all subsets of } X\}.$$

Moreover, clearly, if $\mathcal{B}$ is any algebra of subsets of X then $\{\emptyset, X\} \subset \mathcal{B} \subset 2^X$. So, $\{\emptyset, X\}$ is the smallest and 2^X is the largest of all algebras of subsets of X.

In the statement that follows, $\mathcal{I}$ is an arbitrary set whose sole use is to index the elements of the family of σ-algebras.

Proposition A.1.5. *Consider any non-empty family $\{\mathcal{B}_i : i \in \mathcal{I}\}$ of σ-algebras of subsets of the same set X. Then the intersection $\mathcal{B} = \bigcap_{i \in \mathcal{I}} \mathcal{B}_i$ is also a σ-algebra of subsets of X.*

Given any family $\mathcal{E}$ of subsets of X, we may apply Proposition A.1.5 to the family of all σ-algebras that contain $\mathcal{E}$. Note that this family is non-empty, since it contains the σ-algebra 2^X of all subsets of X. According to the previous proposition, the intersection of all these σ-algebras is also a σ-algebra. By construction, this σ-algebra contains $\mathcal{E}$ and is contained in every σ-algebra that contains $\mathcal{E}$. In other words, it is the smallest σ-algebra that contains $\mathcal{E}$. This leads to the following definition:

Definition A.1.6. The *σ-algebra generated* by a family $\mathcal{E}$ of subsets of X is the smallest σ-algebra $\sigma(\mathcal{E})$ that contains $\mathcal{E}$ or, in other words, the intersection of all the σ-algebras that contain $\mathcal{E}$.

Recall that a *topological space* is a pair (X, τ) where X is a set and τ is a family of subsets of X that contains $\{\emptyset, X\}$ and is closed under finite intersections and arbitrary unions. Such a family τ is called a *topology* and its elements are called *open subsets* of X. In this book we take all topological spaces to be *Hausdorff*, that is, such that for any pair of distinct points there exists a pair of disjoint open subsets each of which contains one of the points.

Definition A.1.7. The *Borel σ-algebra* of a topological space is the σ-algebra $\sigma(\tau)$ generated by the topology τ, that is, the smallest σ-algebra that contains all the open subsets of X. The elements of $\sigma(\tau)$ are called *Borel subsets* of X. The *closed* subsets of X, being the complements of the open subsets, are also in the Borel σ-algebra.

Analogously to Proposition A.1.5, the intersection of any non-empty family $\{\tau_i : i \in \mathcal{I}\}$ of topologies of the same set X is also a topology of X. Then, by the same argument as we used before for σ-algebras, given any family $\mathcal{E}$ of subsets of X there exists a smallest topology $\tau(\mathcal{E})$ that contains $\mathcal{E}$. We call it the *topology generated* by $\mathcal{E}$.

Example A.1.8. Let $(X, \mathcal{B})$ be a measurable space. The *limit superior* of a sequence of sets $E_n \in \mathcal{B}$ is the set $\limsup_n E_n$ formed by the points $x \in X$ such that $x \in E_n$ for infinitely many values of n. Analogously, the *limit inferior* of $(E_n)_n$ is the set $\liminf_n E_n$ of points $x \in X$ such that $x \in E_n$ for every value of n sufficiently large. In other words,

$$\liminf_n E_n = \bigcup_{n \geq 1} \bigcap_{m \geq n} E_m \quad \text{and} \quad \limsup_n E_n = \bigcap_{n \geq 1} \bigcup_{m \geq n} E_m.$$

Observe that $\liminf_n E_n \subset \limsup_n E_n$ and both sets are in $\mathcal{B}$.

Example A.1.9. The *extended line* $\bar{\mathbb{R}} = [-\infty, \infty]$ is the union of the real line $\mathbb{R} = (-\infty, +\infty)$ with the two points $\pm\infty$ at infinity. This space has a natural topology, generated by the intervals $[-\infty, b)$ and $(a, +\infty]$, with $a, b \in \mathbb{R}$. It is easy to see that the extended line is homeomorphic to a compact interval in the real line: for example, the function $\arctan : \mathbb{R} \to (-\pi/2, \pi/2)$ extends straightforwardly to a *homeomorphism* (that is, a continuous bijection whose inverse is also continuous) between $\bar{\mathbb{R}}$ and $[-\pi/2, \pi/2]$. We always consider on the extended line the Borel σ-algebra associated with this topology.

Of course, the real line $\mathbb{R}$ is a subspace (measurable as well as topological) of the extended line. The Borel subsets of the real line constitute a large family and one might even be led to think that every subset of $\mathbb{R}$ is a Borel subset. However, this is not true: a counterexample is constructed in Exercise A.1.4.

A.1.2 Measure spaces

Let $(X, \mathcal{B})$ be a measurable space. The following notions have a central role in this book:

Definition A.1.10. A *measure* on $(X, \mathcal{B})$ is a function $\mu : \mathcal{B} \to [0, +\infty]$ such that $\mu(\emptyset) = 0$ and

$$\mu\left(\bigcup_{j=1}^{\infty} A_j\right) = \sum_{j=1}^{\infty} \mu(A_j)$$

for any countable family of pairwise disjoint sets $A_j \in \mathcal{B}$. This last property is called *countable additivity* or *σ-additivity*. Then the triple $(X, \mathcal{B}, \mu)$ is called a *measure space*. If $\mu(X) < \infty$ then we say that μ is a *finite* measure and if $\mu(X) = 1$ then we call μ a *probability measure*. In this last case, $(X, \mathcal{B}, \mu)$ is called a *probability space*.

Example A.1.11. Let X be an arbitrary set, endowed with the σ-algebra $\mathcal{B} = 2^X$. Given any $p \in X$, consider the function $\delta_p : 2^X \to [0, +\infty]$ defined by:

$$\delta_p(A) = \begin{cases} 1 & \text{if } p \in A \\ 0 & \text{if } p \notin A. \end{cases}$$

It is easy to see that δ_p is a measure. It is usually called the *Dirac measure*, or *Dirac mass* at p.

Definition A.1.12. We say that a measure μ is *σ-finite* if there exists a sequence $A_1, \dots, A_n, \dots$ of subsets of X such that $\mu(A_i) < \infty$ for every $i \in \mathbb{N}$ and

$$X = \bigcup_{i=1}^{\infty} A_i.$$

We say that a function $\mu : \mathcal{B} \to [0, +\infty]$ is *finitely additive* if

$$\mu\left(\bigcup_{j=1}^{N} A_j\right) = \sum_{j=1}^{N} \mu(A_j)$$

for any finite family $A_1,\dots,A_N \in \mathcal{B}$ of pairwise disjoint subsets. Note that if μ is σ-additive then it is also finitely additive. Moreover, if μ is finitely additive and is not constant equal to $+\infty$ then $\mu(\emptyset) = 0$.

The main tool for constructing measures is the following theorem:

Theorem A.1.13 (Extension). *Let $\mathcal{A}$ be an algebra of subsets of X and let $\mu_0 : \mathcal{A} \to [0,+\infty]$ be a σ-additive function with $\mu_0(X) < \infty$. Then there exists a unique measure μ defined on the σ-algebra $\mathcal{B}$ generated by $\mathcal{A}$ that is an extension of μ_0, meaning that it satisfies $\mu(A) = \mu_0(A)$ for every $A \in \mathcal{A}$.*

Theorem A.1.13 remains valid for σ-finite measures. Moreover, there is a version for finitely additive functions: if μ_0 is finitely additive then it admits a finitely additive extension to σ-algebra $\mathcal{B}$ generated by $\mathcal{A}$. However, in this context the extension need not be unique.

The most useful criterion for proving that a given function is σ-additive is provided by the following theorem:

Theorem A.1.14 (Continuity at the empty set). *Let $\mathcal{A}$ be an algebra of subsets of X and $\mu : \mathcal{A} \to [0,+\infty)$ be a finitely additive function with $\mu(X) < \infty$. Then μ is σ-additive if and only if*

$$\lim_n \mu(A_n) = 0 \tag{A.1.1}$$

for every sequence $A_1 \supset \cdots \supset A_j \supset \cdots$ of elements of $\mathcal{A}$ with $\bigcap_{j=1}^{\infty} A_j = \emptyset$.

The proof of this theorem is proposed in Exercise A.1.7. Exercise A.1.9 deals with some variations of the statement.

Definition A.1.15. We say that an algebra $\mathcal{A}$ is *compact* if any decreasing sequence $A_1 \supset \cdots \supset A_n \supset \cdots$ of non-empty elements of $\mathcal{A}$ has non-empty intersection.

An *open cover* of a topological space is a family of open subsets whose union is the whole of K. A *subcover* is just a subfamily of elements of a cover whose union is still the whole space. A topological space is *compact* if every open cover admits some finite subcover. A subset K of a topological space X is *compact* if the topology of X restricted to K turns the latter into a compact topological space. Every closed subset of a compact space is compact. Conversely, (assuming X is a Hausdorff space) then every compact subset is closed. Another important fact is that the intersection $\bigcap_n K_n$ of any decreasing sequence $K_1 \supset \cdots \supset K_n \supset \cdots$ of compact subsets is non-empty.

Example A.1.16. It follows from what we have just said that if X is a (Hausdorff) topological space and every element of the algebra $\mathcal{A}$ is compact then $\mathcal{A}$ is a compact algebra.

It follows from Theorem A.1.14 that if $\mathcal{A}$ is a compact algebra then every finitely additive function $\mu : \mathcal{A} \to [0,+\infty)$ with $\mu(X) < \infty$ is σ-additive.

Hence, by Theorem A.1.13, μ extends uniquely to a measure defined on the σ-algebra generated by $\mathcal{A}$.

Definition A.1.17. We say that a non-empty family $\mathcal{C}$ of subsets of X is a *monotone class* if $\mathcal{C}$ contains X and is closed under countable monotone unions and intersections:

- if $A_1 \subset A_2 \subset \cdots$ are in $\mathcal{C}$ then $\bigcup_{n\geq 1} A_n \in \mathcal{C}$, and
- if $A_1 \supset A_2 \supset \cdots$ are in $\mathcal{C}$ then $\bigcap_{n\geq 1} A_n \in \mathcal{C}$.

Clearly, the two families $\{\emptyset, X\}$ and 2^X are monotone classes. Moreover, if $\{\mathcal{C}_i : i \in \mathcal{I}\}$ is any family of monotone classes then the intersection $\bigcap_{i\in\mathcal{I}} \mathcal{C}_i$ is a monotone class. Thus, for every subset $\mathcal{A}$ of 2^X there exists the smallest monotone class that contains $\mathcal{A}$.

Theorem A.1.18 (Monotone class). *The smallest monotone class that contains an algebra $\mathcal{A}$ coincides with the σ-algebra $\sigma(\mathcal{A})$ generated by $\mathcal{A}$.*

Another important result about σ-algebras that will be useful later states that every element of a σ-algebra $\mathcal{B}$ generated by an algebra $\mathcal{A}$ is approximated by the elements of $\mathcal{A}$, in the sense that the measure of the symmetric difference

$$A \Delta B = (A \setminus B) \cup (B \setminus A) = \big(A \cup B\big) \setminus \big(A \cap B\big)$$

can be made arbitrarily small. More precisely:

Theorem A.1.19 (Approximation). *Let $(X, \mathcal{B}, \mu)$ be a probability space and $\mathcal{A}$ be an algebra $\mathcal{A}$ of subsets of X that generates the σ-algebra $\mathcal{B}$. Then, for every $\varepsilon > 0$ and every $B \in \mathcal{B}$ there exists $A \in \mathcal{A}$ such that $\mu(A \Delta B) < \varepsilon$.*

Definition A.1.20. A measure space is *complete* if every subset of a measurable set with zero measure is also measurable.

It is possible to transform any measure space $(X, \mathcal{B}, \mu)$ into a complete space, as follows. Let $\bar{\mathcal{B}}$ be the family of all subsets $A \subset X$ such that there exist $B_1, B_2 \in \mathcal{B}$ with $B_1 \subset A \subset B_2$ and $\mu(B_2 \setminus B_1) = 0$. Then $\bar{\mathcal{B}}$ is a σ-algebra and it contains $\mathcal{B}$. Consider the function $\bar{\mu} : \bar{\mathcal{B}} \to [0, +\infty]$ defined by $\bar{\mu}(A) = \mu(B_1) = \mu(B_2)$, for any $B_1, B_2 \in \mathcal{B}$ as before. The function $\bar{\mu}$ is well defined, it is a measure on $\bar{\mathcal{B}}$ and its restriction to $\mathcal{B}$ coincides with μ. By construction, $(X, \bar{\mathcal{B}}, \bar{\mu})$ is a complete measure space. It is called the *completion* of $(X, \mathcal{B}, \mu)$.

Given subsets $\mathcal{U}_1$ and $\mathcal{U}_2$ of the σ-algebra $\mathcal{B}$, we say that $\mathcal{U}_1 \subset \mathcal{U}_2$ *up to measure zero* if for every $B_1 \in \mathcal{U}_1$ there exists $B_2 \in \mathcal{U}_2$ such that $\mu(B_1 \Delta B_2) = 0$. By definition, $\mathcal{U}_1 = \mathcal{U}_2$ up to measure zero if $\mathcal{U}_1 \subset \mathcal{U}_2$ up to measure zero and $\mathcal{U}_2 \subset \mathcal{U}_1$ up to measure zero. We say that a set $\mathcal{U} \subset \mathcal{B}$ generates the σ-algebra $\mathcal{B}$ up to measure zero if the σ-algebra generated by $\mathcal{U}$ is equal to $\mathcal{B}$ up to measure zero. Equivalently, $\mathcal{U}$ generates $\mathcal{B}$ up to measure zero if the completion of the σ-algebra generated by $\mathcal{U}$ coincides with the completion of $\mathcal{B}$.

By definition, a measure takes values in $[0,\infty]$. Whenever it is convenient to stress that fact, we speak of *positive measure* instead. But it is possible to weaken that requirement and, indeed, such generalizations are useful for our purposes.

We call a *signed measure* on a measurable space $(X,\mathcal{B})$ any σ-additive function $\mu:\mathcal{B}\to[-\infty,\infty]$ such that $\mu(\emptyset)=0$. More precisely, μ may take either the value $-\infty$ or the value $+\infty$, but not both; this is to avoid the "indetermination" $\infty-\infty$ in the additivity condition.

Theorem A.1.21 (Hahn decomposition)**.** *If μ is a signed measure then there exist measurable sets $P,N\subset X$ such that $P\cup N=X$ and $P\cap N=\emptyset$, and*

$$\mu(E)\geq 0 \text{ for every } E\subset P \quad \text{and} \quad \mu(E)\leq 0 \text{ for every } E\subset N.$$

This means that we may write $\mu=\mu^{+}-\mu^{-}$, where μ^{+} and μ^{-} are the (positive) measures defined by

$$\mu^{+}(E)=\mu\big(E\cap P\big) \quad \text{and} \quad \mu^{-}(E)=-\mu\big(E\cap N\big).$$

In particular, the sum $|\mu|=\mu^{+}+\mu^{-}$ is also a positive measure; it is called the *total variation* of the signed measure μ.

If μ takes values in $(-\infty,\infty)$ only, we call it a *finite signed measure*. In this case, the measures μ^{+} and μ^{-} are finite. The set $\mathcal{M}(X)$ of finite signed measures is a real vector space and the function $\|\mu\|=|\mu|(X)$ is a complete norm in this space (see Exercise A.1.10). In other words, $(\mathcal{M}(X),\|\cdot\|)$ is a real Banach space. When X is a compact metric space, this Banach space is isomorphic to the dual of the space $C^{0}(X)$ of continuous real functions X (theorem of Riesz–Markov).

More generally, we call a *complex measure* on a measurable space $(X,\mathcal{B})$ any σ-additive function $\mu:\mathcal{B}\to\mathbb{C}$. Observe that $\mu(\emptyset)$ is necessarily zero. Clearly, we may write $\mu=\Re\mu+i\Im\mu$, where the real part $\Re\mu$ and the imaginary part $\Im\mu$ are finite signed measures. The *total variation* of μ is the finite measure defined by

$$|\mu|(E)=\sup_{\mathcal{P}}\sum_{P\in\mathcal{P}}|\mu(P)|,$$

where the supremum is taken over all countable partitions of the measurable set E into measurable subsets (this definition coincides with the one we gave previously in the special case when μ is real). The function $\|\mu\|=|\mu|(X)$ defines a norm in the vector space of complex measures on X, which we also denote as $\mathcal{M}(X)$. Moreover, this norm is complete. When X is a compact metric space, the complex Banach space $(\mathcal{M}(X),\|\cdot\|)$ is isomorphic to the dual of the space $C^{0}(X)$ of continuous complex functions on X (theorem of Riesz–Markov).

A.1.3 Lebesgue measure

The notion of Lebesgue measure corresponds to the notion of volume of subsets of the Euclidean space $\mathbb{R}^d$. It is defined as follows.

Let $X = [0,1]$ and $\mathcal{A}$ be the family of all subsets of the form $A = I_1 \cup \cdots \cup I_N$ where $I_1, \ldots, I_N$ are pairwise disjoint intervals. It is easy to check that $\mathcal{A}$ is an algebra of subsets of X. Let $m_0 : \mathcal{A} \to [0,1]$ be the function defined on this algebra by

$$m_0\big(I_1 \cup \cdots \cup I_N\big) = |I_1| + \cdots + |I_N|,$$

where $|I_j|$ represents the length of each interval I_j. Note that $m_0(X) = 1$. In Exercise A.1.8 we ask the reader to show that m_0 is σ-additive.

Note that the σ-algebra $\mathcal{B}$ generated by $\mathcal{A}$ coincides with the Borel σ-algebra of X, since every open subset is a countable union of pairwise intervals. So, by Theorem A.1.13, there exists a unique probability measure m defined on $\mathcal{B}$ that is an extension of m_0. It is called the *Lebesgue measure* on $[0,1]$.

More generally, one defines the Lebesgue measure m on the *cube* $X = [0,1]^d$ of any dimension $d \geq 1$, in the following way. First, we call a *rectangle* in X any subset of the form $R = I_1 \times \cdots \times I_d$ where the I_j are intervals. Then we define:

$$m_0(R) = |I_1| \times \cdots \times |I_d|.$$

Next, we consider the algebra $\mathcal{A}$ of subsets of X of the form $A = R_1 \cup \cdots \cup R_N$, where $R_1, \ldots, R_N$ are pairwise disjoint rectangles, and we define

$$m_0(A) = m_0(R_1) + \cdots + m_0(R_N)$$

for every A in that algebra. The σ-algebra generated by $\mathcal{A}$ coincides with the Borel σ-algebra of X. The *Lebesgue measure* on the cube $X = [0,1]^d$ is the extension of m_0 to that σ-algebra.

In order to define the Lebesgue measure on the whole Euclidean space $\mathbb{R}^d$, we decompose the space into cubes of unit size:

$$\mathbb{R}^d = \bigcup_{k_1 \in \mathbb{Z}} \cdots \bigcup_{k_d \in \mathbb{Z}} [k_1, k_1 + 1) \times \cdots \times [k_d, k_d + 1).$$

Each cube $[k_1, k_1 + 1) \times \cdots \times [k_d, k_d + 1)$ may be identified with $[0,1)^d$ through the translation $T_{k_1,\ldots,k_d}(x) = x - (k_1, \ldots, k_d)$ that maps $(k_1, k_2, \ldots, k_d)$ to the origin. That allows us to define a measure $m_{k_1,k_2,\ldots,k_d}$ on C, by setting

$$m_{k_1,k_2,\ldots,k_d}(B) = m_0\big(T_{k_1,\ldots,k_d}(B)\big)$$

for every measurable set $B \subset C$. Finally, given any measurable set $B \subset \mathbb{R}^d$, define:

$$m(B) = \sum_{k_1 \in \mathbb{Z}} \cdots \sum_{k_d \in \mathbb{Z}} m_{k_1,\ldots,k_d}\big(B \cap [k_1, k_1 + 1) \times \cdots \times [k_d, k_d + 1)\big).$$

Note that this measure m is σ-finite but not finite.

Example A.1.22. It is worthwhile outlining a classical alternative construction of the Lebesgue measure (see Chapter 2 of Royden [Roy63] for details). We call the *Lebesgue exterior measure* of an arbitrary set $E \subset \mathbb{R}^d$ the number

$$m^*(E) = \inf \sum_k m_0(R_k),$$

where the infimum is taken over all countable covers $(R_k)_k$ of E by open rectangles. The function $E \mapsto m(E)$ is defined for every $E \subset \mathbb{R}^d$, but is not finitely additive (although it is countably subadditive). We say that E is a *Lebesgue measurable set* if

$$m^*(A) = m^*\left(A \cap E\right) + m^*\left(A \cap E^c\right) \quad \text{for every } A \subset \mathbb{R}^d.$$

Every rectangle R is a Lebesgue measurable set and satisfies $m^*(R) = m_0(R)$. The family $\mathcal{M}$ of all Lebesgue measurable sets is a σ-algebra. Moreover, the restriction of m^* to $\mathcal{M}$ is σ-additive and, hence, a measure. By the previous observation, $\mathcal{M}$ contains every Borel set of $\mathbb{R}^d$. The restriction of m^* to the Borel σ-algebra $\mathcal{B}$ of $\mathbb{R}^d$ coincides with the Lebesgue measure on $\mathbb{R}^d$.

Actually, $\mathcal{M}$ coincides with the completion of the Borel σ-algebra of $\mathbb{R}^d$ with respect to the Lebesgue measure. This and other related properties are part of Exercise A.1.13.

Example A.1.23. Let $\phi : [0,1] \to \mathbb{R}$ be a positive continuous function. Given any interval I, with endpoints $0 \le a < b \le 1$, define

$$\mu_\phi(I) = \int_a^b \phi(x)\,dx \qquad \text{(Riemann integral).}$$

Next, extend the definition of μ_ϕ to the algebra $\mathcal{A}$ formed by the finite unions $A = I_1 \cup \cdots \cup I_k$ of pairwise disjoint intervals, through the relation

$$\mu_\phi(A) = \sum_{j=1}^{k} \mu_\phi(I_j).$$

The basic properties of the Riemann integral ensure that μ_ϕ is finitely additive. We leave it to the reader to check that the measure μ_ϕ is σ-additive in the algebra $\mathcal{A}$ (see Exercise A.1.7). Moreover, $\mu_\phi(\emptyset) = 0$ and $\mu_\phi([0,1]) < \infty$, because ϕ is continuous and, hence, bounded. With the help of Theorem A.1.13, we may extend μ_ϕ to the whole Borel σ-algebra of $[0,1]$.

The measure μ_ϕ that we have just constructed has the following special property: if a set $A \subset [0,1]$ has Lebesgue measure zero then $\mu_\phi(A) = 0$. This property is called *absolute continuity* (with respect to the Lebesgue measure) and is studied in a lot more depth in Appendix A.2.4.

Here is an example of a measure that is positive on any open set but is not absolutely continuous with respect to Lebesgue measure:

Example A.1.24. Fix any enumeration $\{r_1, r_2, \dots\}$ of the set $\mathbb{Q}$ of rational numbers. Consider the measure μ defined on $\mathbb{R}$ by

$$\mu(A) = \sum_{r_i \in A} \frac{1}{2^i}.$$

On the one hand, the measure of any non-empty open subset of the real line is positive, for such a subset must contain some r_i. On the other hand, the measure of $\mathbb{Q}$ is

$$\mu(\mathbb{Q}) = \sum_{r_i \in \mathbb{Q}} \frac{1}{2^i} = 1.$$

Since $\mathbb{Q}$ has Lebesgue measure zero (because it is a countable set), this implies that μ is not absolutely continuous with respect to the Lebesgue measure.

This example also motivates the concept of the support of a measure on a topological space (X, τ), which we introduce next. For that, we must recall a few basic ideas from topology.

A subset τ' of the topology τ is a *basis of the topology*, or a *basis of open sets*, if for every $x \in X$ and every open set U containing x there exists $U' \in \tau'$ such that $x \in U' \subset U$. We say that the topological space admits a *countable basis of open sets* if such a subset τ' may be chosen to be countable. A set $V \subset X$ is a *neighborhood* of a point $x \in X$ if there exists some open set U such that $x \in U \subset V$. Thus, a subset X is open if and only if it is a neighborhood of each one of its points. A family υ' of subsets of X is a *basis of neighborhoods* of a point $x \in X$ if for every neighborhood V there exists some $V' \in \upsilon'$ such that $x \in V' \subset V$. We say that x admits a *countable basis of neighborhoods* if υ' may be chosen to be countable. If the topological space admits a countable basis of open sets then every $x \in X$ admits a countable basis of neighborhoods, namely, the family of elements of the countable basis of open sets that contain x.

Definition A.1.25. Let (X, τ) be a topological space and μ be a measure on the Borel σ-algebra of X. The *support* of the measure μ is the set $\operatorname{supp}\mu$ formed by the points $x \in X$ such that $\mu(V) > 0$ for any neighborhood V of x.

It follows immediately from the definition that the support of a measure is a closed set. In Example A.1.24 above, the support of μ is the whole real line, despite the fact that $\mu(\mathbb{Q}) = 1$.

Proposition A.1.26. *If X is a topological space with a countable basis of open sets and μ is a non-zero measure on X, then the support* $\operatorname{supp}\mu$ *is non-empty.*

Proof. If $\operatorname{supp}\mu$ is empty then for each point $x \in X$ we may find an open neighborhood V_x such that $\mu(V_x) = 0$. Let $\{A_j : j = 1, 2, \dots\}$ be a countable basis of the topology of X. Then, for each $x \in X$ we may choose $i(x) \in \mathbb{N}$ such that $x \in A_{i(x)} \subset V_x$. Hence,

$$X = \bigcup_{x \in X} V_x = \bigcup_{x \in X} A_{i(x)}$$

and so

$$\mu(X) = \mu\left(\bigcup_{x \in X} A_{i(x)}\right) \leq \sum_{i=1}^{\infty} \mu(A_i) = 0.$$

This is a contradiction, and so $\operatorname{supp} \mu$ cannot be empty.

A.1.4 Measurable maps

Measurable maps play a role in measure theory similar to the role of continuous maps in topology: measurability corresponds to the idea that the map preserves the family of measurable subsets, just as continuity means that the family of open subsets is preserved by the map.

Definition A.1.27. Given measurable spaces $(X, \mathcal{B})$ and $(Y, \mathcal{C})$, we say that a map $f : X \to Y$ is *measurable* if $f^{-1}(C) \in \mathcal{B}$ for every $C \in \mathcal{C}$.

In general, the family of sets $C \in \mathcal{C}$ such that $f^{-1}(C) \in \mathcal{B}$ is a σ-algebra. So, to prove that f is measurable it suffices to show that $f^{-1}(C_0) \in \mathcal{B}$ for every set C_0 in some family $\mathcal{C}_0 \subset \mathcal{C}$ that generates the σ-algebra $\mathcal{C}$. See also Exercise A.1.1.

Example A.1.28. A function $f : X \to [-\infty, \infty]$ is measurable if and only if the set $f^{-1}((c, +\infty])$ belongs to $\mathcal{B}$ for every $c \in \mathbb{R}$. This follows from the previous observation, since the family of intervals $(c, +\infty]$ generates the Borel σ-algebra of the extended line (recall Example A.1.9). In particular, if a function f takes values in $(-\infty, +\infty)$ then it is measurable if and only if $f^{-1}((c, +\infty))$ belongs to $\mathcal{B}$ for every $c \in \mathbb{R}$.

Example A.1.29. If X is a topological space and $\mathcal{B}$ is the corresponding Borel σ-algebra, then every continuous function $f : X \to \mathbb{R}$ is measurable. Indeed, continuity means that the pre-image of every open subset of $\mathbb{R}$ is an open subset of X and, hence, is in $\mathcal{B}$. Since the family of open sets generates the Borel σ-algebra of $\mathbb{R}$, it follows that the pre-image of every Borel subset of the real line is also in $\mathcal{B}$.

Example A.1.30. The *characteristic function* $\mathcal{X}_B : X \to \mathbb{R}$ of a set $B \subset X$ is defined by:

$$\mathcal{X}_B(x) = \begin{cases} 1, & \text{if } x \in B; \\ 0, & \text{otherwise.} \end{cases}$$

Observe that the function $\mathcal{X}_B$ is measurable if and only if B is a measurable subset: indeed, $\mathcal{X}_B^{-1}(A) \in \{\emptyset, B, X \setminus B, X\}$ for any $A \subset \mathbb{R}$.

Among the basic properties of measurable functions, let us highlight:

Proposition A.1.31. *Let $f,g : X \to [-\infty,+\infty]$ be measurable functions and let $a,b \in \mathbb{R}$. Then the following functions are also measurable:*

$$(af+bg)(x) = af(x)+bg(x) \quad \textit{and} \quad (f\cdot g)(x) = f(x)\cdot g(x).$$

Moreover, if $f_n : X \to [-\infty,+\infty]$ is a sequence of measurable functions, then the following functions are also measurable:

$$s(x) = \sup\{f_n(x) : n \geq 1\} \quad \textit{and} \quad i(x) = \inf\{f_n(x) : n \geq 1\},$$

$$f^*(x) = \limsup_n f_n(x) \quad \textit{and} \quad f_*(x) = \liminf_n f_n(x).$$

In particular, if $f(x) = \lim f_n(x)$ exists then f is measurable.

The linear combinations of characteristic functions form an important class of measurable functions:

Definition A.1.32. We say that a function $s : X \to \mathbb{R}$ is *simple* if there exist constants $\alpha_1,\dots,\alpha_k \in \mathbb{R}$ and pairwise disjoint measurable sets $A_1,\dots,A_k \in \mathcal{B}$ such that

$$s = \sum_{j=1}^{k} \alpha_j \mathcal{X}_{A_j}, \tag{A.1.2}$$

where $\mathcal{X}_A$ is the characteristic function of the set A.

Note that every simple function is measurable. In the converse direction, the result that follows asserts that every measurable function is the limit of a sequence of simple functions. This fact will be very useful in the next appendix, when defining the Lebesgue integral.

Proposition A.1.33. *Let $f : X \to [-\infty,+\infty]$ be a measurable function. Then there exists a sequence $(s_n)_n$ of simple functions such that $|s_n(x)| \leq |f(x)|$ for every n and*

$$\lim_n s_n(x) = f(x) \textit{ for every } x \in X.$$

If f takes values in $\mathbb{R}$, we may take every s_n with values in $\mathbb{R}$. If f is bounded, the sequence $(s_n)_n$ may be chosen such that the convergence is uniform. If f is non-negative, we may take $0 \leq s_1 \leq s_2 \leq \cdots \leq f$.

In Exercise A.1.16 the reader is invited to prove this proposition.

A.1.5 Exercises

A.1.1. Let X be a set and $(Y,\mathcal{C})$ be a measurable space. Show that, for any transformation $f : X \to Y$ there exists some σ-algebra $\mathcal{B}$ of subsets of X such that the transformation is measurable with respect to the σ-algebras $\mathcal{B}$ and $\mathcal{C}$.

A.1.2. Let X be a set and consider the family of subsets

$$\mathcal{B}_0 = \{A \subset X : A \text{ is finite or } A^c \text{ is finite}\}.$$

Show that $\mathcal{B}_0$ is an algebra. Moreover, $\mathcal{B}_0$ is a σ-algebra if and only if the set X is finite. Show also that, in general,

$$\mathcal{B}_1 = \{A \subset X : A \text{ is finite or countable or } A^c \text{ is finite or countable}\}$$

is the σ-algebra generated by the algebra $\mathcal{B}_0$.

A.1.3. Prove Proposition A.1.5.

A.1.4. The purpose of this exercise is to exhibit a *non-Borel subset* of the real line. Let α be any irrational number. Consider the following relation on $\mathbb{R} : x \sim y \Leftrightarrow$ there are $m, n \in \mathbb{Z}$ such that $x - y = m + n\alpha$. Check that $\sim$ is an equivalence relation and every equivalence class intersects $[0,1)$. Let E_0 be a subset of $[0,1)$ containing exactly one element of each equivalence class (the existence of such a set is a consequence of the Axiom of Choice). Show that E_0 is not a Borel set.

A.1.5. Let $(X, \mathcal{B}, \mu)$ be a measure space. Show that if $A_1, A_2, \ldots$ are in $\mathcal{B}$ then

$$\mu\Big(\bigcup_{j=1}^{\infty} A_j\Big) \le \sum_{j=1}^{\infty} \mu(A_j).$$

A.1.6. (Lemma of Borel–Cantelli). Let $(E_n)_n$ be a countable family of measurable sets. Let F be the set of points that belong to E_n for infinitely many values of n, that is, $F = \limsup_n E_n = \bigcap_{k=1}^{\infty} \bigcup_{n=k}^{\infty} E_n$. Show that if $\sum_n \mu(E_n) < \infty$ then $\mu(F) = 0$.

A.1.7. Prove Theorem A.1.14.

A.1.8. Let $\mathcal{A}$ be the collection of subsets of $X = [0,1]$ that may be written as finite unions of pairwise disjoint intervals. Check that $\mathcal{A}$ is an algebra of subsets of X. Let $m_0 : \mathcal{A} \to [0,1]$ be the function defined on this algebra by

$$m_0(I_1 \cup \cdots \cup I_N) = |I_1| + \cdots + |I_N|,$$

where $|I_j|$ represents the length of I_j. Show that m_0 is σ-additive.

A.1.9. Let $\mathcal{B}$ be an algebra of subsets of X and $\mu : \mathcal{B} \to [0, +\infty)$ be a finitely additive function with $\mu(X) < \infty$. Show that μ is σ-additive if and only if any one of the following conditions holds:

(a) $\lim_n \mu(A_n) = \mu(\bigcap_{j=1}^{\infty} A_j)$ for any decreasing sequence $A_1 \supset \cdots \supset A_j \supset \cdots$ of elements of $\mathcal{B}$;

(b) $\lim_n \mu(A_n) = \mu(\bigcup_{j=1}^{\infty} A_j)$ for any increasing sequence $A_1 \subset \cdots \subset A_j \subset \cdots$ of elements of $\mathcal{B}$.

A.1.10. Show that $\|\mu\| = |\mu|(X)$ defines a complete norm in the vector space of finite signed measures on a measurable space $(X, \mathcal{B})$.

A.1.11. Let $X = \{1, \ldots, d\}$ be a finite set, endowed with the discrete topology, and let $M = X^{\mathcal{I}}$ with $\mathcal{I} = \mathbb{N}$ or $\mathcal{I} = \mathbb{Z}$.

(a) Check that (A.2.7) defines a distance on M and that the topology defined by this distance coincides with the product topology on M. Describe the open balls and the closed balls around any point $x \in X^{\mathcal{I}}$.

(b) Without using the theorem of Tychonoff, show that (M, d) is a compact space.

(c) Let $\mathcal{A}$ be the algebra generated by the elementary cylinders of M. Show that every additive function $\mu : \mathcal{A} \to [0,1]$ with $\mu(M) = 1$ extends to a probability measure on the Borel σ-algebra of M.

A.1.12. Let $K \subset [0,1]$ be the *Cantor set*, that is, $K = \bigcap_{n=0}^{\infty} K_n$ where $K_0 = [0,1]$ and each K_n is the set obtained by removing from each connected component C of K_{n-1} the open interval whose center coincides with the center of C and whose length is one third of the length of C. Show that K has Lebesgue measure equal to zero.

A.1.13. Given a set $E \subset \mathbb{R}^d$, prove that the following conditions are equivalent:

(a) E is a Lebesgue measurable set.

(b) E belongs to the completion of the Borel σ-algebra relative to the Lebesgue measure, that is, there exist Borel sets $B_1, B_2 \subset \mathbb{R}^d$ such that $B_1 \subset E \subset B_2$ and $m(B_2 \setminus B_1) = 0$.

(c) (Approximation from above by open sets) Given $\varepsilon > 0$ we can find an open set A such that $E \subset A$ and $m^*(A \setminus E) < \varepsilon$.

(d) (Approximation from below by closed sets) Given $\varepsilon > 0$ we can find a closed set F such that $F \subset E$ and $m^*(E \setminus F) < \varepsilon$.

A.1.14. Prove Proposition A.1.31.

A.1.15. Let $g_n : M \to \mathbb{R}$, $n \geq 1$ be a sequence of measurable functions such that $f(x) = \sum_{n=1}^{\infty} g_n(x)$ converges at every point. Show that the sum f is a measurable function.

A.1.16. Prove Proposition A.1.33.

A.1.17. Let $f : X \to X$ be a measurable transformation and ν be a measure on X. Define $(f_*\nu)(A) = \nu(f^{-1}(A))$. Show that $f_*\nu$ is a measure and note that it is finite if and only if ν itself is finite.

A.1.18. Let $\omega_5 : [0,1] \to [0,1]$ be the function assigning to each $x \in [0,1]$ the *upper frequency of the digit* 5 in the decimal expansion of x. In other words, writing $x = 0.a_0a_1a_2\ldots$ with $a_i \neq 9$ for infinitely many values of i,

$$\omega_5(x) = \limsup_n \frac{1}{n}\#\{0 \leq j \leq n-1 : a_j = 5\}.$$

Prove that the function ω_5 is measurable.

A.2 Integration in measure spaces

In this appendix we define the Lebesgue integral of a measurable function with respect to a measure. This generalizes the notion of Riemann integral that is usually presented in calculus or introductory analysis courses to a much broader class of functions. Indeed, the Riemann integral is not defined for many useful functions, for example the characteristic functions of arbitrary measurable sets (see Example A.2.5 below). In contrast, the Lebesgue integral makes sense for the whole class of measurable functions, which, as we have seen in Proposition A.1.31, is closed under all the main operations in analysis.

Also in this appendix, we state some important results about the behavior of the (Lebesgue) integral under limits of sequences. Moreover, we describe the product of any finite family of finite measures; for probability measures we even extend this construction to countable families. Near the end, we discuss the related notions of absolute continuity and Lebesgue derivation.

A.2.1 Lebesgue integral

Throughout this section, we always take $(X, \mathcal{B}, \mu)$ to be a measure space. We are going to introduce the notion of Lebesgue integral in a certain number of steps. The first one deals with the integral of a simple function:

Definition A.2.1. Let $s = \sum_{j=1}^{k} \alpha_j \mathcal{X}_{A_j}$ be a simple function. The *integral* of s is given by:

$$\int s \, d\mu = \sum_{j=1}^{k} \alpha_j \mu(A_j).$$

It is easy to check (Exercise A.2.1) that this definition is consistent: if two different linear combinations of characteristic functions define the same function then the values of the integrals obtained from those two linear combinations are equal.

The next step is to define the integral of a non-negative measurable function. The idea is to approximate the function by a monotone sequence of simple functions, using Proposition A.1.33:

Definition A.2.2. Let $f : X \to [0, \infty]$ be a non-negative measurable function. Then

$$\int f d\mu = \lim_n \int s_n d\mu,$$

where $s_1 \le s_2 \le \ldots$ is a non-decreasing sequence of simple functions such that $\lim_n s_n(x) = f(x)$ for every $x \in X$.

It is not difficult to check (Exercise A.2.2) that this definition is consistent: the value of the integral does not depend on the choice of the sequence $(s_n)_n$.

Next, to extend the definition of integral to an arbitrary measurable function, let us observe that given any function $f : X \to [-\infty, +\infty]$ we can always write $f = f^+ - f^-$ with

$$f^+(x) = \max\{f(x), 0\} \quad \text{and} \quad f^-(x) = \max\{-f(x), 0\}.$$

It is clear that the functions f^+ and f^- are non-negative. Moreover, by Proposition A.1.31, they are measurable whenever f is measurable.

Definition A.2.3. Let $f : X \to [-\infty, +\infty]$ be a measurable function. Then

$$\int f \, d\mu = \int f^+ \, d\mu - \int f^- \, d\mu,$$

as long as at least one of the integrals on the right-hand side is finite (with the usual conventions that $(+\infty) - a = +\infty$ and $a - (+\infty) = -\infty$ for every $a \in \mathbb{R}$).

Definition A.2.4. A function $f : X \to [-\infty, +\infty]$ is *integrable* if it is measurable and its integral is a real number. We denote the set of all integrable functions as $\mathcal{L}^1(X, \mathcal{B}, \mu)$ or, simply, as $\mathcal{L}^1(\mu)$.

Given a measurable function $f : X \to [-\infty, \infty]$ and a measurable set E, we define the *integral of f over E* to be

$$\int_E f d\mu = \int f \mathcal{X}_E \, d\mu,$$

where $\mathcal{X}_E$ is the characteristic function of the set E.

Example A.2.5. Consider $X = [0,1]$ endowed with the Lebesgue measure m. Let $f = \mathcal{X}_B$, where B is the subset of rational numbers. Then $m(B) = 0$ and so, using Definition A.2.2, the Lebesgue integral of f is equal to zero. On the other hand, a direct calculation shows that every lower Riemann sum of f is equal to 0, while every upper Riemann sum of f is equal to 1. So, the Riemann integral of f does not exist. Indeed, more generally, the Riemann integral of the characteristic function of a measurable set exists if and only if the boundary of the set has zero Lebesgue measure. Note that in the present case the boundary is the whole of $[0,1]$, which has positive Lebesgue measure.

Example A.2.6. Let $x_1, \ldots, x_m \in X$ and $p_1, \ldots, p_m > 0$ with $p_1 + \cdots + p_m = 1$. Let μ be the probability measure μ defined on 2^X by

$$\mu = \sum_{i=1}^{m} p_i \delta_{x_i} \quad \text{where } \delta_{x_i} \text{ is the Dirac mass at } x_i.$$

In other words, $\mu(A) = \sum_{x_i \in A} p_i$ for every subset A of X. Then, for any function $f : X \to [-\infty, +\infty]$,

$$\int f \, d\mu = \sum_{i=1}^{m} p_i f(x_i).$$

Proposition A.2.7. *The set $\mathcal{L}^1(\mu)$ of all real integrable functions is a real vector space. Moreover, the map $I : \mathcal{L}^1(\mu) \to \mathbb{R}$ given by $I(f) = \int f \, d\mu$ is a positive linear functional:*

(1) $\int af + bg \, d\mu = a \int f \, d\mu + b \int g \, d\mu$, and
(2) $\int f \, d\mu \geq \int g \, d\mu$ if $f(x) \geq g(x)$ for every x.

In particular, $|\int f \, d\mu| \leq \int |f| \, d\mu$ if $|f| \in \mathcal{L}^1(\mu)$. Moreover, $|f| \in \mathcal{L}^1(\mu)$ if and only if $f \in \mathcal{L}^1(\mu)$.

The notion of the Lebesgue integral may be extended to an even broader class of functions, in two different ways. On the one hand, we may consider complex functions $f : X \to \mathbb{C}$. In this case, we say that f is integrable if and only if the real part $\Re f$ and the imaginary part $\Im f$ are both integrable. Then, by definition,

$$\int f \, d\mu = \int \Re f \, d\mu + i \int \Im f \, d\mu.$$

On the other hand, we may consider functions that are not necessarily measurable but coincide with some measurable function on a subset of the domain with total measure. To explain this, we need the following notion, which is used frequently throughout the text:

Definition A.2.8. We say that a property holds *at μ-almost every point* (or *μ-almost everywhere*) if the subset of points of X for which it does not hold is contained in some zero measure set.

For example, we say that a sequence of functions $(f_n)_n$ converges to some function at μ-almost every point if there exists some measurable set $N \subset X$ with $\mu(N) = 0$ such that $f(x) = \lim_n f_n(x)$ for every $x \in X \setminus N$. Analogously, we say that two functions f and g are equal at μ-almost every point if there exists a measurable set $N \subset X$ with $\mu(N) = 0$ such that $f(x) = g(x)$ for every $x \in X \setminus N$. Clearly, this is an equivalence relation in the space of functions defined on X. Moreover, assuming that the two functions are integrable, it implies that the two integrals coincide:

$$\int f\,d\mu = \int g\,d\mu \quad \text{if } f = g \text{ at } \mu\text{-almost every point.}$$

This observation permits the definition of the integral for any function f, possibly non-measurable, that coincides at μ-almost every point with some measurable function g: it suffices to take $\int f\,d\mu = \int g\,d\mu$.

To close this section, let us observe that the notion of integral may also be extended to signed measures and even complex measures, as follows. Let μ be a signed measure and $\mu = \mu^+ - \mu^-$ be its Hahn decomposition. We say that a function ϕ is integrable with respect to μ if it is integrable with respect to both μ^+ and μ^-. Then we define:

$$\int \phi\,d\mu = \int \phi\,d\mu^+ - \int \phi\,d\mu^-.$$

Similarly, let μ be a complex measure. By definition, a function ϕ is integrable with respect to μ if it is integrable with respect to both the real part $\Re\mu$ and the imaginary part $\Im\mu$. Then we define:

$$\int \phi\,d\mu = \int \phi\,d\Re\mu - \int \phi\,d\Im\mu.$$

A.2.2 Convergence theorems

Next, we mention three important results concerning the convergence of functions under the integral sign. The first one deals with monotone sequences of functions:

Theorem A.2.9 (Monotone convergence). *Let $f_n : X \to [-\infty, +\infty]$ be a non-decreasing sequence of non-negative measurable functions. Consider the* *function* $f : X \to [-\infty, +\infty]$ *defined by* $f(x) = \lim_n f_n(x)$. *Then*

$$\lim_n \int f_n \, d\mu = \int f(x) \, d\mu.$$

The next result applies to much more general sequences, not necessarily monotone:

Theorem A.2.10 (Lemma of Fatou)**.** *Let* $f_n : X \to [0, +\infty]$ *be a sequence of non-negative measurable functions. Then the function* $f : X \to [-\infty, +\infty]$ *defined by* $f(x) = \liminf_n f_n(x)$ *is integrable and satisfies*

$$\int \liminf_n f_n(x) \, d\mu \le \liminf_n \int f_n \, d\mu.$$

The most powerful of the results in this section is the dominated convergence theorem, which asserts that we may take the limit under the integral sign whenever the sequence of functions is bounded by some integrable function:

Theorem A.2.11 (Dominated convergence)**.** *Let* $f_n : X \to \mathbb{R}$ *be a sequence of measurable functions and assume that there exists some integrable function* $g : X \to \mathbb{R}$ *such that* $|f_n(x)| \le |g(x)|$ *for* μ*-almost every* x *in* X. *Assume moreover that the sequence* $(f_n)_n$ *converges at* μ*-almost every point to some function* $f : X \to \mathbb{R}$. *Then* f *is integrable and satisfies*

$$\lim_n \int f_n \, d\mu = \int f \, d\mu.$$

In Exercise A.2.7 we invite the reader to deduce the dominated convergence theorem from the Lemma of Fatou.

A.2.3 Product measures

Let $(X_j, \mathcal{A}_j, \mu_j), j = 1, \dots, n$ be finite measure spaces, that is, such that $\mu_j(X_j) < \infty$ for every j. One can endow the Cartesian product $X_1 \times \cdots \times X_n$ with the structure of a finite measure space in the following way. Consider on $X_1 \times \cdots \times X_n$ the σ-algebra generated by the family of all subsets of the form $A_1 \times \cdots \times A_n$ with $A_j \in \mathcal{A}_j$. This is called the *product σ-algebra* and is denoted by $\mathcal{A}_1 \otimes \cdots \otimes \mathcal{A}_n$.

Theorem A.2.12. *There exists a unique measure* μ *on the measurable space* $(X_1 \times \cdots \times X_n, \mathcal{A}_1 \otimes \cdots \otimes \mathcal{A}_n)$ *such that* $\mu(A_1 \times \cdots \times A_n) = \mu_1(A_1) \cdots \mu_n(A_n)$ *for every* $A_1 \in \mathcal{A}_1, \dots, A_n \in \mathcal{A}_n$. *In particular,* μ *is a finite measure.*

The proof of this result (see Theorem 35.B in Halmos [Hal50]) combines the extension theorem (Theorem A.1.13) with the monotone convergence theorem (Theorem A.2.9). The measure μ in the statement is the *product* of the measures $\mu_1, \dots, \mu_n$ and is denoted by $\mu_1 \times \cdots \times \mu_n$. In this way one defines the *product measure space*

$$(X_1 \times \cdots \times X_n, \mathcal{A}_1 \otimes \cdots \otimes \mathcal{A}_n, \mu_1 \times \cdots \times \mu_n).$$

Theorem A.2.12 remains valid when the measures μ_j are just σ-finite, except that in this case the product measure μ is also only σ-finite.

Next, we describe the product of a *countable* family of measure spaces. Actually, for now we restrict ourselves to the case of probability spaces. Let $(X_j, \mathcal{B}_j, \mu_j), j \in \mathcal{I}$ be probability measure spaces with $\mu_j(X_j) = 1$ for every $j \in \mathcal{I}$. What follows holds for both $\mathcal{I} = \mathbb{N}$ and $\mathcal{I} = \mathbb{Z}$. Consider the Cartesian product

$$\Sigma = \prod_{j \in \mathcal{I}} X_j = \{(x_j)_{j \in \mathcal{I}} : x_j \in X_j\}. \tag{A.2.1}$$

We call *cylinders* of Σ all subsets of the form

$$[m; A_m, \dots, A_n] = \{(x_j)_{j \in \mathcal{I}} : x_j \in A_j \text{ for } m \le j \le n\}, \tag{A.2.2}$$

where $m \in \mathcal{I}$ and $n \ge m$ and $A_j \in \mathcal{B}_j$ for each $m \le j \le n$. Note that X itself is a cylinder: we may write $X = [1; X_1]$, for example. By definition, the *product σ-algebra* on Σ is the σ-algebra $\mathcal{B}$ generated by the family of all cylinders. The family $\mathcal{A}$ of all finite unions of pairwise disjoint cylinders is an algebra and it generates the product σ-algebra $\mathcal{B}$.

Theorem A.2.13. *There exists a unique measure μ on $(\Sigma, \mathcal{B})$ such that*

$$\mu([m; A_m, \dots, A_n]) = \mu_m(A_m) \cdots \mu_n(A_n) \tag{A.2.3}$$

for every cylinder $[m; A_m, \dots, A_n]$. In particular, μ is a probability measure.

The proof of this theorem (see Theorem 38.B in Halmos [Hal50]) uses the extension theorem (Theorem A.1.13) together with the theorem of continuity at the empty set (Theorem A.1.14). The probability measure μ is called the *product* of the measures μ_j and is denoted as $\prod_{j \in \mathcal{I}} \mu_j$. The probability space $(\Sigma, \mathcal{B}, \mu)$ is called the *product* of the spaces $(X_j, \mathcal{B}_j, \mu_j), j \in \mathcal{I}$.

An important special case is when the spaces $(X_i, \mathcal{B}_i, \mu_i)$ are all equal to a given $(X, \mathcal{C}, \nu)$. The corresponding product space may be used to model a sequence of identical random experiments such that the outcome of each experiment is independent of all the others. To explain this, take X to be the set of possible outcomes of each experiment and let ν be the probability distribution of those outcomes. In this context, the measure $\mu = \nu^{\mathcal{I}} = \prod_{j \in \mathcal{I}} \nu$ is usually called the *Bernoulli measure* defined by ν. Property (A.2.3) corresponds to the identity

$$\mu([m; A_m, \dots, A_n]) = \prod_{j=m}^{n} \nu(A_j), \tag{A.2.4}$$

which may be read in the following way: the probability of any composite event $\{x_m \in A_m, \dots, x_n \in A_n\}$ is equal to the product of the probabilities of the individual events $x_i \in A_i$. So, (A.2.4) does reflect the assumption that the successive experiments are mutually independent.

We have a special interest in the case when X is a finite set, endowed with the σ-algebra $\mathcal{C} = 2^X$ of all its subsets. In this case, it is useful to consider the *elementary cylinders*

$$[m; a_m, \dots, a_n] = \{(x_j)_{j \in \mathcal{I}} \in X : x_m = a_m, \dots, x_n = a_n\}, \tag{A.2.5}$$

corresponding to subsets A_j consisting of a single point a_j. Observe that every cylinder is a finite union of pairwise disjoint elementary cylinders. In particular, the σ-algebra generated by the elementary cylinders coincides with the σ-algebra generated by all the cylinders, and the same is true for the generated algebra. Moreover, the relation (A.2.4) may be written as

$$\mu([m; a_m, \dots, a_n]) = p_{a_m}, \cdots p_{a_n} \quad \text{where } p_a = \nu(\{a\}) \text{ for } a \in X. \tag{A.2.6}$$

Consider the finite set X endowed with the discrete topology. The product topology on $\Sigma = X^{\mathcal{I}}$ coincides with the topology generated by the elementary cylinders. Moreover (see Exercise A.1.11), it coincides with the topology associated with the distance defined by

$$d\big((x_i)_{i \in \mathcal{I}}, (y_i)_{i \in \mathcal{I}}\big) = \theta^N, \tag{A.2.7}$$

where $\theta \in (0,1)$ is fixed and $N = N((x_i)_{i \in \mathcal{I}}, (y_i)_{i \in \mathcal{I}}) \geq 0$ is the largest integer such that $x_i = y_i$ for every $i \in \mathcal{I}$ with $|i| < N$.

A.2.4 Derivation of measures

Let m be the Lebesgue measure on $\mathbb{R}^d$. Given a measurable subset A of $\mathbb{R}^d$, we say that $a \in \mathbb{R}^d$ is a *density point* of A if the subset A occupies most of every small neighborhood of a, in the following sense:

$$\lim_{\delta \to 0} \frac{m(B(a,\delta) \cap A)}{m(B(a,\delta))} = 1. \tag{A.2.8}$$

Theorem A.2.14. *Let A be a measurable subset of $\mathbb{R}^d$ with Lebesgue measure $m(A)$ positive. Then m-almost every $a \in A$ is a density point of A.*

In Exercise A.2.11 we propose a proof of this result. It is also a direct consequence of the theorem that we state next. We say that a function $f : \mathbb{R}^d \to \mathbb{R}$ is *locally integrable* if the product $f\mathcal{X}_K$ is integrable for every compact set $K \subset \mathbb{R}^d$.

Theorem A.2.15 (Lebesgue derivation). *Let $X = \mathbb{R}^d$ and $\mathcal{B}$ be a Borel σ-algebra and m be the Lebesgue measure on $\mathbb{R}^d$. Let $f : X \to \mathbb{R}$ be a locally integrable function. Then*

$$\lim_{r \to 0} \frac{1}{m(B(x,r))} \int_{B(x,r)} |f(y) - f(x)|\, dm = 0 \quad \textit{at } m\textit{-almost every point } x.$$

In particular,

$$\lim_{r\to 0}\frac{1}{m(B(x,r))}\int_{B(x,r)} f(y)dm = f(x) \quad \text{at } m\text{-almost every point } x.$$

The crucial ingredient in the proof of these results is the following geometric fact:

Theorem A.2.16 (Lemma of Vitali). *Let m be the Lebesgue measure on $\mathbb{R}^d$ and suppose that for every $x \in \mathbb{R}$ one is given a sequence $(B_n(x))_n$ of balls centered at x with radii converging to zero. Let $A \subset \mathbb{R}^d$ be a measurable set with $m(A) > 0$. Then, for every $\varepsilon > 0$ there exist sequences $(x_j)_j$ in $\mathbb{R}$ and $(n_j)_j$ in $\mathbb{N}$ such that*

1. *the balls $B_{n_j}(x_j)$ are pairwise disjoint;*
2. $m\Big(\bigcup_j B_{n_j}(x_j) \setminus A\Big) < \varepsilon$ *and* $m\Big(A \setminus \bigcup_j B_{n_j}(x_j)\Big) = 0$.

This theorem remains valid if, instead of balls, we take for $(B_n(x))_n$ any sequence of sets such that $\bigcap_n B_n(x) = \{x\}$ and

$$\sup_{x,n}\frac{\sup\{d(x,y) : y \in B_n(x)\}}{\inf\{d(x,z) : z \notin B_n(x)\}} < \infty.$$

The set of measures defined on the same measurable space possesses a natural partial order relation:

Definition A.2.17. Let μ and ν be two measures in the same measurable space $(X,\mathcal{B})$. We say that ν is *absolutely continuous* with respect to μ if every measurable set E that satisfies $\mu(E) = 0$ also satisfies $\nu(E) = 0$; then we write $\nu \ll \mu$. We say that μ and ν are *equivalent* if each one of them is absolutely continuous with respect to the other; then we write $\mu \sim \nu$. In other words, two measures are equivalent if they have exactly the same zero measure sets.

Another very important result, known as the theorem of Radon–Nikodym, asserts that if $\nu \ll \mu$ then the measure ν may be seen as the product of μ by some measurable function ρ:

Theorem A.2.18 (Radon–Nikodym). *If μ and ν are finite measures such that $\nu \ll \mu$ then there exists a measurable function $\rho : X \to [0,+\infty]$ such that $\nu = \rho\mu$, meaning that*

$$\int \phi\, d\nu = \int \phi\rho\, d\mu \quad \text{for any bounded measurable function } \phi : X \to \mathbb{R}. \tag{A.2.9}$$

In particular, $\nu(E) = \int_E \rho\, d\mu$ for every measurable set $E \subset X$. Moreover, ρ is essentially unique: any two functions satisfying (A.2.9) *coincide at μ-almost every point.*

We call ρ the *density*, or *Radon–Nikodym derivative*, of ν relative to μ and we write

$$\rho = \frac{d\nu}{d\mu}.$$

Definition A.2.19. Let μ and ν be two measures in the same measurable space $(X,\mathcal{B})$. We say that μ and ν are *mutually singular* if there exist disjoint measurable subsets A and B such that $A\cup B = X$ and $\mu(A)=0$ and $\nu(B)=0$. Then we write $\mu \perp \nu$.

The Lebesgue decomposition theorem states that, given any two finite measures μ and ν in the same measurable space, we may write $\nu = \nu_a + \nu_s$ where ν_a and ν_s are finite measures such that $\nu_a \ll \mu$ and $\nu_s \perp \mu$. Combining this with the theorem of Radon–Nikodym, we get:

Theorem A.2.20 (Lebesgue decomposition). *Given any finite measures μ and ν, there exist a measurable function $\rho : X \to [0,+\infty]$ and a finite measure η such that $\nu = \rho\mu + \eta$ and $\eta \perp \mu$.*

A.2.5 Exercises

A.2.1. Prove that the integral of a simple function is well defined: if two linear combinations of characteristic functions define the same function, then the values of the integrals obtained from the two combinations coincide.

A.2.2. Show that if $(r_n)_n$ and $(s_n)_n$ are non-decreasing sequences of non-negative functions converging at μ-almost every point to the same function $f : M \to [0,+\infty)$, then $\lim_n \int r_n\, d\mu = \lim_n \int s_n\, d\mu$.

A.2.3. Prove Proposition A.2.7.

A.2.4. (Tchebysheff–Markov inequality) Let $f : M \to \mathbb{R}$ be a non-negative function integrable with respect to a finite measure μ. Then, given any real number $a > 0$,

$$\mu\big(\{x \in M : f(x) \geq a\}\big) \leq \frac{1}{a}\int_X f\, d\mu.$$

In particular, if $\int |f|\, d\mu = 0$ then $\mu\big(\{x \in X : f(x) \neq 0\}\big) = 0$.

A.2.5. Let f be an integrable function. Show that for every $\varepsilon > 0$ there exists $\delta > 0$ such that $|\int_E f\, d\mu| < \varepsilon$ for every measurable set E with $\mu(E) < \delta$.

A.2.6. Let $\psi_1, \dots, \psi_N : M \to \mathbb{R}$ be bounded measurable functions defined on a probability space $(M,\mathcal{B},\mu)$. Show that for any $\varepsilon > 0$ there exist $x_1,\dots,x_s \in M$ and positive numbers $\alpha_1,\dots,\alpha_s$ such that $\sum_{j=1}^{s} \alpha_j = 1$ and

$$\left|\int \psi_i\, d\mu - \sum_{j=1}^{s} \alpha_j \psi_i(x_j)\right| < \varepsilon \quad \text{for every } i = 1,\dots,N.$$

A.2.7. Deduce the dominated convergence theorem (Theorem A.2.11) from the Lemma of Fatou (Theorem A.2.10).

A.2.8. A set $\mathcal{F}$ of measurable functions $f : M \to \mathbb{R}$ is said to be *uniformly integrable* with respect to a probability measure μ if for every $\alpha > 0$ there exists $C > 0$ such that $\int_{\{|f|>C\}} |f|\, d\mu < \alpha$ for every $f \in \mathcal{F}$. Show that

(a) $\mathcal{F}$ is uniformly integrable with respect to μ if and only if there exists $L > 0$ and for every $\varepsilon > 0$ there exists $\delta > 0$ such that $\int |f|\, d\mu < L$ and $\int_A |f|\, d\mu < \varepsilon$ for every $f \in \mathcal{F}$ and every measurable set A with $\mu(A) < \delta$.

(b) If there exists a function $g : M \to \mathbb{R}$ integrable with respect to μ such that $|f| \le |g|$ for every $f \in \mathcal{F}$ (we say that $\mathcal{F}$ is *dominated* by g) then the set $\mathcal{F}$ is uniformly integrable with respect to μ.

(c) If the set $\mathcal{F}$ is uniformly integrable with respect to μ then $\lim_n \int f_n\, d\mu = \int \lim f_n\, d\mu$ for any sequence $(f_n)_n$ in $\mathcal{F}$ such that $\lim_n f_n$ exists at μ-almost every point.

A.2.9. Show that a is a density point of a set $A \subset \mathbb{R}^d$ if and only if

$$\liminf_{\delta \to 0} \left\{ \frac{m(B \cap A)}{m(B)} : B \text{ a ball with } a \in B \subset B(a,\delta) \right\} = 1. \tag{A.2.10}$$

A.2.10. Let $\mathcal{P}_n$, $n \ge 1$ be a sequence of countable partitions of $\mathbb{R}^d$ into measurable subsets. Assume that the diameter $\operatorname{diam} \mathcal{P}_n = \sup\{\operatorname{diam} P : P \in \mathcal{P}_n\}$ converges to zero when $n \to \infty$. Show that, given any measurable set $A \subset \mathbb{R}^d$ with positive Lebesgue measure, it is possible to choose sets $P_n \in \mathcal{P}_n$, $n \ge 1$ in such a way that $m(A \cap P_n)/m(P_n) \to 1$ when $n \to \infty$.

A.2.11. Prove Theorem A.2.14.

A.2.12. Consider $x_1, x_2 \in M$ and $p_1, p_2, q_1, q_2 > 0$ with $p_1 + p_2 = q_1 + q_2 = 1$. Let μ and ν be the probability measures given by

$$\mu(A) = \sum_{x_i \in A} p_i, \qquad \nu(A) = \sum_{x_i \in A} q_i,$$

that is, $\mu = p_1 \delta_{x_1} + p_2 \delta_{x_2}$ and $\nu = q_1 \delta_{x_1} + q_2 \delta_{x_2}$. Check that $\nu \ll \mu$ and $\mu \ll \nu$ and calculate the corresponding Radon–Nikodym derivatives.

A.2.13. Construct a probability measure μ on $[0,1]$ absolutely continuous with respect to the Lebesgue measure m and such that there exists a measurable set $K \subset [0,1]$ with $\mu(K) = 0$ and $m(K) = 1/2$. In particular, m is not absolutely continuous with respect to μ. Could we require that $m(K) = 1$?

A.2.14. Assume that $f : X \to X$ is such that there exists a countable cover of M by measurable sets B_n, $n \ge 1$, such that the restriction of f to each B_n is a bijection onto its image, with measurable inverse. Let η be a probability measure on M such that $A \subset B_n$ and $\eta(A) = 0$ implies $\eta(f(A)) = 0$. Show that there exists a function $J_\eta : X \to [0, +\infty]$ such that

$$\int_{f(B_n)} \psi \, d\eta = \int_{B_n} (\psi \circ f) J_\eta \, d\eta$$

for every bounded measurable function $\psi : X \to \mathbb{R}$ and every n. Moreover, J_η is essentially unique.

A.2.15. Let $\mu = \mu^+ - \mu^-$ be the Hahn decomposition of a finite signed measure μ. Show that there exist functions $\rho^\pm$ and $\tau^\pm$ such that $\mu^+ = \rho^+ |\mu| = \tau^+ \mu$ and $\mu^- = \rho^- |\mu| = \tau^- \mu$. Which functions are these?

A.2.16. Let $(\mu_n)_n$ and $(\nu_n)_n$ be two sequences of measures such that $\mu = \sum_n \mu_n$ and $\nu = \sum_n \nu_n$ are finite measures. Let $\hat{\mu}_n = \sum_{i=1}^n \mu_i$ and $\hat{\nu}_n = \sum_{i=1}^n \nu_i$. Show that if $\hat{\mu}_n \ll \hat{\nu}_n$ for every n then $\mu \ll \nu$ and

$$\frac{d\mu}{d\nu} = \lim_n \frac{d\hat{\mu}_n}{d\hat{\nu}_n} \quad \text{at } \nu\text{-almost every point.}$$

A.3 Measures in metric spaces

In this appendix, unless stated otherwise, μ is a Borel probability measure on a metric space M, that is, a probability measure defined on the Borel σ-algebra of M. Most of the results extend immediately to finite Borel measures, in fact.

Recall that a *metric space* is a pair (M,d) where M is a set and d is a *distance* in M, that is, a function $d: M \times M \to \mathbb{R}$ satisfying:

1. $d(x,y) \geq 0$ for any x,y and the equality holds if and only if $x = y$;
2. $d(x,y) = d(y,x)$ for any x,y;
3. $d(x,y) \leq d(x,z) + d(z,y)$ for any x,y,z.

We denote $B(x,r) = \{y \in M : d(x,y) < r\}$ and call it the *ball* of center $x \in M$ and radius $r > 0$.

Every metric space has a natural structure of a topological space where the family of balls centered at each point is a basis of neighborhoods for that point. Equivalently, a subset of M is open if and only if it contains some ball centered at each one of its points. In the converse direction, one says that a topological space is *metrizable* if its topology can be defined in this way, from some distance function.

A.3.1 Regular measures

A first interesting fact is that any probability measure on a metric space is completely determined by the values it takes on the open subsets (or the closed subsets) of the space.

Definition A.3.1. A (Borel) measure μ on a topological space is *regular* if for every measurable subset B and every $\varepsilon > 0$ there exists a closed set F and an open set A such that $F \subset B \subset A$ and $\mu(A \setminus F) < \varepsilon$.

Proposition A.3.2. *Any probability measure on a metric space is regular.*

Proof. Let $\mathcal{B}_0$ be the family of all Borel subsets B for which the condition in the definition holds, that is, such that for every $\varepsilon > 0$ there exist a closed set F and an open set A satisfying $F \subset B \subset A$ and $\mu(A \setminus F) < \varepsilon$. Begin by noting that $\mathcal{B}_0$ contains all the closed subsets of M. Indeed, let B be any closed set and let B^δ denote the (open) set of points whose distance to B is less than δ.

By Theorem A.1.14, we have that $\mu(B^\delta \setminus B) \to 0$ when $\delta \to 0$. Hence, we may take $F = B$ and $A = B^\delta$ for $\delta > 0$ sufficiently small.

It is immediate that the family $\mathcal{B}_0$ is closed under taking the complement, that is, $B^c \in \mathcal{B}_0$ whenever $B \in \mathcal{B}_0$. Furthermore, consider any countable family B_n, $n = 1,2,\dots$ of elements of $\mathcal{B}_0$ and let $B = \bigcup_{n=1}^\infty B_n$. By hypothesis, for every $n \in \mathbb{N}$ and $\varepsilon > 0$ there exist a closed set F_n and an open set A_n satisfying $F_n \subset B_n \subset A_n$ and $\mu(A_n \setminus F_n) < \varepsilon/2^{n+1}$. The union $A = \bigcup_{n=1}^\infty A_n$ is an open set and any finite union $F = \bigcup_{n=1}^m F_n$ is a closed set. Fix m large enough that

$$\mu\Big(\bigcup_{n=1}^{\infty} F_n \setminus F\Big) < \varepsilon/2$$

(recall Theorem A.1.14). Then $F \subset B \subset A$ and

$$\mu(A \setminus F) \le \sum_{n=1}^{\infty} \mu(A_n \setminus F_n) + \mu\Big(\bigcup_{n=1}^{\infty} F_n \setminus F\Big) < \sum_{n=1}^{\infty} \frac{\varepsilon}{2^{n+1}} + \frac{\varepsilon}{2} = \varepsilon.$$

This shows that $B \in \mathcal{B}_0$. In this way, we have shown that $\mathcal{B}_0$ is a σ-algebra. Hence, $\mathcal{B}_0$ contains all the Borel subsets of M.

It follows that, as stated above, the values that the probability measure μ takes on the closed subsets of M determine μ completely: if ν is another probability measure such that $\mu(F) = \nu(F)$ for every closed set F then, taking the complement, $\mu(A) = \nu(A)$ for every open set A and, using the theorem, $\mu(B) = \nu(B)$ for every Borel set B. In other words, $\mu = \nu$. The same argument shows that the values of μ on the open sets also determine the measure completely.

The proposition that we state and prove next implies that the values of the integrals of the bounded continuous functions also determine the probability measure completely. Indeed, the same is true for the (smaller) set of bounded Lipschitz functions.

Recall that a map $h: M \to N$ is *Lipschitz* if there exists some constant $C > 0$ such that $d(h(x), h(y)) \le C d(x,y)$ for every $x, y \in M$. When it is necessary to specify the constant, we say that the function h is C-Lipschitz. More generally, we say that h is *Hölder* if there exist $C, \theta > 0$ such that $d(h(x), h(y)) \le C d(x,y)^\theta$ for every $x, y \in M$. Then we also say that h is θ-Hölder or even (C,θ)-Hölder.

Proposition A.3.3. *If μ and ν are probability measures on a metric space M with $\int \varphi\, d\mu = \int \varphi\, d\nu$ for every bounded Lipschitz function $\varphi: M \to \mathbb{R}$ then $\mu = \nu$.*

Proof. We are going to use the following simple topological fact:

Lemma A.3.4. *Given any closed subset F of M and any $\delta > 0$, there exists a Lipschitz function $g_\delta: M \to [0,1]$ such that $g_\delta(x) = 1$ for every $x \in F$ and $g_\delta(x) = 0$ for every $x \in M$ such that $d(x,F) \ge \delta$.*

Proof. Consider the function $h : \mathbb{R} \to [0,1]$ given by $h(s) = 1$ if $s \le 0$ and $h(s) = 0$ if $s \ge 1$ and $h(s) = 1 - s$ if $0 \le s \le 1$. Define

$$g : M \to [0,1], \quad g(x) = h\Big(\frac{1}{\delta} d(x,F)\Big).$$

Note that g is Lipschitz, since it is a composition of Lipschitz functions. The other properties in the lemma follow immediately from the definition.

Now we may finish the proof of Proposition A.3.3. Let F be any closed subset of M and, for every $\delta > 0$, let $g_\delta : M \to [0,1]$ be a function as in the lemma above. By assumption,

$$\int g_\delta \, d\mu = \int g_\delta \, d\nu \quad \text{for every } \delta > 0.$$

Moreover, by the dominated convergence theorem (Theorem A.2.11),

$$\lim_{\delta \to 0} \int g_\delta \, d\mu = \mu(F) \quad \text{and} \quad \lim_{\delta \to 0} \int g_\delta \, d\nu = \nu(F).$$

This shows that $\mu(F) = \nu(F)$ for every closed subset F. As pointed out before, the latter implies $\mu = \nu$.

As observed in Example A.1.29, continuous maps are automatically measurable relative to the Borel σ-algebra. The result that we prove next asserts that, under a simple condition on the metric space, there is a kind of converse: measurable maps are continuous, restricted to certain subsets with almost full measure.

A subset of a topological space M is *dense* if it intersects every open subset of M. We say that the space M is *separable* if it admits some countable dense subset. In the special case of metric spaces this is equivalent to saying that the topology admits a countable basis of open sets (Exercise A.3.1).

Theorem A.3.5 (Lusin). *Let $\varphi : M \to N$ be a measurable map with values in some separable metric space N. Given any $\varepsilon > 0$, there exists a closed set $F \subset M$ such that $\mu(M \setminus F) < \varepsilon$ and the restriction of φ to F is continuous.*

Proof. Let $\{x_n : n \in \mathbb{N}\}$ be a countable dense subset of N and, for every $k \ge 1$, let $B_{n,k}$ be the ball of center x_n and radius $1/k$. Fix $\varepsilon > 0$. By Proposition A.3.2, for every (n,k) we may find an open set $A_{n,k} \subset M$ containing $\varphi^{-1}(B_{n,k})$ and satisfying $\mu(A_{n,k} \setminus \varphi^{-1}(B_{n,k})) < \varepsilon / 2^{n+k+1}$. Define

$$E = \bigcap_{n,k=1}^{\infty} \big(\varphi^{-1}(B_{n,k}) \cup A_{n,k}^c\big).$$

On the one hand,

$$\mu(M \setminus E) \le \sum_{n,k=1}^{\infty} \mu(A_{n,k} \setminus \varphi^{-1}(B_{n,k})) < \sum_{n,k=1}^{\infty} \frac{\varepsilon}{2^{n+k+1}} = \frac{\varepsilon}{2}.$$

On the other hand, every $\varphi^{-1}(B_{n,k})$ is an open subset of $\varphi^{-1}(B_{n,k}) \cup A^c_{n,k}$, since the complement is the closed set $A^c_{n,k}$. Consequently, $\varphi^{-1}(B_{n,k})$ is open in E for every (n,k). This shows that the restriction of φ to the set E is continuous. To conclude the proof it suffices to use Proposition A.3.2 once more to find a closed set $F \subset E$ such that $\mu(E \setminus F) < \varepsilon/2$.

A.3.2 Separable complete metric spaces

Next, we discuss another important property of measures on metric spaces that are both separable and complete. Recall that the latter means that every Cauchy sequence converges.

Definition A.3.6. A (Borel) measure μ on a topological space is *tight* if for every $\varepsilon > 0$ there exists a compact subset K such that $\mu(K^c) < \varepsilon$.

Since every closed subset of a compact metric space is also compact, it follows immediately from Proposition A.3.2 that every probability measure on a compact metric space is tight. However, this conclusion is a lot more general:

Proposition A.3.7. *Every probability measure on a separable complete metric space is tight.*

Proof. Let $\{p_k : k \in \mathbb{N}\}$ be a countable dense subset of M. Then, for every $n \geq 1$, the closed balls $\bar{B}(p_k, 1/n)$, $k \in \mathbb{N}$ form a countable cover of M. Given $\varepsilon > 0$ and $n \geq 1$, fix $k(n) \geq 1$ in such a way that the (closed) set

$$L_n = \bigcup_{k=1}^{k(n)} \bar{B}(p_k, 1/n)$$

satisfies $\mu(L_n) > 1 - \varepsilon/2^n$. Take $K = \bigcap_{n=1}^{\infty} L_n$. Note that K is closed and

$$\mu(K^c) \leq \mu\left(\bigcup_{n=1}^{\infty} L_n^c\right) < \sum_{n=1}^{\infty} \frac{\varepsilon}{2^n} = \varepsilon.$$

It remains to check that K is compact. For that, it is enough to show that every sequence $(x_i)_i$ in K admits some Cauchy subsequence (since M is complete, this subsequence converges). Such a subsequence may be found as follows. Since $x_i \in L_1$ for every i, there exists $l(1) \leq k(1)$ such that the set of indices

$$\mathcal{I}_1 = \{i \in \mathbb{N} : x_i \in B(p_{l(1)}, 1)\}$$

is infinite. Let $i(1)$ be the smallest element of $\mathcal{I}_1$. Next, since $x_i \in L_2$ for every i, there exists $l(2) \leq k(2)$ such that

$$\mathcal{I}_2 = \{i \in \mathcal{I}_1 : x_i \in B(p_{l(2)}, 1/2)\}$$

is infinite. Let $i(2)$ be the smallest element of $\mathcal{I}_2 \setminus \{i(1)\}$. Repeating this procedure, we construct a decreasing sequence $\mathcal{I}_n$ of infinite subsets of $\mathbb{N}$, and an increasing sequence $i(1) < i(2) < \cdots < i(n) < \cdots$ of integers such that

$i(n) \in \mathcal{I}_n$ and all the x_i, $i \in \mathcal{I}_n$ are contained in the same closed ball of radius $1/n$. In particular,

$$d(x_{i(a)}, x_{i(b)}) \leq 2/n \quad \text{for every} \quad a, b \geq n.$$

This shows that the subsequence $(x_{i(n)})_n$ is indeed Cauchy.

Corollary A.3.8. *Assume that M is a separable complete metric space and μ is a probability measure on M. For every $\varepsilon > 0$ and every Borel set $B \subset M$ there exists a compact set $L \subset B$ such that $\mu(B \setminus L) < \varepsilon$.*

Proof. By Proposition A.3.2, we may find some closed set $F \subset B$ such that $\mu(B \setminus F) < \varepsilon/2$. By Theorem A.3.5, there exists a compact subset $K \subset M$ such that $\mu(M \setminus K) < \varepsilon/2$. Take $L = F \cap K$. Then L is compact and $\mu(B \setminus L) < \varepsilon$.

Analogously, when the metric space M is separable and complete we can improve the statement of Lusin's theorem, replacing "closed" with "compact" in the conclusion:

Theorem A.3.9 (Lusin). *Assume that M is a separable complete metric space and μ is a probability measure on M. Let $\varphi : M \to N$ be a measurable map with values in a separable metric space N. Then, given any $\varepsilon > 0$ there exists a compact set $K \subset M$ such that $\mu(M \setminus K) < \varepsilon$ and the restriction of φ to K is continuous.*

We close this section with another important fact about measures on separable complete metric spaces. A measure μ is called *atomic* if there exists some point x such that $\mu(\{x\}) > 0$; any such point is called an *atom*. Otherwise, the measure μ is said to be non-atomic.

The next theorem states that every non-atomic probability measure on a separable complete metric space is equivalent to the Lebesgue measure in the interval. The proof is given in Section 8.5.

Theorem A.3.10. *Let M be a separable complete metric space and μ be a non-atomic probability measure on M. Then there exists a measurable map $\psi : M \to [0,1]$ such that ψ is a bijection with measurable inverse, restricted to a subset with full measure, and $\psi_* \mu$ is the Lebesgue measure on $[0,1]$.*

A.3.3 Space of continuous functions

Let M be a compact metric space. We are going to describe some important properties of the vector space $C^0(M)$ of continuous functions, real or complex, defined on M. We consider on this space the norm of uniform convergence, given by:

$$\|\phi\| = \sup\{|\phi(x)| : x \in M\}.$$

This norm is complete and, hence, endows $C^0(M)$ with the structure of a Banach space.

The conclusions of the previous sections hold in this setting, since every compact metric space is separable and complete. Another useful fact about compact metric spaces is that every open cover admits some *Lebesgue number*, that is, some number $\rho > 0$ such that for every $x \in M$ there exists some element of the cover that contains the ball $B(x,\rho)$.

A linear functional $\Phi : C^0(M) \to \mathbb{C}$ is said to be *positive* if $\Phi(\varphi) \geq 0$ for every function $\varphi \in C^0(M)$ with $\varphi(x) \geq 0$ for every $x \in M$. The theorem of Riesz–Markov (see Theorem 6.19 in Rudin [Rud87]) shows that the only positive linear functionals on $C^0(M)$ are the integrals:

Theorem A.3.11 (Riesz–Markov)**.** *Let M be a compact metric space. Consider any positive linear functional $\Phi : C^0(M) \to \mathbb{C}$. Then there exists a unique finite Borel measure μ on M such that*

$$\Phi(\varphi) = \int \varphi \, d\mu \quad \text{for every } \varphi \in C^0(M).$$

Moreover, μ is a probability measure if and only if $\Phi(1) = 1$.

The next result, which is also known as the theorem of Riesz–Markov, gives an analogous representation for *continuous* linear functionals in $C^0(M)$, not necessarily positive. Recall that the *norm* of a linear functional $\Phi : C^0(M) \to \mathbb{C}$ is defined by

$$\|\Phi\| = \sup\{\frac{|\Phi(\varphi)|}{\|\varphi\|} : \varphi \neq 0\} \tag{A.3.1}$$

and that Φ is continuous if and only if the norm is finite.

Theorem A.3.12 (Riesz–Markov)**.** *Let M be a compact metric space. Consider any continuous linear functional $\Phi : C^0(M) \to \mathbb{C}$. Then there exists some complex Borel measure μ on M such that*

$$\Phi(\varphi) = \int \varphi \, d\mu \quad \text{for every } \varphi \in C^0(M).$$

The norm $\|\mu\| = |\mu|(X)$ of the measure μ coincides with the norm $\|\Phi\|$ of the functional Φ. Moreover, μ takes values in $[0,\infty)$ if and only if Φ is positive and μ takes values in $\mathbb{R}$ if and only if $\Phi(\varphi) \in \mathbb{R}$ for every real function φ.

In other words, this last theorem asserts that *the dual space of $C^0(M)$ is isometrically isomorphic to $\mathcal{M}(M)$*. Theorems A.3.11 and A.3.12 extend to locally compact topological spaces, with suitable assumptions on the behavior of the functions at infinity. In this context the measure μ is still regular, but not necessarily finite.

We also use the fact that the space $C^0(M)$ has countable dense subsets (Exercise A.3.6 is a particular instance):

Theorem A.3.13. *If M is a compact metric space then $C^0(M)$ is separable.*

Proof. We treat the case of real functions; the complex case is entirely analogous. Every compact metric space is separable. Let $\{x_k : k \in \mathbb{N}\}$ be a countable dense subset of M. For each $k \in N$, consider the function $f_k : M \to \mathbb{R}$ defined by $f_k(x) = d(x, x_k)$. Represent by $\mathcal{A}$ the set of all functions $f : M \to \mathbb{R}$ of the form

$$f = c + \sum_{k_1,\dots,k_s} c_{k_1,\dots,k_s} f_{k_1} \cdots f_{k_s} \tag{A.3.2}$$

with $c \in \mathbb{R}$ and $c_{k_1,\dots,k_s} \in \mathbb{R}$ for every $k_1, \dots, k_s \in \mathbb{N}$. It is clear that $\mathcal{A}$ contains all constant functions. Observe also that $\mathcal{A}$ is an *algebra of functions*, that is, it is closed under the operations of addition and multiplication (including multiplication by any constant). Moreover, $\mathcal{A}$ *separates* the points of M, in the sense that for any $x \neq y$ there exists some $f \in \mathcal{A}$ such that $f(x) \neq f(y)$. To see that, fix $\varepsilon > 0$ such that $d(x,y) > 2\varepsilon$, consider $k \in \mathbb{N}$ such that $d(x, x_k) < \varepsilon$ and then take $f = f_k$. Note that $f(x) = d(x, x_k) < \varepsilon$ while, by the triangle inequality, $f(y) = d(y, x_k) \geq d(x,y) - d(x, x_k) > \varepsilon$. So, the algebra of functions $\mathcal{A}$ is separating, as we claimed. Now, the theorem of Stone–Weierstrass (see [DS57, Theorem 4.6.16]) asserts that *every separating subalgebra of the space of continuous functions that contains the constant function* 1 *is dense in* $C^0(M)$. The previous observations show that this applies to $\mathcal{A}$. It follows that the (countable) set of functions of the form (A.3.2) with $c \in \mathbb{Q}$ and $c_{k_1,\dots,k_s} \in \mathbb{Q}$ is also dense in $C^0(M)$.

A.3.4 Exercises

A.3.1. Let M be a metrizable topological space. Justify that every point of M admits a countable basis of neighborhoods. Check that M is separable if and only if it admits a countable basis of open sets. Give examples of separable metric spaces and non-separable metric spaces.

A.3.2. Let μ be a finite measure on a metric space M. Show that for every closed set $F \subset M$ there exists some finite or countable set $E \subset (0, \infty)$ such that

$$\mu(\{x \in M : d(x,F) = r\}) = 0 \quad \text{for every } r \in (0,\infty) \setminus E.$$

A.3.3. Let μ be a finite measure on a separable metric space M. Show that for every $\varepsilon > 0$ there exists a countable partition of M into measurable subsets with diameter less than ε and whose boundaries have measure zero.

A.3.4. Let μ be a probability measure on $[0,1]$ and $\phi : [0,1] \to [0,1]$ be the function given by $\phi(x) = \mu([0,x])$. Check that ϕ is continuous if and only if μ is non-atomic. Check that ϕ is absolutely continuous if and only if μ is absolutely continuous with respect to the Lebesgue measure.

A.3.5. Let μ be a probability measure on some metric space M. Show that for every integrable function $\psi : M \to \mathbb{R}$ there exists a sequence $\psi_n : M \to \mathbb{R}$, $n \geq 1$ of uniformly continuous functions converging to ψ at μ-almost every point. Moreover, if ψ is bounded then we may choose the sequence in such a way that $\sup|\psi_n| \le \sup|\psi|$ for every n. Do these claims remain true if we require convergence at *every* point?

A.3.6. Without using Theorem A.3.13, show that the space $C^0([0,1]^d)$ of continuous functions, real or complex, on the compact unit cube is separable, for every $d \ge 1$.

A.4 Differentiable manifolds

In this appendix we review some fundamental notions and facts from differential topology and Riemannian geometry.

A.4.1 Differentiable manifolds and maps

A *differentiable manifold* of dimension d is a (Hausdorff) topological space M endowed with a *differentiable atlas* of dimension d, that is, a family of homeomorphisms $\varphi_\alpha : U_\alpha \to X_\alpha$ such that

1. each U_α is an open subset of M and each X_α is an open subset of $\mathbb{R}^d$ and $M = \bigcup_\alpha U_\alpha$;
2. the map $\varphi_\beta \circ \varphi_\alpha^{-1} : \varphi_\alpha(U_\alpha \cap U_\beta) \to \varphi_\beta(U_\alpha \cap U_\beta)$ is differentiable, for any α and β such that $U_\alpha \cap U_\beta \neq \emptyset$.

More generally, instead of $\mathbb{R}^d$ we may consider any Banach space E. Then we say that M is a differentiable manifold *modelled* on the space E.

The homeomorphisms φ_α are called *local charts*, or *local coordinates*, and the transformations $\varphi_\beta \circ \varphi_\alpha^{-1}$ are called *coordinate changes*. Exchanging the roles of α and β, we see that the inverse $(\varphi_\beta \circ \varphi_\alpha^{-1})^{-1} = \varphi_\alpha \circ \varphi_\beta^{-1}$ is also differentiable. So, the definition of a differentiable manifold requires the coordinate changes to be diffeomorphisms between open subsets of Euclidean space.

Unless explicitly stated otherwise, we only consider manifolds such that M admits a countable basis of open sets and is *connected*. The latter means that no subset of M is both open and closed, except for M and $\emptyset$.

Let $r \in \mathbb{N} \cup \{\infty\}$. If every coordinate change is of class C^r (that is, all its partial derivatives up to order r exist and are continuous), we say that the manifold M (and the atlas $\{\varphi_\alpha : U_\alpha \to X_\alpha\}$) are of class C^r. Clearly, every manifold of class C^r is also of class C^s for every $s \le r$.

Example A.4.1. The following are manifolds of class C^∞ and dimension d:

Euclidean space $\mathbb{R}^d$: consider the atlas consisting of a unique map, namely, the identity map $\mathbb{R}^d \to \mathbb{R}^d$.

Sphere $S^d = \{(x_0,x_1,\dots,x_d) \in \mathbb{R}^{d+1} : x_0^2 + x_1^2 + \cdots + x_d^2 = 1\}$: consider the atlas formed by the two *stereographic projections*:

$$S^d \setminus \{(1,0,\dots,0)\} \to \mathbb{R}^d, \quad (x_0,x_1,\dots,x_d) \mapsto (x_1,\dots,x_d)/(1-x_0)$$
$$S^d \setminus \{(-1,0,\dots,0)\} \to \mathbb{R}^d, \quad (x_0,x_1,\dots,x_d) \mapsto (x_1,\dots,x_d)/(1+x_0).$$

Torus $\mathbb{T}^d = \mathbb{R}^d/\mathbb{Z}^d$: consider the atlas formed by the inverses of the maps $g_z : (0,1)^d \to \mathbb{T}^d$, defined by $g_z(x) = z + x \mod \mathbb{Z}^d$ for every $z \in \mathbb{R}^d$.

Example A.4.2 (Grassmannian manifolds). Given $0 \le k \le d$, denote by $\operatorname{Gr}(k,d)$ the set of all vector subspaces of dimension k of the Euclidean space $\mathbb{R}^d$. For each $j_1 < \cdots < j_k$, denote by $\operatorname{Gr}(k,d,j_1,\dots,j_k)$ the subset of elements of $\operatorname{Gr}(k,d)$ that are transverse to $\{(x_j)_j \in \mathbb{R}^d : x_{j_1} = \cdots = x_{j_k} = 0\}$. For every $V \in \operatorname{Gr}(k,d,j_1,\dots,j_k)$ there exists a unique matrix $(u_{i,j})_{i,j}$ with $(d-k)$ rows and k columns such that

$$V = \big\{(x_j)_j \in \mathbb{R}^d : x_i = u_{i,j_1}x_{j_1} + \cdots + u_{i,j_k}x_{j_k} \text{ for every } i \notin \{j_1,\dots,j_k\}\big\}.$$

The maps $\operatorname{Gr}(k,d,j_1,\dots,j_k) \to \mathbb{R}^{(d-k)k}$ associating with each V the corresponding matrix $(u_{i,j})_{i,j}$ constitute an atlas of class C^∞ for $\operatorname{Gr}(k,d)$. So, every $\operatorname{Gr}(k,d)$ is a manifold of class C^∞ and dimension $(d-k)k$.

Let M be a manifold of dimension d and $\mathcal{A} = \{\varphi_\alpha : U_\alpha \to X_\alpha\}$ be the corresponding atlas. Let S be a subset of M. We say that S is a *submanifold* of dimension $k < d$ if there exists some atlas $\mathcal{B} = \{\psi_\beta : V_\beta \to Y_\beta\}$ of M such that

(i) $\mathcal{A}$ and $\mathcal{B}$ are *compatible*: the coordinate changes $\psi_\beta \circ \varphi_\alpha^{-1}$ and $\varphi_\alpha \circ \psi_\beta^{-1}$ are differentiable in their domains, for every α and every β;
(ii) for every β, the local chart ψ_β maps $V'_\beta = V_\beta \cap S$ onto an open subset Y'_β of $\mathbb{R}^k \times \{0^{d-k}\}$.

Identifying $\mathbb{R}^k \times \{0^{d-k}\} \simeq \mathbb{R}^k$, we get that the family formed by the restrictions $\psi_\beta : V'_\beta \to Y'_\beta$ constitutes an atlas for S. Hence, S is a manifold of dimension k. If M is a manifold of class C^r and the atlases $\mathcal{A}$ and $\mathcal{B}$ are C^r-compatible, that is, if all the coordinate changes in (i) are of class C^r, then S is a (sub)manifold of class C^r.

We say that a map $f : M \to N$ between two manifolds is *differentiable* if

$$\psi_\beta \circ f \circ \varphi_\alpha^{-1} : \varphi_\alpha(U_\alpha \cap f^{-1}(V_\beta)) \to \psi_\beta\Big(V_\beta \cap f(U_\alpha)\Big) \tag{A.4.1}$$

is a differentiable map for every local chart $\varphi_\alpha : U_\alpha \to X_\alpha$ of M and every local chart $\psi_\beta : V_\beta \to Y_\beta$ of N with $f(U_\alpha) \cap V_\beta \neq \emptyset$. Moreover, we say that f is *of class C^r* if M and N are manifolds of class C^r and every map $\psi_\beta \circ f \circ \varphi_\alpha^{-1}$ in (A.4.1) is of class C^r. A *diffeomorphism* $f : M \to N$ is a bijection between two manifolds such that both f and f^{-1} are differentiable. If both maps are of class C^r then we say that the diffeomorphism is *of class C^r*.

Let $C^r(M,N)$ be the space of maps of class C^r between two manifolds M and N. We are going to introduce in this space a certain topology, called the *C^r topology*, for which two maps are close if and only if they are uniformly close and the same is true for their derivatives up to order r. The definition may be given in a very broad context (see Section 2.1 of Hirsch [Hir94]), but we restrict ourselves to the case when M and N are compact. In this case, the C^r topology may be defined in the following way.

Fix finite families of local charts $\varphi_i : U_i \to X_i$ of M and $\psi_j : V_j \to Y_j$ of N, such that $\bigcup_i U_i = M$ and $\bigcup_j V_j = N$. Let $\delta > 0$ be a Lebesgue number for the open cover $\{U_i \cap f^{-1}(V_j)\}$ of M. For each pair (i,j) such that $U_i \cap f^{-1}(V_j) \neq \emptyset$, let $K_{i,j}$ be the set of points whose distance to the complement of $U_i \cap f^{-1}(V_j)$ is greater than or equal to δ. Then $K_{i,j}$ is a compact set contained in $U_i \cap f^{-1}(V_j)$ and the union $\bigcup_{i,j} K_{i,j}$ is the whole M. Consider

$$\mathcal{U}(f) = \{g \in C^r(M,N) : g(K_{i,j}) \subset V_j \text{ for any } i,j\}.$$

It is clear that $f \in \mathcal{U}(f)$. For each $g \in \mathcal{U}(f)$ and each pair (i,j) such that $K_{i,j}$ is non-empty, denote by $g_{i,j}$ the restriction of $\psi_j \circ g \circ \varphi_i^{-1}$ to the set $\varphi_i(K_{i,j})$. For each $r \in \mathbb{N}$ and $\varepsilon > 0$, define

$$\mathcal{U}^r(f,\varepsilon) = \{g \in \mathcal{U}(f) : \sup_{s,x,i,j} \|D^s f_{i,j}(x) - D^s g_{i,j}(x)\| < \varepsilon\}, \tag{A.4.2}$$

where the supremum is over every $s \in \{1,\dots,r\}$, every $x \in \varphi_i(K_{i,j})$ and every pair (i,j) such that $K_{i,j} \neq \emptyset$. By definition, the family $\{\mathcal{U}^r(f,\varepsilon) : \varepsilon > 0\}$ is a basis of neighborhoods of each $f \in C^r(M,N)$ relative to the C^r topology. Also by definition, the family $\{\mathcal{U}^r(f,\varepsilon) : \varepsilon > 0 \text{ and } r \in \mathbb{N}\}$ is a basis of neighborhoods of $f \in C^\infty(M,N)$ relative to the C^∞ topology.

The C^r topology has very nice properties: in particular, it admits a countable basis of open sets and is *completely metrizable*, that is, it is generated by some complete distance. An interesting consequence is that $C^r(M,N)$ is a *Baire space*: every intersection of a countable family of open dense subsets is dense in the space. The set $\operatorname{Diffeo}^r(M)$ of diffeomorphisms of class C^r is an open subset of $C^r(M,M)$ relative to the C^r topology.

A.4.2 Tangent space and derivative

Let M be a manifold. For each $p \in M$, consider the set $\mathcal{C}(p)$ of all the curves $c : I \to M$ whose domain is some open interval I containing $0 \in \mathbb{R}$, such that $c(0) = p$ and c is differentiable at the point 0. The latter means that the map $\varphi_\alpha \circ c$ is differentiable at the point 0 for every local chart $\varphi_\alpha : U_\alpha \to X_\alpha$ with $p \in U_\alpha$. We say that two curves $c_1, c_2 \in \mathcal{C}(p)$ are *equivalent* if $(\varphi_\alpha \circ c_1)'(0) = (\varphi_\alpha \circ c_2)'(0)$ for every local chart $\varphi_\alpha : U_\alpha \to X_\alpha$ with $p \in U_\alpha$. Actually, if the equality holds for some local chart then it holds for all the other charts as well. We denote by $[c]$ the equivalence class of any curve $c \in \mathcal{C}(p)$.

The *tangent space* to the manifold M at the point p is the set of such equivalence classes. We denote this set by T_pM. For any fixed local chart $\varphi_\alpha : U_\alpha \to X_\alpha$ with $p \in U_\alpha$, the map

$$D\varphi_\alpha(p) : T_pM \to \mathbb{R}^d, \quad [c] \mapsto (\varphi_\alpha \circ c)'(0)$$

is well defined and is a bijection. We may use this bijection to identify T_pM with $\mathbb{R}^d$. In this way, the tangent space acquires the structure of a vector space, transported from $\mathbb{R}^d$ via $D\varphi_\alpha(p)$. Although this identification $D\varphi_\alpha(p)$ depends on the choice of the local chart, the vector space structure on T_pM does not. That is because, for any other local chart $\varphi_\beta : U_\beta \to X_\beta$ with $p \in U_\beta$, the corresponding map $D\varphi_\beta(p)$ is given by

$$D\varphi_\beta(p) = D\big(\varphi_\beta \circ \varphi_\alpha^{-1}\big)(\varphi_\alpha(p)) \circ D\varphi_\alpha(p).$$

Since $D\big(\varphi_\beta \circ \varphi_\alpha^{-1}\big)(\varphi_\alpha(p))$ is a linear isomorphism, it follows that the vector space structures transported from Euclidean space to T_pM by $D\varphi_\alpha(p)$ and $D\varphi_\beta(p)$ coincide, as we stated.

If $f : M \to N$ is a differentiable map, its *derivative* at a point $p \in M$ is the linear map $Df(p) : T_pM \to T_{f(p)}N$ defined by

$$Df(p) = D\psi_\beta(f(p))^{-1} \circ D\big(\psi_\beta \circ f \circ \varphi_\alpha^{-1}\big)(\varphi_\alpha(p)) \circ D\varphi_\alpha(p),$$

where $\varphi_\alpha : U_\alpha \to X_\alpha$ is a local chart of M with $p \in U_\alpha$ and $\psi_\beta : V_\beta \to Y_\beta$ is a local chart of N with $f(p) \in V_\beta$. The definition does not depend on the choice of these local charts.

The *tangent bundle* to M is the (disjoint) union $TM = \bigcup_{p\in M} T_pM$ of all the tangent spaces to M. For each local chart $\varphi_\alpha : U_\alpha \to X_\alpha$, consider the union $T_{U_\alpha}M = \bigcup_{p\in U_\alpha} T_pM$ and the map

$$D\varphi_\alpha : T_{U_\alpha}M \to X_\alpha \times \mathbb{R}^d$$

that associates with each $[c] \in T_{U_\alpha}M$ the pair

$$((\varphi_\alpha \circ c)(0), (\varphi_\alpha \circ c)'(0)) \in X_\alpha \times \mathbb{R}^d.$$

We consider on TM the (unique) topology that turns every $D\varphi_\alpha$ into a homeomorphism. Assuming that the atlas $\{\varphi_\alpha : U_\alpha \to X_\alpha\}$ of the manifold M is of class C^r, the coordinate change

$$D\varphi_\beta \circ D\varphi_\alpha^{-1} : \varphi_\alpha\big(U_\alpha \cap U_\beta\big) \times \mathbb{R}^d \to \varphi_\beta\big(U_\alpha \cap U_\beta\big) \times \mathbb{R}^d$$

is a map of class C^{r-1} for any α and β such that $U_\alpha \cap U_\beta \neq \emptyset$. So, the tangent bundle TM is endowed with the structure of a manifold of class C^{r-1} and dimension $2d$.

The *derivative* $Df : TM \to TN$ of a differentiable map $f : M \to N$ is the map whose restriction to each tangent space T_pM is given by $Df(p)$. If f is of class C^r then Df is of class C^{r-1}, relative to the manifold structure on the tangent bundles TM and TN that we introduced in the previous paragraph. For example, the canonical projection $\pi : TM \to M$, associating with each $v \in TM$ the unique point $p \in M$ such that $v \in T_pM$, is a map of class C^{r-1} (Exercise A.4.9).

A *vector field* on a manifold M is a map that associates with each point $p \in M$ an element $X(p)$ of the tangent space T_pM, that is, a map $X : M \to TM$ such that $\pi \circ X = \mathrm{id}$. We say that the vector field is of class C^k, with $k \le r-1$, if this map is of class C^k.

Assuming that $k \ge 1$, we may apply the theorem of existence and uniqueness of solutions of ordinary differential equations to conclude that for every point $p \in M$ there exists a unique curve $c_p : I_p \to M$ such that

- $c_p(0) = p$ and $c_p'(t) = X(c(t))$ for every $t \in I_p$, and
- I_p is the largest open interval where such a curve can be defined.

If M is compact then $I_p = \mathbb{R}$ for any $p \in M$. Moreover, the maps $f^t : M \to M$ defined by $f^t(p) = c_p(t)$ are diffeomorphisms of class C^k, with $f^0 = \mathrm{id}$ and $f^s \circ f^t = f^{s+t}$ for any $s,t \in \mathbb{R}$. The family $\{f^t : t \in \mathbb{R}\}$ is called the *flow* of the vector field X.

A.4.3 Cotangent space and differential forms

The *cotangent space* T_p^*M to a manifold M at a point p is the dual of the tangent space T_pM, that is, the space of linear functionals $\xi : T_pM \to \mathbb{R}$. For any local chart $\varphi_\alpha : U_\alpha \to X_\alpha$ with $p \in U_\alpha$, the isomorphism $D\varphi_\alpha(p) : T_pM \to \mathbb{R}^d$ induces an isomorphism

$$D\varphi_\alpha^*(p) : T_p^*M \to \mathbb{R}^d$$

as follows. For each $i = 1,\dots,d$, let $dx_i = \pi_i \circ D\varphi_\alpha(p)$, where $\pi_i : \mathbb{R}^d \to \mathbb{R}$ is the projection to the i-th coordinate. Then $dx_i \in T_p^*M$ and, in fact, the family $\{dx_1,\dots,dx_d\}$ is a basis of T_p^*M. For each $\xi \in T_p^*M$, define

$$D\varphi_\alpha^*(p)\xi = (\xi_1,\dots,\xi_d) \quad \Leftrightarrow \quad \xi = \sum_{i=1}^{d} \xi_i dx_i.$$

The *cotangent bundle* of M is the (disjoint) union $T^*M = \bigcup_{p\in M} T_p^*M$ of all the cotangent spaces to M. For each local chart $\varphi_\alpha : U_\alpha \to X_\alpha$, consider the union $T_{U_\alpha}^*M = \bigcup_{p\in U_\alpha} T_p^*M$ and the map

$$D\varphi_\alpha^* : T_{U_\alpha}^*M \to X_\alpha \times \mathbb{R}^d$$

defined by $D\varphi_\alpha^*\xi = (\varphi_\alpha(p), D\varphi_\alpha^*(p)\xi)$ if $\xi \in T_p^*M$. It is clear that this is a bijection. We consider on T^*M the unique topology that turns every $D\varphi_\alpha^*$ into a homeomorphism. If $\{\varphi_\alpha : U_\alpha \to X_\alpha\}$ is an atlas of class C^r for the manifold M then

$$\{D\varphi_\alpha^* : T_{U_\alpha}^*M \to X_\alpha \times \mathbb{R}^d\}$$

is an atlas of class C^{r-1} for T^*M. So, the cotangent bundle T^*M is also endowed with the structure of a manifold of class C^{r-1} and dimension $2d$.

Moreover, the canonical map $\pi^* : T^*M \to M$ defined by $\pi^* \mid T_p^*M = p$ is of class C^{r-1}.

A *differential* 1*-form* in M is a differentiable map $\theta : M \to T^*M$ such that $\pi^* \circ \theta = \mathrm{id}$. In other words, θ assigns to each point $p \in M$ a linear functional (or *linear form*) $\theta_p : T_pM \to \mathbb{R}$ that depends differentiably on the point.

More generally, for any $0 \le k \le d$, an *alternate k-linear form* in T_pM is a map[1]

$$\theta_p : (T_pM)^k \to \mathbb{R}, \quad (v_1, \dots, v_k) \mapsto \theta_p(v_1, \dots, v_k)$$

such that θ_p is linear on each variable v_i and

$$\theta_p(v_1, \dots, v_i, v_{i+1}, \dots, v_k) = -\theta_p(v_1, \dots, v_{i+1}, v_i, \dots, v_k)$$

for any $1 \le i < k$ and any $(v_1, \dots, v_k) \in (T_pM)^k$.

Let $\{dx_1, \dots, dx_d\}$ be the basis of the cotangent space associated with a local chart $\varphi_\alpha : U_\alpha \to X_\alpha$ and let $\{\partial/\partial x_1, \dots, \partial/\partial x_d\}$ be the dual basis of T_pM, defined by

$$dx_i(\partial/\partial x_j) = \begin{cases} 1 & \text{if } i = j \\ 0 & \text{if } i \neq j. \end{cases}$$

If $i_1 \dots, i_k \in \{1, \dots, d\}$ are all distinct, there exists a unique alternate k-linear form $dx_{i_1} \wedge \cdots \wedge dx_{i_k}$ such that

- $dx_{i_1} \wedge \cdots \wedge dx_{i_k}(\partial/\partial x_{i_1}, \dots, \partial/\partial x_{i_k}) = 1$, and
- $dx_{i_1} \wedge \cdots \wedge dx_{i_k}(\partial/\partial x_{j_1}, \dots, \partial/\partial x_{j_k}) = 0$ when $\{i_1, \dots, i_k\} \neq \{j_1, \dots, j_k\}$.

The family $\{dx_{i_1} \wedge \cdots \wedge dx_{i_k} : 1 \le i_1 < \cdots < i_k \le d\}$ is a basis of the vector space of alternate k-linear forms in T_pM.

A *differential k-form* in M is a map θ assigning to each point $p \in M$ an alternate k-linear form in the tangent space T_pM that depends differentiably on the point. In local coordinates, this may be written as

$$\theta_p = \sum_{1 \le i_1 < \cdots < i_k \le d} a_{i_1, \dots, i_k}(p) dx_{i_1} \wedge \cdots \wedge dx_{i_k}.$$

The differentiability condition means that the coefficients $a_{i_1, \dots, i_k}(p)$ depend differentiably on the point p.

Assuming that $k < d$, the *exterior derivative* of θ is the differential $(k+1)$-form $d\theta$ determined by

$$d\theta_p = \sum_{1 \le i_1 < \cdots < i_k \le d} \sum_j \frac{\partial a_{i_1, \dots, i_k}}{\partial x_j}(p) dx_j \wedge dx_{i_1} \wedge \cdots \wedge dx_{i_k},$$

where the second sum is over all $j \notin \{i_1, \dots, i_k\}$; one can check that the expression on the right-hand side does not depend on the choice of the local chart. A differential k-form θ is *closed* if $d\theta = 0$ (or else $k = d$) and it is *exact* if

[1] An alternate 0-linear form is just a real number.

there exists some $(k-1)$-form η such that $d\eta = \theta$ (or else $k = 0$). Every exact differential form is closed.

For much more information on the subject of differential forms, see the book of Henri Cartan [Car70].

A.4.4 Transversality

The result that we state next is an important tool for constructing new manifolds. We say that $y \in N$ is a *regular value* of a differentiable map $f : M \to N$ if the derivative $Df(x) : T_xM \to T_yN$ is surjective for every $x \in f^{-1}(y)$. Note that this holds, automatically, if y is not in the image of f, that is, if $f^{-1}(y)$ is the empty set. On the other hand, in order that some point $y \in f(M)$ is a regular value of f it is necessary that $\dim M \geq \dim N$.

Theorem A.4.3. *Let $f : M \to N$ be a map of class C^r and $y \in f(M)$ be a regular value of f. Then $f^{-1}(y)$ is a submanifold (not necessarily connected) of class C^r of M, with dimension equal to* $\dim M - \dim N$.

Example A.4.4. For any $d \geq 1$, the space of square matrices of dimension d with real coefficients is isomorphic to the Euclidean space $\mathbb{R}^{(d^2)}$ and, hence, it is a manifold of dimension d^2 and class C^∞. The *linear group* $\mathrm{GL}(d,\mathbb{R})$ of invertible matrices is an open subset of that space and, hence, it is also a manifold of dimension d^2 and class C^∞. The function $\det : \mathrm{GL}(d,\mathbb{R}) \to \mathbb{R}$ that maps each matrix to its determinant is of class C^∞ and $y = 1$ is a regular value (see Exercise A.4.5). Using Theorem A.4.3, it follows that the *special linear group* $\mathrm{SL}(d,\mathbb{R})$ formed by the matrices with determinant equal to 1 is a submanifold of class C^∞ of $\mathrm{GL}(d,\mathbb{R})$, with dimension equal to $d^2 - 1$.

It is possible to generalize Theorem A.4.3, using the notion of transversality. We say that a submanifold S of N is *transverse* to f if

$$Df(x)\big(T_xM\big) + T_{f(x)}S = T_{f(x)}N \quad \text{for every } x \in f^{-1}(S). \tag{A.4.3}$$

For example, if S is a submanifold of dimension zero, that is, if it consists of a unique point, then S is transverse to f if and only if that point is a regular value of f. Therefore, the following statement generalizes Theorem A.4.3:

Theorem A.4.5. *Let $f : M \to N$ be a map of class C^r and let S be a submanifold of class C^r of N transverse to f. Then $f^{-1}(S)$ is a submanifold (not necessarily connected) of class C^r of M, with dimension equal to* $\dim M - \dim N + \dim S$.

The next theorem asserts that, for every map $f : M \to N$ of class C^r with r sufficiently high, "almost all" points $y \in N$ are regular values. We say that a set $X \subset N$ is *residual* if it contains some countable intersection of open and dense subsets. Every residual set is dense in the manifold, because manifolds are Baire spaces. We say that a set $Z \subset N$ has *volume zero* if for every local chart $\psi_\beta : V_\beta \to Y_\beta$ the image $\psi_\beta(Z \cap V_\beta)$ is a subset of the Euclidean space with volume zero, that is, it may be covered by balls in such a way that the sum of the volumes of those balls is arbitrarily small.

Theorem A.4.6 (Sard). *Assume that $f : M \to N$ is a map of class C^r with $r > \max\{0, \dim M - \dim N\}$. Then the set of regular points of f is a residual subset of N and its complement has volume zero.*

A.4.5 Riemannian manifolds

A *Riemannian metric* on a manifold M is a map that associates with each point $p \in M$ an inner product in the tangent space T_pM, that is, a symmetric bilinear map

$$\cdot_p : T_pM \times T_pM \to \mathbb{R}$$

such that $v \cdot_p v > 0$ for every non-zero vector $v \in T_pM$. As part of the definition, this inner product is required to vary in a differentiable way with the point p, in the following sense. Consider any local chart $\varphi_\alpha : U_\alpha \to X_\alpha$ of M. As explained previously, for every $p \in U_\alpha$ we may identify T_pM with $\mathbb{R}^d$, through the map $D\varphi_\alpha(p)$. Thus, we may view $\cdot_p$ as an inner product in the Euclidean space. Let $e_1, \dots, e_d$ be a basis of $\mathbb{R}^d$. Then the functions $g_{\alpha,i,j}(p) = e_i \cdot_p e_j$ are required to be differentiable, for every pair (i,j) and any choice of the local chart φ_α and the basis $e_1, \dots, e_d$.

We call a *Riemannian manifold* any manifold endowed with a Riemannian metric. Every submanifold S of a Riemannian manifold M inherits the structure of a Riemannian manifold, given by the restriction of the inner product $\cdot_p$ of M to the tangent subspace T_pS of each point $p \in S$. Every compact manifold admits (infinitely many) Riemannian metrics. That follows from the theorem of Whitney (see Section 1.3 of Hirsch [Hir94]), according to which every compact manifold may be realized as a submanifold of some Euclidean space. Actually, this remains true in the much larger class of *paracompact* manifolds (which we do not define here): every paracompact manifold of dimension d is diffeomorphic to some submanifold of $\mathbb{R}^{2d}$. In particular, paracompact manifolds are always metrizable.

Starting from the Riemannian metric, we may define the *length* of a differentiable curve $\gamma : [a,b] \to M$, by

$$\operatorname{length}(\gamma) = \int_a^b \|\gamma'(t)\|_{\gamma(t)}\, dt, \quad \text{where } \|v\|_p = (v \cdot_p v)^{1/2}.$$

This also allows us to define on the manifold M the following *distance associated with the Riemannian metric*: the distance $d(p,q)$ between two points $p, q \in M$ is the infimum of the lengths of all the differentiable curves connecting the two points. We say that a differentiable curve $\gamma : [a,b] \to M$ is *minimizing* if it realizes the distance between its endpoints, that is, if

$$\text{length}(\gamma) = d(\gamma(a), \gamma(b)).$$

Any two points $p, q \in M$ are connected by some minimizing curve; in other words, the infimum in the definition of $d(p,q)$ is always realized.

A differentiable curve $\gamma : I \to M$ defined on an open interval I is called a *geodesic* if it is locally minimizing, in the following sense: for every $c \in I$ there exists $\delta > 0$ such that the restriction of γ to the interval $[c-\delta, c+\delta]$ is minimizing. Every minimizing curve is a geodesic, but the converse is not true: for example, the great circles are geodesics on the sphere S^2, but closed curves cannot be minimizing. An important fact is that if γ is a geodesic then the norm $\|\gamma'(t)\|_{\gamma(t)}$ is constant on the domain I. The theory of ordinary differentiable equations may be used to show that for every $p \in M$ and every $v \in T_pM$ there exists a unique geodesic $\gamma_{p,v} : I_{p,v} \to M$ such that $\gamma_{p,v}(0) = p$, $\gamma'_{p,v}(0) = v$ and $I_{p,v}$ is a maximal interval such that $\gamma_{p,v}$ is locally minimizing.

If the manifold M is compact then $I_{p,v} = \mathbb{R}$ for every $p \in M$ and every $v \in T_pM$. Then we define the *exponential map* at each point $p \in M$:

$$\exp_p : T_pM \to M, \quad v \mapsto \gamma_{p,v}(1).$$

This is a differentiable map and its derivative at $v = 0$ is the identity transformation on the tangent space T_pM. We also define the *geodesic flow* on the tangent bundle:

$$f^t : TM \to TM, \quad (p,v) \mapsto (\gamma_{p,v}(t), \gamma'_{p,v}(t)).$$

Most of the time, one considers the restriction of the geodesic flow to the *unit tangent bundle* $T^1M = \{(p,v) \in TM : \|v\|_p = 1\}$. This is well defined since, as we mentioned before, the norm of the velocity vector of any geodesic is constant.

A.4.6 Exercises

A.4.1. Check that every set X with the cardinality of $\mathbb{R}$ may be endowed with the structure of a differentiable manifold of class C^∞ and dimension d, for any $d \geq 1$.

A.4.2. Consider the differentiable manifolds $M = (\mathbb{R}, \mathcal{A})$ and $N = (\mathbb{R}, \mathcal{B})$, where $\mathcal{A}$ is the atlas consisting of the map $\phi(x) = x$ and $\mathcal{B}$ is the atlas consisting of the map $\psi(x) = x^3$. Is the map $f : M \to N$ defined by $f(x) = x$ a diffeomorphism between these manifolds?

A.4.3. A topological space is *path connected* if any two points are connected by some continuous curve. Show that every (connected) manifold is path connected.

A.4.4. For each $d \geq 2$, the *projective space of dimension d* is the set $\mathbb{P}^d$ of all subspaces of $\mathbb{R}^{d+1}$ with dimension 1. Equivalently, $\mathbb{P}^d$ is the quotient space of $\mathbb{R}^{d+1} \setminus \{0\}$ for the equivalence relation defined by:

$$(x_0, \ldots, x_d) \sim (y_0, \ldots, y_d) \Leftrightarrow \text{there exists } c \neq 0 \text{ such that } x_i = cy_i \text{ for every } i.$$

Show that the family of maps $\varphi_i : U_i \to \mathbb{R}^d$, $i = 0,\dots,d$ defined by

$$U_i = \{[x_0 : \cdots : x_d] \in \mathbb{P}^d : x_i \neq 0\}$$

(where $[x_0 : \cdots : x_d]$ denotes the equivalence class of $(x_0,\dots,x_d)$) and

$$\varphi_i([x_0 : \cdots : x_d]) = \left(\frac{x_0}{x_i},\dots,\frac{x_{i-1}}{x_i},\frac{x_{i+1}}{x_i},\dots,\frac{x_d}{x_i}\right),$$

constitutes an atlas of class C^∞ and dimension d for $\mathbb{P}^d$.

A.4.5. Check the claims in Example A.4.4.

A.4.6. Let M and N be two compact (connected) manifolds with the same dimension. A map $f : M \to N$ of class C^1 is a *local diffeomorphism* if the derivative $Df(x) : T_xM \to T_{f(x)}N$ is an isomorphism for every $x \in M$. Show that in that case there exists an integer $k \geq 1$ such that every $y \in M$ has exactly k pre-images:

$$\#f^{-1}(y) = k \quad \text{for every } y \in N.$$

[Observation: The number k is called the *degree* of f and is denoted degree(f).]

A.4.7. Consider on $\mathbb{R}_+ = \{x \in \mathbb{R} : x > 0\}$ the Riemannian metric defined by $u \cdot_x v = uv/x^2$. Calculate the distance $d(a,b)$ between any two points $a, b \in \mathbb{R}_+$.

A.4.8. Let M and N be submanifolds of $\mathbb{R}^{m+n}$ with $\dim M = m$ and $\dim N = n$. Show that there exists a set $Z \subset \mathbb{R}^{m+n}$ with volume zero such that, for every v in the complement of Z, the translate $M + v$ is *transverse* to N:

$$T_x(M+v) + T_xN = \mathbb{R}^d \quad \text{for every } x \in (M+v) \cap N.$$

A.4.9. Show that if M is a manifold of class C^r then the canonical projection $\pi : TM \to M$ is a map of class C^{r-1}.

A.5 $L^p(\mu)$ spaces

In this appendix we review certain Banach spaces formed by functions with special integrability properties. Throughout, $(X, \mathcal{B}, \mu)$ is a measure space. Recall that a Banach space is a vector space endowed with a norm relative to which the space is complete. We also state some properties of the norms in these spaces.

A.5.1 $L^p(\mu)$ spaces with $1 \leq p < \infty$

Given any $p \in [1, \infty)$, we say that a function $f : X \to \mathbb{C}$ is *p-integrable* with respect to μ if the function $|f|^p$ is integrable with respect to μ. For $p = 1$ this is the same as saying that the function f is integrable (Definition A.2.4 and Proposition A.2.7).

Definition A.5.1. We denote by $L^p(\mu)$ the set of all complex functions p-integrable with respect to μ, modulo the equivalence relation that identifies any two functions that are equal at μ-almost every point.

Note that if the measure μ is finite, which is the case in most of our examples, then all bounded measurable functions are in $L^p(\mu)$:

$$\int |f|^p \, d\mu \leq (\sup |f|)^p m(X) < \infty.$$

In particular, if X is a compact topological space then every continuous function is in $L^p(\mu)$. In other words, the space $C^0(X)$ of all continuous functions is contained in $L^p(\mu)$ for every $p \geq 1$.

For every function $f \in L^p(\mu)$, define the *L^p-norm* of f by:

$$\|f\|_p = \left(\int |f|^p \, d\mu \right)^{\frac{1}{p}}.$$

The next theorem asserts that $\|\cdot\|_p$ turns $L^p(\mu)$ into a Banach space:

Theorem A.5.2. *The set $L^p(\mu)$ is a complex vector space. Moreover, $\|\cdot\|_p$ is a norm in $L^p(\mu)$ and this norm is complete.*

The most interesting part of the proof of this theorem is to establish the triangle inequality, which in this context is known as the *Minkowski inequality*:

Theorem A.5.3 (Minkowski inequality). *Let $f, g \in L^p(\mu)$. Then:*

$$\left(\int |f+g|^p \, d\mu \right)^{\frac{1}{p}} \leq \left(\int |f|^p \, d\mu \right)^{\frac{1}{p}} + \left(\int |g|^p \, d\mu \right)^{\frac{1}{p}}.$$

In Exercises A.5.2 and A.5.5 we invite the reader to prove the Minkowski inequality and to complete the proof of Theorem A.5.2.

A.5.2 Inner product in $L^2(\mu)$

The case $p = 2$ deserves special attention. The reason is that the norm $\|\cdot\|_2$ introduced in the previous section arises from an (Hermitian) inner product. Indeed, consider:

$$f \cdot g = \int f \bar{g} \, d\mu. \tag{A.5.1}$$

It follows from the properties of the Lebesgue integral that this expression does define an inner product on $L^2(\mu)$. Moreover, this product gives rise to the norm $\|\cdot\|_2$ through:

$$\|f\|_2 = (f \cdot f)^{1/2}.$$

In particular, we have the Cauchy–Schwarz inequality:

Theorem A.5.4 (Cauchy–Schwarz Inequality). *For every $f, g \in L^2(\mu)$ we have that $f\bar{g} \in L^1(\mu)$ and*

$$\left| \int f \bar{g} \, d\mu \right| \leq \int |f \bar{g}| \, d\mu \leq \left(\int |f|^2 \, d\mu \right)^{1/2} \left(\int |g|^2 \, d\mu \right)^{1/2}.$$

This inequality has the following interesting consequence. Assume that the measure μ is finite and consider any $f \in L^2(\mu)$. Then, taking $g \equiv 1$,

$$\int |f|\,d\mu = \int |f\bar{g}|\,d\mu \le \Big(\int |f|^2\,d\mu\Big)^{1/2}\Big(\int 1\,d\mu\Big)^{1/2} < \infty. \qquad \text{(A.5.2)}$$

This proves that every function in $L^2(\mu)$ is also in $L^1(\mu)$. In fact, when the measure μ is finite one has $L^p(\mu) \subset L^q(\mu)$ whenever $p \ge q$ (Exercise A.5.3).

The next result is a generalization of the Cauchy–Schwarz inequality for all values of $p > 1$:

Theorem A.5.5 (Hölder inequality). *Given $1 < p < \infty$, consider $q > 1$ defined by the relation $\frac{1}{p} + \frac{1}{q} = 1$. Then, for every $f \in L^p(\mu)$ and every $g \in L^q(\mu)$, we have that $f\bar{g} \in L^1(\mu)$ and*

$$\int |f\bar{g}|\,d\mu \le \Big(\int |f|^p\,d\mu\Big)^{\frac{1}{p}}\Big(\int |g|^q\,d\mu\Big)^{\frac{1}{q}}.$$

A.5.3 Space of essentially bounded functions

Next, we extend the definition of $L^p(\mu)$ to the case $p = \infty$. For that we need the following notion. We say that a function $f : X \to \mathbb{C}$ is *essentially bounded* with respect to μ if there exists some constant $K > 0$ such that $|f(x)| \le K$ at μ-almost every point. Then the infimum of all such constants K is called the *essential supremum* of f and is denoted by $\operatorname{supess}_\mu(f)$.

Definition A.5.6. We denote by $L^\infty(\mu)$ the set of all complex functions essentially bounded with respect to μ, identifying any two functions that coincide at μ-almost every point.

We endow $L^\infty(\mu)$ with the following norm:

$$\|f\|_\infty = \operatorname{supess}_\mu(f).$$

The conclusion of Proposition A.5.2 remains valid for $p = \infty$ (Exercise A.5.5): the space $L^\infty(\mu)$ is a Banach space for the norm $\|\cdot\|_\infty$. Clearly, if μ is a finite measure then $L^\infty(\mu) \subset L^p(\mu)$ for any $p \ge 1$.

The *dual* of a complex Banach space E is the space E^* of all continuous linear functionals $\phi : E \to \mathbb{C}$, endowed with the norm

$$\|\phi\| = \sup\Big\{\frac{|\phi(v)|}{\|v\|} : v \in E \setminus \{0\}\Big\}. \qquad \text{(A.5.3)}$$

The Hölder inequality (Theorem A.5.5) leads to the following explicit characterization of the dual space of $L^p(\mu)$ for every $p < \infty$:

Theorem A.5.7. *For each $p \in [1,\infty)$ consider $q \in (1,\infty]$ defined by the relation $\frac{1}{p} + \frac{1}{q} = 1$. The map $L^q(\mu) \to L^p(\mu)^*$ defined by $g \mapsto \big[f \mapsto \int fg\,d\mu\big]$ is an isomorphism and an isometry between $L^q(\mu)$ and the dual space of $L^p(\mu)$.*

This statement is false for $p = \infty$: in general, the dual space of $L^\infty(\mu)$ is *not* isomorphic to $L^1(\mu)$.

A.5.4 Convexity

We say that a function $\phi : I \to \mathbb{R}$ defined on an interval I of the real line is *convex* if

$$\phi(tx + (1-t)y) \le t\phi(x) + (1-t)\phi(y)$$

for every $x, y \in I$ and $t \in [0,1]$. Moreover, we say that ϕ is *concave* if $-\phi$ is convex. For functions that are twice differentiable we have the following practical criterion (Exercise A.5.1): ϕ is convex if $\phi''(x) \ge 0$ for every $x \in I$ and it is concave if $\phi''(x) \le 0$ for every $x \in I$.

Theorem A.5.8 (Jensen inequality). *Let $\phi : I \to \mathbb{R}$ be a convex function. If μ is a probability measure on X and $f \in L^1(\mu)$ is such that $\int f\, d\mu \in I$, then:*

$$\phi\left(\int f\, d\mu\right) \le \int \phi \circ f\, d\mu.$$

Example A.5.9. For any probability measure μ and any integrable positive function f, we have

$$\log \int f\, d\mu \ge \int \log f\, d\mu.$$

Indeed, this corresponds to the Jensen inequality for the function $\phi : (0,\infty) \to \mathbb{R}$ given by $\phi(x) = -\log x$. Note that ϕ is convex: $\phi''(x) = 1/x^2 > 0$ for every x.

Example A.5.10. Let $\phi : \mathbb{R} \to \mathbb{R}$ be a convex function, $(\lambda_i)_i$ be a sequence of non-negative real numbers satisfying $\sum_{i=1}^\infty \lambda_i \le 1$ and $(a_i)_i$ be a bounded sequence of real numbers. Then

$$\phi\left(\sum_{i=1}^\infty \lambda_i a_i\right) \le \sum_{i=1}^\infty \lambda_i \phi(a_i). \tag{A.5.4}$$

This may be seen as follows. Consider $X = [0,1]$ endowed with the Lebesgue measure μ. Let $f : [0,1] \to \mathbb{R}$ be a function of the form $f = \sum_{i=1}^\infty a_i \mathcal{X}_{E_i}$, where the E_i are pairwise disjoint measurable sets such that $\mu(E_i) = \lambda_i$. The Jensen inequality applied to this function f gives precisely the relation (A.5.4).

A.5.5 Exercises

A.5.1. Consider any function $\varphi : (a,b) \to \mathbb{R}$. Show that if φ is twice differentiable and $\phi'' \ge 0$ then φ is convex. Show that if φ is convex then it is continuous.

A.5.2. Consider $p, q > 1$ such that $1/p + 1/q = 1$. Prove:
- (a) The Young inequality: $ab \le a^p/p + a^q/q$ for every $a, b > 0$.
- (b) The Hölder inequality (Theorem A.5.5).
- (c) The Minkowski inequality (Theorem A.5.3).

A.5.3. Show that if μ is a finite measure then we have $L^q(\mu) \subset L^p(\mu)$ for every $1 \le p < q \le \infty$.

A.5.4. Let μ be a finite measure and $f \in L^\infty(\mu)$ be different from zero. Show that

$$\|f\|_\infty = \lim_n \frac{\int |f|^{n+1}\, d\mu}{\int |f|^n\, d\mu}.$$

A.5.5. Show that a normed vector space $(V, \|\cdot\|)$ is complete if and only if every series $\sum_k v_k$ that is absolutely summable (meaning that $\sum_k \|v_k\|$ converges) is convergent. Use this fact to show that if μ is a probability measure then $\|\cdot\|_p$ is a complete norm on $L^p(\mu)$ for every $1 \le p \le \infty$.

A.5.6. Show that if μ is a finite measure and $1/p + 1/q = 1$ with $1 \le p < \infty$ then the map $\Phi : L^q(\mu) \to L^p(\mu)^*$, $\Phi(g)f = \int fg\, d\mu$ is an isomorphism and an isometry.

A.5.7. Show that if X is a metric space then, given any Borel probability measure μ, the set $C^0(X)$ of all continuous functions is dense in $L^p(\mu)$ for every $1 \le p \le \infty$. Indeed, the same holds for the subset of all uniformly continuous bounded functions.

A.5.8. Let $f, g : X \to \mathbb{R}$ be two positive measurable functions such that $f(x)g(x) \ge 1$ for every x. Show that $\int f\, d\mu \int g\, d\mu \ge 1$ for every probability measure μ.

A.6 Hilbert spaces

Let H be a vector space, real or complex. An (Hermitian) *inner product* on H is a map $(u, v) \mapsto u \cdot v$ from $H \times H$ to the scalar field ($\mathbb{R}$ or $\mathbb{C}$, respectively) satisfying: for any $u, v, w \in H$ and any scalar λ,

1. $(u + w) \cdot v = u \cdot v + w \cdot v$ and $u \cdot (v + w) = u \cdot v + u \cdot w$;
2. $(\lambda u) \cdot v = \lambda (u \cdot v)$ and $u \cdot (\lambda v) = \bar{\lambda}(u \cdot v)$;
3. $u \cdot v = \overline{v \cdot u}$;
4. $u \cdot u \ge 0$ and $u \cdot u = 0$ if and only if $u = 0$.

Then we can define the *norm* of a vector $u \in H$ to be $\|u\| = (u \cdot u)^{1/2}$.

A *Hilbert space* is a vector space endowed with an inner product whose norm $\|\cdot\|$ is complete: relative to $\|\cdot\|$ every Cauchy sequence is convergent. Thus, in particular, $(H, \|\cdot\|)$ is a Banach space. A standard example of a Hilbert space is the space $L^2(\mu)$ of square-integrable functions that we introduced in Appendix A.5.2.

Given $v \in H$ and any family $(v_\alpha)_\alpha$ of vectors of H, we say that $v = \sum_\alpha v_\alpha$ if for every $\varepsilon > 0$ there exists a finite set I such that

$$\left\| v - \sum_{\beta \in J} v_\beta \right\| \le \varepsilon \quad \text{for every finite set } J \supset I.$$

Given any family $(H_\alpha)_\alpha$ of subspaces of H, the set of all vectors of the form $v = \sum_\alpha v_\alpha$ with $v_\alpha \in H_\alpha$ for every α is a subspace of H (see Exercise A.6.2). It is called the *sum* of the family $(H_\alpha)_\alpha$ and it is denoted by $\sum_\alpha H_\alpha$.

A.6.1 Orthogonality

Let H be a Hilbert space. Two vectors $u, v \in H$ are said to be *orthogonal* if $u \cdot v = 0$. We call a subset of H *orthonormal* if its elements have norm 1 and are pairwise orthogonal.

A *Hilbert basis* of H is an orthonormal subset $B = \{v_\beta\}$ such that the set of all (finite) linear combinations of elements of B is dense in H. For example, the *Fourier basis*

$$\{x \mapsto e^{2\pi ikx} : k \in \mathbb{Z}\} \tag{A.6.1}$$

is a Hilbert basis of the space $L^2(m)$ of all measurable functions on the unit circle whose square is integrable with respect to the Lebesgue measure.

A Hilbert basis $B = \{v_\beta\}$ is usually not a basis of the vector space in the usual sense (Hammel basis): it is usually not true that every vector of H is a finite linear combination of the elements of B. However, every $v \in H$ may be written as an *infinite* linear combination of the elements of the Hilbert basis:

$$v = \sum_\beta (v \cdot v_\beta) v_\beta \quad \text{and, moreover,} \quad \|v\|^2 = \sum_\beta |v \cdot v_\beta|^2.$$

In particular, $v \cdot v_\beta = 0$ except, possibly, for a countable subset of values of β.

Every orthonormal subset of H may be extended to a Hilbert basis. In particular, Hilbert bases always exist. Moreover, any two Hilbert bases have the same cardinal, which is called the *Hilbert dimension* of H. The Hilbert dimension depends monotonically on the space: if H_1 is a subspace of H_2 then $\dim H_1 \le \dim H_2$. We say that two Hilbert spaces are *isometrically isomorphic* if there exists some isomorphism between the two that also preserves the inner product. A necessary and sufficient condition is that the two spaces have the same Hilbert dimension.

A Hilbert space is said to be *separable* if it admits some countable subset that is dense for the topology defined by the norm. This happens if and only if the Hilbert dimension is either finite or countable. In particular, all separable Hilbert spaces with infinite Hilbert dimension are isometrically isomorphic. For this reason, one often finds in the literature (especially in the area of mathematical physics) mentions of *the* Hilbert space, as if there were only one.

Given any family $(H_\alpha)_\alpha$ of Hilbert spaces, we denote by $\bigoplus_\alpha H_\alpha$ their *orthogonal direct sum*, that is, the vector space of all $(v_\alpha)_\alpha \in \prod_\alpha H_\alpha$ such that $\sum_\alpha \|v_\alpha\|_\alpha^2 < \infty$ (this implies that $v_\alpha = 0$ except, possibly, for a countable set of values of α), endowed with the inner product

$$(v_\alpha)_\alpha \cdot (w_\alpha)_\alpha = \sum_\alpha v_\alpha \bar{w}_\alpha.$$

The *orthogonal complement* of a subset S of a Hilbert space H is the set $S^\perp$ of all the vectors of H that are orthogonal to every vector of S. It is easy to see that $S^\perp$ is a closed subspace of H (Exercise A.6.7). If S itself is a closed subspace of H then $S = (S^\perp)^\perp$ and every vector $v \in H$ may be decomposed as a sum $v = s + s^\perp$ of some $s \in S$ and some $s^\perp \in S^\perp$. Moreover, this decomposition is unique and the vectors s and $s^\perp$ are the elements of S and $S^\perp$, respectively, that are closest to v.

A.6.2 Duality

A *linear functional* on a Hilbert space H (or, more generally, on a Banach space) is a linear map from H to the scalar field ($\mathbb{R}$ or $\mathbb{C}$). It is said to be *bounded* if

$$\|\phi\| = \sup\left\{\frac{|\phi(v)|}{\|v\|} : v \neq 0\right\} < \infty.$$

This is equivalent to saying that the linear functional is continuous, relative to the topology defined by the norm of H (see Exercise A.6.3). The *dual space* of a Hilbert space H is the vector space H^* formed by all the bounded linear functionals. The function $\phi \mapsto \|\phi\|$ is a complete norm on H^* and, hence, it endows the dual with the structure of a Banach space. The map

$$h : H \to H^*, \quad w \mapsto \left[v \mapsto v \cdot w\right] \tag{A.6.2}$$

is a bijection between the two spaces and it preserves the norms. In particular, h is a homeomorphism. Moreover, it satisfies $h(w_1 + w_2) = h(w_1) + h(w_2)$ and $h(\lambda w) = \bar{\lambda} h(w)$.

The *weak* topology in H is the smallest topology relative to which all the linear functionals $v \mapsto v \cdot w$ are continuous. In terms of sequences, it can be characterized as follows:

$$(w_n)_n \to w \text{ weakly} \quad \Leftrightarrow \quad (v \cdot w_n)_n \to v \cdot w \text{ for every } v \in H.$$

The *weak** topology in the dual space H^* is the smallest topology relative to which $\phi \mapsto \phi(v)$ is continuous for every $v \in H$.

It is known from the theory of Banach spaces (theorem of Banach–Alaoglu) that every bounded closed subset of the dual space is compact for the weak* topology. In the special case of Hilbert spaces, the weak topology in the space H is homeomorphic to the weak* topology in the dual space H^*: the map h in (A.6.2) is also a homeomorphism for these topologies. Since h preserves the class of bounded sets, it follows that the weak topology in the space H itself enjoys the property in the theorem of Banach–Alaoglu:

Theorem A.6.1 (Banach–Alaoglu). *Every bounded closed subset of a Hilbert space H is compact for the weak topology in H.*

A linear operator $L : H_1 \to H_2$ between two Hilbert spaces is *continuous* (or *bounded*) if

$$\|L\| = \sup\left\{\frac{|L(v)|}{\|v\|} : v \neq 0\right\}$$

is finite. The *adjoint* of a continuous linear operator is the linear operator $L^*: H_2 \to H_1$ defined by

$$v \cdot Lw = L^*v \cdot w \quad \text{for every } v, w \in H.$$

The adjoint operator is continuous, with $\|L^*\| = \|L\|$ and $\|L^*L\| = \|LL^*\| = \|L\|^2$. Moreover, $(L^*)^* = L$ and $(L_1 + L_2)^* = L_1^* + L_2^*$ and $(\lambda L)^* = \bar{\lambda} L^*$ (in Exercise A.6.5 we invite the reader to prove these facts).

A continuous linear operator $L: H \to H$ is *self-adjoint* if $L = L^*$. More generally, L is *normal* if it satisfies $L^*L = LL^*$. We are especially interested in the case when L is *unitary*, that is, $L^*L = \mathrm{id} = LL^*$. We call *linear isometry* to every linear operator $L: H \to H$ such that $L^*L = \mathrm{id}$. Hence, the unitary operators are the linear isometries that are also normal operators.

A.6.3 Exercises

A.6.1. Let H be a Hilbert space. Prove:
- (a) That every ball (either open or closed) is a convex subset of H.
- (b) The *parallelogram identity*: $\|v+w\|^2 + \|v-w\|^2 = \|v\|^2 + \|w\|^2$ for any $v, w \in H$.
- (c) The *polarization identity*: $4(v \cdot w) = \|v+w\|^2 - \|v-w\|^2$ (real case) or $4(v \cdot w) = (\|v+w\|^2 - \|v-w\|^2) + i(\|v+iw\|^2 - \|v-iw\|^2)$ (complex case).

A.6.2. Show that, given any family $(H_\alpha)_\alpha$ of subspaces of a Hilbert space H, the set of all the vectors of the form $v = \sum_\alpha v_\alpha$ with $v_\alpha \in H_\alpha$ for every α is a vector subspace of H.

A.6.3. Show that a linear operator $L: E_1 \to E_2$ between two Banach spaces is continuous if and only if there exists $C > 0$ such that $\|L(v)\|_2 \le C\|v\|_1$ for every $v \in E_1$, where $\|\cdot\|_i$ denotes the norm in the space E_i (we say that L is a *bounded* operator).

A.6.4. Consider the Hilbert space $L^2(\mu)$. Let V be the subspace formed by the constant functions. What is the orthogonal complement of V? Determine the (orthogonal) projection to V of an arbitrary function $g \in L^2(\mu)$.

A.6.5. Prove that if $L: H \to H$ is a bounded operator on a Hilbert space H then the adjoint operator L^* is also bounded and $\|L^*\| = \|L\|$ and $\|L^*L\| = \|LL^*\| = \|L\|^2$ and $(L^*)^* = L$.

A.6.6. Show that if K is a closed convex subset of a Hilbert space then for every $z \in H$ there exists a unique $v \in K$ such that $\|z - v\| = d(z, K)$.

A.6.7. Let S be a subspace of a Hilbert space H. Prove that:
- (a) The orthogonal complement $S^\perp$ of S is a closed subspace of H and it coincides with the orthogonal complement of the closure $\bar{S}$. Moreover, $(S^\perp)^\perp = \bar{S}$.
- (b) Every $v \in H$ may be written, in a unique fashion, as a sum $v = s + s^\perp$ of some $s \in \bar{S}$ and some $s^\perp \in S^\perp$. The two vectors s and $s^\perp$ are the elements of S and $S^\perp$ that are closest to v.

A.6.8. Let E be a closed subspace of a Hilbert space H. Show that E is also closed in the weak topology. Moreover, $U(E)$ is a closed subspace of H, for every isometry $U : H \to H$.

A.6.9. Show that a linear operator $L : H \to H$ on a Hilbert space H is an isometry if and only if $\|L(v)\| = \|v\|$ for every $v \in H$. Moreover, L is a unitary operator if and only if L is an isometry and is invertible.

A.7 Spectral theorems

Let H be a complex Hilbert space. The *spectrum* of a continuous linear operator $L : H \to H$ is the set $\operatorname{spec}(L)$ of all numbers $\lambda \in \mathbb{C}$ such that $L - \lambda \operatorname{id}$ is not an isomorphism. The spectrum is closed and it is contained in the closed disk of radius $\|L\|$ around $0 \in \mathbb{C}$. In particular, $\operatorname{spec}(L)$ is a compact subset of the complex plane. When H has finite dimension, $\operatorname{spec}(L)$ consists of the eigenvalues of L, that is, the complex numbers λ such that $L - \lambda \operatorname{id}$ is not injective. In general, the spectrum is strictly larger than the set of eigenvalues (see Exercise A.7.2).

A.7.1 Spectral measures

By definition, a *projection* in H is a continuous linear operator $P : H \to H$ that is idempotent ($P^2 = P$) and self-adjoint ($P^* = P$). Then the image and the kernel of P are closed subspaces of H and they are orthogonal complements to each other. In fact, the image coincides with the set of all fixed points of P.

Consider any map E associating with each measurable subset of the plane $\mathbb{C}$ a projection in H. Such a map is called a *spectral measure* if it satisfies $E(\mathbb{C}) = \operatorname{id}$ and

$$E\Big(\bigcup_{n \in \mathbb{N}} B_n\Big) = \sum_{n \in \mathbb{N}} E(B_n)$$

whenever the B_n are pairwise disjoint (σ-additivity). Then, given any $v, w \in H$, the function

$$Ev \cdot w : B \mapsto E(B)v \cdot w \tag{A.7.1}$$

is a complex measure in $\mathbb{C}$. Clearly, it depends on the pair (v, w) in a bilinear fashion.

We call the *support* of a spectral measure E the set $\operatorname{supp} E$ of all the points $z \in \mathbb{C}$ such that $E(V) \neq 0$ for every neighborhood V of z. Note that the support is always a closed set. Moreover, the support of the complex measure $Ev \cdot w$ is contained in $\operatorname{supp} E$ for every $v, w \in H$.

Example A.7.1. Consider $\{\lambda_1, \dots, \lambda_s\} \subset \mathbb{C}$ and let $V_1, \dots, V_s$ be a finite family of subspaces of $\mathbb{C}^d$, pairwise orthogonal and such that $\mathbb{C}^d = V_1 \oplus \cdots \oplus V_s$. For each set $J \subset \{1, \dots, s\}$, denote by P_J the projection in $\mathbb{C}^d$ whose image is

$\bigoplus_{j \in J} V_j$. For each measurable set $B \subset \mathbb{C}$ define

$$E(B) : \mathbb{C}^d \to \mathbb{C}^d, \quad E(B) = P_{J(B)},$$

where $J(B)$ is the set of all $j \in \{1, \ldots, s\}$ such that $\lambda_j \in B$. The function E is a spectral measure.

Example A.7.2. Let μ be a probability measure in $\mathbb{C}$ and $H = L^2(\mu)$ be the space of all complex functions whose square is integrable with respect to μ. For each measurable set $B \subset \mathbb{C}$, let

$$E(B) : L^2(\mu) \to L^2(\mu), \quad \varphi \mapsto \mathcal{X}_B \varphi.$$

Each $E(B)$ is a projection and the function E is a spectral measure.

The next lemma collects a few simple properties of the spectral measures:

Lemma A.7.3. *Let E be a spectral measure and A, B be measurable subsets of $\mathbb{C}$. Then:*

1. $E(\emptyset) = 0$ *and* $E(\operatorname{supp} E) = \mathrm{id}$;
2. *if* $A \subset B$ *then* $E(A) \le E(B)$ *and* $E(B \setminus A) = E(B) - E(A)$;
3. $E(A \cup B) + E(A \cap B) = E(A) + E(B)$;
4. $E(A)E(B) = E(A \cap B) = E(B)E(A)$.

In what follows we always assume that E is a spectral measure with compact support. Then the support of every complex measure $Ev \cdot w$ is also compact. Consequently, the integral $\int z d(E(z)v \cdot w)$ is well defined and it is a bilinear function of (v, w). Hence, there exists a bounded linear operator $L : H \to H$ such that

$$Lv \cdot w = \int z d(E(z)v \cdot w) \quad \text{for every } v, w \in H. \tag{A.7.2}$$

We write, in shorter form:

$$L = \int z dE(z). \tag{A.7.3}$$

More generally, given any bounded measurable function ψ in the support of the spectral measure E, there exists a bounded linear operator $\psi(L) : H \to H$ that is characterized by

$$\psi(L)v \cdot w = \int \psi(z) d(E(z)v \cdot w) \quad \text{for every } v, w \in H. \tag{A.7.4}$$

We write

$$\psi(L) = \int \psi(z) dE(z). \tag{A.7.5}$$

Lemma A.7.4. *Let E be a spectral measure with compact support. Given bounded measurable functions φ, ψ and numbers $\alpha, \beta \in \mathbb{C}$,*

(1) $\int (\alpha\varphi + \beta\psi)(z) dE(z) = \alpha \int \varphi \, dE(z) + \beta \int \psi \, dE(z)$;
(2) $\int \bar{\varphi}(z) \, dE(z) = \left(\int \varphi(z) \, dE(z) \right)^*$;

(3) $\int (\varphi\psi)(z)\,dE(z) = \left(\int \varphi(z)\,dE(z)\right) \circ \left(\int \psi(z)\,dE(z)\right)$.

In particular, by part (3) of this lemma,

$$L^j = \left(\int z\,dE(z)\right)^j = \int z^j\,dE(z) \quad \text{for every } j \in \mathbb{N}. \tag{A.7.6}$$

Analogously, using also part (2) of the lemma,

$$\begin{aligned} LL^* &= \left(\int z\,dE(z)\right)\left(\int \bar{z}\,dE(z)\right) = \int |z|^2\,dE(z) \\ &= \left(\int \bar{z}\,dE(z)\right)\left(\int z\,dE(z)\right) = L^*L. \end{aligned} \tag{A.7.7}$$

Consequently, the linear operator defined by (A.7.3) is normal. Conversely, the spectral theorem asserts that every normal operator may be written in this way:

Theorem A.7.5 (Spectral). *For every normal operator $L: H \to H$ there exists a spectral measure E such that $L = \int z\,dE(z)$. This measure is unique and its support coincides with the spectrum of L. In particular, L is unitary if and only if* $\operatorname{supp} E$ *is contained in the unit circle* $\{z \in \mathbb{C} : |z| = 1\}$.

Example A.7.6 (Spectral theorem in finite dimension). Let H be a complex Hilbert space with finite dimension. Then for every normal operator $L: H \to H$ there exists a basis of H formed by eigenvectors of L. Let $\lambda_1, \dots, \lambda_s$ be the eigenvalues of L. The eigenspaces $V_j = \ker(L - \lambda_j \mathrm{id})$ are pairwise orthogonal, because L is normal. Moreover, by Theorem A.7.5, the direct sum $\bigoplus_{j=1}^s V_j$ is the whole of H. So

$$L = \sum_{j=1}^{s} \lambda_j \pi_j$$

where π_j denotes the orthogonal projection to V_j. In other words, the spectral measure E of the operator L is given by $E(\{\lambda_j\}) = \pi_j$ for every $j = 1, \dots, s$ and $E(B) = 0$ if B contains no eigenvalue of L.

Example A.7.7. Let $(\sigma_\alpha)_{\alpha \in \mathcal{A}}$ be any family of finite measures in the unit circle $\{z \in \mathbb{C} : |z| = 1\}$. Consider $H = \bigoplus_{\alpha \in \mathcal{A}} L^2(\sigma_\alpha)$ and the linear operator

$$L: H \to H, \quad (\varphi_\alpha)_\alpha \mapsto (z \mapsto z\varphi_\alpha(z))_\alpha.$$

Consider the spectral measure E given by

$$E(B): H \to H, \quad (\varphi_\alpha)_\alpha \mapsto (\mathcal{X}_B \varphi_\alpha)_\alpha$$

(compare with Example A.7.2). Then, $L = \int z\,dE(z)$. Indeed, the definition of E gives that $E\varphi \cdot \psi = \sum_\alpha \varphi_\alpha \bar{\psi}_\alpha \sigma_\alpha$ for every $\varphi = (\varphi_\alpha)_\alpha$ and $\psi = (\psi_\alpha)_\alpha$ in the space H. Then,

$$L\varphi \cdot \psi = \sum_\alpha \int z\varphi_\alpha(z)\bar{\psi}_\alpha(z)\,d\sigma_\alpha(z) = \int z\,d(E(z)\varphi \cdot \psi) \tag{A.7.8}$$

for every φ, ψ.

We say that $\lambda \in \mathbb{C}$ is an *atom* of the spectral measure if $E(\{\lambda\}) \neq 0$ or, equivalently, if there exists some non-zero vector $\omega \in H$ such that $E(\{\lambda\})\omega \neq 0$. The proof of the next proposition is outlined in Exercise A.7.4.

Proposition A.7.8. *Every eigenvalue of L is an atom of the spectral measure E. Conversely, if λ is an atom of E then λ is an eigenvalue of the operator L and every non-zero vector of the form $v = E(\{\lambda\})\omega$ is an eigenvector.*

A.7.2 Spectral representation

Theorem A.7.5 shows that normal linear operators on a Hilbert space are essentially the same thing as spectral measures in that space. Theorems of this type, establishing a kind of dictionary between two classes of objects that a priori do not seem to be related, are among the most fascinating results in mathematics. Of course, just how useful such a dictionary is to study one of those classes (normal linear operators, say) depends on to what extent we are capable of understanding the other one (spectral measures, in this case). In the present situation this is handled, in a most satisfactory way, by the next result, which exhibits a canonical form (inspired by Example A.7.2) in which every normal linear operator may be written.

As before, we use $\oplus$ to denote the orthogonal direct sum of Hilbert spaces. Given any cardinal χ, finite or infinite, and a Hilbert space V, we denote by V^χ the orthogonal direct sum of χ copies of V.

Theorem A.7.9 (Spectral representation). *Let $L : H \to H$ be a normal linear operator. Then there exist mutually singular finite measures $(\sigma_j)_j$ with support in the spectrum of L, there exist cardinals $(\chi_j)_j$ and there exists a unitary operator $U : H \to \bigoplus_j L^2(\sigma_j)^{\chi_j}$, such that the conjugate $ULU^{-1} = T$ is given by:*

$$T : \bigoplus_j L^2(\sigma_j)^{\chi_j} \to \bigoplus_j L^2(\sigma_j)^{\chi_j}, \quad (\varphi_{j,l})_{j,l} \mapsto \big(z \mapsto z\varphi_{j,l}(z)\big)_{j,l}. \qquad \text{(A.7.9)}$$

We call (A.7.9) the *spectral representation* of the normal operator L. Let us point out that the measures σ_j in Theorem A.7.9 are not uniquely determined. However, the spectral representation is unique, in the following sense.

Call the *multiplicity function* of the operator L the function associating with each finite measure θ in $\mathbb{C}$ the smallest cardinal χ_j such that the measures θ and σ_j are not mutually singular. One can prove that this function is uniquely determined by the operator L, that is, it does not depend on the choice of the measures σ_j in the statement. Moreover, two normal operators are conjugate by some unitary operator if and only if they have the same multiplicity function.

Example A.7.10 (Spectral representation in finite dimension). Let us go back to the setting of Example A.7.6. For each $j = 1,\dots,s$, let σ_j be the Dirac mass at the eigenvalue λ_j and χ_j be the dimension of the eigenspace V_j. Note that the space $L^2(\sigma_j)$ has dimension 1. Hence, we may choose a unitary operator

$U_j : V_j \to L^2(\sigma_j)$, for each $j = 1,\dots,s$. Since $L = \lambda_j \mathrm{id}$ restricted to V_j, we have that $T_j = U_j L U_j^{-1} = \lambda_j \mathrm{id}$, that is,

$$T_j : ((\varphi_\alpha)_\alpha) \mapsto \big(z \mapsto \lambda_j \varphi_\alpha(z)\big)_\alpha = \big(z \mapsto z\varphi_\alpha(z)\big)_\alpha.$$

In this way, we have found a unitary operator

$$U : \mathbb{C}^d \to \bigoplus_{j=1}^{s} L^2(\sigma_j)^{\chi_j}$$

such that $T = ULU^{-1}$ is a spectral representation of L.

A.7.3 Exercises

A.7.1. Let $T : E \to E$ be a Banach space isomorphism, that is, a continuous linear bijection whose inverse is also continuous. Show that $T + H$ is a Banach space isomorphism for every continuous linear map $H : E \to E$ such that $\|H\|\,\|T^{-1}\| < 1$. Use this fact to prove that the spectrum of every continuous linear operator $L : E \to E$ is a closed set and is contained in the closed disk of radius $\|L\|$ around the origin.

A.7.2. Show that if $L : H \to H$ is a linear operator in a Hilbert space H with finite dimension then $\operatorname{spec}(L)$ consists of the eigenvalues of L, that is, the complex numbers λ for which $L - \lambda \mathrm{id}$ is not injective. Give an example, in infinite dimension, such that the spectrum is strictly larger than the set of eigenvalues.

A.7.3. Prove Lemma A.7.3.

A.7.4. Prove Proposition A.7.8, along the following lines:

(a) Assume that $Lv = \lambda v$ for some $v \neq 0$. Consider the functions

$$\varphi_n(z) = \begin{cases} (z-\lambda)^{-1} & \text{if } |z-\lambda| > 1/n \\ 0 & \text{otherwise.} \end{cases}$$

Show that $\varphi_n(L)(L - \lambda \mathrm{id}) = E(\{z : |z-\lambda| > 1/n\})$ for every n. Conclude that $E(\{\lambda\})v = v$ and, consequently, λ is an atom of E.

(b) Assume that there exists $w \in H$ such that $v = E(\{\lambda\})w$ is non-zero. Show that, given any measurable set $B \subset \mathbb{C}$,

$$E(B)v = \begin{cases} v & \text{if } \lambda \in B \\ 0 & \text{if } \lambda \notin B. \end{cases}$$

Conclude that $Lv = \lambda v$ and, consequently, λ is an eigenvalue of L.

A.7.5. Let $(\sigma_j)_j$ be the family of measures given in Theorem A.7.9. Given any measurable set $B \subset \mathbb{C}$, check that $E(B) = 0$ if and only if $\sigma_j(B) = 0$ for every j. Therefore, given any measure η in $\mathbb{C}$, we have that $E \ll \eta$ if and only if $\sigma_j \ll \eta$ for every j.

Hints or solutions for selected exercises

1.1.2. Use Exercise A.3.5 to approximate characteristic functions by continuous functions.

1.2.5. Show that if $N > 1/\mu(A)$, then there exists $j \in V_A$ with $0 \le j \le N$. Adapting the proof of the previous statement, conclude that if K is a set of non-negative integers with $\#K > 1/\mu(A)$, then we may find $k_1, k_2 \in K$ and $n \in V_A$ such that $n = k_1 - k_2$. That is, the set $K - K = \{k_1 - k_2; k_1, k_2 \in K\}$ intersects V_A. To conclude that S is syndetic assume, by contradiction, that for every $n \in \mathbb{N}$ there is some number l_n such that $\{l_n, l_n+1, \dots, l_n+n\} \cap V_A = \emptyset$. Consider an element $k_1 \notin V_A$ and construct, recursively, a sequence $k_{j+1} = l_{k_j} + k_j$. Prove that the set $K = \{k_1, \dots, k_N\}$ is such that $(K-K) \cap V_A = \emptyset$.

1.2.6. Otherwise, there exists $k \ge 1$ and $b > 1$ such that the set $B = \{x \in [0,1] : n|f^n(x) - x| > b \text{ for every } n \ge k\}$ has positive measure. Let $a \in B$ be a density point of B. Consider $E = B \cap B(a,r)$, for r small. Get a lower estimate for the return time to E of any point of $x \in E$ and use the Kač theorem to reach a contradiction.

1.3.5. Consider the sequence $\log_{10} a_n$, where $\log_{10}$ denotes the base 10 logarithm, and observe that $\log_{10} 2$ is an irrational number.

1.3.12. Consider orthonormal bases $\{v_1, \dots, v_d\}$, at x, and $\{w_1, \dots, w_d\}$, at $f(x)$, such that v_1 and w_1 are orthogonal to H_c. Check that $\operatorname{grad} H(f(x)) \cdot Df(x)v = \operatorname{grad} H(x) \cdot v$ for every v. Deduce that the matrix of $Df(x)$ with respect to those bases has the form

$$Df(x) = \begin{pmatrix} \alpha & 0 & \cdots & 0 \\ \beta_2 & \gamma_{2,2} & \cdots & \gamma_{2,d} \\ \vdots & \vdots & \ddots & \vdots \\ \beta_d & \gamma_{d,2} & \cdots & \gamma_{d,d} \end{pmatrix},$$

with $\|\operatorname{grad} H(f(x))\| \, |\alpha| = \|\operatorname{grad} H(x)\|$. Note that $\Gamma = (\gamma_{i,j})_{i,j}$ is the matrix of $D(f \mid H_c)$ and observe that $|\det \Gamma| = \|\operatorname{grad} H(x)\| / \|\operatorname{grad} H(f(x))\|$. Using the formula of change of variables, conclude that $f \mid H_c$ preserves the measure $ds/\|\operatorname{grad} H\|$.

1.4.4. Choose a set $E \subset M$ with measure less than ε/n and, for each $k \ge 1$, let E_k be the set of points $x \in E$ that return to E in exactly k iterates. Take for B the union of the sets E_k, with $k \ge n$, of the n-th iterates of the sets E_k with $k \ge 2n$, and so on. For the second part, observe that if (f,μ) is aperiodic then μ cannot have atoms.

1.4.5. By assumption, $f^\tau(y) \in H_{n-\tau(y)}$ whenever $y \in H_n$ with $n > \tau(y)$. Therefore, $T(y) \in H$ if $y \in H$. Consider $A_n = \{1 \le j \le n : x \in H_j\}$ and $B_n = \{l \ge 1 : \sum_{i=0}^{l} \tau(T^i(x)) \le n\}$. Show, by induction, that $\#A_n \le \#B_n$ and deduce that $\limsup_n \#B_n/n \ge \theta$. Now suppose that $\liminf_k (1/k)\sum_{i=0}^{k-1} \tau(T^i(x)) > (1/\theta)$. Show that there exists $\theta_0 < \theta$ such that $\#B_n < \theta_0 n$, for every n sufficiently large. This contradicts the previous conclusion.

1.5.5. Observe that the maps $f, f^2, \dots, f^k$ commute with each other and then use the Poincaré multiple recurrence theorem.

1.5.6. By definition, the complement of $\Omega(f_1,\dots,f_q)^c$ is an open set. The Birkhoff multiple recurrence theorem ensures that the non-wandering set is non-empty.

2.1.6. Consider the image $V_*\mu$ of the measure μ under V. Check that $V_*\mu((a,b]) = F(b) - F(a)$ for every $a < b$. Consequently, $V_*\mu(\{b\}) = F(b) - \lim_{a\to b} F(a)$. Therefore, $(-\infty, b]$ is a continuity set for $V_*\mu$ if and only if b is a continuity point for F. Using Theorem 2.1.2, it follows that if $(V_{k*}\mu)_k$ converges to $V_*\mu$ in the weak* topology then $(V_k)_k$ converges to V in distribution. Conversely, if $(V_k)_k$ converges to V in distribution then $V_{k*}\mu((a,b]) = F_k(b) - F_k(a)$ converges to $F(b) - F(a) = V_*\mu((a,b])$, for any continuity points $a < b$ of F. Observing that such intervals $(a,b]$ generate the Borel σ-algebra of the real lines, conclude that $(V_{k*}\mu)_k$ converges to $V_*\mu$ in the weak* topology.

2.1.8. (Billingsley [Bil68]) Use the hypothesis to show that if $(U_n)_n$ is an increasing sequence of open subsets of M such that $\bigcup_n U_n = M$ then, for every $\varepsilon > 0$ there exists n such that $\mu(U_n) \ge 1 - \varepsilon$ for every $\mu \in \mathcal{K}$. Next, imitate the proof of Proposition A.3.7.

2.2.2. For the first part of the statement use induction in q. The case $q = 1$ corresponds to Theorem 2.1. Consider continuous transformations $f_i : M \to M$, $1 \le i \le q+1$ commuting with each other. By the induction hypothesis, there exists a probability ν invariant under f_i for $1 \le i \le q$. Define $\mu_n = (1/n)\sum_{j=0}^{n-1} (f_{q+1})^j_*(\nu)$. Note that $(f_i)_*\mu_n = \mu_n$ for every $1 \le i \le q$ and every n. Hence, every accumulation point of $(\mu_n)_n$ is invariant under every f_i, $1 \le i \le q$. By compactness, there exists some accumulation point $\mu \in \mathcal{M}_1(M)$. Check that μ is invariant under f_{q+1}. For the second part, denote by $M_q \subset \mathcal{M}_1(M)$ the set of probability measures invariant under f_i, $1 \le i \le q$. Then, $(M_q)_q$ is a non-increasing sequence of closed non-empty subsets of $\mathcal{M}_1(M)$. By compactness, the intersection $\bigcap_q M_q$ is non-empty.

2.2.6. Define μ in each iterate $f^j(W)$, $j \in \mathbb{Z}$ by letting $\mu(A) = m(f^{-j}(A))$ for each measurable set $A \subset f^j(W)$.

2.3.2. Clearly, convergence in norm implies weak convergence. To prove the converse, assume that $(x^k)_k$ converges to zero in the weak topology but not in the norm topology. The first condition implies that, for every fixed N, the sum $\sum_{n=0}^{N} |x_n^k|$ converges to zero when $k \to \infty$. The second condition means that, up to restricting to a subsequence, there exists $\delta > 0$ such that $\|x^k\| > \delta$

for every k. Then, there exists some increasing sequence $(l_k)_k$ such that

$$\sum_{n=0}^{l_{k-1}} |x_n^k| \leq \frac{1}{k} \quad \text{but} \quad \sum_{n=0}^{l_k} |x_n^k| \geq \|x^k\| - \frac{1}{k} \geq \delta - \frac{1}{k} \quad \text{for every } k.$$

Take $a_n = \overline{x_n^k}/|x_n^k|$ for each $l_{k-1} < n \leq l_k$. Then, for every k,

$$\left|\sum_{n=0}^{\infty} a_n x_n^k\right| \geq \sum_{l_{k-1}<n\leq l_k} |x_n^k| - \sum_{n\leq l_{k-1}} |x_n^k| - \sum_{n>l_k} |x_n^k| \geq \|x^k\| - \frac{4}{k} \geq \delta - \frac{4}{k}.$$

This contradicts the hypotheses. Now take $x_n^k = 1$ if $k = n$ and $x_n^k = 0$ otherwise. Given any $(a_n)_n \in c_0$, we have that $\sum_n a_n x_n^k = a_k$ converges to zero when $k \to \infty$. Therefore, $(x^k)_k$ converges to zero in the weak* topology. But $\|x^k\| = 1$ for every k, hence $(x^k)_k$ does not converge to zero in the norm topology.

2.3.6. Take $W = U(H)^\perp$ and $V = (\bigoplus_{n=0}^\infty U^n(W))^\perp$.

2.3.7. Suppose that there exist tangent functionals T_1 and T_2 with $T_1(v) > T_2(v)$ for some $v \in E$. Show that $\phi(u+tv) + \phi(u-tv) - 2\phi(u) \geq t(T_1(u) - T_2(u))$ for every t and deduce that ϕ is not differentiable in the direction of v.

2.4.1. Consider the set $\mathcal{P}$ of all probability measures on $X \times M$ of the form $\nu^{\mathbb{Z}} \times \eta$. Note that $\mathcal{P}$ is compact in the weak* topology and is invariant under the operator F_*.

2.4.2. The condition $\hat{p} \circ g = \hat{f} \circ \hat{p}$ entails $\hat{f}^n \circ \hat{p} = \hat{p} \circ g^n$ for every $n \in \mathbb{Z}$. Using $\pi \circ \hat{p} = p$, it follows that $\pi \circ \hat{f}^n \circ \hat{p} = p \circ g^n$ for every $n \leq 0$. Therefore, $\hat{p}(y) = \left(p(g^n(y))\right)_{n\leq 0}$. This proves the existence and uniqueness of $\hat{p}$. Now suppose that p is surjective. The hypotheses of compactness and continuity ensure that

$$\left(g^{-n}(p^{-1}(\{x_n\}))\right)_{n\leq 0}$$

is a nested sequence of compact sets, for every $(x_n)_{n\leq 0} \in \hat{M}$. Take y in the intersection and note that $\hat{p}(y) = (x_n)_{n\leq 0}$.

2.5.2. Fix q and l. Assume that for every $n \geq 1$ there exists a partition $\{S_1^n, \dots, S_l^n\}$ of the set $\{1, \dots, n\}$ such that no subset of S_j^n contains an arithmetic progression of length q. Consider the function $\phi_n : \mathbb{N} \to \{1, \dots, l\}$ given by $\phi_n(i) = j$ if $i \in S_j$ and $\phi_n(i) = l$ if $i > n$. Take $(n_k)_k \to \infty$ such that the subsequence $(\phi_{n_k})_k$ converges at every point to some function $\phi : \mathbb{N} \to \{1, \dots, l\}$. Consider $S_j = \phi^{-1}(j)$ for $j = 1, \dots, l$. Some S_j contains some arithmetic progression of length q. Then $S_j^{n_k}$ contains that arithmetic progression for every k sufficiently large.

2.5.4. Consider $\Sigma = \{1, \dots, l\}^{\mathbb{N}^k}$ with the distance $d(\omega, \omega') = 2^{-N}$ where $N \geq 0$ is largest such that $\omega(i_1, \dots, i_k) = \omega'(i_1, \dots, i_k)$ for every $i_1, \dots, i_k < N$. Note that Σ is a compact metric space. Given $q \geq 1$, let $F_q = \{(a_1, \dots, a_k) : 1 \leq a_i \leq q \text{ and } 1 \leq i \leq k\}$. Let $e_1, \dots, e_m$ be an enumeration of the elements of F_q. For each $j = 1, \dots, m$, consider the shift map $\sigma_j : \Sigma \to \Sigma$ given by $(\sigma_j \omega)(n) = \omega(n + e_j)$ for $n \in \mathbb{N}^k$. Consider the point $\omega \in \Sigma$ defined by $\omega(n) = i \Leftrightarrow n \in S_i$. Let Z be the closure of $\{\sigma_1^{l_1} \cdots \sigma_m^{l_m}(\omega) : l_1, \dots, l_m \in \mathbb{N}\}$. Note that Z is invariant under the shift maps σ_j. By the Birkhoff multiple recurrence theorem, there exist $\zeta \in Z$ and $s \geq 1$ such that $d(\sigma_j^s(\zeta), \zeta) < 1$ for every $j = 1, \dots, m$. Let $e = (1, \dots, 1) \in \mathbb{N}^k$. Then $\zeta(e) = \zeta(e + se_1) = \cdots = \zeta(e + se_m)$. Consider $\sigma_1^{l_1} \cdots \sigma_m^{l_m}(\omega)$ close enough to ζ that $\omega(b) = \omega(b + se_1) = \cdots = \omega(b + se_m)$, where $b = e + l_1 e_1 +$

$\cdots + l_m e_m$. It follows that if $i = \omega(b)$, then $b + sF_q \subset S_i$. Given that there are only finitely many sets S_i, some of them must contain infinitely many sets of the type $b + sF_q$, with q arbitrarily large.

3.1.1. Mimic the proof of Theorem 3.1.6.

3.1.2. Suppose that for every $k \in \mathbb{N}$ there exists $n_k \in \mathbb{N}$ such that $\mu(A \cap f^{-j}(A)) = 0$ for every $n_k + 1 \le j \le n_k + k$. It is no restriction to assume that $(n_k)_k \to \infty$. Take $\varphi = \mathcal{X}_A$. By Exercise 3.1.1, $(1/k)\sum_{j=n_k+1}^{n_k+k} \varphi \cdot \varphi \circ f^j \to \varphi \cdot P(\varphi)$. The left-hand side is identically zero and the right-hand side is equal to $\|P(\varphi)\|^2$. Hence, the time average $P(\varphi) = 0$ and so $\mu(A) = \int P(\varphi)\, d\mu = 0$.

3.2.3. (a) Consider $\varepsilon = 1$ and let $C = \sup\{|\varphi(l)| : |l| \le L(1)\}$. Given $n \in \mathbb{Z}$, fix $s \in \mathbb{Z}$ such that $sL(1) < n \le (s+1)L(1)$. By hypothesis, there exists $\tau \in \{sL(1) + 1, \dots, (s+1)L(1)\}$ such that $|\varphi(k+\tau) - \varphi(k)| < 1$ for every $k \in \mathbb{Z}$. Take $k = n - \tau$ and observe that $|k| \le L(1)$. It follows that $|\varphi(n)| < 1 + C$. (b) Take $\rho\varepsilon > 2L(\varepsilon)\sup|\varphi|$. For every $n \in \mathbb{Z}$ there exists some ε-quasi-period $\tau = n\rho + r$ with $1 \le r \le L(\varepsilon)$. Then,

$$\left| \sum_{j=n\rho+1}^{(n+1)\rho} \varphi(j) - \sum_{j=1-r}^{\rho-r} \varphi(j) \right| < \rho\varepsilon \quad \text{and} \quad \left| \sum_{j=1-r}^{\rho-r} \varphi(j) - \sum_{j=1}^{\rho} \varphi(j) \right| \le 2r\sup|\varphi| < \rho\varepsilon.$$

(c) Given $\varepsilon > 0$, take ρ as in part (b). For each $n \ge 1$, write $n = s\rho + r$, with $1 \le r \le \rho$. Then,

$$\frac{1}{n}\sum_{j=1}^{n} \varphi(j) = \frac{\rho}{s\rho + r}\sum_{i=0}^{s-1} \frac{1}{\rho} \sum_{l=i\rho+1}^{(i+1)\rho} \varphi(l) + \frac{1}{n}\sum_{l=s\rho+1}^{s\rho+r} \varphi(l).$$

For s large, the first term on the right-hand side is close to $(1/\rho)\sum_{j=0}^{\rho-1}\varphi(j)$ (by part (b)) and the last term is close to zero (by part (a)). Conclude that the left-hand side of the identity is a Cauchy sequence. (d) Observe that

$$\left| \frac{1}{n}\sum_{j=1}^{n} \varphi(x+k) - \frac{1}{n}\sum_{j=1}^{n} \varphi(j) \right| \le \frac{2|x|}{n} \sup|\varphi|$$

and use parts (a) and (c).

3.3.3. Let μ be a probability measure invariant under a flow $f^t : M \to M$, $t \in \mathbb{R}$ and let $(\varphi_s)_{s>0}$ be a family of functions, indexed by the positive real numbers, such that $\varphi_{s+t} \le \varphi_t + \varphi_s \circ f^t$ and the function $\Phi = \sup_{0<s<1} \varphi_s^+$ is in $L^1(\mu)$. Then, $(1/T)\varphi_T$ converges at μ-almost every point to a function φ such that $\varphi^+ \in L^1(\mu)$ and $\int \varphi\, d\mu = \lim_{T\to\infty}(1/T)\int \varphi_T\, d\mu$. To prove this, take $\varphi = \lim_n (1/n)\varphi_n$ (Theorem 3.3.3). For $T > 0$ non-integer, write $T = n + s$ with $N \in \mathbb{N}$ and $s \in (0,1)$. Then,

$$\varphi_T \le \varphi_n + \varphi_s \circ f^n \le \varphi_n + \Phi \circ f^n \quad \text{and} \quad \varphi_T \ge \varphi_{n+1} - \varphi_{1-s} \circ f^T \ge \varphi_n - \Phi \circ f^T.$$

Using Lemma 3.2.5, the first inequality shows that $\limsup_{T\to\infty}(1/T)\varphi_T \le \varphi$. Analogously, using the version of Lemma 3.2.5 for continuous time, the second inequality above gives that $\liminf_{T\to\infty}(1/T)\varphi_T \ge \varphi$. It also follows that $\lim_{T\to\infty}(1/T)\int \varphi_T\, d\mu$ coincides with $\lim_n (1/n)\int \varphi_n\, d\mu$. By Theorem 3.3.3, this last limit is equal to $\int \varphi\, d\mu$.

3.3.6. Since $\log^+ \|\phi\| \in L^1(\mu)$, for every $\varepsilon > 0$ there exists $\delta > 0$ such that $\mu(B) < \delta$ implies $\int_B \log^+ \|\theta\| \, d\mu < \varepsilon$. Using that $\log^+ \|\phi^n\| \le \sum_{j=0}^{n-1} \log^+ \|\theta\| \circ f^j$, one gets that

$$\mu(E) < \delta \Rightarrow \frac{1}{n} \int_E \log^+ \|\phi^n\| \, d\mu \le \frac{1}{n} \sum_{j=0}^{n-1} \int_{f^{-j}(E)} \log^+ \|\theta\| \, d\mu \le \varepsilon.$$

3.4.3. Consider local coordinates $x = (x_1, x_2, \ldots, x_d)$ such that Σ is contained in $\{x_1 = 0\}$. Write $\nu = \psi(x)\, dx_1 dx_2 \ldots dx_d$. Then $\nu_\Sigma = \psi(y)\, dx_2 \ldots dx_d$ with $y = (0, x_2, \ldots, x_d)$. Given $A \subset \Sigma$ and $\delta > 0$, the map $\xi : (t, y) \mapsto g^t(y)$ is a diffeomorphism from $[0,\delta] \times A$ to A_δ. Therefore, $\nu(A_\delta) = \int_{[0,\delta]\times A} (\psi \circ \xi) |\det D\xi| \, dt dx_2 \ldots dx_d$ and, consequently,

$$\lim_{\delta \to 0} \frac{\nu(A_\delta)}{\delta} = \int_A \psi(y) |\det D\xi|(y)\, dx_2 \ldots dx_d.$$

Next, note that $|\det D\xi|(y) = |X(y) \cdot (\partial/\partial t)| = \phi(y)$ for every $y \in \Sigma$. It follows that the flux of ν coincides with the measure $\eta = \phi \nu_\Sigma$. In particular, η is invariant under the Poincaré map.

4.1.2. Use the theorem of Birkhoff and the dominated convergence theorem.

4.1.8. Assume that $U_f \varphi = \lambda \varphi$. Since U_f is an isometry, $|\lambda| = 1$. If $\lambda^n = 1$ for some n then $\varphi \circ f^n = \varphi$ and, by ergodicity, φ is constant almost everywhere. Otherwise, given any $c \neq 0$, the sets $\varphi^{-1}(\lambda^{-k} c)$, $k \ge 0$ are pairwise disjoint. Since they all have the same measure, this measure must be zero. Finally, the set $\varphi^{-1}(c)$ is invariant under f and, consequently, its measure is either zero or total.

4.2.4. Let K be such a set. We may assume that K contains an infinite sequence of periodic orbits $(\mathcal{O}_n)_n$ with period going to infinity. Let $Y \subset K$ be the set of accumulation points of that sequence. Show that Y cannot consist of a single point. Let $p \neq q$ be periodic points in Y and z be a *heteroclinic point*, that is, such that $\sigma^n(z)$ converges to the orbit of p when $n \to -\infty$ and to the orbit of q when $n \to +\infty$. Show that $z \in Y$ and deduce the conclusion of the exercise.

4.2.10. Let $J_k = (0, 1/k)$, for each $k \ge 1$. Check that the continued fraction expansion of x is of bounded type if and only if there exists $k \ge 1$ such that $G^n(x) \notin J_k$ for every n. Observe that $\mu(J_k) > 0$ for every k. Deduce that for every k and μ-almost every x there exists $n \ge 1$ such that $G^n(x) \in J_k$. Conclude that $\mathcal{L}$ has zero Lebesgue measure.

4.2.11. For each $L \in \mathbb{N}$, consider $\varphi_L(x) = \min\{\phi(x), L\}$. Then, $\varphi_L \in L^1(\mu)$ and, by ergodicity, $\tilde{\varphi}_L = \int \varphi_L \, d\mu$ at μ-almost every point. To conclude, observe that $\tilde{\phi} \ge \tilde{\phi}_L$ for every L and $\int \phi_L \, d\mu \to +\infty$.

4.3.7. Let $M = \{0,1\}^{\mathbb{N}}$ and, for each n, let μ_n be the invariant measure supported on the periodic orbit $\alpha^n = (\alpha^n_k)_k$, with period $2n$, defined by $\alpha^n_k = 0$ if $0 \le k < n$ and $\alpha^n_k = 1$ if $n \le k < 2n$. Show that $(\mu_n)_n$ converges to $(\delta_0 + \delta_1)/2$, where 0 and 1 are the fixed points of the shift map.

4.3.9. (a) Take $k \ge 1$ such that every cylinder of length k has diameter less than δ. Take $y = (y_j)$ defined by $y_{j+n_i} = x^i_j$ for each $0 \le j < m_i + k$. (b) Take $\delta > 0$ such that $d(z, w) < \delta$ implies $|\varphi(z) - \varphi(w)| < \varepsilon$ and consider $k \ge 1$ given by part (a). Choose m_i, $i = 1, \ldots, s$ such that $m_i/n_s \approx \alpha_i$ for every i. Then take y as in part (a). (c) By the ergodic theorem, $\int \varphi \, d\mu = \int \tilde{\varphi} \, d\mu$. Take $x^1, \ldots, x^s \in \Sigma$ and

$\alpha^1,\dots,\alpha^s$ such that $\int \tilde{\varphi}\,d\mu \approx \sum_i \alpha_i \tilde{\varphi}(x^i)$. Note that $\tilde{\varphi}(y) = \int \varphi\,d\nu_y$, where ν_y is the invariant measure supported on the orbit of y. Recall Exercise 4.1.1.

4.4.3. On each side of the triangle, consider the foot of the corresponding height, that is, the orthogonal projection of the opposite vertex. Show that the trajectory defined by those three points is a periodic orbit of the billiard.

4.4.5. Using (4.4.10) and the twist condition, we get that for each $\theta \in \mathbb{R}$ there exists exactly one number $\rho_\theta \in (a,b)$ such that $\Theta(\theta,\rho_\theta) = \theta$. The function $\theta \mapsto \rho_\theta$ is continuous and periodic, with period 1. Consider its graph $\Gamma = \{(\theta,\rho_\theta) : \theta \in S^1\}$. Every point in $\Gamma \cap f(\Gamma)$ is fixed under f: if $(\theta,\rho_\theta) = f(\gamma,\rho_\gamma) = \big(\Theta(\gamma,\rho_\gamma),R(\gamma,\rho_\gamma)\big)$ then, since $\Theta(\gamma,\rho_\gamma) = \gamma$, it follows that $\theta = \gamma$ and so $\rho_\theta = \rho_\gamma$. Since f preserves the area measure, none of the connected components of $A \setminus \Gamma$ may be mapped inside itself. This implies that $f(\Gamma)$ intersect Γ at no less than two points.

4.4.7. Taking inspiration from Example 4.4.12, show that the billiard map in Ω extends to a *Dehn twist* in the annulus $A = S^1 \times [-\pi/2,\pi/2]$, that is, a homeomorphism $f : A \to A$ that coincides with the identity on both boundary components but is homotopically non-trivial: actually, f admits a lift $F : \mathbb{R} \times [-\pi/2,\pi/2] \to \mathbb{R} \times [-\pi/2,\pi/2]$ such that $F(s,-\pi/2) = (s-2\pi,-\pi/2)$ and $F(s,\pi/2) = (s,\pi/2)$ for every s. Consider rational numbers $p_n/q_n \in (-2\pi,0)$ with $q_n \to \infty$. Use Exercise 4.4.6 to show that g has periodic points of period q_n. One way to ensure that these periodic points are all distinct is to take the q_n mutually prime.

5.1.7. The statement does not depend on the choice of the ergodic decomposition, since the latter is essentially unique. Consider the construction in Exercise 5.1.6. The set M_0 is saturated by the partition $\mathcal{W}^s$, that is, if $x \in M_0$ then $\mathcal{W}^s(x) \subset M_0$. Moreover, the map $y \mapsto \mu_y$ is constant on each $\mathcal{W}^s(x)$. Since the partition $\mathcal{P}$ is characterized by $\mathcal{P}(x) = \mathcal{P}(y) \Leftrightarrow \mu_x = \mu_y$, it follows that $\mathcal{P} \prec \mathcal{W}^s$ restricted to M_0.

5.2.1. Consider the canonical projections $\pi_{\mathcal{P}} : M \to \mathcal{P}$ and $\pi_{\mathcal{Q}} : M \to \mathcal{Q}$, the quotient measures $\hat{\mu}_{\mathcal{P}} = (\pi_{\mathcal{P}})_*\mu$ and $\hat{\mu}_{\mathcal{Q}} = (\pi_{\mathcal{Q}})_*\mu$ and the disintegrations $\mu = \int \mu_P\,d\hat{\mu}_{\mathcal{P}}(P)$ and $\mu = \int \mu_Q\,d\hat{\mu}_{\mathcal{Q}}(Q)$. Moreover, for each $P \in \mathcal{P}$, consider $\hat{\mu}_{P,\mathcal{Q}} = (\pi_{\mathcal{Q}})_*\mu_P$ and the disintegration $\mu_P = \int \mu_{P,Q}\,d\hat{\mu}_{P,\mathcal{Q}}(Q)$. Observe that $\int \hat{\mu}_{P,\mathcal{Q}}\,d\hat{\mu}_{\mathcal{P}}(P) = \hat{\mu}_{\mathcal{Q}}$: given any $B \subset \mathcal{Q}$,

$$\int \hat{\mu}_{P,\mathcal{Q}}(B)\,d\hat{\mu}_{\mathcal{P}}(P) = \int \mu_P(\pi_{\mathcal{Q}}^{-1}(B))\,d\hat{\mu}_{\mathcal{P}}(P) = \mu(\pi_{\mathcal{Q}}^{-1}(B)) = \hat{\mu}_{\mathcal{Q}}(B).$$

To check that $\mu_{\pi(Q),Q}$ is a disintegration of μ with respect to $\mathcal{Q}$: (a) $\mu_{P,Q}(Q) = 1$ for $\hat{\mu}_{P,\mathcal{Q}}$-almost every Q and $\hat{\mu}_{\mathcal{P}}$-almost every P. Moreover, $\mu_{P,Q} = \mu_{\pi(Q),Q}$ for $\hat{\mu}_{P,\mathcal{Q}}$-almost every Q and $\hat{\mu}_{\mathcal{P}}$-almost every P, because $\mu_P(P) = 1$ for $\hat{\mu}_P$-almost every P. By the previous observation, it follows that $\mu_{\pi(Q),Q}(Q) = 1$ for $\hat{\mu}_{\mathcal{Q}}$-almost every Q. (b) $P \mapsto \mu_P(E)$ is measurable, up to measure zero, for every Borel set $E \subset M$. By construction (Section 5.2.3), there exists a countable generating algebra $\mathcal{A}$ such that $\mu_{P,Q}(E) = \lim_n \mu_P(E \cap Q_n)/\mu_P(Q_n)$ for every $E \in \mathcal{A}$ (where Q_n is the element of $\mathcal{Q}_n$ that contains Q). Deduce that $P \mapsto \mu_{\pi(Q),Q}(E)$ is measurable, up to measure zero, for every $E \in \mathcal{A}$. Extend this conclusion to every Borel set E, using the monotone class argument in

Section 5.2.3. (c) Note that $\mu = \int \mu_P \, d\hat{\mu}_{\mathcal{P}}(P) = \int\int \mu_{P,Q} \, d\hat{\mu}_{P,\mathcal{Q}}(Q) d\hat{\mu}_{\mathcal{P}}(P) = \int\int \mu_{\pi(Q),Q} \, d\hat{\mu}_{P,\mathcal{Q}}(Q) d\hat{\mu}_{\mathcal{P}}(P) = \int \mu_{\pi(Q),Q} \, d\hat{\mu}_{\mathcal{Q}}(Q)$.

5.2.2. Argue that the partition $\mathcal{Q}$ of the space $\mathcal{M}_1(M)$ into points is measurable. Given any disintegration $\{\mu_P : P \in \mathcal{P}\}$, consider the measurable map $M \mapsto \mathcal{M}_1(M)$, $x \mapsto \mu_{P(x)}$. The pre-image of $\mathcal{Q}$ under this map is a measurable partition. Check that this pre-image coincides with $\mathcal{P}$ on a subset with full measure.

6.1.3. The function φ is invariant.

6.2.5. Denote by X the closure of the orbit of x. If X is minimal, for each $y \in X$ there exists $n(y) \geq 1$ such that $d(f^{n(y)}(y),x) < \varepsilon$. Then, by continuity, y admits an open neighborhood $V(y)$ such that $d(f^{n(y)}(z),x) < \varepsilon$ for every $z \in V(y)$. Take $y_1,\dots,y_s$ such that $X \subset \bigcup_i V(y_i)$ and let $m = \max_i n(y_i)$. Given any $k \geq 1$, take i such that $f^k(x) \in V(y_i)$. Then, $d(f^{k+n_i}(x),x) < \varepsilon$, that is, $k + n_i \in R_\varepsilon$. This proves that, given any $m+1$ consecutive integers, at least one of them is in R_ε. Hence, R_ε is syndetic. Now assume that X is not minimal. Then, there exists a non-empty, closed invariant set F properly contained in X. Note that $x \notin F$ and so, for every ε sufficiently small, there exists an open set U that contains F and does not intersect $B(x,\varepsilon)$. On the other hand, since R_ε is syndetic, there exists $m \geq 1$ such that for any $k \geq 1$ there exists $n \in \{k,\dots,k+m\}$ satisfying $f^n(x) \in B(x,\varepsilon)$. Take k such that $f^k(x) \in U_1$, where $U_1 = U \cap f^{-1}(U) \cap \cdots \cap f^{-m}(U)$, and find a contradiction.

6.2.6. By Exercise 6.2.5, the set $R_\varepsilon = \{n \in \mathbb{N} : d(x,f^n(x)) < \varepsilon\}$ is syndetic for every $\varepsilon > 0$. If y is close to x then $\{n \in \mathbb{N} : d(f^n(x),f^n(y)) < \varepsilon\}$ contains blocks of consecutive integers with arbitrary length, no matter the choice of $\varepsilon > 0$. Let U_1 be any neighborhood of x. It follows from the previous observations that there exist infinitely many values of $n \in \mathbb{N}$ such that $f^n(x), f^n(y)$ are in U_1. Fix n_1 with this property. Next, consider $U_2 = U_1 \cap f^{-n_1}(U_1)$. By the previous step, there exists $n_2 > n_1$ such that $f^{n_2}(x), f^{n_2}(y) \in U_2$. Continuing in this way, construct a non-increasing sequence of open sets U_k and an increasing sequence of natural numbers n_k such that $f^{n_k}(U_{k+1}) \subset U_k$ and $f^{n_k}(x), f^{n_k}(y) \in U_k$. Check that $f^{n_{i_1}+\cdots+n_{i_k}}(x)$ and $f^{n_{i_1}+\cdots+n_{i_k}}(y)$ are in U_1 for any $i_1 < \cdots < i_k$, $k \geq 1$.

6.2.7. Consider the shift map $\sigma : \Sigma \to \Sigma$ in $\Sigma = \{1,2,\dots,q\}^{\mathbb{N}}$. The partition $\mathbb{N} = S_1 \cup \cdots \cup S_q$ defines a certain element $\alpha = (a_n) \in \Sigma$, given by $a_n = i$ if and only if $n \in S_i$. Consider β in the closure of the orbit of α such that α and β are near and the closure of the orbit of β is a minimal set. Apply Exercise 6.2.6 with $x = \beta$, $y = \alpha$ and $U = [0; a_0]$ to obtain the result.

6.3.6. Write $g = (a_{11}, a_{12}, a_2, a_{22})$. Then,

$$E_g(x_{11},x_{12},x_{21},x_{22}) = (a_{11}x_{11}+a_{12}x_{21}, a_{11}x_{12}+a_{12}x_{22}, a_{21}x_{11}+a_{22}x_{21}, a_{21}x_{12}+a_{22}x_{22}).$$

Write the right-hand side as $(y_{11}, y_{12}, y_{21}, y_{22})$. Use the formula of change of variables, observing that $\det(y_{11},y_{12},y_{21},y_{22}) = (\det g)\det(x_{11},x_{12},x_{21},x_{22})$ and

$$dy_{11}dy_{12}dy_{21}dy_{22} = (\det g)^2 dx_{11}dx_{12}dx_{21}dx_{22}.$$

In the complex case, take

$$\int_{\mathrm{GL}(2,\mathbb{R})} \varphi \, d\mu = \int \frac{\varphi(z_{11},z_{12},z_{21},z_{22})}{|\det(z_{11},z_{12},z_{21},z_{22})|^4} dx_{11}dy_{11}dx_{12}dy_{12}dx_{21}dy_{21}dx_{22}dy_{22},$$

where $z_{jk} = x_{jk} + y_{jk}i$. [Observation: Generalize these constructions to any dimension!]

6.3.9. Given $x \in M$, there exists a unique number $0 \le r < 10^k$ such that $f^r(x) \in [b_0, \dots, b_{k-1}]$. Moreover, $f^n(x) \in [b_0, \dots, b_{k-1}]$ if and only if $n - r$ is a multiple of 10^k. Use this observation to conclude that

$$\tau([b_0, \dots, b_{k-1}], x) = 10^{-k} \quad \text{for every } x \in M.$$

Conclude that if f admits an ergodic probability measure μ then $\mu([b_0, \dots, b_{k-1}]) = 10^{-k}$ for every $b_0, \dots, b_{k-1}$. This determines μ uniquely. To conclude, show that μ is well defined and invariant.

6.3.11. Consider the sequence of words w_n defined inductively by $w_1 = \alpha$ and $s(w_{n+1}) = w_n$ for $n \ge 1$. Decompose the word $s(\alpha) = w_2 = \alpha r_1$ and prove, by induction, that w_{n+1} may be decomposed as $w_{n+1} = w_n r_n$, for some word r_n with length greater than or equal to n, such that $s(r_n) = r_{n+1}$. Define $w = \alpha r_1 r_2 \cdots$ and note that $s(w) = s(\alpha)s(r_1)s(r_2)\cdots = \alpha r_1 r_2 r_3 \cdots = w$. This proves existence. To prove uniqueness, let $\gamma \in \Sigma$ be a sequence starting with α and such that $S(\gamma) = \gamma$. Decompose γ as $\gamma = \alpha\gamma_1\gamma_2\gamma_3\cdots$, in such a way that γ_i and r_i have the same length. Note that $S(\alpha) = \alpha\gamma_1 = \alpha r_1$, and so $\gamma_1 = r_1$. Conclude by induction.

6.4.2. Given any $0 \le \alpha < \beta \le 1$, we have that $\sqrt{n} \in (\alpha, \beta)$ in the circle if and only if there exists some integer $k \ge 1$ such that $k^2 + 2k\alpha + \alpha^2 < n < k^2 + 2k\beta + \beta^2$. For each k the number of values of n that satisfy this inequality is equal to the integer part of $2k(\beta - \alpha) + (\beta^2 - \alpha^2)$. Therefore,

$$\#\{1 \le n < N^2 : \sqrt{n} \in (\alpha, \beta)\} \le \sum_{k=1}^{N-1} 2k(\beta - \alpha) + (\beta^2 - \alpha^2)$$

and the difference between the term on the right and the one on the left is less than N. Hence,

$$\lim \frac{1}{N^2} \#\{1 \le n < N^2 : \sqrt{n} \in (\alpha, \beta)\} = \beta - \alpha.$$

A similar calculation shows that the sequence $(\log n \mod \mathbb{Z})_n$ is not equidistributed in the circle. [Observation: But it does admit a continuous (non-constant) limit density. Calculate that density!]

6.4.3. Define $\phi_n = a^n + (-1/a)^n$. Check that $(\phi_n)_n$ is the Fibonacci sequence and, in particular, $\phi_n \in \mathbb{N}$ for every $n \ge 1$. Now observe that $(-1/a)^n$ converges to zero. Hence, $\{n \ge 1 : a^n \mod \mathbb{Z} \in I\}$ is finite, for any interval $I \subset S^1$ whose closure does not contain zero.

7.1.1. It is clear that the condition is necessary. To see that it is sufficient: Given A, consider the closed subspace $\mathcal{V}$ of $L^2(\mu)$ generated by the functions 1 and $\mathcal{X}_{f^{-k}(A)}$, $k \in \mathbb{N}$. The hypothesis ensures that $\lim_n U_f^n(\mathcal{X}_A) \cdot \mathcal{X}_{f^{-k}(A)} = (\mathcal{X}_A \cdot 1)(\mathcal{X}_{f^{-k}(A)} \cdot 1)$ for every k. Conclude that $\lim_n U_f^n(\mathcal{X}_A) \cdot \phi = (\mathcal{X}_A \cdot 1)(\phi \cdot 1)$ for every $\phi \in \mathcal{V}$. Given a measurable set B, write $\mathcal{X}_B = \phi + \phi^\perp$ with $\phi \in \mathcal{V}$ and $\phi^\perp \in \mathcal{V}^\perp$ to conclude that $\lim_n U_f^n(\mathcal{X}_A) \cdot \mathcal{X}_B = (\mathcal{X}_A \cdot 1)(\mathcal{X}_B \cdot 1)$.

7.1.2. Assuming that E exists, decompose $(1/n)\sum_{j=0}^{n-1} |a_j|$ into two terms, one over $j \in E$ and the other over $j \notin E$. The hypotheses imply that the two terms converge to zero. Conversely, assume that $(1/n)\sum_{j=0}^{n-1} |a_j|$ converges to zero. Define $E_m =$

$\{j \geq 0 : |a_j| \geq (1/m)\}$ for each $m \geq 1$. The sequence $(E_m)_m$ is increasing and each E_m has density zero; in particular, there exists $\ell_m \geq 1$ such that $(1/n)\#\big(E_m \cap \{0,\dots,n-1\}\big) < (1/m)$ for every $n \geq \ell_m$. Choose $(\ell_m)_m$ increasing and define $E = \bigcup_m (E_m \cap \{\ell_m,\dots,\ell_{m+1}-1\})$. For the second part of the exercise, apply the first part to both sequences, $(a_n)_n$ and $(a_n^2)_n$.

7.1.6. (Pollicott and Yuri [PY98]) It is enough to treat the case when $\int \varphi_j \, d\mu = 0$ for every j. Use induction on the number k of functions. The case $k = 1$ is contained in Theorem 3.1.6. Use the inequalities

$$\frac{1}{N}\sum_{n=1}^{n} a_n \leq \frac{1}{N}\sum_{n=1}^{N-m+1}\Big(\frac{1}{m}\sum_{j=0}^{m-1} a_{n+j}\Big) + \frac{m}{N}\big(\max_{1\leq i\leq m}|a_i| + \max_{N-m\leq i\leq N}|a_i|\big)$$

$$\Big(\frac{1}{N}\sum_{n=1}^{N} b_n\Big)^2 \leq (1/N)\sum_{n=1}^{N}|b_n|^2$$

to conclude that $\int \big|(1/N)\sum_{j=0}^{N-1}(\varphi_1 \circ f^n)\cdots(\varphi_k \circ f^{kn})\big|^2 d\mu$ is bounded above by

$$\frac{1}{N}\sum_{n=1}^{N}\Big(\int \Big|\frac{1}{m}\sum_{j=0}^{m-1}(\varphi_1 \circ f^{n+j})\cdots(\varphi_k \circ f^{k(n+j)})\Big|^2 d\mu + \Big(\frac{2m}{N} + \frac{m^2}{N^2}\Big)\Big(\max_{1\leq i\leq k} \operatorname{supess}|\varphi_i|\Big)^2.$$

The integral is equal to

$$\sum_{i=0}^{m-1}\sum_{j=0}^{m-1}\int \prod_{l=1}^{k}\big(\varphi_l(\varphi_l \circ f^{l(j-i)})\big) \circ f^{l(n+i)} \, d\mu.$$

By the induction hypothesis,

$$\frac{1}{N}\sum_{n=1}^{N}\prod_{l=2}^{k}\big(\varphi_l(\varphi_l \circ f^{l(j-i)})\big) \circ f^{l(n+i)} \to \prod_{l=2}^{k}\int \varphi_l(\varphi_l \circ f^{l(j-i)}) \, d\mu$$

in $L^2(\mu)$, when $N \to \infty$. Therefore,

$$\frac{1}{N}\sum_{n=1}^{N}\int \prod_{l=1}^{k}\big(\varphi_l(\varphi_l \circ f^{l(j-i)})\big) \circ f^{l(n+i)} \, d\mu \to \prod_{l=1}^{k}\int \varphi_l(\varphi_l \circ f^{l(j-i)}) \, d\mu$$

in $L^2(\mu)$, when $N \to \infty$. Combining these estimates,

$$\limsup_N \int \Big|\frac{1}{N}\sum_{n=1}^{N}(\varphi_1 \circ f^n)\cdots(\varphi_k \circ f^{kn})\Big|^2 d\mu \leq \frac{1}{m^2}\sum_{i=0}^{m-1}\sum_{j=0}^{m-1}\prod_{l=1}^{k}\int \varphi_l(\varphi_l \circ f^{l(j-i)}) \, d\mu.$$

Since (f,μ) is weak mixing, $\int \varphi_l(\varphi_l \circ f^{lr}) \, d\mu$ converges to 0 when $r \to \infty$, restricted to a set of values of l with density 1 at infinity (recall Exercise 7.1.2). Therefore, the expression on the right-hand side is close to zero when m is large.

7.2.5. The first statement is analogous to Exercise 7.2.1. The definition ensures that μ_k has memory k. Given $\varepsilon > 0$ and any (uniformly) continuous function $\varphi : \Sigma \to \mathbb{R}$, there exists $\kappa \geq 1$ such that $|\int_C \varphi \, d\eta - \varphi(x)\eta(C)| \leq \varepsilon\eta(C)$ for every $x \in C$, every cylinder C of length $l \geq \kappa$ and every probability measure η. Since $\mu = \mu_k$ for cylinders of length k, it follows that $|\int \varphi \, d\mu_k - \int \varphi \, d\mu| \leq \varepsilon$ for every $k \geq \kappa$. This proves that $(\mu_k)_k$ converges to μ in the weak* topology.

7.2.6. (a) Use that $P^{n_1+n_2}_{i,i} = \sum_j P^{n_1}_{i,j} P^{n_2}_{j,i}$. All the terms in this expression are non-negative and the term corresponding to $j=i$ is positive. (b) Up to replacing R by R/κ, we may suppose that $\kappa = 1$. Start by showing that if $S \subset \mathbb{Z}$ is closed under addition and subtraction then $S = a\mathbb{Z}$, where a is the smallest positive element of S. Use that fact to show that if $a_1, \dots, a_s$ are positive integers with greatest common divisor equal to 1 then there exist integers $b_1, \dots, b_s$ such that $b_1 a_1 + \cdots + b_s a_s = 1$. Now take $a_1, \dots, a_s \in R$ such that their greatest common divisor is equal to 1. Using the previous observation, and the hypothesis that R is closed under addition, conclude that there exists $p, q \in R$ such that $p - q = 1$. To finish, show that R contains every integer $n \geq pq$. (c) Consider any $i, j \in X$ and let κ_i, κ_j be the greatest common divisors of $R(i), R(j)$, respectively. By irreducibility, there exist $k, l \geq 1$ such that $P^k_{i,j} > 0$ and $P^l_{j,i} > 0$. Deduce that if $n \in R(i)$ then $n + k + l \in R(j)$. In view of (b), this is possible only if $\kappa_i \geq \kappa_j$. Exchanging the roles of i and j, it also follows that $\kappa_i \leq \kappa_j$. If $\kappa \geq 2$ then, given any i, we have $P^n_{i,i} = 0$ for n arbitrarily large and so P cannot be aperiodic. Now suppose that $\kappa = 1$. Then, using (b) and the hypothesis that X is finite, there exists $m \geq 1$ such that $P^n_{i,i} > 0$ for every $i \in X$ and every $n \geq m$. Then, since P is irreducible and X is finite, there exists $k \geq 1$ such that for any i, j there exists $l \leq k$ such that $P^l_{i,j} > 0$. Deduce that $P^{m+k}_{i,j} > 0$ for every i, j and so P is aperiodic. (d) Fix any $i \in X$ and, for each $r \in \{0, \dots, \kappa - 1\}$, define $X_r = \{j \in X : \text{ there exists } n \equiv r \mod \kappa \text{ such that } P^n_{i,j} > 0\}$. Check that these sets X_r cover X and are pairwise disjoint. Show that the restriction of P^κ to each of them is aperiodic.

7.3.1. By the theorem of Darboux, there exist coordinates (x_1, x_2) in the neighborhood of any point of S such that $\omega = dx_1 \wedge dx_2$. Consider the expression of the vector field in those coordinates: $X = X_1(\partial/\partial x_1) + X_2(\partial/\partial x_2)$. Show that $\beta = X_1 dx - 2 - X_2 dx_1$ and so $d\beta = (\operatorname{div} X)\, dx_1 \wedge dx_2$. Hence, β is closed if and only if the divergent of X vanishes.

7.3.5. Observe that f is invertible and if A is a d-adic interval of level $r \geq 1$ (that is, an interval of the form $A = [id^{-r}, (i+1)d^{-r}])$, then there exists $s \geq r$ such that $f(A)$ consists of d^{s-r} d-adic intervals of level s. Deduce that f preserves the Lebesgue measure. Show also that if A and B are d-adic intervals then, since σ has no periodic points, $m(f^k(A) \cap B) = m(A)m(B)$ for every large k.

7.4.2. (a) Given $y^1, y^2 \in M$, write $f^{-1}(y^i) = \{x^i_1, \dots, x^i_d\}$ with $d(x^1_j, x^2_j) \leq \sigma^{-1} d(y^1, y^2)$. Then,

$$|\mathcal{L}\varphi(y^1) - \mathcal{L}\varphi(y^2)| = \frac{1}{d} \sum_{j=1}^{d} |\varphi(x^1_j) - \varphi(x^2_j)| \leq K_\theta(\varphi) \sigma^{-\theta} d(y^1, y^2)^\theta.$$

(b) It follows that $\|\mathcal{L}\varphi\| \leq \sup|\varphi| + \sigma^{-\theta} K_\theta(\varphi) \leq \|\varphi\|$ for every $\varphi \in \mathcal{E}$, and the identity holds if and only if φ is constant. Hence, $\|\mathcal{L}\| = 1$. (c) Let $J_n = [\inf \mathcal{L}^n\varphi, \sup \mathcal{L}^n\varphi]$. By part (a), the sequence $(J_n)_n$ is decreasing and the diameter of J_n converges to zero exponentially fast. Take ν_φ to be the point in the intersection and note that $\|\mathcal{L}^n\varphi - \nu_\varphi\| = \sup|\mathcal{L}^n\varphi - \nu_\varphi| + K_\theta(\mathcal{L}^n\varphi)$. (d) The constant functions are eigenvectors of $\mathcal{L}$, associated with the eigenvalue $\lambda = 1$. It follows that $\nu_{\varphi+c} = \nu_\varphi + c$ for every $\varphi \in \mathcal{E}$ and every $c \in \mathbb{R}$. Then, $H = \{\varphi : \nu_\varphi = 0\}$ is a hyperplane of $\mathcal{E}$ transverse to the line of constant functions. This hyperplane is invariant under $\mathcal{L}$ and, by part (c), the spectral radius of $\mathcal{L} \mid H$ is less or equal than $\sigma^{-\theta} < 1$. (e) By part (b), $\|\mathcal{L}^n\varphi - \mathcal{L}^n\psi\| \le \|\mathcal{L}^k\varphi - \mathcal{L}^k\psi\|$ for every $n \ge k \ge 1$. Making $n \to \infty$, we get that $|\nu_\varphi - \nu_\psi| \le \|\mathcal{L}^k\varphi - \mathcal{L}^k\psi\|$ for every $k \ge 1$. Using part (a) and making $k \to \infty$, we get that $|\nu_\varphi - \nu_\psi| \le \sup|\varphi - \psi|$. Therefore, the linear operator $\psi \mapsto \nu_\psi$ is continuous, relative to the norm in the space $C^0(M)$.

8.1.2. Denote $X_i = X \cap [0;i]$ and $p_i = \mu([0;i])$, for $i = 1,\dots,k$. Since μ is a Bernoulli measure, $\mu(X_i) = p_i\mu(f(X_i))$. Hence, $\sum_i p_i\mu(f(X_i)) = 1$. Since $\sum_i p_i = 1$, it follows that $\mu(f(X_i)) = 1$ for every i. Consequently, $\bigcap_i f(X_i)$ has full measure. Take x in that intersection. If (f,μ) and (g,ν) are ergodically equivalent, there exists a bijection $\phi : X \to Y$ between full measure invariant subsets such that $\phi \circ f = g \circ \phi$. Take $x \in X$ with k pre-images $x_1,\dots,x_k$ in X. The points $\phi(x_i)$ are pre-images of $\phi(x)$ for the transformation g. Hence, $k \le l$; by symmetry, we also have that $l \le k$.

8.2.5. Assume that (f,μ) is not weak mixing. By Theorem 8.2.1, there exists a non-constant function φ such that $U_f\varphi = \lambda\varphi$ for some $\lambda = e^{2\pi i\theta}$. By ergodicity, the absolute value of φ is constant μ-almost everywhere. Using that f^n is ergodic for every n (Exercise 4.1.8), θ is irrational and any set where φ is constant has measure zero. Given $\alpha < \beta$ in $[0,2\pi]$, consider $A = \{x \in \mathbb{C} : \alpha \le \arg(\varphi(x)) \le \beta\}$. Show that for every $\varepsilon > 0$ there exists n such that $\mu(f^{-n}(A) \setminus A) < \varepsilon$. Show that, by choosing $|\beta - \alpha|$ sufficiently small, one gets to contradict the inequality in the statement.

8.2.7. Note that $f_{n+1}(x) = f_n(x)$ for every $x \in J_n$ that is not on the top of $\mathcal{S}_n$. Hence, (for example, arguing as in Exercise 6.3.10), $f(x) = f_n(x)$ for every $x \in [0,1)$ and every n sufficiently large; moreover, f preserves the Lebesgue measure. Let $a_n = \#\mathcal{S}_n$ be the height of each pile $\mathcal{S}_n$. Denote by $\{I^e, I^c, I^d\}$ the partition of each $I \in \mathcal{S}_n$ into subintervals of equal length, ordered from left to right. (a) If A is a set with $m(A) > 0$ then for every $\varepsilon > 0$ there exists $n \ge 1$ and some interval $I \in \mathcal{S}_n$ such that $m(A \cap I) \ge (1-\varepsilon)m(I)$. If A is invariant, it follows that $m(A \cap J) \ge (1-\varepsilon)m(J)$ for every $J \in \mathcal{S}_n$. (b) Assume that $U_f\varphi = \lambda\varphi$. Since U_f is an isometry, $|\lambda| = 1$. By ergodicity, $|\varphi|$ is constant almost everywhere; we may suppose that $|\varphi| \equiv 1$. Initially, assume that there exists n and some interval $I \in \mathcal{S}_n$ such that the restriction of φ to I is constant. Take $x \in I^e$ and $y \in I^c$ and $z \in I^d$. Then, $\varphi(x) = \varphi(y) = \varphi(z)$ and $\varphi(y) = \lambda^{a_n}\varphi(x)$ and $\varphi(z) = \lambda^{a_n+1}\varphi(y)$. Hence, $\lambda = 1$ and, by ergodicity, φ is constant. In general, use the theorem of Lusin (Theorems A.3.5–A.3.9) to reach the same conclusion. (c) A is a union of intervals I_j in the pile $\mathcal{S}_n$ for each $n \ge 2$. Then, $f^{a_n}(I_j^e) = I_j^c$ for every j. Hence, $m(f^{a_n}(A) \cap A) \ge m(A)/3 = 2/27$.

8.3.1. Let $\{v_j : j \in \mathcal{I}\}$ be a basis of H formed by eigenvectors with norm 1 and λ_j be the eigenvalue associated with each eigenvector v_j. The hypothesis ensures that we may consider $\mathcal{I} = \mathbb{N}$. Show that for every $\delta > 0$ and every $k \ge 1$ there exists $n \ge 1$ such that $|\lambda_j^n - 1| \le \delta$ for every $j \in \{1,\dots,k\}$ (use the pigeonhole principle). Decompose $\varphi = \sum_j c_j v_j$, with $c_j \in \mathbb{C}$. Observe that $U_f^n\varphi = \sum_{j\in\mathbb{N}} c_j\lambda_j^n v_j$, and so

$$\|U_f^n\varphi - \varphi\|_2^2 \le \sum_{j=1}^{k} |c_j(\lambda_j^n - 1)|^2 + \sum_{j=k+1}^{\infty} 2|c_j|^2 \le \delta^2\|\varphi\|_2^2 + \sum_{j=k+1}^{\infty} 2|c_j|^2.$$

Given $\varepsilon > 0$, we may choose δ and k in such a way that each one of the terms on the right-hand side is less than $\varepsilon/2$.

8.4.3. Let $U : H \to H$ be a non-invertible isometry. Recalling Exercise 2.3.6, show that there exist closed subspaces V and W of H such that $U : H \to H$ is unitarily conjugate to the operator $U_1 : V \oplus W^{\mathbb{N}} \to V \oplus W^{\mathbb{N}}$ given by $U_1 \mid V = U \mid V$ and $U_1 \mid W^{\mathbb{N}} = \mathrm{id}$. Let $U_2 : V \oplus W^{\mathbb{Z}} \to V \oplus W^{\mathbb{Z}}$ be the linear operator defined by $U_1 \mid V = U \mid V$ and $U_1 \mid W^{\mathbb{Z}} = \mathrm{id}$. Check that U_2 is a unitary operator such that $U_2 \circ \jmath = \jmath \circ U_1$, where $\jmath : V \oplus W^{\mathbb{N}} \to V \oplus W^{\mathbb{Z}}$ is the natural inclusion. Show that if $E \subset V \oplus W^{\mathbb{N}}$ satisfies the conditions in the definition of Lebesgue spectrum for U_1 then $j(E)$ satisfies those same conditions for U_2. Conclude that the rank of U_1 is well defined.

8.4.6. The lemma of Riemann–Lebesgue ensures that F takes values in c_0. The operator F is continuous: $\|F(\varphi)\| \le \|\varphi\|$ for every $\varphi \in L^1(\lambda)$. Moreover, F is injective: if $F(\varphi) = 0$ then $\int \varphi(z)\psi(z)\, d\lambda(z) = 0$ for every linear combination $\psi(z) = \sum_{|j|\le l} a_j z^j$, $a_j \in \mathbb{C}$. Given any interval $I \subset S^1$, the sequence $\psi_N = \sum_{|n|\le N} c_n z^n$, $c_n = \int_I z^{-n}\, d\lambda(z)$ of partial sums of the Fourier series of the characteristic function $\mathcal{X}_I$ is bounded (see [Zyg68, page 90]). Using the dominated convergence theorem, it follows that $F(\varphi) = 0$ implies $\int_I \varphi(z)\, d\lambda(z) = 0$, for any interval I. Hence, $\varphi = 0$. If F were bijective then, by the open mapping theorem, its inverse would be a continuous linear operator. Then, there would be $c > 0$ such that $\|F(\varphi)\| \ge c\|\varphi\|$ for every $\varphi \in L^1(\lambda)$. But that is false: consider $D_N(z) = \sum_{|n|\le N} z^n$ for $N \ge 0$. Check that $F(D_N) = (a_n^N)_n$ with $a_n^N = 1$ if $|n| \le N$ and $a_n^N = 0$ otherwise. Hence, $\|F(D_N)\| = 1$ for every N. Writing $z = e^{2\pi i t}$, check that $D_N(z) = \sin((2N+1)\pi t)/\sin(\pi t)$. Conclude that $\|D_N\| = \int |D_N(z)|\, d\lambda(z)$ converges to infinity when $N \to \infty$. [Observation: One can also give explicit examples. For instance, if $(a_n)_n$ converges to zero and satisfies $\sum_{n=1}^{\infty} a_n/n = \infty$ then the sequence $(\alpha_n)_n$ given by $\alpha_n = a_n/(2i)$ for $n \ge 1$ and $\alpha_n + \alpha_{-n} = 0$ for every $n \ge 0$ may not be written in the form $\alpha_n = \int z^n\, d\nu(z)$. See Section 7.3.4 of Edwards [Edw79].]

8.5.3. By Exercise 8.5.2, $\tilde f$ is always injective. Conclude that if $\tilde f$ is surjective then it is invertible: there exists a homomorphism of measure algebras $h : \tilde{\mathcal{B}} \to \tilde{\mathcal{B}}$ such that $h \circ \tilde f = \tilde f \circ h = \mathrm{id}$. Use Proposition 8.5.6 to find $g : M \to M$ such that $g \circ f = f \circ g$ at μ-almost every point. The converse is easy: if (f, μ) is invertible at almost every point then the homomorphism of measure algebras $\tilde g$ associated with $g = f^{-1}$ satisfies $\tilde g \circ \tilde f = \tilde f \circ \tilde g = \mathrm{id}$; in particular, $\tilde f$ is surjective.

8.5.6. Check that the unions of elements of $\bigcup_n \mathcal{P}_n$ are pre-images, under the inclusion $\imath$, of an open subset of K. Use that fact to show that if the chains have measure zero then for each $\delta > 0$ there exists an open set $A \subset K$ such that $m(A) < \delta$ and every point outside A is in the image of the inclusion: in other words, $K \setminus \imath(M_{\mathcal{P}}) \subset A$. Conclude that $\imath(M_{\mathcal{P}})$ is a Lebesgue measurable set and its complement in K has measure zero. For the converse, use the fact that (a) implies (c) in Exercise A.1.13.

9.1.1. $H_\mu(\mathcal{P}/\mathcal{R}) \le H_\mu(\mathcal{P} \vee \mathcal{Q}/\mathcal{R}) = H_\mu(\mathcal{Q}/\mathcal{R}) + H_\mu(\mathcal{P}/\mathcal{Q} \vee \mathcal{R}) \le H_\mu(\mathcal{Q}/\mathcal{R}) + H_\mu(\mathcal{P}/\mathcal{Q})$.

9.1.3. Let $g = f^k$. Then $H_\mu(\bigvee_{i=0}^{k-1} f^{-i}(\mathcal{P}) / \bigvee_{j=k}^{n} f^{-j}(\mathcal{P})) = H_\mu(\mathcal{P}^k / \bigvee_{i=1}^{n-k} g^{-i}(\mathcal{P}^k))$. By Lemma 9.1.12, this expression converges to $h_\mu(g, \mathcal{P}_k)$. Now use Lemma 9.1.13.

9.2.5. Write $\mathcal{Q}^n = \bigvee_{j=0}^{n-1} f^{-j}(\mathcal{Q})$ for each n and let $\mathcal{A}$ be the σ-algebra generated by $\bigcup_n \mathcal{Q}^n$. Check that f is measurable with respect to the σ-algebra $\mathcal{A}$. Show that the hypothesis implies that $\mathcal{P} \subset \mathcal{A}$. By Corollary 9.2.4, it follows that $H_\mu(\mathcal{P}/\mathcal{Q}^n)$ converges to zero. By Lemmas 9.1.11 and 9.1.13, we have that $h_\mu(f,\mathcal{P}) \le h_\mu(f,\mathcal{Q}) + H_\mu(\mathcal{P}/\mathcal{Q}^n)$ for every n.

9.2.7. The set $\mathcal{A}$ of all finite disjoint of rectangles $A_i \times B_i$, with $A_i \subset M$ and $B_i \subset N$, is an algebra that generates the σ-algebra of $M \times N$. Given partitions $\mathcal{P}$ and $\mathcal{Q}$ of M and N, respectively, the family $\mathcal{P} \times \mathcal{Q} = \{P \times Q : P \in \mathcal{P} \text{ and } Q \in \mathcal{Q}\}$ is a partition of $M \times N$ contained in $\mathcal{A}$ and such that $h_{\mu\times\nu}(f \times g, \mathcal{P} \times \mathcal{Q}) = h_\mu(f,\mathcal{P}) + h_\nu(g,\mathcal{Q})$. Conversely, given any partition $\mathcal{R} \subset \mathcal{A}$ of $M \times N$, there exist partitions $\mathcal{P}$ and $\mathcal{Q}$ such that $\mathcal{R} \prec \mathcal{P} \times \mathcal{Q}$ and so $h_{\mu\times\nu}(f \times g, \mathcal{R}) \le h_\mu(f,\mathcal{P}) + h_\nu(g,\mathcal{Q})$. Conclude using Exercise 9.2.6.

9.3.3. It is clear that $B(x,n,\varepsilon) \subset B(f(x),n-1,\varepsilon)$. Hence, $h_\mu(f,x) \ge h_\mu(f,f(x))$ for μ-almost every x. On the other hand, $\int h_\mu(f,x)\,d\mu(x) = \int h_\mu(f,f(x))\,d\mu(x)$ since the measure μ is invariant under f.

9.4.2. Use the following consequence of the Jordan canonical form: there exist numbers $\rho_1,\dots,\rho_l > 0$, there exists an A-invariant decomposition $\mathbb{R}^d = E_1 \oplus \cdots \oplus E_l$ and, given $\alpha > 0$, there exists an inner product in $\mathbb{R}^d$ relative to which the subspaces E_j are orthogonal and satisfy $e^{-\alpha}\rho_j\|v\| \le \|Av\| \le e^{\alpha}\rho_j\|v\|$ for every $v \in E_j$. Moreover, the ρ_i are the absolute values of the eigenvalues of A and they satisfy $\sum_{i=1}^d \log^+|\lambda_i| = \sum_{j=1}^l \dim E_j \log^+ \rho_j$.

9.4.3. Consider any countable partition $\mathcal{P}$ with $\{B,B^c\} \prec \mathcal{P}$. Let $\mathcal{Q}$ be the restriction of $\mathcal{P}$ to the set B. Write $\mathcal{P}^n = \bigvee_{j=0}^{n-1} f^{-j}(\mathcal{P})$ and $\mathcal{Q}^k = \bigvee_{j=0}^{k-1} g^{-j}(\mathcal{Q})$. Check that, for every $x \in B$ and $k \ge 1$, there exists $n_k \ge 1$ such that $\mathcal{Q}^k(x) = \mathcal{P}^{n_k}(x)$. Moreover, by ergodicity, $\lim_k k/n_k = \tau(B,x) = \mu(B)$ for almost every x. By the theorem of Shannon–McMillan–Breiman,

$$h_\nu(g,\mathcal{Q},x) = \lim_k -\frac{1}{k}\log\nu(\mathcal{Q}^k(x)) \quad\text{and}\quad h_\mu(f,\mathcal{P},x) = \lim_k -\frac{1}{n_k}\log\mu(\mathcal{P}^{n_k}(x)).$$

Conclude that $h_\nu(f,\mathcal{Q},x) = \mu(B)h_\nu(g,\mathcal{Q},x)$ for almost every $x \in B$. Varying $\mathcal{P}$, deduce that $h_\nu(f) = \mu(B)h_\nu(g)$.

9.5.2. Consider $A \in \bigcap_n f^{-n}(\mathcal{B})$ with $m(A) > 0$. Then, for each n there exists $A_n \in \mathcal{B}$ such that $A = f^{-n}(A_n)$. Consider the intervals $I_{j,n} = \big((j-1)/10^n, j/10^n\big)$. Then,

$$\frac{m(A \cap I_{j,n})}{m(I_{j,n})} = \frac{m(A_n)}{m((0,1))} = m(A) \quad \text{for every } 1 \le j \le 10^n.$$

Making $n \to \infty$, conclude that A^c has no points of density. Hence, $m(A^c) = 0$.

9.5.6. Assume that $h_\mu(f,\mathcal{P}) = 0$. Use Lemma 9.5.4 to show that $\mathcal{P} \prec \bigvee_{j=1}^\infty f^{-jk}(\mathcal{P})$ for every $k \ge 1$. Deduce that, up to measure zero, $\mathcal{P}$ is contained in $f^{-k}(\mathcal{B})$ for every $k \ge 1$. Conclude that the partition $\mathcal{P}$ is trivial.

9.6.1. Uniqueness is immediate. To prove the existence, consider the functional Ψ defined by $\Psi(\psi) = \int\big(\int \psi\,d\eta\big)\,dW(\eta)$ in the space of bounded measurable functions $\psi : M \to \mathbb{R}$. Note that Ψ is linear and non-negative and satisfies $\Psi(1) = 1$. Use the monotone convergence theorem to show that if B_n, $n \ge 1$ are pairwise disjoint measurable subsets of M then $\Psi(\mathcal{X}_{\bigcup_n B_n}) = \sum_n \Psi(\mathcal{X}_{B_n})$. Conclude that $\xi(B) = \Psi(\mathcal{X}_B)$ defines a probability measure in the σ-algebra of measurable subsets of M. Show that $\int \psi\, d\xi = \Psi(\psi)$ for every bounded measurable function. Take $\mathrm{bar}(W) = \xi$.

9.6.5. For the penultimate identity one would need to know that $n^{-1}\log\mu_P(\mathcal{Q}^n(x))$ is a dominated sequence (for example).

9.7.4. By Exercise 9.7.3, given any bounded measurable function ψ,

$$\int (\psi \circ f)\, d\eta = \int \psi(x)\Big(\sum_{z\in f^{-1}(x)} \frac{1}{J_\eta f}(z)\Big)\, d\eta(x).$$

Deduce the first part of the statement. For the second part, note that if η is invariant then $\eta(f(A)) = \eta(f^{-1}(f(A))) \geq \eta(A)$ for every domain of invertibility A.

9.7.7. The "if" part of the statement is easy: we may exhibit the ergodic equivalence explicitly. Assume that the two systems are ergodically equivalent. The fact that $k = l$ follows from Exercise 8.1.2. To prove that p and q are permutations of one another, use the fact that the Jacobian is invariant under ergodic equivalence (Exercise 9.7.6), together with the expressions of the Jacobians given by Example 9.7.1.

10.1.6. Note that $\psi(M)$ is compact and the inverse $\psi^{-1} : \psi(M) \to M$ is (uniformly) continuous. Hence, given $\varepsilon > 0$ there exists $\delta > 0$ such that if $E \subset M$ is (n,ε)-separated for f then $\psi(E) \subset N$ is (n,δ)-separated for g. Conclude that $s(f,\varepsilon,M) \leq s(g,\delta,N)$ and deduce that $h(f) \leq h(g)$. [Observation: The statement remains valid in the non-compact case, as long as we assume the inverse $\psi^{-1} : \psi(M) \to M$ to be uniformly continuous.]

For the second part, consider the distance defined in Σ by

$$d\big((x_n)_n, (y_n)_n\big) = \sum_{n\in\mathbb{Z}} 2^{-|n|}|x_n - y_n|.$$

Consider a discrete set $A \subset [0,1]$ with n elements. Check that the restriction to $A^{\mathbb{Z}}$ of the distance of $[0,1]^{\mathbb{Z}}$ is uniformly equivalent to the distance defined in (9.2.15). Using Example 10.1.2, conclude that the topological entropy of σ is at least $\log n$, for any n.

10.1.10. (Carlos Gustavo Moreira) Let $\theta_1 = 0$, $\theta_2 = 01$ and, for $n \geq 2$, $\theta_{n+1} = \theta_n\theta_{n-1}$. We claim that, for every $n \geq 1$, there exists a word τ_n such that $\theta_n\theta_{n+1} = \tau_n\alpha_n$ and $\theta_{n+1}\theta_n = \tau_n\beta_n$, where $\alpha_n = 10$ and $\beta_n = 01$ if n is even and $\alpha_n = 01$ and $\beta_n = 10$ if n is odd. That holds for $n = 1$ with $\tau_1 = 0$ and for $n = 2$ with $\tau_2 = 010$. If it holds for a given n, then $\theta_{n+1}\theta_{n+2} = \theta_{n+1}\theta_{n+1}\theta_n = \theta_{n+1}\tau_n\beta_n = \tau_{n+1}\alpha_{n+1}$ and also $\theta_{n+2}\theta_{n+1} = \theta_{n+1}\theta_n\theta_{n+1} = \theta_{n+1}\tau_n\alpha_n = \tau_{n+1}\beta_{n+1}$, as long as we take $\tau_{n+1} = \theta_{n+1}\tau_n$. This proves the claim. It follows that the last letters of θ_n and θ_{n+1} are distinct.

Now, we claim that $\theta = \lim_n \theta_n$ is not pre-periodic. Indeed, suppose that θ were pre-periodic and let m be its period. Since the length of θ_n is F_{n+1} (where F_k is the k-th Fibonacci number), we may take n large such that m divides F_{n+1} and the pre-period (that is, the length of the non-periodic part) of θ is less than F_{n+2}. Then, θ starts with $\theta_{n+3} = \theta_{n+2}\theta_{n+1} = \theta_{n+1}\theta_n\theta_{n+1}$. However, since the length F_{n+1} of θ_n is a multiple of the period m, the F_{m+2}-th letter of θ, which is the last letter of θ_{n+1}, must coincide with the $(F_{m+2} + F_{n+1})$-th letter of θ, which is the last letter of θ_n. This would contradict the conclusion of the previous paragraph.

Next, we claim that $c_{k+1}(\theta) > c_k(\theta)$ for every k. Indeed, suppose that $c_{k+1}(\theta) = c_k(\theta)$ for some k. Then, every subword of length k can have only one continuation of length $k+1$. Hence, we have a transformation in the set of subwords of length k, assigning to each subword its unique continuation, without the first letter. Since the domain is finite, all the orbits of this transformation are pre-periodic. In particular, θ is also pre-periodic, which contradicts the conclusion in the previous paragraph.

Since $c_1(\theta) = 2$, it follows that $c_k(\theta) \geq k+1$ for every k. We claim that $c_{F_{n+1}}(\theta) \leq F_{n+1} + 1$ for every $n > 1$. To prove that fact, note that θ may be written as a concatenation of words belonging to $\{\theta_n, \theta_{n+1}\}$ because (by induction) every θ_r with $r \geq n$ may be written as a concatenation of words belonging to $\{\theta_n, \theta_{n+1}\}$. Thus, any subword of θ of length F_{n+1} (which is the length of θ_n) is a subword of $\theta_n\theta_{n+1}$ or $\theta_{n+1}\theta_n$. Since $\theta_n\theta_{n+1} = \theta_n\theta_n\theta_{n-1}$, is a subword of $\theta_n\theta_n\theta_{n-1}\theta_{n-2} = \theta_n\theta_n\theta_n$, there are at most $|\theta_n| = F_{n+1}$ subwords of length $|\theta_n| = F_{n+1}$ of $\theta_n\theta_n\theta_n$ and, hence, of $\theta_n\theta_{n+1}$. Since $\theta_n\theta_{n+1} = \tau_n\alpha_n$ and $\theta_{n+1}\theta_n = \tau_n\beta_n$, and $\theta_{n+1}\theta_n$ ends with θ_n and $|\beta_n| = 2$, the unique subword of $\theta_{n+1}\theta_n$ of length $|\theta_n| = F_{n+1}$ that may not be a subword of $\theta_n\theta_{n+1}$ is the subword that ends with the first letter of β_n (that is, one position before the end of $\theta_{n+1}\theta_n$). Hence, $c_{F_{n+1}}(\theta) \leq F_{n+1} + 1$ as stated.

We are ready to obtain the statement of the exercise. Assume that $c_k(\theta) > k+1$ for some k. Taking n such that $F_{n+1} > k$, we would have $c_{F_{n+1}}(\theta) - c_k(\theta) < F_{n+1} + 1 - (k+1) = F_{n+1} - k$ and that would imply that $c_{m+1}(\theta) \leq c_m(\theta)$ for some m with $k \leq m < F_{n+1}$. This would contradict the conclusion in the previous paragraph.

10.2.4. By Proposition 10.2.1, $h(f) = g(f,\delta,M)$ whenever f is ε-expansive and $\delta < \varepsilon/2$. Show that if $d(f,h) < \delta/3$ then $g(h,\delta/3,M) \leq g(f,\delta,M)$. Deduce that if $(f_k)_k$ converges to f then $\limsup_k h(f_k) = \limsup_k g(f_k,\delta/3,M) \leq g(f,\delta,M) = h(f)$.

10.2.8. (Bowen [Bow72]) Write $a = g_*(f,\varepsilon)$. Observe that if E is an (n,δ)-generating set of M, with $\delta < \varepsilon$, then $M = \bigcup_{x\in E} B(x,n,\varepsilon)$. Combining this fact with the result of Bowen, show that $g_n(f,\delta,M) \leq \#E e^{c+(a+b)n}$. Take $b \to 0$ to conclude the inequality.

10.3.3. (a) The hypothesis implies that for every n and every subcover δ of β^n there exists a subcover γ of α^n such that $\gamma \prec \delta$. Taking γ minimal, $\#\gamma \leq \#\delta$ and $\sum_{U\in\gamma} \inf_{x\in U} e^{\phi_n(x)} \leq \sum_{V\in\delta} \inf_{y\in V} e^{\phi_n(y)}$. It follows that $Q_n(f,\phi,\alpha) \leq Q_n(f,\phi,\beta)$ for every n. (b) Lemma 10.1.11 gives that α^{n+k-1} is a subcover of $(\alpha^k)^n$. A variation of the argument in part (a) gives that $Q_n(f,\phi,\alpha^k) \leq e^{(k-1)\sup|\phi|} Q_{n+k-1}(f,\phi,\alpha)$ for every n. Hence, $Q^{\pm}(f,\phi,\alpha^k) \leq Q^{\pm}(f,\phi,\alpha)$. [Observation: Analogously, $P(f,\phi,\alpha^k) \leq P(f,\phi,\alpha)$.] By the second part of Lemma 10.1.11, for every subcover β of $(\alpha^k)^n$ there exists a subcover γ of α^{n+k-1} such that $\gamma \prec \beta$, $\#\gamma \leq \#\beta$ and $\sum_{U\in\gamma} \inf_{x\in U} e^{\phi_{n+k-1}(x)} \leq e^{(k-1)\sup|\phi|} \sum_{V\in\beta} \inf_{y\in V} e^{\phi_n(y)}$ (taking γ minimal). Deduce that $Q_{n+k-1}(f,\phi,\alpha) \leq e^{(k-1)\sup|\phi|} Q_n(f,\phi,\alpha^k)$. Hence, $Q^{\pm}(f,\phi,\alpha) \leq Q^{\pm}(f,\phi,\alpha^k)$. (c) Follows from part (b) and Corollary 10.3.3. (d) If the elements of α are disjoint then $(\alpha^k)^n = \alpha^{n+k-1}$ and so

$$P_n(f,\phi,\alpha^k) = \inf\{\sum_{U\in\gamma} \sup_{x\in U} e^{\phi_n(x)} : \gamma \subset (\alpha^k)^n\} = \inf\{\sum_{U\in\gamma} \sup_{x\in U} e^{\phi_n(x)} : \gamma \subset \alpha^{n+k-1}\}.$$

It follows that $e^{-(k-1)\sup|\phi|}P_n(f,\phi,\alpha^k) \le P_{n+k-1}(f,\phi,\alpha) \le e^{(k-1)\sup|\phi|}P_n(f,\phi,\alpha^k)$. (e) Follows from part (d) and the definition of pressure (Lemma 10.3.1). (f) Note that $\alpha^{\pm k} = f^k(\alpha^{2k})$ and use Exercise 10.3.2.

10.3.8. Show that given $\varepsilon > 0$ there exists $\kappa \ge 1$ such that every dynamical ball $B(x,n,\varepsilon)$ has diameter equal to $\varepsilon 2^{-n}$ and contains some periodic point p_x^n of period $n+\kappa$. Show that given $C,\theta > 0$ there exists $K > 0$ such that $|\phi_n(y) - \phi_n(p_x^n)| \le K$ for every $y \in B(x,n,\varepsilon)$, every $n \ge 1$ and every (C,θ)-Hölder function $\phi : S^1 \to \mathbb{R}$. Use this fact to replace generating (or separated) sets by sets of periodic points in the definition of pressure.

10.3.10. Since $\xi \mid \Lambda^c = \eta \mid \Lambda^c$,

$$\left|\mathcal{E}(\xi,\eta) - E_\Lambda(\xi \mid \Lambda) + E_\Lambda(\eta \mid \Lambda)\right| = \Bigg| \sum_{(k,l)\in\Lambda\times\Lambda^c\bigcup\Lambda^c\times\Lambda} \Psi(k-l,\xi_k,\xi_l) - \Psi(k-l,\eta_k,\eta_l)\Bigg|$$
$$\le \sum_{(k,l)\,\sum_{(k,l)\in\Lambda\times\Lambda^c\bigcup\Lambda^c\times\Lambda}} 2Ke^{-\theta|k-l|}.$$

Recalling that Λ is an interval, the cardinal of $\{(k,l) \in \Lambda \times \Lambda^c \cup \Lambda^c \times \Lambda : |k-l| = n\}$ is less than or equal to $4n$ for every $n \ge 1$. Hence,

$$\left|\mathcal{E}(\xi,\eta) - E_\Lambda(\xi \mid \Lambda) + E_\Lambda(\eta \mid \Lambda)\right| \le \sum_{n=1}^{\infty} 8Kne^{-\theta n} < \infty.$$

The second part of the statement is an immediate consequence of the first one.

10.4.4. Consider the shift map σ in the space $\Sigma = \{0,1\}^{\mathbb{N}}$. Consider the function $\phi : \Sigma \to \mathbb{R}$ defined by $\phi(x) = 0$ if $x_0 = 0$ and $\phi(x) = 1$ if $x_0 = 1$. Let N be the set of points $x \in \Sigma$ such that the time average in the orbit of x does not converge. Check that N is invariant under σ and is non-empty: for each finite sequence $(z_0,\dots,z_k)$ one can find $x \in N$ with $x_i = z_i$ for $i = 0,\dots,k$. Deduce that the topological entropy of the restriction $f \mid N_\phi$ is equal to $\log 2$. Justify that N does not support any probability measure invariant under f.

10.4.5. Consider the open cover ξ of K whose elements are $K \cap [0,a]$ and $K \cap [1-\beta,1]$. Check that $P(f,\phi) = P(f,\phi,\xi)$ for every potential ϕ. Moreover,

$$P_n(f,-t\log g',\xi) = \sum_{U\in\alpha^n} [(g^n)']^{-t}(U) = (\alpha^t + \beta^t)^n.$$

Conclude that $\psi(t) = \log(\alpha^t + \beta^t)$. Check that $\psi' < 0$ and $\psi'' > 0$ (convexity also follows from Proposition 10.3.7). Moreover, $\psi(0) > 0 > \psi(1)$. By the variational principle, the last inequality implies that $h_\mu(f) - \int \log g'\,d\mu < 0$.

10.5.3. The Gibbs property gives that $\lim_n (1/n)\log\mu(C^n(x)) = \tilde{\varphi}(x) - P$, where $C^n(x)$ is the cylinder of length n that contains x. Combine this identity with the theorem of Brin–Katok (Theorem 9.3.3) and the theorem of Birkhoff to get the first claim. Now assume that μ_1 and μ_2 are two ergodic Gibbs states with the same constant P. Observe that there exists C such that $C^{-1}\mu_1(A) \le \mu_2(A) \le C\mu_1(A)$ for every A in the algebra formed by the finite disjoint unions of cylinders. Using the monotone class theorem (Theorem A.1.18), deduce that $C^{-1}\mu_1(A) \le \mu_2(A) \le C\mu_1(A)$ for any measurable set A. This implies that μ_1 and μ_2 are equivalent measures. Using Lemma 4.3.1, it follows that $\mu_1 = \mu_2$.

10.5.5. By Proposition 10.3.7, the pressure function is convex. By Exercise A.5.1, it follows that it is also continuous. By the smoothness theorem of Mazur (recall Exercise 2.3.7), there exists a residual subset $\mathcal{R} \subset C^0(M)$ such that the pressure function is differentiable at every $\varphi \in \mathcal{R}$. Apply Exercise 10.5.4.

11.1.3. Adapt the arguments in Section 9.4.2, as follows. Start by checking that the iterates of f have bounded distortion: there exists $K > 1$ such that

$$\frac{1}{K} \le \frac{|Df^n(x)|}{|Df^n(y)|} \le K,$$

for every $n \ge 1$ and any points x, y with $\mathcal{P}^n(x) = \mathcal{P}^n(y)$. Consider the sequence $\mu_n = (1/n)\sum_{j=0}^{n-1} f_*^j m$ of averages of the iterates of the Lebesgue measure m. Show that the Radon–Nikodym derivatives $d\mu_n/dm$ are uniformly bounded and are Hölder, with uniform Hölder constants. Deduce that every accumulation point μ of that sequence is an invariant probability measure absolutely continuous with respect to the Lebesgue measure. Show that the Radon–Nikodym derivative $\rho = d\mu/dm$ is bounded from zero and infinity (in other words, $\log\rho$ is bounded). Show that ρ and $\log\rho$ are Hölder.

11.1.5. Check that $J_\mu f = (\rho \circ f)|f'|/\rho$ and use the Rokhlin formula (Theorem 9.7.3).

11.2.4. Take $\Lambda = \{2^{-n} : n \ge 0\} \mod \mathbb{Z}$. The restriction $f : \Lambda \to \Lambda$ cannot be an expanding map because $1/2$ is an isolated point in Λ but $1 = f(1/2)$ is not. [Observation: Note that $\Lambda = S^1 \setminus \bigcup_{n=0}^{\infty} f^{-n}(I)$, where $I = (1/2, 1) \mod \mathbb{Z}$. Modifying suitably the choice of I, one finds many other examples, possibly with Λ uncountable.]

11.3.3. Let $a = \int \varphi\, d\mu_1$ and $b = \int \varphi\, d\mu_2$. Assume that $a < b$ and write $r = (b-a)/5$. By the ergodic decomposition theorem, we may assume that μ_1 and μ_2 are ergodic. Then, there exist x_1 and x_2 such that $\tilde\varphi(x_1) = a$ and $\tilde\varphi(x_2) = b$. Using the hypothesis that f is topologically exact, construct a pseudo-orbit $(z_n)_{n\ge0}$ alternating (long) segments of the orbits of x_1 and x_2 in such a way that the sequence of time averages of φ along the pseudo-orbit $(z_n)_n$ oscillates from $a+r$ to $b-r$ (meaning that $\liminf \le a+r$ and $\limsup \ge b-r$). Next, use the shadowing lemma to find $x \in M$ whose orbit shadows this pseudo-orbit. Using that φ is uniformly continuous, conclude that the sequence of time averages of φ along the orbit of x oscillates from $a+2r$ to $b-2r$.

12.1.2. Theorem 2.1.5 gives that $\mathcal{M}_1(M)$ is weak* compact and it is clear that it is convex. Check that the operator $\mathcal{L} : C^0(M) \to C^0(M)$ is continuous and deduce that its dual $\mathcal{L}^* : \mathcal{M}(M) \to \mathcal{M}(M)$ is also continuous. If $(\eta_n)_n \to \eta$ in the weak* topology then $(\int \mathcal{L}1\, d\eta_n)_n \to \int \mathcal{L}1\, d\eta$. Conclude that the operator $G : \mathcal{M}_1(M) \to \mathcal{M}_1(M)$ is continuous. Hence, by the Tychonoff–Schauder theorem, G has some fixed point ν. This means that $\mathcal{L}^*\nu = \lambda\nu$, where $\lambda = \int \mathcal{L}1\, d\nu$. Since $\lambda > 0$, this proves that ν is a reference measure. Using Corollary 12.1.9, check that $\lambda = \limsup_n \sqrt[n]{\|\mathcal{L}^n 1\|}$ and deduce that λ is the spectral radius of $\mathcal{L}$.

12.2.4. Fix in S^1 the orientation induced by $\mathbb{R}$. Consider the fixed point $p_0 = 0$ of f and let $p_1, \dots, p_d$ be its pre-images, ordered cyclically, with $p_d = p_0$. Analogously, let q_0 be a fixed point of g and $q_1, \dots, q_d$ be its pre-images, ordered cyclically, with $q_d = q_0$. Note that f maps each $[p_{i-1}, p_i]$ and g maps each $[q_{i-1}, q_i]$ onto S^1. Then, for each sequence $(i_n)_n \in \{1, \dots, d\}^{\mathbb{N}}$ there exists exactly one point $x \in S^1$ and one point $y \in S^1$ such that $f^n(x) \in [p_{i_n-1}, p_{i_n}]$ and $g^n(y) \in [q_{i_n-1}, q_{i_n}]$ for every n. Clearly, the maps $(i_n)_n \mapsto x$ and $(i_n)_n \mapsto y$ are surjective. Consider two sequences $(i_n)_n$ and $(j_n)_n$ to be equivalent if there exists $N \in \mathbb{N} \cup \{\infty\}$ such that (1) $i_n = j_n$ for every $n \le N$ and either (2a) $i_n = 1$ and $j_n = d$ for every $n > N$ or (2b) $i_n = d$ and $j_n = 1$ for every $n > N$. Show that the points x corresponding to $(i_n)_n$ and $(j_n)_n$ coincide if and only if the two sequences are equivalent and a similar fact holds for the points y corresponding to the two sequences. Conclude that the map $\phi : x \mapsto y$ is well defined and is a bijection in S^1 such that $\phi(f(x)) = g(\phi(x))$ for every x. Observe that ϕ preserves the orientation of S^1 and, thus, is a homeomorphism.

12.2.5. (a) $\Rightarrow$ (b): Trivial. (b) $\Rightarrow$ (c): Let μ_a be the absolutely continuous invariant probability measure and μ_m be the measure of maximum entropy of f; let ν_a and ν_m be the corresponding measures for g. Show that $\mu_a = \mu_m$. Let $\phi : S^1 \to S^1$ be a topological conjugacy. Show that $\nu_m = \phi_* \mu_m$ and $\nu_a = \phi_* \mu_a$ if ϕ is absolutely continuous. Use Corollary 12.2.4 to conclude that in the latter case $|(g^n)'(x)| = k^n$ for every $x \in \mathrm{Fix}(f^n)$. (c) $\Rightarrow$ (a): The hypothesis implies that $\nu_a = \nu_m$ and so $\nu_a = \phi_* \mu_a$. Recall (Proposition 12.1.20) that the densities $d\mu_a/dm$ and $d\nu_a/dm$ are continuous and bounded from zero and infinity. Conclude that ϕ is differentiable, with $\phi' = (d\mu/dm)/(d\nu/dm) \circ \phi$.

12.3.2. Consider $A = (a,1)$, $P = (p,1)$, $Q = (q,1)$, $B = (b,1)$, $O = (0,0) \in \mathbb{C} \times \mathbb{R}$. Let A' (respectively, B') be the point where the line parallel to OQ (respectively, OP) passing through P (respectively, Q) intersects the boundary of C. Note that all these points belong to the plane determined by P, Q and O; note also that $A' \in OA$ and $B' \in OB$. By definition, $\alpha(P,Q) = |B'Q|/|OP|$ and $\beta(P,Q) = |OQ|/|A'P|$. Check that $|AP|/|AQ| = |A'P|/|OQ|$ and $|BQ|/|BP| = |B'Q|/|OP|$. Hence,

$$\theta(P,Q) = \log \frac{\beta(P,Q)}{\alpha(P,Q)} = \log \frac{|OQ|\,|OP|}{|A'P|\,|B'Q|} = \log \frac{|AQ|\,|BP|}{|AP|\,|BQ|}.$$

In other words, $d(p,q) = \log(|aq|\,|bp|)/(|ap|\,|bq|) = \Delta(p,q)$, for any $p, q \in \mathbb{D}$.

12.3.4. Consider the cone C_0 of positive continuous functions in M. The corresponding projective distance θ_0 is given in Example 12.3.5. Check that θ_1 is the restriction of θ_0 to the cone C_1. Consider a sequence of positive differentiable functions converging uniformly to a (continuous but) non-differentiable function g_0 . Show that $(g_n)_n$ converges to g_0 with respect to the distance θ_0 and, thus, is a Cauchy sequence for θ_0 and θ_1. Argue that $(g_n)_n$ cannot be convergent for θ_1.

12.3.8. (a) It is clear that $\log g$ is (b,β)-Hölder and $\sup g/\inf g$ is close to 1 if the norm $\|v\|_{\beta,\rho}$ is small; this will be implicit in all that follows. Then, $g \in C(b,\beta,R)$. To estimate $\theta(1,g)$, use the expression given by Lemma 12.3.8. Observe that

$$\beta(1,g) = \sup \left\{ g(x), \frac{\exp(b\delta)g(x) - g(y)}{\exp(b\delta) - 1} : x \neq y, d(x,y) < \rho \right\} \quad \text{where } \delta = d(x,y)^\beta.$$

Clearly, $g(x) \le 1 + \sup|v|$. Moreover,

$$\frac{\exp(b\delta)g(x) - g(y)}{\exp(b\delta) - 1} \le \frac{\exp(b\delta)g(y) + \exp(b\delta)H_{\beta,\rho}(v)\delta - g(y)}{\exp(b\delta) - 1}$$
$$= g(y) + \frac{\delta\exp(b\delta)}{\exp(b\delta) - 1}H_{\beta,\rho}(v).$$

Take $K_1 > K_2 > 0$, depending only on b, β, ρ, such that $K_1 \ge \exp(bs)s/(\exp(bs) - 1) \ge K_2$ for every $s \in [0, \rho^\beta]$. Then, the term on the right-hand side of the previous inequality is bounded by $1 + \sup|v| + K_1 H_{\beta,\rho}(v)$. Hence, $\log\beta(1,g) \le \log(1 + \sup|v| + K_1 H_{\beta,\rho}(v)) \le K_1' \|v\|_{\beta,\rho}$, where $K_1' = \max\{K_1, 1\}$. Varying x and y in the previous arguments, we also find that $\beta(1,g) \ge 1 + \sup|v|$ and $\beta(1,g) \ge 1 - \sup|v| + K_2 H_{\beta,\rho}(v)$. Deduce that

$$\log\beta(1,g) \ge \max\left\{\log(1 + \sup|v|), \log(1 - \sup|v| + K_2 H_{\beta,\rho}(v))\right\} \ge K_2' \|v\|_{\beta,\rho},$$

where the constant K_2' depends only on K_2, β and ρ. Analogously, there exist constants $K_3' > K_4' > 0$ such that $-K_3' \|v\|_{\beta,\rho} \le \log\alpha(1,g) \le -K_4' \|v\|_{\beta,\rho}$. Fixing

$$K \ge \max\{(K_1 + K_3), 1/(K_2 + K_4)\},$$

it follows that $K^{-1}\|v\|_{\beta,\rho} \le \theta(1,g) \le K\|v\|_{\beta,\rho}$. (b) It is no restriction to assume that $\|v\|_{\beta,\rho} < r$. Note that $\mathcal{P}^n g = 1 + \mathcal{P}^n v$ for every n. Corollary 12.3.12 gives that

$$\theta(\mathcal{P}^{kN} g, 1) \le \Lambda_0^k \theta(1,g) \quad \text{for every } k,$$

with $\Lambda_0 < 1$. By part (a), it follows that $\|\mathcal{P}^{kN} v\| \le K^2 \Lambda_0^k$ for every k. This yields the statement, with $\tau = \Lambda_0^{1/k}$ and $C = K^2 \|\mathcal{P}\|^N \Lambda_0^{-1}$.

12.4.4. Consider $0 < \delta \le \rho$. For every cover $\mathcal{U}$ of A with diameter less than δ, we have

$$\sum_{U\in\mathcal{U}} (\operatorname{diam} U)^d \ge \sum_{U\in\mathcal{U}} K^{-1}\mu(U) \ge K^{-1}\mu(A).$$

Taking the infimum over $\mathcal{U}$, we get that $m_d(A,\delta) \ge K^{-1}\mu(A)$. Making $\delta \to 0$, we find that $m_d(A) > K^{-1}\mu(A)$; hence, $d(A) \ge d$.

12.4.7. Consider $\ell = 1$. Then, $D, D_1, \ldots, D_N$ (Section 12.4.3) are compact intervals. It is no restriction to assume that $D = [0,1]$. Write $D_{i^n} = h_{i_0} \circ \cdots \circ h_{i_{n-1}}(D)$ for each $i^n = (i_0, \ldots, i_{n-1})$ in $\{1, \ldots, N\}^n$. Starting from the bounded distortion property (Proposition 12.4.5), prove that there exists $c > 0$ such that, for every i^n and every n,

(i) $c \le |(f^n)'(x)| \operatorname{diam} D_{i^n} \le c^{-1}$ for every $x \in D_{i^n}$;
(ii) $d(D_{i_n}, D_{j^n}) \ge c \operatorname{diam} D_{i^n}$ for every $j^n \ne i^n$;
(iii) $\operatorname{diam} D_{i^{n+1}} \ge c \operatorname{diam} D_{i^n}$ for every i_n, where $i^{n+1} = (i_0, \ldots, i_{n-1}, i_n)$.

Let ν be the reference measure of the potential $\varphi = -d_0 \log|f'|$. Since $P(f,\varphi) = 0$, it follows from Lemma 12.1.3 and Corollary 12.1.15 that $J_\nu f = |f'|^{d_0}$. Deduce that $c \le |(f^n)'(x)|^{d_0} \nu(D_{i^n}) \le c^{-1}$ for any $x \in D_{i^n}$ and, using (i) once more, conclude that

$$c^2 \le \frac{\operatorname{diam}(D_{i^n})^{d_0}}{\nu(D_{i^n})} \le c^{-2} \quad \text{for every } i^n \text{ and every } n.$$

It follows that $\sum_{i^n} \operatorname{diam}(D_{i^n})^{d_0} \le c^{-2} \sum_{i^n} \nu(D_{i^n}) = c^{-2}$. Since the diameter of D_{i^n} converges uniformly to zero when $n \to \infty$, this implies that $m_{d_0}(\Lambda) \le c^{-2}$.

For the lower estimate, let us prove that ν satisfies the hypothesis of the mass distribution principle (Exercise 12.4.4). Given any U with $\operatorname{diam} U < c\min\{\operatorname{diam} D_1,\dots,\operatorname{diam} D_N\}$, there exist $n \geq 1$ and i^n such that D_{i^n} intersects U and $c\operatorname{diam} D_{i^n} > \operatorname{diam} U$. By (ii), we have that $\nu(U) \leq \nu(D_{i^n}) \leq c^{-2}\operatorname{diam} D_{i^n}^{d_0}$. Take n maximum. Then, using (iii), $\operatorname{diam} U \geq c\operatorname{diam} D_{i^{n+1}} \geq c^2 \operatorname{diam} D_{i^n}$ for some choice of i_n. Combining the two inequalities, we get $\nu(U) \leq c^{-2-2d_0}(\operatorname{diam} U)^{d_0}$. Then, by the mass distribution principle, $m_{d_0}(\Lambda) \geq c^{2+2d_0}$. Finally, extend these arguments to any dimension $\ell \geq 1$.

A.1.9. Given $A_1 \supset \cdots \supset A_i \supset \cdots$, take $A = \bigcap_{i=1}^{\infty} A_i$. For $j \geq 1$, consider $A_j' = A_j \setminus A$. By Theorem A.1.14, we have that $\mu(A_j') \to 0$ and so $\mu(A_j) \to \mu(A)$. Given $A_1 \subset \cdots \subset A_i \subset \cdots$, take $A = \bigcup_{i=1}^{\infty} A_i$. For each j, consider $A_j' = A \setminus A_j$. By Theorem A.1.14, we have that $\mu(A_j') \to 0$, that is, $\mu(A_j) \to \mu(A)$.

A.1.13. (Royden [Roy63]) (b) $\Rightarrow$ (a) Assume that there exist Borel sets B_1, B_2 such that $B_1 \subset E \subset B_2$ and $m(B_2 \setminus B_1) = 0$. Deduce that $m^*(E \setminus B_1) = 0$, hence $E \setminus B_1$ is a Lebesgue measurable set. Conclude that E is a Lebesgue measurable set. (a) $\Rightarrow$ (c) Let E be a Lebesgue measurable set such that $m^*(E) < \infty$. Given $\varepsilon > 0$, there exists a cover by open rectangles $(R_k)_k$ such that $\sum_k m^*(R_k) < m^*(E) + \varepsilon$. Then, $A = \bigcup_k R_k$ is an open set containing E and such that $m^*(A) - m^*(E) < \varepsilon$. Using that E is a Lebesgue measurable set, deduce that $m^*(A \setminus E) < \varepsilon$. For the general case, write E as a disjoint union of Lebesgue measurable sets with finite exterior measure. (c) $\Leftrightarrow$ (d) It is clear that E is a Lebesgue measurable set if and only if its complement is. (c) and (d) $\Rightarrow$ (b) For each $k \geq 1$, consider a closed set $F_k \subset E$ and an open set $A_k \supset E$ such that $m^*(E \setminus F_k)$ and $m^*(A_k \setminus E)$ are less than $1/k$. Then, $B_1 = \cup F_k$ and $B_2 = \bigcap_k A_k$ are Borel sets such that $B_1 \subset E \subset B_2$ and $m^*(E \setminus B_1) = m^*(B_2 \setminus E) = 0$. Conclude that $m(B_2 \setminus B_1) = m^*(B_2 \setminus B_1) = 0$.

A.1.18. Show that $x \mapsto \frac{1}{n}\#\{0 \leq j \leq n-1 : a_j = 5\}$ is a simple function for each $n \geq 1$. By Proposition A.1.31, it follows that ω_5 is measurable.

A.2.8. (a) Assume that $\mathcal{F}$ is uniformly integrable. Consider $C > 0$ corresponding to $\alpha = 1$ and take $L = C + 1$. Check that $\int |f|\,d\mu < L$ for every $f \in \mathcal{F}$. Given $\varepsilon > 0$, consider $C > 0$ corresponding to $\alpha = \varepsilon/2$ and take $\delta = \varepsilon/(2C)$. Check that $\int_A |f|\,d\mu < \varepsilon$ for every $f \in \mathcal{F}$ and every set with $\mu(A) < \delta$. Conversely, given $\alpha > 0$, take $\delta > 0$ corresponding to $\varepsilon = \alpha$ and let $C = L/\delta$. Show that $\int_{|f|>C} |f|\,d\mu < \alpha$. (b) Applying Exercise A.2.5 to the function $|g|$, show that $\mathcal{F}$ satisfies the criterion in (a). (c) Let us prove three facts about $f = \lim_n f_n$. (i) *f is finite at almost every point*: Consider L as in (a). Note that $\mu(\{x : |f_n(x)| \geq k\}) \leq L/k$ for every $n, k \geq 1$ (Exercise A.2.4) and deduce that $\mu(\{x : |f(x)| \geq k\}) \leq L/k$ for every $k \geq 1$. (ii) *f is integrable*: Fix $K > 0$. Given any $\varepsilon > 0$, take δ as in (a). Take n sufficiently large that $\mu(\{x : |f_n(x) - f(x)| > \varepsilon\}) < \delta$. Note that

$$\int_{|f|\leq K} |f|\,d\mu \leq \int_{|f_n - f|\leq\varepsilon} |f|\,d\mu + \int_{|f|\leq K, |f_n - f|>\varepsilon} |f|\,d\mu \leq (L+\varepsilon) + K\delta.$$

Deduce that $\int_{|f|\leq K} |f|\,d\mu \leq L$ for every K and $\int |f|\,d\mu \leq L$. (iii) *$(f_n)_n$ converges to f in $L^1(\mu)$*: Show that given $\varepsilon > 0$ there exists $K > 0$ such that $\int_{|f|>K} |f|\,d\mu < \varepsilon$ and $\int_{|f|>K} |f_n|\,d\mu < \varepsilon$ for every n. Take δ as in part (a) and n large enough that

$\mu(\{x : |f_n(x) - f(x)| > \varepsilon\}) < \delta$. Then

$$\int_{|f|\le K} |f_n - f| \, d\mu \le \int_{|f_n - f|\le\varepsilon} |f_n - f| \, d\mu + \int_{|f_n - f|>\varepsilon} |f_n| \, d\mu + \int_{|f_n|\le K, |f_n - f|>\varepsilon} |f| \, d\mu.$$

The right-hand side is bounded above by $2\varepsilon + K\delta$. Combining these inequalities, $\int |f_n - f| \, d\mu < 4\varepsilon + K\delta$ for every n sufficiently large.

A.2.14. It is no restriction to assume that the B_n are pairwise disjoint. For each n, consider the measure η_n defined in B_n by $\eta_n(A) = \eta(f(A))$. Then, $\eta_n \ll (\eta \mid B_n)$ and, by the theorem of Radon–Nikodym, there exists $\rho_n : B_n \to [0, +\infty]$ such that $\int_{B_n} \phi \, d\eta_n = \int_{B_n} \phi \rho_n \, d\eta$ for every bounded measurable function $\phi : B_n \to \mathbb{R}$. Define $J_\eta \mid B_n = \rho_n$. The essential uniqueness of J_η is a consequence of the essential uniqueness of the Radon–Nikodym derivative.

A.3.5. Given any Borel set $B \subset M$, use Proposition A.3.2 and Lemma A.3.4 to construct Lipschitz functions $\psi_n : M \to [0,1]$ such that $\mu(\{x \in M : \psi_n(x) \neq \mathcal{X}_B(x)\}) \le 2^{-n}$ for every n. Conclude that the claim in the exercise is true for every simple function. Extend the conclusion to every bounded measurable function, using the fact that it is a uniform limit of simple functions. Finally, for any integrable function, use the fact that the positive part and the negative part are monotone pointwise limits of bounded measurable functions. Now consider $M = [0,1]$ and assume that there exists a sequence of continuous functions $\psi_n : M \to \mathbb{R}$ converging to the characteristic function ψ of $M \cap \mathbb{Q}$ at every point. Consider the set $R = \bigcap_m \bigcup_{n>m} \{x \in M : \psi_n(x) > 1/2\}$. On the one hand, $R = \mathbb{Q} \cap M$; on the other hand, R is a residual subset of M; this is a contradiction.

A.4.6. By the inverse function theorem, for every $x \in M$ there exist neighborhoods $U(x) \subset M$ of x and $V(x) \subset N$ of $f(x)$ such that f maps $U(x)$ diffeomorphically onto $V(x)$. This implies that the function $y \mapsto \#f^{-1}(y)$ is lower semi-continuous. Moreover, this function is bounded. Indeed, if there were $y_n \in N$ with $\#f^{-1}(y_n) \ge n$ for every $n \ge 1$ then, since M is compact, we could find $x_n, x_n' \in f^{-1}(y_n)$ distinct with $d(x_n, x_n') \to 0$. Let x be any accumulation point of either sequence. Then f would not be injective in the neighborhood of x, contradicting the hypothesis. Let k be the maximum value of $\#f^{-1}(y)$. The set B_k of points $y \in N$ such that $\#f^{-1}(y) = k$ is open, closed and non-empty. Since N is connected, it follows that $B_k = M$.

A.4.9. Consider local charts $\varphi_\alpha : U_\alpha \to X_\alpha, x \mapsto \varphi_\alpha(x)$ of M and $\varphi_\alpha : T_{U_\alpha} M \to X_\alpha \times \mathbb{R}^d$, $(x, v) \mapsto (\varphi_\alpha, D\varphi_\alpha(x)v)$ of TM. Note that $\varphi_\alpha \circ \pi \circ D\varphi_\alpha^{-1}$ is the canonical projection $X_\alpha \times \mathbb{R}^d \to X_\alpha$, which is infinitely differentiable. Since M is of class C^r and TM is of class C^{r-1}, it follows that π is of class C^{r-1}.

A.5.2. (a) Use the fact that the exponential function is convex. (b) Starting from the Young inequality, show that $\int |f\bar{g}| \, d\mu \le 1$ whenever $\|f\|_p = \|g\|_q = 1$. Deduce the general case of the Hölder inequality. (c) Start by noting that $|f + g|^p \le |f||f+g|^{p-1} + |g||f+g|^{p-1}$. Apply the Hölder inequality to each of the terms on the right-hand side of this inequality to obtain the Minkowski inequality.

A.5.6. (Rudin [Rud87, Theorem 6.16]) Note that $\Phi(g) \in L^p(\mu)^*$ and $\|\Phi(g)\| \le \|g\|_q$: for $q < \infty$, that follows from the Hölder inequality; the case $q = \infty$ is immediate. It is clear that Φ is linear. To see that it is injective, given g such that $\Phi(g) = 0$, consider a function β with values on the unit circle such that $\beta g = |g|$. Then, $\phi(g)\beta = \int |g| \, d\mu = 0$, hence $g = 0$. We are left to prove that for every

$\phi \in L^p(\mu)^*$ there exists $g \in L^q(\mu)$ such that $\phi = \Phi(g)$ and $\|g\|_q = \|\phi\|$. For each measurable set $B \subset M$, define $\eta(B) = \phi(\mathcal{X}_B)$. Check that η is a complex measure (to prove σ-additivity one needs $p < \infty$) and observe that $\eta \ll \mu$. Consider the Radon–Nikodym derivative $g = (d\eta/d\mu)$. Then, $\phi(\mathcal{X}_B) = \int_B g\,d\mu$ for every B; conclude that $\phi(f) = \int fg\,d\mu$ for every $f \in L^\infty(\mu)$. In the case $p = 1$, this construction yields $|\int_B g\,d\mu| \le \|\phi\|\,\mu(B)$ for every measurable set. Deduce that $\|g\|_\infty \le \|\phi\|$. Now suppose that $1 < p < \infty$. Take $f_n = \mathcal{X}_{B_n}\beta|g|^{q-1}$, where $B_n = \{x : |g(x)| \le n\}$. Observe that $f_n \in L^\infty(\mu)$ and $|f_n|^p = |g|^q$ in the set B_n and

$$\int_{B_n} |g|^q\,d\mu = \int f_n g\,d\mu = \phi(f_n) \le \|\phi\|\left(\int |f_n|^p\,d\mu\right)^{1/p} \le \|\phi\|\left(\int_{B_n} |g|^q\,d\mu\right)^{1/p}.$$

This yields $\int_{B_n} |g|^q\,d\mu \le \|\phi\|^q$ for every n and, thus, $\|g\|_q \le \|\phi\|$. Finally, $\phi(f) = \int fg\,d\mu$ for every $f \in L^p(\mu)$, since the two sides are continuous functionals and they coincide on the dense subset L^∞.

A.6.5. By definition, $u \cdot Lv = L^*u \cdot v$ and $u \cdot L^*v = (L^*)^*u \cdot v$ for any u and v. Hence, $v \cdot (L^*)^*u = L^*v \cdot u$ for any u and v. Reversing the roles of u and v, we see that $L = (L^*)^*$. Note that $\|L^*u \cdot v\| \le \|L\|\,\|u\|\,\|v\|$ for every u and v. Taking $v = L^*u$, it follows that $\|L^*u\| \le \|L\|\,\|u\|$ for every u and so $\|L^*\| \le \|L\|$. Since $L = (L^*)^*$, it follows that $\|L\| \le \|L^*$, hence the two norms coincide. Since the operator norm is submultiplicative, $\|L^*L\| \le \|L\|^2$. On the other hand, $u \cdot L^*Lu = \|Lu\|^2$ and so $\|L^*L\|\,\|u\|^2 \ge \|Lu\|^2$, for every u. Deduce that $\|L^*L\| \ge \|L\|^2$ and so the two expressions coincide. Analogously, $\|LL^*\| = \|L\|^2$.

A.6.8. Assume that $v \in H$ and $(u_n)_n$ is a sequence in E such that $u_n \cdot v \to u \cdot v$ for every $v \in H$. Considering $v \in E^\perp$, conclude that $v \in (E^\perp)^\perp$. By Exercise A.6.7, it follows that $u \in E$. Therefore, E is closed in the weak topology. Now consider any sequence $(v_n)_n$ in $U(E)$ converging to some $v \in H$. For each n, take $u_n = h^{-1}(v_n) \in E$. Since h is an isometry, $\|u_m - u_n\| = \|v_m - v_n\|$ for any m, n. It follows that $(u_n)_n$ is a Cauchy sequence in E and so it admits a limit $u \in E$. Hence, $v = h(u)$ is in $U(E)$.

A.7.1. The inverse of $T + H$ is given by the equation $(T + H)(T^{-1} + J) = \mathrm{id}$, which may be rewritten as a fixed point equation $J = -L^{-1}HL^{-1} + L^{-1}HJ$. Use the hypothesis to show that this equation admits a (unique) solution. Hence, $T+H$ is an isomorphism. Deduce that $L - \lambda\mathrm{id}$ whenever $\lambda > \|L\|$. Therefore, the spectrum of L is contained in the disk of radius $\|L\|$. It also follows from the previous observation that if $L - \lambda\mathrm{id}$ is an isomorphism then the same is true for $L - \lambda'\mathrm{id}$ if λ' is sufficiently close to λ.

A.7.4. (a) Observe that $L - \lambda\mathrm{id} = \int (z - \lambda)\,dE(z)$ and use Lemma A.7.4. By the continuity from below property (Exercise A.1.9), $E(\{\lambda\}) = \lim_n E(\{z : |z - \lambda| \le 1/n\})$. It follows that $E(\{\lambda\})v = v$. (b) It follows from Exercise A.7.3 that $E(B)E(\{\lambda\}) = E(\{\lambda\})$ if $\lambda \in B$ and $E(B)E(\{\lambda\}) = E(\emptyset) = 0$ otherwise. Since $L = \int z\,dE(z)$, we get that $Lv = \lambda E(\{\lambda\})v = \lambda v$.

References

[Aar97] J. Aaronson. *An introduction to infinite ergodic theory*, volume 50 of *Mathematical Surveys and Monographs*. American Mathematical Society, 1997.

[AB] A. Avila and J. Bochi. Proof of the subadditive ergodic theorem. Preprint www.mat.puc-rio.br/~jairo/docs/kingbirk.pdf.

[AF07] A. Avila and G. Forni. Weak mixing for interval exchange transformations and translation flows. *Ann. Math.*, 165:637–664, 2007.

[AKM65] R. Adler, A. Konheim and M. McAndrew. Topological entropy. *Trans. Amer. Math. Soc.*, 114:309–319, 1965.

[AKN06] V. Arnold, V. Kozlov and A. Neishtadt. *Mathematical aspects of classical and celestial mechanics*, volume 3 of *Encyclopaedia of Mathematical Sciences*. Springer-Verlag, third edition, 2006. [Dynamical systems. III], Translated from the Russian original by E. Khukhro.

[Ano67] D. V. Anosov. Geodesic flows on closed Riemannian manifolds of negative curvature. *Proc. Steklov Math. Inst.*, 90:1–235, 1967.

[Arn78] V. I. Arnold. *Mathematical methods of classical mechanics*. Springer-Verlag, 1978.

[AS67] D. V. Anosov and Ya. G. Sinai. Certain smooth ergodic systems. *Russian Math. Surveys*, 22:103–167, 1967.

[Bal00] V. Baladi. *Positive transfer operators and decay of correlations*. World Scientific Publishing Co. Inc., 2000.

[BDV05] C. Bonatti, L. J. Díaz and M. Viana. *Dynamics beyond uniform hyperbolicity*, volume 102 of *Encyclopaedia of Mathematical Sciences*. Springer-Verlag, 2005.

[Bil68] P. Billingsley. *Convergence of probability measures*. John Wiley & Sons Inc., 1968.

[Bil71] P. Billingsley. *Weak convergence of measures: Applications in probability*. Society for Industrial and Applied Mathematics, 1971. Conference Board of the Mathematical Sciences Regional Conference Series in Applied Mathematics, No. 5.

[Bir13] G. D. Birkhoff. Proof of Poincaré's last Geometric Theorem. *Trans. Amer. Math. Soc.*, 14:14–22, 1913.

[Bir67] G. Birkhoff. *Lattice theory*, volume 25. A.M.S. Colloq. Publ., 1967.

[BK83] M. Brin and A. Katok. On local entropy. In *Geometric dynamics (Rio de Janeiro, 1981)*, volume 1007 of *Lecture Notes in Math.*, pages 30–38. Springer-Verlag, 1983.

[BLY] D. Burguet, G. Liao and J. Yang. Asymptotic h-expansiveness rate of C^∞ maps. arxiv:1404.1771.

[Bos86] J.-B. Bost. Tores invariants des systèmes hamiltoniens. *Astérisque*, 133–134:113–157, 1986.

[Bos93] M. Boshernitzan. Quantitative recurrence results. *Invent. Math.*, 113(3): 617–631, 1993.

[Bow71] R. Bowen. Entropy for group endomorphisms and homogeneous spaces. *Trans. Amer. Math. Soc.*, 153:401–414, 1971.

[Bow72] R. Bowen. Entropy expansive maps. *Trans. Am. Math. Soc.*, 164:323–331, 1972.

[Bow75a] R. Bowen. *Equilibrium states and the ergodic theory of Anosov diffeomorphisms*, volume 470 of *Lect. Notes in Math.* Springer-Verlag, 1975.

[Bow75b] R. Bowen. A horseshoe with positive measure. *Invent. Math.*, 29:203–204, 1975.

[Bow78] R. Bowen. Entropy and the fundamental group. In *The Structure of Attractors in Dynamical Systems*, volume 668 of *Lecture Notes in Math.*, pages 21–29. Springer-Verlag, 1978.

[BS00] L. Barreira and J. Schmeling. Sets of "non-typical" points have full topological entropy and full Hausdorff dimension. *Israel J. Math.*, 116:29–70, 2000.

[Buz97] J. Buzzi. Intrinsic ergodicity for smooth interval maps. *Israel J. Math*, 100:125–161, 1997.

[Car70] H. Cartan. *Differential forms*. Hermann, 1970.

[Cas04] A. A. Castro. *Teoria da medida*. Projeto Euclides. IMPA, 2004.

[Cla72] J. Clark. *A Kolmogorov shift with no roots*. ProQuest LLC, Ann Arbor, MI, 1972. PhD. Thesis, Stanford University.

[dC79] M. do Carmo. *Geometria riemanniana*, volume 10 of *Projeto Euclides*. Instituto de Matemática Pura e Aplicada, 1979.

[Dei85] K. Deimling. *Nonlinear functional analysis*. Springer-Verlag, 1985.

[Din70] E. Dinaburg. A correlation between topological entropy and metric entropy. *Dokl. Akad. Nauk SSSR*, 190:19–22, 1970.

[Din71] E. Dinaburg. A connection between various entropy characterizations of dynamical systems. *Izv. Akad. Nauk SSSR Ser. Mat.*, 35:324–366, 1971.

[dlL93] R. de la Llave. Introduction to K.A.M. theory. In *Computational physics (Almuñécar, 1992)*, pages 73–105. World Sci. Publ., 1993.

[DS57] N. Dunford and J. Schwarz. *Linear operators I: General theory*. Wiley & Sons, 1957.

[DS63] N. Dunford and J. Schwarz. *Linear operators II: Spectral theory*. Wiley & Sons, 1963.

[Dug66] J. Dugundji. *Topology*. Allyn and Bacon Inc., 1966.

[Edw79] R. E. Edwards. *Fourier series. A modern introduction. Vol. 1*, volume 64 of *Graduate Texts in Mathematics*. Springer-Verlag, second edition, 1979.

[ET36] P. Erdös and P. Turán. On some sequences of integers. *J. London. Math. Soc.*, 11:261–264, 1936.

[Fal90] K. Falconer. *Fractal geometry: Mathematical foundations and applications*. John Wiley & Sons Ltd., 1990.

[Fer02] R. Fernandez. *Medida e integração*. Projeto Euclides. IMPA, 2002.

[FFT09] S. Ferenczi, A. Fisher and M. Talet. Minimality and unique ergodicity for adic transformations. *J. Anal. Math.*, 109:1–31, 2009.

[FO70] N. Friedman and D. Ornstein. On isomorphism of weak Bernoulli transformations. *Advances in Math.*, 5:365–394, 1970.

[Fri69] N. Friedman. *Introduction to ergodic theory*. Van Nostrand, 1969.

[Fur61] H. Furstenberg. Strict ergodicity and transformation of the torus. *Amer. J. Math.*, 83:573–601, 1961.

[Fur77] H. Furstenberg. Ergodic behavior and a theorem of Szemerédi on arithmetic progressions. *J. d'Analyse Math.*, 31:204–256, 1977.

[Fur81] H. Furstenberg. *Recurrence in ergodic theory and combinatorial number theory*. Princeton University Press, 1981.

[Goo71a] T. Goodman. Relating topological entropy and measure entropy. *Bull. London Math. Soc.*, 3:176–180, 1971.

[Goo71b] G. Goodwin. Optimal input signals for nonlinear-system identification. *Proc. Inst. Elec. Engrs.*, 118:922–926, 1971.

[GT08] B. Green and T. Tao. The primes contain arbitrarily long arithmetic progressions. *Ann. of Math.*, 167:481–547, 2008.

[Gur61] B. M. Gurevič. The entropy of horocycle flows. *Dokl. Akad. Nauk SSSR*, 136:768–770, 1961.

[Hal50] P. Halmos. *Measure Theory*. Van Nostrand, 1950.

[Hal51] P. Halmos. *Introduction to Hilbert space and the theory of spectral multiplicity*. Chelsea Publishing Company, 1951.

[Hay] N. Haydn. Multiple measures of maximal entropy and equilibrium states for one-dimensional subshifts. Preprint, Penn State University.

[Hir94] M. Hirsch. *Differential topology*, volume 33 of *Graduate Texts in Mathematics*. Springer-Verlag, 1994. Corrected reprint of the 1976 original.

[Hof77] F. Hofbauer. Examples for the nonuniqueness of the equilibrium state. *Trans. Amer. Math. Soc.*, 228:223–241, 1977.

[Hop39] E. F. Hopf. Statistik der geodätischen Linien in Mannigfaltigkeiten negativer Krümmung. *Ber. Verh. Sächs. Akad. Wiss. Leipzig*, 91:261–304, 1939.

[HvN42] P. Halmos and J. von Neumann. Operator methods in classical mechanics. II. *Ann. Math.*, 43:332–350, 1942.

[Jac60] K. Jacobs. *Neuere Methoden und Ergebnisse der Ergodentheorie*. Ergebnisse der Mathematik und ihrer Grenzgebiete. N. F., Heft 29. Springer-Verlag, 1960.

[Jac63] K. Jacobs. *Lecture notes on ergodic theory, 1962/63. Parts I, II.* Matematisk Institut, Aarhus Universitet, Aarhus, 1963.

[Kal82] S. Kalikow. T, T^{-1} transformation is not loosely Bernoulli. *Ann. Math.*, 115:393–409, 1982.

[Kat71] Yi. Katznelson. Ergodic automorphisms of T^n are Bernoulli shifts. *Israel J. Math.*, 10:186–195, 1971.

[Kat80] A. Katok. Lyapunov exponents, entropy and periodic points of diffeomorphisms. *Publ. Math. IHES*, 51:137–173, 1980.

[Kea75] M. Keane. Interval exchange transformations. *Math. Zeit.*, 141:25–31, 1975.

[KM10] S. Kalikow and R. McCutcheon. *An outline of ergodic theory*, volume 122 of *Cambridge Studies in Advanced Mathematics*. Cambridge University Press, 2010.

[Kok35] J. F. Koksma. Ein mengentheoretischer Satz über die Gleichverteilung modulo Eins. *Compositio Math.*, 2:250–258, 1935.

[KR80] M. Keane and G. Rauzy. Stricte ergodicité des échanges d'intervalles. *Math. Zeit.*, 174:203–212, 1980.

[Kri70] W. Krieger. On entropy and generators of measure-preserving transformations. *Trans. Amer. Math. Soc.*, 149:453–464, 1970.

[Kri75] W. Krieger. On the uniqueness of the equilibrium state. *Math. Systems Theory*, 8:97–104, 1974/75.

[KSS91] A. Krámli, N. Simányi and D. Szász. The K-property of three billiard balls. *Ann. Math.*, 133:37–72, 1991.

[KSS92] A. Krámli, N. Simányi and D. Szász. The K-property of four billiard balls. *Comm. Math. Phys.*, 144:107–148, 1992.

[KW82] Y. Katznelson and B. Weiss. A simple proof of some ergodic theorems. *Israel J. Math.*, 42:291–296, 1982.

[Lan73] O. Lanford. Entropy and equilibrium states in classical statistical mechanics. In *Statistical mechanics and mathematical problems*, volume 20 of *Lecture Notes in Physics*, page 1–113. Springer-Verlag, 1973.

[Led84] F. Ledrappier. Propriétés ergodiques des mesures de Sinaï. *Publ. Math. I.H.E.S.*, 59:163–188, 1984.

[Lin77] D. Lind. The structure of skew products with ergodic group actions. *Israel J. Math.*, 28:205–248, 1977.

[LS82] F. Ledrappier and J.-M. Strelcyn. A proof of the estimation from below in Pesin's entropy formula. *Ergod. Th & Dynam. Sys*, 2:203–219, 1982.

[LVY13] G. Liao, M. Viana and J. Yang. The entropy conjecture for diffeomorphisms away from tangencies. *J. Eur. Math. Soc. (JEMS)*, 15(6):2043–2060, 2013.

[LY85a] F. Ledrappier and L.-S. Young. The metric entropy of diffeomorphisms. I. Characterization of measures satisfying Pesin's entropy formula. *Ann. Math.*, 122:509–539, 1985.

[LY85b] F. Ledrappier and L.-S. Young. The metric entropy of diffeomorphisms. II. Relations between entropy, exponents and dimension. *Ann. Math.*, 122:540–574, 1985.

[Man75] A. Manning. Topological entropy and the first homology group. In *Dynamical Systems, Warwick, 1974*, volume 468 of *Lecture Notes in Math.*, pages 185–190. Springer-Verlag, 1975.

[Mañ85] R. Mañé. Hyperbolicity, sinks and measure in one-dimensional dynamics. *Comm. Math. Phys.*, 100:495–524, 1985.

[Mañ87] R. Mañé. *Ergodic theory and differentiable dynamics*. Springer-Verlag, 1987.

[Mas82] H. Masur. Interval exchange transformations and measured foliations. *Ann. Math*, 115:169–200, 1982.

[Mey00] C. Meyer. *Matrix analysis and applied linear algebra*. Society for Industrial and Applied Mathematics (SIAM), 2000.

[Mis73] M. Misiurewicz. Diffeomorphim without any measure of maximal entropy. *Bull. Acad. Pol. Sci.*, 21:903–910, 1973.

[Mis76] M. Misiurewicz. A short proof of the variational principle for a Z_+^N action on a compact space. *Asterisque*, 40:147–187, 1976.

[MP77a] M. Misiurewicz and F. Przytycki. Entropy conjecture for tori. *Bull. Pol. Acad. Sci. Math.*, 25:575–578, 1977.

[MP77b] M. Misiurewicz and F. Przytycki. Topological entropy and degree of smooth mappings. *Bull. Pol. Acad. Sci. Math.*, 25:573–574, 1977.

[MP08] W. Marzantowicz and F. Przytycki. Estimates of the topological entropy from below for continuous self-maps on some compact manifolds. *Discrete Contin. Dyn. Syst. Ser.*, 21:501–512, 2008.

[MT78] G. Miles and R. Thomas. Generalized torus automorphisms are Bernoullian. *Advances in Math. Supplementary Studies*, 2:231–249, 1978.

[New88] S. Newhouse. Entropy and volume. *Ergodic Theory Dynam. Systems*, 8*(Charles Conley Memorial Issue):283–299, 1988.

[New90] S. Newhouse. Continuity properties of entropy. *Ann. Math.*, 129:215–235, 1990. Errata in Ann. Math. 131:409–410, 1990.

[NP66] D. Newton and W. Parry. On a factor automorphism of a normal dynamical system. *Ann. Math. Statist.*, 37:1528–1533, 1966.

[NR97] A. Nogueira and D. Rudolph. Topological weak-mixing of interval exchange maps. *Ergod. Th. & Dynam. Sys.*, 17:1183–1209, 1997.

[Orn60] D. Ornstein. On invariant measures. *Bull. Amer. Math. Soc.*, 66:297–300, 1960.

[Orn70] D. Ornstein. Bernoulli shifts with the same entropy are isomorphic. *Advances in Math.*, 4:337–352 (1970), 1970.

[Orn72] Donald S. Ornstein. On the root problem in ergodic theory. In *Proceedings of the Sixth Berkeley Symposium on Mathematical Statistics and Probability (Univ. California, Berkeley, Calif., 1970/1971), Vol. II: Probability theory*, pages 347–356. Univ. California Press, 1972.

[Orn74] D. Ornstein. *Ergodic theory, randomness, and dynamical systems*. Yale University Press, 1974. James K. Whittemore Lectures in Mathematics given at Yale University, Yale Mathematical Monographs, No. 5.

[OS73] D. Ornstein and P. Shields. An uncountable family of K-automorphisms. *Advances in Math.*, 10:63–88, 1973.

[OU41] J. C. Oxtoby and S. M. Ulam. Measure-preserving homeomorphisms and metrical transitivity. *Ann. Math.*, 42:874–920, 1941.

[Par53] O. S. Parasyuk. Flows of horocycles on surfaces of constant negative curvature. *Uspehi Matem. Nauk (N.S.)*, 8:125–126, 1953.

[Pes77] Ya. B. Pesin. Characteristic Lyapunov exponents and smooth ergodic theory. *Russian Math. Surveys*, 324:55–114, 1977.

[Pes97] Ya. Pesin. *Dimension theory in dynamical systems: Contemporary views and applications*. University of Chicago Press, 1997.

[Pet83] K. Petersen. *Ergodic theory*. Cambridge University Press, 1983.

[Phe93] R. Phelps. *Convex functions, monotone operators and differentiability*, volume 1364 of *Lecture Notes in Mathematics*. Springer-Verlag, second edition, 1993.

[Pin60] M. S. Pinsker. *Informatsiya i informatsionnaya ustoichivostsluchainykh velichin i protsessov*. Problemy Peredači Informacii, Vyp. 7. Izdat. Akad. Nauk SSSR, 1960.

[PT93] J. Palis and F. Takens. *Hyperbolicity and sensitive-chaotic dynamics at homoclinic bifurcations*. Cambridge University Press, 1993.

[PU10] F. Przytycki and M. Urbański. *Conformal fractals: Ergodic theory methods*, volume 371 of *London Mathematical Society Lecture Note Series*. Cambridge University Press, 2010.

[PW72a] W. Parry and P. Walters. Errata: "Endomorphisms of a Lebesgue space". *Bull. Amer. Math. Soc.*, 78:628, 1972.

[PW72b] W. Parry and P. Walters. Endomorphisms of a Lebesgue space. *Bull. Amer. Math. Soc.*, 78:272–276, 1972.

[PY98] M. Pollicott and M. Yuri. *Dynamical systems and ergodic theory*, volume 40 of *London Mathematical Society Student Texts*. Cambridge University Press, 1998.

[Qua99] A. Quas. Most expanding maps have no absolutely continuous invariant mesure. *Studia Math.*, 134:69–78, 1999.

[Que87] M. Queffélec. *Substitution dynamical systems—spectral analysis*, volume 1294 of *Lecture Notes in Mathematics*. Springer-Verlag, 1987.

[Rok61] V. A. Rokhlin. Exact endomorphisms of a Lebesgue space. *Izv. Akad. Nauk SSSR Ser. Mat.*, 25:499–530, 1961.

[Rok62] V. A. Rokhlin. On the fundamental ideas of measure theory. *A. M. S. Transl.*, 10:1–54, 1962. Transl. from *Mat. Sbornik* 25 (1949), 107–150. First published by the A. M. S. in 1952 as Translation Number 71.

[Rok67a] V. A. Rokhlin. Lectures on the entropy theory of measure-preserving transformations. *Russ. Math. Surv.*, 22(5):1–52, 1967. Transl. from *Uspekhi Mat. Nauk.* 22(5) (1967), 3–56.

[Rok67b] V. A. Rokhlin. Metric properties of endomorphisms of compact commutative groups. *Amer. Math. Soc. Transl.*, 64:244–252, 1967.

[Roy63] H. L. Royden. *Real analysis*. Macmillan, 1963.

[RS61] V. A. Rokhlin and Ja. G. Sinaĭ. The structure and properties of invariant measurable partitions. *Dokl. Akad. Nauk SSSR*, 141:1038–1041, 1961.

[Rud87] W. Rudin. *Real and complex analysis*. McGraw-Hill, 1987.

[Rue73] D. Ruelle. Statistical mechanics on a compact set with Z^ν action satisfying expansiveness and specification. *Trans. Amer. Math. Soc.*, 186:237–251, 1973.

[Rue78] D. Ruelle. An inequality for the entropy of differentiable maps. *Bull. Braz. Math. Soc.*, 9:83–87, 1978.

[Rue04] D. Ruelle. *Thermodynamic formalism: The mathematical structures of equilibrium statistical mechanics*. Cambridge Mathematical Library. Cambridge University Press, second edition, 2004.

[RY80] C. Robinson and L. S. Young. Nonabsolutely continuous foliations for an Anosov diffeomorphism. *Invent. Math.*, 61:159–176, 1980.

[SC87] Ya. Sinaĭ and Nikolay Chernov. Ergodic properties of some systems of two-dimensional disks and three-dimensional balls. *Uspekhi Mat. Nauk*, 42:153–174, 256, 1987.

[Shu69] M. Shub. Endomorphisms of compact differentiable manifolds. *Amer. Journal of Math.*, 91:129–155, 1969.

[Shu74] M. Shub. Dynamical systems, filtrations and entropy. *Bull. Amer. Math. Soc.*, 80:27–41, 1974.

[Sim02] N. Simányi. The complete hyperbolicity of cylindric billiards. *Ergodic Theory Dynam. Systems*, 22:281–302, 2002.

[Sin63] Ya. Sinaĭ. On the foundations of the ergodic hypothesis for a dynamical system of statistical mechanics. *Soviet. Math. Dokl.*, 4:1818–1822, 1963.

[Sin70] Ya. Sinaĭ. Dynamical systems with elastic reflections. Ergodic properties of dispersing billiards. *Uspehi Mat. Nauk*, 25:141–192, 1970.

[Ste58] E. Sternberg. On the structure of local homeomorphisms of Euclidean *n*-space – II. *Amer. J. Math.*, 80:623–631, 1958.

[SW75] M. Shub and R. Williams. Entropy and stability. *Topology*, 14:329–338, 1975.

[SX10] R. Saghin and Z. Xia. The entropy conjecture for partially hyperbolic diffeomorphisms with 1-D center. *Topology Appl.*, 157:29–34, 2010.

[Sze75] S. Szemerédi. On sets of integers containing no k elements in arithmetic progression. *Acta Arith.*, 27:199–245, 1975.

[vdW27] B. van der Waerden. Beweis eibe Baudetschen Vermutung. *Nieuw Arch. Wisk.*, 15:212–216, 1927.

[Vee82] W. Veech. Gauss measures for transformations on the space of interval exchange maps. *Ann. of Math.*, 115:201–242, 1982.

[Ver99] Alberto Verjovsky. *Sistemas de Anosov*, volume 9 of *Monographs of the Institute of Mathematics and Related Sciences*. Instituto de Matemática y Ciencias Afines, IMCA, Lima, 1999.

[Via14] M. Viana. *Lectures on Lyapunov exponents*. Cambridge University Press, 2014.

[VO14] M. Viana and K. Oliveira. *Fundamentos da Teoria Ergódica*. Coleção Fronteiras da Matemática. Sociedade Brasileira de Matemática, 2014.

[Wal73] P. Walters. Some results on the classification of non-invertible measure preserving transformations. In *Recent advances in topological dynamics (Proc. Conf. Topological Dynamics, Yale Univ., New Haven, Conn., 1972; in honor of Gustav Arnold Hedlund)*, pages 266–276. Lecture Notes in Math., Vol. 318. Springer-Verlag, 1973.

[Wal75] P. Walters. A variational principle for the pressure of continuous transformations. *Amer. J. Math.*, 97:937–971, 1975.

[Wal82] P. Walters. *An introduction to ergodic theory*. Springer-Verlag, 1982.

[Wey16] H. Weyl. Uber die Gleichverteilungen von Zahlen mod Eins. *Math. Ann.*, 77:313–352, 1916.

[Yan80] K. Yano. A remark on the topological entropy of homeomorphisms. *Invent. Math.*, 59:215–220, 1980.

[Yoc92] J.-C. Yoccoz. Travaux de Herman sur les tores invariants. *Astérisque*, 206:Exp. No. 754, 4, 311–344, 1992. Séminaire Bourbaki, Vol. 1991/92.

[Yom87] Y. Yomdin. Volume growth and entropy. *Israel J. Math.*, 57:285–300, 1987.

[Yos68] K. Yosida. *Functional analysis*. Second edition. Die Grundlehren der mathematischen Wissenschaften, Band 123. Springer-Verlag, 1968.

[Yuz68] S. A. Yuzvinskii. Metric properties of endomorphisms of compact groups. *Amer. Math. Soc. Transl.*, 66:63–98, 1968.

[Zyg68] A. Zygmund. *Trigonometric series: Vols. I, II*. Second edition, reprinted with corrections and some additions. Cambridge University Press, 1968.

Index of notation

Index